THE DEATH OF MASSIVE STARS:
SUPERNOVAE AND GAMMA-RAY BURSTS

IAU SYMPOSIUM No. 279

COVER ILLUSTRATION: DEATH OF A MASSIVE STAR

An illustration of the death of a rapidly rotating massive star shortly after its core collapse. A highly relativistic jet is launched along the rotation axis.

INTERNATIONAL ASTRONOMICAL UNION
UNION ASTRONOMIQUE INTERNATIONALE

THE DEATH OF MASSIVE STARS: SUPERNOVAE AND GAMMA-RAY BURSTS

PROCEEDINGS OF THE 279th SYMPOSIUM OF THE INTERNATIONAL ASTRONOMICAL UNION
HELD IN NIKKO, JAPAN
MARCH 12–16, 2012

Edited by

PETER W. A. ROMING
Southwest Research Institute, San Antonio, TX, USA

NOBUYUKI KAWAI
Tokyo Institute of Technology, Tokyo, JAPAN

and

ELENA PIAN
Scuola Normale Superiore di Pisa, Pisa, ITALY

CAMBRIDGE UNIVERSITY PRESS
The Edinburgh Building, Cambridge CB2 2RU, UnitedKingdom
32 Avenue of the Americas, New York, NY 10013 2473, USA
10 Stamford Road, Oakleigh, Melbourne 3166, Australia

First published 2012

Printed in the UK by MPG Books Ltd

Typeset in System LaTeX 2ε

A catalogue record for this book is available from the British Library

Library of Congress Cataloguing in Publication data

This journal issue has been printed on FSC-certified paper and cover board. FSC is an independent, non-governmental, not-for-profit organization established to promote the responsible management of the worlds forests. Please see www.fsc.org for information.

ISBN 9781107019799 hardback
ISSN 1743-9213

Table of Contents

Section E. Environments and Host Galaxies of GRBs & Supernovae

Chair: Annalisa De Cia

Section F. Massive Star Formation and Cosmological Implications

Chair: Nobu Kawai

Part 2. POSTERS

Part 3. ABSTRACTS

Preface

IAU Symposium 279 took place in Nikko, in the Tochigi Prefecture of Japan. Its science motivation centers around the death of stars that are larger than eight solar masses. These massive stars end their lives in a fiery explosion and are manifest as core-collapse supernovae or gamma-ray bursts. In rare cases, a highly stripped massive star explodes and exhibits properties of both core-collapse supernovae and gamma-ray bursts: the prototype of this class is the nearby (35 Mpc) core-collapse explosion of supernova SN 1998bw, accompanied by the gamma-ray burst of 25 April 1998, in the direction of the Telescopium constellation. In contrast, there are clear cases in which no bright supernova is found to be associated with a gamma-ray burst, and vice versa. The quest in understanding supernovae and gamma-ray bursts, and the connection between them, has raised many questions. Since the elements synthesized in the explosion of massive stars are the building blocks for much of the visible Universe, it is important to understand the life cycle of these massive stars.

This symposium brought together international leaders, in both theory and observation, who study core-collapse supernovae and gamma-ray bursts, to discuss the range of activities in the field. These include: stellar evolution and explosion; progenitors, environments and host galaxies; astroparticle physics; as well as multi-wavelength observations of these objects and their use as cosmological probes, particularly in the very early Universe. The symposium, opened by the welcoming address of Nobu Kawai, was divided into eight sessions, 62 talks, and 82 posters. The 158 participants came from 25 countries with 28 invited and 34 contributed speakers, of which five and six were women, respectively. A very large fraction of participants were younger scientists. We had the honor and privilege of having with us for the whole meeting Thierry Montmerle, Assistant General Secretary of the IAU Executive Committee, who also delivered a talk on the future organization of the IAU.

Our understanding of the lives and death of stars with masses greater than eight solar masses are beginning to expand thanks to increasingly powerful diagnostic tools, models, and numerical simulations that have become available. These resources are helping identify the evolutionary channels and eventual fates of massive stars, as well as investigating how a fraction of them are able to produce high-energy emission and jets, and possibly accelerate cosmic rays. The talks during the meeting focused on twelve primary themes:

- What are the differing models relating to the death of massive stars telling us;
- X-ray and optical properties of all classes of supernovae including the superluminous supernovae (sometimes referred to as "Quimbies," derived from the last name of Robert Quimby, a Symposium participant who has critically contributed to the discovery and investigation of these sources);
- What we are learning from multi-wavelength observations of the prompt and afterglow phases of gamma-ray bursts;
- What we are learning from X-ray to radio observations of core-collapse supernovae and their remnants, including the nearest and by far best studied SN 1987A in the Large Magellanic Cloud, that had just celebrated its Silver Jubilee at the time of the Symposium;
- The challenges associated with observing and constraining the progenitors of gamma-ray bursts and supernovae;

- Current thoughts and scenarios on supernova theory and the role of asymmetry in the explosion;
- Gravitational waves and gamma-ray bursts;
- Host galaxies and the local environments, particularly the metallicities, of gamma-ray bursts and core-collapse supernovae;
- Current theories in early Universe star formation including Population III stars;
- Using gamma-ray bursts as probes of the early Universe and looking into correlations that may turn them into standard candles;
- Understanding the shock break out of supernovae, i.e. the blast driven by the rebouncing of the collapse stall, that produces a bright multi-wavelength early outburst;
- The possible connection between short gamma-ray bursts and magnetars, highly magnetized and rapidly spinning neutron stars.

Some speakers tackled the above fundamental issues by focussing on one particular source of specific relevance, a sub-set, or an entire class. Every one did an excellent job in underlining the merits of any particular observing campaign or technique, or of dedicated models and/or simulations. The picture that emerged is that much more data are necessary to develop reliable interpretative paradigms. Thanks to the systematic, sensitive, wide-field optical and radio surveys that are already taking place or are soon to be on line, the near-term future looks promising with regards to data accumulation. It is also highly desirable that space-based assets keep pace by ensuring continuity in the X- and gamma-ray regime in the follow-up and monitoring of supernovae and gamma-ray bursts.

All talks generated lively debates that often continued during the coffee break sessions or were resumed during lunches and dinners. The symposium was concluded by Shri Kulkarni, who summarized the content of the meeting as well as included some of his own thoughts about our current understanding in the field. One invited talk was not given: Chris Fryer had a last minute emergency that prevented him from giving his talk on Stellar Collapse and Gamma-Ray Burst Explosion Mechanisms.

No more appropriate setting could be chosen for the Symposium than the little and lovely city of Nikko. Its magic atmosphere fascinated the participants, who cherished the opportunity of discovering this Japanese hidden gemstone, and to discuss science in this pure, quiet and inspiring mountain environment.

We would like to thank all the people, entities and authorities who helped this event happen. We are grateful to the participants, speakers, session chairpersons, and Science Organizing Committee who ensured a high scientific standard and lively forum for the exchange of new ideas and concepts. We are also grateful to the secretaries, assistants, and Local Organizing Committee whose active support made possible the realization of the numerous details always associated with such a symposium. A special thanks goes to Takeo Minezaki, whose help as an assistant editor was invaluable, and to Tanja Karthaus (IAC) and Sarah Tayler (SRON) for their careful transcription of the discussion sheets.

Elena Pian, Nobu Kawai, and Pete Roming, co-chairs SOC,
Keiichi Maeda, chair LOC
June 6, 2012

THE ORGANIZING COMMITTEE

Scientific

Z.-G. Dai (China)
J. Fynbo (Denmark)
N. Kawai (co-chair, Japan)
M. Modjaz (USA)
K. Nomoto (Japan)
E. Pian (co-chair, Italy)
S. Savaglio (Germany)
S. Smartt (UK)
S. Yamada (Japan)
M. Della Valle (Italy)
N. Gehrels (USA)
S. McBreen (Ireland)
E. Nakar (Israel)
P. O'Brien (UK)
P. Roming (co-chair, USA)
B. Schmidt (Australia)
A. Soderberg (USA)

Local

K. Maeda (chair)
A. Bamba
T. Minezaki
T. Tamagawa
Y. Yatsu
K. Asano
K. Kotake
T. Ohkawa
M. Tanaka

Acknowledgements

The symposium is supported by IAU Divisions VIII (Galaxies & the Universe) and XI (Space & High Energy Astrophysics); and by IAU Commission No. 44 (Space & High Energy Astrophysics).

Funding by the
International Astronomical Union,
Ministry of Education, Culture, Sport, Science & Technology of Japan,
Japan Society for Promotion of Science,
Institute for Physics & Mathematics of the Universe, University of Tokyo,
Global COE Program "Quantum Physics & Nanoscience," Tokyo Tech,
Southwest Research Institute,
Grant-in-Aid for Priority Research Area "Deciphering the Ancient Universe
with Gamma-Ray Bursts,"
and
Astronomical Society of Japan,
is gratefully acknowledged.

Participants

Lorenzo **Amati**, INAF - IASF Bologna, Bologna, Italy amati@iasfbo.inaf.it
Joseph **Anderson**, Universidad de Chile, Santiago, Chile anderson@das.uchile.cl
Yu **Aoki**, Tokyo Institute of Technology, Tokyo, Japan aoki@hp.phys.titech.ac.jp
Iair **Arcavi**, Weizmann Institute of Science, Rehovot, Israel iair.arcavi@weizmann.ac.il
Katsuaki **Asano**, Tokyo Institute of Technology, Tokyo, Japan asano@phys.titech.ac.jp
Tri L. **Astraatmadja**, Nikhef Amsterdam, Amsterdam, The Netherlands trias@nikhef.nl
Andrey **Baranov**, LAPTH, Universite de Savoie, ANNECY-LE-VIEUX, France a.hc.baranov@gmail.com
Aldo **Batta**, IA UNAM, Ciudad de México, D.F., Mexico abatta@astro.unam.mx
Franz **Bauer**, Pontificia Universidad Catolica de Chile / Space Science Institute, Santiago, Chile fbauer@astro.puc.cl
Sagi **Ben-Ami**, Weizmann Institute of Science, Rehovot, Israel sagiba.ph@gmail.com
Melina **Bersten**, IPMU, Kashiwa, Chiba, Japan melina.bersten@ipmu.jp
Joanne **Bibby**, American Museum of Natural History, New York, NY, USA jbibby@amnh.org
Omer **Bromberg**, The Hebrew university of Jerusalem, Jerusalem, Israel omerb@phys.huji.ac.il
Niccolo' **Bucciantini**, INAF Osservatorio di Arcetri, Firenze, Italy niccolo@arcetri.astro.it
Andrew **Bunker**, University of Oxford, Oxford, United Kingdom a.bunker1@physics.ox.ac.uk
David **Burrows**, Penn State University, PA, USA burrows@astro.psu.edu
Alberto J. **Castro-Tirado**, IAA-CSIC, Granada, Spain ajct@iaa.es
Sayan **Chakraborti**, Tata Institute of Fundamental Research, Mumbai, India sayan@tifr.res.in
Pascal **Chardonnet**, Universite de Savoie, Chambery, Savoie, France chardonnet@lapp.in2p3.fr
Ryan **Chornock**, Harvard-Smithsonian Center for Astrophysics, Cambridge, MA, USA rchornock@cfa.harvard.edu
Bethany **Cobb**, George Washington University, Washington, DC, USA bcobb@gwu.edu
Alessandra **Corsi**, California Institute of Technology, Pasadena, CA, USA corsi@caltech.edu
Sean **Couch**, University of Chicago, Chicago, IL, USA smc@flash.uchicago.edu
Paul **Crowther**, University of Sheffield, Sheffield, United Kingdom Paul.Crowther@sheffield.ac.uk
Maria **Dainotti**, Stanford University, Santa Clara, CA, USA dainotti@stanford.edu
Annalisa **De Cia**, University of Iceland, Reykjavik, Iceland annalisa@raunvis.hi.is
Massimo **Della Valle**, Capodimonte Astronomical Observatory, INAF-Naples, Napoli, Italy dellavalle@na.astro.it
Takahiro **Enomoto**, Tokyo Institute of Technology, Tokyo, Japan enomoto@hp.phys.titech.ac.jp
Romanus **Eze**, JAXA/ISAS, Sagamihara, Kanagawa, Japan romanus.eze@gmail.com
Thierry **Foglizzo**, CEA-Saclay, Saclay, France foglizzo@cea.fr
Gaston **Folatelli**, IPMU, University of Tokyo, Kashiwa, Chiba, Japan gaston.folatelli@ipmu.jp
Shin-ichiro **Fujimoto**, Kumamoto National College of Technology, Goshi, Kumamoto, Japan fujimoto@ec.knct.ac.jp
Shun **Furusawa**, Waseda University, Tokyo, Japan furusawa.shun@gmail.com
Johan **Fynbo**, Dark Cosmology Centre, Copenhagen, Denmark jfynbo@dark-cosmology.dk
Avishay **Gal-Yam**, Weizmann Institute of Science, Rehovot, Israel avishay.gal-yam@weizmann.ac.il
Viktoriya **Giryanskaya**, Institute of Nuclear Physics, Tashkent, Uzbekistan moroz_vs@yahoo.com
Tomer **Goldfriend**, The Hebrew University of Jerusalem, Jerusalem, Israel tomergf@gmail.com
John **Graham**, STScI & Johns Hopkins University, Baltimore, MD, USA graham@pha.jhu.edu
Stephan **Hachinger**, INAF - Osservatorio Astronomico di Padova, Padova, Italy s.hachinger@gmx.de
Paul **Hancock**, The Univerity of Sydney, Sydney, NSW, Australia Paul.Hancock@sydney.edu.au
Tetsuya **Hashimoto**, National Astronomical Observatory of Japan, Mitaka, Tokyo, Japan tetsuya.hashimoto@nao.ac.jp
Mayumi **Hayashi**, Tokyo Institute of Technology, Tokyo, Japan hayashi@hp.phys.titech.ac.jp
Andrew **Howell**, LCOGT / UCSB, Santa Barbara, CA, USA ahowell@lcogt.net
Stefan **Immler**, NASA/CRESST/GSFC, Greenbelt, MD, USA stefan.m.immler@nasa.gov
Susumu **Inoue**, ICRR, University of Tokyo, Kashiwa, Chiba, Japan sinoue@icrr.u-tokyo.ac.jp
Tsuyoshi **Inoue**, Aoyama-Gakuin University, Sagamihara, Kanagawa, Japan inouety@phys.aoyama.ac.jp
Kunihito **Ioka**, KEK Theory Center, Tsukuba, Ibaraki, Japan kunihito.ioka@kek.jp
Wakana **Iwakami-Nakano**, Tohoku University, Sendai, Miyagi, Japan wakana@dragon.ifs.tohoku.ac.jp
Natsuko **Izutani**, Univ. of Tokyo, Tokyo, Japan izutani@astron.s.u-tokyo.ac.jp
Palli **Jakobsson**, University of Iceland, Reykjavik, Iceland pja@raunvis.hi.is
Samuel **Jones**, Keele University, Staffordshire, UK swj@astro.keele.ac.uk
Jun **Kakuwa**, Hiroshima Univ., Higashihiroshima, Hiroshima, Japan junkakuwa@hiroshima-u.ac.jp
Yasuomi **Kamiya**, University of Tokyo/IPMU, Kashiwa, Chiba, Japan yasuomi.kamiya@ipmu.jp
Boaz **Katz**, Institute for Advanced Study, Princeton, NJ, USA boazka@ias.edu
Nobuyuki **Kawai**, Tokyo Institute of Technology, Tokyo, Japan nkawai@phys.titech.ac.jp
Kosuke **Kawakami**, Tokyo Institute of Technology, Tokyo, Japan kawakami@hp.phys.titech.ac.jp
Sylvio **Klose**, Thueringer Landessternwarte Tautenburg, Germany, Tautenburg, Germany klose@tls-tautenburg.de
Kei **Kotake**, NAOJ, Mitaka, Tokyo, Japan kkotake@th.nao.ac.jp
Thomas **Kruehler**, Dark Cosmology Centre, Copenhagen, Denmark tom@dark-cosmology.dk
S. **Kulkarni**, Caltech, Pasadena, CA, USA srk@astro.caltech.edu
Hanin **Kuncarayakti**, Institute of Astronomy, University of Tokyo, Mitaka, Tokyo, Japan hanin@ioa.s.u-tokyo.ac.jp
Giorgos **Leloudas**, Dark Cosmology Centre, Copenhagen, Denmark giorgos@dark-cosmology.dk
Marie **Lemoine-Busserolle**, Gemini Observatory, Hilo, HI, USA mbusserolle@gemini.edu
Andrew **Levan**, University of Warwick, Coventry, United Kingdom A.J.Levan@warwick.ac.uk
Emily **Levesque**, University of Colorado at Boulder, Boulder, CO, USA Emily.Levesque@colorado.edu
Amy **Lien**, NASA/GSFC/ORAU, Greenbelt, MD, USA amy.y.lien@nasa.gov
Heuijin **Lim**, Institute of the Early Universe, Ewha Womans University, Seoul, Korea heuijin.lim@gmail.com
Andrew **MacFadyen**, New York University, New York, NY, USA macfadyen@nyu.edu
Yoshitomo **Maeda**, ISAS/JAXA, Sagamihara, Kanagawa, Japan ymaeda@astro.isas.jaxa.jp
Keiichi **Maeda**, IPMU, U. Tokyo, Kashiwa, Chiba, Japan keiichi.maeda@ipmu.jp
Jirong **Mao**, Korea Astronomy and Space Science Institute, Daejeon, Korea jirongmao@kasi.re.kr
Paolo **Mazzali**, INAF-Padova / MPA-Garching, Garching, Germany mazzali@mpa-garching.mpg.de
Brittany **McDonald**, McMaster University, Ontario, Canada mcdonb5@mcmaster.ca
Brian **Metzger**, Princeton University, Princeton, NJ, USA bmetzger@astro.princeton.edu
Takeo **Minezaki**, Institute of Astronomy, University of Tokyo, Mitaka, Tokyo, Japan minezaki@ioa.s.u-tokyo.ac.jp
I. Felix **Mirabel**, IAFE-CONICET & CEA-IRFU, Buenos Aires, Argentina fmirabel@eso.org
Akira **Mizuta**, KEK, Tsukuba, Ibaraki, Japan mizuta@post.kek.jp
Maryam **Modjaz**, New York University, New York, NY, USA mmodjaz@nyu.edu
Sergey **Moiseenko**, Space Research Institute, Moscow, Russia moiseenko@iki.rssi.ru
Thierry **Montmerle**, Institut d'Astrophysique de Paris, Paris, France montmerle@iap.fr
Mikio **Morii**, Tokyo Institute of Technology, Tokyo, Japan morii@hp.phys.titech.ac.jp
Takashi **Moriya**, IPMU, University of Tokyo, Kashiwa, Chiba, Japan takashi.moriya@ipmu.jp
Nidia **Morrell**, Las Campanas Observatory, La Serena, Chile nmorrell@lco.cl
Hiroki **Nagakura**, Kyoto University and Waseda University, Tokyo, Japan hiroki@heap.phys.waseda.ac.jp
Yujin **Nakagawa**, Waseda University, Tokyo, Japan yujin@aoni.waseda.jp
Ko **Nakamura**, National Astronomical Observatory of Japan, Mitaka, Tokyo, Japan nakamura.ko@nao.ac.jp
Ehud **Nakar**, Tel Aviv University, Tel Aviv, Israel udini@wise.tau.ac.il
Ken'ichiro **Nakazato**, Department of Physics, Tokyo University of Science, Noda, Chiba, Japan nakazato@rs.tus.ac.jp
Ana Maria **Nicuesa Guelbenzu**, Thueringer Landessternwarte Tautenburg, Tautenburg, Germany ana@tls-tautenburg.de

Yuu **Niino**, NAOJ, Mitaka, Tokyo, Japan yuu.niino@nao.ac.jp
Kenichi **Nishikawa**, UAH/CSPAR, Huntsville, AL, USA ken-ichi.nishikawa-1@nasa.gov
Ken'ichi **Nomoto**, IPMU, University of Tokyo, Kashiwa, Chiba, Japan nomoto@astron.s.u-tokyo.ac.jp
Takaya **Nozawa**, IPMU, University of Tokyo, Kashiwa, Chiba, Japan takaya.nozawa@ipmu.jp
Paul **O'Brien**, University of Leicester, Leicester, United Kingdom pto@star.le.ac.uk
Evan **O'Connor**, California Institute of Technology, Pasadena, CA, USA evanoc@tapir.caltech.edu
Felipe **Olivares**, Max-Planck-Institute for extraterrestrial Physics, Garching, Germany foe@mpe.mpg.de
Kazu **Omukai**, Kyoto U., Kyoto, Kyoto, Japan omukai@tap.scphys.kyoto-u.ac.jp
Oded **Papish**, Technion, Haifa, Israel papish@physics.technion.ac.il
Daniel **Perley**, Caltech, Pasadena, CA, USA dperley@astro.caltech.edu
Elena **Pian**, Scuola Normale Superiore di Pisa, Pisa, Italy elena.pian@sns.it
Graziella **Pizzichini**, IASF/INAF, Bologna, Italy pizzichini@iasfbo.inaf.it
Antonio de Ugarte **Postigo**, IAA-CSIC, DARK-NBI, Granada, Spain adeugartepostigo@gmail.com
Tyler **Pritchard**, Pennsylvania State University, PA, USA tapritchard@astro.psu.edu
Robert **Quimby**, IPMU, Kashiwa, Chiba, Japan robert.quimby@ipmu.jp
Itay **Rabinak**, Weizmann Institute of Science, Rehovot, Israel itay.rabinak@weizmann.ac.il
Jakub **Ripa**, Institute for the Early Universe, Ewha Womans University, Seoul, Korea ripa@ewha.ac.kr
Pete **Roming**, SwRI, San Antonio, TX, USA proming@swri.edu
Yoshihiko **Saito**, Tokyo Institute of Technology, Tokyo, Japan saitoys@hp.phys.titech.ac.jp
Takanori **Sakamoto**, NASA/CRESST/GSFC, Greenbelt, MD, USA Taka.Sakamoto@nasa.gov
Nir **Sapir**, Weizmann Institute of Science, Rehovot, Israel nir.sapir@weizmann.ac.il
Sandra **Savaglio**, Max-Planck Institute for Extraterrestrial Physics, Garching, Germany savaglio@mpe.mpg.de
Patricia **Schady**, MPE, Garching, Germany pschady@mpe.mpg.de
Yuichiro **Sekiguchi**, Yukawa Institute for Theoretical Physics, Kyoto, Kyoto, Japan sekig@yukawa.kyoto-u.ac.jp
Motoko **Serino**, RIKEN, Wako, Saitama, Japan motoko@crab.riken.jp
Sanshiro **Shibata**, Konan University, Kobe, Hyogo, Japan mn021006@center.konan-u.ac.jp
Joshua **Shiode**, UC Berkeley, Berkeley, CA, USA jhshiode@berkeley.edu
Brendan **Sinnott**, McMaster University, Ontario, Canada sinnotbp@physics.mcmaster.ca
Michael **Smith**, ESAC / Vega Space, Madrid, Spain Michael.Smith@sciops.esa.int
Nathan **Smith**, U. Arizona, Tucson, AZ, USA nathans@as.arizona.edu
Jesper **Sollerman**, Stockholm University, Stockholm, Sweden jesper@astro.su.se
Seongdeng **Song**, Tokyo Institute of Technology, Tokyo, Japan song@hp.phys.titech.ac.jp
Rhaana **Starling**, University of Leicester, UK, Leicester, United Kingdom rlcs1@star.le.ac.uk
Christopher **Stockdale**, Marquette University, Milwaukee, WI, USA chris.stockdale@mu.edu
Kohsuke **Sumiyoshi**, Numazu College of Technology, Numazu, Shizuoka, Japan sumi@numazu-ct.ac.jp
Yudai **Suwa**, Kyoto University, Kyoto, Japan suwa@yukawa.kyoto-u.ac.jp
Akihiro **Suzuki**, RESCEU, University of Tokyo, Tokyo, Japan suzuki@resceu.s.u-tokyo.ac.jp
Gilad **Svirski**, Tel Aviv University, Tel Aviv, Israel giladsv@gmail.com
Tamas **Szalai**, University of Szeged, Szeged, Hungary szaszi@titan.physx.u-szeged.hu
Francesco **Taddia**, Stockholm University, Stockholm, Sweden ftadd@astro.su.se
Koh **Takahashi**, University of Tokyo, Tokyo, Japan s102006@mail.ecc.u-tokyo.ac.jp
Hiroyuki **Takahashi**, National Astronomical Observatory of Japan, Mitaka, Tokyo, Japan takahashi@cfca.jp
Tomoya **Takiwaki**, National Astronomical Observatory of Japan, Mitaka, Tokyo, Japan takiwaki.tomoya@nao.ac.jp
Masaomi **Tanaka**, National Astronomical Observatory of Japan, Mitaka, Tokyo, Japan masaomi.tanaka@nao.ac.jp
Makoto **Tashiro**, Saitama University, Saitama, Saitama, Japan tashiro@phy.saitama-u.ac.jp
Christina **Thoene**, IAA Granada, Granada, Spain christina.thoene@gmail.com
Takahiro **Toizumi**, Tokyo Institute of Technology, Tokyo, Japan toizumi@hp.phys.titech.ac.jp
Kazuki **Tokoyoda**, Tokyo Institute of Technology, Tokyo, Japan tokoyoda@hp.phys.titech.ac.jp
Nozomu **Tominaga**, Konan University, Kobe, Hyogo, Japan tominaga@konan-u.ac.jp
Tomonori **Totani**, Dept. Astronomy, Kyoto University, Kyoto, Kyoto, Japan totani@kusastro.kyoto-u.ac.jp
Rachel **Tunnicliffe**, University of Warwick, Coventry, United Kingdom r.l.tunnicliffe@warwick.ac.uk
Ryuichi **Usui**, Tokyo Institute of Technology, Tokyo, Japan usui@hp.phys.titech.ac.jp
Schuyler **Van Dyk**, Spitzer Science Center/Caltech, Pasadena, CA, USA vandyk@ipac.caltech.edu
Nicolas **Vasquez**, ObservatorioAstronomico de Quito, Quito, Ecuador vasqpaz@gmail.com
Jorick **Vink**, Armagh Observatory, Armagh, United Kingdom jsv@arm.ac.uk
Roni **Waldman**, Hebrew University, Jerusalem, Israel roni181066@gmail.com
Joseph **Walmswell**, Institute of Astronomy, University of Cambridge, Cambridge, United Kingdomjjw49@cam.ac.uk
Daniel **Whalen**, Carnegie Mellon University, Pittsburgh, PA, USA dwhalen@lanl.gov
Annop **Wongwathanarat**, Max-Planck Institute for Astrophysics, Garching, Germany annop@mpa-garching.mpg.de
Chao **Wu**, NAOC,CAS, Beijing, China cwu@bao.ac.cn
Shoichi **Yamada**, Waseda University, Tokyo, Japan shoichi@waseda.jp
Kazutaka **Yamaoka**, Aoyama Gakuin University, Sagamihara, Kanagawa, Japan yamaoka@phys.aoyama.ac.jp
Ryo **Yamazaki**, Aoyama Gakuin University, Sagamihara, Kanagawa, Japan ryo@phys.aoyama.ac.jp
Yoichi **Yatsu**, Tokyo Institute of Technology, Tokyo, Japan yatsu@hp.phys.titech.ac.jp
Daisuke **Yonetoku**, Kanazawa University, Kanazawa, Ishikawa, Japan yonetoku@astro.s.kanazawa-u.ac.jp
Takashi **Yoshida**, University of Tokyo, Tokyo, Japan tyoshida@astron.s.u-tokyo.ac.jp
Patrick **Young**, Arizona State University, Tempe, AZ, USA patrick.young.1@asu.edu
Norhasliza **Yusof**, University of Malaya, Kuala Lumpur, Malaysia norhasliza@siswa.um.edu.my
Bing **Zhang**, University of Nevada Las Vegas, Las Vegas, NV, USA zhang@physics.unlv.edu

Post Symposium Address by the Local Organizing Committee

Dear colleagues,

It is a great pleasure to observe that the IAU symposium 279 has been a great success. As mentioned by Prof. Kawai at the opening address, we suffered from the huge earthquake which hit the northern part of Japan on March 11, 2011, that created many tragic stories. While the Kanto-area including Nikko and Tokyo was not seriously damaged, many side effects have been unavoidably there. Because of this, the symposium organizing efforts suffered from various uncertainties, including the postponement from the original schedule (April, 2011) to March, 2012. The difficulties have been overcome by the help and devotion of many people involved in the organization, and I want to take this opportunity to thank all them on behalf of the Local Organizing Committee.

We have very efficient, self-sacrificing people in the LOC: Katsuaki Asano, Yoichi Yatsu, Takuya Ohkawa from Tokyo Tech.; Masaomi Tanaka and Kei Kotake from NAOJ; Aya Bamba from Aoyama Univ.; Takeo Minezaki from U. Tokyo; and Toru Tamagawa from RIKEN. They all actively worked on the LOC, and without their great efforts the symposium would never have taken place. Pete Roming, Elena Pian, and Nobu Kawai, as SOC chairs, have provided great supervision of the LOC efforts.

There are many people who are not listed in the organizing committee despite their great contributions to the symposium. You saw many students (from Tokyo Tech. and IPMU) assisting during the symposium sessions and coffee breaks. I'd like to thank the secretaries from these institutes and the company Quality Management – especially Yasuko Konagai, Rie Ujita, Yasuhiro Kato, and Noriko Kagaya who provided the critical help necessary for the symposium. I also want to express my gratitude to the IAU secretaries for their continuous help, especially on how to proceed with the symposium organization after the earthquake on March, 2011. I am happy to see that Thierry Montmerle from the IAU, whose advice was most critical for the reorganization, made it to the symposium. There are many institutions who supported the symposium, as you can see in the preface and the opening address of this proceedings edition.

I also want to emphasize that the people in Nikko area were very helpful during the symposium organization. The hotel Nikko Senhime Monogatari provided the beautiful venue, and the staff was very flexible and helpful in providing the warm and stimulating atmosphere. As you might have noticed, everywhere you went people in the Nikko area were all friendly and always willing to provide help - beautiful landscapes and people. I am sure that with all that Nikko has to offer, that shortly the city will fully recover from the economic damage they suffered as a place of tourism after the earthquake - I hope that the symposium had a good effect on this respect as well.

Last but not least, I want to remark that the symposium was successful thanks to the excellent talks and active discussions from all the participants. I hope that every participant feels the same way as I do and that they enjoyed the symposium very much - thanks very much to all of you, and I look forward to seeing you again.

Keiichi Maeda, LOC chair
20 June 2012

Welcome Address

I would like to welcome all of you to IAU Symposium 279 titled "Death of Massive Stars." The idea of having a symposium in Japan that deals with supernovae and gamma-ray bursts together was originally suggested by Pete Roming, who is here to chair the session today. I enthusiastically agreed with his idea, and suggested to have it here in Nikko. This is actually my home town; I spent my childhood here. Some of my old friends still live here, and the owner of this hotel is actually my classmate from the primary school.I am really happy to see the impressive program in this symposium thanks to the SOC. In particular Elena Pian did a superb job in assembling the ideas. I hope you enjoy the discussions here, and gain some new research ideas during this meeting.

As all of you are aware, this symposium was originally scheduled in April of last year, but was postponed due to the huge earthquake that hit northern Japan exactly one year ago yesterday. Because of the earthquake, and the subsequent Tsunami, which is the largest in Japanese history, lives of 20,000 people were lost.Here, I would like to propose a moment of silence for them with our eyes closed.

Thank you.

Immediately after the earthquake, eastern Japan, including Tokyo and Nikko was in severe confusion. In two weeks or so, the situation in Tokyo and Nikko was more or less back to normal, except for some inconvenience to save electric power. However, we realized that the accident at the Fukushima Nuclear power plant was a very strong concern in and outside Japan, and we ourselves did not know how it was going to be settled. After discussion with the IAU, we decided to postpone the symposium to now. This is rather unusual, because IAU symposia are selected on a yearly basis, and this symposium had been selected for 2011. IAU allocates some yearly budgets for symposia and moving a symposium to the next year causes non-trivial arrangements at IAU. I would like to thank the IAU for supporting this symposium, and also all the efforts they made to make it happen despite all of the trouble.

Here in Japan, we also obtained funding for supporting this symposium, and it also was approved for the fiscal year 2011, which ends this month. That is why we set the new date in March. The weather in March in Nikko is not so nice as in April, as you may have already noticed. Furthermore, we have some snow today It is actually, uncommon even here in Nikko in March. However, I hope that the shrines here, that are selected as UNESCO's World Cultural Heritage, look more beautiful with snow, and I hope you have time to see them during this week. Wednesday afternoon is allocated to relax and see this beautiful and interesting area.

Lastly, I thank the support from the IAU; the Ministry of Education, Culture, Sport, Science and Technology of Japan; Japan Society for Promotion of Science; Institute for Physics and Mathematics of the Universe of the University of Tokyo; Global COE Program "Quantum Physics and Nanoscience," Tokyo Institute of Technology; and Southwest Research Institute. It is also supported by the Astronomical Society of Japan.

Please enjoy the meeting.

Nobuyuki Kawai, co-chair SOC
Nikko, 12 March 2012

Death of Massive Stars: Supernovae and Gamma-Ray Bursts
Proceedings IAU Symposium No. 279, 2012
P. Roming, N. Kawai & E. Pian, eds.

doi:10.1017/S1743921312012604

Final Fates of Massive Stars

Ken'ichi Nomoto

Kavli Institute for the Physics and Mathematics of the Universe, University of Tokyo,
5-1-5 Kashiwanoha, Kashiwa, Chiba 277-8583, Japan
email: nomoto@astron.s.u-tokyo.ac.jp

Abstract. Massive stars are thought to play important roles in the early evolution of the Universe. In this paper, we first classify the final fates of massive stars into 7 cases according to their mass ranges. These variations of the final fate may correspond to the observed large diversities of supernova properties, such as extremely faint and extremely luminous (superluminous) supernovae, and the extremely energetic hypernovae. We then focus on the properties of the peculiar superluminous Type Ic supernova 1999as. We examine radioactive decay models, magnetar models, and circumstellar interaction models for the light curve of SN 1999as. We find that these models are not quite successful, and thus it is crucially important to improve these models to clarify the final fates of massive stars.

Keywords. supernova, hypernova, superluminous, nucleosynthesis

1. Final Fates of Massive Stars

The final stages of massive star evolution, supernova properties, and their chemical yields depend on the progenitor's main-sequence masses M (e.g., Arnett 1996, Filippenko 1997, Nomoto *et al.* 2006, Smartt 2009). Here we call some specific supernovae (SNe) as follows. In terms of the kinetic explosion energy E, we use "Hypernovae" for such energetic SNe as $E_{51} = E/10^{51}\mathrm{erg} > 10$. In terms of brightness, we use "Faint SNe" for low luminosity SNe, and "Superluminous SNe (SLSNe)" for SNe brighter than -21 mag at maximum (Quimby in this volume).

The following mass ranges are set by various types of criteria, based on some combinations of observations and models. But the criteria and critical masses are not quite systematic yet, and should still be regarded as working hypothesis (Nomoto *et al.* 2009, 2010).

(1) $M_{\rm up}$ **- 10** $M_\odot$ **stars: Faint supernovae**: These stars become electron capture SNe because their degenerate O+Ne+Mg cores collapse due to electron capture. $M_{\rm up} \sim 9\pm0.5 M_\odot$ depending on the mass loss rate on the super-AGB phase thus on the metallicity (e.g., Pumo *et al.* 2009).

(2) **10 - 13** $M_\odot$ **stars: Faint Supernovae**: These stars undergo Fe-core collapse to form a neutron star (NS) after the phase of strong Neon shell-flashes (Nomoto & Hashimoto 1988). Their Fe core is relatively small, and the resultant SNe tend to be faint (Smartt 2009).

(3) **13** $M_\odot$ **-** $M_{\rm BN}$ **stars: Normal Supernovae**: These stars undergo Fe-core collapse to form a NS, and produce significant amount of heavy elements from α-elements and Fe-peak elements. The boundary mass between the NS and black hole (BH) formation, $M_{\rm BN} \sim 25 M_\odot$, is only tentative.

(4) $M_{\rm BN}$ **- 80** $M_\odot$ **stars: Hypernovae and Faint Supernovae**: These stars undergo Fe-core collapse to form a BH. SNe seem to bifurcate into two branches, Hypernovae and Faint SNe. If the BH has little angular momentum, little mass ejection would take

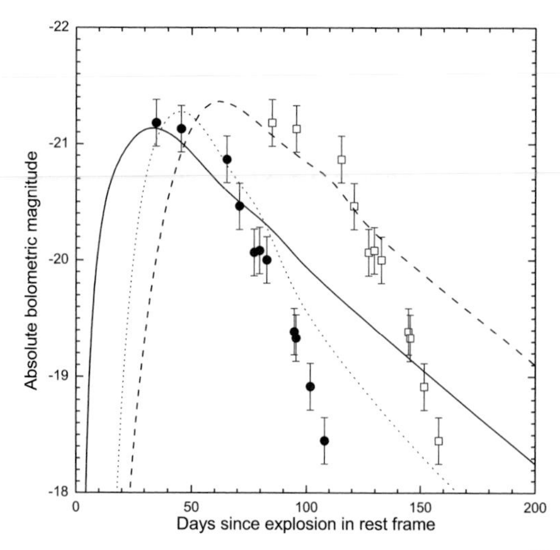

Figure 1. Model LCs with 30 $M_\odot$ ejecta and a kinetic energy of 3×10^{52} erg, compared with the observed bolometric LC of SN 1999as, with the first data epoch of $t_0 = 35$ days (*filled circles*) and of $t_0 = 85$ days (*open squares*) are shown (Deng *et al.* 2009). The model LC to approximately fit the former was synthesized assuming $\sim 7\ M_\odot$ ^{56}Ni (*solid line*), while the latter $\sim 9.5\ M_\odot$ (*dashed line*).

place and be observed as Faint SNe. On the other hand, a rotating BH could eject a matter in a form of jets to make a Hypernova. The latter explosions produce a large amount of heavy elements from α-elements and Fe-peak elements. Nucleosynthesis in these jet-induced explosions is in good agreement with the abundance patterns observed in extremely metal-poor stars.

(5) **80 - 140 $M_\odot$ stars: Superluminous SNe (SLSNe)**: These massive stars undergo nuclear instabilities and associated pulsations (ϵ-mechanism) at various nuclear burning stages depending on the mass loss and thus metallicity. Eventually, these stars undergo Fe-core collapse. Depending on the angular momentum, Hypernova-like energetic SNe could occur to produce large amount ^{56}Ni. (Because of the large ejecta mass, the expansion velocities may not be high enough to form a broad line features.)

Thanks to the large E and ^{56}Ni mass, SNe in this mass range could be SLSNe. The upper limit of ^{56}Ni mass has been estimated to be $\sim 10 M_\odot$. The possible presence of circumstellar matter (CSM) leads to an energetic SN IIn. Pulsation could also cause superluminous events.

(6) **140 - 300 $M_\odot$ stars: SLSNe**: If these very massive stars (VMS) do not lose much mass, they become pair-instability supernovae (PISN). The star is completely disrupted without forming a BH and thus ejects a large amount of heavy elements, especially ^{56}Ni of mass up to $\sim 40 M_\odot$. Radioactive decays could produce SLSNe (Gal-Yam in this volume).

(7) **Stars with $M \gtrsim 300 M_\odot$: SLSNe**: These VMSs are too massive to be disrupted by PISN but undergo core collapse (CVMS), forming intermediate-mass black holes (IMBHs). Some mass ejection could be possible, associated with the possible jet-induced explosion, which could form a superluminous SNe (SLSNe).

Here we focus on the light curve models for superluminous Type Ic supernova 1999as (Deng *et al.* 2009, Nugent *et al.* 2009) to understand the explosion mechanisms of very massive stars. We find that the light curve models for SN 1999as are not quite successful and need further improvement.

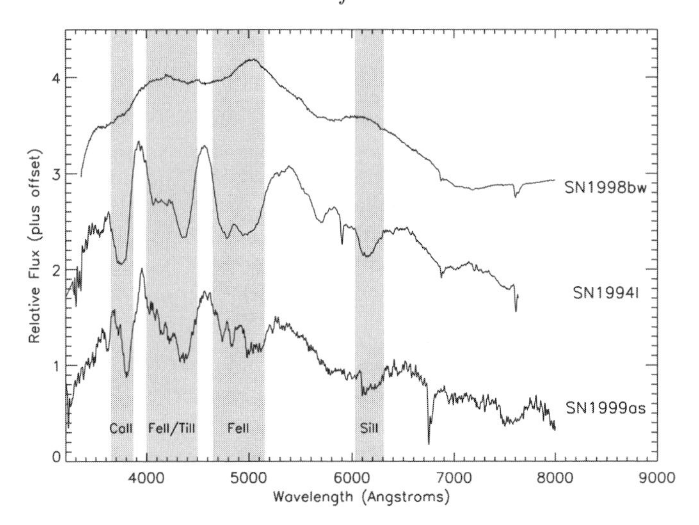

Figure 2. The spectrum of SN 1999as 25 days after discovery compared to SN 1998bw and SN 1994I (Kasen 2004).

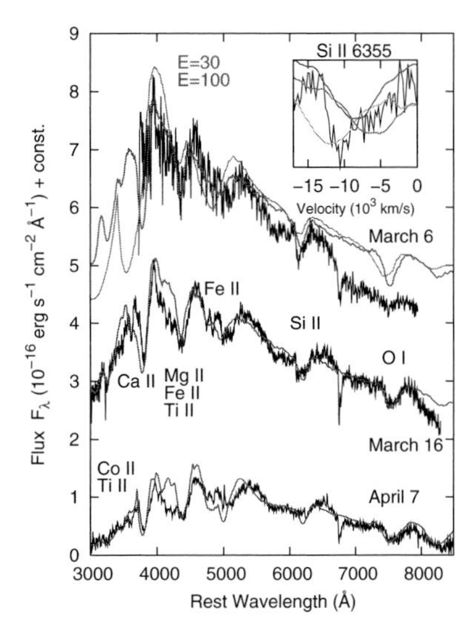

Figure 3. Synthetic spectra compared with the observed spectra taken on 1999 March 6, 16 and April 7 (from *top* to *bottom*). The *red* lines show synthetic spectra for the model with 30 $M_\odot$ ejecta and a kinetic energy of 3×10^{52} ergs (Deng *et al.* 2009).

2. Superluminous Type Ic Supernova 1999as

Supernova (SN) 1999as, discovered in a galaxy of $z = 0.127$ (Knop *et al.* 1999), had long been the most luminous SN in terms of absolute brightness (Richardson *et al.* 2002, Hatano *et al.* 2001). With a peak magnitude of $M_{\rm V} \sim -21.5$ (Nugent *et al.* 2009; $m - M = 38.8$), it is ~ 8 times brighter than GRB-supernova 1998bw, ~ 3 times brighter than Type Ia/IIn (IIa) SN 2002ic (e.g., Deng *et al.* 2004), and ~ 6 times brighter than normal Type Ia SNe. On the other hand, SN 1999as shows overall spectral similarities to normal Type Ic SN (SN Ic) 1994I (Fig. 2).

3. Radioactive Decay Models

The optical LC of SN 1999as (Fig. 1), like other Type Ib/Ic SNe, was possibly powered by γ-rays and positrons released in the radioactive decays of ^{56}Ni, i.e., ^{56}Ni$\rightarrow^{56}$Co$\rightarrow^{56}$Fe.

3.1. *Mass of Radioactive ^{56}Ni*

SN 1999as had a peak $M_{\rm bol} \sim -21.2$ (Nugent *et al.* 2009), while the spectral modeling can constrain its peak epoch to be $\sim 35-85$ days after explosion. In comparison, GRB-SN 1998bw had a peak $M_{\rm bol} \sim -18.7$ and $t \sim 15$ days at light maximum, and its ^{56}Ni mass has been modeled to be ~ 0.4 M$_\odot$ (Nakamura *et al.* 2001). Then one gets $M(^{56}{\rm Ni}) \sim 5-9$ M$_\odot$ for SN 1999as.

From the viewpoint of explosive nucleosynthesis, such a large amount of fresh ^{56}Ni clearly indicates that the SN explosion was extremely energetic. In a core-collapse SN, most ^{56}Ni are the outcome of complete Si-burning in the region where temperatures can reach $\gtrsim 5 \times 10^9$K (i.e., $T_{\rm Si}$) as the SN shock wave propagates through it (Arnett 1996). The region behind the shock front is radiation dominated and has roughly the same temperature (e.g., Shigeyama *et al.* 1988). Thus, complete Si-burning is confined below a radius $R_{\rm Ni}$ in the pre-explosion star for a given explosion energy, E, and $E \sim (4\pi/3)R_{\rm Ni}^3 a T_{\rm Si}^4$, or $R_{\rm Ni} \sim 3700(E/10^{51}{\rm ergs})^{1/3}$ km.

As the model for SN 1999as, the explosion of the C+O core of the $M_{\rm MS} = 80$ M$_\odot$ star leads to a final ejecta mass of ~ 30 M$_\odot$. As the energy rises from 1×10^{52} to 8×10^{52} ergs, $M(^{56}{\rm Ni})$ increases from ~ 1 to ~ 3 M$_\odot$ for $M_{\rm MS} = 50$ M$_\odot$, and from ~ 1.5 to ~ 7 M$_\odot$ for $M_{\rm MS} = 80$ M$_\odot$.

For $M_{\rm MS} = 100$M$_\odot$ star, Type Ic SN ejecta mass is ~ 40 M$_\odot$. This model synthesize ~ 7 M$_\odot$ ^{56}Ni for only $E \sim 4 \times 10^{52}$ ergs. The synthetic spectra will be similar for the similar $E/M_{\rm ej}$.

3.2. *Synthetic Spectra of SN 1999as*

The spectra of SN 1999as, which were taken on 1999 March 6, 16 and April 17 (Nugent *et al.* 2009), are calculated for the ~ 30 M$_\odot$ ejecta produced from the explosion of the C+O core of the 80 M$_\odot$ star with a kinetic energy of 3×10^{52} ergs.

The synthesized spectra are shown in Figure 3, and compared with the observed SN spectra. The overall fitness is satisfactory. All the absorption features are identified, being attributed to the P-Cygni lines of O I, Mg II, Si II, Ca II, Ti II, and Fe II.

The Si abundance is higher than the solar value, while Fe and Co are ~ 10 times more abundant than the solar values. This is evidence of explosive-synthesis products being mixed out into the ejecta above $8,000$ km s^{-1}.

The supernova explosion date is also a key parameter which was estimated through spectrum modeling. The observed bolometric luminosity can be taken as approximately proportional to $(v_{\rm ph}t)^2 T_{\rm eff}^4$, where t is the rest-frame epoch with respect to explosion and $T_{\rm eff}$ the effective temperature. $v_{\rm ph}$ can be measured by fitting the absorption shapes of the observed P-Cygni lines (Jeffery & Branch 1990), so t is derived. A rest-frame epoch is estimated as $t_1 \sim 50-100$ days after explosion for the March 6 spectrum. The best-fitting model spectra discussed above correspond to $t_1 = 80$ days.

3.3. *Light Curve Models*

Both the LC assuming $t_0 = 35$ days after explosion and that of $t_0 = 85$ days are shown in Figure 1. The relative flatness of LC during the first month suggests that the SN was possibly discovered not far from its actual light maximum.

The bolometric LCs of 30 M$_\odot$ ejecta with 3×10^{52} ergs are shown in Figure 1. To fit the slowest LC case, i.e., $t_0 = 85$ days, a total amount of ~ 9.5 M$_\odot$ ^{56}Ni is required, which

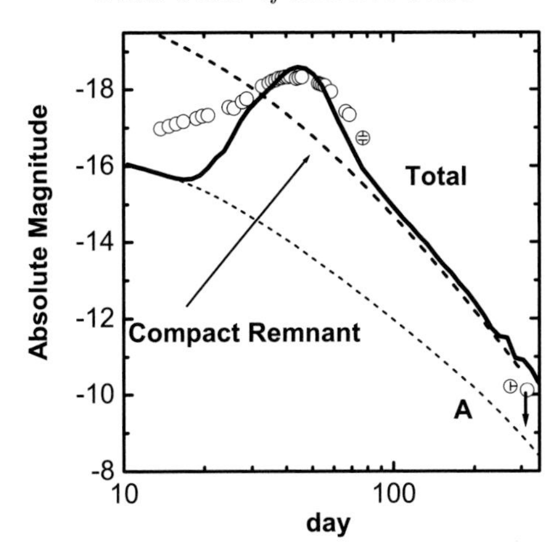

Figure 4. Synthetic light curve with the putative energy input from the central remnant (thick solid) (Maeda *et al.* 2007). The input luminosity of the remnant (thick dashed) and the contribution from the blob (thin dashed) are also shown.

were assumed by us to fill the inner part of the ejecta up to ~ 7000 km s^{-1}. The LC (*dashed line*) peaks ~ 10 days before t_0, attributable to the existence of ^{56}Ni at relatively high velocities.

3.4. *Fast Decline of the Light Curve*

The above radioactive decay models fail to reproduce the fast LC drop after $\sim t_0 + 60$ days. The observed luminosity decayed at a rate of $\gtrsim 0.06$ mag/day, fast as compared with other Type Ic SNe during similar epochs ($\lesssim 0.02$ mag/day; Tomita *et al.* 2006). In the CSM-interacting Type Ib SN 2006jc, dust formation shifted the bulk of the bolometric luminosity to the infrared (Tominaga *et al.* 2008, Nozawa *et al.* 2008). However, a pronounced red continuum should have been seen, which characterized the optical spectra of SN 2006jc at corresponding epochs (Smith, Foley, & Filippenko 2008).

4. Magnetar Models

We note that a fast late LC drop has also been observed in the peculiar Type Ib SN 2005bf, which shows two peaks in the optical light curve (Fig. 4 and references therein). The first peak luminosity is reproduced if $\sim 0.07 M_\odot$ of ^{56}Ni is mixed out to outer layers (as a blob), and the second peak (at ~ 40 days) can be reproduced by the radioactive decay of $\sim 0.32 M_\odot$ ^{56}Ni (Tominaga *et al.* 2005).

At ~ 270 days since the explosion, however, the absolute R-band magnitude of SN 2005bf is ~ -10.2 (Maeda *et al.* 2007). It is very faint as compared to other SNe Ib/c, at least by 2 magnitudes (e.g., by 3 magnitudes fainter than SN 1998bw at a similar epoch). If the R magnitude is close the bolometric magnitude, ^{56}Ni required to fit the luminosity is only $\sim 0.03 - 0.08 M_\odot$. Why this is much smaller than $\sim 0.32 M_\odot$ that reproducing the early phase peak luminosity is a serious question.

Encountered by this difficulty in the ^{56}Ni heating model, Maeda *et al.* (2007) have adopted an alternative scenario in which the heating source is a newly born, strongly magnetized neutron star (a magnetar). A synthetic light curve is shown in Figure 4 (Maeda *et al.* 2007). The input luminosity of the remnant (thick dashed) and the contribution from the ^{56}Ni blob (thin dashed) for the 1st peak are shown.

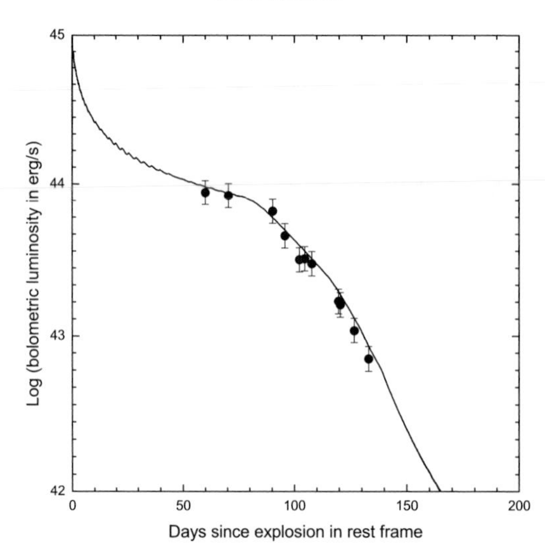

Figure 5. The LC of a CSM-interaction model (*solid line*) compared with the observed LC (*filled circles*; assuming a first data epoch of $t_0 = 60$ days). The model ejecta have $\sim 10\ M_\odot$ and a kinetic energy of 1×10^{52} erg, while the model CSM have a power-law density profile index of $n = 2$ for the inner $\sim 0.90\ M_\odot$ and of $n = 5$ for the outer $\sim 0.25\ M_\odot$ (Deng *et al.* 2009).

The large second peak luminosity and the rapid decline after the second peak can qualitatively be explained. The model parameters are: $L_0 = 8 \times 10^{43}$ erg s^{-1}, $t_0 = 60^{\rm d}$, $\beta = 4$, which corresponds to a pulsar with $B_{\rm mag} \sim 3 \times 10^{14}(P_0/10\ {\rm ms})^2\sqrt{0.1/f_{\rm rad}}$ gauss, using the dipole radiation formula. Here P_0 is the initial spin period, t_0 is the characteristic time scale of the high energy input, and $f_{\rm rad}$ is a fraction of energy going into the radiation. A relatively large breaking index $\beta = 4$ is required to reproduce the large contrast between the peak and the tail. The total energy injection with these parameters is $\sim 7 \times 10^{50}(f_{\rm rad}/0.1)^{-1}$ erg, a fraction of which might be consumed to increase the kinetic energy of the SN ejecta to $E_{51} = 1 - 1.5$ and/or to develop the pulsar nebula in the early phase. For the LC of SN 1999as, it would be interesting to apply the magnetar model (also Kasen & Bildsten 2010, Woosley 2010).

5. Circumstellar Interaction Model

As seen in Figure 2, narrow spectral features are observed around 4200 and 4800 Å in SN 1999as (line width ~ 1000 km s^{-1}). These features seem to be high-velocity Fe lines with $v \sim 12,000$ km s^{-1}. We note that a jet-like explosion may result in a high-velocity shell of explosive-synthesized Fe-group elements along the jet directions (Khokhlov *et al.* 1999, Maeda *et al.* 2002). Multi-dimensional radiative transfer calculations (e.g., Höflich *et al.* 1996, Tanaka *et al.* 2007) are required to investigate such scenarios.

Kasen (2004) modeled high-velocity narrow Fe II features by assuming the existence of a thin, detached shell, and suggested circumstellar interactions for the LC of SN 1999as. Deng *et al.* (2009) calculated the circumstellar interaction model for the LC of SN 1999as in the same way as done by Suzuki & Nomoto (1995). The total ejecta mass is $\sim 10\ {\rm M}_\odot$. The collision starts at a distance of $r_0 = 2 \times 10^{14}$ cm, where the CSM density is ρ_0, and the CSM has a power-law density profile $\rho = \rho_0(r/r_0)^{-n}$.

High CSM densities and a large explosion energy are required to reproduce the brightness of SN 1999as. The density in the shocked ejecta and CSM is so high that the reverse and forward shocks are radiative to form a dense cooling shell. The regions heated by the forward and reverse shocks emit X-rays, a large part of which is converted into UV and optical light.

The best-fitting model LC is shown in Figure 5 (*dotted line*), where the observed LC is drawn adopting an epoch of 60 days for the first data point. The model explosion energy is $\sim 1 \times 10^{52}$ ergs. The CSM parameters are $\rho_0 \sim 3.5 \times 10^{-13}$ g cm^{-3}, $n = 2$ for the inner 0.90 $M_\odot$, and $n = 5$ for the outer 0.25 $M_\odot$. The steep outer density profile is required to reproduce the late fast LC drop. The total interacting-CSM mass is then $\sim 1.15\ M_\odot$. The corresponding mass-loss rate would be unusually large, e.g., $\sim 0.03\ M_\odot$ yr^{-1} within ~ 30 years prior to the SN explosion if adopting a wind velocity of ~ 100 km s^{-1}.

Further refinement is necessary by considering that the CSM formed with such an extensive mass loss could be optically thick. The optical thickness would improve the early part of the model LC.

6. Discussion

The optical spectra and light curve of the superluminous Type Ic SN 1999as are modeled. First we apply the radioactive model, which requires $M_{\rm ej} \sim 30$ M$_\odot$ ($M_{\rm MS} = 80$ M$_\odot$) and $E \sim 3 \times 10^{52}$ ergs. The high explosion energy is consistent with ~ 6 M$_\odot$ ^{56}Ni that are required to power the exceptional brightness of the SN. These model values of $M_{\rm ej}$ and E are larger than the GRB-supernovae, i.e., SN 1998bw (~ 10 M$_\odot$ and $\sim 5 \times 10^{52}$ ergs; Nakamura *et al.* 2001), SN 2003dh (~ 7 M$_\odot$ and $\sim 3.5 \times 10^{52}$ ergs; Deng *et al.* 2005), and SN 2003lw (~ 13 M$_\odot$ and $\sim 6 \times 10^{52}$ ergs; Mazzali *et al.* 2006), although the energy value is more or less comparable.

It has been demonstrated that the core-collapse explosions of massive metal-poor stars (Umeda & Nomoto 2008) can synthesize the required large amount of ^{56}Ni of several M$_\odot$. Nugent *et al.* (2009) did find that the host galaxy of SN 1999as has a relatively low metallicity.

Pair-instability SNe from $M_{\rm MS} \sim 140 - 270$ M$_\odot$ (Barkat, Rakavy, & Sack 1967, Bond, Arnett, & Carr 1984, Umeda & Nomoto 2002, Heger & Woosley 2002) have very large $M_{\rm ej}/E$, which results in very late LC peak epochs, very slow brightness decline after the peak, and relatively narrow-lined spectra.

However, the above radioactive decay models fail to reproduce the fast decline of the light curve of SN 1999as. We thus need to explore possible alternative models.

The magnetar model has parameters to fit to the observed light curve. However, the progenitor's main-sequence mass should be smaller than ~ 25 M$_\odot$ to produce a magnetic neutron star rather than a black hole.

The strong circumstellar interaction models for SN 1999as would require the existence of an optically-thick, non-Hydrogen shell. It's configuration would be non-spherical to avoid the contamination of the Type Ic spectral features which dominated the observed spectra. However, the presence of non-Hydrogen CSM is not trivial. The collision of massive stars (in particular, Wolf-Rayet stars) in a dense stellar cluster can make a massive C+O star surrounded by a massive non-Hydrogen CSM (Suzuki *et al.* 2007).

The above models are not quite successful in explaining the properties of SN 1999as, in particular, the late fast LC drop and some narrow spectral features. It is crucially important to improve these models to clarify the final fates of massive stars.

I would like to thank Peter Nugent, Jinsong Deng, and many collaborators for the work on SN 1999as. This research has been supported in part by World Premier International Research Center Initiative, MEXT, and by the Grant-in-Aid for Scientific Research of the JSPS (23540262) and MEXT (22012003, 23105705), Japan.

References

Aldering, G., *et al.* 2006, *ApJ*, 650, 510
Arnett, W. D. 1996, *Nucleosynthesis and Supernovae* (Princeton: Princeton Univ. Press)
Barkat, Z., Rakavy, G., & Sack, N. 1967, *Phys. Rev. Letters*, 18, 379
Bond, J. R., Arnett, W. D., & Carr, B. J. 1984, *ApJ*, 280, 825
Deng, J., *et al.* 2002, *ApJ*, 605, L37
Deng, J., Tominaga, N., Mazzali, P. A., Maeda, K., & Nomoto, K. 2005, *ApJ*, 624, 898
Deng, J., *et al.* 2009, in preparation
Filippenko, A. V. 1997, *ARAA*, 35, 30
Hatano, K., Maeda, K., Deng, J., Nomoto, K., *et al.* 2001, in ASPC Conf. Proc. 251., New Century of X-ray Astronomy, ed. H. Inoue & H. Kunieda (San Francisco: ASPC), 244
Herger, A. & Woosley, S. E. 2002, *ApJ*, 567, 532
Höflich, P., Wheeler, J. C., Hines, D. C., & Trammell, S. R. 1996, *ApJ*, 459, 307
Jeffery, D. J. & Branch, D. 1990, in *Jerusalem Winter School for Theor. Phy. V. 6, Supernovae*, ed. J. C. Wheeler, T. Piran, & S. Weinberg (Singapore: World Scientific), 149
Kasen, D. 2004, Ph.D. Thesis, Univ. California at Berkeley
Kasen, D. & Bildsten, L. 2010, *ApJ*, 717, 245
Khokhlov, A. M., Höflich, P. A., Oran, E. S., *et al.* 1999, *ApJ*, 524, L107
Knop, R., *et al.* 1999, *IAUC*, 7128
Maeda. K., Nakumura, T., Nomoto, K., *et al.* 2002, *ApJ*, 565, 405
Maeda, K. *et al.* 2007, *ApJ*, 666, 1069
Mazzali, P. A., *et al.* 2006, *ApJ*, 645, 1323
Nakamura, T, Mazzali, P. A., Nomoto, K., & Iwamoto, K. 2001, *ApJ*, 550, 991
Nomoto, K. & Hashimoto, M 1988, *Phys. Rep.*, 163, 13
Nomoto, K., *et al.* 2006, *Nuclear Phys A* 777, 424 (astro-ph/0605725)
Nomoto, K., *et al.* 2009, in *IAU Symp. 254, The Galaxy Disk in Cosmological Context*, ed. J. Andersen, *et al.* (Cambridge: Cambridge Univ. Press), 355 (arXiv: 0901.4536)
Nomoto, K., *et al.* 2010, in *IAU Symp. 265, Chemical Abundances in the Universe: Connecting First Stars to Planet*, ed. K. Cunha, *et al.* (Cambridge: Cambridge Univ. Press), 34
Nozawa, T., *et al.* 2008, *ApJ*, 684, 1343
Nugent, P., *et al.* 2009, private communication
Pumo, M. L., *et al.* 2009, *ApJ*, 705, L138
Quimby, R. M., Aldering, C., Wheeler, J. C., *et al.* 2007, *ApJ*, 668, L99
Richardson, D., Branch, D., Casebeer, D., *et al.* 2002, *ApJ*, 123, 745
Shigeyama, T., Nomoto, K., & Hashimoto, M. 1988, *A&A*, 196, 141
Smartt, S. J. 2009, *ARAA*, 47, 63
Smith, N., Foley, R. J., & Filippenko, A. V., 2008, *ApJ*, 680, 568
Suzuki, T. & Nomoto, K. 1995, *ApJ*, 455, 658
Suzuki, T. K., Nakasato, N., Baumgardt, H., *et al.* 2007, *ApJ*, 668, L19
Tanaka, M., Maeda, K., Mazzali, P. A., & Nomoto, K. 2007, *ApJ*, 668, L19
Tomita, H., *et al.* 2006, *ApJ*, 644, 400
Tominaga, N., *et al.* 2005, *ApJ*, 633, L97
Tominaga, N., *et al.* 2008, *ApJ*, 687, 1208
Umeda, H. & Nomoto, K. 2002, *ApJ*, 565, 385
Umeda, H. & Nomoto, K. 2008, *ApJ*, 673, 1014
Woosley, S. E., 2010, *ApJ*, 719, L204

Death of Massive Stars: Supernovae and Gamma-Ray Bursts
Proceedings IAU Symposium No. 279, 2012
P. Roming, N. Kawai & E. Pian, eds.

doi:10.1017/S1743921312012616

Environments of massive stars and the upper mass limit

Paul A. Crowther

Department of Physics and Astronomy, University of Sheffield,
Hounsfield Road, Sheffield, United Kingdom, S3 7RH
email: Paul.Crowther@sheffield.ac.uk

Abstract. The locations of massive stars ($\geqslant 8M_{\odot}$) within their host galaxies is reviewed. These range from distributed OB associations to dense star clusters within giant H II regions. A comparison between massive stars and the environments of core-collapse supernovae and long duration Gamma Ray Bursts is made, both at low and high redshift. We also address the question of the upper stellar mass limit, since very massive stars (VMS, $M_{\rm init} \gg 100M_{\odot}$) may produce exceptionally bright core-collapse supernovae or pair instability supernovae.

Keywords. stars: early-type, stars: supernovae: general, stars: Wolf-Rayet, ISM: HII regions, galaxies: star clusters, galaxies: ISM

---

1. Environments of Massive Stars

Massive star formation in the Milky Way spans a broad spectrum, from dispersed, low intensity OB associations to concentrated, high intensity starbursts. Within a few hundred parsec of the Sun, high mass stars ($M_{\rm init} \geqslant 8M_{\odot}$) are rather distributed, typically located in loose, spatially extended OB associations (de Zeeuw *et al.* 1999). A notable exception is Orion OB1, which hosts the Orion Nebula Cluster (ONC), responsible for our closest H II region. Further afield, large numbers of massive stars are associated with relatively intense bursts of star formation such as the high mass, compact clusters (Trumpler 14, 16) within the Carina Nebula giant H II region.

1.1. *Star clusters*

It is generally accepted that the majority of stars form within star clusters (Lada & Lada 2003), although recent evidence suggests star formation occurs in a continuum of stellar densities (e.g. Evans *et al.* 2009). Nevertheless, given their short-lifetimes (3–50 Myr) only a few percent of massive stars appear genuinely 'isolated' (de Wit *et al.* 2005) such that they either tend to be associated with their natal cluster or are plausible runaways from it†.

According to Weidner & Kroupa (2006), there is a tight relation between cluster mass, and the most massive star formed within the cluster, although this remains controversial (Calzetti *et al.* 2010, Eldridge 2012). Examples of well known star clusters spanning a range of masses are shown in Table 1, all of which are sufficiently young ($< 1 - 2$ Myr) that the most massive stars have yet to end their lives. We include the most massive star in each cluster, which increases towards the highest mass clusters.

If there is a relation between a cluster and its most massive star, the galaxy-wide stellar initial mass function (IMF) will also depend upon the cluster mass function and

† Runaways may be ejected from their cluster either dynamically during the formation process or at a later stage after receiving a kick following a supernova explosion in a close binary system.

Table 1. Selected young star clusters spanning a range of masses, $M_{\rm cl}$, for which (initial) masses of the highest mass stars, $M_{*,init}$, have been determined.

Cluster	$M_{\rm cl}/M_{\odot}$	Ref	Star	$M_{*,init}/M_{\odot}$	Ref
ρ Oph	$\sim 10^2$	a	ρ Oph Source 1	9	a
ONC	1.8×10^3	b	θ^1 Ori C	39±6	c
NGC 3603 (HD 97950)	$\sim 10^4$	d	NGC 3603-B	166±20	e
R136 (HD 38268)	5×10^4	e	R136a1	320^{+100}_{-40}	e

(a) Wilking *et al.* (1989); (b) Hillenbrand & Hartmann (1998); (c) Simón-Díaz *et al.* (2006)
(d) Harayama *et al.* (2008); (e) Crowther *et al.* (2010)

the range of cluster masses, as set out by Pflamm-Altenburg *et al.* (2007). In normal star-forming galaxies the cluster mass distribution follows a power law with index −2, albeit this is truncated at high mass depending upon the rate of star formation (Gieles 2009). Consequently, similar absolute numbers of stars are formed in low mass ($M_{\rm cl} \sim 10^2 M_{\odot}$), intermediate mass ($\sim 10^3 M_{\odot}$) and high mass ($\sim 10^4 M_{\odot}$) clusters, while high mass stars should be rare in the former. This is not always the case, since star formation in some nearby dwarf irregular starbursts is strongly biased towards a few very high mass clusters (e.g. NGC 1569, Hunter *et al.* 2000).

1.2. *H II regions and star formation rates*

Because of the (universal?) Salpeter IMF slope, the overall statistics of massive stars in galaxies will be heavily biased towards 8–20 $M_{\odot}$ (early B-type) stars. However, the most frequently used indicator of active star formation is nebular hydrogen emission (e.g. Hα) from gas associated with young, massive stars. The Lyman continuum ionizing output from hot, young stars is a very sensitive function of temperature (stellar mass), such that one O3 dwarf ($\sim$75 $M_{\odot}$) will emit more ionizing photons than 25,000 B2 dwarfs ($\sim$9 $M_{\odot}$, Conti *et al.* 2008). Therefore, H II regions are biased towards high mass (O-type) stars with >20 $M_{\odot}$ since B stars will produce extremely faint H II regions.

Beyond several Mpc, current sensitivies limit detections of H II regions to relatively bright examples, involving several ionizing early O-type stars (Pflamm-Altenburg *et al.* 2007). Still, the Hα luminosity of bright H II regions can be converted into the corresponding number of Lyman continuum ionizing photons, for which the number of equivalent O7 dwarf stars, N(O7V), serves as a useful reference (Vacca & Conti 1992), as indicated in Table 2. Kennicutt *et al.* (1989) have studied the behaviour of the H II region luminosity function in nearby spirals and irregular galaxies. Early-type (Sa-Sb) spirals possess a steep luminosity function, with the bulk of massive star formation occurring in small regions ionized by one of a few O stars, plus a low cut-off to the luminosity function. Late-type spirals and irregulars possess a shallower luminosity function, in which most of the massive stars form within large H II regions/OB complexes, for which 30 Doradus in the LMC serves as a useful template. For example, although the LMC contains considerably fewer H II regions than M31 (SAb), it contains ten H II regions more luminous than any counterpart in M31 (Kennicutt *et al.* 1989).

The integrated nebular Hα luminosity of a galaxy is widely used as a proxy for the rate of (near-instantaneous) star formation (Kennicutt 1998), although conversions into total star formation rates (SFR) rely upon the adopted stellar mass function and evolutionary models for single and binary stars (e.g. Leitherer 2008). In addition, since the youngest star forming regions are deeply embedded, the combination of gas (Hα) and dust (24μm continuum) provide a more complete SFR indicator (Calzetti *et al.* 2007), although the

Table 2. Examples of nearby H II regions, spanning a range of luminosities (adapted from Kennicutt 1984), for an assumed O7V Lyman continuum ionizing flux of 10^{49} ph/s).

Region	Type	galaxy	Distance (kpc)	Diameter (pc)	L(Hα) (erg s^{-1}	N(O7V)
Orion (M42)	Classical	Milky Way	0.5	5	1×10^{37}	<1
Rosette (NGC 2244)	Classical	Milky Way	1.5	50	9×10^{37}	7
N66	Giant	SMC	60	220:	6×10^{38}	50
Carina (NGC 3372)	Giant	Milky Way	2.3	300:	1.5×10^{39}	120
NGC 604	Giant	M33	800	400	4.5×10^{39}	320
30 Doradus	(Super)giant	LMC	50	370	1.5×10^{40}	1100
NGC 5461	(Super)giant	M101	6400	1000:	7×10^{40}	5000

situation is more complicated for galaxies with low SFR (e.g. Pflamm-Altenburg *et al.* 2007). In addition, Hα-derived star formation rates differ from FUV continuum diagnostics for dwarf galaxies (Lee *et al.* 2009b), while FUV indicators closely match the local ccSNe rate (Botticella *et al.* 2012).

1.3. *30 Doradus: Template extragalactic giant H II region*

30 Doradus, the brightest star forming complex within the Local Group, provides a useful template for extragalactic 'supergiant' H II regions (Kennicutt *et al.* 1995, Table 2). The 30 Dor nebula is shown in Fig. 1 and spans an angular size of $\sim 15' \times 15'$, corresponding to a linear scale of 220×220 pc at the distance of the LMC. Consequently, individual stars may be studied in detail (e.g. Evans *et al.* 2011). Walborn & Blades (1997) identified five distinct spatial structures within 30 Dor, (i) the central 1–2 Myr cluster R136; (ii) a surrounding triggered generation embedded in dense knots (< 1 Myr); (iii) OB supergiants spread throughout the region (4–6 Myr); (iv) an OB association to the southeast surrounding R143 (~5 Myr); (v) an older cluster containing red supergiants to the northwest (10–20 Myr). 30 Dor would only subtend 1.5″ at a distance of 30 Mpc, so care should be taken for nebular-derived ages of stars within extra-galactic H II regions (e.g. Leloudas *et al.* 2011).

2. Environments of supernovae and gamma-ray bursts

2.1. *H II regions and core-collapse SNe*

Turning to studies of the environments of supernovae, locally neither type II nor type Ib/c supernovae are associated with ongoing star formation. Specifically, Smartt (2009) examined the host environment of a volume limited ($cz < 2{,}000$ km/s), statistically complete sample of ccSNe, of which 0 from 20 type II SN were located in bright H II regions. A number of type II SN were located in loose associations, with two in older clusters (e.g. SN2004am, II-P, in M82), while only 1 of 10 type Ib/c SN from Smartt (2009) was in a large star forming region (SN2007gr, Ic, in NGC 1058), albeit spatially offset from regions of H II emission.

Anderson & James (2008) took a different approach, studying the association between ccSNe and H II regions within (mostly) bright spirals, whose recession velocities extended up to $cz = 10{,}000$ km/s. In common with Smartt (2009), Anderson & James (2008) did not find type II SNe associated with H II regions, concluding that the *"type II progenitor population does not trace the underlying star formation"*. In contrast, Anderson & James (2008) found that type Ib, and especially Ic ccSNe were spatially coincident with (presumably bright) H II regions.

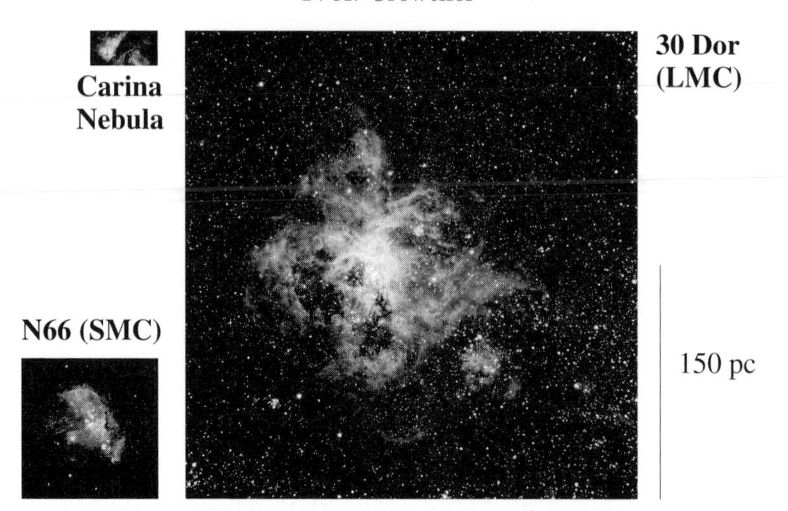

Figure 1. Three Local Group giant H II regions shown on the same physical scale: Carina Nebula (Milky Way, ESO/WFI, $60' \times 30'$), 30 Doradus (LMC, ESO/TRAPPIST, $20' \times 20'$). N66 (SMC, HST/ACS, $4.7' \times 4.7'$). 30 Doradus hosts multiple stellar generations (Walborn & Blades 1997) but would only subtend $1.5''$ at a distance of 30 Mpc, so care should be taken for characteristic ages of stars within extra-galactic H II regions.

Let us consider the typical duration of the H II phase in young, isolated clusters. Walborn (2010) compared the properties of young star clusters, revealing an association with a H II region only for the first ~2–3 Myr, after which the gas has been dispersed (e.g. Westerlund 1, Clark *et al.* 2005). Therefore, one would *not* expect ccSNe to be spatially coincident with *isolated* H II regions unless the mass of the progenitor was sufficiently short for its lifetime to be comparable to the gas dispersion timescale. This is illustrated in Figure 2(a) where we compare the lifetime of the most massive stars in clusters (masses according to Eqn. 10 from Pflamm-Altenberg *et al.* 2007), adopting stellar lifetimes from Ekström *et al.* (2012), with an estimate of the duration of isolated H II regions (adapted from Walborn 2010). This naturally explains the lack of any association between type II ccSNe and H II regions for both Smartt (2009) and Anderson & James (2008).

How, then, can one explain the *empirical* association between type Ib/c SNe and H II regions in the Anderson & James (2008) study? These either arise from very massive stars, which would be inconsistent with Smartt (2009), or more likely we have to appreciate that not all massive star formation occurs within isolated, compact star clusters.

Late-type spirals and irregulars, which form the majority of Anderson & James' host galaxy sample, host large star forming complexes, up to several hundred parsec in size, involving (super)giant H II regions. These are ionized by successive generations of star clusters, separated by a few Myr (Table 2), with a total duty cycle of $\geqslant$10 Myr. Therefore, a massive star exploding within such an environment as a SN after 5–10+ Myr would still be associated with a bright H II region, as illustrated in Fig. 2(b), even if its natal star cluster had cleared the gas from its immediate vicinity. Resolving the location of the ccSNe within the region would be especially difficult at larger distances. Recall that the average distance of galaxies within the Anderson & James (2008) sample was ~32 Mpc, and that their study was based upon moderate resolution ground-based Hα imaging. Typically higher spatial resolution datasets were employed by Smartt (2009), which together with a lower host distance (~27 Mpc maximum for an adopted $H_0 = 75$ km/s/Mpc) enabled a higher spatial inspection of the SN environment (recall Fig. 1).

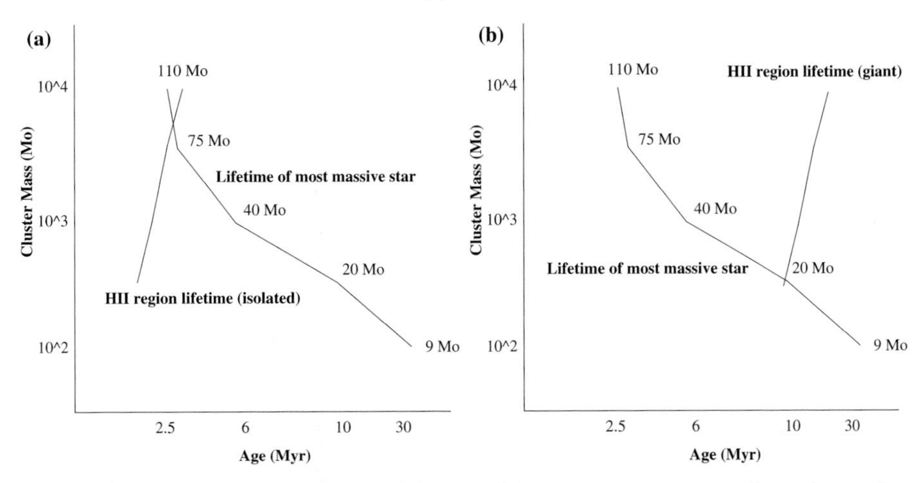

Figure 2. (a) Schematic comparing the lifetime of the most massive star in a cluster (according to Pflamm-Eltenberg *et al.* 2007) and isolated H II regions (adapted from Walborn 2010). Core-collapse SNe should only be associated with isolated H II regions for very massive progenitors; (b) as (a) except for (super)giant H II regions, whose 10–20 Myr lifetimes should imply an association with ccSNe except for only relatively low mass, long lived (type II-P) progenitors.

2.2. *Wolf-Rayet stars and ccSNe*

Main-sequence O stars may preceed ccSNe by up to 3–10 Myr, whereas Wolf-Rayet stars, their evolved descendents, should preceed (type Ib/c) SNe by a timescale that is an order of magnitude shorter (Crowther 2007). Therefore, comparisons between the environment of Wolf-Rayet stars and ccSNe provide information upon whether the former are plausibly the parent population of the latter (e.g. Leloudas *et al.* 2010), since lower mass close binaries might dominate type Ib/c SNe statistics (Smith *et al.* 2011). In addition, we can also compare the environment of Wolf-Rayet stars in their host galaxies (e.g. associated H II regions) with those of ccSNe.

Since the Wolf-Rayet content of the Milky Way is highly incomplete due to foreground interstellar dust, let us instead turn to local external galaxies. From the LMC Wolf-Rayet catalogue of Breysacher *et al.* (1999), 58% (78/134) lie within OB associations, while 83% (112/134) lie within catalogued H II regions. Of course, the LMC is not particularly representative of the local star forming galaxy population, since star formation is largely confined to several giant H II regions (Kennicutt *et al.* 1995) and faint H II regions would not necessarily be identified in distant galaxies. Neugent & Massey (2011) have re-assessed the Wolf-Rayet content of M33 (Scd), revealing a total of 206 stars which they argue is complete to $\sim$5%. As for the LMC, the majority of WR stars (80%) reside in OB associations. Further afield, a comparison between the recent Wolf-Rayet photometric survey of the Scd spiral NGC 5068 (Bibby & Crowther 2012) with Hα images reveals that 50% of the Wolf-Rayet candidates lie in bright or giant H II regions, while 25% are associated with faint H II regions and 25% lie away from any nebulosity.

Recalling Sect. 2.1, type Ib/c SN are rarely associated with H II regions. This suggests that most do *not* result from young, massive Wolf-Rayet stars, arising instead from lower mass close binaries in which the ccSNe arise from the H-deficient, mass-losing primary (Fryer *et al.* 2007). In contrast, the preference of type Ic SNe for H II regions suggest that massive Wolf-Rayet stars *are* realistic progenitors in such instances. Close binaries in such a scenario ought to mimic the masses, and in turn, the lifetimes of type II ccSNe progenitors, which positively shy away from H II regions.

Table 3. Summary of expected association between H II regions and ccSNe/long GRBs in different host galaxies (following Kennicutt *et al.* 1989, Gieles 2009).

Host	SFR	Cluster range ($M_\odot$)	Characteristic H II region	SN-H II association?	Example
Spiral (Sab)	Low	10^{2-4}	Isolated	No (all types)	M31
Spiral (Scd)	High	10^{2-6}	Giant	Yes (Ib/c), No (II-P)	M101
Irr	Low	10^{2-4}	Isolated	No (all types)	SMC
Irr	High	10^{2-6}	Giant	Yes (Ib/c, GRB), No (II-P)	NGC 1569

2.3. *ccSNe, long GRBs and host galaxy types*

Mindful of the spatial resolution issue, let us turn to the Kelly *et al.* (2008) study of SNe locations with respect to the continuum g'-band light from their low redshift ($z < 0.06$) host galaxies. Kelly *et al.* revealed that Ic SNe are much more likely to be found in the brightest regions of their hosts than Ib or II SNe. An earlier analysis of high redshift galaxies by Fruchter *et al.* (2006) revealed that long GRBs ($< z > = 1.25$) were also strongly biased towards the brightest pixel of their hosts, in contrast to core-collapse SNe ($< z > = 0.63$, presumably mostly type II-P) which merely traced the light from their hosts. Kelly *et al.* (2008) concluded that if the brightest locations correspond to the largest star-forming regions, type Ic SNe (and long GRBs) are restricted to the most massive stars, while type Ib and especially type II-P SNe are drawn from stars with more moderate masses, results in common with Anderson & James (2008).

However, one significant difference between the low-redshift SN Hα study of Anderson & James (2008) and the high-redshift GRB study of Fruchter *et al.* (2006) is that hosts of the former are relatively high mass, metal-rich spirals, while those of the latter are low mass, metal-poor dwarfs. In normal disk galaxies the number of stars forming across the mass distribution of star clusters is relatively flat, albeit with a cut-off linked to the star formation intensity (Gieles 2009). The star cluster mass function is repeated in nearby dwarf galaxies (Cook *et al.* 2012), but galaxy-wide triggers may induce intense, concentrated bursts of star formation, leading to disproportionately numerous massive star clusters (Billett *et al.* 2002)†. We have attempted to set out the potential association between H II regions, ccSNe and long GRBs in Table 3 for star forming spirals and irregulars, based upon the above arguments, although exceptions are anticipated (and subject to uncertainties regarding the main progenitors of type Ib/c SNe).

Relatively massive, metal-rich galaxies would represent the primary site of all star formation for the sample of Fruchter *et al.* (2006), resulting in (type II-P) ccSNe unassociated with the brightest regions in their hosts. Yet, when localised starburst activity does occur, it is very intense (Billett *et al.* 2002), leading to very massive clusters, and in turn large numbers of high mass, metal-poor stars, a subset of which would be progenitors of the long GRBs witnessed by Fruchter *et al.* (2006).

3. Upper Mass Limit

The lower limit to the mass of stars is relatively well known (e.g. Burrows *et al.* 1993), yet establishing whether there is a corresponding upper mass limit has proved elusive (Massey 2011). In part, this is because obtaining robust masses for VMS is extremely challenging, and in part because of the scarcity of star clusters that are sufficiently

† Of course, not all dwarf galaxies are starbursting. Within the local volume (<11 Mpc) only a quarter of the star formation from dwarf galaxies is formed during starbursts (Lee *et al.* 2009a)

nearby, young and massive for their most massive stars to be studied in detail. Up until recently, a mass limit of $\sim 150 M_{\odot}$ has been commonly adopted, based upon a near-IR photometric study of the Arches cluster (Figer 2005). However, it is well known that the temperature of hot, massive stars is rather insensitive to optical/IR photometry. Spectroscopic analysis is required for robust temperatures and in turn luminosities, from which stellar masses are derived.

3.1. *R136 stars*

The situation is especially difficult for the brightest main-sequence members of the most massive young clusters, which possess unusual (emission line) spectral morphologies, reminiscent of Wolf-Rayet stars (e.g. Drissen *et al.* 1995). The mass-luminosity relationship for main-sequence VMS is relatively flat, $L \propto M^{1.5}$ (e.g. Crowther *et al.* 2012), so inferred masses are particularly sensitive to temperature, $M \propto T_{\rm eff}^{8/3}$. Recent advances in atmospheric models for stars with dense stellar winds has led to an upward revision to the temperatures of such stars, by ~25%, corresponding to as much as an 80% increase in the resulting mass. Fortunately, several very massive, double-lined eclipsing binaries have been identified within the past few years, including the Wolf-Rayet binary NGC 3603 A1 (Schnurr *et al.* 2008), permitting an independent check on spectroscopic results for similar systems.

R136, the central ionizing cluster of 30 Dor, has both a very high stellar mass (~55,000 $M_{\odot}$) and a sufficiently young age (1–2 Myr) for its most massive stars not to have undergone core-collapse. Previous estimates of their stellar masses, based on conventional O star calibrations, implied 120 – 155 $M_{\odot}$ (Massey & Hunter 1998). Schnurr *et al.* (2009) searched for close binaries among the visually brightest members, but none revealed radial velocities, with the possible exception of R136c. Still, their near-IR integral field datasets provided spatially resolved spectroscopy of individual stars within R136, which, together with archival UV/optical spectroscopy and AO-assisted photometry permitted a reassessment of their stellar masses. Spectroscopic analyses together with new evolutionary models for VMS enabled Crowther *et al.* (2010) to revise their (current) stellar masses upward to 135–265 $M_{\odot}$. Initial masses of 165–320 $M_{\odot}$ were inferred, adopting standard main-sequence mass-loss rates for VMS (Vink *et al.* 2001) which closely matched spectroscopically-derived values, and were reinforced by the close agreement between spectroscopic and dynamical masses obtained for NGC 3603-A1. Overall, R136 supports the trend that higher (initial) mass stars reside within the most massive star clusters set out by Weidner & Kroupa (2006). However, statistics of high mass clusters for which accurate stellar masses have been determined remain very poor.

3.2. *Pair instability supernovae*

Based upon their re-assessment of the most massive stars in R136 and other young, high mass clusters (Arches, NGC 3603), Crowther *et al.* (2010) concluded that their stellar content was consistent with a revised upper mass limit of $\sim 300 M_{\odot}$. Regardless of the physical origin of this limit, such high initial masses raise the prospect of extremely luminous core-collapse SNe (Waldman 2008) or even pair-instability SNe (Heger & Woosley 2002). Models have recently been calculated for the post-main sequence evolution of VMS spanning a range of metallicities (N. Yusof, these proc.). From these, it would appear that the VMS in R136 will end their lives as core-collapse SNe, with lower metallicity (SMC-like) required to reduce mass-loss rates sufficiently for pair-instability SNe, as has been proposed for SN 2007bi (Gal-Yam *et al.* 2009). However, details remain very sensitive to mass-loss prescriptions for the post-main sequence evolution (e.g. Crowther *et al.* 2012).

Acknowledgements

I am grateful to financial support from the Royal Society, IAU and local organizers, enabling participation in the Symposium. Thanks also to Raphael Hirschi and Lisa Yusof for providing results of evolutionary models for VMS prior to publication, plus Mark Gieles for helpful discussions.

References

Anderson, J. P. & James, P. A. 2008, *MNRAS*, 390, 1527
Bibby, J. L. & Crowther, P. A. 2012, *MNRAS*, 420, 3091
Billett, O. H., Hunter, D. A., & Elmegreen B. G., 2002, *AJ* 123, 1454
Botticella, M. T., Smartt, S. J., Kennicutt, R. C. Jr. *et al.* 2012, *A&A*, 537, A132
Breysacher, J., Azzopardi, M., & Testor, G. 1999, *A&AS*, 137, 117
Burrows, A., Hubbard, W. B., Saumon, D., & Lunine, J. I. 1993, *ApJ*, 406, 158
Calzetti, D., Kennicutt, R. C., Engelbracht C. W. *et al.* 2007, *ApJ*, 666, 870
Calzetti, D., Chandar, R., Lee, J. C. *et al.* 2010, *ApJ*, 719, L158
Clark, J. S., Negueruela, I., Crowther, P. A., & Goodwin, S. P. 2005, *A&A*, 434, 949
Conti, P. S., Crowther, P. A., & Leitherer, C., 2008, *From Luminous Hot Stars to Starburst Galaxies* (Cambridge: CUP)
Cook, D. O., Seth, A. C., Dale, D. A. *et al.* 2012, *ApJ*, in press (arXiv:1203.4826)
Crowther, P. A., 2007, *ARA&A*, 45, 177
Crowther, P. A., Schnurr, O., Hirschi R. *et al.* 2010, *MNRAS*, 408, 731
Crowther, P. A., Hirschi, R., & Walborn, N. R. 2012, in: L. Drissen (ed.), *Four Decades of Research on Massive Stars* (San Francisco: ASP), ASP Conf. Ser, in press
de Wit, W. J., Testi, L., Palla, F., & Zinnecker, H. 2005, *A&A*, 437, 247
de Zeeuw, P. T., Hoogerwerf, R., de Bruijne, J. H. J. *et al.* 1999, *AJ*, 117, 354
Drissen, L., Moffat, A. F. J., Walborn, N. R., & Shara, M. M. 1995, *AJ*, 110, 2235
Ekström, S., Georgy, C., Eggenberger, P. *et al.* 2012, *A&A*, 537, A146
Eldridge, J. J., 2012, *MNRAS*, in press (arXiv:1106.4311)
Evans, N. J. II, Dunham, M. M., & Jørgensen, J. K., 2009, *ApJS*, 181, 321
Evans C. J., Taylor, W. D., Henault-Brunet, V. *et al.* 2011, *A&A*, 530, A108
Figer D. F., 2005, *Nat*, 434, 192
Fruchter, A. S., Levan, A. J., Strolger, L. *et al.* 2006, *Nat*, 441, 463
Fryer, C. L., Mazzali, P. A., Prochaska, J. *et al.* 2007, *PASP*, 119, 861
Gal-Yam, A., Mazzali, P., Ofek, E. O. *et al.* 2009, *Nat*, 462, 624
Gieles, M. 2009, *MNRAS*, 394, 2113
Harayama, Y., Eisenhauer, F., & Martins, F., 2008, *ApJ*, 675, 1319
Heger, A. & Woosley, S. E., 2002, *ApJ*, 567, 532
Hillebrand, L. A., & Hartmann, L. W., 1998, *ApJ*, 492, 540
Hunter D. A., O'Connell, R. W., Gallagher, J. S., & Smecker-Hane, T. A., 2000, *AJ* 120, 2383
Kelly P. L., Kirschner, R. P., & Pahre M., 2008, *ApJ*, 687, 1201
Kennicutt, R. C. Jr., 1984, *ApJ*, 287, 116
Kennicutt, R. C. 1998, *ARA&A*, 36, 189
Kennicutt, R. C. Jr., Edgar, B. K., & Hodge, P. W., 1989, *ApJ*, 337, 761
Kennicutt R. C. Jr, Bresolin, F., Bomans, D. J., Bothun, G. D., & Thompson, I. B. 1995, *AJ*, 109, 594
Lada, C. J. & Lada, E. A., 2003, *ARA&A*, 41, 57
Lee J. C., Kennicutt, R. C. Jr., Funes, J. G. *et al.* 2009a, *ApJ*, 692, 1305
Lee J. C., Gil de Paz, A., Tremonti, C. *et al.* 2009b, *ApJ*, 706, 599
Leitherer C., 2008, in: L.K. Hunt, S. Madden & R. Schneider (eds.), *Low-Metallicity Star Formation: From the First Stars to Dwarf Galaxies, Proc. IAU Symp. 255* (Cambridge: CUP), p. 305
Leloudas, G., Sollerman, J., Levan, A. J. *et al.*, 2010, *A&A*, 518, A29
Leloudas, G., Gallazzi, A., Sollerman, J., *et al.*, 2011, *A&A*, 530, A95

Massey, P. & Hunter, D. A., 1998, *ApJ*, 493, 180
Massey P., 2011, in: M. Treyer, T.K. Wyder, J.D. Neill *et al.* (eds.), *UP2010: Have Observations Revealed a Variable Upper End of the Initial Mass Function?* (San Francisco: ASP), ASP Conf. Ser 440, p. 29
Neugent, K. F., & Massey, P., 2011, *ApJ*, 733, 123
Pflamm-Altenburg, J., Weidner, C., & Kroupa, P., 2007, *ApJ*, 671, 1550
Schnurr O., Casoli, J., Chené, A.-N. *et al.* 2008, *MNRAS*, 389, L38
Schnurr, O., Chené, A.-N., Casoli, J. *et al.* 2009, *MNRAS*, 397, 2049
Simón-Díaz, S. , Herrero, A., Esteban, C., & Najarro, F., 2006, *A&A*, 448, 351
Smartt, S. J. 2009, *ARA&A*, 47, 63
Smith, N., Li, W., Filippenko, A. V., & Chornock, R., 2011, *MNRAS*, 412, 1522
Vacca, W. D. & Conti, P. S., 1992, *ApJ*, 401, 543
Vink, J. S., de Koter, A., & Lamers, H. J. G. L. M. 2001, *A&A*, 369, 574
Walborn N. R., 2010, in: C. Leitherer, P. D. Bennett, P. W. Morris & J. Th. van Loon (eds.), *Hot and Cool: Bridging Gaps in Massive-Star Evolution* (San Francisco: ASP), ASP Conf. Ser 425, p. 45
Walborn N. R. & Blades J. C., 1997, *ApJS*, 112, 457
Waldman, R. 2008, *ApJ*, 685, 1103
Weidner, C. & Kroupa, P., 2006, *MNRAS*, 365, 1333
Wilking, B. A., Lada, C. J., & Young, E. T., 1989, *ApJ*, 340, 823

Discussion

MODJAZ: Could you comment on the agreement of spectroscopically derived mass loss rate with theoretically predicted ones for R136?

CROWTHER: Spectroscopically derived (clumped) mass-loss rates for the R136 very massive stars match the main-sequence Vink *et al.* (2001) predictions fairly well, but these rates are expected to increase for the post-main sequence phase as the star approaches the Eddington limit.

MODJAZ: If you were to use the light-weighted average of the spatially-resolved cluster, which ones are the dominating clusters?

CROWTHER: The integrated light from 30 Doradus is strongly biased towards the youngest high mass cluster R136a so an average characteristic age of $\sim$ 3 Myr would be expected despite OB stars spanning 0 - 20 Myr within this region.

BROMBERG: Are most of the radio loud Ib/c SNe connected with HII regions or not?

CROWTHER: Natal gas is observed to be removed from clusters within only a few Myrs so I would not expect the SN environment to be affected by the ISM/HII region in general, although a high density ISM may exist in some circumstances such as the central region of starbursts such as M82.

Death of Massive Stars: Supernovae and Gamma-Ray Bursts
Proceedings IAU Symposium No. 279, 2012
P. Roming, N. Kawai & E. Pian, eds.

doi:10.1017/S1743921312012628

A supergiant progenitor for SN 2011dh

Melina C. Bersten[1], Omar Benvenuto[2] and Ken Nomoto[1]

[1]Kavli Institute for the Physics and Mathematics of the Universe, Todai Institutes for Advanced Study, the University of Tokyo, Kashiwa, Japan 277-8583 (Kavli IPMU, WPI)
email: `melina.bersten@ipmu.jp`

[2]Facultad de Ciencias Astronómicas y Geofísicas, Universidad Nacional de La Plata, Paseo del Bosque S/N, B1900FWA La Plata, Argentina
email: `obenvenu@fcaglp.unlp.edu.ar`

Abstract. A set of hydrodynamical models based on stellar evolutionary progenitors is used to study the nature of SN 2011dh. Our modeling suggests that a large progenitor star —with $R \sim 200\ R_\odot$— is needed to reproduce the early light curve (LC) of SN 2011dh. This is consistent with the suggestion that the progenitor is a yellow super-giant star detected at the location of the SN in deep pre-explosion images. From the main peak of the bolometric light curve (LC) and expansion velocities we constrain the mass of the ejecta to be $\approx 2\ M_\odot$, the explosion energy to be $E = 8 \times 10^{50}$ erg, and the ^{56}Ni mass to be 0.063 $M_\odot$. The progenitor star is composed of a helium core of $\approx 4\ M_\odot$ and a thin hydrogen envelope, and it had a main-sequence mass of $\approx 13\ M_\odot$. Our models rule out progenitors with helium-core masses larger than 8 $M_\odot$, which correspond to $M_{\rm ZAMS} \gtrsim 25\ M_\odot$. This suggests that a single evolutionary scenario for SN 2011dh is highly unlikely.

Keywords. hydrodynamics, stellar evolution, supernovae, SN 2011dh

---

1. Introduction

SN 2011dh was discovered in the nearby spiral galaxy M51 and classified as a type IIb supernova (SN IIb) because of the early presence of H lines in the spectra that were later dominated by He lines. Soon after discovery, a source was identified as the possible progenitor of SN 2011dh in archival, multi-band HST images (Maund *et al.* 2011; Van Dyk *et al.* 2011). Photometry of the source was compatible with a yellow super-giant (YSG) star with $R \sim 270 R_\odot$. However, some authors have suggested a more compact progenitor ($R \sim R_\odot$), claiming that the YSG star detected in the pre-SN images may be its binary companion or even an unrelated object (Arcavi *et al.* 2011; Van Dyk *et al.* 2011; Soderberg *et al.* 2011). Here we use hydrodynamical models applied to stellar evolutionary progenitors that aim to elucidate the compact or extended nature of the progenitor of SN 2011dh, as well as the main physical parameters of the explosion.

2. Hydrodynamical Models

Our supernova models were computed using a one-dimensional Lagrangian hydrodynamic code with flux-limited radiation diffusion including $\gamma-$ray transfer in gray approximation for any distribution of ^{56}Ni (Bersten *et al.* 2011). As initial model we adopted progenitors with different He core mass from stellar evolution calculations by Nomoto & Hashimoto (1988). The external envelope of the models was removed by hand without specifying the responsible mechanism of such mass loss, thus leaving a compact structure, similar to a Wolf-Rayet star. To take into account the thin hydrogen envelope, we

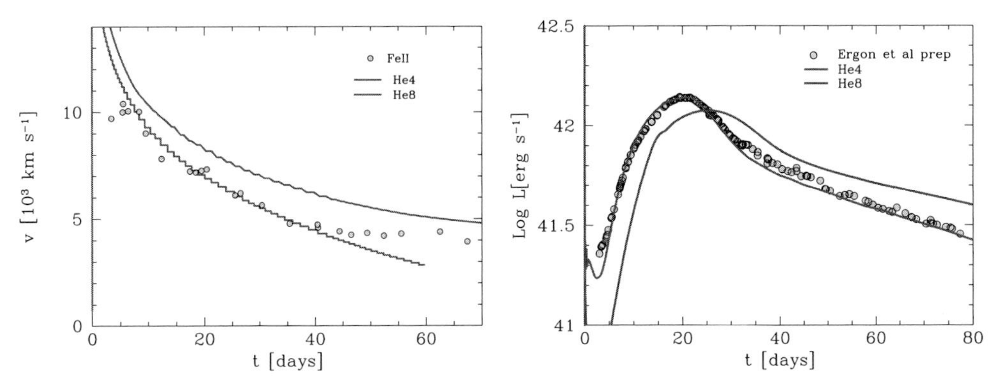

Figure 1. Evolution of the photospheric velocity (**left panel**) and bolometric LC (**right panel**) for our optimal model, He4, with a He core mass of $\approx 4M_\odot$ (red line) and for a model with He core mass of $8M_\odot$ (He8; blue line) as compared with the observed bolometric LC and Fe II line velocities of SN 2011dh (cyan dots). Note that the model with large helium core mass is not compatible with the observations.

smoothly attached a low-mass H-rich envelope in hydrostatic and thermal equilibrium to the He core. The presence of the external envelope significantly modified the progenitor radius but not the mass of the progenitor.

LCs of SNe IIb show two distinct phases (1) the early cooling phase with strong dependence on progenitor radius (see §2.1) and (2) the second peak mainly powered by radioactive decay but also dependent on explosion energy and ejecta mass. Comparing our models with the observed bolometric LC during the second peak and with line expansion velocities we found that a progenitor with He core mass of $\approx 4M_\odot$, an explosion energy of 8×10^{50} erg, and a ^{56}Ni mass of 0.063 $M_\odot$ reproduce very well the observations. This optimal model (He4) is presented in Figure 1 and it is consistent with a main-sequence mass of $\approx$13 $M_\odot$.

2.1. *Compact versus extended progenitor*

In order to test the effect of the progenitor radius on the LC we attached a H/He envelope to the optimal model He4, in hydrostatic and thermal equilibrium. The envelope mass and composition was fit so that the luminosity and effective temperature matched those of a YSG star. Figure 2 (left panel) shows a comparison of our models with the g'-band observations of Arcavi *et al.* (2011) for a compact ($R \approx 2\ R_\odot$; blue line) and an extended ($R \approx 270\ R_\odot$; red line) progenitor. The latter model (He4R270) produces a pronounced spike, in agreement with the observations, while the compact progenitor shows a much weaker bump. From the figure it is clear that obtaining data during the early adiabatic cooling phase is critical to elucidate the progenitor radius.

The right panel of Figure 2 shows the evolution of the effective temperature for the compact and extended models along with a measurement obtained from a spectrum by Arcavi *et al.* (2011). Although the models show large differences at $t \lesssim 2$ days, the evolution is remarkably similar thereafter. That is why the observation obtained at 2.4 days is not a good discriminant of the progenitor radius. Also shown for comparison are the analytic expressions of Rabinak & Waxman (2011) for compact and extended progenitors which do show large temperature differences.

From this analysis we conclude that an extended progenitor, similar to the YSG star detected in pre-SN images, is favored by the early-time observations.

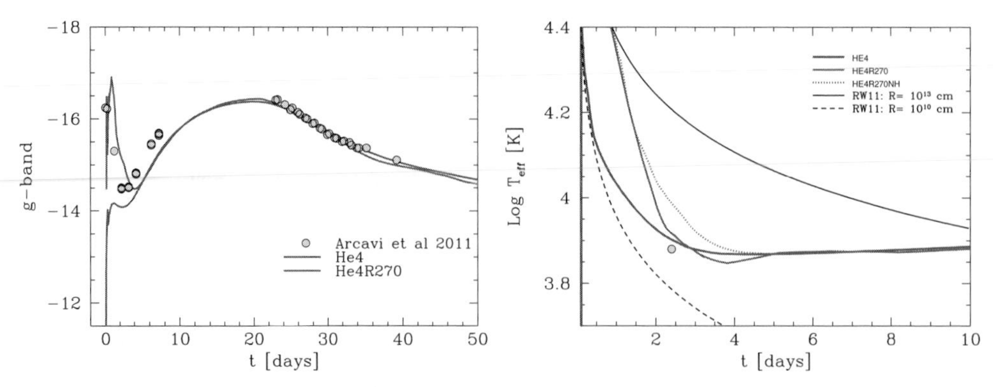

Figure 2. Observed and modeled g'-band LCs **(left panel)** and effective temperature evolution **(right panel)** for the compact model He4 and the extended He4R270 model. The effective temperature calculated using the analytic expression of Rabinak & Waxman(2011) with $R \sim 10^{13}$ cm (solid black line) and $R \sim 10^{10}$ cm (dashed black line) are also shown. The black-body temperature (cyan dot) estimated from a spectrum of SN 2011dh obtained at 2.4 days is included for comparison.

2.2. *Single versus binary progenitor*

Fig1 shows that progenitors with He core masses larger than 8 $M_\odot$ (model He8) which correspond to $M_{\rm ZAMS} \gtrsim 25 M_\odot$ are ruled out. Considering the limitations of single stars of such stellar masses to almost entirely expel the H-rich envelope via stellar winds, as required for SNe IIb, this result is indicative of a binary origin for SN 2011dh.

To test the plausibility of a binary system compatible with the pre-SN observations of SN 2011dh we performed binary evolution calculations with mass transfer using a code developed by Benvenuto & De Vito (2003). Figure 3 shows the evolutionary tracks in the H-R diagram for a system with 16 $M_\odot$ + 10 $M_\odot$ and an initial period of 150

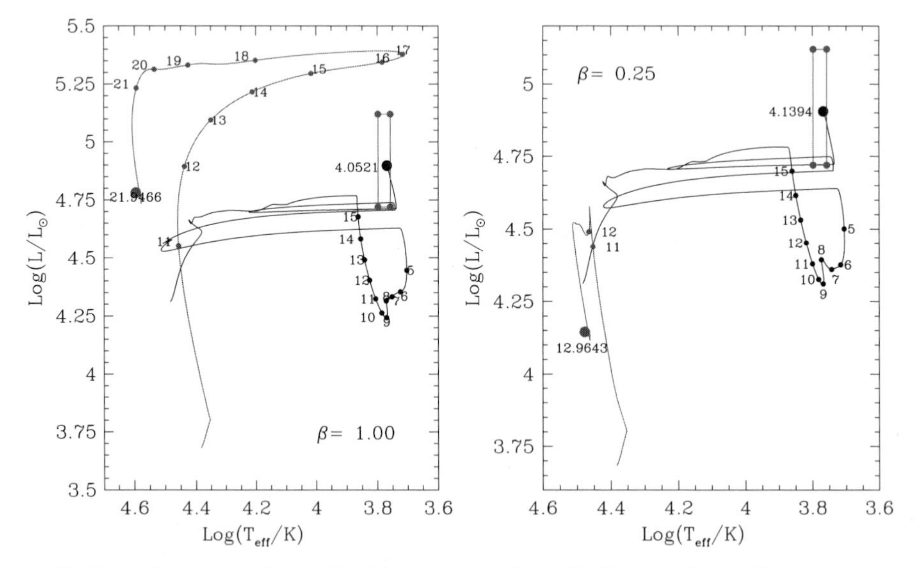

Figure 3. Evolutionary tracks in the Hertzsprung-Russell diagram for both components of a binary system with solar composition. The stars have $M_{\rm ZAMS}$ of 16 $M_\odot$ and 10 $M_\odot$ and an initial period of 150 days. **Left panel:** assuming conservative mass transfer ($\beta = 1$). **Right panel:** non-conservative mass transfer ($\beta = 0.25$). Labeled dots along the tracks indicate the masses of the stars (in solar units) while mass-transfer by Roche-Lobe overflow occurs.

days. In this configuration the primary star ends its evolution at the right position as compared with the YSG star detected in the pre-SN images. Furthermore, the final mass of the primary is $\approx 4M_\odot$, which is consistent with our hydrodynamical modeling. The calculations further predict that a hydrogen envelope with a mass of $\approx 4 \times 10^{-3} M_\odot$ is retained by the primary star, which is required to produce a SN IIb.

The binary scenario was further tested by estimating the effect of the putative companion star on the pre-explosion photometry and comparing this with the observations. Because the secondary star is predicted to be much hotter than the primary star, we found that the largest effect appears in the blue and UV ranges. The contribution of the secondary to the flux in the F336W band, however, is marginal, at the 1.5-σ level for a conservative mass-transfer case, and at the 0.6-σ level for a non-conservative case. The existence of the binary companion can be tested in a few years time by a search for a blue object at the location of the SN.

References

Arcavi, I., Gal-Yam, A., Yaron, O., *et al.* 2011, *ApJL*, 742, L18
Bersten, M. C., Benvenuto, O., & Hamuy, M. 2011, *ApJ*, 729, 61
Benvenuto, O. G. & De Vito, M. A. 2003, *MNRAS*, 342, 50
Maund, J. R., Fraser, M., Ergon, M., *et al.* 2011, *ApJL*, 739, L37
Nomoto, K. & Hashimoto, M. 1988, Phys. Rep., 163, 1
Rabinak, I. & Waxman, E. 2011, *ApJ*, 728, 63
Soderberg, A. M., Margutti, R., Zauderer, B. A., *et al.* 2011, arXiv:1107.1876
Van Dyk, S. D., Li, W., Cenko, S. B., *et al.* 2011, *ApJL*, 741, L28

Discussion

N. Kawai: What is the exact radiation process during the early spike of the LC?

Bersten: The early spike is produced by the shock breakout (the arrival of the shock wave to the object surface). This is also observed for compact progenitors but it is less noticeable because a larger fraction of the shock energy is lost to expand adiabatically a more compact object.

A. Gal-Yam: Would this model fit SN 1993J?

Bersten: I did not calculate a model for SN 1993J in detail but I did some tests and the model seems consistent. In any case the progenitor of SN 1993J was a RSG with a radius ~2 times larger than the radius of the YSG star detected in pre-explosion images of SN 2011dh.

I. Rabinak: What is the mass you have outside of the He-core?

Bersten: The mass of the envelope is a free parameter which is fit when attaching the envelope to the He core for values of L and $T_{\rm eff}$ consistent with a YSG star. The envelope mass in this case is $\approx 0.1M_\odot$.

Death of Massive Stars: Supernovae and Gamma-Ray Bursts
Proceedings IAU Symposium No. 279, 2012
P. Roming, N. Kawai & E. Pian, eds.

doi:10.1017/S174392131201263X

Superluminous Supernovae

Robert M. Quimby

Kavli Institute for the Physics and Mathematics of the Universe
Todai Institutes for Advanced Study, University of Tokyo
5-1-5 Kashiwa-no-Ha, Kashiwa City, Chiba 277-8583, Japan
email: robert.quimby@ipmu.jp

Abstract. Not long ago the sample of well studied supernovae, which were gathered mostly through targeted surveys, was populated exclusively by events with absolute peak magnitudes fainter than about −20. Modern searches that select supernovae not just from massive hosts but from dwarfs as well have produced a new census with a surprising difference: a significant percentage of supernovae found in these flux limited surveys peak at −21 magnitude or brighter. The energy emitted by these superluminous supernovae in optical light alone rivals the total explosion energy available to typical core collapse supernovae ($> 10^{51}$ erg). This makes superluminous supernovae difficult to explain through standard models. Adding further complexity to this picture are the distinct observational properties of various superluminous supernovae. Some may be powered in part by interactions with a hydrogen-rich, circumstellar material but others appear to lack hydrogen altogether. Some appear to be powered by large stores of radioactive material, while others fade quickly and have stringent limits on 56-Ni production. In this talk I will discuss the current observational constrains on superluminous supernova and the prospects for revealing their origins.

Keywords. supernovae: general

Until nearly the turn of the century, the sample of supernovae discoveries with peak absolute magnitudes brighter than about −19 in the optical was dominated by Type Ia events, only a handful of supernovae connected to the deaths of massive stars were observed to cross this threshold, and there were no well studied supernovae brighter than −21 magnitude. Richardson *et al.* (2002) studied the peak magnitude distributions of supernovae including events listed in the Asiago Supernova Catalog (Barbon *et al.* 1999) for which only minimal discovery reports were available. A few events were noted as possibly brighter than −21 magnitude, although most of these can likely be explained by calibration errors†.

The absolute magnitude distribution expected for a local, *volume* limited survey is shown in Fig. 1. None of the supernovae in the volume limited samples of the Lick Observatory Supernova Search (LOSS) have peak magnitudes brighter than −20 (Li *et al.* 2011). A flux limited survey was conducted by the Supernova Cosmology Project in the Spring of 1999, which perhaps netted the first high luminosity ($M < -21$) supernova discovery, SN 1999as (Deng *et al.* 2001). However, it would be several more years before the first unambiguous superluminous supernova (SLSN) was detected and confirmed by multiple research groups.

SN 2006gy proved to be a rather surprising event. At just 72 Mpc, this supernova was observed to peak at about 14th magnitude–bright enough to be studied with even small aperture telescopes or photon hungry instruments on the larger ones–or an absolute

† For example, the reported magnitude for SN 1988O places it near −22 mag absolute, but it was spectroscopically connected to the subluminous Type Ia class (Branch *et al.* 1993), and the magnitude given in IAUC 4601 was likely incorrect (J. Mueller, priv. comm.).

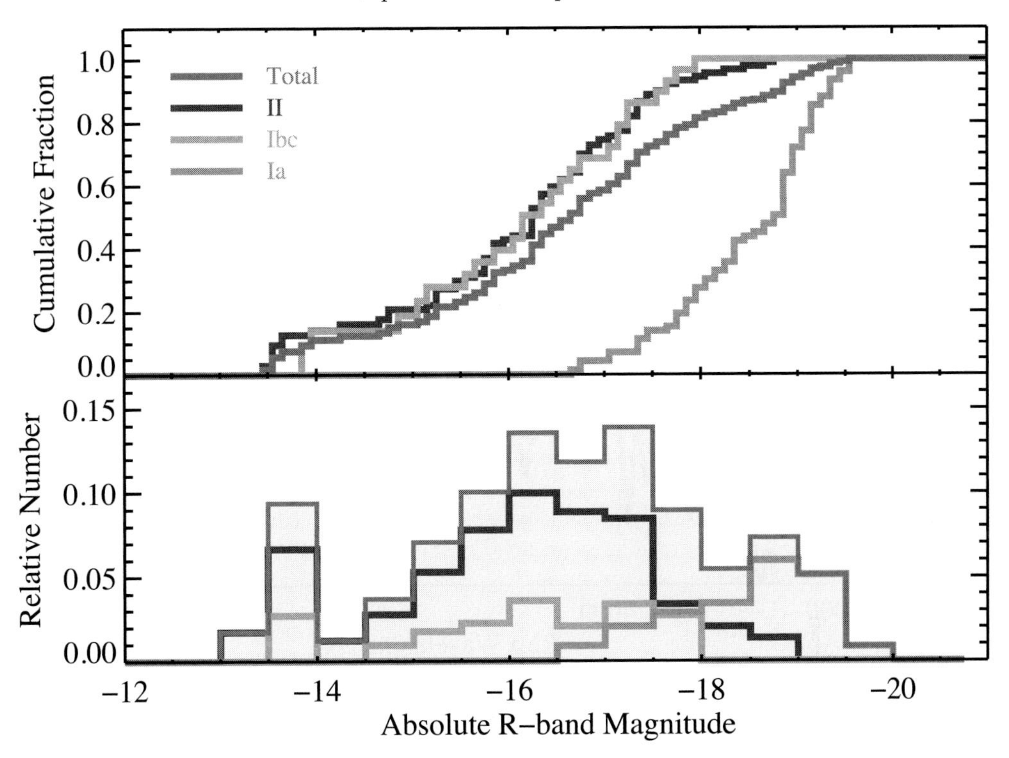

Figure 1. Peak R-band magnitude distribution for supernovae in an local, volume limited sample. Based on data from the Lick Observatory Supernova Search (Li *et al.* 2011).

magnitude of nearly -22 after extinction correction. And SN 2006gy did not just peak at 100 times the optical luminosity of a typical supernova, it remained brighter than normal Type Ia supernovae for 5 months, and brighter than any other well studied supernovae for 3 months (e.g. Smith *et al.* 2007; Ofek *et al.* 2007). Integration of the light curve shows the energy radiated in photons exceeds 10^{51} erg. Just how this energy was generated and how such a massive star (likely $M_{\rm MS} > 100 M_{\odot}$) came to be in the core of a giant, metal rich galaxy, remain active research topics.

In any case, SN 2006gy clearly demonstrated that high luminosity supernovae can exist in the local universe, and this realization helped in the discovery of additional SLSNe. In particular, SN 2005ap had (obviously) been detected earlier, but it was difficult to accept the consequences of its spectroscopically determined redshift. At $z = 0.28$, SN 2005ap was possibly even more luminous than SN 2006gy (Quimby *et al.* 2007; a definitive statement remains difficult given the unfiltered nature of the light curve). The spectra (see Fig. 2) are similar to SN 2006gy in the sense that the strong P-Cygni profiles evident in normal supernova are lacking, leaving mainly featureless blue continua, but with an important difference: SN 2006gy shows a complex Hα profile (Smith *et al.* 2010), while SN 2005ap appears to lack hydrogen all together. SN 2005ap also shows a series of broad absorption dips near rest frame 4200Å (identified as OII; Quimby *et al.* 2011), which are not present in the spectra of SN 2006gy.

Another SLSN detected by the Robotic Optical Transient Search Experiment (ROTSE-IIIb; Akerlof *et al.* 2003) supernovae searches shows further spectroscopic differences. Spectra of SN 2008es shows a broad HeII λ4686 line early on and broad Hα develops later, but the narrow emission lines such as those present in SN 2006gy are not observed

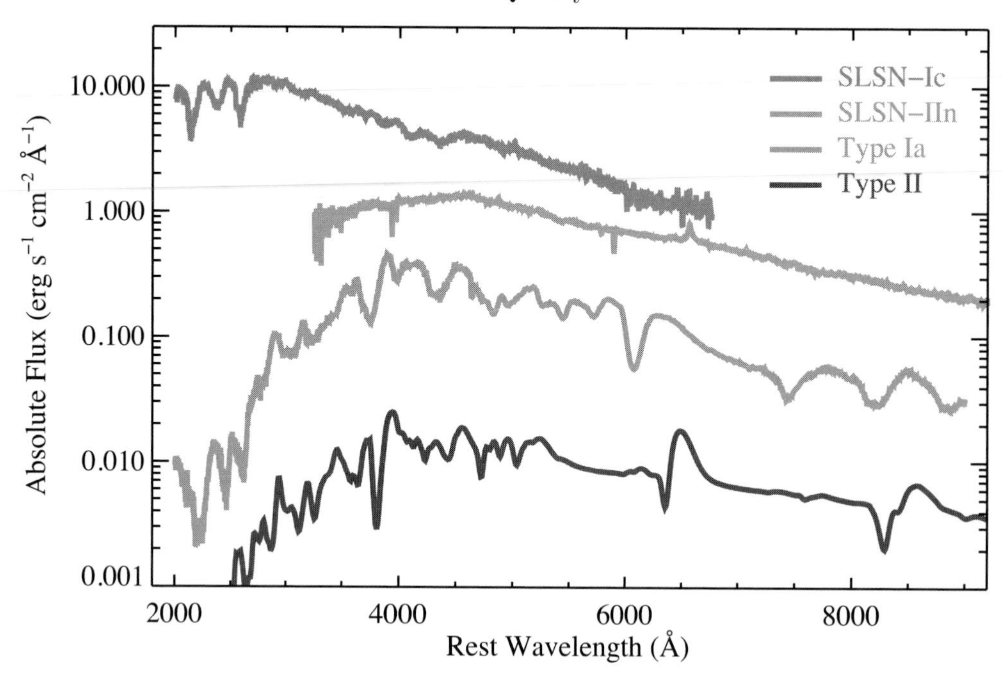

Figure 2. Spectra of normal and superluminous supernovae taken near peak (optical) brightness. The SLSN-Ic spectrum is a composite of SCP06F6 (Barbary *et al.* 2009), PTF09cnd (Quimby *et al.* 2011), and SN2005ap (Quimby *et al.* 2007), the SLSN-IIn is SN 2006gy from Smith *et al.* (2007), the Type Ia is a combination of SN 1992A (Kirshner *et al.* 1993) and SN 2003hv (Leloudas *et al.* 2009), and the Type II is a Nugent template (see supernova.lbl.gov/~nugent/nugent_templates.html). Flux values have been scaled to typical values for each class. SLSNe are about 10 times brighter than typical Type Ia supernovae in the optical, but in the UV, they can be a thousand times more luminous.

(Miller *et al.* 2009; Gezari *et al.* 2009). SN 2008es was also the first SLSNe to be studied in the UV with the *Swift* (Gehrels *et al.* 2004) Ultraviolet/Optical Telescope (UVOT; Roming *et al.* 2005). These data constrain the broad band spectral energy distribution and its time evolution, which can be fit relatively well with a cooling black body. The temperatures derived from this fit show that SN 2008es stayed rather hot for over a month even as the photosphere expanded at $\sim 10000\,\mathrm{km\,s^{-1}}$. The lack of adiabatic cooling thus suggest a large initial radius at (or beyond) the limits of super red giant envelopes.

The Palomar Transient Factory (PTF; Law *et al.* 2009; Rau *et al.* 2009) began discovering a steady supply of SLSNe from its commissioning in 2009. Some of the first of these, PTF09atu, PTF09cnd, SN 2009jh, and SN 2010gx were found to be hydrogen poor SLSNe similar to SN 2005ap (SLSN-Ic; Quimby *et al.* 2011; see also Pastorello *et al.* 2010). These supernovae were also followed-up in the UV using *Swift*. Like SN 2008es, there is only slow cooling over time even as the ejecta expand at $> 10000\,\mathrm{km\,s^{-1}}$. Unlike SN 2008es, however, these events lack hydrogen. There are no stripped envelop progenitor systems known (to the author at least) with bound, hydrogen poor material distributed at $\sim 10^{15}$ cm, which suggests an unbound configuration.

Prior to the start of the PTF survey proper, a "dry run" was conducted to test the follow-up paradigm that would come to characterize the future survey. This resulted in the intensive follow-up of SN 2007bi (Gal-Yam *et al.* 2009). The spectra of SN 2007bi lack hydrogen (and strong SiII $\lambda6355$), and it can thus be classified as a SLSN-Ic. However,

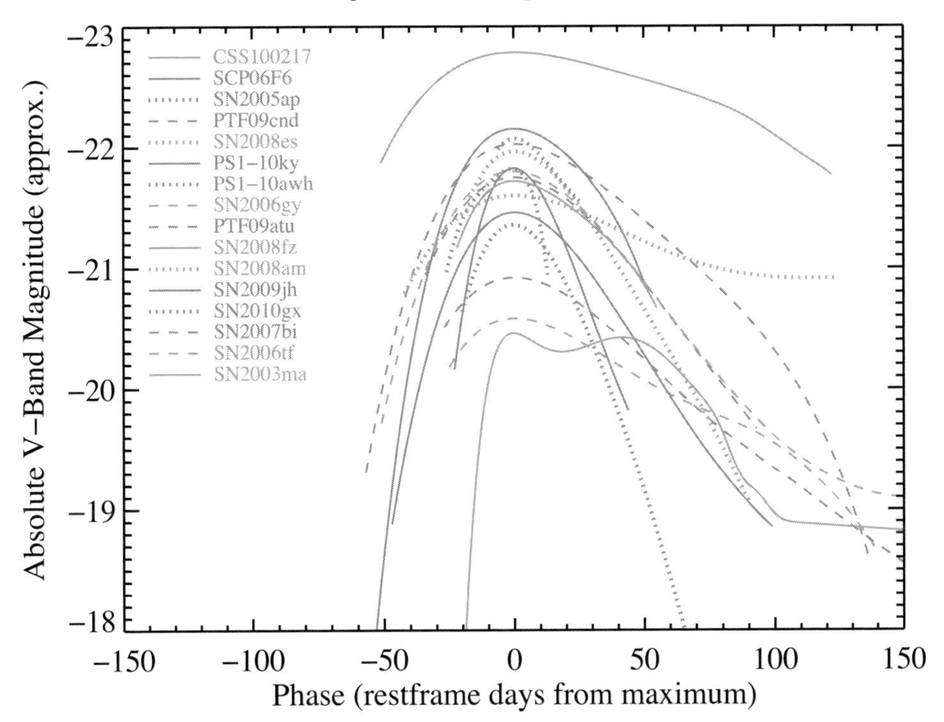

Figure 3. Approximate light curves in the rest frame, absolute V-band magnitudes for a collection of published SLSNe, including the SLSNe candidate, CSS100217. Data adapted from Drake *et al.* 2011, Barbary *et al.* 2009, Quimby *et al.* 2007, Quimby *et al.* 2011, Gezari *et al.* 2009, Miller *et al.* 2009, Chomiuk *et al.* 2011, Smith *et al.* 2007, Drake *et al.* 2010, Pastorello *et al.* 2010, Gal-Yam *et al.* 2009, Smith *et al.* 2008, and Rest *et al.* 2011. The light curve of SN 2006tf is supplemented with unpublished ROTSE-IIIb observations. Note that for some objects (e.g. SN 2005ap) only unfiltered data are available, and for others (e.g. SCP06F6) the available pass bands do not sample the rest frame V-band. This and the simplistic k-corrections used may lead to systematic errors of up to a few tenths of a magnitude.

there is a key difference from the SN 2005ap-like SLSNe-Ic: SN 2007bi shows photometric and spectroscopic evidence for a large ^{56}Ni yield while this is excluded for the SN 2005ap-like events (Quimby *et al.* 2011). In particular, the light curve of SN 2007bi fades by about 1 magnitude every 100 days, as expected from the decay of ^{56}Ni's daughter product, ^{56}Co into ^{56}Fe. Other SLSN-Ic like PTF09cnd fade more rapidly and limit the production of ^{56}Ni to $\lesssim 1 M_{\odot}$. The light curves of various SLSN-Ic and hydrogen rich SLSN-II are shown in Fig. 3.

Some SLSNe, like SN 2007bi for example, may be powered mainly by radioactive decay ^{56}Ni, but others (e.g. PTF09cnd) reach peak luminosities too great to be explained by the ^{56}Ni production allowed by their late time photometric limits. Some SLSNe, like SN 2006gy for example, may be powered mainly by interactions of the SN ejecta with pre-SN winds, but others (e.g. SN 2008es) show no obvious signs of such ongoing interactions. It may therefore be that there are fundamentally different engines powering these observationally distinct events. On the other hand, the principle sources of power may yet be related and it is stochastic differences in the final years of the progenitors that affect the observations.

A possible process connecting the engines powering at least some SLSNe is the conversion of kinetic energy in the supernova ejecta into radiant energy via interaction with slower moving material. This is most clearly evident in SLSNe-IIn such as SN 2006gy, where the narrow emission features require slow moving material in the vicinity of the explosion (fast moving material would give rise to broad, not narrow emission features). It is possible that events like SN 2008es also derive some of their power from ejecta/CSM interactions, but in this case the distribution of CSM must be fast moving or truncated such that the slow moving material has mostly been overtaken by the SN ejecta by the time the spectroscopic observations begin. Extending this model further, if the CSM was depleted of its hydrogen (for example, if the progenitor was striped of its hydrogen long before the SN explosion), the ejecta/CSM interaction could in principle provide a similar transfer of kinetic energy into photons. A possible source for such hydrogen poor CSM may be material cast off by instabilities in the cores of very massive stars in their final years (e.g. Woosley *et al.* 2007; Umeda & Nomoto 2008).

Another possibility is that the high luminosities are achieved by thermalization of energy deposited into an expanding SN envelope by a compact remnant that formed as a result of the core-collapse. In the magnetar model (Kasen & Bildsten 2010; Woosley 2010), rotational energy from the nascent neutron star is transferred (by an unspecified process) to the ejecta mass. Kasen & Bildsten 2010 show that such models can reproduce at least the light curves of events like SN 2008es and even SN 2007bi with plausible initial rotation periods and magnetic field strengths. In this case, the progenitors could be of more modest initial masses.

We can get some insights into the progenitors by studying the broader environments in which these SLSNe explode (e.g. their hosts). Neill *et al.* 2011 have studied the NUV-Optical color vs. Optical magnitude distribution of a number of high luminosity supernovae and they find a preference for fainter, bluer hosts when compared to the broader population of GALEX to SDSS matched galaxies. However, the sample studied is still consistent with the giant to dwarf host distribution of normal luminosity core-collapse supernovae from PTF (Arcavi *et al.* 2010).

Looking at the SLSNe samples from ROTSE-IIIb and PTF, there is not an obvious preference among SLSNe-IIn for dwarf or giant hosts (both surveys find SLSNe-IIn in hosts of various luminosities, faint to bright). However, the SLSN-Ic do appear to prefer dwarf host galaxies. There is possibly only one SLSN-Ic out of more than a dozen discoveries that is hosted by a giant.

The rates of SLSNe can also offer some constraints on the progenitor systems when the birth rates of such progenitors are known. Based on the ROTSE-IIIb sample, *preliminary* results (Quimby *et al.* in prep.) suggest that there is one SLSN (of any type) for about every 1000 core-collapse supernova in the local ($z \sim 0.2$) universe. If we assume that the distribution of SLSN-Ic is skewed brighter than the SLSN-II distribution, as hinted at by the ROTSE-IIIb sample at least, then this would imply that most SLSN have hydrogen, since ROTSE-IIIb has found a roughly equal number of each while the SLSN-Ic would be drawn from a larger volume in this case. The SLSN-Ic rate would then be about one for every 10^4 core-collapse supernovae.

This is still a sufficiently high rate, and the SLSNe are luminous enough that it should be possible with existing instruments to detect SLSNe out to very high redshifts (e.g. $z \sim 4$ with Subaru). This opens the possibility of using SLSNe to glean insights into the distant universe. First of all, if the SLSNe are connected to the most massive stars (as seems to be the case at least for SN 2006gy), then their rates should evolve with redshift with the cosmic star formation history. If there are changes in the IMF such as a "top-heavy" IMF at higher redshifts, then we could expect more SLSNe per unit star formation. Thus checking if the distant to local SLSN rate differs from the distant to local SFR could be one way to search for evolution in the IMF. Additionally, absorption features imprinted in the otherwise smooth continua of SLSNe could carry information about the chemistry of distant stellar nurseries. Considering the coming suite of wide field cameras on moderate to large aperture telescopes, the future of SLSNe research appears, well, bright.

References

Akerlof, C. W., Kehoe, R. L., McKay, T. A., *et al.* 2003, *PASP*, 115, 132

Arcavi, I., Gal-Yam, A., Kasliwal, M., *et al.* 2010, *ApJ*, 721, 777

Barbary, K., Dawson, K. S., Tokita, K., *et al.* 2009, *ApJ*, 690, 1358

Barbon, R., Buondí, V., Cappellaro, E., & Turatto, M. 1999, *A&A* (Supplement), 139, 531

Branch, D., Fisher, A., & Nugent, P. 1993, *AJ*, 106, 2383

Chatzopoulos, E., Wheeler, J. C., Vinko, J., *et al.* 2011, *ApJ*, 729, 143

Chomiuk, L., Chornock, R., Soderberg, A. M., *et al.* 2011, *ApJ*, 743, 114

Deng, J. S., Hatano, K., Nakamura, T., *et al.* 2001, in Astronomical Society of the Pacific Conference Series, Vol. 251, New Century of X-ray Astronomy, ed. H. Inoue & H. Kunieda, 238

Drake, A. J., Djorgovski, S. G., Mahabal, A., *et al.* 2011, *ApJ*, 735, 106

Drake, A. J., Djorgovski, S. G., Prieto, J. L., *et al.* 2010, *ApJ*, 718, L127

Gal-Yam, A., Mazzali, P., Ofek, E. O., *et al.* 2009, *Nature*, 462, 624

Gehrels, N., Chincarini, G., Giommi, P., *et al.* 2004, *ApJ*, 611, 1005

Gezari, S., Halpern, J. P., Grupe, D., *et al.* 2009, *ApJ*, 690, 1313

Kasen, D. & Bildsten, L. 2010, *ApJ*, 717, 245

Kirshner, R. P., Jeffery, D. J., Leibundgut, B., *et al.* 1993, *ApJ*, 415, 589

Law, N. M., Kulkarni, S. R., Dekany, R. G., *et al.* 2009, *PASP*, 121, 1395

Li, W., Leaman, J., Chornock, R., *et al.* 2011, *MNRAS*, 413

Miller, A. A., Chornock, R., Perley, D. A., *et al.* 2009, *ApJ*, 690, 1303

Neill, J. D., Sullivan, M., Gal-Yam, A., *et al.* 2011, *ApJ*, 727, 15

Ofek, E. O., Cameron, P. B., Kasliwal, M. M., *et al.* 2007, *ApJ*, 659, L13

Pastorello, A., Smartt, S. J., Botticella, M. T., *et al.* 2010, *ApJ*, 724, L16

Quimby, R. M., Aldering, G., Wheeler, J. C., *et al.* 2007, *ApJ*, 668, L99

Quimby, R. M., Kulkarni, S. R., Kasliwal, M. M., *et al.* 2011, *Nature*, 474, 487

Rau, A., Kulkarni, S. R., Law, N. M., *et al.* 2009, *PASP*, 121, 1334

Rest, A., Foley, R. J., Gezari, S., *et al.* 2011, *ApJ*, 729, 88

Richardson, D., Branch, D., Casebeer, D., *et al.* 2002, *AJ*, 123, 745

Roming, P. W. A., Kennedy, T. E., Mason, K. O., *et al.* 2005, *Space Sci. Revs*, 121, 95

Smith, N., Chornock, R., Li, W., *et al.* 2008, *ApJ*, 686, 467

Smith, N., Chornock, R., Silverman, J. M., Filippenko, A. V., & Foley, R. J. 2010, *ApJ*, 709, 856

Smith, N., Li, W., Foley, R. J., *et al.* 2007, *ApJ*, 666, 1116

Umeda, H. & Nomoto, K. 2008, *ApJ*, 673, 1014

Woosley, S. E. 2010, *ApJ*, 719, L204

Woosley, S. E., Blinnikov, S., & Heger, A. 2007, *Nature*, 450, 390

Discussion

KATZ: What are the upper limits for X-ray emission from the SLSNe?

QUIMBY: The only SLSN I know of with a reported X-ray detection is SN 2006gy (Smith *et al.* 2007), and not everyone agrees that this is actually a detection. Upper limits exist for an additional number of SLSN-IIn (e.g. Chatzopoulos *et al.* 2011) and SLSN-Ic (e.g. Quimby *et al.* 2011), but these are not often all that constraining given the typically large distances: the limits near optical maximum are typically around 10^{43} erg s^{-1} in the *Swift*/XRT band, which is lower than the optical luminosities.

GAL-YAM: I know of no detections in X-ray, including, in my opinion, SN 2006gy.

METZGER: How do the host galaxies of SLSN-Ic compare to those of GRB hosts?

QUIMBY: The sample of SLSN-Ic hosts is still small, but it would appear that SLSN-Ic may favor even lower luminosity hosts than do GRBs.

Death of Massive Stars: Supernovae and Gamma-Ray Bursts
Proceedings IAU Symposium No. 279, 2012
P. Roming, N. Kawai & E. Pian, eds.

doi:10.1017/S1743921312012641

Mass loss and fate of the most massive stars

Jorick S. Vink

Armagh Observatory, College Hill, BT61 9DG, Armagh, United Kingdom
email: jsv@arm.ac.uk

Abstract. The fate of massive stars up to $300 M_{\odot}$ is highly uncertain. Do these objects produce pair-instability explosions, or normal Type Ic supernovae? In order to address these questions, we need to know their mass-loss rates during their lives. Here we present mass-loss predictions for very massive stars (VMS) in the range of 60-$300 M_{\odot}$. We use a novel method that simultaneously predicts the wind terminal velocities v_{∞} and mass-loss rate $\dot{M}$ as a function of the stellar parameters: (i) luminosity/mass Γ, (ii) metallicity Z, and (iii) effective temperature $T_{\rm eff}$. Using our results, we evaluate the likely outcomes for the most massive stars.

Keywords. stars: mass loss, stars: evolution, stars: Wolf-Rayet, supernovae: general

---

1. Introduction

Mass loss is the decisive parameter for predicting final stellar masses and the types of supernova (SN) explosion. Do the most massive stars disrupt as pair-instability SNe (PISNs), or do they produce normal SNe Ic? When does this occur in conjunction with a long gamma-ray burst (GRB)? Is low metallicity Z simply in *favour* due to lower mass-loss rates $\dot{M}$, or is it even a *stringent* requirement? Furthermore, the formation of intermediate mass-black holes (IMBHs) and the stellar black-hole mass distribution are determined by Z-dependent $\dot{M}$ (Heger *et al.* 2003; Eldridge & Vink 2006).

Another relevant issue concerns the stellar upper-mass limit. Until recently many researchers accepted a $150 M_{\odot}$ limit. Crowther *et al.* (2010) recently agued for much higher luminosities – with masses twice as high – for the WNh objects in dense clusters. A potential issue with the Crowther *et al.* luminosities is that these objects are clustered, involving a non-negligible chance of photon pollution from line-of-sight objects.

We have found a new WNh star VFTS 682 in 30 Dor (Evans *et al.* 2011; Bestenlehner *et al.* 2011). It is a near-identical twin of one of the 'Crowther' stars, R136a3. Surprisingly, VFTS 682 is in apparent isolation from the R136 cluster (see Bestenlehner *et al.* for a discussion on isolated formation or a "slow runaway" status). This enables a check on the reliability of the luminosities derived for the core stars. Our finding of $\log(L/L_{\odot}) = 6.5 \pm 0.2$ for VFTS 682 provides support for high luminosities and masses, as the chance of line-of-sight pollution is small for this isolated star. Mass-loss rates for VMS up to $300 M_{\odot}$ are needed to establish their fate. VMS are extremely close to the Eddington limit $\Gamma = g_{\rm rad}/g_{\rm grav} = \kappa L/(4\pi c G M)$.

2. Method: Monte Carlo mass-loss predictions

Stellar winds from massive stars are driven by radiation pressure on spectral lines (Castor *et al.* 1975, CAK), predominantly on Fe. The approach we use to compute $\dot{M}$ for VMS is similar to the Monte Carlo method used to predict $\dot{M}$ for normal OB stars (Vink *et al.* 2000). Until 2008 our methodology was semi-empirical, as we assumed a velocity law that reached a certain empirical v_{∞}. Müller & Vink (2008) suggested a new line-force

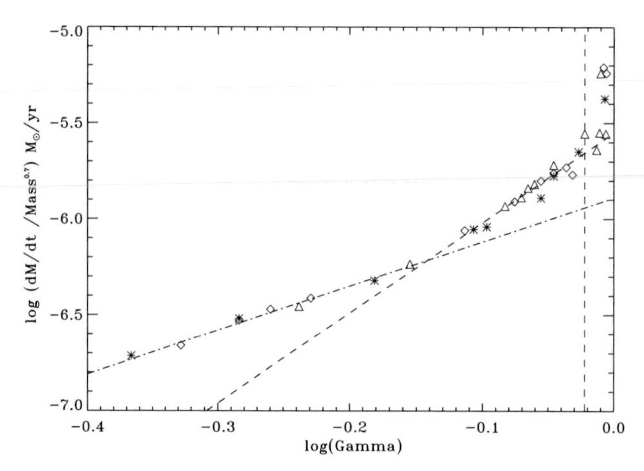

Figure 1. Mass-loss predictions versus the Eddington parameter Γ – divided by $M^{0.7}$. Symbols correspond to models of different mass ranges (Vink *et al.* 2011a).

parametrization that explicitly depends on radius (rather than the velocity gradient, as in CAK theory). We predicted v_∞ within ~25% of the observations. In Muijres *et al.* (2012) we tested the Müller & Vink approach by comparison to hydrodynamical models. As both methods gave similar results, we use the Müller & Vink approach for VMS.

Nugis & Lamers (2002) and Gräfener & Hamann (2008) studied radiative driving due to Fe-peak opacities in deep photospheric layers of Wolf-Rayet (WR) winds. As Γ crosses unity in deep layers, the sonic point is located at high optical depth, leading to the initiation of optically-thick winds. In the Monte Carlo models, one traces the driving over the entire wind, and as the bulk of the energy is transferred in the supersonic portion of the wind, one is less susceptible to the details of the photospheric region. Our strategy allows us to explore the transition from optically thin O-star winds to optically thick WR winds.

3. $\dot{M} - \Gamma$ dependence - Do PISNs exist at $Z_\odot$?

In Figure 1 we show mass-loss predictions for VMS as a function of the Eddington parameter Γ (see Vink *et al.* 2011a for details). Most notable is the presence of a *kink* in the relation. For O-type stars with "low" Γ and optically-thin winds, the $\dot{M} \propto \Gamma^x$ relationship is shallow, with $x \simeq 2$. There is a steepening at higher Γ, where x becomes $\simeq 5$. Here the objects show optically thick WR-like winds, with optical depths and wind efficiencies above unity.

Gräfener *et al.* (2011) recently provided empirical evidence for our predicted steep exponent ($x \simeq 5$), but note that there are still issues with our v_∞ values for the high Γ range. For now we employ the Vink *et al.* (2000) mass-loss recipe for our assessment of the fate of the most massive stars. These mass-loss rates agree extremely well with the rates discussed by Crowther *et al.* (2010) for the 30 Dor R136 core stars. We have recently also calibrated the Vink *et al.* rates using an analytic method and applied it to the most massive stars in the Arches cluster (Vink & Gräfener 2012).

Using Vink *et al.* (2000) rates for a star starting with $300 M_\odot$ we find $\dot{M} = 10^{-4.2}$ $M_\odot \mathrm{yr}^{-1}$. For a lifetime of 2.5 Myrs, this leads to a total main-sequence mass lost of $\simeq 150 M_\odot$. Additional mass loss during the core helium WR phase should further "evaporate" the object. Our results indicate that there is little room for substantial additional mass loss in luminous blue variable (LBV) eruptions. Our results also imply that IMBHs and pair-instability explosions are unlikely. Unless we go to lower Z environments.

4. $\dot{M} - Z$ dependence - Are GRBs confined to low Z?

The issue of mass loss and evolution at low Z has gained attention due to the issue of the progenitors of long GRBs. Within Woosley's collapsar model, GRB progenitors require two key properties: (i) a rapidly rotating core, and (ii) the absence of a hydrogen envelope. Therefore, GRB progenitors are thought to be rotating WR stars. The potential problem with this is that WR star have high mass loss which should remove the angular momentum before the core collapses.

In the rapidly rotating stellar models of Yoon & Langer (2005), the objects evolve "quasi-homogeneously". The stars are subject to a strong magnetic coupling between the core and envelope. If the rapid rotation can be maintained due to low main-sequence mass loss in low Z galaxies, the objects may avoid slow-down in a red supergiant (RSG) or LBV phase, and directly become rapidly rotating WR stars. If the WR winds also depend on Fe driving (Vink & de Koter 2005), the WR stars can maintain rapid rotation towards the very end, making GRBs – but *only* at low Z.

GRB data presented at this meeting suggest that GRBs are not restricted to low Z, but there seems to be a need for a GRB channel at high Z. We have recently identified a sub-group of rotating Galactic WR stars – allowing for a potential solution to this problem (Vink *et al.* 2011b; Gräfener *et al.* 2012b). Spectropolarimetry surveys show that the majority of WR stars have spherically symmetric winds indicative of slow rotation, but a small minority display signatures of a spinning stellar surface. We found this spinning sub-group to be surrounded by ejecta nebulae, which are thought to be ejected during a recent RSG/LBV phase, which suggests that these WR stars are still young and rotating.

If the core-surface coupling were strong enough, the cores would not be expected to rotate rapidly enough to make a GRB, but if the core-envelope coupling is less efficient, they may have the required angular momentum in their cores to make GRBs. In most high Z cases these stars would nonetheless still be expected to spin down due to mass loss, but within our post-RSG/LBV scenario one would not exclude the possibility of a high Z GRB. Yet, low Z environments are still preferred due to weaker WR winds.

5. $\dot{M} - T_{\rm eff}$ dependence - Do Luminous Blue Variables (LBVs) explode?

The stellar winds of O supergiants are fast ($\simeq$2000-4000 km/s) and transparent, whilst those emanating from lower $T_{\rm eff}$ B supergiants are much slower ($\simeq$100-1000 km/s). This is because O star winds are driven by high Fe ionization states, whilst those of B and later sub-types are driven by lower ones. This is wind bi-stability (BS).

LBVs increase their radii continuously on timescales of $\sim$10 yrs. These S Dor excursion across the HR diagram lead to winds with variable v_∞ and $\dot{M}$. If the LBV wind changes instantaneously at the BS-jump, we can explain the double-troughed Hα absorptions seen in LBV spectra (Groh & Vink 2011). Intriguingly such double-troughed Hα line profiles have also been seen in the luminous IIn SN 2005gj, which was for this reason suggested to have an LBV progenitor (Trundle *et al.* 2008). The same BS jump was also used to first suggest the LBV-SNe II link (Kotak & Vink 2006).

Even if $\dot{M}$ varies as a result of LBV radius changes, we still do not understand *why* LBVs change their radii (see Vink 2009 for a recent review). One possibility would be that the sub-photospheric outer envelopes of the stars become "inflated" as a result of the proximity to the Eddington limit (see Fig. 2). Ishii *et al.* (1999) first studied the

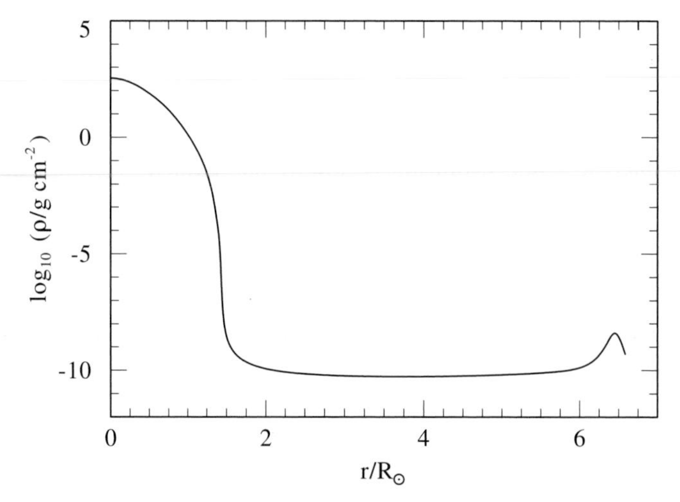

Figure 2. Density vs. radius for a $23M_\odot$ helium model of Gräfener *et al.* (2012a) showing a density inversion. This leads to an inflation of the outer envelope.

outer envelope inflation from stellar evolution models, and in Gräfener *et al.* (2012a) we developed an analytic explanation for how such an envelope inflation would occur. We described the radial inflation as a function of a dimensionless parameter W, which largely depends on the topology of the Fe-opacity peak. For $W > 1$, we discovered an instability limit for which the stellar envelope becomes unbound. Within our framework, we are in principle able to explain LBV S Dor variations. Stellar temperatures could be strongly affected, and there could be important implications of the *radii* of WR and LBV progenitors prior to collapse, as SN with different sub-types II, SN Ibc, and GRBs.

References

Bestenlehner, J. M., Vink, J. S., Gräfener, G., *et al.*, 2011, *A&A* 530L, 14
Castor, J. I., Abbott, D. C., & Klein, R. I., 1975, *ApJ* 195, 157
Crowther, P. A., Schnurr, O., Hirschi, R., *et al.*, 2010, *MNRAS* 408, 731
Eldridge, J. J. & Vink, J. S., 2006, *A&A* 452, 295
Evans, C. J., Taylor, W. D., Henault-Brunet, V., *et al.* 2011, *A&A* 530, 108
Gräfener, G. & Hamann, W.-R. 2008, *A&A* 482, 945
Gräfener, G., Vink, J. S., de Koter, A., & Langer, N., 2011, *A&A* 535, 56
Gräfener, G., Owocki, S. P., & Vink, J. S., 2012a, *A&A* 538, 40
Groh, J. H. & Vink, J. S., 2011, *A&A* 531L, 10
Heger, A., Fryer, C. L., Woosley, S. E., *et al.*, 2003, *ApJ* 591, 288
Ishii, M., Ueno, M., & Kato, M., 1999, PASJ 51, 417
Kotak, R. & Vink, J. S., 2006, *A&A* 460L, 5
Muijres, L., Vink, J. S., de Koter, A., *et al.*, 2012, *A&A* 537, 37
Müller, P. E. & Vink, J. S., 2008, *A&A* 492, 493
Nugis, T., & Lamers, H. J. G. L. M., 2002, *A&A* 389, 162
Trundle C., Kotak R., Vink J. S., & Meikle, W. P. S., 2008, *A&A* 483L, 47
Vink, J. S., 2009, astro-ph/0905.3338
Vink, J. S. & de Koter. A., 2005, *A&A* 442, 587
Vink, J. S., de Koter, A., & Lamers, H. J. G. L. M., 2000, *A&A* 362, 295
Vink, J. S., Muijres, L. E., Anthonisse, B., *et al.*, 2011a, *A&A* 531, 132
Vink, J. S., Gräfener, G., & Harries, T. J., 2011b, *A&A* 536L, 10
Yoon, S.-C. & Langer, N., 2005, *A&A* 443, 643

Discussion

OMUKAI: $300 M_\odot$ stars are unstable to pulsation by the epsilon mechanism. Did you include this effect in evaluating the mass-loss rate?

VINK: No, we didn't. The epsilon mechanism is thought to grow too slowly, and is usually not considered all that relevant.

OMUKAI: You said that the WR envelope has a density inversion during the inflation phase. Is it hydrodynamically stable?

VINK: The Gräfener *et al.* (2012) models are static, and until we have studied the hydrodynamic case we cannot be 100% sure. However, the suggested structure might not be all that unstable. Note that there is a lot of supporting radiation pressure!

KULKARNI: Angular momentum will only be efficiently removed from a mass-losing star if the core is coupled to the envelope. Could you comment on our current understanding of this coupling?

VINK: There is a debate regarding the magnetic coupling of the core and the envelope. Some massive star evolution modellers include magnetic fields, which results in a strong core-envelope coupling (e.g. Brott *et al.* 2011, *A&A* 530, 115), as this seems to be favoured when regarding the spins of neutron stars (Langer/Bonn argument). The Geneva models do not include magnetic fields, leading to less coupling.

Death of Massive Stars: Supernovae and Gamma-Ray Bursts
Proceedings IAU Symposium No. 279, 2012
P. Roming, N. Kawai & E. Pian, eds.

doi:10.1017/S1743921312012653

The Flavours of SN II Light Curves

Iair Arcavi

Department of Particle Physics and Astrophysics, The Weizmann Institute of Science, Rehovot 76100, Israel
email: iair.arcavi@weizmann.ac.il

Abstract. We present R-Band light curves of Type II supernovae (SNe) from the Caltech Core Collapse Program (CCCP). With the exception of interacting (Type IIn) SNe and rare events with long rise times, we find that most light curve shapes belong to one of three distinct classes: plateau, slowly declining and rapidly declining events. The latter class is composed solely of Type IIb SNe which present similar light curve shapes to those of SNe Ib, suggesting, perhaps, similar progenitor channels. We do not find any intermediate light curves, implying that these subclasses are unlikely to reflect variance of continuous parameters, but rather might result from physically distinct progenitor systems, strengthening the suggestion of a binary origin for at least some stripped SNe. We find a large plateau luminosity range for SNe IIP, while the plateau lengths seem rather uniform at approximately 100 days. We present also host galaxy trends from the Palomar Transien Factory (PTF) core collapse SN sample, which augment some of the photometric results.

Keywords. supernovae: general

---

1. Introduction

Type II supernovae (SNe) are widely recognized as the end stages of massive H-rich stars. Together with Type Ib/c events, they represent the bulk of observed core collapse SNe (see Filippenko 1997 for a review of SN classifications). Several sub-types of Type II SNe have been observed. Those showing a plateau in their light curve are known as Type IIP events, while those showing a linear decline from peak magnitude are classified as IIL. A third class of events, Type IIb, characterized by its spectral rather than its photometric properties, develops prominent He features at late times. Finally, Type IIn SNe display narrow lines in their spectra, indicative of interaction between the SN ejecta and a dense circumstellar medium.

Red supergiants (RSGs) have been directly identified as the progenitors of Type IIP SNe (see Smartt 2009 for a review). Such stars have thick hydrogen envelopes that are ionized by the explosion shock wave. As the shocked envelope expands and cools, it recombines, releasing radiation at a roughly constant rate, thus producing a plateau in the light curve (e.g. Popov 1993; Kasen & Woosley 2009). It follows that SNe IIL might be the explosions of stars with less massive H envelopes that can not support a plateau in their light curve. SN IIb progenitors, then, would contain an even smaller H mass.

However, if SNe IIP-IIL-IIb progenitors represent merely a sequence of decreasing H-envelope mass, one would expect the properties of these SNe to behave as a continuum. Specifically, a gradual transition in light curve shape should be observed when examining a homogeneous sample of events.

The Caltech Core Collapse Program (CCCP) is a large observational survey which made use of the robotic 60-inch (P60) and Hale 200-inch telescopes at Palomar Observatory to obtain optical $BVRI$ photometry and spectroscopy of 48 nearby core collapse SNe.

Light curves of Type Ib/c SNe from CCCP have been presented and analyzed by Drout *et al.* (2011). Type IIn CCCP events are treated by Kiewe *et al.* (2012). Here we present the R-Band photometry of 21 non-interacting Type II SNe with well observed light curves collected through CCCP.

2. Photometry

Our light curves are produced using image subtraction via the CPM method (Gal-Yam *et al.* 2008a) for PSF matching. Due to incomplete data for three of the events, we use photometry published in the literature. The light curve of SN2004fx is taken from Hamuy *et al.* (2006), that of SN2005ay from Gal-Yam *et al.* (2008b) and that of SN2005cs from Pastorello *et al.* (2009).

3. Results and Discussion

We plot the R-Band light curves of 15 Type II events normalized to peak magnitude in Figure 1. Rather than forming a continuum, we find that the light curves group into three distinct sub-classes: plateau, slowly declining (1-2 Mag/100 days) and initially rapidly declining (5-6 Mag/100 days) events. We note that the three rapidly declining events are all Type IIb and that they display similar light curve shapes to those of Type Ib/c SNe (Drout *et al.* 2011).

Three events (SN2004ek, SN2005ci and SN2005dp; Figure 2) display prolonged rising periods in their light curves. They do not show signs of interaction in their spectra and may be explosions of compact blue supergiant progenitors (Kleiser *et al.* 2011; Pastorello *et al.* 2012).

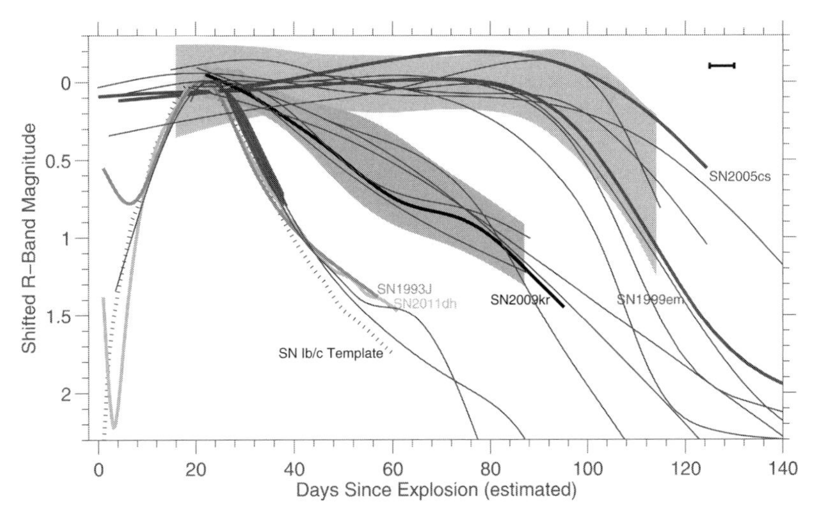

Figure 1. R-band light curves of 15 Type II SNe from CCCP, normalized in peak magnitude (SN2004fx data taken from Hamuy *et al.* 2006; SN2005ay data taken from Gal-Yam *et al.* (2008b); SN2005cs data taken from Pastorello *et al.* 2009). Reference SNe are shown for comparison (SN1999em from Leonard *et al.* 2002; SN2009kr from Fraser *et al.* 2010; SN1993J from Richmond *et al.* 1994; SN2011dh from Arcavi *et al.* 2011). We also overplot the SN Ib/c template derived by Drout *et al.* (2011). The data have been interpolated with spline fits (except for SN2005by, where a polynomial fit provided a better trace to the data). The shaded regions denote the average light curve $\pm 2\sigma$ of each subclass. The typical 5-day uncertainty in determining the explosion times is illustrated by the interval in the top right corner.

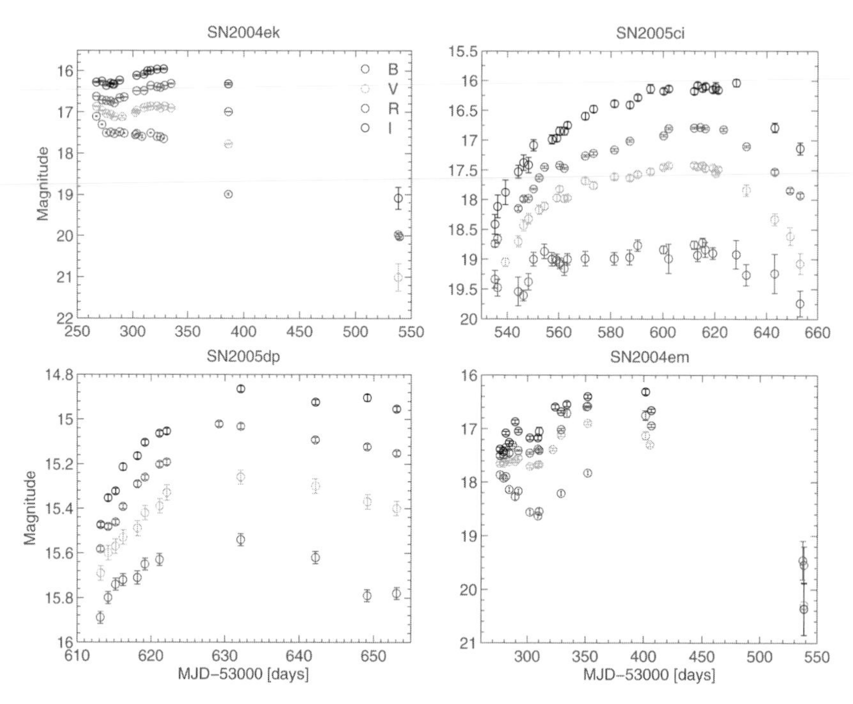

Figure 2. *BVRI* light curves of the CCCP events not included in Figure 1. Three events (SN2004ek, SN2005ci, SN2005dp) show long rise times, while one peculiar event (SN2004em) changes behavior from flat to rising around three weeks after explosion.

Finally, one event (SN2004em; Figure 2) displays a very peculiar photometric behavior. For the first few weeks it is similar to a Type IIP SN, while around day 25 it suddenly changes behavior to resemble a SN1987A-like event.

3.1. *Declining SNe*

Aside from establishing a different rate of decline for SNe IIb compared to SNe IIL, Figure 1 suggests that the IIP, IIL and IIb subtypes do not span a continuum of physical parameters, such as H envelope mass. Rather, additional factors should be considered. Specifically, Type IIb events might arise from binary systems (as suggested also by recent progenitor studies for SN1993J, Maund *et al.* 2004; SN2008ax, Crockett *et al.* 2008; SN2011dh, Arcavi *et al.* 2011, Van-Dyk *et al.* 2011). The similarity of the Type IIb light curves to those of Type Ib events (in addition to the known spectral similarities at late times) suggests that these two types of events might come from similar progenitor systems. Metallicity might be an important factor driving some of the systems to explode as Ib SNe and others as IIb's (Fig. 3; Arcavi *et al.*, in prep.).

3.2. *Plateau SNe*

The *R*-Band light curves of the Type IIP SNe, on an absolute magnitude scale, can be seen in Figure 4. We find a wide range of plateau luminosities, but do not have enough statistics to test whether they form a continuous distribution or if there are two distinct underlying types (bright and faint), as previously suggested (Pastorello *et al.* 2004). The

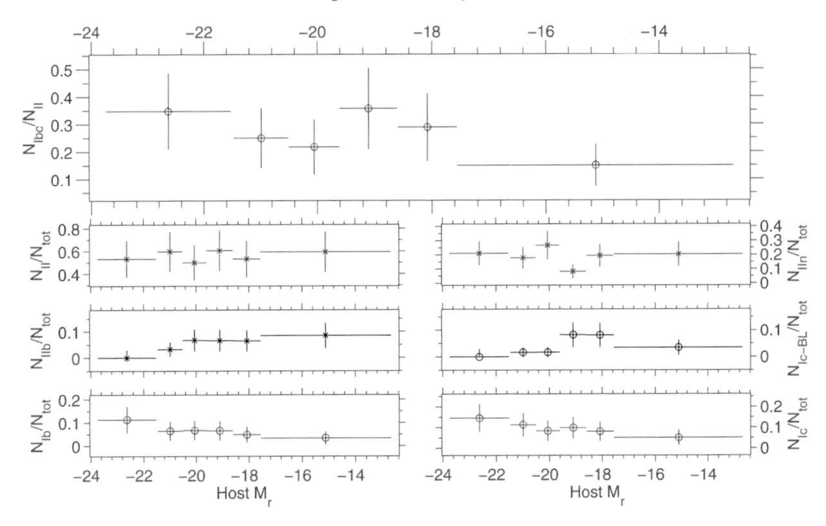

Figure 3. Core collapse SN fractions as a function of host galaxy luminosity, considered as a proxy for metallicity (Tremonti *et al.* 2004), for 369 events discovered and followed through the Palomar Transient Factory (PTF; Rau *et al.* 2009, Law *et al.* 2009).

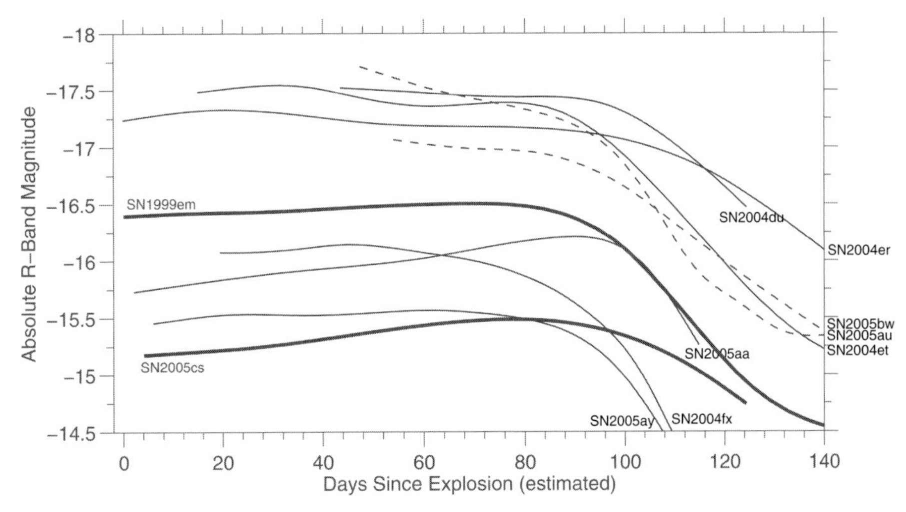

Figure 4. *R*-band light curves of 9 Type IIP SNe from CCCP (SN2004fx data taken from Hamuy *et al.* 2006; SN2005ay data taken from Gal-Yam *et al.* (2008b); SN2005cs data taken from Pastorello *et al.* 2009) with respect to their estimated explosion time (except for SN2005au and SN2005bw, marked by dashed lines, for which the explosion date is not known to good accuracy). SN1999em (Leonard *et al.* 2002) is shown for comparison. Spline fits were applied to the data

plateau lengths, however, seem rather uniform at ~ 100 days (with the sole exception of SN2004fx, displaying a shorter plateau)†.

The scarcity of observed short plateaus and the possibly related sharp distinction between IIP and IIL light curves is evident. Such a pronounced absence of intermediate

† note that SN2005au and SN2005bw are plotted only to show their plateau luminosity, their plateau lengths are unknown due the lack of sufficient constraints on their explosion time

events might suggest that Type IIL SNe are powered by a different mechanism than that associated with SNe IIP (e.g., Kasen & Bildsten 2010).

References

Arcavi, I., Gal-Yam, A., Yaron, O., *et al.* 2011, *ApJL*, 742, L18
Crockett, R. M., Eldridge, J. J., Smartt, S. J., *et al.* 2008, *MNRAS*, 391, L5
Drout, M. R., Soderberg, A. M., Gal-Yam, A., *et al.* 2011, *ApJ*, 741, 97
Filippenko, A. V. 1997, *ARA&A*, 35, 309
Fraser, M., Takáts, K., Pastorello, A., *et al.* 2010, *ApJL*, 714, L280
Gal-Yam, A., Maoz, D., Guhathakurta, P., & Filippenko, A. V. 2008a, *ApJ*, 680, 550
Gal-Yam, A., Bufano, F., Barlow, T. A., *et al.* 2008b, *ApJL*, 685, L117
Hamuy, M., Folatelli, G., Morrell, N. I., *et al.* 2006, *PASP*, 118, 2
Kasen, D. & Bildsten, L. 2010, *ApJ*, 717, 245
Kasen, D. & Woosley, S. E. 2009, *ApJ*, 703, 2205
Kiewe, M., Gal-Yam, A., Arcavi, I., *et al.* 2012, *ApJ*, 744, 10
Kleiser, I. K. W., Poznanski, D., Kasen, D., *et al.* 2011, *MNRAS*, 415, 372
Law, N. M., Kulkarni, S. R., Dekany, R. G., *et al.* 2009, *PASP*, 121, 1395
Leonard, D. C., Filippenko, A. V., Gates, E. L., *et al.* 2002, *PASP*, 114, 35
Maund, J. R., Smartt, S. J., Kudritzki, R. P., Podsiadlowski, P., & Gilmore, G. F. 2004, *Nature*, 427, 129
Pastorello, A., Zampieri, L., Turatto, M., *et al.* 2004, *MNRAS*, 347, 74
Pastorello, A., Valenti, S., Zampieri, L., *et al.* 2009, *MNRAS*, 394, 2266
Pastorello, A., Pumo, M. L., Navasardyan, H., *et al.* 2012, *A&A*, 537, A141
Popov, D. V. 1993, *ApJ*, 414, 712
Rau, A., Kulkarni, S. R., Law, N. M., *et al.* 2009, *PASP*, 121, 1334
Richmond, M. W., Treffers, R. R., Filippenko, A. V., *et al.* 1994, *AJ*, 107, 1022
Smartt, S. J. 2009, *ARA&A*, 47, 63
Tremonti, C. A., Heckman, T. M., Kauffmann, G., *et al.* 2004, *ApJ*, 613, 898
Van Dyk, S. D., Li, W., Cenko, S. B., *et al.* 2011, *ApJL*, 741, L28

Discussion

CHORNOCK: Can you comment on the peak luminosities of the objects you have classified as SNe IIL / slow decliners?

ARCAVI: We do not have the peaks resolved for all these events, but they seem to be confined to the −16 to −17 magnitude range.

CHORNOCK: What about SN 1979C?

ARCAVI: SN 1979C, known as the canonical Type IIL, does not fit into any of the three subtypes we identify in Fig. 1. SN 1979C was exceptionally bright and it is still debated whether it was an interacting event. We therefore choose not to include it in the analysis.

SMITH: You showed different behavior of SNe IIb vs. Ib with host galaxy luminosity (and presumably metallicity). Have you looked for a similar trend with the location within the host galaxy?

ARCAVI: The location within the host galaxy can serve as a proxy for progenitor mass, and is therefore and excellent complementary paramater to check. We hope to be able to conduct this analysis in the future, especially for the very faint host galaxies in the PTF sample (most of which require the resolution of *HST*).

KATZ: Is there a continuous spectrum between SNe Ib and IIb?

ARCAVI: Yes, in fact at late times, the two events have identical spectra (which is why multi-epoch spectroscopy is required to correctly classify them).

Additional discussion points where incorporated into the text.

Death of Massive Stars: Supernovae and Gamma-Ray Bursts
Proceedings IAU Symposium No. 279, 2012
P. Roming, N. Kawai & E. Pian, eds.

doi:10.1017/S1743921312012665

GRB Prompt X-ray Emission

Takanori Sakamoto[1,2,3]

[1]Center for Research and Exploration in Space Science and Technology (CRESST), NASA Goddard Space Flight Center, Greenbelt, MD 20771, U.S.A.

[2]Joint Center for Astrophysics, University of Maryland, Baltimore County, 1000 Hilltop Circle, Baltimore, MD 21250, U.S.A.

[3]NASA Goddard Space Flight Center, Greenbelt, MD 20771, U.S.A.
email: Taka.Sakamoto@nasa.gov

Abstract. I present the observational properties of the prompt emission of supernova associated GRBs (SN-GRBs) focusing on temporal and spectral characteristics. I compare the properties of SN-GRBs with typical long GRBs to see whether there is a distinct difference or not. Furthermore, I present our attempt to search for hard X-ray emission prior to the discovery date from optically identified type Ibc supernovae using *Swift* BAT survey data.

Keywords. gamma rays: bursts

1. Introduction

After the first discovery of the supernova associated gamma-ray burst (SN-GRB) GRB 980425 detected by BATSE and *Beppo*SAX (Galama *et al.* 1998), the number of SN-GRBs has been increasing thanks to a rapid and accurate GRB position notice by *HETE-2* (Ricker *et al.* 2003), *INTEGRAL* (Winkler *et al.* 2003), and *Swift* (Gehrels *et al.* 2004). On the other hand, we have started to see a large diversity in the characteristics of the prompt emission of SN-GRBs (e.g., Kaneko *et al.* 2007). I will review the current observational status of the prompt emission properties of spectroscopically identified SN-GRBs by comparing them to typical long GRBs.

2. Sample of SN-GRBs

Table 1 shows the list of the spectroscopically identified SN-GRB samples in this work. Three GRBs – 030329, 060218, and 100316D (bold fonts in the table) – have secure associations with supernovae since a spectral evolution from a non-thermal power-law to a supernova-like thermal feature is clearly seen in the optical spectroscopic observations (e.g., Stanek *et al.* 2003, Hjorth *et al.* 2003 for GRB 030329; e.g., Pian *et al.* 2006 for GRB 060218; e.g., Chornock *et al.* 2011 for GRB 100316D).

Table 1. Spectroscopically identified SN-GRB sample.

GRB	SN	Mission	Redshift	SN-Type
980425	1998bw	BATSE/BSAX	0.0085	Ic
030329	2003dh	HETE-2	0.1685	Ic
031203	2003lw	INTEGRAL	0.1005	Ibc
060218	2006aj	Swift	0.0331	Ic
091127	2009nz	Swift	0.490	Ic
100316D	2010bh	Swift	0.059	Ic
101219B	2010ma	Swift	0.55	Ic

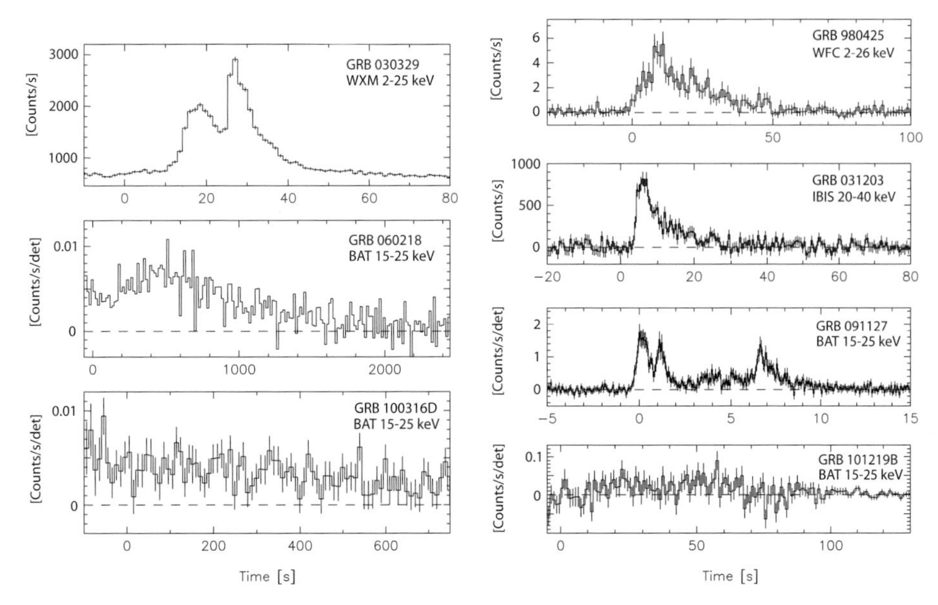

Figure 1. Prompt emission light curves of the SN-GRBs.

3. Temporal properties

3.1. *Light curves*

Figure 1 shows the prompt emission light curves of the SN-GRBs observed by various GRB instruments. As seen in the figure, there are large varieties in the light curves. For example, there are SN-GRBs which are composed of several bright overlapping pulses such as GRB 030329 and GRB 091127. On the other hand, GRB 060218, GRB 100316D and GRB 101219B show a very smooth and a long duration profile in their light curves.

3.2. *Durations*

The observed durations of SN-GRBs show diversity. While GRB 060218 and GRB 100316D show a duration of several thousands of seconds (Campana *et al.* 2006; Starling *et al.* 2011), SN-GRBs with a duration of a several tens to hundreds of seconds are also common. Although the durations of GRB 060218 and GRB 100316D are exceptionally long, the rest of the sample is well within the duration distribution of typical long GRBs (Figure 2).

3.3. *Lag-Luminosity relation*

Norris *et al.* (2000) found a correlation between the spectral lag and the peak luminosity (the so called lag-luminosity relation) of the prompt emission from the BATSE long GRBs. It is also known that short GRBs and several low luminous GRBs do not follow this relation (Norris & Bonnell 2006). The lag-luminosity relation is further confirmed by the *HETE-2* (Arimoto *et al.* 2010) and the *Swift* (Ukwatta *et al.* 2010) GRB samples. Figure 3 shows the BATSE, the *HETE-2* and the *Swift* long GRBs and the SN-GRBs in a peak luminosity versus lag plane. The lag and the peak luminosity values of GRB 980425, GRB 031203 and GRB 060218 are extracted from Norris *et al.* (2000), Sazonov, Lutovinov & Sunyaev (2004) and Liang *et al.* (2006), respectively. The values of the rest of the SN-GRBs are derived in this work. GRB 091127, GRB 030329 and possibly GRB 060218 are consistent with the lag-luminosity relations. According to Liang *et al.* (2006), the lag value of GRB 060218 has a large uncertainty since it requires an extrapolation

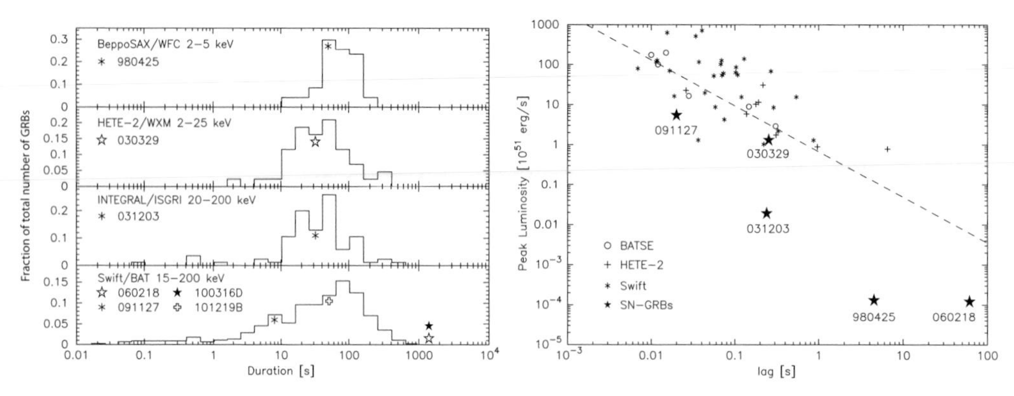

Figure 2. Observed duration distribution of *Beppo*SAX/WFC (2-5 keV), *HETE-2*/WXM (2-25 keV), *INTEGRAL*/ISGRI (20-200 keV) and *Swift*/BAT (15-200 keV) from top to bottom panel. The durations of the SN-GRBs are overlaid in the histograms.

Figure 3. Lag-Luminosity relation of the BATSE (Norris *et al.* 2000), the *HETE-2* (Arimoto *et al.* 2010) and the *Swift* (Ukwatta *et al.* 2010) long GRBs. The dotted line is the lag-luminosity relation, $L_{\rm iso} \propto lag^{-1.14}$, originally proposed by Norris *et al.* (2000). The SN-GRBs are shown as stars.

from the original lag measurement between the BAT (Barthelmy *et al.* 2005) and the XRT (Burrows *et al.* 2005) band to that of the BATSE standard band to be able to plot on the lag-luminosity plane. Based on their analysis, GRB 060218 is within the 2σ confidence region of the relation. However, as noted by various previous works, GRB 980425 and GRB 031203 do not follow the relation. The relatively low peak luminosity for both GRB 980425 and GRB 031203 makes them outliers to this relationship. Note that since the low luminosity GRB 060218 seems consistent with the relation, not all low luminosity GRBs are outliers of the lag-luminosity relation.

4. Spectral properties

4.1. *Band function parameters*

The prompt emission spectra of SN-GRBs which have reported broad-band spectral parameters are well fitted with the Band function (Band *et al.* 1993). To compare the best fit parameters of the Band function between the SN-GRBs and typical long GRBs, the histograms of the low-energy photon index, α, the high-energy photon index, β, and the spectral peak energy, $E_{\rm peak}$, of the long GRBs are shown in Figure 4. α and β values are from the BATSE long GRBs (Goldstein *et al.* 2010). $E_{\rm peak}$ values are from Goldstein *et al.* (2010) for the BATSE, Sakamoto *et al.* (2005) and Pélangeon *et al.* (2008) for the *HETE-2*, and Sakamoto *et al.* (2011) for the *Swift* GRBs. The values of the best fit Band function parameters of the SN-GRBs are marked in the histograms. All the Band function parameters of the SN-GRBs are consistent with typical long GRBs.

4.2. *Additional blackbody component in X-ray spectrum*

The existence of an additional blackbody component in the prompt X-ray spectrum of GRB 060218 (Campana *et al.* 2006) and GRB 100316D (Starling *et al.* 2011) has been reported. The blackbody temperature is stable in the range of 0.1-0.2 keV up to several hundreds of seconds after the trigger, and then, decreases to < 0.01 keV. I also found an additional blackbody component for GRB 101219B. The *Swift* XRT spectrum extracted from 152 s to 300 s after the trigger shows a significant improvement in the fit with a blackbody plus a power-law model (χ^2/d.o.f. $= 85.8/100$) over a power-law model

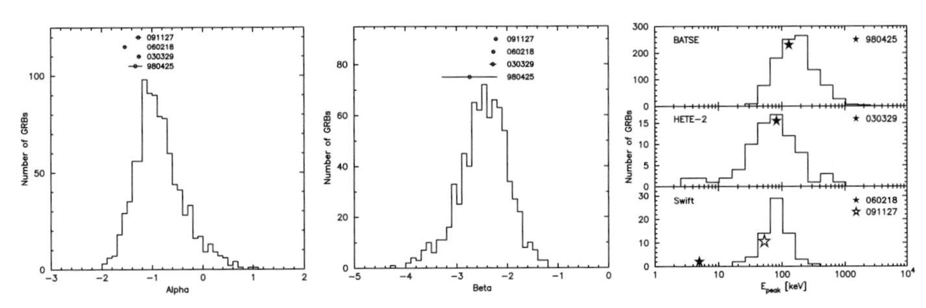

Figure 4. Distributions of the low-energy photon index α (left), the high-energy photon index β (middle) and E_{peak} (right). The E_{peak} histograms of the BATSE, the *HETE-2* and the *Swift* long GRBs are shown from top to bottom.

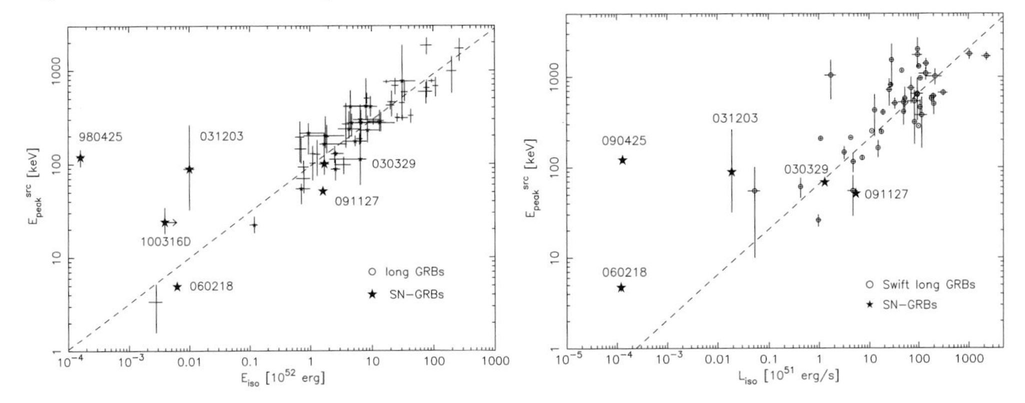

Figure 5. E_{peak}-E_{iso} relation. The long GRB samples are from Amati *et al.* 2006 and Sakamoto *et al.* 2011.

Figure 6. E_{peak}-L_{iso} relation. The long GRB samples are from Nava *et al.* 2012.

(χ^2/d.o.f. = 104.1/102). The F-test probability is 6.3×10^{-5} between these two fits. The best fit blackbody temperature is 0.12 ± 0.02 keV which is similar to the case of GRB 060218 and GRB 100316D. Since most of the XRT spectra of long GRBs can well be fit by a simple power-law model, this additional blackbody component could be a unique spectral feature of SN-GRBs.

4.3. E_{peak}-E_{iso} *and* E_{peak}-L_{iso} *relations*

The empirical spectral relations between E_{peak} and the isotropically radiated γ-ray energy E_{iso} (E_{peak}-E_{iso} relation; Amati *et al.* 2002) and between E_{peak} and the peak isotropic luminosity L_{iso} (E_{peak}-L_{iso} relation; Yonetoku *et al.* 2004) are well discussed relationships of the prompt emission. As shown in Figure 5 and 6, most of the long GRBs are consistent with the E_{peak}-E_{iso} and the E_{peak}-L_{iso} relations including the SN-GRBs except GRB 980425.

5. Searching for hard X-ray emission in supernova

We searched for hard X-ray emission using the *Swift* BAT survey data for 123 type Ibc and IIp supernovae (< 100 Mpc) discovered between 2005 and 2010. The BAT data starting from $\sim$ 28 days prior to the discovery up to the discovery date are processed. No significant hard X-ray emission was found for ordinary type Ibc and IIp supernovae. Our result is consistent with a radio survey of ordinary type Ibc supernovae (Soderberg *et al.* 2006), and might indicate that a fundamentally different mechanism is required to produce a GRB from a supernova.

6. Summary

- Burst duration, spectral lag and spectral parameters of SN-GRBs are consistent with typical long GRBs.
- In addition to the Band function spectrum, GRB 060218 and GRB 100316D (and possibly GRB 101219B) show kT = 0.1-0.2 keV blackbody component in the prompt emission spectrum.
- Most of the SN-GRBs follow the empirical relations between lag and luminosity, E_{peak}-E_{iso} and E_{peak}-L_{iso}. However, GRB 980425 and possibly GRB 031203 are outliers of those relations.
- No hard X-ray emission is found from ordinary type Ibc and IIp supernovae.

References

Amati, L., *et al.* 2002, *A&A*, 390, 81
Amati, L., *et al.* 2006, *MNRAS*, 372, 233
Arimoto, M., *et al.* 2010, *PASJ*, 62, 487
Band, D. L., *et al.* 1993, *ApJ*, 413, 281
Barthelmy, S. D., *et al.* 2005, *Space Sci. Revs*, 121, 143
Burrows, D. N., *et al.* 2005, *Space Sci. Revs*, 121, 165
Campana, S., *et al.* 2006, *Nature*, 442, 1008
Chornock, R., *et al.* 2011, *ApJ* submitted (astro-ph/arXiv:1004.2262)
Galama, T. J., *et al.* 1998, *Nature*, 434, 1104
Gehrels, N., *et al.* 2004, *ApJ*, 611, 1005
Goldstein, A., *et al.* 2010, http://gammaray.msfc.nasa.gov/ goldstein/
Hjorth, J., *et al.* 2003, *Nature*, 423, 847
Kaneko, Y., *et al.* 2007, *ApJ*, 654, 385
Liang, E.-W., *et al.* 2006, *ApJ*, 653, L81
Nava, L., *et al.* 2012, *MNRAS*, 421, 1256
Norris, J. P., *et al.* 2000, *ApJ*, 534, 248
Norris, J. P. & Bonnell, J. T. 2006, *ApJ*, 643, 266
Pélangeon, A., *et al.* 2008, *A&A*, 491, 157
Pian, E., *et al.* 2006, *Nature*, 442, 1011
Ricker, G. R., *et al.* 2003, *AIP-CP*, 662, 3
Sakamoto, T. *et al.* 2005, *ApJ*, 629, 311
Sakamoto, T., *et al.* 2011, *ApJS*, 195, 1
Sazonov, S. Y., Lutovinov, A. A., & Sunyaev, R. A. 2004, *Nature*, 430, 646
Soderberg, A. M., *et al.* 2006, *ApJ*, 638, 930
Starling, R. L. C., *et al.* 2011, *MNRAS*, 411, 2792
Stanek, K. Z., *et al.* 2003, *ApJ*, 591, L17
Ukwatta, T. N., *et al.* 2010, *ApJ*, 711, 1073
Winkler, C., *et al.* 2003, *A&A*, 411, 1
Yonetoku, D., *et al.* 2004, *ApJ*, 609, 935

Discussion

BURROWS: How does beaming affect your conclusions about the lack of hard X-ray emission from type Ibc SNe?

SAKAMOTO: We believe that beaming is playing a major role in generating a hard X-ray prompt emission between SN-GRBs and ordinary type Ibc SNe.

KULKARNI: As yourself noted the SN-GRB sample has strong selection biases. The events have to be at low redshift. This selection effect may strongly affect the detection of very soft (BB) components.

KULKARNI: It was my impression that the volumetric rate of SN-GRBs is an order of magnitude larger than those of cosmological GRBs. If so, the selection biases are even more severe.

Death of Massive Stars: Supernovae and Gamma-Ray Bursts
Proceedings IAU Symposium No. 279, 2012
P. Roming, N. Kawai & E. Pian, eds.

doi:10.1017/S1743921312012677

Optical and near-infrared flares in GRB afterglows

Thomas Krühler

Dark Cosmology Centre, Niels Bohr Institute, University of Copenhagen, Juliane Maries Vej 30, 2100 Copenhagen, Denmark.
email: tom@dark-cosmology.dk

Abstract. Among the diversities in the very early evolution of GRB afterglows are bright optical/near-infrared flares before or superimposed onto an otherwise smoothly decaying afterglow light curve. A lot has been learned about GRBs by using an optical flare or lack thereof as a diagnostic of the emission mechanisms and outflow conditions. In this contribution I will review the observational properties of rising and decaying light-curves in GRB afterglows, discuss their possible physical origins, and highlight in which way they help in understanding GRB and afterglows physics.

Keywords. gamma rays: bursts

1. Introduction

The launch of the *Swift* satellite (Gehrels *et al.* 2004) in 2004 opened a new field of gamma-ray burst (GRB) afterglow physics. With its precise localization by the Burst Alert Telescope, rapid slewing capabilities and early follow-up with two instruments in the X-ray and ultraviolet/optical regime, studies of the early afterglow phase were possible for the first time with larger sample statistics of around 100 per year.

The optical/near-infrared (NIR) afterglow shows significant diversity in its early evolution. While the late phases ($t \gtrsim 1$ day) of the afterglow of long GRBs are generally well described with smoothly decaying power-laws with temporal indices between $t^{-0.5}$ to $t^{-2.5}$, and additional components from associated supernovae or host galaxies, the early afterglow ($t \lesssim 1$ day) often shows epochs of flaring, or more generally, rise and decay (see Fig. 1).

2. Reverse shocks and lack thereof

After the prompt internal shock phase, the optical afterglow light curve is composed of two different emission components. The reverse shock (RS) propagating into the ejecta and the forward shock (FS) travelling into the surrounding medium (e.g., Zhang *et al.* 2003). Rapid optical observations of the early transition phase between prompt and afterglow emission can constrain the nature of the outflow (e.g., Nakar & Piran 2004). Baryonic ejecta are expected to produce an optical flash, that can be associated with a RS. The characteristic observational signature of a reverse shock is a relatively steep decline ($\propto t^{-2}$), that is too fast for the standard afterglow emission. A Poynting flux dominated afterglow, however, should preferentially show the FS emission.

The hydrodynamical calculations from the fireball model have succeeded in describing the generic afterglow of GRBs from several minutes to days post burst. However, the majority of bursts does not show bright optical flashes and apparently lack a strong RS

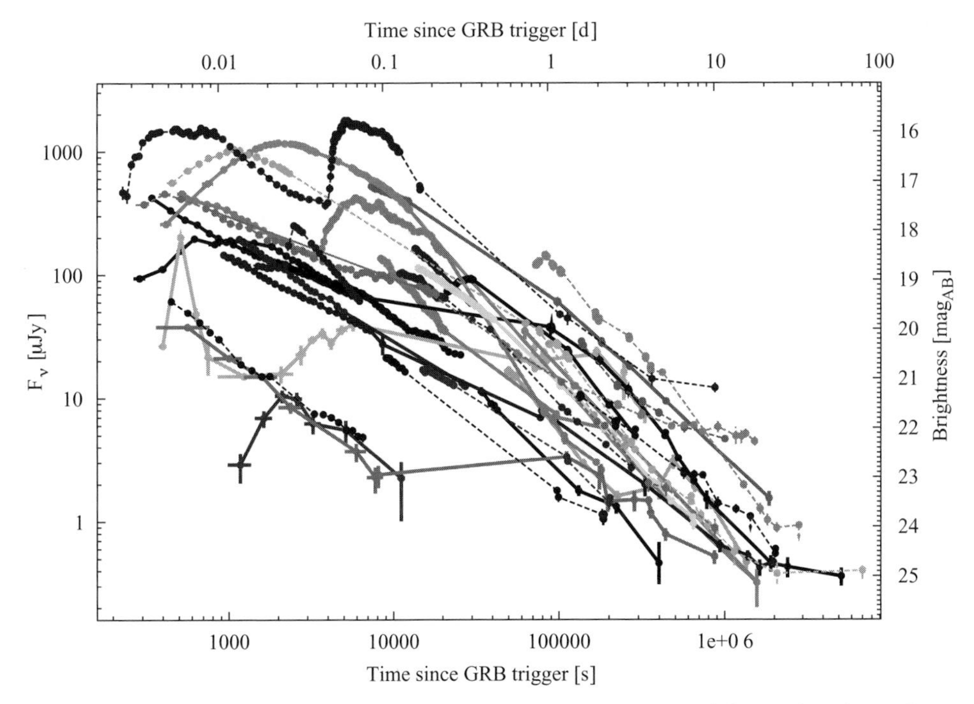

Figure 1. Selection of observed optical light-curves for a number of GRB afterglows obtained with the Gamma-ray Optical Near-infrared Detector (GROND, Greiner *et al.* 2008). Note the considerable variety in the early evolution of optical/NIR afterglows.

emission component (Roming *et al.* 2006, Kann *et al.* 2010). This fact provides observational support that the strength of the optical emission from the RS may be weaker than previously calculated, or that the RS is not radiated in the optical wavelength regime. The properties of the magnetic fields in the ejecta arguably play a crucial role, and might shift the typical synchrotron emission of the reverse shock out of the optical wavelength bands (e.g., Beloborodov 2005).

3. The onset of the afterglow

The optical/NIR light curve is in many cases dominated by an early increase in brightness (e.g., Molinari *et al.* 2007). This early rise is achromatic (see Fig. 2). Thus a movement of the characteristic synchrotron frequency through the observed optical bands, as well as dust destruction are readily ruled out as the origin of the initial rise (see e.g., Krühler *et al.* 2009b, Perley *et al.* 2010). The observational characteristics of this afterglow onset are a temporal rise of approximately $\propto t^{0.5-3}$ and a very smooth turnover to the subsequent decay (see Fig. 2).

The early rise in the optical afterglow light-curve is generally attributed to the onset of the forward shock emission. This happens when the swept up medium efficiently decelerates the ejecta. From the time of the light curve peak (typically at few tens to hundreds of seconds), physical parameters of the outflow, such as the initial bulk Lorentz factor Γ_0 or the deceleration radius can be constrained. The early optical afterglow hence provides a robust measurement and confirmation of the ultra-relativistic nature of the GRB phenomenon, yielding values of $100 \lesssim \Gamma_0 \lesssim 500$, and deceleration radii of 10^{16} to 10^{17} cm with only a weak dependence on the uncertain micro-physical parameters.

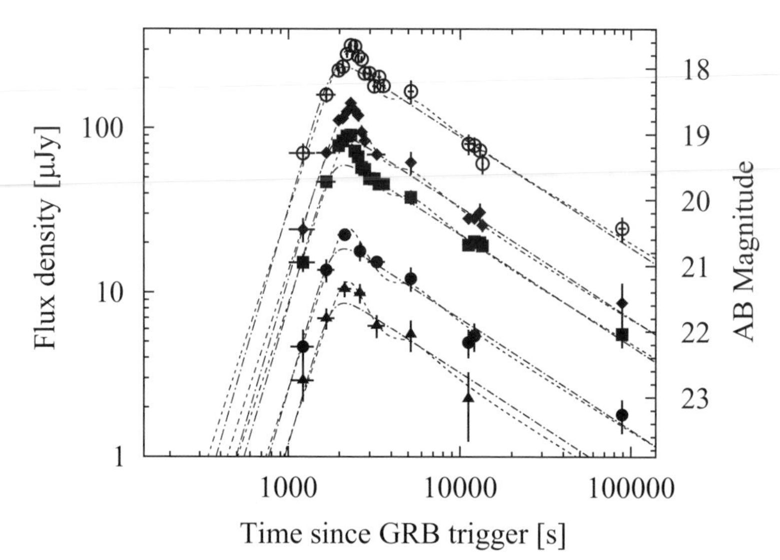

Figure 2. Optical/NIR light-curve of the afterglow of GRB 070802. Five different filters from the GROND instrument ($rzJHK$) are shown, illustrating the achromatic nature of the rise and decay. Adapted from Krühler *et al.* (2008).

Even higher Lorentz-factors approaching or possibly exceeding $\Gamma_0 \sim 1000$ arguably exist. Optical follow-up observations are in those cases however not rapid enough to catch the peak of the afterglow emission at a few seconds after the initial γ-ray trigger.

Alternatively, an initial achromatic rise of the afterglow might be caused by an off-axis location of the observer with respect to the outflow geometry. Because of the relativistic beaming of the decelerating ejecta, an observer located off-axis to the central jet will see a rising optical afterglow light curve at early times. The steepness of the rise would then be characteristic of the off-axis angle and the jet structure: the farther the observer is located from the central emitting cone or the faster the energy per solid angle decreases outside the jet, the shallower is the observed rise in a structured jet model (Panaitescu & Vestrand 2008). This scenario is particularly appealing for soft or sub-energetic events such as X-ray flashes, or X-ray rich bursts. A unified picture that attributes both, prompt and early afterglow emission to the observer's viewing angle could be obtained in this scenario.

4. True flares - fast rise and fast decays

In some cases, early optical observations reveal a light-curve morphology that is remarkably similar to the X-ray flares, which are detected in approximately 50% of all X-ray afterglows (e.g., Burrows *et al.* 2005). These morphologies are characterized by a very fast rise and similarly fast decay (see Fig. 3). Furthermore, these optical flares show strong spectral evolution and time lags, and are in many cases correlated with contemporaneous X-ray flares. There can also be multiple optical/NIR flares for a single event (Krühler *et al.* 2009a).

This strongly suggests late central engine activity as the common origin. The optical/NIR flares with fast rise and fast decay would thus be the soft tail of emission correlated with late internal shocks. This connection provides additional evidence that

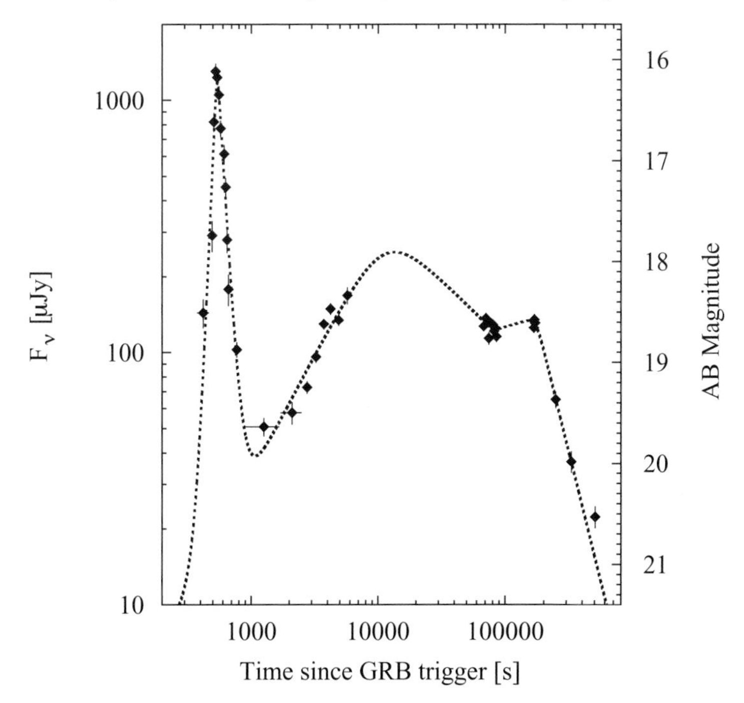

Figure 3. The optical flare of GRB 080129 before its rising afterglow as obtained with the Gamma-ray Optical Near-infrared Detector (GROND). The light-curve is parametrized with a series of smoothly connected power-laws (dashed lines). Super-imposed to the overall afterglow evolution peaking at 10 ks, a strong flare with fast rise and fast decay at $t \sim 500$ s is apparent. Adapted from Greiner *et al.* (2009).

inner engine activity may last or be revived on a timescale of hours or days at least for some bursts.

The early optical afterglow light curve is however not as often and not as strongly affected as the X-rays by flaring episodes. This is readily explained with the spectrum of the flares (peaking in the sub keV to few keV range), and a bright forward shock component typically dominating the optical emission. Thus, if the emission in the flares is not strong enough with respect to the underlying forward-shock emission, a bright afterglow can easily outshine flare signatures in the optical bands even for very bright X-ray flares.

5. Jumps - fast rise followed by slow decays

Remarkably different from the previous light-curve morphologies (fast rising and decaying flares from Section 4, as well as the smooth evolution of the onset of the afterglow from Section 3), is a class of objects characterized by a very fast rise, followed by a sharp turnover to a slow decay. Exemplary of these objects is GRB 081029 (Nardini *et al.* 2011) or GRB 100621A (Krühler *et al.* 2011). The excellent coverage in both time and frequency domain for these two events provided by GROND gives a detailed observational picture of their step-like afterglow light-curves.

5.1. *Temporal evolution*

The overall temporal evolution is dominated by the very fast rise (up to $F_\nu \propto t^{12}$), and can be divided into three phases. The first phase (before $t = 3000$ s in Fig. 4) shows

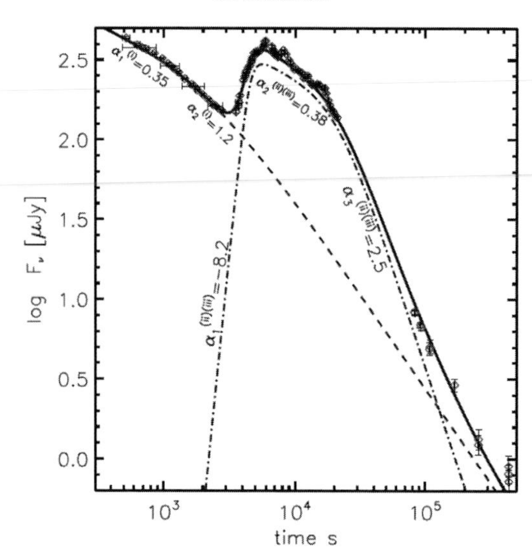

Figure 4. Afterglow light-curve for GRB 081029 as observed with GROND and parametrized using an empirical model consisting of different power-law segments. Taken from Nardini *et al.* (2011).

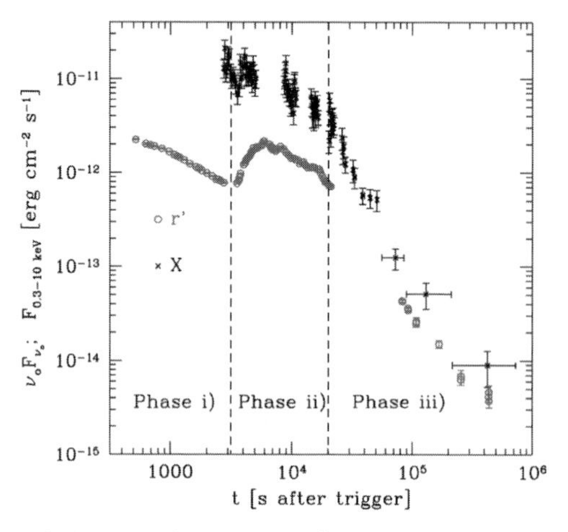

Figure 5. Broad-band behavior of the optical/X-ray afterglow GRB 081029 taken with GROND and the XRT onboard *Swift*. Taken from Nardini *et al.* (2011)

a canonical afterglow behavior characterized by a smoothly decaying light-curve with a possible break. Phase 2 displays a sudden rebrightening in all optical/NIR bands, followed by a shallow decay phase. Small scale variations are superimposed onto the shallow decay. The last phase of the afterglow shows a relatively steep $F_\nu \propto t^{-2.5}$ decay. An in depth discussion on the specific properties of the afterglow of GRB 081029 can be found in Nardini *et al.* (2011). This temporal evolution is a common characteristic for the events with a jump feature in their light-curve.

5.2. *Spectral evolution & broad-band behavior*

Figure 5 shows the optical r-band light-curve together with the X-ray data as observed with the XRT. While in the late phases, the optical and X-ray afterglow seem to track each other well, there is a strong discrepancy during the optical jump: there is hardly any evolution apparent in the high-energy bands. This is consistent with a strong blue-to-red evolution in the optical/NIR bands during the steep rise. This color evolution observed in the optical/NIR bands is an intrinsic feature of the emission component and is not related to changes of the intervening dust properties. This spectral evolution and broad-band behavior seems to be common for all events. In many cases, however, the data are not as detailed as obtained with GROND, and any intrinsic color evolution can not be tested with high accuracy.

5.3. *The class of jumping afterglows*

The literature provides several optical light-curves with similar light curve morphologies, fast rises followed by a shallow decay. They are shown in Figure 6, shifted to rest-frame time, and scaled to $z = 1$ for a direct comparison. The shift to $z = 1$ includes a rough K-correction and intrinsic dust correction. Due to the very different redshifts (from $z = 0.5$ to $z = 4$) the observed R-band, where most of the follow-up is performed, is probing very different intrinsic wavelengths.

It is immediately clear, that the jump can appear at very different times, that the rise time is different, and there is only one prominent jump in a single light-curve. The brightness of the underlying afterglow varies over several orders of magnitudes, and the increase in brightness during the jump is typically between one and two magnitudes (corresponding to a factor of 3 to 6 in flux). Selection effects however play a crucial role: the jump component needs to be prominent enough for a burst to enter the sample. The light-curves shown are hence the most extreme examples, and very likely many more cases with less pronounced variability exist.

The apparent dependence on the afterglow is in striking contrast to the true flares as discussed in Section 4, where the flare amplitude is not related to the underlying afterglow. Hence there seems to be a connection between afterglow and the jump components. Furthermore, the steepness of the rise varies significantly, and is not correlated with the time when the jump occurs. The canonical afterglow is typically well established before the rise, and the post-jump decay is shallow and comparable for all the events. In many cases, and if the photometric monitoring is dense and accurate enough, small scale variabilities exist in the shallow decay phase after the steep rise.

5.4. *Jumps at face with theoretical models*

Several theoretical interpretations have been proposed to explain variability in afterglow light-curves (see e.g., Ioka *et al.* 2005 and references therein). Possible physical explanations for afterglow variability apart from the earlier discussed reverse shocks (Section 2), afterglow onset (Section 3) and flares (Section 4) include inhomogeneities in the circum-burst medium (e.g., Wang & Loeb 2000) or the angular distribution of the energy in the jet (patchy shell model, e.g., Kumar & Piran 2000) or late energy injection by refreshed shocks (e.g., Rees & Meszaros 1998) for later flares. Alternatively, it is possible to obtain a chromatic and variable light-curve when decoupling the X-ray and optical afterglow, e.g., through 2-component jets (e.g., Peng *et al.* 2005) or late prompt models (e.g., Ghisellini *et al.* 2007, Nardini *et al.* 2010).

All models, however, have difficulties explaining the very fast rise with a temporal index of up to $F_\nu \propto t^{12}$, without invoking a shift of T_0, i.e., a restarting of the inner engine (Nardini *et al.* 2011, and references therein). A possible explanation might be present in refreshed shock models, when a late-emitted shell catches up with the decelerating afterglow, and a new forward and reverse shock are formed. These two-shell collisions (Vlasis *et al.* 2011) could in principle account for the observed features: the steep rise would then be caused by the interaction of the late-and-slow shell with the shocked ISM.

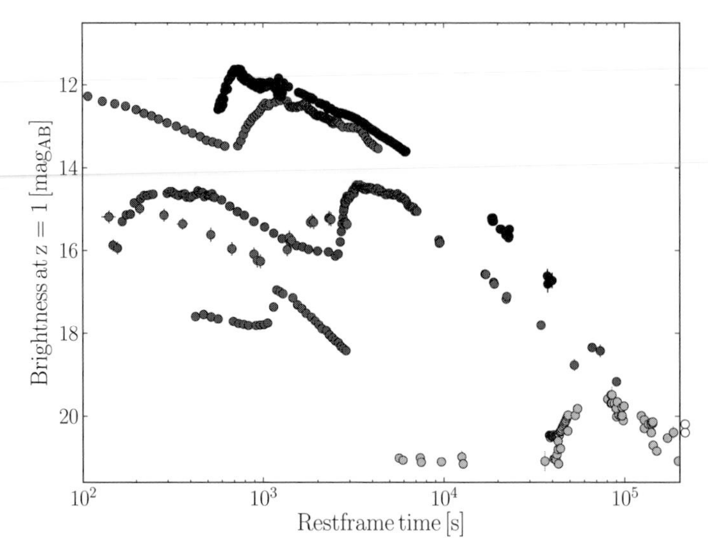

Figure 6. Afterglow light-curves for the sample of jumping afterglows. Data collected from the literature (Pedersen *et al.* 1998, Stanek *et al.* 2007, Cenko *et al.* 2009, Krühler *et al.* 2011, Nardini *et al.* 2011) or from the GROND archive.

The morphology of the rise is in this scenario a combination of two emission processes: the increase in energy and Lorentz-factor due to the propagation of the forward shock and the reverse shock traveling back into the late shell. Different steepnesses and amplitudes of the jumps are obtained for different jet geometries, i.e., opening angles, and energy ratios between early and late shells (Vlasis *et al.* 2011).

6. Summary

Despite the large diversity in the morphologies of early optical light curves, they are excellent tools for probing the jet physics. Early optical light curves provide, for example, detailed measurements of the initial Lorentz-factor via the peak of the forward shock. This directly probes the ultra-relativistic nature of the outflow as well as the emission radius of the afterglow. Furthermore, early optical data provide constraints on the role of magnetic fields via the signatures of reverse shocks or lack thereof.

Similar as in many X-ray afterglows, there are also optical/NIR flares superimposed onto the canonical afterglow component. They appear, however, much less frequently in the optical than in the X-ray regime. This is readily explained with the flare spectrum peaking in the few keV range, and the bright afterglow forward shock, dominating the optical emission. These flares indicate, that the inner engine may be active, or revived, up to several hours after the initial burst.

A distinct class of afterglow light-curves is revealed through a very fast rise in brightness followed by a shallow decay phase. Unlike flares, the emission does not drop back immediately to the already established afterglow level. These jumps with rise indices approaching t^{12} present a challenge to canonical models of afterglow variability, and indicate an impulsive release of energy very late after the prompt γ-ray emission has ceased. This might be caused by refreshed shocks of late or slow shells, catching up with the decelerating forward shock.

I acknowledge support by the EU under the Marie Curie Programme. I am also very grateful to the GROND team, in particular J. Greiner, M. Nardini and D. A. Kann for constant support and highly valuable input and discussion.

References

Beloborodov, A. M. 2005, *ApJL*, 618, L13
Burrows, D. N. *et al.* 2005, *Science*, 309, 1833
Gehrels, N. *et al.* 2004, *ApJ*, 611, 1005
Cenko, S. B., *et al.* 2009, *ApJ*, 693, 1484
Ghisellini, G. *et al.* 2007, *ApJ*, 685, 75
Greiner, J. *et al.* 2008, *PASP*, 120, 405
Greiner, J. *et al.* 2009, *ApJ*, 693, 1912
Ioka, K., *et al.* 2005, *ApJ*, 631, 429
Kann, D. A., *et al.* 2010, *ApJ*, 720, 1513
Krühler, J. *et al.* 2008, *ApJ*, 685, 376
Krühler, J. *et al.* 2009a, *ApJ*, 697, 758
Krühler, J. *et al.* 2009b, *A&A*, 508, 593
Krühler, J. *et al.* 2011, *A&A*, 534, 108
Kumar, P. & Piran, T. 2000, *ApJ*, 535, 152
Molinari, E. *et al.* 2007, *A&A*, 469, 13
Nakar, E. & Piran, T. 2004, *MNRAS*, 353, 647
Nardini, M. *et al.* 2010, *MNRAS*, 403, 1131
Nardini, M. *et al.* 2011, *A&A*, 535, 57
Panaitescu, A. & Vestrand, W., T. 2008, *MNRAS*, 387, 497
Pedersen, H. *et al.* 1998, *ApJ*, 496, 311
Peng, F. *et al.* 2005, *ApJ*, 626, 966
Perley, D. *et al.* 2010, *MNRAS*, 406, 2473
Rees, M. J. & Meszaros, P. 1998, *ApJL*, 496, 1
Roming, P. W. A. *et al.* 2006, *ApJ*, 652, 1416
Stanek, K. Z. *et al.* 2007, *ApJ*, 654, 21
Vlasis, A. *et al.* 2011, *MNRAS*, 415, 279
Wang, X. & Loeb, A. 2000, *ApJ*, 535, 788
Zhang, B., Kobayashi, S., & Mészáros, P. 2003, *ApJ*, 595, 950

Discussion

PERLEY: Can you comment on the energetics of the refreshed shock scenario ? It looks to me that in order to increase the brightness by a factor of 3 or more, the radiated energy integrated over time must increase by a similar factor.

KRÜHLER: In the shown examples, arguably the most extreme ones, the energy carried by the second shell is few times larger than in the initial shell.

IOKA: What is the time-seperation for the two shells ? The usual refreshed shock model with simultaneous ejection would be impossible to reproduce in the early fast rise.

KRÜHLER: The time separation is of order several thousand seconds, so also the refreshed-shock scenario needs a late time activity of the central engine. The fast rise is indeed challenging to explain in canonical models of afterglow variability without invoking a late engine activity.

Death of Massive Stars: Supernovae and Gamma-Ray Bursts
Proceedings IAU Symposium No. 279, 2012
P. Roming, N. Kawai & E. Pian, eds.

doi:10.1017/S1743921312012689

Ultraviolet-Bright Type IIP Supernovae from Massive Red Supergiants

Takashi J. Moriya[1], Nozomu Tominaga[2], Sergei I. Blinnikov[3], Petr V. Baklanov[3], and Elena I. Sorokina[4]

[1]Kavli IPMU, University of Tokyo, Kashiwanoha 5-1-5, Kashiwa, Chiba 277-8583, Japan
email: takashi.moriya@ipmu.jp

[2]Department of Physics, Faculty of Science and Engineering, Konan University, 8-9-1 Okamoto, Kobe, Hyogo 658-8501, Japan

[3]Institute for Theoretical and Experimental Physics, Bolshaya Cheremushkinskaya 25, 117218 Moscow, Russia

[4]Sternberg Astronomical Institute, Moscow University, Universitetski pr. 13, 119992 Moscow, Russia

Abstract. Red supergiants (RSGs) are progenitors of Type IIP supernovae (SNe). It is suggested that RSGs can experience a mass loss with a very high mass-loss rate (even as high as 0.01 $M_{\odot}$ yr^{-1}) due to, e.g., dynamical instabilities of their envelopes (e.g., Yoon & Cantiello (2010)). Because of the extensive mass loss, RSGs can have very dense circumstellar medium (CSM) around them. If a SN explosion occurs soon after the extensive mass loss of a RSG, the SN ejecta will collide with the dense CSM. Due to the collision, the kinetic energy of the ejecta is converted to radiation energy and such SNe with collision can be brighter than usual Type IIP SNe. By performing one-dimensional multi-group radiation hydrodynamical calculations, we investigate the effects of the collision on Type IIP SN LCs. We show that if RSGs explode within a dense CSM, the SN will be very bright, especially in ultraviolet, at early epochs. We also compare our models with the ultraviolet-bright Type IIP SN 2009kf and show that the progenitor of SN 2009kf can be a massive RSG which experienced extensive mass loss just before its explosion. We conclude that this is evidence that massive RSGs experience extensive mass loss and the existence of such mass loss can actually be the cause of the contradiction between theoretical and observational mass ranges of Type IIP SN progenitors.

Keywords. circumstellar matter, stars: mass loss, supernovae: general, supernovae: individual (SN 2009kf)

---

1. Introduction

SN 2009kf is a Type IIP supernova (SN) discovered by the Pan-STARRS 1 survey (Botticella *et al.* (2010)). The bolometric luminosity of the plateau obtained by just adding the optical light curves (LCs) exceeded 4×10^{42} erg s^{-1} and SN 2009kf is one of the most luminous Type IIP SNe. It was also observed by the GALEX satellite in the NUV band at the early epochs. Surprisingly, the NUV luminosity stayed almost constant for about 10 days during the GALEX observations. Ultraviolet (UV) LCs of usual Type IIP SNe decline rapidly due to the adiabatic cooling of the SN ejecta after the shock breakout. The very slow decline of UV luminosity observed in one of the most luminous Type IIP, SNe SN 2009kf, is difficult to interpret only with the adiabatic cooling of SN ejecta.

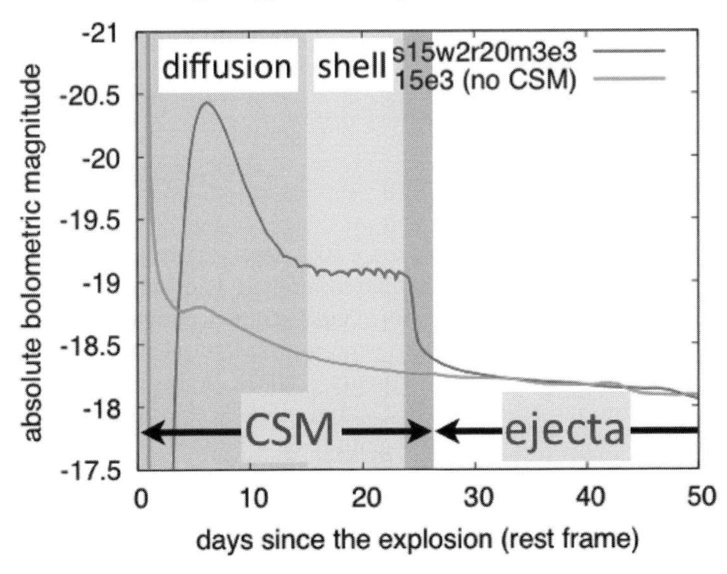

Figure 1. Early LCs from the explosion of an RSG with and without dense CSM.

2. UV-Bright Type IIP SNe

One possible cause of the slowly-declining high UV luminosity is the existence of a dense circumstellar medium (CSM) around the progenitor. With the typical mass loss history of RSGs, we do not expect the existence of a dense enough CSM around RSGs to affect the optical LCs of the subsequent SN. However, some RSGs are observationally known to have much higher mass-loss rates (e.g., VY CMa (Smith *et al.* (2009)), IRAS 05280-6910 (Boyer *et al.* (2010))) and it is also theoretically suggested that some RSGs experience extensive mass loss just before their explosions (e.g., Yoon & Cantiello (2010)). We investigate the effect of the dense CSM around the progenitors of RSGs on their LCs.

As is shown in Moriya *et al.* (2011) and Figure 1, the dense CSM affects the early LCs of the explosions of RSGs. The models without RSGs show a quick rise in the LCs because of the shock breakout. If there is a dense CSM around the progenitor, however, the shock breakout signal is extended because of the diffusion in the dense CSM. In addition, because of the interaction, the LC becomes brighter than those without the dense CSM, especially in UV.

Looking into the explosions with dense CSM, the photosphere is located in the CSM and photons diffuse in the CSM at first. Then, the photosphere stays in the cold dense shell between the dense CSM and SN ejecta for a while and the luminosity stays constant at these epochs. When the shock goes through the dense CSM or the remaining CSM becomes small enough, the LC suddenly drops and the effect of the CSM on it is no longer apparent.

The UV-bright LC of SN 2009kf is found to be consistent with an explosion of an RSG with 3×10^{51} erg within a dense CSM from $10^{-2}\ M_{\odot}\ \mathrm{yr}^{-1}$ ($0.6\ M_{\odot}$), 2×10^{15} cm (Moriya *et al.* (2011)). Unfortunately, the spectra of SN 2009kf were only taken during the plateau phase long after the UV-bright phase and all the dense CSM is presumed to be shocked away at these epochs. We expect that if we take the spectra of SN 2009kf-like SNe during the UV-bright phase, the spectral type would be Type IIn, which shows the narrow lines due to the dense CSM, and then transforms to Type II spectral type without the effect of the dense CSM.

3. Implications

The plateau luminosity of SN 2009kf ($\simeq 4 \times 10^{42}$ erg s^{-1}) is one of the brightest among the currently known Type IIP SNe. This high luminosity and the very large line velocities obtained from the spectra indicate that SN 2009kf is one of the most energetic explosions of RSGs currently observed. There are known empirical implications that more massive RSGs explode with higher energy (e.g., Hamuy (2003)). If this is true, the progenitor of SN 2009kf is one of the most massive RSGs and close to the upper mass limit of Type IIP SN progenitors. The fact that the progenitor of a Type IIP SN is close to the upper mass limit indicates that the upper mass limit of Type IIP SNe may be determined by the extensive mass loss which is currently not known well. This kind of extensive mass loss by massive RSGs can reduce the upper mass of Type IIP SN progenitors and will be a key to solve the so called 'red supergiant problem' (Smartt *et al.* (2009)).

References

Botticella, M. T., *et al.* 2010, *ApJ* (Letters), 717, L52
Boyer, M. L., *et al.* 2010, *A&A* (Letters), 518, L142
Hamuy, M. 2010, *ApJ* , 582, 905
Moriya, T., *et al.* 2011, *MNRAS*, 415, 199
Smartt, S. J., *et al.* 2009, *MNRAS*, 395, 1409
Smith, N., *et al.* 2009, *AJ*, 137, 3558
Utrobin, V. P., *et al.* 2010, *ApJ* (Letters), 723, L89
Yoon, S.-C. & Cantiello, M. 2010, *ApJ* (Letters), 717, L62

Discussion

KATZ: Can you separate the effects of UV and total brightness due to the two changes: energy and CSM?

MORIYA: The UV LC of the model without CSM with 3×10^{51} erg, which is already rather a high explosion energy, is more than two magnitudes fainter than that of SN 2009kf. You can increase the UV luminosity by increasing the energy but then the duration of the UV LC will also decrease and it will be difficult to explain the long duration of the UV LC of SN 2009kf. If you also increase the RSG radius, you can get the LC of SN 2009kf without CSM but the required values are extreme (2000 $R_\odot$ and 2.2×10^{52} erg, Utrobin *et al.* (2010)).

CHAKRABORTI: Do you predict X-ray emission?

MORIYA: No, because I expect X-rays are absorbed by the massive CSM.

CHAKRABORTI: Did you observe SN 2009kf, Stefan?

IMMLER: No, I didn't.

SMITH: A progenitor mass-loss rate of 10^{-2} $M_{\odot}$ yr^{-1} is very high, probably requiring an explosive pre-SN outburst instead of a wind. We would usually expect that to make a Type IIn spectrum. Do you have early-time spectra during the CSM-interacting phase to show that SN 2009kf was not a Type IIn supernova?

MORIYA: We do expect that the early time spectra will be those of Type IIn. We expect that the spectral type changes from Type IIn to Type II when the entire dense CSM is shocked away. Unfortunately, the early time spectra of SN 2009kf were not taken.

Death of Massive Stars: Supernovae and Gamma-Ray Bursts
Proceedings IAU Symposium No. 279, 2012
P. Roming, N. Kawai & E. Pian, eds.

doi:10.1017/S1743921312012690

Multiwavelength observations of GRB afterglows

Alberto J. Castro-Tirado

Instituto de Astrofísica de Andalucía (IAA-CSIC), Glorieta de las Astronomía s/n, E-18080 Granada, Spain
email: ajct@iaa.es

Abstract. Multiwavelength observations of gamma-ray burst afterglows are presented, in particular those in the optical and millimetre wavelengths. I will focus on the observations mostly carried out at Spanish ground-based observatories (mainly the 10.4m GTC) and at the Plateau de Bure Interferometer in the French Alps. The importance of global networks of robotic telescopes (like BOOTES, established worldwide) for early time observations in order to put constraints on the physical mechanisms of the GRB early time emission phase is also discussed. The overall observational efforts provide additional clues for a better understanding of the reverse and forward shock. Finally I will report on the *Lomonosov*/UFFO-p capabilities taking into account its launch in 2012.

Keywords. gamma-rays: bursts, accretion disks, acceleration of particles, black hole physics

1. Introduction

Since their discovery in gamma-rays in 1967 and the detection of the first counterparts at other wavelengths in 1997, thanks to the *BeppoSax* improvement on the localization accuracy leading to the first X-ray afterglow discovery (Costa *et al.* 1997), now we know that Gamma-Ray Bursts (GRBs) originate at cosmological distances with energy releases of 10^{51}–10^{53} ergs. The multiwavelength emission that follows the gamma-ray emission (the "afterglow") satisfies the predictions of the "standard" relativistic fireball model. See Castro-Tirado (2001) for a review of the field prior to the year 2000.

Since the launch of the *Swift* mission in 2004, more than 500 GRBs have been detected, with 85% of them being recorded in follow-up X-ray observations, 60% with optical detection and nearly 200 redshifts have been measured since 1997 (including 41 prior to *Swift*). *Fermi* was launched in 2008, to record the higher energy (MeV) population.

For the record, it is worth mention that a likely X-ray afterglow was pinpointed 5 yr before the *BeppoSAX* detection of GRB 970228. Such was the case for GRB 920723B (Castro-Tirado 1994), and the word "afterglow" was indeed used by Terekhov *et al.* (1993). Moreover, the first optical afterglow was already serendipitously imaged in 1992 for GRB 920925c (Hurley *et al.* 2000) being reported 15 yr later (Denisenko & Terekhov 2007).

2. Afterglow science in long-duration GRBs

Since the initial works by Paczynski (1986) and Goodman (1986), the physics of relativistic ejections by a compact source was further developed. The standard model was developed by Meszaros & Rees (1997) and also by Paczynski & Rhoads (1993). Hereafter, we will consider a few selected results on both the reverse and forward shock emission.

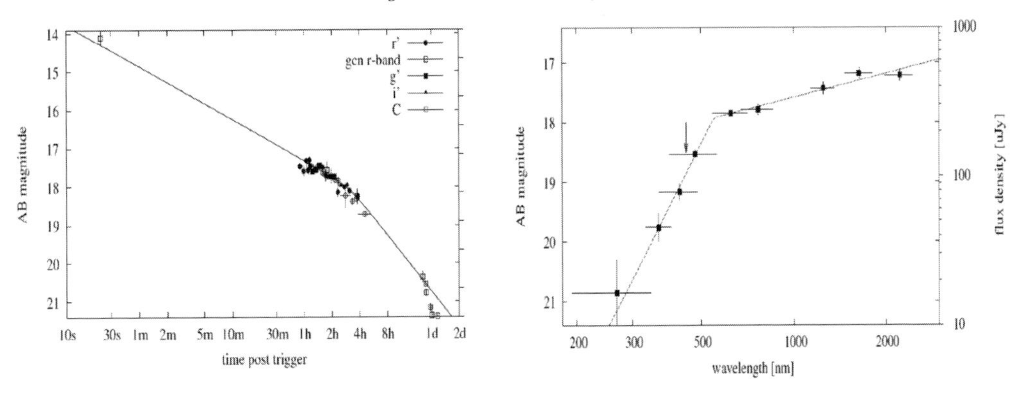

Figure 1. Left panel: The overall lightcurve of the GRB 080603B afterglow including BOOTES data. Right panel: The spectral energy distribution of the optical afterglow in the rest frame. The small arrow marks the Ly-α position for z = 2.69. From Jelinek *et al.* (2012).

2.1. *The reverse shock emission*

Due to the interaction with the surrounding medium, a reverse shock propagates within the ejecta. It is usually believed to contribute to the prompt emission (optical flash) and/or the early afterglow. For a uniform shell, the reverse shock is short-lived.

An important assumption is that the reverse shock microphysics parameters are equivalent to the internal shock microphysics parameters and different (in both cases) to the forward shock ones. If the outflow is variable, the reverse shock can be much more complicated than this simple picture and can even be long-lived. Thus, the strength of the reverse shock depends on magnetization content of the ejecta (Zhang, Kobayashi & Meszaros 2003, Gomboc *et al.* 2009).

For instance, for the case of GRB 060117 (Jelinek *et al.* 2006) and GRB 090902B (Pandey *et al.* 2010), a bright afterglow and a decay slope are suggestive of a reverse shock origin.

2.2. *The forward shock emission*

The afterglow is usually interpreted as the signature of the deceleration of the relativistic outflow by the external medium. The forward shock can be modelled under the frame of the standard fireball model, by means of the synchrotron spectrum moving towards longer wavelengths as function of time.

In the case of GRB 060117 (Jelinek *et al.* 2006), the peak time of the rising optical afterglow lightcurves could be used for determining the initial Lorentz factor following Molinari *et al.* (2007). The rising lightcurves ($F \propto t^{\alpha}$) are also important to understand the onset of the afterglow (Sari *et al.* 1999): α ~2 ($\nu_c < \nu_{optical}$) or α ~3 ($\nu_c < \nu_{optical}$) in the case of ISM or α ~0.5 for a WIND density profile. It allows also off-axis and structured jet models to be constrained (Panaitescu *et al.* 1998).

Prior to *Swift*, there were very promising results (multiwavelength fits) but even then problems were noticed, like the jet breaks, first detected in GRB 990123 (Castro-Tirado *et al.* 1999, Kulkarni *et al.* 1999; see Fig. 1).

After the initial *Swift* results, the picture has turned out to be more complicated: the canonical *Swift* X-ray afterglow lightcurve presents five distinct regions (Nousek *et al.* 2005): i) a steep decline; ii) a shallow slope; iii) the classical afterglow; iv) a jet break/late plateau; and v) flares (mostly in X-rays).

In many occasions, energy injections have to be taken into account to properly fit $F(\lambda, t)$. Thus, GRB 021004 (z = 2.33) was modelled by multiple energy injections (de

Ugarte Postigo *et al.* 2005), as was GRB 050730 (z = 3.97) and GRB 051028 (unknown-z) (Castro-Tirado *et al.* 2006). GRB 030329/SN 2003dh (z = 0.168) is also modelled by multiple energy injections but the initial phase cannot be properly accounted for (Guziy *et al.* 2012). GRB 060904B is an example of a GRB displaying all features on its lightcurve (Jelinek *et al.* 2012).

2.3. *The GRB-highly energetic SN connection*

The first hint for the long duration GRB - highly energetic supernova (SN) connection occurred in 1998, when GRB 980425 took place in the same region of the sky (also coincident in time) with SN 1998bw (Galama *et al.* 1998, Pian *et al.* 1999). It exhibited all the characteristics of a core-collapse SN Ic, expect for its kinetic energy ($\sim 10^{52}$ erg) and its high luminosity (Iwamoto *et al.* 1998), with an inferred progenitor mass of $\sim$30 $M_{\odot}$ and a remnant mass consistent with that of a black hole. Later on, a rebrightening seen in GRB 980326 was attributed to an underlying SN (Castro-Tirado & Gorosabel 1999, Bloom *et al.* 1999) being properly modelled by the contribution of a SN at $z \sim 1$ (Bloom *et al.* 1999).

Indeed, the smoking gun took place in 2003, when a multiwavelength campaign for the nearby GRB 030329 afterglow (at z = 0.168) led to the detection of prominent broad emission lines similar to the ones seen in GRB 980425/SN 1998bw (Stanek *et al.* 2003, Hjorth *et al.* 2003). The optical afterglow spectrum showed some evolution starting from the first night after the burst, and the beginning of spectral changes were seen as early as $\sim$ 10-12 h after the GRB. The onset of the spectral changes for t < 1 day indicated that the contribution from a Type Ic supernova (SN) -dubbed SN 2003dh later on- into the optical afterglow flux could be detected much earlier (Sokolov *et al.* 2003).

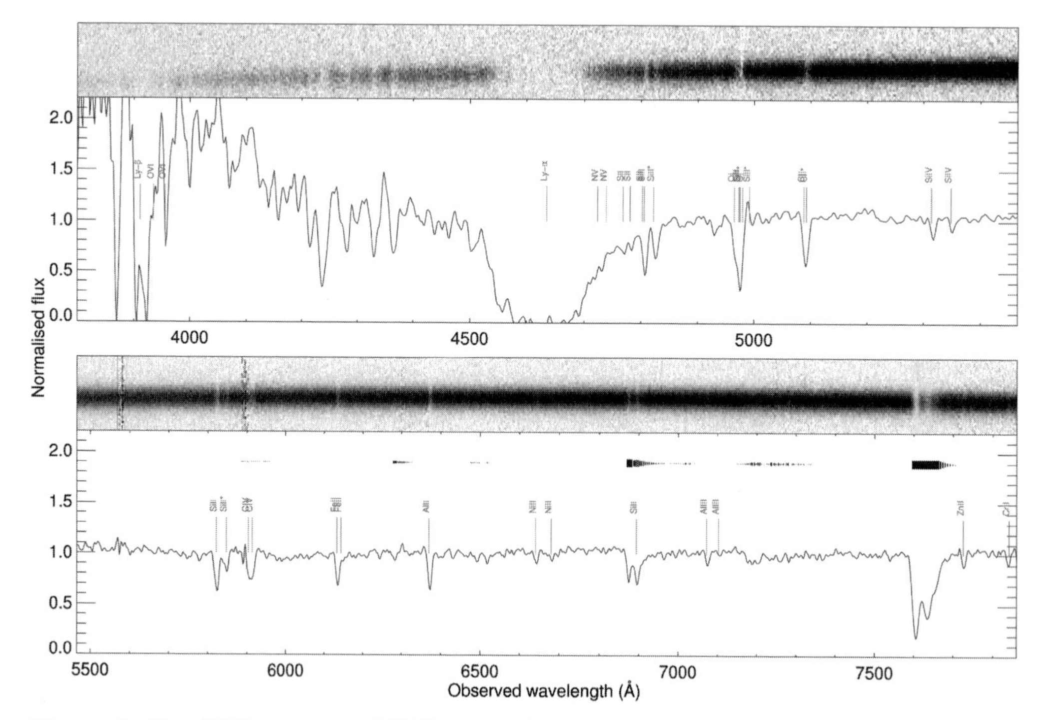

Figure 2. The GTC spectrum of GRB 100327A obtained $\sim$ 2.33 hr after the trigger with the 10.4m GTC telescope (+OSIRIS). A strong DLA system (with $N_H \sim 10^{22}$ cm^{-2} is detected, besides several absorption lines, at a common redshift of 2.813. From Castro-Tirado *et al.* (2012c).

Rebrightenings have been observed for many long-duration GRBs at epochs consistent with a rising light curve from a SN exploding within 1 day of the GRB time, a few of them being confirmed by spectroscopy and changes of colours around $\sim 15\ (1+z)$ days after the GRB trigger. Thus, SNe/GRB should be reachable with the Spanish 10.4m GTC up to z $\sim$ 1. In fact, GTC has allowed detection of the underlying SNe for both GRB 091127 (SN 2009nz) at z = 0.490 (Vergani *et al.* 2011a; see also Cobb *et al.* 2010, Berger *et al.* 2011) and GRB 111211A (de Ugarte Postigo *et al.* 2012).

2.4. *Spectroscopic observations of GRB afterglows*

Besides helping to determine the distance scale, spectroscopic observations of nearby events (z < 0.3) may reveal on a few cases unequivocal SN signatures similar to those of SN 1998bw, which coupled to the lightcurve monitoring will help to estimate the energy output, ^{56}Ni mass and remnant mass of the associated SN.

Besides this, optical (and near-IR) spectroscopy is most essential for understanding the GRB environment (abundances, metallicities, etc). As an example, in GRB 021004 (z = 2.33), high resolution spectroscopy revealed several high velocity systems in the range 200-3000 km/s (Starling *et al.* 2005, Castro-Tirado *et al.* 2010, Vergani *et al.* 2011b). In GRB 021004, the high velocity systems were naturally explained by multiple shells formed by stellar winds of a WR progenitor after passing through a LBV phase after reaching the Eddington limit. A $\sim$60 $M_\odot$ ZAMS progenitor is suggested following the accepted evolutionary track: O - LBV - WR - SN (Castro-Tirado *et al.* 2010).

We have also used the 10.4m GTC (+OSIRIS) to determine redshifts for a handful of events (GRB 100316A, 110503, 110801A, 110918A and 120326A; Sánchez-Ramírez *et al.* 2012, Tello-Salas *et al.* 2012) and confirmed many others (like GRB 100816A, 110422A, 110918A and 120327A; see Fig. 2).

2.5. *Polarimetric observations of GRB afterglows and hypernovae*

Optical light of the afterglows should be polarized because the magnetic field of the ISM, and also the central source, are amplified in the shocked front, so the electrons emit polarized synchrotron light. The ejecta concentrates the radiation in a beam of width $\theta \sim 1/\Gamma(t)$, where $\Gamma(t)$ is the Lorentz factor. The observer line of sight must be within the beam (otherwise the GRB would be undetectable) but in general not perfectly aligned with the jet. This breaks the source-observer axi-symmetry, avoiding the cancellation of electric vectors created in different emitting regions (Lazzati *et al.* 2004).

Observations performed so far show varying levels of polarized emission in the jet and the geometry. Polarized optical emission has been claimed for a few classical GRBs, usually P = 1-2% (Gorosabel *et al.* 2004) and rarely up to 10% (Steele *et al.* 2009). P $\sim$ 1.5 % was measured for SN 2003dh / GRB 030329 (Greiner *et al.* 2003).

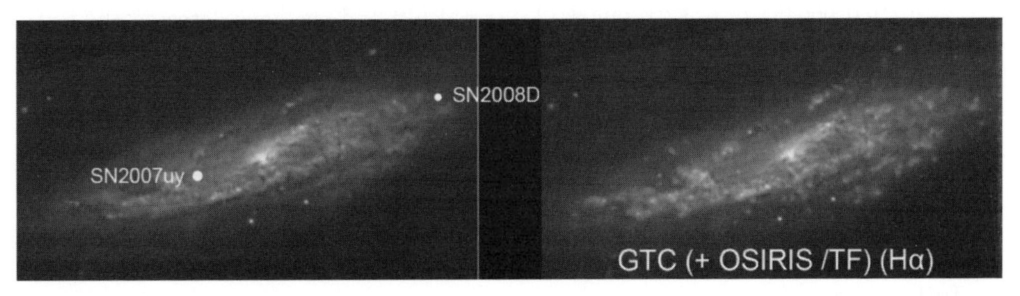

Figure 3. Left panel: NGC 2770 in broad-band optical filters showing the location of SN2007uy and SN 2008D. Right panel: The same field imaged in H-α thanks to the OSIRIS tunable filters in the GTC. From Gorosabel *et al.* (2011).

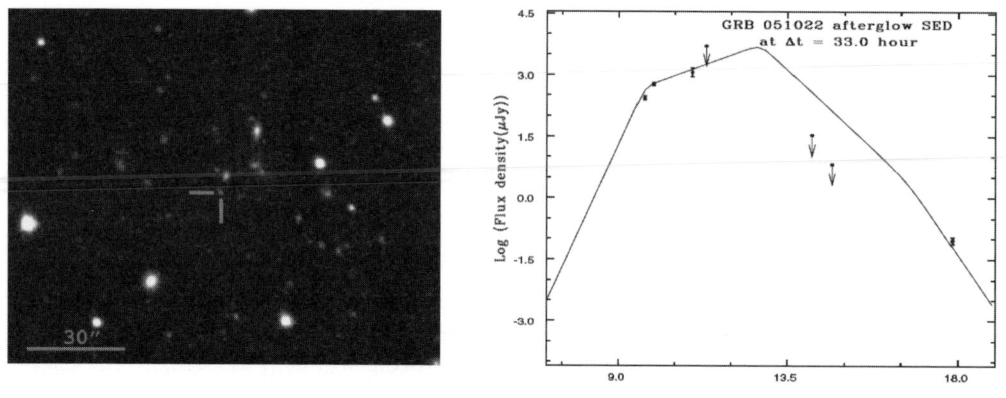

Figure 4. Left panel: the GRB 051022 mm afterglow was coincident with a blue galaxy at redshift z = 0.809. Right panel: the SED of the dark GRB 051022 afterglow. From Castro-Tirado *et al.* (2010).

The optical light of hypernovae can also be polarized as in the case of normal supernovae: Thomson scattering in the expanding stellar atmosphere. If the expansion is perfectly spherical, the polarization coming from different stellar parts would be canceled so no net polarization would be detected, as in normal SNe. However, in hypernovae, the expansion is likely to be ellipsoidal, so the symmetry around the line of sight is broken, producing non-zero polarization.

For SN 2006aj (type Ic) / XRF 060218A there was evidence for aspherical expansion with P = 4 % decreasing to 1.4 % plus a 100° rotation in the position angle (Gorosabel *et al.* 2006). For SN 2008D / XRT 080109 in NGC 2270 there was evidence for aspherical expansion with P = 1% (Gorosabel *et al.* 2010). Neither SN 2008D nor XRT 080109 are associated with any given giant H-II region in NGC 2770 (Fig. 3).

2.6. *Afterglows at mm wavelengths*

Millimetre observations are not affected by high-z or extinction and usually can lead to sampling the synchrotron peak in the spectrum as well as detecting the tail of the prompt emission and the forward shock. Since the first mm afterglow ever detected by us (Bremer *et al.* 1998), the number of follow-ups amounted to 60, with 54 events being observed at 3mm and 20 of them being detected so far. Redshifts for the GRB afterglows lie in the range z = 0.03-9.2, with measured flux densities (at 3 mm) varying between 0.25 and 60 mJy (but usually < 1.5 mJy) with the first observations taking place around 1-2 days after the GRB. We have detected 2 (out of 3) GRBs at z > 6: GRB 050904 (z = 6.3) and GRB 090423 (z = 8.3). On the lower redshift end, we have detected the low-z, faint GRB 080109 / SN 2008D (z = 0.0065). Amongst the long-duration class, we have detected two "dark" GRBs: one associated with a z = 0.809 galaxy (GRB 051022, Fig. 4) and another one (GRB 090404) whose host galaxy is not properly identified. Altogether, this implies a $\sim$ 42% success rate (for this non-blind sample). A more in-depth analysis is be presented in Castro-Tirado *et al.* (2012a).

3. Automated and Robotic Telescopes and their usage for GRB follow-ups

The (B)urst (O)bserver and (O)ptical (T)ransient (E)xploring (S)ystem (BOOTES), is a set of instruments that was conceived in 1995 and nowadays comprises four astronomical

Figure 5. Left panel: an image obtained by the CASANDRA-3 all-sky camera showing the silhouette of the 0.6m diameter YA telescope against the Milky Way at the BOOTES-3 robotic astronomical station in Blenheim (New Zealand). Right panel: the enclosure of 0.6m diameter MET telescope at the BOOTES-4 robotic astronomical station in Lijiang (China).

stations in Spain (two), New Zealand and China (Fig. 5). The main scientific goal is the follow-up of GRBs by means of fast ultralight weight 60cm diameter telescopes (Castro-Tirado *et al.* 2012b).

BOOTES-1 follow-ups have been partially summarized by Jelinek *et al.* (2011). The BOOTES network has allowed the discovery of a handful of optical afterglows (GRB 080330, GRB 080603B, GRB 080605, GRB 080606, GRB 101112A), observing many other at early times. BOOTES-2 and BOOTES-3, with larger diameter telescopes, have also contributed to some detailed studies on the forward shock evolution, as in the case of GRB 080603B, GRB 080605 and GRB 080606 (Jelinek *et al.* 2012) including the discovery of the optical afterglows of GRB 091029 and GRB 091208B. On the aggregate, BOOTES-1 have followed $\sim$ 40 events, detecting 10 afterglows, BOOTES-2/TELMA have followed up 9 events, detecting 3 afterglows, whereas BOOTES-3/YA have followed up 17 events, detecting 5 afterglows. BOOTES-4/MET has just followed up two GRBs, as it was officially opened in Mar 2012.

4. UFFO-p onboard Lomonosov

GRB afterglows can also be monitored by doing the follow-up using the triggering satellite itself, besides sending the position to the Earth (*BeppoSAX* in 6-8 hr, *Swift* in 1 min). Early follow-up (within $\sim$1 hr) is only available to *Swift* so far (sometimes even very early, with response times of $\sim$1 min) due to the slewing time of the entire spacecraft. Is it possible to beat this 1 min barrier from space? Indeed, in UFFO-p (UFFO-pathfinder) the optical path moves, not the spacecraft itself. This is achieved thanks to a fast slewing motorized mirror (new concept) in the Slewing Mirror Telescope (SMT) instrument (a mirror of 10 cm diameter) and an intensified CCD at the Cassegrain focus to achieve less than 1 s optical response after X-ray trigger. The second instrument is the UFFO Burst Alert & Trigger telescope (UBAT) for detecting the GRB locations. The UBAT employs a coded-mask γ/X-ray camera with a wide field of field of $\sim$ 1.8 sterad ($90^\circ \times 90^\circ$) and is comprised of three parts: a coded mask, a hopper, and a detector module. The UBAT DM consists of a LYSO scintillator crystal array sensitive to $\sim$ 200 keV (effective area

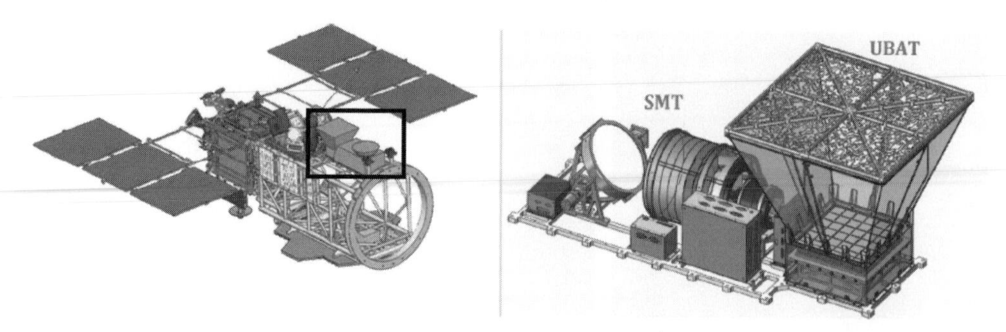

Figure 6. The integrated UFFO-pathfinder, to be launched aboard the *Lomonosov* spacecraft. The SMT enclosure is not shown in this drawing. Adapted from Chen *et al.* (2011).

of $\sim$ 190 cm^2), multi-anode photo multipliers, and analog and digital readout electronics (Park *et al.* 2009, Chen *et al.* 2011; see Fig. 6).

5. Summary

Afterglow emission can be detected in all the electromagnetic range, in all timescales from seconds to months (the later only in some cases). A variety of features can be studied by different techniques (photometry, spectroscopy, polarimetry) to gain insight into the underlying hypernovae and progenitors, environments, abundances, metallicities, host galaxies, etc. The overall observational efforts provide additional clues for a better understanding of the reverse and forward shock in the framework of the existing theoretical models, which should also accommodate the last observational results (not always easy).

In order to shed additional light in the field, technological developments are also required, e.g. new instrumentation at large facilities (like GTC) and at robotic telescope networks (like BOOTES). Future space missions, like *Lomonosov*/UFFO-p (to be launched in 2012) should be well suited for studying GRB emission at very early stages.

6. Acknowledgements

I appreciate the contribution over the years of the members of my ARAE Group (arae.iaa.es) as well as others belonging to the BOOTES Collaboration (bootes.iaa.es), UFFO-p Team and additional long-term collaborators worldwide. I also acknowledge support from CSIC, INTA, Junta de Andalucía and the Spanish Ministry of Science and Innovation through Project AYA 2009-14000-C03-01 (including FEDER funds).

References

Berger, E. *et al.* 2011, *ApJ*, 743, 204
Bloom, J. *et al.* 1999, *Nat*, 401, 453
Bremer, M. *et al.* 1998, *A&A*, 332, L13
Castro-Tirado, A. J. 1994, *Ph.D. Thesis*, Copenhagen Univ.
Castro-Tirado, A. J & Gorosabel, J. 1999, *A&AS*, 138, 449
Castro-Tirado, A. J. *et al.* 1999, *Sci*, 283, 2069
Castro-Tirado, A. J. 2001, *Fourth INTEGRAL Workshop: Exploring the Gamma-ray Universe*, ESA-SP Conf. Proc., SP-459, p. 367
Castro-Tirado, A. J. *et al.* 2006, *A&A*, 459, 763

Castro-Tirado, A. J. *et al.* 2010, *A&A*, 517, A61
Castro-Tirado, A. J. *et al.* 2012a, *A&A*, in press
Castro-Tirado, A. J. *et al.* 2012b, *BASI*, in press
Castro-Tirado, A. J. *et al.* 2012c, *A&A*, in preparation
Cobb, B. *et al.* 2010, *ApJ*, 718, L150
Costa, E. *et al.* 1998, *Nat*, 387, 783
Chen, P. *et al.* 2011, *Proc. of the 32nd International Conference on Cosmic Rays (ICRC)*, in press (arXiv1106.3929C)
Denisenko & Terekhov 2007, *GCNC*, 6155
Galama, T. *et al.* 1998, *Nat*, 395, 670
Gomboc, A. *et al.* 2009, *GRB: Sixth Huntsville Symposium, AIP Conf. Proc.*, 133, 145
Goodman, J. 1986, *ApJ*, 308, L47
Gorosabel, J. *et al.* 2004, *A&A*, 422, 113
Gorosabel, J. *et al.* 2006, *A&A*, 459, L33
Gorosabel, J. *et al.* 2010, *A&A*, 522, A14
Greiner, J. *et al.* 2003, *Nat*, 426, 157
Guziy, S. *et al.* 2012, *A&A*, in preparation
Hjorth, J. *et al.* 2003, *Nat*, 423, 847
Hurley, K. *et al.* 2000, *ApJS*, 128, 549
Iwamoto, K. *et al.* 1998, *Nat*, 395, 672
Jelinek, M. *et al.* 2006, *A&A*, 454, L119
Jelinek, M. *et al.* 2012, *A&A*, in preparation
Kasen, X. *et al.* 2003, *ApJ*, 593, 788
Kulkarni, S. R. *et al.* 1999, *Nat*, 398, 389
Lazzati, D. *et al.* 2004, *A&A*, 422, 121
Meszaros, P. & Rees, M. J. 1997, *ApJ*, 428, L29
Molinari, E. *et al.* 2007, *ApJ*, 469, L13
Nousek, J. A. *et al.* 2005, *ApJ*, 642, 389
Paczynski 1986, *ApJ*, 494, L45
Paczynski, B. & Rhoads, J. E. 1986, *ApJ*, 418, L5
Panaitescu, A., Meszaros, P.& Rees, M. J. 1998, *ApJ*, 503, 314
Pandey, S. B. 2010, *ApJ*, 714, 799
Park, I. *et al.* 2009, unpublished (arXiv0912.0773)
Pian, E. *et al.* 2000, *ApJ*, 536, 778
Sánchez-Ramírez, R. *et al.* 2012, *A&A*, in preparation
Sari, R., Piran, T., & Narayan, R. 1998, *ApJ*, 497, L17
Sokolov, V. V. *et al.* 2003, *BSAO*, 56, 5 [Stanek *et al.* 2003]Stanek03 Stanek, K. *et al.* 2003, *ApJ*, 591, L17
Starling, R. L. *et al.* 2005, *MNRAS*, 360, 305
Steele, I. *et al.* 2009, *Nat*, 462, 767
Tello-Salas, J. C. *et al.* 2012, *A&A*, in preparation
Kasen, X. *et al.* 1993, *Pisma Astron. Zh.*, 19, 686
de Ugarte Postigo, A. *et al.* 2005, *A&A*, 443, 841
de Ugarte Postigo, A. *et al.* 2012, *GCNC*, 12802
Vergani, S. *et al.* 2011a, *A&A*, 535, A127
Vergani, S. *et al.* 2011b, *AN*, 332, 292
Zhang, B., Kobayashi, S., & Meszaros, P. 1993, *ApJ*, 595, 950

Discussion

L. AMATTI: What about the UFFO Mission?

A. J. CASTRO-TIRADO: The UFFO-p is scheduled for launch this year, aboard the *Lomonosov* spacecraft, in a low-Earth orbit. See also the contribution by H. Lim *et al.* (in these proceedings).

B. ZHANG: Can you say a few words about using ALMA for studying GRB afterglows?

A. J. CASTRO-TIRADO: ALMA, thanks to its superb sensitivity, will opening a new era in the study of GRB afterglows detecting most of them if observed closed to the passage of the peak synchrotron emission. See also the contribution by A. de Ugarte Postigo *et al.* (in these proceedings).

Death of Massive Stars: Supernovae and Gamma-Ray Bursts
Proceedings IAU Symposium No. 279, 2012
P. Roming, N. Kawai & E. Pian, eds.

doi:10.1017/S1743921312012707

Ultra High Energy Cosmic Rays from Engine-driven Relativistic Supernovae

Sayan Chakraborti

Tata Institute of Fundamental Research,
1 Homi Bhabha Road, Mumbai, India
email: sayan@tifr.res.in

Abstract. The sources of the highest energy cosmic rays remain an enigma half a century after their discovery. Understanding their origin is a crucial step in probing new physics at energies unattainable by terrestrial accelerators. They must be accelerated in the local universe as otherwise interaction with cosmic background radiations would severely deplete the flux of protons and nuclei at energies above the Greisen-Zatsepin-Kuzmin (GZK) limit. Hypernovae, nearby GRBs, AGNs and their flares have all been suggested and debated in the literature as possible sources. Type Ibc supernovae have a local sub-population with mildly relativistic ejecta which are known to be sub-energetic GRBs or X-Ray Flashes for sometime and more recently as those with radio afterglows but without detected GRB counterparts, such as SN 2009bb. In this talk we present the size-magnetic field evolution, baryon loading and energetics of SN 2009bb using its radio spectra obtained with VLA and GMRT. We show that the engine-driven SNe lie above the Hillas line and they can explain the characteristics of post-GZK UHECRs.

Keywords. cosmic rays, supernovae: individual (SN 2009bb, SN 2012ap), gamma rays: bursts

1. Introduction

The Pierre Auger Collaboration (2007) show that the ultra high energy cosmic rays (UHECRs) have such a large energy and a low flux. They were first detected by Linsley (1963) only by air showers where according to Bhabha & Heitler (1937) the Earth's atmosphere acts as the active medium. UHECRs beyond the GZK limit have been invoked by Coleman, S. & Glashow (1999) to propose tests of known physical laws and symmetries. Understanding their origin is important for their use as probes of new physics. The magnetic rigidity of these particles are such that the magnetic fields in our galaxy are neither strong enough to contain them nor bend them sufficiently. However UHECR protons at energies above 60 EeV can interact with Cosmic Microwave Background (CMB) photons via the Δ resonance. The cross section of this process is such that only local extragalactic cosmic ray sources within 200 Mpc of the Earth can contribute significantly to the flux of UHECRs above the GZK limit found by Greisen (1966), Zatsepin & Kuz'min (1966), The High Resolution Fly'S Eye Collaboration (2008). Since particles of such high energy could not have traveled to Earth from cosmological distances, unless Lorentz invariance breaks down at these energies, as proposed by Coleman, S. & Glashow (1999), their detection encourages the search for potential cosmic ray accelerators in the local Universe. Accordingly Waxman (1995), Milgrom & Usov (1995), Murase *et al.* (2006) suggested nearby GRBs, Wang *et al.* (2007), Budnik *et al.* (2008) Hypernovae, The Pierre Auger Collaboration (2007) AGNs and Farrar & Gruzinov (2009) their flares, as possible sources.

SNe with relativistic ejecta have been detected until recently, exclusively through associated Long GRBs like GRB 980425 or its twin GRB 031203. Soderberg *et al.* (2006)

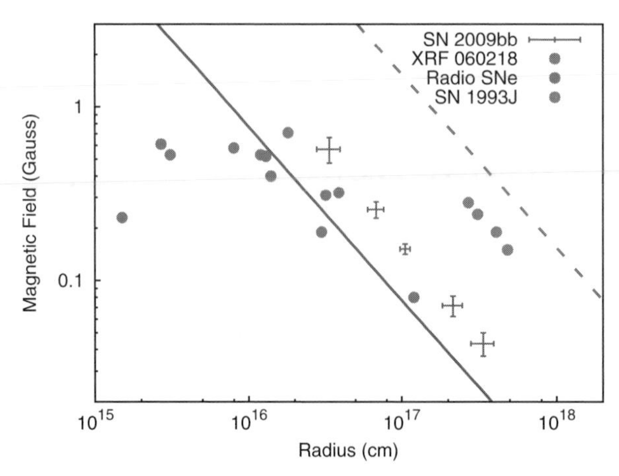

Figure 1. Hillas Diagram: Mildly relativistic sources ($\beta/\Gamma \sim 1$) must lie above the *solid line*, to accelerate Iron nuclei to 60 EeV by diffusive shock acceleration, according to $E_Z \lesssim \beta eZBR/\Gamma$. Non-relativistic SNe ($\beta/\Gamma \sim 0.05$) must lie above the *dashed line* to reach the same energies. Radius and magnetic field of SN 2009bb (crosses, at 5 epochs, determined here from radio observations with VLA and GMRT assuming equipartition) and XRF 060218 lie above the solid red line. Other balls denote other radio SNe from Chevalier (1998). For SN 1993J only, the magnetic fields are obtained by Chandra, Ray, & Bhatnagar (2004) without assuming equipartition. All non-relativistic SNe including SN 1993J lie below the dashed line and are unable to produce UHECRs unlike the mildly relativistic SN 2009bb and XRF 060218 which lie above the solid line.

showed that mildly relativistic SNe, like SN 2006aj associated with XRF 060218, are hundred times less energetic but thousand times more common (in their isotropic equivalent rate, the relevant rate for UHECRs reaching the observer) than classical GRBs. Soderberg *et al.* (2010) now discovered the presence of a radio emitting engine driven outflow from SN 2009bb, without a detected GRB. These mildly relativistic SNe, a subset of SNe Ibc detected either using X-Ray Flashes (XRFs) or radio afterglows, are far more abundant at low redshifts required for the UHECR sources, than the classical GRBs. Their mildly relativistic nature makes them have the most favorable combination of $\beta/\Gamma \sim 1$, unlike both non-relativistic SNe and ultra-relativistic classical Long GRBs. In Chakraborti *et al.* (2010) we measured the size and magnetic field of the prototypical SN 2009bb at several epochs. Such engine-driven supernovae are placed above the Hillas line and we demonstrate that they may accelerate cosmic rays beyond the GZK threshold. Together with the rates and energetics of such events, we establish that they readily explain the post-GZK UHECRs.

2. Magnetic field evolution with expanding radius

In order to derive the highest energy upto which these relativistic SNe can accelerate cosmic rays, we determine the evolution of the size and the magnetic field in the blastwave. Radii and magnetic fields were measured from VLA and GMRT data at 5 epochs, for plotting on the Hillas diagram (Figure 1).

With a set of assumptions for the electron energy distribution and magnetic fields the radius of the forward shock wave at the time of the synchrotron self-absorption peak was given by Chevalier (1998) as

$$R \simeq 4.0 \times 10^{14} \alpha^{-1/19} \left(\frac{f}{0.5}\right)^{-1/19} \left(\frac{F_{op}}{\mathrm{mJy}}\right)^{9/19} \left(\frac{D}{\mathrm{Mpc}}\right)^{18/19} \left(\frac{\nu}{5\ \mathrm{GHz}}\right)^{-1} \mathrm{cm}, \quad (2.1)$$

where $\alpha = \epsilon_e/\epsilon_B$ is the ratio of relativistic electron energy density to magnetic energy density, f is the fraction of the spherical volume occupied by the radio emitting region, F_{op} is the observed peak flux, and D is the distance. Using the same variables, the magnetic field is given by

$$B \simeq 1.1\alpha^{-4/19}\left(\frac{f}{0.5}\right)^{-4/19}\left(\frac{F_{op}}{\mathrm{mJy}}\right)^{-2/19}\left(\frac{D}{\mathrm{Mpc}}\right)^{-4/19}\left(\frac{\nu}{5\ \mathrm{GHz}}\right)\ \mathrm{G}. \qquad (2.2)$$

The radio spectrum of SN 2009bb at all epochs from discovery, paper by Soderberg *et al.* (2010) and Chakraborti *et al.* (2010), as obtained from observations using the Very Large Array (VLA) and the Giant Metrewave Radio Telescope (GMRT), is well fit by the SSA model. Thus we can measure the size and magnetic field of a candidate accelerator, instead of indirect arguments connecting luminosity with the Poynting flux.

SN 2009bb and XRF 060218 can both confine UHECRs and accelerate them to the highest energies seen experimentally. At the time of the earliest radio observations (Soderberg *et al.* (2010)) the combination of $\beta/\Gamma \sim 1$ for SN 2009bb shows that it could have accelerated nuclei of atomic number Z to an energy of $\sim 6.5 \times Z$ EeV. Thus the source could have accelerated protons, Neon, and Iron nuclei to 6.4, 64 and 166 EeV, respectively. Here the highest energy particles are likely to be nuclei heavier than protons, consistent with the latest results from The Pierre Auger Collaboration (2010) indicating an increasing average rest mass of primary UHECRs with energy. Therefore, our results support the Auger collaboration's claimed preference of heavier UHECRs at the highest energies.

3. Rates of Engine-driven Supernovae

We require the rate of relativistic SNe to estimate whether there are enough of them to explain the target objects associated with the ~ 60 detected UHECRs. SNe Ibc occur at a rate of $\sim 1.7 \times 10^4$ Gpc^{-3} yr^{-1}. The fraction that have relativistic outflows is still somewhat uncertain, estimated to be around $\sim 0.7\%$ (Soderberg *et al.* (2010)). Hence the rate of SN 2009bb-like mildly relativistic SNe is $\sim 1.2 \times 10^{-7}$ Mpc^{-3} yr^{-1}, comparable to the rate of mildly relativistic SNe detected as sub-energetic GRBs or XRFs of $\sim 2.3 \times 10^{-7}$ Mpc^{-3} yr^{-1}. This leads to ~ 4 (or 0.5) such objects within a distance of 200 (or 100) Mpc every year. Since SN 2009bb is still a unique object, only a systematic radio survey can establish their cosmic rate and statistical properties. However, cosmic rays of different energies have different travel delays due to deflections by magnetic fields. For a conservative mean delay (Farrar & Gruzinov (2009)) of $\langle\tau_{delay}\rangle \approx 10^5$ yrs we may receive cosmic rays from any of 4 (or 0.5) $\times 10^5$ possible sources at any given time. Since a direct association between a detected UHECR and its source is unlikely (Kashti & Waxman (2008)), most workers have focused on the constraints placed on plausible sources (Hillas (1984), Waxman & Loeb (2009)). Our arguments above show that this new class of objects satisfy all such constraints.

4. Energy Injection and Energy Budget

The required energy injection rate per logarithmic interval in UHECRs is $\Gamma_{inj} = (0.7 - 20) \times 10^{44}$ erg Mpc^{-3} yr^{-1}. With the volumetric rate of mildly relativistic SNe in the local universe, if all UHECR energy injection is provided by local mildly relativistic SNe, then each of them has to put in around $E_{SN} = (0.3 - 9) \times 10^{51}$ ergs. This is comparable to the kinetic energy in even a normal SN and can easily be supplied by a collapsar model. The mildly relativistic outflow of SN 2009bb is in nearly free expansion for ~ 1 year. Our measurements of this expansion show that this relativistic outflow, without a detected GRB, is significantly baryon loaded and the energy carried by the relativistic baryons is $E_{Baryons} \gtrsim 3.3 \times 10^{51}$ ergs (Chakraborti *et al.* (2010)).

5. Conclusion

We have shown that the newly found subset of nearby SNe Ibc, with engine-driven mildly relativistic outflows detected as sub-energetic GRBs, XRFs or solely via their strong radio emission, can be a source of UHECRs with energies beyond the GZK limit. Our study shows, for the first time, a new class of objects, which satisfy the constraints which any proposed accelerator of UHECRs has to satisfy. The author thanks Alicia Soderberg, Alak Ray, Abraham Loeb and Poonam Chandra for collaboration.

References

Bhabha, H. J. & Heitler, W. 1937, *Royal Society of London Proceedings Series A*, 159, 432
Budnik, R., Katz, B., MacFadyen, A., & Waxman, E. 2008, *ApJ*, 673, 928
Chakraborti, S., *et al.* 2010, *Nature Communications*, 2, 175
Chandra, P., Ray, A., & Bhatnagar, S. 2004, *ApJ*, 612, 974
Chevalier, R. A. 1998, *ApJ*, 499, 810
Coleman, S. & Glashow, S. L. 1999, *Phys. Rev. D*, 59, 116008
Farrar, G. R. & Gruzinov, A. 2009, *ApJ*, 693, 329
Greisen, K. 1966, *Phys. Rev. Lett.*, 16, 748
Hillas, A. M. 1966, *ARAA*, 22, 425
Horiuchi, S., Murase, K., Ioka, K., & Meszaros, P. 2012, *ArXiv e-prints*, astro-ph.HE:1203.0296
Kashti, T. & Waxman, E. 2008, *Journal of Cosmology and Astro-Particle Physics*, 5, 6
Linsley, J. 1963, *Phys. Rev. Lett.*, 10, 146
Milgrom, M. & Usov, V. 1995, *Ap. Lett.*, 449, 37
Murase, K., Ioka, K., Nagataki, S., & Nakamura, T. 2006, *Ap. Lett.*, 651, 5
Soderberg, A. M., Kulkarni, S. R., Nakar, E., *et al.* 2006, *Nature*, 442, 1014
Soderberg, A. M., Chakraborti, S., Pignata, G., *et al.* 2010, *Nature*, 463, 513
The High Resolution Fly'S Eye Collaboration 2008, *Phys. Rev. Lett.*, 100, 101101
The Pierre Auger Collaboration 2007, *Science*, 318, 938
The Pierre Auger Collaboration 2010, *Phys. Rev. Lett.*, 104, 091101
Wang, X.-Y., Razzaque, S., Mészáros, P., & Dai, Z.-G. 2007, *Phys. Rev. D*, 76, 083009
Waxman, E. 1995, *Phys. Rev. Lett.*, 75, 386
Waxman, E. & Loeb, A. 2009, *Journal of Cosmology and Astro-Particle Physics*, 8, 26
Zatsepin, G. T. & Kuz'min, V. A. 1966, *Soviet Journal of Experimental and Theoretical Physics Letters*, 4, 78

Discussion

METZER: Is this proposal consistent with the Auger observations of the composition of UHECRs? Chakraborti: The depth of maxima measurements from The Pierre Auger Collaboration (2007) favour a high atomic weight for their highest energy showers. This is consistent with our picture as we favour high atomic numbers such as Iron.

IOKA: If the Cosmic Rays are accelerated at the forward shock, the iron fraction is not high. So the Injection efficiency would have to be higher for iron than protons. Chakraborti: It is possible that the relativistic outflow which originated from near the core of the progenitor may carry some nucleosynthetic products. Whether such nuclei can survive in jets is being investigated by Horiuchi *et al.* (2012).

Death of Massive Stars: Supernovae and Gamma-Ray Bursts
Proceedings IAU Symposium No. 279, 2012
P. Roming, N. Kawai & E. Pian, eds.

doi:10.1017/S1743921312012719

SN1987A: the X-ray remnant at age 25 years

David N. Burrows[1], Sangwook Park[2], Eveline A. Helder[1], Daniel Dewey[3], Richard McCray[4], Svetozar A. Zhekov[5], Judith L. Racusin[6], and Eli Dwek[6]

[1]Dept. of Astronomy & Astrophysics, Penn State University, University Park, PA 16802 USA; email: burrows@astro.psu.edu;
[2]U. Texas-Arlington, Arlington, TX, USA; [3]MIT Kavli Institute, Cambridge, MA, USA
[4]U. Colorado, Boulder, CO, USA; [5]Space Research and Technology Institute, Sofia, Bulgaria
[6]NASA/GSFC, Greenbelt, MD, USA

Abstract. SN1987A is the best-studied core-collapse supernova in the sky. We know what the progenitor was, what the circumstellar environment was, and what the explosion looked like over a broad electromagnetic bandpass and in neutrinos. For over a decade, the Chandra X-ray Observatory has been monitoring SN1987A on a regular basis, obtaining resolved images of the developing interaction with the circumstellar material, as well as high resolution grating spectroscopy of the X-ray emission. We highlight the latest results from this campaign and discuss the overall picture of the remnant's structure that emerges from these observations.

Keywords. X-rays:individual(SN1987A), stars:supernovae:individual(SN1987A)

1. Introduction

Supernova SN1987A is unique among the myriad supernovae discovered in the past century. It was the first naked-eye supernova since 1604 (Kepler). Unlike the Galactic supernova remnants (SNRs), it has an accurately known distance (51.4 ± 1.2 kpc; Panagia *et al.* 1999); a known type (Type IIP), and even a known progenitor (Sk -69 202, a B3I blue supergiant). The explosion time (07:36 UT on 23 February 1987) is known precisely due to the detection of neutrinos from the event (Hirata *et al.* 1987), which also confirmed that it was a core-collapse supernova. Because of its proximity in the Large Magellanic Cloud, it has been observed across the electromagnetic spectrum. It is without a doubt the best-studied supernova to date.

SN1987A was also unique in a number of its characteristics. The fact that the progenitor was a blue supergiant was quite surprising at the time. *HST* began making high-resolution optical observations shortly after its launch in 1990, discovering the unique triple-ring system that was illuminated by the UV flash from the explosion. Subsequent *HST* observations found unexpected "hot spots" on the inner ring (Lawrence *et al.* 2000), and tracked the expansion of the ejecta inside the inner ring (Larsson *et al.* 2011).

Chandra began making high resolution X-ray images shortly after its launch in 1999 and has been tracking the X-ray behavior ever since, in both high-resolution images and high-resolution spectography. Here we highlight recent results from our *Chandra* campaign with the Advanced CCD Imaging Spectrometer (ACIS), spanning the years 1999 – 2011.

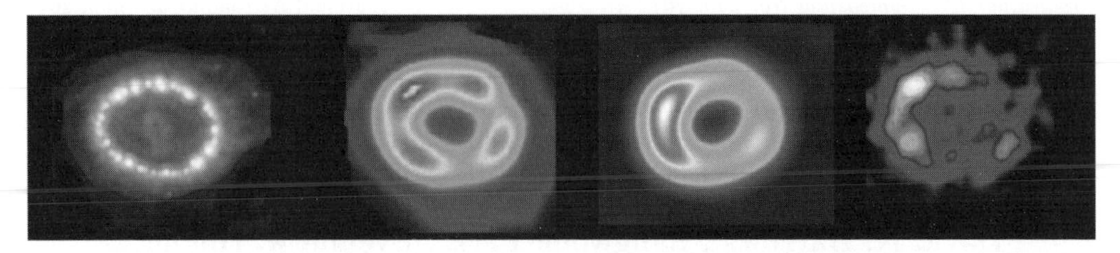

Figure 1. SN1987A ca. 2005. Left to right: *HST*, *Chandra* 0.5-2 keV, ATCA 12 mm (Manchester *et al.* 2005), Gemini 12μm (Bouchet *et al.* 2006).

2. Multiwavelength Overview

We begin with a very brief multiwavelength overview (Fig. 1). From left to right:

Optical. The inner equatorial ring (ER) begins to develop hot spots ca. 1995, and these have now encircled the entire ring. The ejecta in the center are expanding. After dimming until day $\sim$ 5000 due to radioactive decay of ^{44}Ti, they are now brightening again as the soft X-rays from the ER have begun to reheat the ejecta (Larsson *et al.* 2011).

X-ray. The X-ray image shows an expanding ring of emission due to interactions of the shock with the HII region inside of the ER, and with the ER itself. Bright spots in the NE and SE quadrants correspond to optical hot spots #1 (NE) and #2–4 (SE).

Radio. The radio image also shows an expanding shell-like structure, but the eastern emission is distributed somewhat differently than the soft X-rays, while the bright spot in the west aligns well with the western bright spot in the X-ray image.

Infrared. The 12μm image corresponds very well to the X-ray image. *Spitzer* measurements show that the mid-IR flux tracks the soft X-ray flux quite well, providing evidence that the mid-IR component is due to dust heated to $T \sim$ 180K by collisions with electrons in the hot X-ray emitting shocked ER (Dwek *et al.* 2010). *Herschel* observations at 100–350 μm suggest that roughly 0.5 solar masses of dust in the ejecta have been heated to 20K, possibly also by the soft X-rays from the ER (Matsuura *et al.* 2011).

3. X-ray Observations

The *Chandra*/ACIS instrument has monitored SN 1987A roughly every six months since October 1999. The observations have been taken in several instrument configurations to obtain both high resolution images and high resolution spectra. There is space here only to sketch a few key results; for more details please see papers published between 2000 and 2011 by S. Park, S. Zhekov, D. Dewey, J. Racusin, and references therein.

Images. We have produced deconvolved images at each observational epoch. Details of the technique are given in Burrows *et al.* (2000). A few images are shown in Figure 2. The emission is dominated by dense gas in the inner ring heated to $\sim 3 \times 10^7$ K by the forward shock, indicating that SN 1987A is now a supernova remnant.

Light Curve. The *Chandra* soft X-ray light curve of SN 1987A is shown in Figure 3. The flux increased linearly for over five years, from day 6500 to 8500, but then flattened out and has been nearly constant for the last 2.5 years (Park *et al.* 2011). This flattening of the light curve suggests that the shock wave has passed through the peak density of the inner ring and may be about to enter the region outside the visible ring, where it will map out the unknown density profile beyond the ER.

Expansion. The superb angular resolution of *Chandra* has allowed us to measure the expansion of the X-ray remnant, which is shown in Fig. 4. The expansion rate slowed dramatically around day 6000 to about 1780 km/s (Racusin *et al.* 2009).

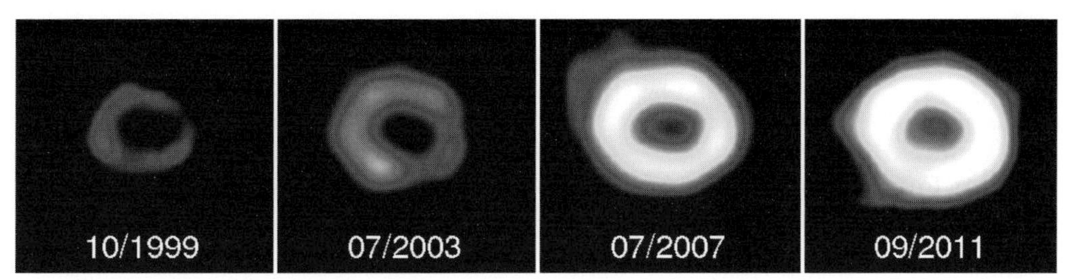

Figure 2. Deconvolved, smoothed images of SN 1987A from 1999 – 2011. The color scale is logarithmic. North is up, East is to the left.

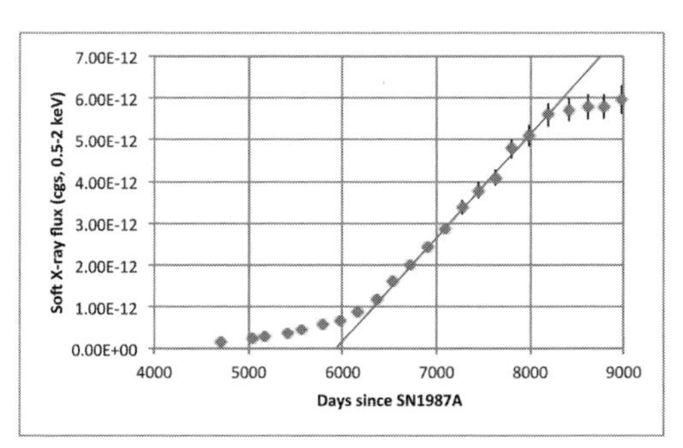

Figure 3. *Chandra* soft X-ray light curve of SN 1987A.

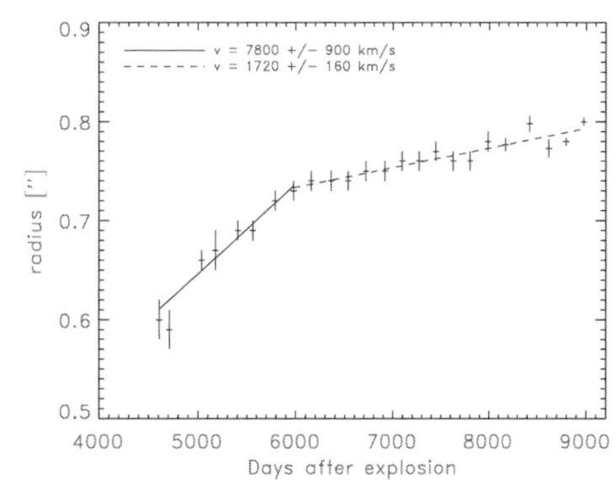

Figure 4. ACIS expansion measurements of SN 1987A.

Compact Remnant. SN 1987A is expected to contain a neutron star buried under optically thick ejecta (Graves *et al.* 2005), which could emerge from the ejecta at any time as the latter continue to expand and thin out. However, our observations show no evidence to date for a central source, with a limit of 1.5×10^{34} erg/s on the observed flux.

4. Physical Model

The physical model for SN 1987A inferred by Dewey *et al.* (2012; hereafter D12) is illustrated in Fig. 5. Analysis of the X-ray images and line spectra indicate that the X-ray emission is produced by 3 components: the shocked, clumpy ER (red regions in Fig. 5), which currently dominates the soft X-ray band; the shocked HII regions inside the ER and off the equatorial plane (blue regions in Fig. 5), which produce very broad X-ray line profiles and which dominate the hard X-ray and radio bands; and the reverse-shocked outer ejecta (green regions in Fig. 5). Zhekov *et al.* (2009, 2010) fit the in-plane components of this model

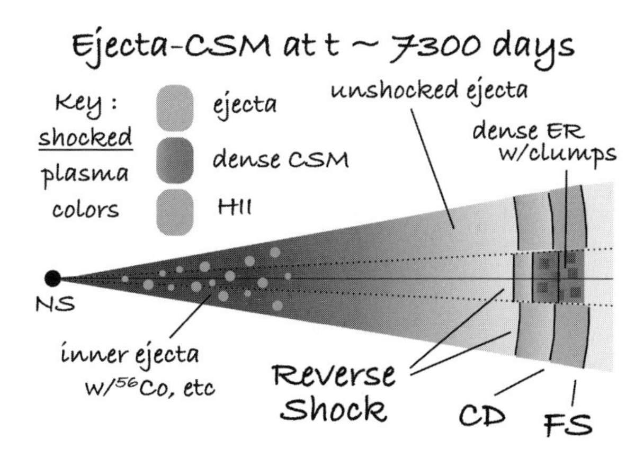

Figure 5. Cross-section (±15° wedge perpendicular to the equatorial plane) of the inferred structure of SN1987A, circa Feb. 2007. CD is the contact discontinuity, FS is the forward shock.

to ACIS light curves and spectra up to 2007. D12 fit this model, including the out-of-plane HII regions, to the measured X-ray fluxes and line profiles to determine the parameters of a set of hydro models that reproduce both the spectra and light curves.

5. The Future of SN1987A

The Forward Shock. The forward shock is expected to pass beyond the dense inner ring in the near future, when it will map out the circumstellar density profile, acting as a "time machine" to measure the mass loss history of Sk -69 202. D12 show that the future X-ray light curve will differ dramatically, depending on whether or not the ER continues beyond the visible ring to larger radii. The same will apply to the IR light curve, which is closely coupled to the X-ray flux.

The Ejecta. Optical observations show that the bulk of the ejecta is rapidly approaching the inner ring (Larsson *et al.* 2011), but both optical images and the hydrodynamical simulations of D12 suggest that the reverse shock has not yet entered this material. The early emergence of X-rays from SN 1987A demonstrated that the supernova explosion ejected some of the newly synthesized ^{56}Co into the outer envelope of the supernova ejecta (McCray 1993). Eventually, clumps of ^{56}Co and other newly synthesized elements in the outer debris will cross the reverse shock, producing dramatic changes in the X-ray spectra. Such events could happen at any time during the next several years. Frequent monitoring of SN 1987A at all wavelengths is required to monitor these expected changes.

The Compact Remnant (?). As the ejecta continue to expand and thin, we expect that the putative neutron star at the center of SN 1987A will become apparent. We have not been able to detect it yet in the *Chandra* bandpass, but it is possible that *NuSTAR*, which will be launched in March 2012, will find evidence at higher energies for a pulsar.

The future should continue to deliver exciting new results, as the forward and reverse shocks probe uncharted territory in the CSM and ejecta, respectively. The continued development of the youngest accessible supernova remnant will undoubtedly continue to puzzle and enlighten us for years to come.

References

Bouchet, P., *et al.* 2006, *ApJ*, 650, 212
Burrows, D. N., *et al.* 2000, *ApJ*, 543, L149
Dewey, D. & Dwarkadas, V. V., *et al.* 2012, *ApJ*, submitted, *arXiv 1111.5314*
Dwek, E., *et al.* 2010, *ApJ*, 722, 425
Graves, G. J. M., *et al.* 2005, *ApJ*, 629, 944
Hirata, K., *et al.* 1987, *Phys. Rev. Lett.*, 58, 1490
Larsson, J., *et al.* 2011, *Nature*, 474, 484
Lawrence, S., S., *et al.* 2000, *ApJ*, 537, L123
Manchester, R., *et al.* 2005, *ApJ*, 628, L131
Matsuura, M., *et al.* 2011, *Science*, 333, 6047
McCray, R. 1993, *ARAA*, 31, 175
Panagia, N. 1999, in: Y.-H. Chu, N. Suntzeff, J. Hesser, & D. Bohlender (eds.), *New Views of the Magellanic Clouds*, Proc. IAU Symposium No. 190, p. 549
Park, S., *et al.* 2011, *ApJ*, 733, L35
Racusin, J. L., *et al.* 2009, *ApJ*, 703, 1752
Zhekov, S. A., *et al.* 2009, *ApJ*, 692, 1190
Zhekov, S. A., *et al.* 2010, *MNRAS*, 407, 1157

Death of Massive Stars: Supernovae and Gamma-Ray Bursts
Proceedings IAU Symposium No. 279, 2012
P. Roming, N. Kawai & E. Pian, eds.

doi:10.1017/S1743921312012720

Supernovae and Gamma-ray Bursts

Paolo A. Mazzali[1,2]

[1]INAF - Osservatorio Astronomico di Padova, Padova, Italy
[2]Max-Planck Institute for Astrophysics, Garching, Germany

Abstract. The properties of the Supernovae discovered in coincidence with long-duration Gamma-ray Bursts and X-Ray Flashes are reviewed, and compared to those of SNe for which GRBs are not observed. The SNe associated with GRBs are of Type Ic, they are brighter than the norm, and show very broad absorption lines in their spectra, indicative of high expansion velocities and hence of large explosion kinetic energies. This points to a massive star origin, and to the birth of a black hole at the time of core collapse. There is strong evidence for gross asymmetries in the SN ejecta. The observational evidence seems to suggest that GRB/SNe are more massive and energetic than XRF/SNe, and come from more massive stars. While for GRB/SNe the collapsar model is favoured, XRF/SNe may host magnetars.

Keywords. supernovae: general, supernovae: individual (SN 1997ef, 1998bw, 2002ap, 2003dh, 2003jd, 2005bf, 2006aj, 2008D), gamma rays: bursts, radiation mechanisms: general, line: formation, nuclear reactions, nucleosynthesis, abundances

---

1. Introduction

The connection between long-duration Gamma-Ray Bursts (GRBs) and a particular class of core-collapse Supernovae (SNe) has been established with the discovery of optically very bright SNe in positional and temporal coincidence with three of the nearest GRBs (Galama *et al.* 1998, Stanek *et al.* 2003, Malesani *et al.* 2004).

The spectra of these SNe are all very similar. They resemble closely those of Type Ic SNe, but are characterised by P-Cygni lines with very broad absorption components, indicative of the presence of material expelled at very high velocities (Fig.1). Type Ic SNe are thought to be the result of the explosion of the carbon-oxygen core of massive stars that had lost their outer hydrogen and helium envelopes prior to core collapse. The broad-lined spectra and the relatively broad light curves of the GRB-SNe suggest that they are all very energetic explosions. GRB/SNe lie at the luminous end of the distribution of SNe Ib/c, indicating a large production of ^{56}Ni ($\sim 0.5 M_{\odot}$, Fig 2). Because of the very large energy and their spectral characteristics, these SNe have also been called "Hypernovae" (HNe). Similarly, a connection between X-ray Flashes (XRF) and SNe Ib/c has been established (Pian *et al.* 2006, Soderberg *et al.* 2008). These SNe are also overenergetic. Here, the properties of HNe are reviewed, as is the evidence that they are aspherical events, which supports the connection with GRBs.

2. Energetics

SN spectra obtained in the early phase reflect mostly the structure of the outer part of the ejecta. The light curves and the spectra of Type Ib/c SNe must be modelled simultaneously in order to obtain an accurate estimate of the properties of the explosion (Arnett 1982). We use a Montecarlo radiation transport code (Mazzali & Lucy 1993, Lucy 1999, Mazzali 2000). The results indicate that GRB-SNe are powerful explosions

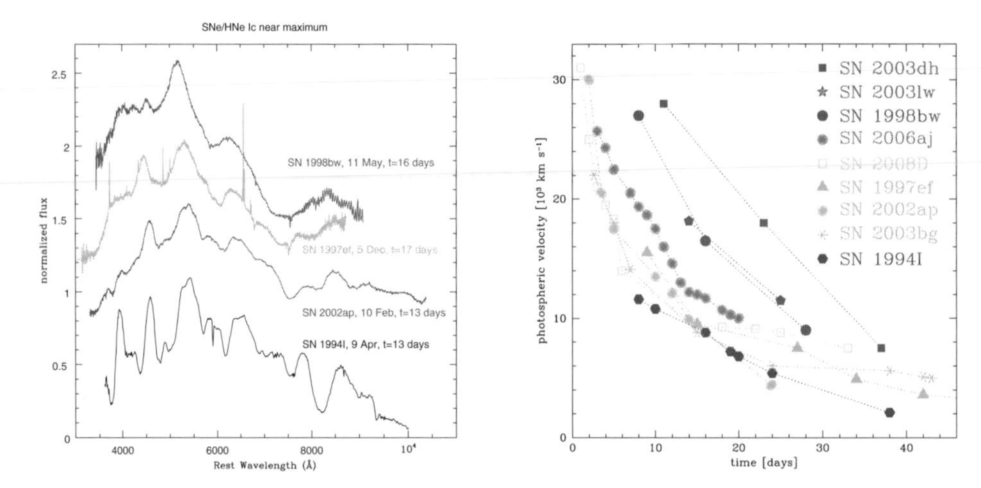

Figure 1. Left: Near-maximum spectra of SNe Ic. The increasing width of the spectral lines marks the transition to Hypernovae. **Right**: Photospheric expansion velocities of SNe Ib/c derived from spectral modelling. GRB-SNe are characterised by the highest velocities.

which ejected large quantities of matter: in the case of the prototypical SN 1998bw (associated with GRB980425) the spherically symmetric explosion kinetic energy derived from modelling is $E \approx 5 \times 10^{52}$ erg, i.e. about 50 times larger than in normal core-collapse SNe, and the ejected mass is $\sim 10 M_{\odot}$ (Iwamoto *et al.* 1998). Other GRB/SNe such as 2003dh/GRB030329 and 2003lw/GRB031203 yield similar values (Mazzali *et al.* 2003, Mazzali *et al.* 2006a, respectively), justifying the name Hypernovae. As an example, the 'prototypical' SN Ic 1994I has $M_{\rm ej} \sim 1.2 M_{\odot}$ and $E_{\rm K} \sim 10^{51}$ erg (Sauer *et al.* 2006).

The large ejecta masses indicate that the progenitor stars were very massive: including the compact remnant (most likely a black hole), the mass of the CO cores must have been $\sim 12 - 15 M_{\odot}$, which points to a zero-age main sequence mass of the progenitor stars of $\sim 40 - 50 M_{\odot}$. A very massive star origin for the SNe connected with GRBs suggests that the ejection of matter at relativistic velocities that is responsible for the emission of the GRB is linked to the formation of the black hole, and supports the scenario envisioned in the so-called "collapsar" model (McFadyen & Woosley 1999).

A number of broad-lined SNe Ic have been discovered that were not associated with GRBs. These are relatively nearby events, so the SNe were discovered optically. The two best observed such events are SNe 1997ef and 2002ap.

SN 1997ef had spectra very similar to those of SN 1998bw, and was analysed to be the very energetic explosion ($E_{\rm K} \sim 2 \times 10^{52}$ erg) of a very massive star ($M_{ZAMS} \sim 35 M_{\odot}$), ejecting however only $\sim 0.15 M_{\odot}$ of ^{56}Ni (Mazzali, Iwamoto, & Nomoto 2000).

The nearby SN 2002ap was also characterized at early times by very broad lines, suggesting again that this was a HN (Fig. 1). Just like SN 1997ef, however, SN 2002ap never became really luminous (Fig. 2). It produced a ^{56}Ni mass of $\sim 0.1 M_{\odot}$. The spectra and the rapidly evolving light curve were modelled as the explosion of a star of relatively small mass, $\sim 23 M_{\odot}$, collapsing to a black hole and ejecting $\sim 2.5 M_{\odot}$ of material with $E_{\rm K} \sim 4 \times 10^{51}$ erg (Mazzali *et al.* 2002).

3. Asphericity

A GRB is thought to be a highly beamed phenomenon, while SNe are traditionally viewed as spherical events, although this is almost certainly an oversimplification

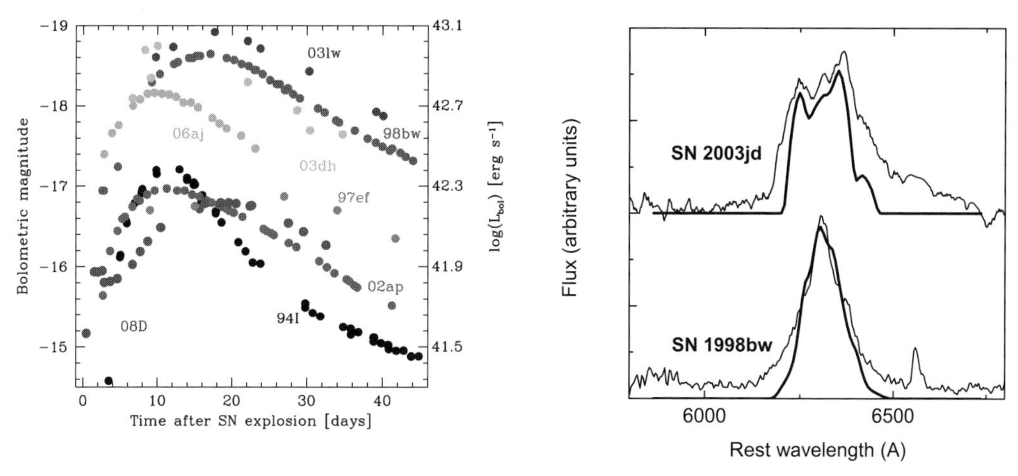

Figure 2. Left: Bolometric light curves of SNe Ib/c. GRB-SNe are at the bright end of the distribution. **Right**: The [O I] 6300Å line in the spectra of SNe 2003jd (top) and 1998bw (bottom), compared to two synthetic lines computed using the same two-dimensional model (Maeda *et al.* 2002). The viewing angle is 15 deg from the polar axis for SN 1998bw, 70 deg for SN 2003jd.

(Leonard *et al.* 2006). Although the measurement of some polarisation suggests that the ejecta may deviate from spherical symmetry, this is not easy to quantify. A much deeper view into the SN is offered by spectra obtained in the late, nebular phase. At this time, the SN nebula is optically thin. The gas is heated by the deposition of the γ-rays and fast positrons emitted in the decay chain ^{56}Ni $\rightarrow$ ^{56}Co $\rightarrow$ ^{56}Fe. Collisions excite the gas, which is then cooled by the emission of radiation in mostly forbidden lines. Fe II lines dominate the spectra of hypernovae in the nebular phase. This testifies to the copius production of ^{56}Ni that makes these SNe so bright. Other strong lines are those of O I and Ca II. These are typical of all SNe Ib/c and indicate a massive star origin for these SNe. Because of the low optical depth, nebular line profiles can be used to map the composition of the SN ejecta down to the lowest velocities, which are located close to the inner core of the explosion, where the black hole is formed. A careful analysis of these lines can therefore provide information about the details of the collapse and the explosion.

A close look at the nebular spectra of the first GRB-SN, 1998bw, provides interesting evidence. The [O I] 6300Å line is very strong, but it has a very sharp profile. This suggests that oxygen is concentrated at the lowest velocities. A uniform distribution would in fact give rise to a parabolic profile. A model based on this assumption (Mazzali *et al.* 2001) can reproduce the [Fe II] lines but produces a synthetic [O I] 6300Å line that is much broader than the observed one. In a massive star, oxygen is located at larger radii than heavier elements. If this mapping is preserved in a spherically symmetric explosion, then the [O I] line is expected to have a broad, flat-topped profile, as oxygen will be ejected at high velocities and it should be absent from the innermost, slow-moving part of the ejecta. This inner part should be dominated by the elements synthesised in the explosion, and in particular by iron. In a spherically symmetric explosion, therefore, Fe lines should be narrower than oxygen lines. In SN 1998bw, however, we see the opposite trend: [Fe II] lines reach velocities of at least 10000 km s^{-1}, and are much broader than [O I] 6300Å, which reaches at most 6,000 km s^{-1}(Mazzali *et al.* 2001).

This surprising observation can most simply be interpreted if we assume that iron (and therefore ^{56}Ni) was ejected at much higher velocities than oxygen. Since ^{56}Ni is synthesised much deeper in the star than oxygen, which is actually a product of the star's previous evolution, the simplest way to explain how it was ejected at a high velocity is to hypotesise that the explosion was highly aspherical. In such a model, kinetic energy was produced mostly along a preferred axis, which may be identified with the star's rotational axis. Consequently, in this region ^{56}Ni was preferentially synthesised. Accordingly, ^{56}Ni would be distributed mostly in a funnel, and after being ejected it remained separate from the bulk of the stellar material, which would be much less nuclearly processed and would be ejected more equatorially and with lower velocities. Thus ^{56}Ni would have a higher expansion velocity than oxygen.

2D hydrodynamic explosion models coupled to nucleosynthesis calculations show how such an explosion can occur. 3D nebular spectra based on such models not only can reproduce the observations, but also allow us to constrain our viewing angle with respect to the SN. For SN 1998bw, we find that the explosion was highly aspherical, with an energy ratio of $\sim 5:1$ in favour of the polar direction, and that an angle of $\sim 15-30$ deg with respect to the axis of the explosion gives the best fit to the observed spectrum (Maeda *et al.* 2002, Maeda, Mazzali, & Nomoto 2006). When the aspherical distribution of the kinetic energy is taken into account, we derive a total kinetic energy for SN 1998bw of $(1-2) \times 10^{52}$erg, which is smaller than the isotropic estimate based on the early-time spectra but still an order of magnitude larger than in classical core-collapse SNe.

Given the connection between SN 1998bw and GRB980425, it is natural to assume that the axis of the explosion is also the direction along which the GRB was emitted. GRB980425 was a rather weak GRB. A slightly off-axis direction for this burst helps to explain its weakness, although it may not be sufficient (Ramirez-Ruiz *et al.* 2005).

If GRB-SNe are intrinsically aspherical, for any GRB-connected SN there should be many more SNe that are not observed to be accompanied by a GRB because their axis of ejection was not pointing towards us. We therefore began a search for the signatures of asphericity in the spectra of SNe Ib/c not connected with GRBs. This limits the sample to sufficiently nearby SNe that can be discovered optically, without the help of the GRB trigger which extends the volume of detection significantly.

We observe these SNe in the nebular phase, looking for signatures of asphericity. One easy prediction of the model discussed in the previous section is that if the axis of ejection was almost perpendicular to our line of sight, the Fe lines would be narrow, and the oxygen line broad. However, because of the disc-like distribution of oxygen, we would expect a double-peaked profile for the [O I] 6300Å line. We use data obtained with Subaru, Keck and the VLT. The most striking result so far was provided by SN 2003jd. This broad-lined SN Ic was almost as bright at peak as SN 1998bw and therefore a very promising candidate. Observations in the nebular phase clearly showed that the [O I] 6300Å line has a double-peak profile, with a large separation between the peaks. The line has a width of $\sim 7000\,\mathrm{km\ s^{-1}}$, and the peak separation is $\sim 5000\,\mathrm{km\ s^{-1}}$. Figure 2 (right) shows that the observed line profile can be reproduced just taking the two-dimensional models that we developed for SN 1998bw and computing emission profiles for an orientation close to the equatorial plane (70 deg, Mazzali *et al.* 2005). The success of this simple test confirms that hypernovae are significantly aspherical events.

Late nebular spectra of SN 1997ef, and of its twin SN 1997dq, show symmetric emission line profiles and do not suggest any major asphericity (Mazzali *et al.* 2004). In the nebular phase, the [O I] line of SN 2002ap was rather sharp, suggesting the presence of a bulk of slow-moving oxygen. Any such material cannot be explained in 1D explosion model and suggests some asphericity in the explosion (Mazzali *et al.* 2007b). The effect is however subtle and it suggests that any asphericity was not as strong as for SN 1998bw. Later, general studies found that asphericity is common among SNe Ib/c, or possibly even the norm (Maeda *et al.* 2008), but indicate that GRB/SNe are on average more aspherical than SNe with lower energy (Maurer *et al.* 2010).

Some evidence that the ejecta include a large mass at low expansion velocity is also provided by the behaviour of the light curve, which is brighter than predicted by standard one-dimensional explosion models at a SN age of 2-3 months. The inclusion of a significant mass of low-velocity material improves the light curve modelling for SNe 1998bw, 1997ef and 2002ap (Maeda *et al.* 2003, 2006). This can be taken as additional indirect evidence for an aspherical distribution of the ejecta.

An inconsistency between the mass derived from the peak of the light curve and the nebular phase is actually the norm for SNe Ic. Even SN 1994I is affected by it. Sauer *et al.* (2006) find that the nebular mass exceeds the mass needed to fit the peak of the light curve by $\sim 0.5 M_{\odot}$. Also, in all SNe Ic significant mixing-out of ^{56}Ni is required to fit the rapid rise of the light curve. This is again not predicted by 1D models and may be the result of some degree of asphericity in the explosion.

4. XRF-SNe and Magnetars

X-Ray Flashes, the soft and weak equivalent of GRBs (Heise *et al.* 2001), were suspected to have a similar origin as GRBs, and to have an associated SN. The first positive discovery of a SN associated with an XRF was the case of XRF060218/SN 2006aj (Pian *et al.* 2006). The SN was of Type Ic, and it had moderately broad lines. It was brighter than normal SNe Ic like SN 1994I, or than non-GRB-HNe such as SNe 2002ap or 1997ef, but not as bright as GRB/SNe. Its derived line velocity was also intermediate between the two groups (Fig. 1, right). Modelling indicates that the explosion that became SN 2006aj was more energetic than normal SNe, but much less so than HNe, with $E_{\rm K} \sim 2\ 10^{51}$ erg. The mass ejected was rather small, $\sim 2 M_{\odot}$, and the mass of ^{56}Ni synthesised was $\sim 0.2 M_{\odot}$. Given these small values, the progenitor star was unlikely to be very massive. Our best estimate is for a progenitor star of $\sim 20 M_{\odot}$. Such a star would probably collapse not to a black hole, but more likely to a neutron star. We therefore suggested that the high explosion energy, as well as the XRF, were the result of a magnetar event (Mazzali *et al.* 2006b). In the late phase, the emission line profiles do now show strong evidence for asphericity (Mazzali *et al.* 2007a).

Perhaps the most intriguing case of a SN which produced a Magnetar is that of the SN Ib 2005bf. This SN reached a first, fairly dim peak, but rather than decline as all other SNe Ib/c, it went through a second, brighter peak phase, reaching a second maximum about one month after the first one (Tominaga *et al.* 2005). This was followed by a sharp decline, and at late time the SN light curve fell on the expected extension of the first peak, suggesting that the first peak was the only one to be powered by ^{56}Ni decay. The second peak may then have been the result of energy injection by a magnetar in an aspherical explosion (Maeda *et al.* 2007).

Another interesting case of a SN associated with an X-ray transient was that of XRF080109/SN 2008D. This SN was discovered following the serendipitous detection of an X-ray burst (Soderberg *et al.* 2008). One of the new aspects of this SN is that it was of Type Ib rather than Ic. Initially, the SN displayed a broad-lined SN Ic spectrum, but later He I lines developed. This is predicted thoretically if a sufficiently large mass of helium is present in the ejecta: He I lines can only develop following non-thermal excitation processes, which require that the ejecta are not very dense (Mazzali & Lucy 1998). The nature of the X-ray transient is debated. Soderberg *et al.* (2008) suggest that it was

Table 1. Properties of GRB-SNe.

GRB/SN	E_K 10^{51} erg	$M(^{56}Ni)$ $M_\odot$	M_{ej} $M_\odot$	M(CO) $M_\odot$	M_{ZAMS} $M_\odot$	Reference
GRB 980425/SN 1998bw	50 ± 5	0.38-0.48	11 ± 1	14 ± 1	35-45	Iwamoto *et al.* 1998
GRB 030329/SN 2003dh	40 ± 10	0.25-0.45	8 ± 2	11 ± 1	30-40	Mazzali *et al.* 2003
GRB 031203/SN 2003lw	60 ± 10	0.45-0.65	13 ± 2	16 ± 1	40-50	Mazzali *et al.* 2006a
XRF 060218/SN 2006aj	2 ± 0.5	0.20-0.25	2 ± 0.5	3.5 ± 0.5	18-22	Mazzali *et al.* 2006b
XRF 080109/SN 2008D (Ib)	7 ± 1	0.07-0.11	7 ± 2	8 ± 2	25-30	Mazzali *et al.* 2008
SN 1997ef	20 ± 4	0.13-0.17	8 ± 2	11 ± 1	30-40	Mazzali *et al.* 2000
SN 2003bg (IIb)	5 ± 1	0.12-0.20	4 ± 1	5 ± 1.5	20-27	Mazzali *et al.* 2009
SN 2002ap	4 ± 1	0.09-0.10	2.5 ± 0.5	5 ± 1	21-25	Mazzali *et al.* 2002
SN 1994I	1 ± 0.2	0.07-0.08	1.2 ± 0.2	2 ± 0.5	14-16	Sauer *et al.* 2006

the result of the breakout of the shock that exploded the star, and that it is a common phenomenon. Mazzali *et al.* (2008) offer a different interpretation. Modelling of the light curve and spectra indicate that SN 2008D was not a typical SN Ib/c, but rather a HN, as indicated by the large $E_K \sim 7 \times 10^{51}$ erg, and the massive ejecta ($M_{ej} \sim 5 M_\odot$; see also Tanaka *et al.* 2009a). The progenitor of SN 2008D may have been a star of $\sim 25 M_\odot$, at the border between black hole and neutron star formation. Therefore the ejection of relativistic material was not unlikely. However, in the case of SN 2008D a relativistic jet may have been weak, and it would also be affected by the presence of the massive ($\sim 2 M_\odot$) He envelope, so that it may only have emerged as a subrelativistic outlow, mimicking the behaviour of the breakout of a shock through the stellar envelope. Such a scenario receives strong support by the nebular spectrum of SN 2008D, which indicates large asphericity (Tanaka *et al.* 2009b).

5. Discussion

The link between energetic, broad-lined type Ic SNe (Hypernovae) and GRBs is established conclusively. The relative rates of GRB and HNe are in good agreement (Podsiadlowski *et al.* 2004), although it is not clear that all HNe make a GRB (Soderberg *et al.* 2006). The exact definition of a HN is not agreed upon. Broad lines and high E_K are an ingredient, but an accompanying GRB is a feature of possibly only the most massive SNe Ic. Presence of a He envelope may quench any jet. Stripping the hydrogen and helium envelopes may require interaction in a binary system.

At lower masses, Type Ic SNe that produce neutron stars may also give rise to an XRF if the neutron star is born spinning rapidly – a magnetar (Mazzali *et al.* 2006b). As in the case of GRBs, this may require the most massive stars that collapse to a neutron star, with ZAMS mass near $20 - 23 M_\odot$.

There is an apparent relation between stellar ZAMS mass, explosion kinetic energy, and luminosity of the SN, as shown in Fig.3 (Nomoto *et al.* 2005).

Nebular spectra can be used to derive the asphericity of the SN explosion and even to determine, albeit only approximately, the direction of the jet axis with respect to our line of sight. This may help us understand the extremely variable properties of the SN-related GRBs, in the face of an amazingly narrow distribution of the properties of the GRB-related SNe (Table 1). Evidence that HNe may occur even in Type IIb SNe (where a thin layer of H has also been preserved) has also been found (Mazzali *et al.* 2009).

A number of questions then arise.

1. Are there minimum requirements for the presence of a GRB in terms of M_{ej}, $M(^{56}Ni)$ and E_K? GRB/SNe are both more energetic and brighter (i.e. they produced more ^{56}Ni) than all other SNe Ib/c (Fig. 3). Recent examples, such as the very broad-lined SN Ic

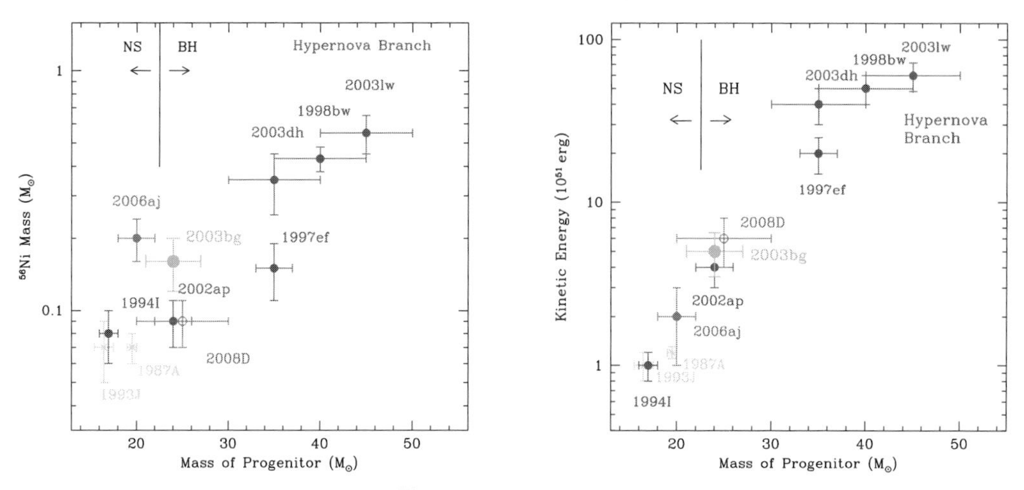

Figure 3. Relation between ^{56}Ni mass, SN kinetic energy and progenitor mass.

SN 2010ah (Corsi *et al.* 2011), or the SN Ic 2010bh, linked with XRF100316D (Cano *et al.* 2011) have not violated this rule. Is this a strict requirement, and if so, why?

2. Where are the off-axis GRB-SNe? Depending on the actual frequency of these events, the volume that we can sample with optically discovered SNe may be too small to include a significant number of GRB-SNe. Additionally, it is possible that not all HNe Ic produced a GRB. Our search indicates that a few SNe Ib/c were aspherical events viewed off-axis. A continued search for the signatures of asphericity in SN explosions on the one hand, and traces of the ejection of material at relativistic velocities on the other are necessary to establish the actual rate of GRB-SNe with respect to that of hypernovae and their fraction relative to all SNe Ic. Alternatively, the jetted nature of the relativistic outflow of nearby GRB/SNe needs to come in question.

3. Why are the GRB-SNe so similar while the SN-GRBs are so different? This may be partly related to orientation, and a study such as that discussed above can also be useful to clarify this apparently puzzling state of affairs. We should also keep in mind that a clear association between GRBs and SNe has only been established for the nearest GRBs. These events may be on average weaker than cosmological ones, and more numerous. All GRB-SNe so far seem to have had progenitors of $\sim 40M_\odot$, while the mass of the SNe associated with an X-Ray Flash is $\sim 20M_\odot$ (Mazzali *et al.* 2006b). As the volume sampled by cosmological GRB is much larger, it is possible that more massive stars contribute to the observed GRBs, which may be intrinsically more powerful.

4. What is the role of Magnetars? Can they contribute kinetic energy to the SN explosion, and can they also help synthesize some ^{56}Ni? It has also been suggested that Magnetars are responsible for most GRB/SNe (Woosley 2010).

References

Arnett, W. D. 1982, *ApJ*, 253, 785
Cano, Z., *et al.* 2011, *ApJ*, 740, 41
Corsi, A., *et al.* 2011, *ApJ*, 741, 76
Galama, T. *et al.* 1998, *Nature*, 395, 670
Heise, J., in't Zand, J., Kippen, R. M., & Woods, P. M. 2001, *GRB in the Afterglow Era*, 16
Iwamoto, K. *et al.* 1998, *Nature*, 395, 672
Leonard, D. *et al.* 2006, *Nature*, 440, 505
Lucy, L. B. 1999, *A&A*, 345, 211

Maeda, K., Mazzali, P. A., & Nomoto, K. 2006, *ApJ*, 645, 1331
Maeda, K. *et al.* 2002, *ApJ*, 565, 405
Maeda, K. *et al.*, 2003, *ApJ*, 593, 931
Maeda, K. *et al.* 2006, *ApJ*, 640, 854
Maeda, K. *et al.* 2007, *ApJ*, 666, 1069
Maeda, K. *et al.* 2008, *Science*, 319, 1220
Malesani, D. *et al.* 2004, *ApJ*, 609, L5
Maurer, J. I. *et al.* 2010, *MNRAS*, 402, 161
Mazzali, P. A. 2000, *A&A*, 363, 705
Mazzali, P. A., Iwamoto K., & Nomoto K. 2000, *ApJ*, 545, 407
Mazzali, P. A. & Lucy, L. B. 1993, *A&A*, 279, 447
Mazzali, P. A. & Lucy, L. B. 1998, *MNRAS*, 295, 4287
Mazzali, P. A., *et al.* 2001, *ApJ*, 559, 1047
Mazzali, P. A., *et al.* 2002, *ApJ*, 572, L61
Mazzali, P. A., *et al.* 2003, *ApJ*, 599, L95
Mazzali, P. A., *et al.* 2004, *ApJ*, 614, 858
Mazzali, P. A., *et al.* 2005, *Science*, 308, 1284
Mazzali, P. A., *et al.* 2006a, *ApJ*, 645, 1323
Mazzali, P. A., *et al.* 2006b, *Nature*, 442, 1018
Mazzali, P. A., *et al.* 2007, *ApJ*, 661, 892
Mazzali, P. A., *et al.* 2007, *ApJ*, 670, 592
Mazzali, P. A., *et al.* 2008, *Science*, 321, 1185
Mazzali, P. A., *et al.* 2009, *ApJ*, 703, 1624
McFadyen, A. & Woosley, S. E. 1999, *ApJ*, 524, 262
Nomoto, K., *et al.* 2005, *ApSS*, 298, 81
Pian, E., *et al.* 2006, *Nature*, 442, 1011
Podsiadlowski, P., *et al.*, 2004, *ApJ*, 607, L.17
Ramirez-Ruiz, E., *et al.* 2005, *ApJ*, 625, L91
Sauer, D. *et al.* 2006, *MNRAS*, 369, 1939
Soderberg, A. M., *et al.* 2006, *ApJ*, 638, 930
Soderberg, A. M., *et al.* 2008, *Nature*, 453, 469
Stanek, K. *et al.* 2003, *ApJ*, 591, L17
Tanaka, M., *et al.* 2009, *ApJ*, 692, 1131
Tanaka, M., *et al.* 2009, *ApJ*, 700, 1680
Tominaga, N., *et al.* 2005, *ApJ*, 633, L97
Woosley, S. E. 2010, *ApJ*, 719, L204

Discussion

O'BRIEN: Where is the mass boundary between progenitors that make a NS or a BH?
P.M.: It is very unclear. The line on my plot is the traditional place to put it.

KULKARNI: Could asphericity be due simply to large convective bubbles rather than a jet? For example, SN 2002ap shows no sign of being a HN.
P.M.: This is possible in the lower energy SNe, but unlikely in HNe. The inferred degree of asphercity is much larger in HNe. Asphericity is higher in the deeper layers, hence SNe Ic tend to be more aspherical than SNe Ib etc. SN 2002ap had enough mass at high velocity to produce broad lines, hence we call it a HN. It however did not have much mass at $v \sim 0.1c$ or higher, as our spectral models showed, and may therefore not have been able to produce a GRB.

Death of Massive Stars: Supernovae and Gamma-Ray Bursts
Proceedings IAU Symposium No. 279, 2012
P. Roming, N. Kawai & E. Pian, eds.

doi:10.1017/S1743921312012732

Recent Observations of GRB-Supernovae

Bethany Elisa Cobb

George Washington University
Department of Physics
725 21st St, NW, Corcoran 105
Washington DC, 20052, USA
email: bcobb@gwu.edu

Abstract. The GRB-SNe connection has been strengthened since 2008 by the detection of 6 additional GRB-SNe at both local and cosmological redshifts. This review summarizes the recent observations of SNe associated with GRBs 081007, 090618, 091127, 100316D, 101219B and 111211A, as well as the observations of SN 2008D, which was associated with a bright X-ray flash (XRF 080109) and may represent a link between "plain" SN and GRB-SNe. It is now clear that most – if not all – long-duration GRBs are produced by the core collapse of massive stars.

Keywords. gamma rays: bursts, supernovae: general, supernovae: individual (2008D, 2008hw, 2009nz, 2010bh, 2010ma)

---

1. Introduction

The connection between long-duration gamma-ray bursts (GRBs) and supernovae (SNe) is supported by indirect evidence including energy considerations (e.g. Woosley 1993), the association of long-duration GRBs with star-forming host galaxies (e.g. Djorgovski *et al.* 1998), and the similarity between the environments in which GRBs occur and the environments in which "plain" SNe-Ic occur (Raskin *et al.* 2008). Between 1998 and 2006, a number of GRBs were also directly connected – both photometrically and spectroscopically – with SNe, including the prime examples of GRB 980425/SN 1998bw, GRB 030329/SN 2003dh, GRB 031203/SN 2003lw, GRB 050525A/SN 2005nc and GRB 060218/2006aj (see Table 1, see also Woosley & Bloom 2006). Interestingly, three of these early, well-studied GRB-SNe (GRBs 980425, 031203 and 060218) had low redshifts ($z \lesssim 0.1$) and low gamma-ray luminosities (isotropic energy $E_{iso} \sim 10^{48-50}$ erg, 2 to 4 orders of magnitude lower than typical for GRBs observed at cosmological distances), showed divergence from the $E_{peak} - E_{iso}$ correlation (Amati relation, e.g. Amati 2006) and produced extremely faint or undetected optical afterglows. Questions naturally arose as to whether or not this group of GRB-SNe represented a special, local population or if all long-duration GRBs are associated with SNe. The detection of additional GRB-SNe, therefore, is crucial in the test of whether or not the GRB-SN connection holds in general for all GRBs, including those with high-redshift and high luminosity.

This review summarizes the recent (since 2008) observations of GRB-related SNe.

2. XRF 080109 – SN 2008D at $z = 0.007$: a special non-GRB SN

The first SN is not actually a GRB-SN, but may represent a link in the continuum between "normal" SNe and GRB-SNe, and therefore merits discussion.

The *Swift* X-ray Telescope (XRT; Burrows *et al.* 2005) serendipitously detected an X-ray flash/transient in NGC 2770 on 2008 January 9 at 13:32:49 UT (Soderberg *et al.* 2008b). This transient does not appear to have been produced by a GRB, as *Swift*

(Gehrels *et al.* 2004) detected no gamma-ray emission from the source despite the fact that the XRF was within the field of view of the Burst Alert Telescope (BAT; Barthelmy *et al.* 2005). While an optical counterpart was not seen in contemporaneous Ultraviolet/Optical Telescope (UVOT; Roming *et al.* 2005) imaging, a brightening counterpart was detected just 1.4 hours after the flash occurred. This bright X-ray flash ($E_X \sim 10^{46}$ erg) was quickly recognized to have been produced by the core collapse of a massive star. Optical imaging and spectra revealed a Helium-rich type Ib/c SN (Deng & Zhu 2008, Soderberg *et al.* 2008a).

The peak optical luminosity of the event fell on the low end of the distribution of GRB-SNe luminosities, but the SN was very similar to other GRB-SNe in having a rest-frame rise time ($\sim$ 18 days) longer than typical for "plain" SN Ib/c (Modjaz *et al.* 2009). In contrast, the spectral features of this SN were narrower than observed in other GRB-SNe (Malesani *et al.* 2008).

The remarkable photometric and spectroscopic temporal coverage of this event allowed for detailed modeling of the SN and its progenitor. Interestingly, the SN appears to have synthesized $\sim 0.09 M_\odot$ of ^{56}Ni, an amount similar to that seen in the broad-lined SN-Ic 2002ap, but much less than that seen in most GRB-SNe, including SN 1998bw which synthesized $\sim 0.7 M_\odot$ of ^{56}Ni (Tanaka *et al.* 2009).

The X-ray flash of SN 2008D may be attributed to a shock break-out from a Wolf-Rayet star, which was slightly delayed by the presence of a dense stellar wind around the progenitor (Soderberg *et al.* 2008b). Alternatively, the X-ray flash may result from the collapse of a $\sim 30 M_\odot$ star that produced a black hole with a weak, mildly relativistic jet (Mazzali *et al.* 2008). In the latter case, the jet is a "failed" relativistic jet – likely failing due to low initial energy or to damping caused by a helium layer, which would not be found around most pre-GRB stars (Mazzali *et al.* 2008). This SN, therefore, may indicate that GRB-like central engine activity may be present in all black-hole-forming SNe Ibc. Even if this SN did not produce a mildly relativistic jet, it may still be an example of a link between He-rich and He-poor explosions (a class which includes the GRB-SNe).

3. GRB 081007 – SN 2008hw at $z = 0.53$

Swift triggered on the long-duration GRB 081007 on 2008 October 7 at 05:23:52 UT (Baumgartner *et al.* 2008). The *Fermi* Gamma-ray Burst Monitor (GBM) also triggered on this GRB (Bissaldi *et al.* 2008). The GRB had a typical isotropic energy release of $E_{iso} \sim 10^{51}$ erg. The burst's X-ray afterglow was detected by the XRT and the burst's optical afterglow was detected by UVOT. Ground-based imaging by the robotic 60-cm Rapid Eye Mount (REM; Chincarini *et al.* 2003) telescope also identified the burst's bright, slowly fading optical/NIR afterglow (Covino *et al.* 2008). Only 73 minutes post-burst, a Gemini Multi-Object Spectrograph (GMOS) observation was started and the redshift of the burst was determined to be $z = 0.53$ (Berger *et al.* 2008).

The SN associated with GRB 081007, SN 2008hw, was first noticed in Gemini-GMOS imaging as an increase in optical afterglow brightness detected between about 17 and 26 days using image subtraction techniques (Soderberg *et al.* 2008c). Analysis of a Very Large Telescope (VLT) spectrum taken 17 days post-burst then revealed (following the subtraction of a template starburst galaxy) the broad bumps indicative of a GRB-SN similar to SN 1998bw (Della Valle *et al.* 2008). Additional photometric follow-up carried out by the Gamma-Ray Burst Optical/Near-Infrared Detector (GROND; Greiner *et al.* 2008) showed a flattening of the afterglow light curve at approximately 7 days (rest-frame) post-burst and a change in the spectral-energy distribution of the afterglow at 10 days (rest-frame) post-burst, consistent with the increased contribution of light from a brightening SN (Della Valle *et al.* 2008).

Of the recent GRB-SNe, GRB 081007/SN 2008hw serves as just the first example of a "typical" energy GRB with a "typical" optical afterglow that is located at "cosmological" distances and yet is undeniably associated with a SN.

4. GRB 090618 – SN2009?? at $z = 0.54$

Swift triggered on the long-duration GRB 090618 on 2009 June 18 at 08:28:29 UT (Schady *et al.* 2009). This was an extremely bright burst that also triggered *Fermi*-GBM (McBreen 2009), *Super-AGILE* (Longo *et al.* 2009), and *Konus-Wind* (Golenetskii *et al.* 2009). The burst's X-ray afterglow was detected by the XRT and the burst's optical afterglow was detected by UVOT. Numerous ground-based telescopes also observed this burst's bright optical afterglow including the Palomar 60-inch (Cenko 2009), the Katzman Automatic Imaging Telescope (KAIT; Perley 2009) and the Robotic Optical Transient Search Experiment (ROTSE)-IIIb (Rujopakarn *et al.* 2009). Just 20 minutes after the burst, the Kast spectrograph on the 3-m Shane telescope at Lick Observatory began obtaining observations, and the redshift of the burst was determined to be $z = 0.54$ (Cenko *et al.* 2009).

The SN associated with GRB 090618 was identified by the presence of a "bump" in the optical afterglow light curve, which was associated with a change in the afterglow's $R_c - i$ color index. The color change is expected to occur as the early, blue afterglow fades and the later, redder SN component rises. The SN accompanying GRB 090618 is similar in brightness and temporal evolution to the prototypical GRB-SN, SN 1998bw. After the host galaxy and afterglow contributions were subtracted, it was determined that the SN peaked around 16 days (rest-frame) and was slightly brighter than SN 1998bw (Cano *et al.* 2011).

GRB 090618 had "typical" gamma-ray burst parameters, including an isotropic equivalent energy release of $E_{iso} \sim 3 \times 10^{53}$ erg (Page *et al.* 2011, Izzo *et al.* 2012). GRB 090618 is, therefore, an excellent example of a normal, cosmological GRB associated with a SN. Unfortunately, no spectrographic observations of this GRB-SN were obtained.

5. GRB 091127 – SN 2009nz at $z = 0.49$

Swift triggered on the long-duration GRB 09117 on 2009 November 27 at 23:25:45 UT (Troja *et al.* 2009). This burst also triggered *Fermi*-GBM (Wilson-Hodge & Preece 2009) and the *Suzaku Wide-band All-sky Monitor* (WAM; Nishioka *et al.* 2009). While no XRT or UVOT observations were initially obtained due to observing constraints, the robotic ground-based 2-m Liverpool Telescope detected a bright ($R \sim 15$) optical afterglow just minutes after the burst occurred (Smith *et al.* 2009). An X-ray afterglow was eventually detected by XRT (Evans *et al.* 2009). The optical afterglow was bright enough to be detected by numerous instruments including UVOT (Immler & Troja 2009), REM (Fugazza *et al.* 2009), GROND (Updike *et al.* 2009), and the 1.3 m Small and Moderate Aperture Research Telescope System (SMARTS; Subasavage *et al.* 2010) telescope (Cobb 2009). The redshift of the burst was measured to be $z = 0.49$ by a Gemini-GMOS spectroscopic observation (Cucchiara *et al.* 2009) and confirmed by an X-shooter (D'Odorico *et al.* 2006) observation at the VLT (Thoene *et al.* 2009).

The SN was detected as a flattening then a brightening of the optical afterglow in a combination of Gemini-GMOS and 1.3m SMARTS imaging. The optical afterglow began to brighten around 10 days (observer frame) post-burst, following an evolution similar to

SN 1998bw (Cobb *et al.* 2010a). The SN was designated SN 2009nz (Cobb *et al.* 2010b). Comparison of SN 2009nz to other well-observed GRB-SNe shows that these events have global similarities, but each one shows individual variations in peak brightness and rise-time (Cobb *et al.* 2010c). While the GRB-SNe form a relatively homogenous group, they are certainly not "clones" of one another.

Spectral identification of SN 2009nz was later published, revealing the broad features typical of GRB-SNe due to their high expansion velocities. Uncovering the SN spectrum required taking a late-time ($\sim$1 year post-burst) spectrum of the host galaxy. This was used to subtract the host's contribution from a spectrum taken near the SN's peak brightness. The spectrum of SN 2009nz was reminiscent of SN 2006aj, and had narrower features than SN 1998bw and SN 2010bh, indicating a somewhat lower expansion velocity (Berger *et al.* 2011).

GRB 091127 had "typical" GRB parameters, including an $E_{iso} \sim 10^{53}$ erg (Stamatikos *et al.* 2009), providing another example of a link between a cosmological long-duration GRB with standard energy to the core-collapse of a massive star. This burst also highlights the difficulty of detecting cosmological GRB-SNe, since both the photometric and spectroscopic observations of this GRB-SN were complicated by the burst's relatively high redshift and light contamination from both the GRB's bright optical afterglow and its relatively bright host galaxy.

6. GRB 100316D – SN 2010bh at $z = 0.059$

Swift triggered on the long-duration GRB 100316D on 2010 March 16 at 12:44:50 UT (Stamatikos *et al.* 2010). An X-ray afterglow was detected, but no UVOT source was observed (Stamatikos *et al.* 2010). An optical source inside the XRT error region was detected in a VLT/X-shooter acquisition image, with a measured redshift of $z = 0.059$ (Vergani *et al.* 2010a). Initially it was unclear whether or not this optical source was associated with the GRB or if it was an extended region of a nearby DSS galaxy. Gemini-GMOS imaging combined with further VLT/X-shooter spectroscopy confirmed that the optical source and the nearby galaxy were at a common redshift and thus likely two parts of a single, morphologically disturbed galaxy (Vergani *et al.* 2010b).

With an isotropic energy release of only $4 - 6 \times 10^{49}$ erg, a redshift of only $z = 0.059$, and an extremely long duration, this burst was not a typically "cosmological" GRB, but was reminiscent of the group of low-redshift, low gamma-ray luminosity bursts detected prior to 2006 (GRBs 980425, 031203 and 060218). Similar to those events, no optical afterglow was detected following the GRB, though the presence of the bright galaxy at the position of the X-ray afterglow may easily have obscured a dim optical afterglow. GRB 100316D shared temporal characteristics with GRB 060218/SN 2006aj; both of these bursts also had very soft gamma-ray spectra (Sakamoto *et al.* 2010a). The notable similarities between these two bursts suggested that observers should be on the lookout for an emergent SN (Sakamoto *et al.* 2010b). In Gemini-GMOS imaging acquired approximately 1.5 days post-burst, a possibly brightening source was noted (Levan *et al.* 2010). This brightening was eventually confirmed by additional Gemini-GMOS imaging (Wiersema *et al.* 2010) and imaging by GROND (Rau *et al.* 2010), indicating a rising SN associated with GRB 100316D: SN 2010bh (Bufano *et al.* 2010a). Spectroscopy obtained on the twin 6.5-m Magellan telescopes and VLT X-shooter and FORS2 then revealed broad photospheric absorption lines similar to SN 1998bw (Chornock *et al.* 2010a, Bufano *et al.* 2010b).

SN 100316D rose to peak brightness in approximately 8 days, which is faster than SN 1998bw but similar to SN 2006aj (Bufano *et al.* 2011, Starling *et al.* 2011). However, SN

100316D showed higher photospheric velocities than SN 1998bw, possibly indicating that SN 100316D was a more energetic explosion than SN 1998bw (Chornock *et al.* 2010b). SN 2010bh is also the faintest of the well-studied GRB-SNe, which suggests it ejected smaller amounts of ^{56}Ni than the other GRB-SNe (Bufano *et al.* 2011).

7. GRB 101219B – SN 2010ma at $z = 0.55$

Swift triggered on the long-duration GRB 101219B on 2010 December 19 at 16:27:53 UT (Gelbord *et al.* 2010). This GRB also triggered *Fermi*-GBM (van der Horst 2010). *Swift* detected both an X-ray and optical afterglow (Gelbord *et al.* 2010); the optical/NIR afterglow was also detected by GROND (Olivares *et al.* 2010). This is another example of a typical energy ($\sim 4 \times 10^{51}$ erg) GRB with a bright afterglow that obeys the "Amati" relation (Sparre *et al.* 2011). The afterglow of this burst was particularly notable in that it had an unusually slow decay and was still detectable by UVOT at 2 weeks post-burst (Kuin & Gelbord 2011).

The first hints of a SN appeared 5 days post-burst when the GROND afterglow light curve began to flatten and change color; at 15 days post-burst, the afterglow resumed its decay (Olivares *et al.* 2011). Several spectra obtained using VLT/X-shooter around 16 days post-burst measured a redshift of the optical source to be $z = 0.55$, with the spectra showing broad SN-like undulations similar to those detected in SN 1998bw and all other GRB-SNe (de Ugarte Postigo *et al.* 2011).

8. GRB 111211A – SN2011?? at $z = 0.478$

Super-AGILE (Tavani *et al.* 2009) triggered on the long-duration GRB 111211A on 2011 December 11 at 22:17:33 UT (Lazzarotto *et al.* 2011). GROND ground-based imaging identified the burst's optical afterglow approximately 8.5 hours post-burst (Kann *et al.* 2011). VLT/X-shooter spectroscopy obtained 1.3 days after the GRB measured the redshift of the burst to be $z = 0.478$ (Vergani *et al.* 2011).

The SN associated with GRB 111211A was first detected in a spectrum obtained using the 10.4m Gran Telescopio Canarias (GTC) equipped with the Ohio State Infrared Imager/Spectrometer (OSIRIS). The spectrum showed that at the position of the optical afterglow, there was a broad-line SN similar to SN 2006aj (associated with GRB 060218) close to its time of maximum (de Ugarte Postigo *et al.* 2012). Additionally, photometric comparison of the spectrum acquisition image to an r-band image taken 5 days prior showed that the afterglow's light curve had flattened, indicating a possible contribution from a supernova component.

9. Conclusion

These recent GRB-SNe observations have significantly added to the "GRB-SNe zoo" (see Table 1). No longer are GRB-SNe just associated with special cases of low-redshift, low-luminosity GRBs. By identifying additional examples of photometrically and spectroscopically confirmed GRB-SNe, it now appears that the GRN-SN connection is proven both for local, low-luminosity GRBs and for typical-energy, cosmological GRBs. Clearly, the evidence suggests that most – if not all – long-duration GRBs are intimately linked with SNe.

Furthermore, the special example of XRF 080109/SN 2008D, while not itself a GRB-SN, might indicate a link between GRB-SNe and non-GRB-SNe. This event helps fill out the continuum between "typical" SNe and the "hyper" SNe associated with GRBs.

Table 1. GRB/SNe

GRB	SN	Redshift
980425	1998bw	0.0085
030329	2003dh	0.16
031203	2003lw	0.11
050525A	2005nc	0.606
060218	2006aj	0.033
XRF 080109	2008D	0.007
081007	2008hw	0.53
090618	2009??	0.54
091127	2009nz	0.49
100316D	2010bh	0.059
101219B	2010ma	0.55
111211A	2011??	0.478

Of course, questions about the GRB-SN connection still remain thanks to two examples of low-redshift GRBs (060505 & 060614) without detected SNe (e.g. Fynbo *et al.* 2006). Possibly the redshifts of these bursts are mis-identified due to the chance alignment of a higher-redshift GRB with a foreground galaxy (e.g. Cobb *et al.* 2006). Alternatively, these bursts might by short-duration bursts and thus not produced by the core collapse of a massive star (e.g. Gehrels *et al.* 2006). However, it is also possible that something about the progenitors of these GRBs was unique and as yet unknown. In any case, it has now become clear that these examples are outliers, and that long-duration GRBs of all redshifts and energies are primarily produced by the core collapse of massive stars.

References

Amati, L. 2006, *MNRAS*, 372, 233
Barthelmy, S. D., *et al.* 2005, *Space Sci. Revs*, 121, 143
Baumgartner, W. H., *et al.* 2008, *GCN*, 8330, 1
Berger, E., Fox, D. B., Cucchiara, A., & Cenko, S. B. 2008, *GCN*, 8335, 1
Berger, E., *et al.* 2011, *ApJ*, 743, 204
Bissaldi, E., McBreen, S., & Connaughton, V. 2008, *GCN*, 8369, 1
Bufano, F., *et al.* 2010, *CBET*, 2227, 1
Bufano, F., *et al.* 2010, *GCN*, 10543, 1
Bufano, F., *et al.* 2011, *arXiv*, arXiv:1111.4527
Burrows, D. N., *et al.* 2005, *Space Sci. Revs*, 121, 165
Cano, Z., *et al.* 2011, *MNRAS*, 413, 669
Cenko, S. B. 2009, *GCN*, 9513, 1
Cenko, S. B., Perley, D. A., Junkkarinen, V., Burbidge, M., Diego, U. S., & Miller, K. 2009, *GCN*, 9518, 1
Chincarini, G., *et al.* 2003, *Messenger*, 113, 40
Chornock, R., Soderberg, A. M., Foley, R. J., Berger, E., Frebel, A., Challis, P., Simon, J. D., & Sheppard, S. 2010, *GCN*, 10541, 1
Chornock, R., *et al.* 2010, *arXiv*, arXiv:1004.2262
Cobb, B. E., Bailyn, C. D., van Dokkum, P. G., & Natarajan, P. 2006, *ApJ*, 651, L85
Cobb, B. E. 2009, *GCN*, 10244, 1
Cobb, B. E., Bloom, J. S., Cenko, S. B., & Perley, D. A. 2010, *GCN*, 10400, 1

Cobb, B. E., Bloom, J. S., Morgan, A. N., Cenko, S. B., & Perley, D. A. 2010, *CBET*, 2288, 1
Cobb, B. E., Bloom, J. S., Perley, D. A., Morgan, A. N., Cenko, S. B., & Filippenko, A. V. 2010, *ApJ*, 718, L150
Covino, S., *et al.* 2008, *GCN*, 8331, 1
Cucchiara, A., Fox, D., Levan, A., & Tanvir, N. 2009, *GCN*, 10202, 1
D'Odorico, S., *et al.* 2006, *SPIE*, 6269, 98
de Ugarte Postigo, A., *et al.* 2011, *GCN*, 11579, 1
de Ugarte Postigo, A., Thoene, C. C., & Gorosabel, J. 2012, *GCN*, 12802, 1
Della Valle, M., *et al.* 2008, *CBET*, 1602, 1
Deng, J. & Zhu, Y. 2008, *GCN*, 7160, 1
Djorgovski, S. G., Kulkarni, S. R., Bloom, J. S., *et al.* 1998, *ApJ Letters*, 508, L17
Evans, P. A., Page, K. L., & Troja, E. 2009, *GCN*, 10201, 1
Fugazza, D., *et al.* 2009, *GCN*, 10194, 1
Fynbo, J. P. U., *et al.* 2006, *Natur*, 444, 1047
Gehrels, N., *et al.* 2004, *ApJ*, 611, 1005
Gehrels, N., *et al.* 2006, *Natur*, 444, 1044
Gelbord, J. M., *et al.* 2010, *GCN*, 11473, 1
Golenetskii, S., *et al.* 2009, *GCN*, 9553, 1
Greiner, J., *et al.* 2008, *PASP*, 120, 405
Immler, S. & Troja, E. 2009, *GCN*, 10193, 1
Izzo, L., *et al.* 2012, *arXiv*, arXiv:1202.4374
Kann, D. A., Greiner, J., Kruehler, T., & Klose, S. 2011, *GCN*, 12668, 1
Kuin, N. P. M. & Gelbord, J. M. 2011, *GCN*, 11516, 1
Lazzarotto, F., *et al.* 2011, *GCN*, 12666, 1
Levan, A. J., Tanvir, N. R., D'Avanzo, P., Vergani, S. D., & Malesani, D. 2010, *GCN*, 10523, 1
Longo, F., *et al.* 2009, *GCN*, 9524, 1
Malesani, D., *et al.* 2008, *GCN*, 7169, 1
Mazzali, P. A., *et al.* 2008, *Sci*, 321, 1185
McBreen, S. 2009, *GCN*, 9535, 1
Modjaz, M., *et al.* 2009, *ApJ*, 702, 226
Nishioka, Y., *et al.* 2009, *GCN*, 10224, 1
Olivares, E. F., Rossi, A., & Greiner, J. 2010, *GCN*, 11478, 1
Olivares, E. F., Schady, P., Kruehler, T., Greiner, J., Klose, S., & Kann, D. A. 2011, *GCN*, 11578, 1
Page, K. L., *et al.* 2011, *MNRAS*, 416, 2078
Perley, D. A. 2009, *GCN*, 9514, 1
Raskin, C., Scannapieco, E., Rhoads, J., & Della Valle, M. 2008, *ApJ*, 689, 358
Rau, A., Nardini, M., Updike, A., Filgas, R., Greiner, J., Kruehler, T., & Klose, S. 2010, *GCN*, 10547, 1
Roming, P. W. A., *et al.* 2005, *Space Sci. Revs*, 121, 95
Rujopakarn, W., Guver, T., Pandey, S. B., & Yuan, F. 2009, *GCN*, 9515, 1
Sakamoto, T., *et al.* 2010, *GCN*, 10511, 1
Sakamoto, T., Barthelmy, S. D., Baumgartner, W. H., Cummings, J. R., Gehrels, N., Markwardt, C. B., Palmer, D. M., & Stamatikos, M. 2010, *GCN*, 10524, 1
Schady, P., *et al.* 2009, *GCN*, 9512, 1
Smith, R. J., Kobayashi, S., Guidorzi, C., & Mundell, C. G. 2009, *GCN*, 10192, 1
Soderberg, A. M., *et al.* 2008, *Natur*, 453, 469
Soderberg, A., Berger, E., Fox, D., Cucchiara, A., Rau, A., Ofek, E., Kasliwal, M., & Cenko, S. B. 2008, *GCN*, 7165, 1
Soderberg, A., Berger, E., & Fox, D. 2008, *GCN*, 8662, 1
Sparre, M., *et al.* 2011, *ApJ*, 735, L24
Stamatikos, M., *et al.* 2009, *GCN*, 10197, 1
Stamatikos, M., *et al.* 2010, *GCN*, 10496, 1
Starling, R. L. C., *et al.* 2011, *MNRAS*, 411, 2792

Subasavage, J. P., Bailyn, C. D., Smith, R. C., Henry, T. J., Walter, F. M., & Buxton, M. M. 2010, *SPIE*, 7737, 77371C
Tanaka, M., *et al.* 2009, *ApJ*, 700, 1680
Tavani, M., *et al.* 2009, *A&A*, 502, 995
Thoene, C. C., *et al.* 2009, *GCN*, 10233, 1
Troja, E., *et al.* 2009, *GCN*, 10191, 1
Updike, A., Rossi, A., Rau, A., Greiner, J., Afonso, P., & Yoldas, A. 2009, *GCN*, 10195, 1
van der Horst, A. J. 2010, *GCN*, 11477, 1
Vergani, S. D., D'Avanzo, P., Levan, A. J., Covino, S., Malesani, D., Hjorth, J., & Antonelli, L. A. 2010, *GCN*, 10512, 1
Vergani, S. D., Levan, A. J., D'Avanzo, P., Covino, S., Malesani, D., Hjorth, J., Tanvir, N. R., & Antonelli, L. A. 2010, *GCN*, 10513, 1
Vergani, S. D., *et al.* 2011, *GCN*, 12677, 1
Wiersema, K., D'Avanzo, P., Levan, A. J., Tanvir, N. R., Malesani, D., & Covino, S. 2010, *GCN*, 10525, 1
Wilson-Hodge, C. A. & Preece, R. D. 2009, *GCN*, 10204, 1
Woosley, S. E. 1993, *ApJ*, 405, 273
Woosley, S. E. & Bloom, J. S. 2006, *ARA&A*, 44, 507

Discussion

SAVAGLIO: In your last table you have 12 GRB-SNe spectroscopically confirmed and 5 are at $z \sim 0.5$, do you have an explanation for this excess?

COBB: There is no exact explanation for this. Of course, it is possible that this is a coincidence resulting from the small number of examples we have. Part of the problem is that most GRBs occur at much higher redshifts ($z \sim 2$ or 3) and GRB-SNe are extremely difficult to detect at redshifts above $z = 0.5$, due to dimness, obscuration by afterglow and host galaxy light, and the fact that the UV line-blanketed spectral regions are redshifted into the optical.

AMATI: The link between long GRBs and SNe is also connected to the issue of classification of GRBs as long or short (e.g. 060614).

COBB: Absolutely. No GRB-SN is expected following short GRBs if they are due to the merger of binary neutron stars. Therefore, if GRB 060505 and GRB 060614 are actually produced via "short-duration" progenitors, then that would naturally explain the lack of GRB-SN associated with them. In that case, they would present no challenge to the GRB-SN paradigm. Time classification is, of course, subject to issues such as detector sensitivity. Classification of bursts as short vs. long may eventually be supplanted by classification by progenitor type.

Death of Massive Stars: Supernovae and Gamma-Ray Bursts
Proceedings IAU Symposium No. 279, 2012
P. Roming, N. Kawai & E. Pian, eds.

doi:10.1017/S1743921312012744

GRB 101225A - a new class of GRBs?

C. C. Thöne[1], A. de Ugarte Postigo[1,2], C. Fryer[3], K. Page[4], J. Gorosabel[1], D. Perley[5], M. Aloy[6], C. Kouveliotou[7] and the Christmas Burst collaboration

[1]IAA - CSIC, Glorieta de la Astronomía s/n, E - 18008, Granada, Spain
email: cthoene@iaa.es

[2]Dark Cosmology Centre, Juliane Maries Vej 30, DK - 2100 Copenhagen, Denmark

[3]Los Alamos National Laboratory, MS D409, CCS-2, Los Alamos, New Mexico 87545, USA

[4]Dep. of Physics and Astronomy, Univ. of Leicester, University Road, Leicester LE1 7RH, UK

[5]Astronomy Department, UC Berkeley, 601 Campbell Hall, Berkeley, California 94720, USA

[6]Departamento de Astronomia y Astrofisica, Universidad de Valencia, 46100 Burjassot, Spain

[7]Science and Technology Office, ZP12, NASA/MSFC, Huntsville, Alabama 35812, USA

Abstract. The Christmas burst, GRB 101225A, was one of the most controversial bursts in the last few years. Its exceptionally long duration but bright X-ray emission showing a thermal component followed by a strange afterglow with a thermal SED lead to two different interpretations. We present here our model ascribing this strange event to a new type of GRB progenitor consisting of a neutron star and an evolved main-sequence star in a very faint galaxy at redshift 0.33 while Campana *et al.* (2011) proposed a Galactic origin. New observations at several wavelengths might resolve the question between the two models in the near future.

Keywords. gamma rays: bursts, supernovae: individual

1. Introduction

Several GRBs in the last years have questioned our original concept of two classes of GRBs first proposed by Kouveliotou *et al.* (1993) and the subsequent connection with two different types of progenitors. Long-duration GRBs are due to the collapse of a massive star (Woosley & Bloom 2006), whereas short GRBs come from the merger of two compact objects. While it seems now established that even intrinsically short GRBs (Levesque *et al.* 2009) or those without SN (Fynbo *et al.* 2006) can be connected to the death of massive stars, some exceptionally long and soft nearby GRBs have been observed which could be connected to a different kind of progenitor.

The standard fireball model predicts the afterglow emission to come from synchrotron radiation and the SED as well as the temporal evolution are subsequently expected to follow power-laws. A few X-ray afterglows, all connected to nearby long GRBs associated with SNe, showed a thermal component (TC) with a contribution of usually around 20% (see e.g. Starling *et al.*, this volume). Surprisingly, most nearby GRB-SNe were subluminous (with the exception of GRB 030329) suggesting they might be a different class than the classical "cosmological" GRB. All SNe connected to those nearby GRBs are luminous broad-line Type Ic SNe. Here we present GRB 101225A, called "The Christmas Burst", whose properties largely deviate from those listed above for usual long GRBs.

2. Observations of GRB 101225A

GRB 101225A was discovered by *Swift* on Dec. 25, 2010 at 18:37:45 UT in an image trigger (Racusin *et al.* 2010). It had an exceptionally long T_{90} of at least 2000 s with indications for emission ongoing for several days (Thöne *et al.* (2011)) and a very soft spectrum. This puts it at the extreme end of the hardness - duration distribution of Swift GRBs, similar to a few other nearby GRB SNe like GRB 060218 and GRB 100316D.

The X-ray counterpart detected by XRT, the brightest at several thousand seconds, had a very shallow decay in the first 0.2 days before rapidly dropping beyond the detection limit with $t^{5/3}$, inconsistent with predictions from synchrotron radiation. Similar behaviour was observed for GRB 060218 and the recent extremely long GRB 111109A. In first two orbits (up to $\sim$ 8 ks) of the X-ray afterglow observation, an additional component to the power-law fit is best modeled with a blackbody (BB) of T$\sim$1 keV and a radius of $\sim 1R_{\odot}$ (assuming a redshift of z=0.33, see below). The BB component contributes around 20% to the total flux and shows no temporal evolution.

An optical counterpart was discovered by AlFOSC/NOT 1.54h after the burst (Xu *et al.* 2010) with nearly constant luminosity over the first 2 days. An extensive UV-optical-IR (UVOIR) follow-up until 2 months after the event revealed a very usual behaviour, both in the light-curve and the SED. Collecting all available UVOIR data, we derive an SED at 7 different epochs from 0.07 to 40 days post burst (Thöne *et al.* 2011). The SED can not be fit with a single power-law as expected from the fireball model. During the first 10 days, it is well modeled with a simple expanding and cooling BB. The radius of the UVOIR BB increases from 13 to 45 AU while the temperature drops from 43,000 to 5,000 K from 0.07 to 18 days. The temperature and radius evolution of X-ray and UVOIR BB are very different suggesting that the emissions come from different processes.

At around 10 days, the simple BB model is no longer valid and an additional, more complicated component becomes evident. At the same time, the light-curve shows a small rise with a maximum at around 30 days before it starts to drop again. This behaviour fits very well to a late, faint supernova associated with the GRB. Spectroscopy at several epochs closely after the GRB and at 41 days did not reveal any absorption or emission lines from the host galaxy nor any broad SN features, therefore, a spectroscopic redshift of the event could not be obtained. We model the SED at 40 days with several SN templates, concluding that a 1998bw-like (Galama *et al.* 1998) spectrum fits the data best, however stretched in time by a factor of 1.25 and only 1/10th of the luminosity. From the SED fit, we obtain a photometric redshift of $z=0.33^{+0.07}_{-0.04}$. At z=0.33, the SN has a luminosity of –16.7 mag, making it the faintest GRB-SN observed.

Preimaging observations of the field from the PAnDAs survey (Richardson *et al.* 2011) showed a possible object with low significance at $g' \sim 27.2$ mag. We reobserved the field 180 days after the burst with OSIRIS/GTC at a total exposure time of 4.1 h. The suggested host candidate from the preimaging is clearly detected, but unresolved, with $g' = 27.21 \pm 0.27$, $r' = 26.90 \pm 0.14$. It shows a blue color, consistent with a star-forming galaxy and lies well above the extrapolation of the light curve (see Fig. 1). Its absolute magnitude of $M_{g'} = -13.7$ mag makes it magnitudes fainter than the host of GRB 060218.

3. Our model

We suggest a new progenitor model to explain all the observed features described in Sect. 2: A merger between an evolved He-star and a neutron star (NS) binary system, which had been suggested as possible GRB progenitor model long ago (Fryer & Woosley (1998)). When the He-star leaves the main sequence, it expands and incorporates the NS, leading to a common envelope phase during which the He-star ejects most of its envelope

Table 1. GRBs with SNe and no observed afterglows

GRB	z	TC^2 X-ray	TC^2 optical	T_{90} (s)	E_{peak} (keV)	E_{iso} (erg)	HR^3	Radio? (mag)	SN M_V [4] (mag)	Host M_B [5]
980425	0.0085	No	?	23.3	55±21	8.1×10^{47}	—	Yes	−19.42	−17.6
031203	0.105	No	?	30	158±51	3×10^{49}	—	Yes	−20.39	−21.0
060218	0.0331	Yes	Yes	∼2100	4.9	6.2×10^{49}	0.835	Yes	−18.76	−15.9
080109*	0.0065	Yes	Yes	∼400	low	2×10^{46}	—	Yes	−16.7	−20.7
100316D	0.059	Yes	?	>1300	—	3.1×10^{49}	0.891	No	−18.62	−18.8
101225A	0.33	Yes	Yes	>2000	38±20	$> 1.4\times10^{51}$	1.06	No	−16.9	−13.7

Notes: [1]*No γ-rays observed, numbers derived from X-rays; [2]TC = thermal component; [3]HR = hardness ratio defined as (50-100 keV)/(25-50 keV); [4]SN M_V SN peak absolute magnitude in V

in form of a thick shell. In our model, this shell must form a broad torus with a narrow opening (funnel) at the rotation axis.

When the two stars merge, an accretion disk and jets are formed leading to a GRB-like event. Only a small part of the jet escapes through the funnel giving rise to the detected γ-ray emission while most of it interacts with the previously ejected material. Backwards scattered material from the inner boundary of the envelope leads to a hot spot causing the X-ray emission. Most of the jet gets thermalized when interacting with the material in the funnel wall. When the material breaks out of the shell, a hot plume is formed responsible for the UVOIR BB emission, which is finally overtaken by the SN. The model predicts only a small amount of radioactive Ni and hence a faint SN.

We investigated if other GRBs in the past showed a similar behaviour, in particular those without "traditional" afterglow (see Tab. 1). Some of them had a TC in X-rays, although GRBs exist that have a TC in X-rays but show the usual power-law afterglow (e.g. GRB 090618 Page *et al.* 2011, Cano *et al.* 2011). Most of these lack early optical data and the onset of the SN is much earlier than for GRB 101225A. The early SEDs of GRB 060218 and XRO 080109, however, both show a TC in the optical during the first days.

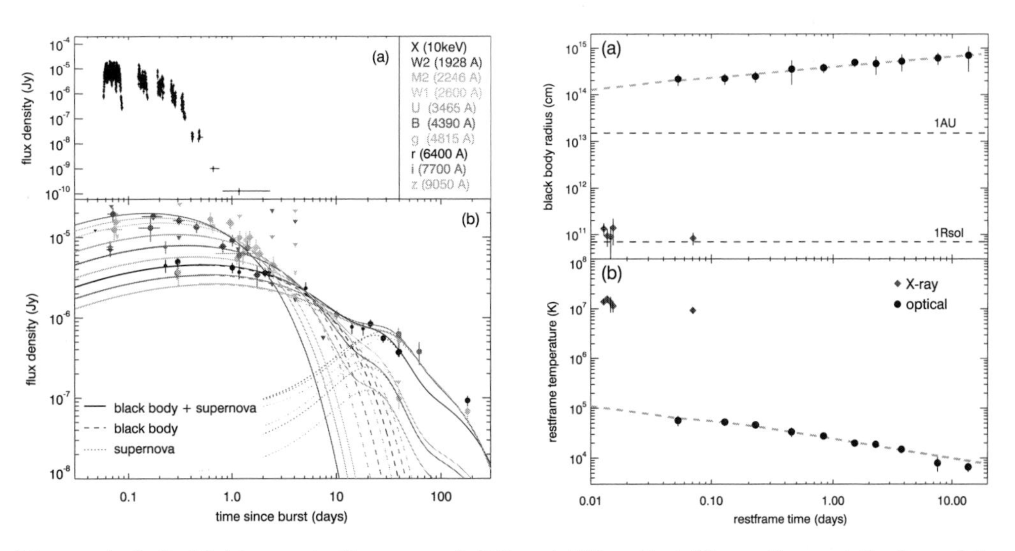

Figure 1. Left: Lightcurve in X-rays, and different UV-optical filters, the contribution of the different emission components described in the text are indicated in the lower panel. Right: Temperature and radius evolution of the X-ray and UVOIR BB emissions together with the predictions from our model (dashed line).

The UVOIR TC of XRO 080109 seems to be just the continuation of the shock-breakout observed in X-rays Soderberg *et al.* (2008). GRB 060218 has a UVOIR BB evolution similar to GRB 101225A and the X-ray and UVOIR BB are also inconsistent with a common origin. The X-ray BB radius however is larger, the SN much brighter and also emerges much earlier, while the GRB itself was subluminous. GRB 060218 could come from a similar event with a somewhat different progenitor.

4. Alternative models and how we can decide between them

Campana *et al.* 2011 suggest an entirely different, Galactic, model for this event namely the tidal disruption of a minor body (e.g. a comet) onto an isolated neutron star at ~3 kpc. The early emission is explained by contributions from the accretion disk of disrupted comet while the late detection (our host candidate) is ascribed to a proto-planetary disk from which the colliding comet had originated. Another recent event, GRB 111209A, with a duration of $>$ 10 ks, showed a remarkably similar X-ray light-curve (Hoversten *et al.* 2011) together with the same conspicuous "dips" still unexplained for GRB 101225A. The afterglow, however, had a normal power-law behaviour. It could be associated with a galaxy at z=0.67 from emission lines of the host, which itself has so far been undetected. This burst was suggested to come from a tidal disruption (TD) event similar to 100328A (A. Levan priv. comm.).

New late time observations might be able to distinguish between these models. The detection of an extended source, e.g. with HST, would clearly suggest an extragalactic origin, since any disk around a Galactic NS at 3 kpc would be unresolved. A late detection in X-rays would rule out our GRB-like progenitor, the same with a late detection in radio since we do not expect any emission except from the host galaxy at late times. The distinction between the TD model and the GRB model can only definitely be answered if the host can be resolved and the afterglow position is clearly offset from the main emission of the galaxy. Those observations, subject to decisions by different time allocation committees, will hopefully reveal the nature of this strange event in the near future.

Acknowledgements: CCT acknowledges support from "Estallidos" under program number AYA2010-21887-C04-01 of the Spanish MEC and by FEDER and generous support from the IAU to attend this conference.

References

Campana, S. *et al.*, 2011, *Nature*, 480, 69
Kouveliotou, C. *et al.* 1993, *ApJ*, 413, L101
Woosley, S. E. & Bloom, J. S. 2006, *ARAA*, 44, 507
Levesque, E. M., *et al.* 2010, *MNRAS*, 401, 963
Fynbo, J. P. U. *et al.* 2006, *Nature*, 444, 1047
Racusin, J. L. *et al.* 2010, *GCN Circ.* 11493
Thöne, C. C. *et al.* 2011, *Nature*, 470, 72
Xu, D., Ilyin, I., & Fynbo, J., 2010, *GCN Circ.*, 11495
Galama, T. J. *et al.* 1998, *Nature*, 395, 670
Richardson, J. C. *et al.* 2011, *ApJ*, 732, 76
Fryer, C. L. & Woosley, S. E. 1998, *ApJ*, 502, L9
Campana, S. *et al.* 2006, *Nature*, 442, 1008
Page, K. L. *et al.* 2011, *MNRAS*, 416, 2078
Cano, Z. *et al.* 2011, *MNRAS*, 413, 669
Soderberg, A. *et al.* 2008, *Nature*, 453, 469
Hoversten, E. A. *et al.* 2011, *GCN Circ.* 12641

Death of Massive Stars: Supernovae and Gamma-Ray Bursts
Proceedings IAU Symposium No. 279, 2012
P. Roming, N. Kawai & E. Pian, eds.

doi:10.1017/S1743921312012756

Constraining gamma-ray burst progenitors

Andrew Levan

Department of Physics, University of Warwick,
Coventry, CV4 9GE, UK,
email: a.j.levan@warwick.ac.uk

Abstract. The past decade has seen great progress towards the unmasking of the progenitors of gamma-ray bursts, starting with the unambiguous detection of a supernova in the light of the long-GRB 030329 almost ten years ago, and the discovery of the first afterglows to short-GRBs in 2005. Here I review progress towards unveiling the progenitors of both long and short-duration GRBs. Furthermore, I examine the diverse broader population of GRBs and high energy transients, and suggest that a full consideration of this parameter space leads to the conclusion that additional progenitor models are likely to be needed, if we are to understand the complete view of GRBs and the transient high-energy sky.

Keywords. gamma-rays : bursts, (stars:) supernovae: general

1. Introduction

The search for the progenitors of gamma-ray bursts has long concentrated on unveiling the nature of two broad classes of gamma-ray bursts which have been known for more than twenty years, and are most cleanly separated by their durations and spectral hardness (Kouveliotou *et al.* 1993). The belief in this pursuit has generally been that each of these classes will ultimately be created by a single set of progenitors, or at least those which are very similar. In this sense, the goal of much GRB research has been to reduce the myriad of models for GRB production to only one or two canonical models.

This search has been extremely successful, we now believe that long-GRBs are created in massive star collapse (see Hjorth & Bloom 2011 for a review). The short duration bursts remain more elusive, but the most promising model for their origin remains in the final merger of compact object binaries (e.g. NS-NS or NS-BH; for a review see Nakar *et al.* 2007). However, while this does indeed reduce the number of progenitor channels remarkably, there are many now disfavoured models that appear physically plausible, and therefore an important question is if any of these can still produce identifiable GRBs? Similarly, events over the past couple of years have shown surprising diversity in GRB populations. A crucial question is if this diversity is intrinsic to the progenitors creating GRBs, or is indicative of additional progenitor systems. In this contribution I will briefly discuss how we know the progenitors of GRBs, and how combinations of different techniques might ultimately allow us to isolate different possible progenitor models.

2. Parameter Space

In studies of optical transients many different classes of objects can be seen in a plot of luminosity against duration (e.g. Rau *et al.* 2009). These diverse events range from faint, fast novae to highly luminous and long-lived SNe, and represent a comparably diverse set of progenitor systems. Figure 1 shows a γ-ray equivalent of this plot, in which various "classes" of γ-ray transient have been marked (note that the borders between these are in many cases rather arbitrary, and in practice likely to be blurred). The classical

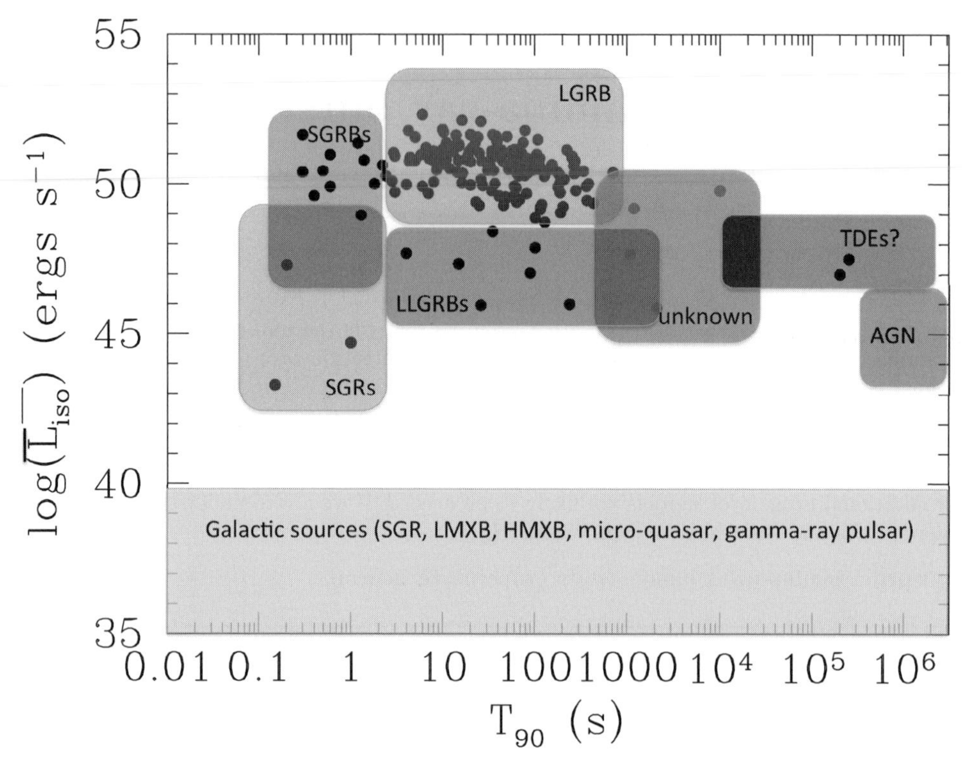

Figure 1. Gamma-ray transient parameter space, expressed as the duration over which 90% of the total fluence is measured (T_{90}), and the average luminosity over this duration (note that the positions of the points within this plot are only approximate, given the differing energy regimes etc). The approximate regions in which various classes of transient are found are marked. Several of these in practice have significant overlap, and so the drawing of a boundary is challenging. A particular note of the last year is the discovery of numerous long duration transients, with durations from ~ 1 hour to several days.

long- and short-duration GRBs (LGRBs and SGRBs, respectively) can clearly be seen, but it is not yet clear if a single progenitor model is sufficient to adequately describe each class. Firstly, some Galactic soft-gamma-repeaters (SGRs) have exhibited flares of sufficient luminosity that they could be seen as SGRBs in external galaxies (Hurley *et al.* 2005) and it is likely that at least some of the SGRB population arise from this mechanism (Tanvir *et al.* 2005; Levan *et al.* 2008). Within the long-GRB distribution there is a population of low luminosity GRBs (LLGRBs) that have typically long durations, and are characterised by strong thermal emission. Aside from associated SN Ic (Pian *et al.* 2006; Starling *et al.* 2011) it is far from clear that these events should be considered as directly scaled down versions of the LGRBs population. Finally, a recently discovered population of transients with extremely long durations pushes the duration distribution of LGRBs even further, and it is not clear what (if any) connection these sources have to the LGRB population at large.

A full characterization of this parameter space is clearly needed, as is the case for the optical sky. Below I will consider some observational approaches that have been applied so far, focussing in particular on the better studied populations of "classical" LGRBs and SGRBs.

3. Observational Techniques

3.1. *Spectroscopy*

Spectroscopic typing of supernovae associated with GRBs has become the gold-standard, and is generally viewed as the clinching proof that GRBs were associated with energetic type Ic supernovae (Hjorth *et al.* 2003; Stanek *et al.* 2003). GRB afterglows can crudely be modeled as power-laws in both time and frequency $F(\nu, t) \propto \nu^{\beta} t^{\alpha}$, where frequently $\alpha \sim \beta \sim -1$. In contrast SN show broadly thermal emission (albeit often heavily modified), and rise slowly. Hence, as the GRB afterglow fades, it is possible to isolate the spectral signature of the SN, as a change in the spectrum, whose template matches that of the class of broad lined SN Ic.

Spectroscopy provides by far the most detail about the SNe associated with GRBs, from detailed typing, to expansion velocities, to evidence for aspherical emission. However, the primary drawback is its limited redshift range, such detailed work is only possible for very low redshift ($z < 0.2$). Beyond this, it is possible to isolate SN signatures (e.g. Sparre *et al.* 2011), but even this is only possible to $z \sim 0.7 - 1$ (e.g. Della Valle *et al.* 2003). However, a potentially larger concern is that it is far from clear that the GRBs in the limited volume open to this technique are representative of GRBs as a class. Of the "best" examples, only GRB 030329 is a typical long duration GRB. The other nearby GRB/SN all show highly atypical prompt properties, often associated with long lived, low luminosity emission, probably belonging to the possibly distinct class of low-luminosity GRBs (LLGRBs; e.g. Campana *et al.* 2006; Pian *et al.* 2006; Starling *et al.* 2011). In this sense, one should be cautious drawing strong inferences from this spectroscopic evidence alone.

Similar strong spectroscopic signatures are not natural to isolate in SGRB progenitors. Spectroscopy of the afterglow may yield signatures of radioactive element production in so called macronovae, (e.g. Metzger & Berger 2012) but such spectroscopy has yet to be successfully obtained. Ultimately spectroscopic observations probing the density of the interstellar (or intergalactic) medium around the SGRB may provide strong constraints on the SGRB environments, but a unambiguous signature is likely to be challenging.

3.2. *Photometry*

Photometric constrains on GRB progenitors largely come from the same principle as the spectroscopic ones. As the afterglow light fades, the fading is slowed, or even reversed by the rising of the supernova light. If the lightcurve is monitored in multiple colours this is normally accompanied by a pronounced reddening of the light (e.g. Zeh *et al.* 2004). Photometric evidence of SNe can be seen out to $z \sim 1$ and beyond (Zeh *et al.* 2004). From this it has been suggested that essentially all long duration GRBs are consistent with the association with a supernova. However, such work is itself difficult. It relies in the successful decomposition of afterglow + supernova + host galaxy. While the latter can be subtracted off (although this is not always done) the first two are complex, both afterglows and supernovae can show variations in brightness and rise/decay rates. Coupled with this additional complications such as refreshing of the afterglow shock, can even reverse the afterglow decay without an associated supernovae. However, the broad picture in which the majority of GRBs are associated with moderately luminous supernovae seems well justified.

For short GRBs, it seems likely that photometric studies are the place in which any macronova emission would be found. Firstly, such emission is likely to be faint, such that spectroscopic observations will be challenging. Secondly, the timescale of this emission is likely to be much more rapid than seen in supernovae, and so identifying a rising source, and obtaining subsequent spectroscopic follow-up will require a fast response.

3.3. *Host environments*

For more distant GRBs, or for an attempt to study the ensemble properties of a large sample, studies of the host galaxies can be extremely powerful. The host galaxies provide information about the stellar scale environments of the progenitors, and as such are a strong route to their identification. In particular, theoretical progress over the past several years suggests GRB progenitors arise from low metallicity progenitors (e.g. Langer *et al.* 2010), probably because magnetic breaking during mass loss slows the rotation of more metal rich progenitors too much for the formation of an accretion disc upon core collapse. This picture is supported from the bulk ensemble properties of GRB hosts, which appear to be smaller and less luminous than those of core collapse supernovae hosts at similar redshifts (Fruchter *et al.* 2006; Svensson *et al.* 2010). In the handful of cases where direct spectroscopic observations of the hosts have yielded reliable metallicity measurements the gas phase metallicities also appear to be low in comparison with the field (Modjaz *et al.* 2008; Levesque *et al.* 2010).

This paints a broadly supportive picture for the collapsar model. However, it should be noted that these insights arise largely (and sometimes exclusively) for bursts where precise positions are available, and these in turn typically come from a sample which exhibit bright optical afterglows. In these cases, extinction along the line of sight within the host galaxy is apparently low, and this may provide a significant bias to the host populations that the GRBs probe. More recent work, focussing on the nature of the host galaxies of dark-GRBs, where the optical afterglow is absent, probably because of heavy extinction in the host galaxy address these concerns. These studies show that the host galaxies of dark GRBs are systematically more luminous, and redder in colour than those of the optically bright host (e.g. Perley *et al.* 2009; Svensson *et al.* 2012). A natural explanation is that they are also more metal rich. It may be that the sightlines to regions of intensive star formation tend to be dust free only in the lower mass galaxies. If correct, and confirmed by further observations this could have important implications for the nature of GRB progenitors.

3.4. *Host locations*

It is not only the bulk properties of the hosts which provide diagnostics towards GRB progenitors. This can also be provided by the locations of GRBs within their host galaxies. LGRBs, because of the short lifetimes of their progenitors, would be expected to lie in the cores of star forming regions within their host galaxies. Indeed, this is what is seen, LGRBs are extremely concentrated on the light of their host galaxies, and are typically found at small radial offsets from their hosts (Bloom *et al.* 2002; Fruchter *et al.* 2006; Svensson *et al.* 2010). They are even more concentrated on their host light than core-collapse supernovae, and this is naturally explained by their production in more massive progenitors (Larsson *et al.* 2007; Raskin *et al.* 2008). However, while these results, and the comparisons they enable are extremely valuable, the fundamental information that can be derived from them is restricted by the resolution of the observations. Even with *HST* we can only probe regions several hundred parsecs across at typical GRB redshifts, and so tying the observed properties back to those of the progenitor stars is a challenging task.

While studies of the locations of LGRBs in their hosts have been extremely influential, they are pivotal for SGRBs, and may represent the best chance of identifying their progenitors prior to the advent of the next generation of gravitational wave detectors (in particular Advanced-LIGO and Virgo, expected in 2015; Abadie *et al.* 2010). If, as thought, SGRBs arise from compact object mergers, then the dynamics of the systems will scatter them well away from the host. Any binary undergoing a merger has survived two supernovae. Each of these SN is likely to have imparted a significant natal kick on the nascent neutron star, while mass loss in the binary at the time of the supernova produces an additional impulse which can propel the binary at several hundred km/s (e.g. Church *et al.* 2011). The typical merger times for the binaries are poorly constrained but are likely to be $> 10^7$ years in most cases (most known NS-NS binaries have merger times $> 10^8$ years). Given this it is likely that SGRBs will be well scattered on their host galaxies, and may escape from low mass systems altogether.

These dynamics are very different from the bulk of the host galaxy stellar population; even intrinsically older systems (e.g white-dwarfs collapsing to neutron stars) would be expected to lie at locations in the host close to stars, the presence of a strongly kicked component would be strong evidence for the compact merger model.

In practice, observations of SGRBs broadly support this model. They are clearly scattered on their hosts, and are inconsistent with either tracing the LGRB distribution, or a linear relationship with the light, at high statistical significance (Fong *et al.* 2010).

4. Oddballs

Perhaps the most remarkable discoveries of the past several years have been of new bursts which don't fit naturally within the standard long/short paradigm. For example, the discovery in 2006 of two bursts (GRB 060505 and GRB 060614) with apparently long duration, but not associated with a supernovae can be taken as evidence that not all long duration GRBs create simultaneous SN (e.g. Fynbo *et al.* 2006; Della Valle *et al.* 2006; Gal-Yam *et al.* 2006). This may represent a new, previously unseen progenitor channel for GRBs, for example the direct collapse to a black hole (Fynbo *et al.* 2006), or could alternatively be indicative of the difficulties in isolating individual progenitor types within any given parameter space, such that these notionally long duration GRBs were in fact created from a progenitor akin to those of short GRBs (Gehrels *et al.* 2006). At the other end of the duration spectrum, gamma-ray transients of particularly long duration are also of great interest, and do not clearly fit within the natural paradigms. These in themselves are a diverse population, ranging in duration from a few hours, up to several days, and it is unclear what relation (if any) they have to the classical GRB population, whose duration distribution peaks at around a minute.

Perhaps most controversial is the burst of Christmas day 2010 (GRB 101225) and its apparent cousin GRB 111209A. These bursts exhibit gamma-ray emission for several hours, accompanied by bright, but apparently uncorrelated UV, and X-ray emission showing prominent dipping behaviour. The first of these two events lacks any redshift measurement, and its host galaxy is unresolved by *HST* (Tanvir *et al.* 2011). Given this, and its moderate Galactic latitude, it was suggested that it could be a Galactic source, perhaps caused by the disruption of a minor body around a neutron star (Campana *et al.* 2011) in a model with remarkable resonance with one initially proposed to describe GRBs prior to the afterglow revolution (e.g. Guseinov *et al.* 1974). A competing proposal is that the source is extragalactic, occurring inside a dense shell created at the latter stages of evolution in a binary progenitor (Thöne *et al.* 2011).

If indeed related then GRB 111209A offers a possible solution to these problems, since it was possible to obtain an absorption line redshift of $z = 0.67$ (Vreeswijk *et al.* 2011), conclusively demonstrating an extragalactic origin. None-the-less, the true origin of these remarkably long bursts remains shrouded in mystery, and further studies of larger examples using the techniques described above are likely to prove highly diagnostic.

At the even more extreme end of the duration distribution is GRB 110328A (now known more commonly, and correctly, as *Swift* 1644+57). This "burst" was detected as a GRB, but had bright, long-lasting emission which clearly marks it apart from the bulk of this population. Its long lived, bright X-ray emission, coupled with an origin in the nucleus of an apparently inactive galaxy at $z = 0.35$ suggest we may well be observing a relativistic jet formed at the time of tidal disruption of a star by the massive black hole in this galaxy (Levan *et al.* 2011; Bloom *et al.* 2011; Burrows *et al.* 2011; Zauderer *et al.* 2011). However, alternative explanations, with much closer resonance to standard LGRB production have also been proposed (Quataert & Kasen 2012; Woosley & Heger 2012).

5. Conclusions

Understanding the nature of GRB progenitors remains one of the central goals of high energy astrophysics. Despite the remarkable success of the past decade our understanding of the progenitors of long GRBs remains rudimentary, and the nature of short-GRBs is still plagued by uncertainty. Coupled with this, ongoing discoveries suggest that the high energy sky is much more diverse that we had previously appreciated. In particular, very long duration GRBs (which at some point possibly should not be called GRBs at all), may point the way to new progenitor channels. This may finally provide evidence for alternative routes that have been suggested for GRB creation, or even open up new, and so far unexplored channels for their production.

References

Abadie, J., Abbott, B. P., Abbott, R., *et al.* 2010, *Classical and Quantum Gravity*, 27, 173001
Bloom, J. S., Kulkarni, S. R., & Djorgovski, S. G. 2002, *AJ*, 123, 1111
Bloom, J. S., Giannios, D., Metzger, B. D., *et al.* 2011, *Science*, 333, 203
Burrows, D. N., Kennea, J. A., Ghisellini, G., *et al.* 2011, *Nature*, 476, 421
Campana, S., Mangano, V., Blustin, A. J., *et al.* 2006, *Nature*, 442, 1008
Campana, S., Lodato, G., D'Avanzo, P., *et al.* 2011, *Nature*, 480, 69
Church, R. P., Levan, A. J., Davies, M. B., & Tanvir, N. 2011, *MNRAS*, 413, 2004
Della Valle, M., Malesani, D., Benetti, S., *et al.* 2003, *A&A*, 406, L33
Della Valle, M., Chincarini, G., Panagia, N., *et al.* 2006, *Nature*, 444, 1050
Fruchter, A. S., Levan, A. J., Strolger, L., *et al.* 2006, *Nature*, 441, 463
Fong, W., Berger, E., & Fox, D. B. 2010, *ApJ*, 708, 9
Fynbo, J. P. U., Watson, D., Thöne, C. C., *et al.* 2006, *Nature*, 444, 1047
Gal-Yam, A., Fox, D. B., Price, P. A., *et al.* 2006, *Nature*, 444, 1053
Gehrels, N., Norris, J. P., Barthelmy, S. D., *et al.* 2006, *Nature*, 444, 1044
Guseinov, O. K. & Vanek, V. 1974, *APSS*, 28, L11
Hjorth, J., Sollerman, J., Møller, P., *et al.* 2003, *Nature*, 423, 847
Hjorth, J. & Bloom, J. S. 2011, arXiv:1104.2274
Hurley, K., Boggs, S. E., Smith, D. M., *et al.* 2005, *Nature*, 434, 1098
Kouveliotou, C., Meegan, C. A., Fishman, G. J., *et al.* 1993, *ApJL*, 413, L101
Langer, N., van Marle, A.-J., & Yoon, S.-C. 2010, *New Astronomy Reviews*, 54, 206
Larsson, J., Levan, A. J., Davies, M. B., & Fruchter, A. S. 2007, *MNRAS*, 376, 1285
Levan, A. J., Tanvir, N. R., Jakobsson, P., *et al.* 2008, *MNRAS*, 384, 541
Levan, A. J., Tanvir, N. R., Cenko, S. B., *et al.* 2011, *Science*, 333, 199
Levesque, E. M., Kewley, L. J., Berger, E., & Zahid, H. J. 2010, *AJ*, 140, 1557
Metzger, B. D. & Berger, E. 2012, *ApJ*, 746, 48
Modjaz, M., Kewley, L., Kirshner, R. P., *et al.* 2008, *AJ*, 135, 1136

Nakar, E. 2007, *Physics Reports*, 442, 166
Perley, D. A., Cenko, S. B., Bloom, J. S., *et al.* 2009, *AJ*, 138, 1690
Pian, E., Mazzali, P. A., Masetti, N., *et al.* 2006, *Nature*, 442, 1011
Quataert, E. & Kasen, D. 2012, *MNRAS*, 419, L1
Raskin, C., Scannapieco, E., Rhoads, J., & Della Valle, M. 2008, *ApJ*, 689, 358
Rau, A., Kulkarni, S. R., Law, N. M., *et al.* 2009, *PASP*, 121, 1334
Sparre, M., Sollerman, J., Fynbo, J. P. U., *et al.* 2011, *ApJL*, 735, L24
Stanek, K. Z., Matheson, T., Garnavich, P. M., *et al.* 2003, *ApJL*, 591, L17
Starling, R. L. C., Wiersema, K., Levan, A. J., *et al.* 2011, *MNRAS*, 411, 2792
Svensson, K. M., Levan, A. J., Tanvir, N. R., Fruchter, A. S., & Strolger, L.-G. 2010, *MNRAS*, 405, 57
Svensson, K. M., Levan, A. J., Tanvir, N. R., *et al.* 2012, *MNRAS*, 421, 25
Tanvir, N. R., Levan, A. J., Fynbo, J. P. U., *et al.* 2011, GRB Coordinates Network, 11564, 1
Thöne, C. C., de Ugarte Postigo, A., Fryer, C. L., *et al.* 2011, *Nature*, 480, 72
Vreeswijk, P., Fynbo, J., & Melandri, A. 2011, *GRB Coordinates Network*, 12648, 1
Woosley, S. E. & heger, A. 2011, arXiv:1110.3842
Zauderer, B. A., Berger, E., Soderberg, A. M., *et al.* 2011, *Nature*, 476, 425
Zeh, A., Klose, S., & Hartmann, D. H. 2004, *ApJ*, 609, 952

Discussion

JESPER SOLLERMAN: Where next for short-GRBs?

ANDREW LEVAN: Short-GRBs are extremely challenging. The most obvious way to make progress is to get a coincident gravitational wave signal, but that is probably several years away (2015 for the first "light" of Advanced-LIGO), given this, the best way to make progress is i) to continue to build larger samples of SGRBs to study the locations of the bursts around their hosts, since this has significant statistical power and ii) to hope that through getting the best monitoring of these events that we can, that we might see an associated macronova.

EHUD NAKAR: How are you selecting your dark GRB redshifts (emission vs spectroscopic)

ANDREW LEVAN: This is a fair concern. A good fraction of optical bright LGRB redshifts come from the afterglow, and so are effectively independent of the host brightness. In contrast, if there is no optical afterglow to take a spectrum of, then we have to look for emission lines in the host galaxy. Although this is possible for very faint galaxies it can introduce a bias, in that it is easier to get such measurements for brighter galaxies, and this does somewhat impact the results that dark-GRB hosts are more luminous than the optically bright hosts. However the results seem to stand when considering this bias.

Death of Massive Stars: Supernovae and Gamma-Ray Bursts
Proceedings IAU Symposium No. 279, 2012
P. Roming, N. Kawai & E. Pian, eds.

doi:10.1017/S1743921312012768

GRB Progenitors and Observational Criteria

Bing Zhang

Department of Physics and Astronomy, University of Nevada, Las Vegas, NV 89154, USA
email: zhang@physics.unlv.edu

Abstract. Phenomenologically, two classes of GRBs (long/soft vs. short/hard) are identified based on their γ-ray properties. The boundary between the two classes is vague. Multi-wavelength observations lead to identification of two types of GRB progenitors: one related to massive stars (Type II), and another related to compact stars (Type I). Evidence suggests that the majority of long GRBs belong to Type II, while at least the majority of nearby short GRBs belong to Type I. Nonetheless, counter examples do exist. Both long-duration Type I and short-duration Type II GRBs have been observed. In this talk, I review the complications in GRB classification and efforts in diagnosing GRB progenitors based on multiple observational criteria. In particular, I raise the caution to readily accept that all short/hard GRBs detected by BATSE are due to compact star mergers. Finally, I propose to introduce "amplitude" as the third dimension (besides "duration" and "hardness") to quantify burst properties, and point out that the "tip-of-the-iceberg" effect may introduce confusion in defining the physical category of GRBs, especially for low-amplitude, high-redshift GRBs.

Keywords. gamma-ray bursts; classification

---

1. Phenomenological vs. physical classification schemes

Observations of GRBs with BATSE led to the identification of two phenomenological classes of GRBs in the duration-hardness (T_{90} − HR) plane: long/soft vs. short/hard (Kouveliotou *et al.* 1993). The boundary between the two classes is vague. The duration separation line is around 2 seconds in the BATSE band (30 keV - 2 MeV). Long and short GRBs roughly comprise 3/4 and 1/4 of the total population.

The main issue of applying the T_{90} criterion to define the class of a GRB is that T_{90} is detector dependent. GRB pulses are typically broader at lower energies. Also a more sensitive detector tends to detect weaker signals which would be otherwise buried in the noise. Indeed, observations carried out with softer detectors such as HETE-2 and *Swift* brought confusion to classifications. Among a total 476 GRBs detected by *Swift* BAT (sensitive in 15 keV - 150 keV) from Dec. 19 2004 to Dec. 21 2009, only 8% have $T_{90} < 2$s (Sakamoto *et al.* 2011), much less than the $\sim 1/4$ fraction of the BATSE sample. An additional 2% of *Swift* GRBs have a short/hard spike typically shorter than or around 2s, but with an extended emission lasting 10's to ~ 100 seconds. These "short GRBs with E.E." (e.g. Norris & Bonnell 2006) have $T_{90} \gg 2$s as observed by *Swift*, but could be "short" GRBs if they were detected by BATSE. So the unfortunate consequence of the T_{90} classification is that the membership in a certain category of the *same* GRB could change when the detector is changed. Nonetheless, the confusion in T_{90} classification only arises in the "grey" area between the two classes. For most GRBs, one can still tell whether they are "long" or "short".

Follow-up afterglow and host galaxy observations of GRBs led to the identification of at least two broad categories of progenitors. Observations led by BeppoSAX, HETE-2, and *Swift* suggest that at least some long GRBs are associated with supernova Type Ic (e.g. Hjorth *et al.* 2003; Campana *et al.* 2006; Pian *et al.* 2006). Most long GRB host galaxies are found to be dwarf star-forming galaxies (Fruchter *et al.* 2006). These facts establish the connection between long GRBs and the death of massive stars (Woosley 1993). The breakthrough led by *Swift* unveiled that some nearby short GRBs (or short GRBs with E.E.) have host galaxies that are elliptical or early-type, with little star formation (Gehrels *et al.* 2005; Barthelmy *et al.* 2005; Berger *et al.* 2005). This points towards another type of progenitor that does not involve massive stars, but is likely related to compact stars, such as NS-NS or NS-BH mergers (e.g. Eichler *et al.* 1992).

The current observational data suggest that the majority of long GRBs belong to the massive star progenitor category (Type II), and at least the majority of nearby short GRBs belong to the compact star category (Type I; see e.g. Zhang 2006 for a discussion on physical classification scheme of GRBs). Nonetheless, the cozy picture of "long GRBs = Type II GRBs", and "short GRBs = Type I GRBs" does not always hold. Long duration Type I GRBs such as GRB 060614 have been observed (e.g. Gehrels *et al.* 2006; Zhang *et al.* 2007), which show deep upper limits on the brightness of any associated supernova, as well as a local galactic environment with low star formation rate (e.g. Gal-Yam *et al.* 2006). Some short-duration Type II GRBs are also observed, including three highest-redshift, "rest-frame short" (i.e. $T_{90}/(1+z) < 2$ s) GRBs, i.e. GRB 080913 at $z = 6.7$ (Greiner *et al.* 2009), GRB 090423 at $z = 8.2$ (Tanvir *et al.* 2009), and GRB 090429B at $z = 9.4$ (Cucchiara *et al.* 2011), and one observer-frame short GRB 090426 at $z = 2.609$ (Levesque *et al.* 2010; Xin *et al.* 2011).

It is then desirable to answer the following challenging question:

2. How can one tell the physical class of a GRB based on the observational data?

In order to address this question, let's remind ourselves what are the observational facts that made us believe in the existence of two distinct classes of progenitors. Zhang *et al.* (2009) summarized 12 multi-wavelength observational criteria that could be connected to the physical nature of a GRB, which include (1) supernova association; (2) specific star formation rate of the host galaxy; (3) position inside the host galaxy; (4) duration; (5) hardness; (6) spectral lag; (7) statistical correlations (e.g. $E_p - E_{\gamma,\rm iso}$, $E_p - L_{\gamma,\rm iso}$, $L-$lag); (8) energetics and beaming; (9) afterglow properties (medium density and spatial profile); (10) redshift distribution; (11) luminosity function; and (12) gravitational wave signal. Except criterion (12), which could be more definitive but is more difficult to carry out, other criteria have been applied to the known GRBs. Criteria (10) and (11) are statistical, relying on a large sample of data. Some interesting results have been obtained, which will be discussed in §3 below. Other criteria can be applied to individual GRBs, and the above particular order of the criteria is based on how closely relevant a particular criterion is to the progenitor. One can see that "duration" and "hardness", which are used in phenomenological classification, are not direct indications of GRB progenitor. This is not surprising, since the bimodal distribution has been known since the BATSE era, but it was after BeppoSAX, HETE-2, and *Swift* when people identified the two broad progenitor types. The first three criteria (SN, host galaxy, and position within the host galaxy) carry much more weight in defining the physical category of a GRB.

A flowchart to diagnose the physical category of a GRB based on multiple observational criteria was proposed in Zhang *et al.* (2009), see Fig. 1. Several noticeable features of the flowchart are the following: 1. Even though duration can be used to roughly separate GRBs, there are several bridges that allow bursts to break the duration separation line.

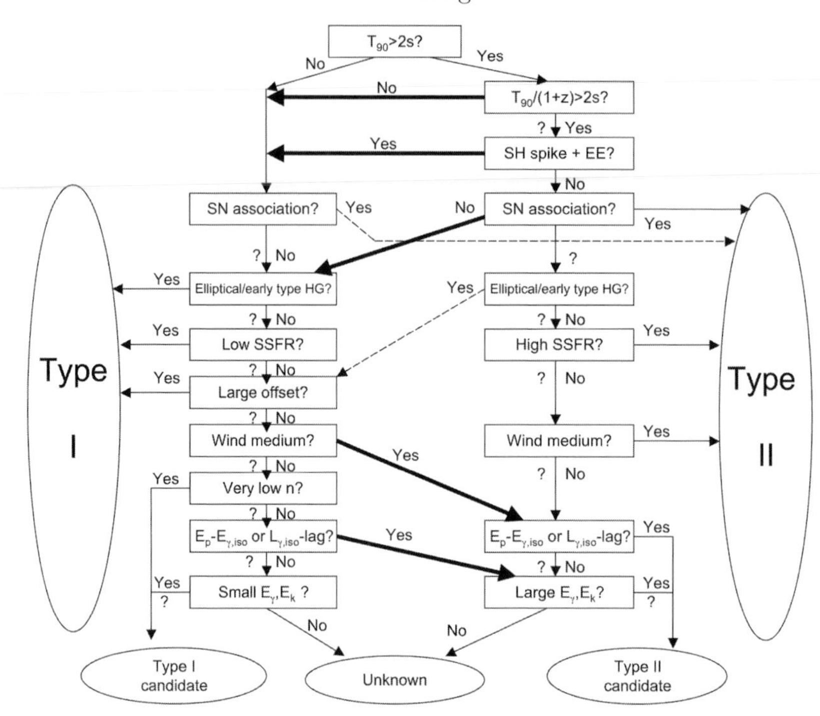

Figure 1. A flowchart using multiple observational criteria to diagnose the physical category of a GRB. After the initial duration criteria (which are used to start with a rough separation), the sequence of the applied criteria are based on their degree of relevance to the progenitor. From Zhang *et al.* (2009).

Noticeable examples include the long-duration Type I GRB 060614, rest-frame short high-z GRBs 080913, 090423, and 090429B, as well as observer-frame short GRB 090426. 2. The upper criteria (SNe and host galaxy properties) carry more weight, and can directly lead a burst to be Type I or Type II; 3. The lower criteria invoking afterglow modeling or empirical correlations carry less weight, and would only lead bursts to be Type I or Type II "candidates".

Such a flowchart has been applied to study currently observed long and short GRBs with afterglow data, and it proves to work well (Kann *et al.* 2010, 2011). In order to apply the flowchart, one needs to have a lot of extra information (afterglow, SN, host galaxy, redshift, etc.) other than prompt γ-ray properties. With γ-ray information only, one cannot determine the physical type of GRBs with high confidence. Nonetheless, one could give a "guess". This is particularly important for a GRB trigger team, since early on there is no afterglow, redshift, SN, and host information. An alert from the team would help the follow-up observers to decide how significant a burst is. The *Swift* team essentially applies the following flowchart to predicts the category of the GRB (Fig. 2).

Some other efforts have been carried out to use fewer parameters to identify the physical category of a GRB. For example, Lü *et al.* (2010) showed that for GRBs with z measurements, the parameter $\varepsilon \equiv E_{\gamma,iso,52}/E_{p,z,2}^{5/3}$ can be a good indicator. The high-ε vs. low-ε categories are found to be more closely related to Type II vs. Type I, respectively, rather than the traditional long vs. short classification.

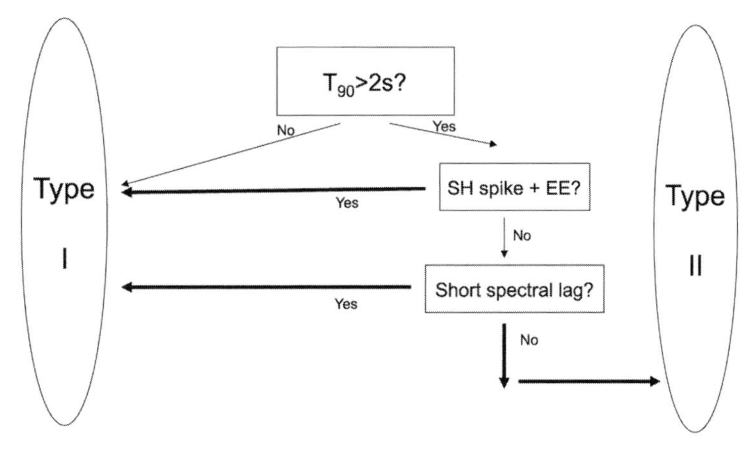

Figure 2. The flowchart used by the *Swift* team.

3. Are all short/hard GRBs of a compact star merger origin?

The leading model for Type I GRBs is NS-NS or NS-BH mergers. An immediate question would be: are all short/hard GRBs as detected by BATSE due to NS-NS or NS-BH mergers?

Most z-known short-duration GRBs are predominantly located at redshifts below 1 (Berger 2009). The peak luminosity spans in a wide range (Virgili *et al.* 2011), from $\sim 7 \times 10^{48}$ erg s^{-1} (for GRB 050509B, Gehrels *et al.* 2005) to $\sim 3.8 \times 10^{52}$ erg s^{-1} (for GRB 090510, De Pasquale *et al.* 2010). An intuitive response to the above question is "perhaps not". Unlike a massive star progenitor that has an extended envelope that can collimate the jet, a NS-NS merger system does not have a natural collimator. A broad jet can be launched, but never highly collimated (e.g. $\sim 30°$, Rezzolla *et al.* 2011). The neutrino annihilation energy power is typically low. Even if the Blandford-Znajek mechanism is invoked, to produce a burst similar to GRB 090510 requires far-stretching of the models. Indeed, Zhang *et al.* (2009) suggested that high-z, high-L short GRB may not be of a Type I origin. Panaitescu (2011) suggested a Type II origin of GRB 090510 based on its wind-like density profile in afterglow modeling.

There are two approaches to address whether all short GRBs are due to compact star mergers. The first approach is to use multiple criteria to identify a Gold sample of Type I GRBs, and check whether it is a good representation of the entire short/hard GRB population. Zhang *et al.* (2009) took this approach and concluded that it may not be. The second approach is to take the observed short GRBs as one population, and statistically compare it with the long GRB population to see the difference. The host galaxy study by Berger (2009) and Fong *et al.* (2010) is of this type. They show that short GRB host galaxies and the afterglow positions inside the hosts are indeed statistically different from those of long GRBs. The sample for such studies is still small, and is dominated by the nearby low-L short GRBs (which are Type I). So the conclusion does not necessarily support that all BATSE short/hard GRBs are Type I.

In Virgili *et al.* (2011), we also performed an analysis using the second approach, but used the gamma-ray data (instead of the host galaxy data). We assume that the z-known short-duration GRBs detected by *Swift* and the z-unknown short/hard GRBs detected by BATSE belong to the same parent sample, and test whether the NS-NS and NS-BH merger models can simultaneously reproduce both daughter samples. As shown in Fig. 3, this is extremely difficult, if not impossible. The reason is that the $L - z$ distribution

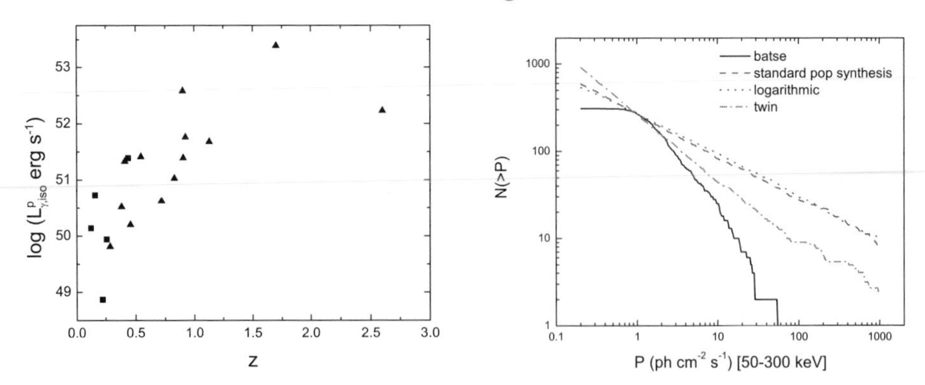

Figure 3. *Left*: the luminosity (L) - redshift (z) distribution of the z-known short GRB population detected by *Swift* (before June 2009). One can see low-z clustering, and lack of significant increase of low-L short GRBs with respect to high-L short GRBs, suggesting a shallow luminosity function; *Right*: Within the compact star merger model, a shallow luminosity function is translated into a shallow peak flux distribution, which is inconsistent with the $\log N - \log P$ distribution of the BATSE short GRBs. This suggests that the compact star merger model cannot simultaneously reproduce both the z-known *Swift* sample and the z-unknown BATSE sample. From Virgili *et al.* (2011).

of the *Swift* sample demands a shallow luminosity function, since one does not see a significant increase of the number of low-L short GRBs with respect to high-L short GRBs. Since merger models predict a low-z clustering due to the merger time delay with respect to star formation, this shallow luminosity function is translated to a shallow peak flux distribution ($\log N - \log P$), which cannot reproduce the BATSE short GRB $\log N - \log P$ data. Notice that even considering the highest fraction of "prompt mergers" (e.g. Belczynski *et al.* 2007), the shallow $\log N - \log P$ cannot be removed. There are two possibilities: 1. the *Swift* sample is not a good manifestation of the BATSE sample. A good fraction of the BATSE short GRBs are different, which may be a mix of Type I and Type II GRBs; 2. All short GRBs are Type I, but the NS-NS and NS-BH merger modes are not the correct one. Other models (e.g. accretion induced collapses) may be needed. In any case, using the short GRB sample to infer gravitational wave detection rate (e.g. Coward *et al.* 2012) may be pre-mature.

4. Introducing "amplitude" as the third dimension and the confusion regimes of GRB classification

In principle, a short GRB can be the "tip-of-the-iceberg" of a long GRB, if the extended longer duration emission is not bright enough to emerge from the detector background. So it is important to introduce the third dimension, i.e. "amplitude", besides "duration" and "hardness" to classify GRBs.

One can define a parameter $f \equiv F_p/F_b$, where F_p is the peak flux, while F_b is the background flux. For a long GRB, if we ideally scale down the flux so that the measured T_{90} becomes shorter than 2 seconds, we will have a dimmer peak flux, F'_p, for which one can define a new parameter $f_{\rm eff} \equiv F'_p/F_b$. For short GRBs, since $T_{90} < 2$ s by definition, one has $f_{\rm eff} = f$. For long GRBs, $f_{\rm eff}$ defines a limit below which the "tip-of-the-iceberg" effect becomes important.

Such an exercise was carried out by Lü *et al.* (2012). Figure 4 shows the distribution of f and $f_{\rm eff}$ for long and short GRBs. The left panel shows both long and short GRBs can have high amplitude spikes. The right panel shows that if only the bright spike of

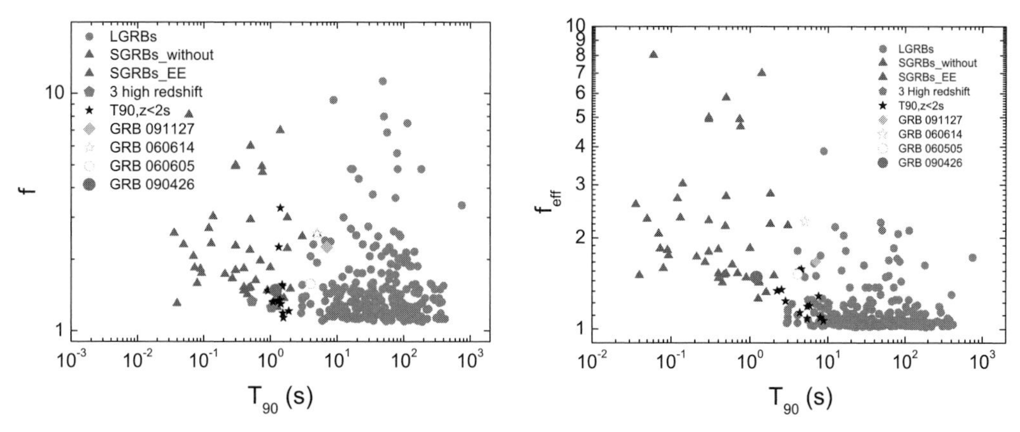

Figure 4. *Left*: The $T_{90} - f$ distribution of long and short GRBs in the *Swift* GRB sample. Both long and short GRBs can have high spikes with large f. *Right*: The $T_{90} - f_{\rm eff}$ distribution of long and short GRBs. Short GRBs can have large f's suggesting that they are intrinsically short. However, some long GRBs can have $f_{\rm eff}$ higher than a few. So short GRBs with low f could be confused with "tip-of-the-iceberg" long GRBs. An example is GRB 090426, which has $f = 1.48$. From H.-J. Lü *et al.*, 2012, in preparation.

a long GRB is above the background, the typical $f_{\rm eff}$ values are substantially dropped. Some short GRBs have a large enough f not to be confused with the bright tip of long GRBs. This means that they are intrinsically short. However, some short GRBs have an f comparable to $f_{\rm eff}$ of some long GRBs, suggesting that they may be confused with a bright spike of long GRBs. It is interesting to note that GRB 090426 is a low f short GRB ($f = 1.48$), which falls into the "confusion" area. Multiple observational criteria suggest that it is a Type II GRB (e.g. Levesque *et al.* 2010; Xin *et al.* 2011). So its short duration is likely due to the "tip-of-the-iceberg" effect.

One can also move a long GRB to progressively higher redshifts until its rest-frame duration is shorter than 2 seconds. Defining $f_{\rm eff,z} \equiv F'_{p,z}/F_b$, one can analyze the range of $f_{\rm eff,z}$ and compare with the f values of the observed rest-frame short GRBs. It turns out that the three rest-frame short GRBs at high redshifts (GRBs 080913, 090423, 090429B) all have f falling into the confusion region, suggesting that they are consistent with being the "tip-of-the-iceberg" of long-duration GRBs (Lü *et al.* 2012).

Acknowledgments

I thank my students/collaborators who carried out the heavy-duty tasks to tackle the complicated problems discussed in this talk, in particular, Hou-Jun Lü, Francisco Virgili, Bin-Bin Zhang, and En-Wei Liang. This work is supported by NSF through grant AST-0908362, and by NASA through grants NNX10AD48G, NNX10AP53G, and NNX11AQ08G.

References

Barthelmy, S. D. *et al.* 2005, *Nature*, 438, 994
Belczynski, K., Stanek, K. Z., & Fryer, C. L. 2007, arXiv:0712.3309
Berger, E. *et al.* 2005, *Nature*, 438, 988
Berger, E. 2009, *ApJ*, 690, 231
Campana, S. *et al.* 2006, *Nature*, 442, 1008
Coward, D. *et al.* 2012, arXiv:1202.2179
Cucchiara, A. *et al.* 2011, *ApJ*, 736, 7

De Pasquale, M. *et al.* 2010, *ApJ*, 709, L146
Eichler, D. *et al.* 1989, *Nature*, 340, 126
Fong, W., Berger, E., & Fox, D. B. 2010, *ApJ*, 708, 9
Fruchter, A. S. *et al.* 2006, *Nature*, 441, 463
Gal-Yam, A. *et al.* 2006, *Nature*, 444, 1053
Gehrels, N. *et al.* 2005, *Nature*, 437, 851
Gehrels, N. *et al.* 2006, *Nature*, 444, 1044
Greiner, J. *et al.* 2009, *ApJ*, 693, 1610
Hjorth, J. *et al.* 2003, *Nature*, 423, 847
Kann, D. A. *et al.* 2010, *ApJ*, 720, 1513
Kann, D. A. *et al.* 2011, *ApJ*, 734, 96
Kouveliotou, C. *et al.* 1993, *ApJ*, 413, L101
Levesque, E. M. *et al.* 2010, *MNRAS*, 410, 963
Lü, H.-J., Liang, E.-W., Zhang, B.-B., & Zhang, B. 2010, *ApJ*, 725, 1965
Lü, H.-J., Zhang, B., Liang, E.-W., & Zhang, B.-B., 2012, in preparation
Norris, J. & Bonnell, J. T. 2006, *ApJ*, 643, 266
Panaitescu, A. 2011, *MNRAS*, 414, 1379
Pian, E. *et al.* 2006, *Nature*, 442, 1011
Rezzolla, L. *et al.* 2011, *ApJ*, 732, L6
Sakamoto, T. *et al.* 2011, *ApJS*, 195, 2
Tanvir, N. *et al.* 2009, *Nature*, 461, 1254
Virgili, F., Zhang, B., O'Brien, P., & Troja, E. 2011, *ApJ*, 727, 109
Woosley, S. E. 1993, *ApJ*, 405, 273
Xin, L.-P. *et al.* 2011, *MNRAS*, 410, 27
Zhang, B. 2006, *Nature*, 444, 1010
Zhang, B. *et al.* 2007, *ApJ*, 655, L25
Zhang, B. *et al.* 2009, *ApJ*, 703, 1696

Discussion

N. Tominaga: In your flowchart, a GRB is moved to Type I if it has no SN component. It is not necessary to have a bright SN, because there is a diversity in SN brightness.

B. Zhang: First, the flowchart is based on what we see not what we believe. Indeed some models can make faint SNe associated with GRBs. However, observationally there is no robust evidence yet for massive star GRBs with no SN association. GRB 060614 is a long GRB. However, it is not that different from the "smoking gun" "short" GRB 050724. GRB 050724 is sitting near the edge of an early type galaxy. However, it is not that short, either. It has an early emission episode with $T_{90} \sim 3$ s, followed by an extended emission tail lasting longer than 100 s (Barthelmy *et al.* 2005). GRB 060614, on the other hand, has an early episode lasting for about ~ 5 s, followed by a softer oscillating tail of ~ 100 s (Gehrels *et al.* 2006). It would be essentially identical to GRB 050724 if it were 8 times less energetic (Zhang *et al.* 2007). Its host galaxy does not have active star formation, and the location of the burst does not track the bright star forming region in the host galaxy. All these suggest that GRB 060614 is not different from GRB 050724, and should be a Type I GRB. Interpreting it as a SN-less Type II GRB is based on the *belief* that all long GRBs are associated with massive stars. Otherwise, most nearby short GRBs (including the actually long-duration "short" GRB 050724) do not have SN associations, either. So why not interpreting them as SN-less core collapses as well? Second, in my flowchart, a long GRB without a SN association is not directly linked to Type I. It is only moved to the left hand side (which becomes a Type I candidate). One needs to go through other criteria to make the final judgment. GRB 060614 is eventually grouped to Type I, but this is because its host galaxy and location within the host galaxy also satisfy the criteria of a Type I GRB. If, in the future, one detects a long duration, SN-less GRB sitting in the bright region of an active star-forming galaxy, having large energetics, and satisfying the empirical correlations of other Type II GRBs, one would move it back to the right hand side of my flowchart, and it will be identified as a Type II candidate. If this indeed happens, then theoretical modeling of GRBs with very faint SNe becomes relevant.

F. MIRABEL: Can you exclude that Doppler boosting dependence on viewing angle does not play any role in GRB classification? Namely, on duration and hardness?

B. ZHANG: The Doppler effect plays an important role in the AGN field, but less in the GRB field. This is mostly because GRBs have higher Lorentz factors ($\Gamma > 100$) and wider jet opening angles ($\theta_j \sim 5°$) than AGNs, so that usually $\Gamma^{-1} \ll \theta_j$ is satisfied. For most geometry, the line of sight is within the jet cone. Observationally, long-duration GRBs typically have multiple peaks, with minimum variability of order milliseconds. So the long duration is the intrinsic property of the central engine, not due to Doppler broadening. Otherwise one should not see much smaller variability time scales. Also early optical afterglows of most GRBs show a decay behavior. This is consistent with the prediction of the on-beam geometry. If the line of sight is outside of the jet cone initially, one would expect to see a rising (or very shallow decay) lightcurve instead. Of course, there is a category of low-luminosity, long-duration GRBs that show one broad peak in the lightcurve. These bursts could be related to bursts viewed at a large off-beam angle with a low Doppler factor. A competitive model is that these are intrinsically different events, probably related to jets that barely breakout from the star, in contrast with high-luminosity long GRBs that have successful jets. The two scenarios can be in principle differentiated with late radio afterglow data. The current data do not support the off-beam jet model. So, to answer your question, even though the viewing angle effect is not fully excluded, one can say that it plays a minor, if any, role in GRB classification.

E. NAKAR: I have a few comments. First, the separation line of $T_{90} = 2$ s is for BATSE. For other detectors (e.g. *Swift*), the separation line can be different. Second, with γ-ray information only, it is impossible to tell which physical category a GRB belongs to. Besides contamination of Type II in short GRBs, there could be contamination of Type I in long GRBs, too.

B. ZHANG: I fully agree with both comments. Regarding T_{90}, I agree that the 2-second separation line is for BATSE only. The *Swift* team currently also uses 2-second as the separation. On the other hand, as is described in the flowchart (Fig. 1), the initial separation is not fundamental. One really needs to go through all the criteria before the final identification of the physical category of a burst is made. There are bridges to connect the two sides. So even if there was inaccuracy in the initial duration criterion, after going through the flowchart, a burst would land in the right physical category. Second, indeed there could be more Type I contaminations in the long duration GRBs. For example, if a long GRB (no short hard spike and no deep SN upper limit) sit in the outskirt of an early type galaxy, one has to be cautious to define its category. In the current flowchart, if this burst also had low energetics, it would be grouped to the "unknown" category. This is a field full of surprises. The flowchart was designed to our best knowledge in 2009, and it still works reasonably well until now (see e.g. Kann *et al.* 2010, 2011). However, it is possible that future observations may suggest that more "bridges" are needed in the flowchart, or even that the global structure of the flowchart has to be modified.

Death of Massive Stars: Supernovae and Gamma-Ray Bursts
Proceedings IAU Symposium No. 279, 2012
P. Roming, N. Kawai & E. Pian, eds.

doi:10.1017/S174392131201277X

Identifying Supernova Progenitors and Constraining the Explosion Channels

Schuyler D. Van Dyk

Spitzer Science Center, Caltech, 220-6, Pasadena, CA USA
email: vandyk@ipac.caltech.edu

Abstract. Connecting the endpoints of massive star evolution with the various types of core-collapse supernovae (SNe) is ultimately the fundamental puzzle to be explored and solved. We can assemble clues indirectly, e.g., from information about the environments in which stars explode and establish constraints on the evolutionary phases of these stars. However, this is best accomplished through direct identification of the actual star that has exploded in pre-supernova imaging, preferably in more than one photometric band, where color and luminosity for the star can be precisely measured. We can then interpret the star's properties in light of expectations from the latest massive stellar evolutionary models, to attempt to assign an initial mass to the progenitor. So far, this has been done most successfully for SNe II-P, for which we now know that red supergiants in a relatively limited initial mass range are responsible. More recently, we have limited examples of the progenitors of SNe II-L, IIn, and IIb. The progenitors of SNe Ib and Ic, however, have been elusive so far; I will discuss the current status of our knowledge of this particular channel.

Keywords. (stars:) supernovae: general, (stars:) supernovae: individual (SN 1987A, SN 1993J, SN 1999A, SN 1999br, SN 1999ev, SN 2002ap, SN 2003ie, SN 2003jg, SN 2004am, SN 2004dj, SN 2004et, SN 2004gt, SN 2005cs, SN 2005gl, SN 2006bc, SN 2007gr, SN 2008bk, SN 2008cn, SN 2009hd, SN 2009jf, SN 2009kr, SN 2010jl, SN 2011dh, SN 2012aw), stars: evolution, (stars:) supergiants, stars: Wolf-Rayet, galaxies: stellar content, (ISM:) dust, extinction

---

1. Direct Identification of SN Progenitors

The most satisfying way of determining which stars explode as which supernovae (SNe) is to directly identify the massive, pre-SN stellar progenitors. We can do this from ground-based imaging data for only the nearest galaxies (distances $d \lesssim 7$ Mpc), the most famous example of which was the identification of Sk $-69°$ 202 as the B3 supergiant progenitor of SN 1987A in the Large Magellanic Cloud (Arnett *et al.* 1989 and references therein). This is the best characterization of a progenitor so far, since not only did we have multi-color photometry for the star (Isserstedt 1975), but we also had an observed spectral type as well (Rousseau *et al.* 1978).

Most of the time we have to identify the progenitor stars in archival, high spatial-resolution, *Hubble Space Telescope* (*HST*) images. This search is typically limited to SNe with $d \lesssim 20$ Mpc, depending on the SN type, the intrinsic luminosity of its progenitor, and the stellar crowding in the SN environment. We have to hope that these pre-SN images contain the SN site and that they might be in more than one band. We initially identify progenitor candidates in the images by comparing with early-time, ground-based SN images. However, inevitably, we need higher-resolution *HST* SN images or images obtained with adaptive optics (AO) on large-aperture, ground-based telescopes.

Once we have the star identified, we attempt to characterize its intrinsic properties via the photometry from the images and estimates of the host galaxy distance, metallicity (Z) at the SN site, and total extinction to the SN. We then compare these properties to theoretical expectations and attempt to map SN types and their progenitors to model predictions. So, what do the latest and greatest theoretical stellar evolutionary tracks predict and explain? For instance, Ekström *et al.* (2012) have produced massive-star models (for $M_{\rm ini} \geqslant 7\ M_\odot$) at solar Z that are either non-rotating or include rotation. The rotating models agree better with the revised Humphreys-Davidson limit on the initial masses of red supergiants (RSGs) at $\sim 25\ M_\odot$ (Levesque *et al.* 2005; Crowther 2007), and therefore are probably more realistic than non-rotating models. Other recent models have also included departures from the standardized mass-loss formulations. Yoon & Cantiello (2010) computed models at solar Z with pulsationally-driven superwinds during the red supergiant (RSG) phase, which strip the star more prior to explosion, resulting in the star being more yellow at terminus. Georgy (2012) also obtained more yellow supergiants (YSGs), also at solar Z, with arbitrarily-increased mass-loss rates used in the Geneva group models for 12–20 $M_\odot$ stars. A special case is the evolution of the most massive stars; that Galactic Wolf-Rayet stars (WRs) have $M_{\rm WR} \lesssim 20\ M_\odot$ (Crowther 2007) may entail continuum-driven luminous blue variable (LBV) eruptions which dramatically shed the star's outer envelope (Humphreys & Davidson 1994; Smith & Owocki 2006). Of course, all of these models are for single-star evolution — binarity has not been taken into account!

2. Progenitors of the Various Supernova Types

2.1. *Type II-P SNe*

What are the progenitors of the most common core-collapse SNe, the Type II-Plateau (II-P)? There are "normal" SNe II-P, such as SN 1999em (Hamuy *et al.* 2001; Leonard *et al.* 2002; Elmhamdi *et al.* 2003; although, it may have been somewhat underluminous, relative to other examples of normal SNe II-P) and low-luminosity SNe II-P, e.g, SN 1999br in NGC 4900 (Pastorello *et al.* 2004) and SN 2005cs in M51 (Pastorello *et al.* 2009). The best example of a low-luminosity SN II-P so far is SN 2008bk in NGC 7793 (Fig. 1), at only 3.4 Mpc distance from us. In Van Dyk *et al.* (2012) we made the "second best" progenitor detection ever, in the form of accurate measurements of the RSG progenitor's spectral energy distribution (SED) at $VRIJHK$ from archival, ground-based Gemini and VLT imaging. We were able to fit this SED with a RSG model stellar atmosphere with $T_{\rm eff} = 3600$ K (Gustafsson *et al.* 2008; and, assuming the extinction to the SN and progenitor were low, $A_V = 0.065$ mag). When this is done, we can estimate $M_{\rm ini} = 8$–$8.5\ M_\odot$ for the progenitor via comparison with stellar tracks at subsolar metallicity. (See also Mattila *et al.* 2008.) The low luminosity from these events may arise from a low ^{56}Ni mass, produced in a shell around the core, rather than the core itself. This mass also intersects with that of the super-AGB stars, and there may be some dependency on low(er) Z in the SN environment, which needs to be further explored.

So, what is the mass for the RSG progenitors of "normal" SNe II-P? This is not yet known. (Actually, at the time of this writing, the mass for the normal SN II-P 2012aw in Messier 95 was being estimated.) Smartt *et al.* (2009) attempted to estimate the initial mass range for all SNe II-P, including low- and high-luminosity ones. The problem is, not all of the SNe II-P assumed to be normal are normal, and not all of the SNe considered were, in all likelihood, SNe II-P. Seven of the SNe II-P considered were of low luminosity. SN 1999ev and, particularly, SN 2004et (Maguire *et al.* 2010) were likely high-luminosity SNe II-P. SN 2003ie was possibly a peculiar SN II-P, similar to SNe 1999A and 1987A (Harutyunyan *et al.* 2008), the latter of which definitely had a BSG as progenitor (not a RSG). SN 2006bc is a probable SN II-Linear (II-L; Gallagher *et al.* 2010). Additionally, the initial mass estimates for both SN 2004dj and 2004am are constraints (not detections), based on the assumed turn-off masses of compact clusters, of which the progenitor stars were presumably members. So, both the $M_{\rm ini}$ for normal SNe II-P and the range of $M_{\rm ini}$ for all SNe II-P is still not known or well constrained.

What may provide some constraint on $M_{\rm ini}$ for normal SNe II-P are the identifications of the progenitors of high-luminosity SNe II-P. Elias-Rosa *et al.* (2009) identified a yellow supergiant (YSG) with $M_{\rm ini} = 15 \pm 2\ M_{\odot}$ for SN 2008cn in pre-SN *HST* images of the host galaxy. This identification is for the most distant SN II-P so far, at $\sim$ 33 Mpc, so some caution should be exercised; Elias-Rosa *et al.* (2009) also consider the possibility that the detected star is really a close binary system. In particular, Van Dyk & Jarrett (in prep.) have revisited the progenitor of SN 2004et in NGC 6946 (at $\sim$5.7 Mpc; much closer than SN 2008cn), for which Li *et al.* (2005) identified a massive YSG progenitor. Crockett *et al.* (2011) confirmed that the progenitor was a YSG, but assigned a far lower $M_{\rm ini}$ for the star. However, Van Dyk & Jarrett have included deep near-infrared JHK_s pre-SN imaging of the host galaxy and find that the star's SED can be fit by a RSG stellar atmosphere with $T_{\rm eff} \simeq 3600$ K, which, at subsolar metallicity in the SN environment, implies that the star had $M_{\rm ini} \simeq 14$–$15\ M_{\odot}$.

2.2. *Type II-L SNe*

What about the progenitors of SNe II-L? Elias-Rosa *et al.* (2010) identified, again, a YSG progenitor for SN 2009kr in NGC 1832, with $M_{\rm ini} = 18$–$24\ M_{\odot}$ (Fraser *et al.* 2010 also consider this star to be a YSG, but assign a much lower $M_{\rm ini} \sim 15\ M_{\odot}$). A criticism one could make is that we have assigned the initial mass range by comparing to normal massive-star evolutionary tracks. However, the $M_{\rm ini} = 20\ M_{\odot}$ track from Yoon & Cantiello (2010) terminates at almost exactly the same bolometric luminosity and

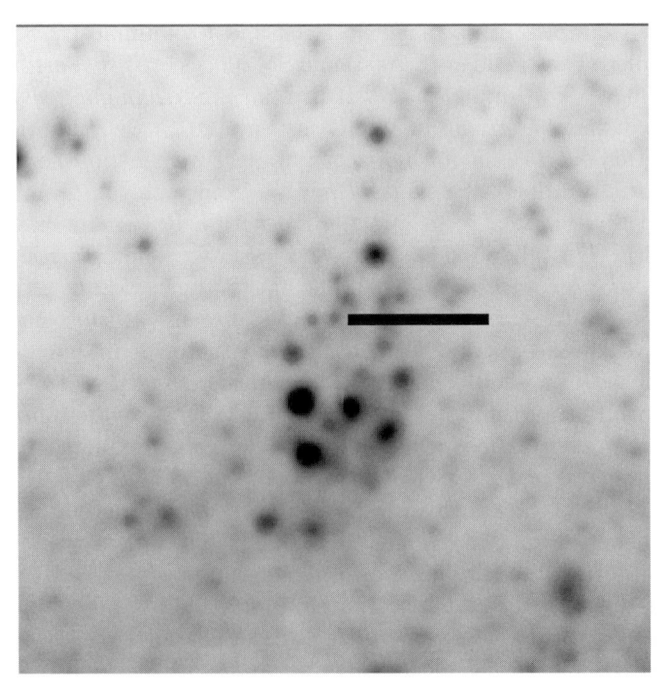

Figure 1. Gemini-S GMOS $g'r'i'$ composite image from 2007, showing the location of the RSG progenitor star of the low-luminosity SN II-P 2008bk in NGC 7793, from Van Dyk *et al.* (2012). We consider this the "second best" SN progenitor detection ever (next to SN 1987A).

effective temperature as the identified progenitor! So, it is quite possible that the yellow color results from enhanced pre-SN mass loss, which strips enough envelope to move the star from the RSG portion of the HR diagram. One other example of a SN II-L progenitor is the nearby SN 2009hd in Messier 66. Elias-Rosa *et al.* (2011) were able to detect the progenitor in pre-SN *HST* F814W images, but not in the corresponding, deep F555W images. This is almost certainly due to the high extinction ($A_V \approx 3.8$ mag). At best, Elias-Rosa *et al.* concluded the star was either a RSG or a YSG, with $M_{\rm ini} \lesssim 20\ M_\odot$.

2.3. *Type IIb SNe*

These are a hybrid of SNe II and SNe Ib, and the difference between SNe IIb and SNe Ib may be skin-deep. The best-known case to date is the SN IIb 1993J in Messier 81. The progenitor was an early K-type supergiant with $M_{\rm ini} \sim 13$–$22\ M_\odot$ (Aldering, Humphreys, & Richmond 1994; Van Dyk *et al.* 2002). The progenitor's B supergiant companion was apparently detected in a very late-time, ground-based, optical spectrum (Maund *et al.* 2004). Another excellent, recent, nearby example is the SN IIb 2011dh in M51. The progenitor star was identified in deep *HST*/ACS multi-band images by both Van Dyk *et al.* (2011) and Maund *et al.* (2011). The latter authors claim that the F-type supergiant identified in the ACS images is the actual progenitor (see also Bersten, this volume). However, the early properties of the SN are most consistent with a *compact* progenitor (Arcavi *et al.* 2011; whereas SN 1993J clearly arose from an extended progenitor). So, Van Dyk *et al.* argue that the progenitor is an unseen, hot star (it was also not detected in *HST*/WFPC2 UV images) in an interacting binary system. When SN 2011dh has sufficiently faded, we can reimage the SN site with *HST* (or, far more likely, JWST!), to determine whether or not the yellow star is still there.

2.4. *Type IIn SNe*

The only relatively certain detection of a progenitor of this very heterogeneous SN type is the case of SN 2005gl in NGC 266 (at $d = 66$ Mpc). Gal-Yam *et al.* (2007) detected an object in a *HST*/WFPC2 F547M image from 1997, confirmed from the ground using Keck AO images of the SN. The astounding aspect is that the object had $M_V \approx -10.3$ mag (!!), which, among stars, only has a counterpart with something like a LBV in eruption. Gal-Yam & Leonard (2009) later very likely confirmed that this object was the progenitor of SN 2005gl, in a *HST*/WFPC2 F547M image they obtained in 2007. The object had vanished! This is the first clear link of at least some SNe IIn to the explosion of LBVs. Other connections are more indirect, such as the modeling of SN 2006gy (Smith *et al.* 2007; Ofek *et al.* 2007) and indications from SNe IIn spectra (Kiewe *et al.* 2011).

Another example of a direct LBV-SN IIn connection may come from identification of the progenitor of SN 2010jl by Smith *et al.* (2011). They identified a luminous, blue object at the SN site, but it was unclear whether this was a single star or a compact star cluster. At the time of this writing, the SN continues to be too bright to determine whether or not the blue object is still at this position.

2.5. *Type Ib/c SNe*

So far, no SN Ib or Ic progenitor has been directly identified. We (Van Dyk *et al.*, in prep.) attempted to make this identification for SN Ib 2009jf in NGC 7479 (at $d = 33.9$ Mpc, with extinction $A_V = 0.53$ mag), in *HST*/WFPC2 F569W and F814W images from 1995. We obtained similar *HST*/ACS F555W and F814W of the SN in 2010, when it had substantially faded. Unfortunately, no clear candidate for the progenitor was located; the SN was simply too distant from us. For this particular object, instead, we made a comparison of the bolometric light curve for SN 2009jf with the models by Dessart *et al.* (2011). The observed light curve is consistent with the model SN resulting from a close binary, with a primary star of $M_{\rm ini} = 18\ M_\odot$ which explodes as a nitrogen-rich WR (WN) star with $M_{\rm final} = 3.79\ M_\odot$; the secondary star in this model has $M_{\rm ini} = 17$–$23\ M_\odot$ (Yoon, Woosley, & Langer 2010).

For SNe Ic, one of the best examples may be SN 2004gt in NGC 4038, which occurred very near a star cluster in the host galaxy, such that both Gal-Yam *et al.* (2005) and Maund, Smartt, & Schweizer (2005) only arrived at not very restrictive limits on the progenitor's nature, based on detection limits of the *HST* images and the inferred star cluster properties. Another similar example is the SN Ic 2007gr in NGC 1058 ($d = 9.3$ Mpc, $A_V \approx 0.3$ mag), which also occurred very near, but still several half-light radii away from, a compact star cluster in the host. Of the other existing cases where progenitors could potentially be identified in *HST* images, many of these are in regions with very high extinction; hence, the progenitor was too extinguished to be identified. An example that we have worked on is the SN Ic 2003jg in NGC 2997, where the total extinction to the SN is estimated at $A_V \approx 4$ mag.

Aggravating the problem further (in addition to the fact that nearby SNe Ib/c are comparatively rare) is that the progenitor is inferred theoretically to be quite blue (particularly, if it in at least some cases it is a single WR star), however, a dearth of (sufficiently deep) images obtained in blue or UV bands for nearby host galaxies exists in the *HST* archive. I illustrate this in Fig. 2, which shows what would be the apparent brightnesses of carbon-rich (WC) WRs at roughly the excitation temperature extremes, subtypes WC4 and WC8 (models from Sander, Hamann, & Todt 2012), at typical distance moduli for host galaxies and possible total extinctions to the SN. The curves are the relevant *HST* instruments, WFPC2, WFC3, and ACS/WFC. What can be seen is that, in the redder bands, little hope exists in detecting WCs, unless the images happen to be very deep (limiting mags ~ 26.5–29), although the ground-based limit at $\sim V$ on the SN Ic 2002ap progenitor (Crockett *et al.* 2007) got very close. The best hope is for bands shortward of ~ 4000 Å. Some potential exists for imaging, e.g., in the F336W band with WFC3, with total exposure times of 1800 s or deeper. However, one would have to match any detection of a progenitor with a detection in a corresponding band, either bluer or redder, to approximately the same depth, to derive color information for the star.

3. Concluding Remarks

We are dealing with small-number statistics here. It is absolutely essential that we continue to add to the numbers of SNe of all types with directly-identified progenitors, so that we can truly map SNe to the end states of stars of all possible initial masses.

Current evolutionary tracks do not adequately predict observed pre-SN stars. Once progenitor stars are directly identified in high spatial-resolution images, it is incumbent upon the theorists to produce models that explain the position of the star, in terms of its intrinsic properties, in the HR diagram. This is especially true for binary models for these progenitors.

I note that, like the monkeys in the famous wood carving in Nikko, I try to "hear no evil, see no evil, and speak no evil." But, hey, this is a competitive game!

Finally, I'd like to acknowledge the contributions of my several collaborators, but, particularly, Nancy Elias-Rosa (IEEC/CSIC, Spain), Alex Filippenko (UC Berkeley), and the late Weidong Li.

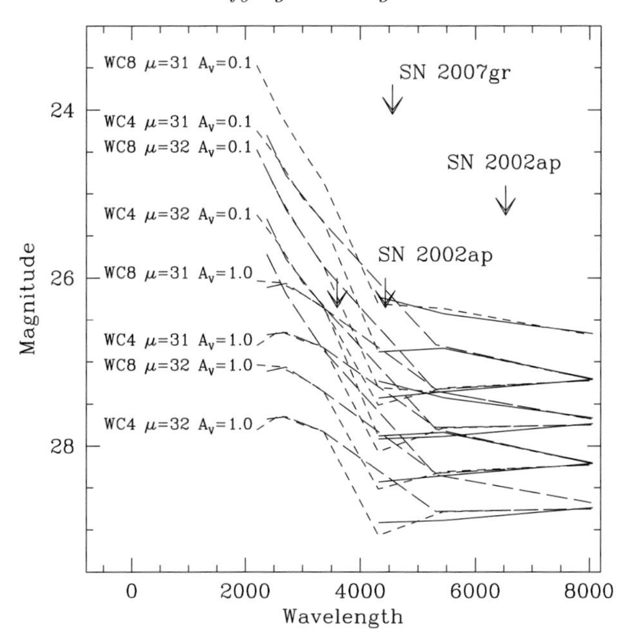

Figure 2. The SEDs of WC4 and WC8 stars from Sander *et al.* (2012) placed at typical distance moduli, μ=31 and 32 mag, for host galaxies of SNe Ic, and further extinguished by $A_V = 0.1$ or 1.0 mag. The various curves represent the available *HST* detectors, WFPC2 (short-dashed lines), WFC3 (long-dashed lines), and ACS/WFC (solid lines). Also shown are upper limits from published searches for SN Ic progenitors, Crockett *et al.* (2007) and Crockett *et al.* (2008), as well as the potential detection threshold for a 1800-s exposure with WFC3 through the F336W band.

References

Aldering, G., Humphreys, R. M., & Richmond, M. 1994, *AJ*, 107, 662

Arcavi, I., Gal-Yam, A., Yaron, O., Sternberg, A., Rabinak, I., Waxman, E., Kasliwal, M. M., Quimby, R. M., Ofek, E. O., & Horesh, A., *et al.* 2011, *ApJ*, 742, L18

Arnett, W. D., Bahcall, J. N., Kirshner, R. P., & Woosley, S. E. 1989, *ARA&A*, 27, 629

Crockett, R. M., Smartt, S. J., Eldridge, J. J., Mattila, S., Young, D. R., Pastorello, A., Maund, J. R., Benn, C. R., & Skillen, I. 2007, *MNRAS*, 381, 835

Crockett , R. M., Maund, J. R., Smartt, S. J., Mattila, S., Pastorello, A., Smoker, J., Stephens, A. W., Fynbo, J., Eldridge, J. J., Danziger, I. J., & Benn, C. R. 2008, *ApJ*, 672, L99

Crockett, R. M., Smartt, S. J., Pastorello, A., Eldridge, J. J., Stephens, A. W., Maund, J. R., & Mattila, S. 2011, *MNRAS*, 410, 2767

Crowther, P. A. 2007, *ARA&A*, 45, 177

Dessart, L., Hillier, D. J., Livne, E., Yoon, S.-C., Woosley, S., Waldman, R., & Langer, N. 2011, *MNRAS*, 414, 2985

Ekström, S., Georgy, C., Eggenberger, P., Meynet, G., Mowlavi, N., Wyttenbach, A., Granada, A., Decressin, T., Hirschi, R., Frischknecht, U., Charbonnel, C., & Maeder, A. 2012, *A&A*, 537, A146

Elias-Rosa, N., Van Dyk, S. D., Li, W., Morrell, N., Gonzalez, S., Hamuy, M., Filippenko, A. V., Cuillandre, J.-C., Foley, R. J., & Smith, N. 2009, *ApJ*, 706, 1174

Elias-Rosa, N., Van Dyk, S. D., Li, W., Miller, A. A., Silverman, J. M., Ganeshalingam, M., Boden, A. F., Kasliwal, M. M., Vinkó, J., Cuillandre, J.-C., *et al.* 2010, *ApJ*, 714, L254

Elias-Rosa, N., Van Dyk, S. D., Li, W., Silverman, J. M., Foley, R. J., Ganeshalingam, M., Mauerhan, J. C., Kankare, E., Jha, S., Filippenko, A. V., *et al.* 2011, *ApJ*, 742, 6

Fraser, M., Takts, K., Pastorello, A., Smartt, S. J., Mattila, S., Botticella, M.-T., Valenti, S., Ergon, M., Sollerman, J., Arcavi, I., *et al.* 2010, *ApJ*, 714, L280

Elmhamdi, A., Danziger, I. J., Chugai, N., Pastorello, A., Turatto, M., Cappellaro, E., Altavilla, G., Benetti, S., Patat, F., & Salvo, M. 2003, *MNRAS*, 338, 939

Gallagher, J. S., Clayton, G., Andrews, J., Sugerman, B., Clem, J., Barlow, M., Ercolano, B., Fabbri, J., Wesson, R., Otsuka, M., & Meixner, M. 2011, *BAAS*, 43, 337.22

Gal-Yam, A., Fox, D. B., Kulkarni, S. R., Matthews, K., Leonard, D. C., Sand, D. J., Moon, D.-S., Cenko, S. B., & Soderberg, A. M. (2005), *ApJ*, 630, L29

Gal-Yam, A., Leonard, D. C., Fox, D. B., Cenko, S. B., Soderberg, A. M., Moon, D.-S., Sand, D. J., Caltech Core Collapse Program, Li, W., Filippenko, A. V., Aldering, G., & Copin, Y. 2007, *ApJ*, 656, 372

Gal-Yam, A. & Leonard, D. C. 2009, *Nature*, 458, 865

Georgy, C. 2012, *A&A*, 538, L8

Gustafsson, B., Edvardsson, B., Eriksson, K., Jørgensen, U. G., Nordlund, A. A., & Plez, B. 2008, *A&A*, 486, 951

Hamuy, M., Pinto, P. A., Maza, J., Suntzeff, N. B., Phillips, M. M., Eastman, R. G., Smith, R. C., Corbally, C. J., Burstein, D., Li, Yong, *et al.* 2001, *ApJ*, 558, 615

Harutyunyan, A. H., Pfahler, P., Pastorello, A., Taubenberger, S., Turatto, M., Cappellaro, E., Benetti, S., Elias-Rosa, N., Navasardyan, H., Valenti, S., *et al.* 2008, *A&A*, 488, 383

Humphreys, R. M. & Davidson, K. 1994, *PASP*, 106, 1025

Isserstedt, J. 1975, *A&AS*, 19, 259

Kiewe, M., Gal-Yam, A., Arcavi, I., Leonard, D. C., Emilio Enriquez, J., Cenko, S. B., Fox, D. B., Moon, D.-S., Sand, D. J., & Soderberg, A. M., The CCCP 2012, *ApJ*, 744, 10

Leonard, D. C., Filippenko, A. V., Gates, E. L., Li, W., Eastman, R. G., Barth, A. J., Bus, S. J., Chornock, R., Coil, A. L., Frink, S., *et al.* 2002, *PASP*, 114, 35

Levesque, E. M., Massey, P., Olsen, K. A. G., Plez, B., Josselin, E., Maeder, A., & Meynet, G. 2005, *ApJ*, 628, 973

Li, W., Van Dyk, S. D., Filippenko, A. V., & Cuillandre, J.-C. 2005, *PASP*, 117, 121

Maguire, K., Di Carlo, E., Smartt, S. J., *et al.* 2010, *MNRAS*, 404, 981

Mattila, S., Smartt, S. J., Eldridge, J. J., Maund, J. R., Crockett, R. M., & Danziger, I. J. 2008, *ApJ*, 688, L91

Maund, J. R., Smartt, S. J., Kudritzki, R. P., Podsiadlowski, P., & Gilmore, G. F. 2004, *Nature*, 427, 129

Maund, J. R., Smartt, S. J., & Schweizer, F. 2005, *ApJ*, 630, L29

Maund, J. R., Fraser, M., Ergon, M., Pastorello, A., Smartt, S. J., Sollerman, J., Benetti, S., Botticella, M.-T., Bufano, F., Danziger, I. J., *et al.* 2011, *ApJ*, 739, L37

Ofek, E. O., Cameron, P. B., Kasliwal, M. M., Gal-Yam, A., Rau, A., Kulkarni, S. R., Frail, D. A., Chandra, P., Cenko, S. B., Soderberg, A. M., & Immler, S. 2007, *ApJ*, 659, L13

Pastorello, A., Zampieri, L., Turatto, M., *et al.* 2004, *MNRAS*, 347, 74

Pastorello, A., Valenti, S., Zampieri, L., *et al.* 2009, *MNRAS*, 394, 2266

Rousseau, J., Martin, N., Prévot, L., *et al.* 1978, *A&AS*, 31, 243

Sander, A., Hamann, W.-R., & Todt, H. 2012, *A&A*, in press (arXiv:1201.6354)

Smartt, S. J., Eldridge, J. J., Crockett, R. M., & Maund, J. R. 2009, *MNRAS*, 395, 1409

Smith, N. & Owocki, S. P. 2006, *ApJ*, 645, L45

Smith, N., Li, W., Foley, R. J., Wheeler, J. C., Pooley, D., Chornock, R., Filippenko, A. V., Silverman, J. M., Quimby, R., Bloom, J. S., & Hansen, C. 2007, *ApJ*, 666, 1116

Smith, N. Li, W., Miller, A. A., Silverman, J. M., Filippenko, A. V., Cuillandre, J.-C., Cooper, M. C., Matheson, T., & Van Dyk, S. D. 2011, *ApJ*, 732, 63

Van Dyk, S. D., Garnavich, P. M., Filippenko, A. V., Höflich, P., Kirshner, R. P., Kurucz, R. L., & Challis, P. 2002, *PASP*, 114, 1322

Van Dyk, S. D., Li, W., Cenko, S. B., Kasliwal, M. M., Horesh, A., Ofek, E. O., Kraus, A. L., Silverman, J. M., Arcavi, I., Filippenko, A. V., *et al.*, 2011, *ApJ*, 741, L28

Van Dyk, S. D., Davidge, T., J., Elias-Rosa, N., Taubenberger, S., Li, W., Levesque, E. M., Howerton, S., Pignata, G., Morrell, N., Hamuy, M., & Filippenko, A. V. 2012, *AJ*, 143, 19

Yoon, S.-C. & Cantiello, M. 2010, *ApJ*, 717, L62

Yoon, S.-C., Woosley, S. E., & Langer, N. 2010, *ApJ*, 725, 940

Discussion

GAL-YAM: We as a community need to push for *HST* imaging of nearby galaxies in the blue-UV bands.

NOMOTO: SNe II-L and II-b are both products of close binary evolution, in my opinion So, the binary evolutionary tracks in the HR diagram should be used for comparison with the observed progenitor candidates. Especially, the binary model for the progenitor of SN IIb 2011dh [M. Bersten's talk] nicely reached the location in the diagram of the observed yellow supergiant.

VAN DYK: I do agree that SNe IIb likely arise from close binaries, and we have observational evidence from the radio emission that at least one SN II-L, 1979C, may have arisen from an interacting, although not necessarily very close, binary system. Our contention with the SN 2011dh system is that the unseen, hot companion exploded, not the detected yellow supergiant.

BERSTEN: How is it possible to determine the $M_{\rm ZAMS}$ for a YSG star using single stellar evolutionary calculations? To obtain a YSG in single stellar calculations, it is necessary for large mass loss, but there is no clear mechanism for that. With hydrodynamic modeling it is possible to explain the early light curve of SN 2011dh, using an extended progenitor with the radius of a YSG ($R \sim 300\ R_\odot$).

VAN DYK: As to the first question, one has to use the tracks with enhanced mass loss, whether due to pulsational instability or some other driver; these are now beginning to get you close to the location of the observed stars in the HR diagram. I also agree that binaries could also potentially get one to the same spot, so such binary models should be calculated to produce YSG progenitors. As to the last point, we feel that the extended progenitor model for SN 2011dh is not consistent with the behavior of the early light curves. Only time will tell, when years from now we can reimage the SN site with *HST* and see if the YSG is still there (in whatever shape it's in) or not.

Death of Massive Stars: Supernovae and Gamma-Ray Bursts
Proceedings IAU Symposium No. 279, 2012
P. Roming, N. Kawai & E. Pian, eds.

doi:10.1017/S1743921312012781

Searching for Wolf-Rayet Stars in M101

J. L. Bibby[1], P. A. Crowther[2], A. F. J. Moffat[3], M. M. Shara[1], D. Zurek[1] and L. Drissen[3]

[1]Dept. of Astrophysics, American Museum of Natural History
Central Park West @ 79th St, New York, NY 10024, USA
email: jbibby@amnh.org

[2]Dept. of Physics & Astronomy, University of Sheffield,
Hounsfield Rd, Sheffield, S3 7RH, UK

[3]Dépt. de physique, Université de Montréal, C.P. 6128,
Succursale Centre-Ville, Montréal, QC, H3C 3J7, Canada

Abstract. Wolf-Rayet (WR) stars are the evolved descendants of massive O-type stars and are considered to be progenitor candidates for Type Ib/c core-collapse supernovae (SNe). Recent results of our HST/WFC3 survey of Wolf-Rayet stars in M101 are summarised based on the detection efficiency of narrow-band optical imaging compared to broad-band methods. We show that on average 42% of WR stars, increasing to ~85% in central regions, are *only* detected in the narrow-band imaging. Hence, the non-detection of a WR star at the location of ~10 Type Ib/c SNe in broad-band imaging is no longer strong evidence for a non-WR progenitor channel.

Keywords. stars: Wolf-Rayet, galaxies: M101, Survey.

---

1. Introduction

Wolf-Rayet (WR) stars are evolved massive O–type stars which are predicted to be the progenitors of Type Ibc core-collapse supernovae (ccSNe). However, to date there has been no direct confirmation of the WR-SNe connection. Pre-SNe broad-band images have failed to reveal the progenitor of ~10 Type Ib/c SNe (Smartt 2009).

WR stars can be classified into two main subtypes, nitrogen–rich (WN) and carbon–rich (WC) stars, which reveal the products of CNO burning and triple alpha reactions, respectively. WR stars exhibit a unique emission line spectrum which is dominated by He IIλ4686 emission lines for WN stars while WC spectra, which also show weaker He IIλ4686, are dominated by C IIIλ4650 and C IVλ5808 (Crowther 2007). These strong emission lines can be 0.2-2.5 magnitudes brighter than the adjacent continuum, making WR stars easy to detect using narrow-band imaging techniques.

Type Ib SNe are hydrogen–poor, while Type Ic SNe are both hydrogen– and helium–poor. The similarities between the observed Type Ib and Ic SN spectra and the chemical composition of the WN and WC stars makes them strong progenitor candidates, respectively.

2. M101

M101 is a grand-design spiral galaxy which lies face-on at a distance of 6.2 Mpc (Shappee & Stanek, 2011). It has a high star-formation rate of at least ~4.5 $M_{\odot} yr^{-1}$ based on the Hα flux or ~6.8 $M_{\odot} yr^{-1}$ from far UV imaging (Lee *et al.* 2009).

There is a wealth of HST/ACS archival data for M101 which was obtained under the legacy program, however the most effective way to identify WR stars is via narrow-band

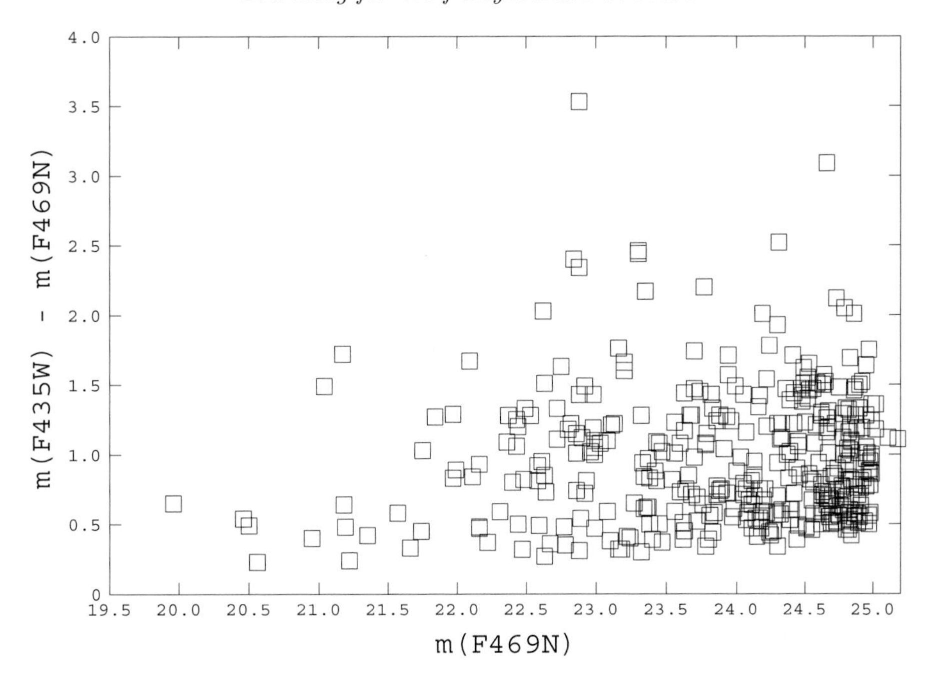

Figure 1. F469N magnitude versus excess magnitude relative to the F435W image for the 372 WR candidates identified in four of the HST/WFC3 pointings.

imaging techniques centered on the WR emission lines. We obtained 18 pointings of M101 using WFC3/F469N over 36 orbits in cycle 17 (PI: Shara).

Identifying WR stars beyond the Local Group with ground-based imaging is challenging since we can only resolve them on scales of ∼25pc. The emission lines of WR stars, although strong, can easily be diluted by strong continuum from other OB stars in the unresolved region (Bibby & Crowther 2010). The high spatial resolution images allow us to resolve sources down to ∼3 pc, decreasing the contamination of the WR emission by continuum sources.

Typically WR stars span an absolute magnitude range of $M_V = -4$ to -8 mag; however, our ground-based studies do not extend fainter than $M_V = -5$ mag. The improved sensitivity of HST allows us to identify the faintest and least massive WR stars, detecting stars to $M_{F469N} = -3.5$ mag and $M_{F435W} = -4$ mag.

3. Identifying candidates

The ACS and WFC3 data were re-drizzled onto the same scale of 0.05 arcsec pix^{-1} using MULTIDRIZZLE. Photometry was performed with the stand-alone DAOPHOT package. Stars that had at least a 3 sigma excess in the F469N filter compared to the F435W and F555W filters were identified as WR candidates. These candidates were then visually inspected using the "blinking" method (Moffat & Shara, 1983).

Four of the WFC3/F469N pointings have currently undergone analysis and have revealed 372 WR candidates with a He II λ4686 excess of at least 3 sigma (Fig. 1). The brightest candidates with m(F435W)-m(F469N) excess of $\leqslant$2.5 mag are likely to be single WR stars, whereas the fainter candidates at m(F469N)∼25 mag with small m(F435W)-m(F469N) excesses $\leqslant$0.6 mag are more likely to host multiple WR stars in unresolved clusters.

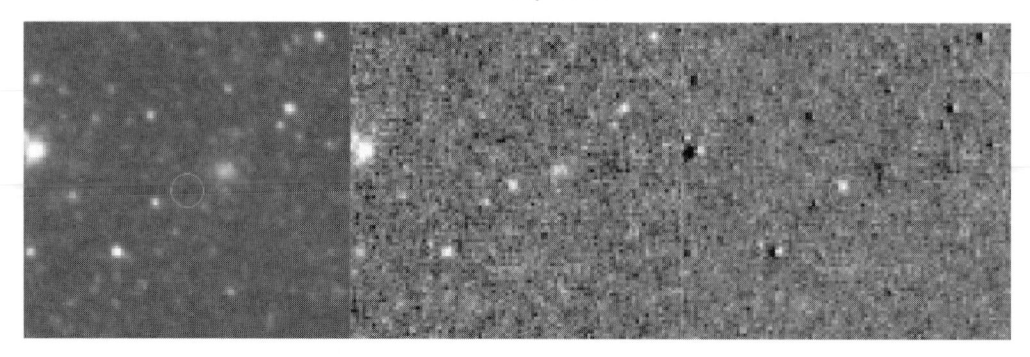

Figure 2. Postage stamp image (~30×30 arcsec) of a WR candidate in M101. The F435W broad-band continuum image (left) and narrow-band F469N image (centre) are subtracted to produce the "net" image (right), revealing He IIλ4686 emission indicating the presence of a WR star.

From the photometric analysis we were also able to identify sources in the F469N image that were not identified in either the F435W or F555W images. This is indicative of a faint WR star with a continuum which lies below the detection threshold of the broad-band images. We identified an additional 269 WR candidates that were only detected in the narrow-band images, which were again checked using the blinking method (Fig. 2).

We are currently awaiting execution of follow-up multi-object spectroscopy with Gemini-North/GMOS for a sample of these candidates (~35%) which will allow us to (i) determine the multiplicity of the sources, (ii) assign a spectral classification of WN or WC to the candidate and (iii) infer the subtype of the remaining candidates for which we did not obtain spectroscopic confirmation.

4. Conclusions & Future Work

We identified 641 WR candidates within four HST/WFC3 pointing of M101, 42% of which are *only* detected in the narrow-band image. This is much higher than the 25% found from ground-based studies of other nearby spiral galaxies, e.g. Bibby & Crowther (2012), which is most likely due to the improved sensitivity and spatial resolution of HST which allows us to detect fainter WR stars. This work highlights the effectiveness of high spatial resolution narrow-band observations in detecting WR stars. If we hope to confirm or, equally important, rule out WR stars as the progenitors of Type Ib/c SNe, then we require a *complete* sample of WR stars in several nearby galaxies.

Once analysis of M101 is complete we will be able to;

(i) Investigate how the number of WR stars varies with the number of O stars and Red Supergiants (RSG) across the galaxy as a function of metallicity and compare the results to predictions from theoretical evolutionary models.

(ii) Degrade our HST imaging and re-run the analysis to determine how many WR stars are not detected, and hence estimate the true completeness of our ground–based surveys.

(iii) Analyse the properties of the regions associated with WR stars to assess whether they are consistent with Type Ib/c SNe.

(iv) Expand our existing catalogue of WR stars which can then be referred to in the event that a Type Ib/c SN occurs in one of these galaxies. This will allow strong, direct observational evidence for, or against, the WR-SNe connection.

References

Bibby, J. L. & Crowther, P. A. 2012, *MNRAS*, 420, 3091
Bibby, J. L. & Crowther, P. A. 2010, *MNRAS*, 405, 2737
Crowther P. A. 2007, *ARA&A*, 54, 177
Crowther, P. A. & Hadfield L. J. 2006, *A&A*, 449, 895
Lee, J. C., Gil de Paz, A., Tremonti, C., Kennicutt, R. C. Jr., Salim, S., Bothwell, M., Calzetti, D., Dalcanton, J., *et al.* 2009, *ApJ*, 706, 599
Moffat A. F. J. & Shara, M. M. 1983, *ApJ*, 273, 544
Shappee, B. J. & Stanek, K. Z. 2011, *ApJ*, 733, 124
Smartt, S. J. 2009, *ARA&A*, 47, 63

Discussion

J. SHIODE: What contaminating sources might masquerade as WR stars?

J. BIBBY: We worried about contamination from planetary nebulae, however our detection limits are not deep enough to detect PNe.

J. SHIODE: Can evolutionary calculations really be considered predictions for N(WR)/N(O) given the uncertainty in very massive star mass-loss histories?

J. BIBBY: N(WR)/N(O) ratios have been relatively consistent with observations where we have a complete WR sample. We should not blame the models until we are sure we are detecting all of the WR population. Our M101 survey will detect the faintest WR stars hence our ratios should be reliable and so if they don't match predictions it gives theoreticians something to work with.

S. CHAKRABORTI: What if some of the stars are just variables?

J.BIBBY: We have not encountered such a problem in our ground-based surveys but if O stars can vary by up to a magnitude on timescales of a few years then this is something we will need to look into. If our follow-up spectroscopy is contaminated by such stars then clearly it is a problem and we need to look at the properties of these stars to see if we can distinguish them from our WR candidates.

T. L. ASTRAATMADJA: Why not look for the WR stars in our Galaxy?

J. BIBBY: There are indeed ongoing WR surveys of the Milky Way. However due to the dust extinction in our galaxy IR rather than optical techniques are used. The overall aim of our work is to identify a future SN progenitor, so surveying many nearby galaxies increases our chances. We select our galaxies based on several criteria, one of which is a face-on orientation to avoid the problem of high extinction.

T. L. ASTRAATMADJA: Do WR stars in other galaxies have the same characteristics?

J. BIBBY: Yes, they still have strong emission lines, however the strength of the lines vary with the metallicity of the environment, i.e. in a low metallicity environment the line flux will be reduced as was demonstrated by Crowther & Hadfield (2006).

P. MAZZALI: Can you distinguish from binary WR stars in your survey?

J. BIBBY: Unfortunately we only obtain one epoch of spectra for a sample of WR stars so we cannot identify which stars are in a binary system from radial velocity motion.

Death of Massive Stars: Supernovae and Gamma-Ray Bursts
Proceedings IAU Symposium No. 279, 2012
P. Roming, N. Kawai & E. Pian, eds.

doi:10.1017/S1743921312012793

H and He in stripped-envelope SNe – how much can be hidden?

S. Hachinger[1,2], P. A. Mazzali[1,2], S. Taubenberger[2], W. Hillebrandt[2], K. Nomoto[3,4] and D. N. Sauer[5,6]

[1]Istituto Nazionale di Astrofisica – OA Padova, vicolo dell'Osservatorio 5, 35122 Padova, Italy
email: stephan.hachinger@oapd.inaf.it

[2]Max-Planck-Institut für Astrophysik, Karl-Schwarzschild-Str. 1, 85748 Garching, Germany

[3]Kavli Institute for the Physics and Mathematics of the Universe, University of Tokyo, 5-1-5 Kashiwanoha, Kashiwa, Chiba 277-8583, Japan

[4]Department of Astronomy, School of Science, University of Tokyo, Bunkyo-ku, Tokyo 113-0033, Japan

[5]Department of Astronomy, Stockholm University, Alba Nova University Centre, SE-10691 Stockholm, Sweden

[6]Meteorologisches Institut, Universität München, Theresienstr. 37, 80333 München, Germany

Abstract. H and He features in photospheric spectra have rarely been used to constrain the structure of Type IIb/Ib/Ic supernovae (SNe IIb/Ib/Ic). The lines have to be modelled with a detailed non-local-thermodynamic-equilibrium (NLTE) treatment, including effects uncommon in stars. Once this is done, however, one obtains valuable hints on the characteristics of progenitors and explosions (composition, explosion energy, ...). We have extended a radiative transfer code to compute synthetic spectra of SNe IIb, Ib and Ic. Here, we discuss our first larger set of models, focusing on the question: How much H/He can be hidden (i.e. remain undetected in photospheric spectra) in SNe Ib/Ic? For the SNe studied (relatively low $M_{\rm ej} = 1...3\ {\rm M}_\odot$), we find a limit of $M_{\rm He} \lesssim 0.1\ {\rm M}_\odot$ in SNe Ic (no unambiguous He lines). Stellar evolution models for single stars normally always yield higher masses. We suggest that low- or moderate-mass SNe Ic result from efficient envelope stripping in binaries. We propose similar studies on H/He in high-mass and extremely aspherical SNe, and observations covering the region of He I λ20581.

Keywords. supernovae: general, supernovae: individual (SN 2008ax, SN 1994I), techniques: spectroscopic, radiative transfer

1. Introduction

Establishing a solid mapping between the spectroscopic appearance of supernovae (SNe) and their physical properties (abundances, explosion energy, progenitor configuration) is a primary objective in SN science. It is achieved by radiative transfer simulations, yielding synthetic observables (e.g. Mazzali 2000; Kasen *et al.* 2004). SN and progenitor models can then be validated by comparison to observations.

In the field of stripped-envelope core-collapse SNe with He (Type IIb/Ib SNe), studies based on synthetic early-phase spectra have rarely been conducted (Utrobin 1996; Mazzali *et al.* 2008). Radiative transfer modelling of their atmospheres requires accurate non-local-thermal-equilibrium (NLTE) calculations to determine the state of the He-rich plasma in the SNe. The excitation/ionisation pattern in this case is strongly influenced by collisions with fast electrons, which result from Compton processes with γ-rays from the decay of ^{56}Ni and ^{56}Co (Lucy 1991). We have enabled a well-established SN spectrum synthesis code (Lucy 1999; Mazzali 2000) to compute the relevant collision rates and solve the stationary NLTE rate equations (Hachinger *et al.* 2012).

Using the upgraded code, we calculated spectra for a sequence of Type IIb/Ib/Ic SNe (SNe IIb/Ib/Ic), varying the envelope mass. In this way, we can determine the mass a (partially stripped) H- or He-envelope of a SN must have for the respective lines to show up. We discuss implications for progenitor models and for future studies on SN spectra.

2. Models

Employing the "abundance tomography" technique (Stehle *et al.* 2005) to infer the abundance stratification, we produced (Hachinger *et al.* 2012) optimum ejecta models for the SN IIb 2008ax (e.g. Chornock *et al.* 2011; Taubenberger *et al.* 2011) and the SN Ic 1994I (e.g. Iwamoto *et al.* 1994; Filippenko *et al.* 1995; Sauer *et al.* 2006). Our models use the density profile of the explosion models "4H47" (Shigeyama *et al.* 1994) and "CO21" (Iwamoto *et al.* 1994), which are based on the same stellar core (with a He/H envelope and without). The synthetic spectra match the observations between 16 d and 41 d past explosion. Furthermore, using the algorithm of Cappellaro *et al.* (1997), we were able to show that the models reproduce the observed bolometric light curves.

A "SN Ic → SN IIb" model sequence (with 40 steps) was obtained from these models by constructing (approximate) "transitional" density and abundance profiles (Hachinger *et al.* 2012). These represent explosions of stars with the same Fe/Si/O/C core, differing only in the envelope. The $E_{\rm kin}$ of all these models is by construction 10^{51} erg, as inferred

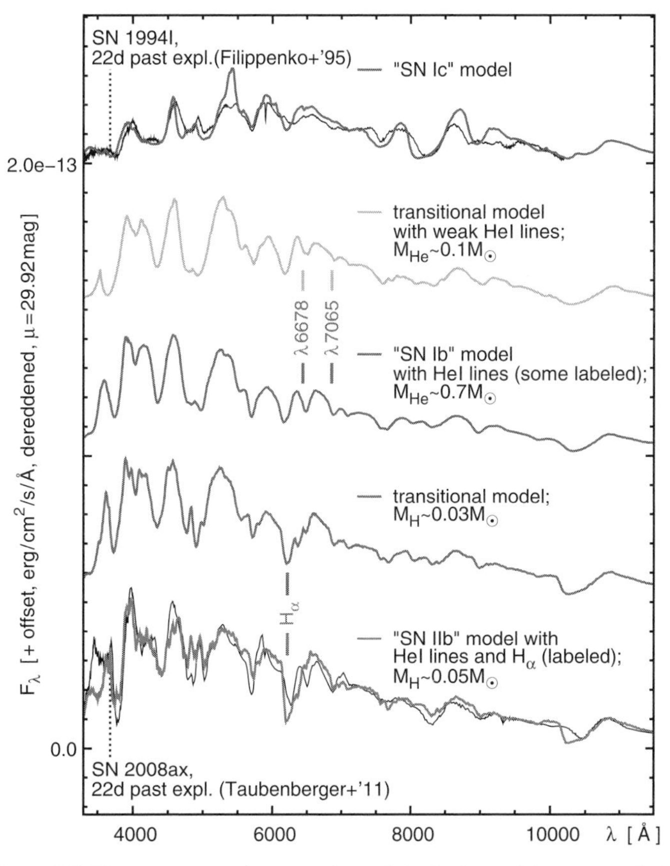

Figure 1. SN Ic → SN IIb sequence of spectral models (coloured/grey graphs) at an epoch of 22 d after explosion. The SN Ic/IIb models have an abundance structure optimised to match observed SNe (black graphs). Identifiers are given for the H/He lines mentioned in the text.

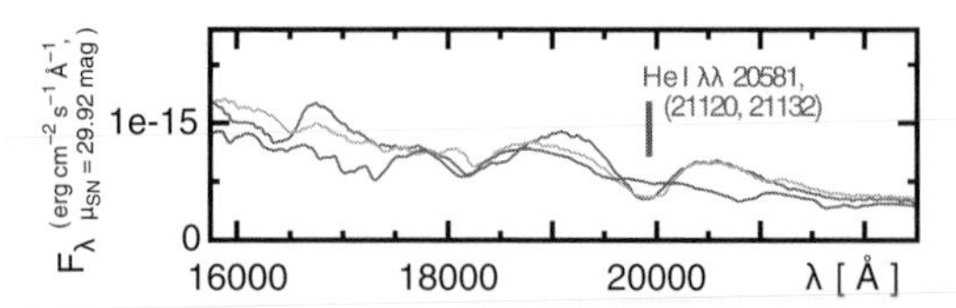

Figure 2. The region around 20000 Å in our SN Ib/c model spectra, the optical part of which is shown in Figure 1. The deepest λ20581 feature is found in the SN Ib model (red), a medium depth in the transitional model (yellow), and no feature in the SN Ic model (green).

for the reference SNe (1994I and 2008ax). The chemical structure of the sequence models was derived from the SN 08ax and 94I models by interpolation at each enclosed M.

Some spectra of the sequence, for an epoch of 22 d after explosion, are shown in Figure 1. We plot the "anchor" models for SN 1994I and SN 2008ax with the respective data, and a few interesting intermediate models.

Going from the SN Ic/1994I model (Fig. 1, top graph) towards more He-rich models (graphs below), one can see that for a He envelope mass of

$$M(\mathrm{He})_{\mathrm{SNIc/Ib}} = 0.1\ \mathrm{M}_\odot,$$

clear lines appear in the optical spectrum ($\lambda\lambda$6678, 7065; these features have sufficiently low contamination by other species). A typical SN Ib is obtained for $M(\mathrm{He}) \sim 0.7\ \mathrm{M}_\odot$.

Similarly, for $M(\mathrm{H})_{\mathrm{SNIb/IIb}} \sim 0.03\ \mathrm{M}_\odot$, the H$\alpha$ feature gets pronounced. For $M(\mathrm{H}) \sim 0.05\ \mathrm{M}_\odot$, we obtain a typical SN IIb spectrum (Fig. 1, bottom). Note that smaller $M(\mathrm{H})$ will be detectable in earlier spectra (cf. Arcavi *et al.* 2011 & questions).

3. Discussion and conclusions

From our sequence of synthetic spectra, we have derived limits on H and He in SNe IIb/Ib/Ic with ejecta masses from $\sim 3\ \mathrm{M}_\odot$ (in 4H47/SN 2008ax) to $\sim 1\ \mathrm{M}_\odot$ (in CO21/ SN 1994I). We find that SNe Ib have $M(\mathrm{H}) \lesssim 0.03\ \mathrm{M}_\odot$, and SNe Ic $M(\mathrm{He}) \lesssim 0.1\ \mathrm{M}_\odot$. While we have focused on the optical, another excellent diagnostic on He is the λ20581 feature (Fig. 2). The well-isolated line reacts very sensitively on the He abundances (as it is strong, but not completely saturated). More observations of He I λ20581 in SNe Ib/Ic would allow for constraining He abundances with unprecedented accuracy.

It is interesting to compare the limits we have derived from the optical spectra with results from stellar evolution calculations for progenitors (Georgy *et al.* 2009; Yoon *et al.* 2010). Stellar evolution studies (involving binarity or not) generally succeed in explaining envelopes as in our SN IIb and SN Ib models, but not the low He masses we infer for SNe Ic. The models presented by Georgy *et al.* (2009), which are single Wolf-Rayet stars, always have $\gtrsim 0.3\ \mathrm{M}_\odot$ of He. In order to explain the total SN Ic rate with their progenitors, they must allow for He masses up to $0.6\ \mathrm{M}_\odot$ in SNe Ic. Also Yoon *et al.* (2010), who simulated binaries with one to three phases of "case A" or "case B" mass transfer, produce only a small number of progenitors with $M(\mathrm{He}) < 0.5\ \mathrm{M}_\odot$. Eldridge *et al.* (2011) find a somewhat larger rate of progenitors with low $M(\mathrm{He})$, but it is unclear whether this suffices. "Exotic" models may be needed to explain SN 1994I-like, low-mass Ic's. Iwamoto *et al.* (1994) and Nomoto *et al.* (1994) proposed for this SN a binary scenario in which the companion is a neutron star in the last mass transfer phase.

It has to be checked whether more He can be hidden in the ejecta of more massive or strongly asymmetric SNe Ic. Accurately constraining the He content of these objects may also help to understand what distinguishes energetic SNe Ic associated with a GRB from similar objects without a GRB (cf. Mazzali *et al.* 2008). While we are further developing our models, we hope that more observations of SNe Ib/Ic (with good time coverage up to 50 d after explosion, and spectra covering the 2μ region) will become available.

Acknowledgements

This contribution has in part been supported by the programme ASI-INAF I/009/10/0. SH thanks the organizers; and IPMU, KN and K. Maeda for hospitality! ST acknowledges support by the TRR33 "The Dark Universe" of the German Research Foundation (DFG).

References

Arcavi, I., *et al.* 2011, *ApJ* (Letters), 742, L18

Cappellaro, E., Mazzali, P. A., Benetti, S., Danziger, I. J., Turatto, M., della Valle, M., & Patat, F. 1997, *A&A*, 328, 203

Chornock, R., *et al.* 2011, *ApJ*, 739, 41

Corsi, A., *et al.* 2012, *ApJ* (Letters), 747, L5

Eldridge, J. J., Langer, N., & Tout, C. A. 2011, *MNRAS*, 414, 3501

Filippenko, A. V., *et al.* 1995, *ApJ* (Letters), 450, L11+

Georgy, C., Meynet, G., Walder, R., Folini, D., & Maeder, A. 2009, *A&A*, 502, 611

Hachinger, S., Mazzali, P. A., Taubenberger, S., Hillebrandt, W., Nomoto, K., & Sauer, D. N. 2012, *MNRAS*, in press, doi: 10.1111/j.1365-2966.2012.20464.x

Iwamoto, K., Nomoto, K., Hoflich, P., Yamaoka, H., Kumagai, S., & Shigeyama, T. 1994, *ApJ* (Letters), 437, L115

Kasen, D., Nugent, P., Thomas, R. C., & Wang, L. 2004, *ApJ*, 610, 876

Lucy, L. B. 1991, *ApJ*, 383, 308

Lucy, L. B. 1999, *A&A*, 345, 211

Mazzali, P. A. 2000, *A&A*, 363, 705

Mazzali, P. A., *et al.* 2008, *Science*, 321, 1185

Nomoto, K., Yamaoka, H., Pols, O. R., van den Heuvel, E. P. J., Iwamoto, K., Kumagai, S., & Shigeyama, T. 1994, *Nature*, 371, 227

Sauer, D. N., Mazzali, P. A., Deng, J., Valenti, S., Nomoto, K., & Filippenko, A. V. 2006, *MNRAS*, 369, 1939

Shigeyama, T., Suzuki, T., Kumagai, S., Nomoto, K., Saio, H., & Yamaoka, H. 1994, *ApJ*, 420, 341

Stehle, M., Mazzali, P. A., Benetti, S., & Hillebrandt, W. 2005, *MNRAS*, 360, 1231

Taubenberger, S., *et al.* 2011, *MNRAS*, 413, 2140

Utrobin, V. P. 1996, *A&A*, 306, 219

Yoon, S., Woosley, S. E., & Langer, N. 2010, *ApJ*, 725, 940

Discussion

CHORNOCK: Why was the first epoch chosen for fitting the models at 16 d? It seems to me that, if you want to model trace amounts of H and He in the outer envelope, you might want to choose an epoch when the photosphere is in the outermost layers.

HACHINGER: The epochs, 16 d to 41 d after explosion, were selected so as to achieve an optimum sensitivity on He, as the He limit seemed more critical. At 16 d there is quite some sensitivity on H, but stricter limits might indeed be derived from earlier spectra.

GAL-YAM: A comment: The result suggesting very little He in SN Ic progenitors is supported by the non-detection in shock-breakout observations by Corsi *et al.* (2012) and Sauer *et al.* (2006) (SN 1994I).

HACHINGER: Thanks! This independent evidence is very interesting!

Death of Massive Stars: Supernovae and Gamma-Ray Bursts
Proceedings IAU Symposium No. 279, 2012
P. Roming, N. Kawai & E. Pian, eds.

doi:10.1017/S174392131201280X

Cutting-edge issues in core-collapse supernova theory

Kei Kotake[1,2]

[1]Division of Theoretical Astronomy, National Astronomical Observatory of Japan, 2-21-1, Osawa, Mitaka, Tokyo, 181-8588, Japan
[2]Center for Computational Astrophysics, National Astronomical Observatory of Japan, 2-21-1, Osawa, Mitaka, Tokyo, 181-8588, Japan

Abstract. Based on our multi-dimensional neutrino-radiation hydrodynamic simulations, we report several cutting-edge issues about the long-veiled explosion mechanism of core-collapse supernovae (CCSNe). In this contribution, we pay particular attention to whether three-dimensional (3D) hydrodynamics and/or general relativity (GR) would or would not help the onset of explosions. Our results from the first generation of full GR 3D simulations including approximate neutrino transport are quite optimistic, indicating that both of the two ingredients can foster neutrino-driven explosions. We give an outlook with a summary of the most urgent tasks to draw a robust conclusion to our findings.

Keywords. supernovae: general, hydrodynamics, relativity

---

1. Introduction

Ever since the first numerical simulation (Colgate & White 1966), the neutrino-heating mechanism of CCSNe, in which a stalled bounce shock could be revived via neutrino absorption on a timescale of several hundred milliseconds after bounce, has been the working hypothesis of supernova theorists for $\sim$ 45 years. However, the simplest, spherically-symmetric (1D) form of this mechanism fails to blow up canonical massive stars (e.g., Liebendörfer *et al.* 2005). Pushed by mounting supernova observations of the blast morphology (e.g., Wang & Wheeler 2008), it is now almost certain that the breaking of the spherical symmetry holds the key to solve the supernova problem (e.g., Kotake *et al.* 2006, Janka *et al.* 2007 for reviews). So far a number of multidimensional (multi-D) hydrodynamic simulations have shown that hydrodynamic motions associated with convective overturn (e.g., Herant *et al.* 1994) and the Standing-Accretion-Shock-Instability (SASI; e.g., Blondin *et al.* 2003, Ohnishi *et al.* 2006, Foglizzo *et al.* 2006, Iwakami *et al.* 2008, and references therein) can help the onset of the neutrino-driven explosion.

In fact, the neutrino-driven explosions have been obtained in the following first-principle two-dimensional (2D) simulations in which the spectral neutrino transport is solved at various levels of approximations (e.g., table 1 in Kotake 2011 for a summary). Among them are the 2D neutrino-radiation-hydrodynamic simulations by Buras *et al.* (2006) and Marek & Janka (2009) who included one of the best available neutrino transfer approximations by the ray-by-ray variable Eddington factor method, by Bruenn *et al.* (2009) who included a ray-by-ray multi-group flux-limited diffusion transport with the best available weak interactions, and by Suwa *et al.* (2010) who employed a ray-by-ray isotropic diffusion source approximation (IDSA; Liebendörfer *et al.* 2009) with a reduced set of weak interactions.

This success, however, is opening further new questions. First of all, the explosion energies obtained in these 2D simulations are typically underpowered by one or two orders of magnitudes to explain the canonical supernova kinetic energy ($\sim 10^{51}$ erg). Moreover, the softer nuclear equation of state (EOS), such as of Lattimer & Swesty EOS with an incompressibility at nuclear densities, K, of 180 MeV, is employed in those simulations. On top of the striking evidence that favors a stiffer EOS based on the nuclear experimental data ($K = 240 \pm 20$ MeV), the soft EOS may not account for the recently observed massive neutron star of $\sim 2M_\odot$ (Demorest *et al.* 2010). Using a stiffer EOS, the explosion energy may be even lower as inferred from Marek & Janka (2009) who did not obtain the neutrino-driven explosion for their model with $K = 263$ MeV†. What else is then missing? The neutrino-driven mechanism would be assisted by other candidate mechanisms such as the acoustic mechanism (Burrows *et al.* 2006) or the magnetohydrodynamic (MHD) mechanism (e.g., Takiwaki & Kotake 2011 for collective references therein).

But before seeking alternative scenarios, it may be of primary importance to investigate how the explosion criteria extensively studied so far in 2D simulations could or could not be changed in 3D simulations. In section 2, we summarize our recent findings obtained in our 3D Newtonian simulations with spectral neutrino transport. In section 3, we move on to report our more recent results based on fully GR simulations including a more approximative neutrino transport.

2. 3D Newtonian simulations with spectral neutrino transport

Going up the ladder beyond previous simulations (Nordhaus *et al.* 2010, Hanke *et al.* 2011) and using a light-bulb scheme to trigger parametric 3D explosions, we reported the first 3D, multigroup (via the IDSA scheme), radiation-hydrodynamic core-collapse simulations for an 11.2 $M_\odot$ progenitor (Takiwaki *et al.* 2012). Firstly we briefly summarize the main results in this section.

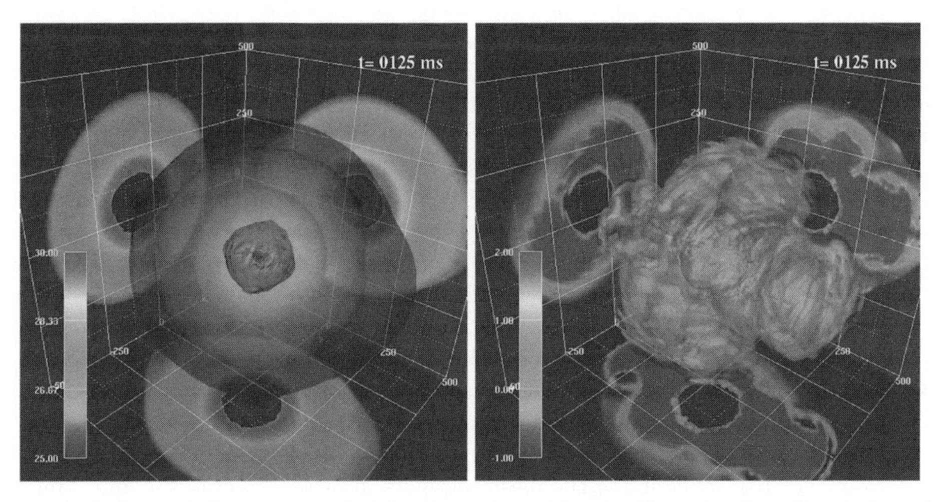

Figure 1. Left panel shows a snapshot for the net neutrino heating rate (logarithmic in unit of erg/cm^3/s at $t = 125$ ms after bounce) for a non-rotating 11.2 $M_\odot$ star (Woosley *et al.* 2002). The right panel shows the ratio of the advection to the neutrino heating timescale.

Figure 1 shows snapshots for the net neutrino heating rate (left panel) and $\tau_{\rm res}/\tau_{\rm heat}$:the ratio of the residency timescale to the neutrino-heating timescale (right panel) for a

† On the other hand, they obtained 2D explosions for Shen EOS ($K = 281$MeV, H.-T. Janka, private communication).

non-rotating 11.2 $M_\odot$ star (Woosley *et al.* 2002; at t = 125 ms after bounce). The left panel shows that there forms the so-called gain region (seen as reddish regions in the wall panels). As shown in the right panel, the condition of $\tau_{\rm adv}/\tau_{\rm heat} \gtrsim 1$ is satisfied behind the aspherical shock, the low-mode deformation of which is characterized by the SASI. The time-scale ratio reaches about two in the gain region, which presents evidence that the shock-revival is driven by the neutrino-heating mechanism.

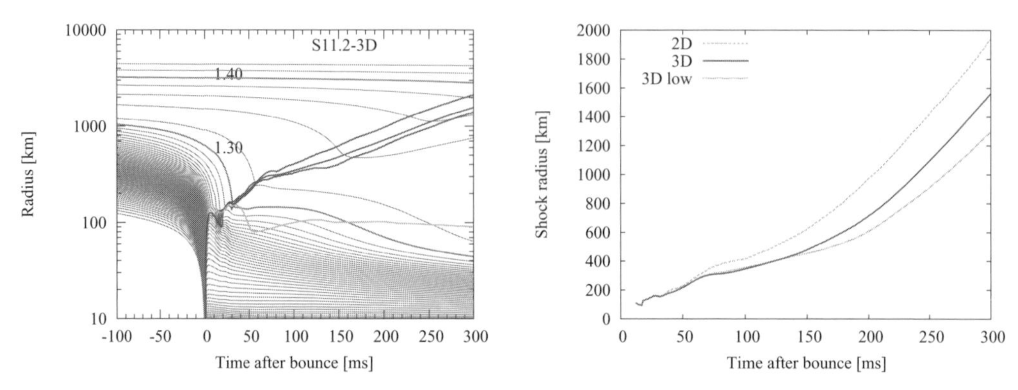

Figure 2. Time evolution of our 3D model of the 11.2 $M_\odot$ star, visualized by mass shell trajectories in thin gray lines (left panel). Thick red lines show the position of shock waves, noting that the maximum (top), average (middle), and the minimum (bottom) shock position are shown, respectively. The green line represents the shock position of the 1D model. "1.30" and "1.40" indicates the mass in unit of $M_\odot$ enclosed inside the mass-shell. Right panel shows the evolution of average shock radii for the 2D (green line), and 3D (red line) models. The "3D low" (pink line) corresponds to the low resolution 3D model, in which the mesh numbers are taken to be half of the standard model (see text).

The left panel of Figure 2 show comparisons of the mass-shell trajectories between the 3D (red lines) and the corresponding 1D model (green line), respectively. At around 300 ms after bounce, the average shock radius for the 3D model exceeds 1000 km in radius. On the other hand, an explosion is not obtained for the 1D model. The right panel of Figure 2 shows a comparison of the average shock radius vs. postbounce time. In the 2D model, the shock expands rather continuously after bounce. These trends in 1D and 2D models are qualitatively consistent with Buras *et al.* (2006) (see, Takiwaki *et al.* 2012 for more detailed comparison).

Comparing the shock evolution between our 2D (green line in the right panel of Figure 2) and 3D model (red line), the shock is shown to expand much faster for 2D. The pink line labeled by "3D low" is for the low resolution 3D model, in which the mesh numbers are taken to be half of the standard model. Note that the 3D computational grid consists of 300 logarithmically spaced, radial zones to cover from the center up to 5000 km and 64 polar (θ) and 32 azimuthal (ϕ) uniform mesh points, which are used to cover the whole solid angle. The low resolution 3D model has one-half of the mesh numbers in the ϕ direction (n_ϕ=16), while fixing the mesh numbers in other directions. Comparing with our standard 3D model (red line), the shock expansion becomes less energetic for the low resolution model (later than $\sim$ 150 ms). The above results indicate that explosions are easiest to obtain in 2D, followed in order by 3D, and 3D (low). At first sight, this may look like a contradiction with the findings obtained in parametric 3D explosion models (e.g., Nordhaus *et al.* 2010) which pointed out that explosions could be more easily obtained in 3D than in 2D. The reason for the discrepancy is summarized shortly as follows.

Figure 3 compares the blast morphology for our 3D (left panel) and 2D (right) model. In the 3D model (left panel), non-axisymmetric structures are clearly seen. By performing a tracer-particle analysis, the maximum residency time of material in the gain region is shown to be longer for 3D than 2D due to the non-axisymmetric flow motions. This is one of advantageous aspects of 3D models to obtain neutrino-driven explosions. On the other hand, our detailed analysis showed that convective matter motions below the gain radius become much more violent in 3D than in 2D, making the neutrino luminosity larger for 3D (e.g., Takiwaki *et al.* 2011). Nevertheless the emitted neutrino energies are made smaller due to the enhanced cooling. Due to these competing ingredients, the neutrino heating timescale becomes shorter for the 3D model, leading to a smaller net-heating rate compared to the corresponding 2D model. Note here that the spectral IDSA scheme, by which the feedback between the mass accretion and the neutrino luminosity can be treated in a self-consistent manner (not like the light-bulb scheme assuming a constant luminosity), sounds quite efficient in the first-generation 3D simulations.

As seen from Figure 2, an encouraging finding is that the shock expansion tends to become more energetic for models with finer resolutions. These results would indicate whether these advantages for driving 3D explosions can or cannot overwhelm the disadvantages in sensitivity to the employed numerical resolutions†. To draw a robust conclusion, 3D simulations with much higher numerical resolutions and more advanced treatment of neutrino transport as well as of gravity are needed, which hopefully will be practical by utilizing forthcoming petaflops-class supercomputers.

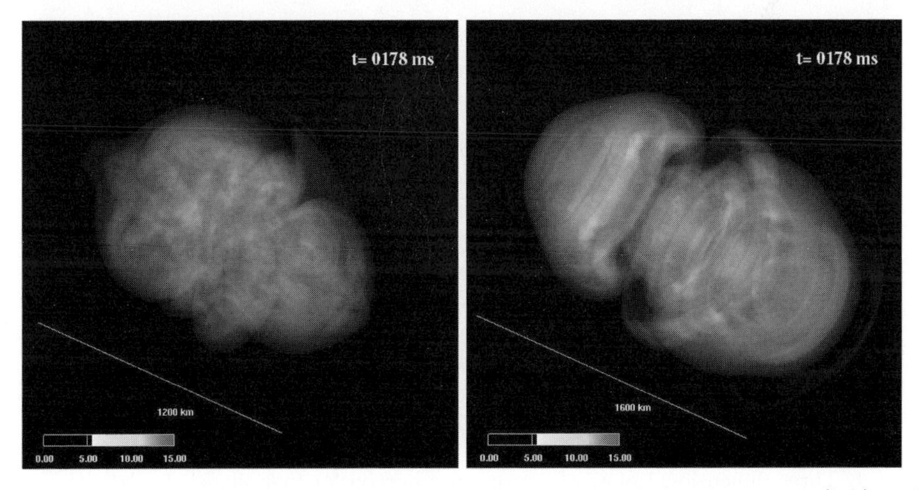

Figure 3. Volume rendering of entropy showing the blast morphology in our 3D (left) and 2D (right) model for the $11.2M_\odot$ progenitor (at t = 178 ms after bounce), respectively. In both panels, the polar axis is tilted (about $\pi/4$). The linear scale is indicated in each panel.

3. 3D full GR simulations with an approximate neutrino transport

In addition to the 3D effects, impacts of GR on the neutrino-driven mechanism stand out among the biggest open questions in supernova theory. Since the late 1990s, the ultimate 1D simulations, in which the GR Boltzmann transport is coupled to 1D GR hydrodynamics, have been made feasible (e.g., Sumiyoshi *et al.* 2005, Liebendörfer *et al.* 2004, and references therein). In these full-fledged 1D simulations, a commonly observed disadvantageous aspect of GR to drive neutrino-driven explosions is that the residency time of material in the gain region becomes shorter due to the stronger gravitational pull.

† It is of crucial importance to conduct a convergence test in which a numerical gridding is changed in a systematic way (e.g. Hanke *et al.* 2011).

In 1D, it is generally agreed that GR works disadvantageously to facilitate the neutrino-driven explosions (e.g., Lentz *et al.* 2011). Although extensive attempts have been made to include microphysics such as by the Y_e formula or by a neutrino leakage scheme in multi-D GR simulations, the effects of neutrino heating have yet to be included in them, which is a main hindrance to studying the GR effects on the multi-D neutrino-driven mechanism‡.

Here we present the first 3D simulations in full GR that include an approximate treatment of neutrino transport (Kuroda *et al.* 2012). The code is a marriage of an adaptive-mesh-refinement (AMR), conservative 3D GR MHD code developed by Kuroda & Umeda (2010), and the approximate neutrino transport code that we recently developed in this work. The spacetime treatment in our full GR code is based on the Baumgarte-Shapiro-Shibata-Nakamura formalism. The hydrodynamics can be solved either in full GR or in special relativity (SR), which allows us to investigate the GR effects on the supernova dynamics. Using a M1 closure scheme with an analytic variable Eddington factor, we solve the energy-independent set of radiation energy and momentum. This part is based on the partial implementation of the Thorne's momentum formalism (Thorne 1981), which has recently been extended by Shibata *et al.* (2011) in a more suitable manner applicable to the neutrino transport problem (see Kuroda *et al.* 2012 for more details).

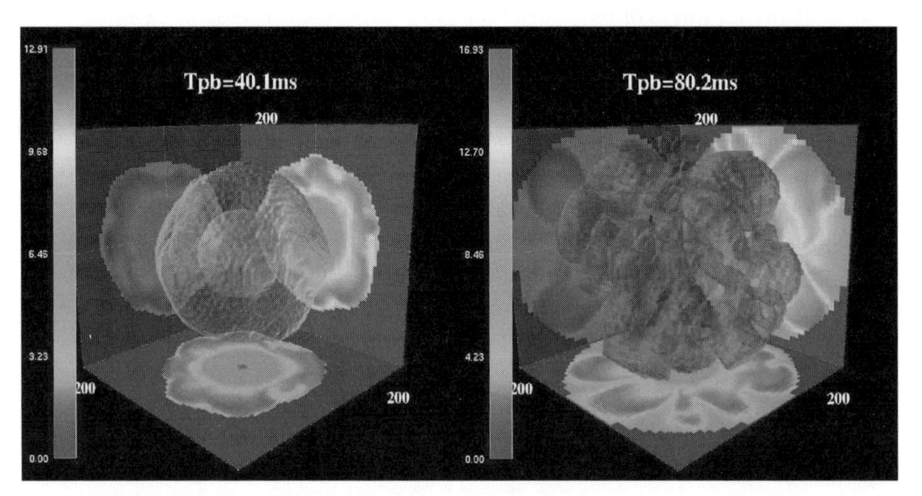

Figure 4. Three dimensional plots of entropy per baryon for four snapshots (left; $t_{\rm pb} = 40$ ms and right; $t_{\rm pb} = 100$ ms) for our 3D-GR model of a 15 $M_\odot$ star. The contours on the cross sections in the $x = 0$ (back right), $y = 0$ (back bottom), and $z = 0$ (back left) planes are, respectively projected on the sidewalls of the graphs to visualize 3D structures. For each snapshot, the arbitrarily chosen iso-entropy surface is shown, and the linear scale is indicated along the axis in units of km.

The left panel of Figure 4 shows a snapshot at $t = 40$ ms postbounce for our 3D GR model†. At this stage, the bounce shock stalls (seen as a greenish sphere) and the gain region forms at $\sim$ 80 ms after bounce in which neutrino heating dominates over neutrino cooling. The neutrino-driven convection gradually develops later on. The entropy

‡ Very recently, Müller *et al.* (2012) reported explosions for 11.2 and 15$M_\odot$ stars based on their 2D GR simulations in CFC with detailed neutrino transport similar to Buras *et al.* (2006).

† The 3D computational domain consists of a cube of 10000^3 km^3 volume in Cartesian coordinates. The maximum refinement AMR level is 5 in the beginning and then increments as the collapse proceeds. The criterion to increment $L_{\rm AMR}$ is set every time the central density exceeds $10^{12,13,13.5}$ g cm^{-3} (see Kuroda & Umeda 2010 for more details), yielding an effective resolution of $\Delta x \sim 600$ m at bounce.

behind the standing shock becomes higher with time due to neutrino-heating. The high entropy bubbles ($s[k_B/\text{baryon}] \gtrsim 10$) rise and sink behind the standing shock. The shock deformation is dominated by unipolar and bipolar modes, which is a characteristic feature of the SASI. The size of the neutrino-heated regions grows bigger with time in a non-axisymmetric way, which is indicated by bubbly structures with increasing entropy (indicated by reddish regions in the right panel).

During our simulation time (100 ms after bounce), the shock radii can reach further out for our 3D-GR model. In contrast, the shock has already shown a trend of recession in other models, namely for 1D-SR, 1D-GR, 3D-SR models, in which "SR" stands for special relativity.

The left panel of Figure 5 shows evolution of the neutrino luminosities (for ν_e and ν_x) for all the computed models. After the neutronization burst ($t_{\rm pb} \sim 10$ ms), the ν_e luminosity for the GR models slightly increases later on, while it stays almost constant for the SR models during the simulation time (green and blue lines). Although the luminosities change with time, the luminosities generally yield to the following order,

for ν_e, 3D-GR $>$ 1DGR, 3D-SR $\sim$ 1D-SR,
for $\bar{\nu}_e$, 3D-GR $>$ 1DGR, 3D-SR $>$ 1D-SR,
for ν_x, 3D-GR $>$ 1DGR, 3D-SR $>$ 1D-SR.

To summarize, both 3D and GR work to raise the neutrino luminosities in the early postbounce phase. As seen from the left panel in Figure 5, GR maximally increases the ν_x luminosity up to $\sim 50\%$ (in 3D), while the maximum increase by 3D is less than $\sim 20\%$ (compare the $\bar{\nu}_e$ luminosity between the 3D-GR and 1D-GR model). These results indicate that compared to the spacial dimensionality, GR holds the key to enhancing the neutrino luminosities. This is also the case for the RMS neutrino energy. The reason for the higher neutrino energy is that the deeper gravitational well of GR produces more compact core structures, and thus hotter neutrino spheres at smaller radii.

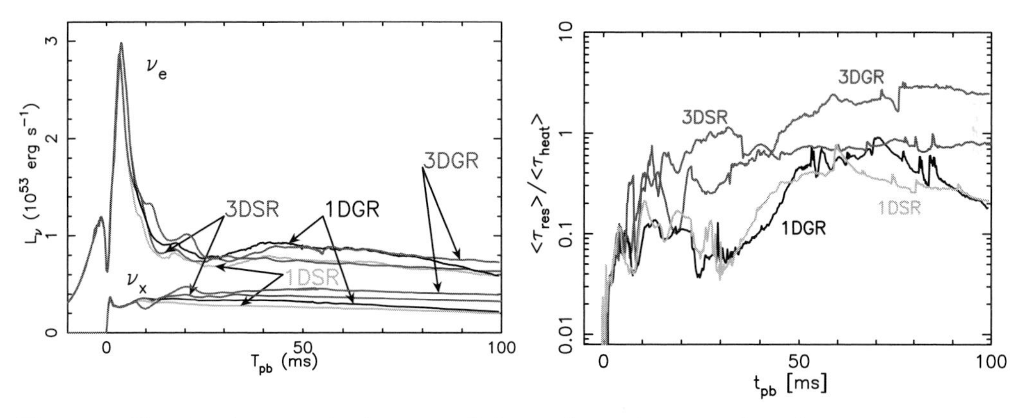

Figure 5. Neutrino luminosities for ν_e, ν_x (left panel) as a function of postbounce time, respectively. Right panel shows the ratio of the residency timescale to the heating timescale for the set of our models as functions of post-bounce time (see Kuroda *et al.* 2012 for the definition of the timescales).

The right panel of Figure 5 shows the ratio of the residency timescale to the neutrino-heating timescale for all the computed models. As seen, the shock revival seems most likely to occur for the 3D-GR model (red line) in our simulation time, which is followed in order by 3D-SR, 1D-SR and 1D-GR models. Thanks to more degrees of freedom, the residency timescale becomes much longer for the 3D models than for the 1D models. In addition, the increase of the neutrino luminosity and RMS energies due to GR enhances the timescale ratio up to the factor of ~ 2 for the 3D-GR model (red line) compared to the SR counterpart (blue line). Therefore our results suggest that the combination of 3D and GR hydrodynamics could provide the most favorable condition to trigger the neutrino-driven explosions.

For the 15 $M_{\odot}$ progenitor employed in our GR simulation, the neutrino-driven explosions are expected to take place later than ~ 200 ms postbounce at the earliest (Bruenn *et al.* 2009) and it could be delayed after ~ 600 ms postbounce (Marek & Janka 2009). The parametric explosion models have shown that the earlier shock revival is good for making the explosion energy larger (e.g., Nordhaus *et al.* 2010). The onset timescale of the neutrino-driven explosions predicted in 2D models (Marek & Janka 2009, Bruenn *et al.* 2009, Suwa *et al.* 2010) could be shorter if the combined effects of GR and 3D would have been included. We anticipate that this can be a possible remedy to turn the relatively underpowered 2D explosions into the powerful ones. It is worth mentioning that the combined effects of GR and 3D should affect not only the supernova dynamics, but also the observational signatures of gravitational-waves (e.g., Kotake *et al.* 2011), neutrino emission (e.g., Abbasi *et al.* 2011), and explosive nucleosynthesis (e.g., Fujimoto *et al.* 2011). To draw a solid conclusion to these important observables, the energy and angle dependence of neutrino transport should be accurately incorporated in our full GR simulations with the use of a more detailed set of weak interactions. This work is only the very first step that leads us to the long and winding road.

Finally, it should be noted that all the numerical results presented in this article should be tested by the next-generation calculations by which more sophistication is made not only in the treatment of multi-D neutrino transport but also in multi-D hydrodynamics including stellar rotation and magnetic fields in full GR. From an optimistic point of view, our understanding of the explosion mechanism can progress in a step-by-step manner at the same pace as our available computational resources will be growing bigger and bigger from now on. Since 2009, several neutrino detectors form the Supernova Early Warning Systems (SNEWS) to broadcast alerts to astronomers to let them know the arrival of neutrinos (Antonioli *et al.* 2004). Currently, Super-Kamiokande, LVD, Borexino, and IceCube contribute to the SNEWS, with a number of other neutrino and GW detectors planning to join in the near future. This is very encouraging news for high-precision multi-messenger astronomy. The interplay between detailed numerical modeling, advancing supercomputing resources, and multi-messenger astronomy, will remain a central issue for advancing our understanding of the theory of massive stellar core-collapse for the future.

References

Abbasi, R., *et al.* 2011, *A&A*, 535, A109
Alcubierre, M. & Brügmann, B. 2001, *Phys. Rev. D.*, 63, 104006
Antonioli, P., *et al.* 2004, *New Journal of Physics*, 6, 114
Blondin, J. M., Mezzacappa, A., & DeMarino, C. 2003, *ApJ*, 584, 971
Bruenn, S. W., Mezzacappa, A., Hix, W. R., Blondin, J. M., Marronetti, P., Messer, O. E. B., Dirk, C. J., & Yoshida, S. 2010, arXiv:1002.4909
Buras, R., Rampp, M., Janka, H.-Th., & Kifonidis, K. 2006, *A&A*, 447, 1049
Burrows, A., Livne, E., Dessart, L., Ott, C. D., & Murphy, J. 2006, *ApJ*, 640, 878
Colgate, S. A. & White, R. H. 1966, *ApJ*, 143, 626
Demorest, P. B., Pennucci, T., Ransom, S. M., Roberts, M. S. E., & Hessels, J. W. T. 2010, *Nature*, 467, 1081
Foglizzo, T., Scheck, L., & Janka, H.-T. 2006, *ApJ*, 652, 1436
Fujimoto, S.-I., Kotake, K., Hashimoto, M.-A., Ono, M., & Ohnishi, N. 2011, *ApJ*, 738, 61

Hanke, F., Marek, A., Mueller, B., & Janka, H.-T. 2011, arXiv:1108.4355
Herant, M., Benz, W., Hix, W. R., Fryer, C. L., & Colgate, S. A. 1994, *ApJ*, 435, 339
Iwakami, W., Kotake, K., Ohnishi, N., Yamada, S., & Sawada, K. 2008, *ApJ*, 678, 1207
Janka, H. & Müller, E. 1996, *A&A*, 306, 167
Janka, H.-T., Langanke, K., Marek, A., Martínez-Pinedo, G., & Müller, B. 2007, *Phy. Rep.*, 442, 38
Kotake, K. 2011, *Comptes Rendus Physique*, accepted (arXiv:1110.5107)
Kotake, K., Sato, K., & Takahashi, K. 2006, *Reports of Progress in Physics*, 69, 971
Kotake, K., Iwakami-Nakano, W., & Ohnishi, N. 2011, *ApJ*, 736, 124
Kuroda, T. & Umeda, H. 2010, *ApJS*, 191, 439
Kuroda, T., Kotake, K., & Takiwaki, T. 2012, *ApJ*, submitted (arXiv:1202.2487)
Lentz, E. J., *et al.* 2011, arXiv:1112.3595
Liebendörfer, M., *et al.* 2004, *ApJS*, 150, 263
Liebendörfer, M., Rampp, M., Janka, H.-T., & Mezzacappa, A. 2005, *ApJ*, 620, 840
Liebendörfer, M., Whitehouse, S. C., & Fischer, T. 2009, *ApJ*, 698, 1174
Marek, A. & Janka, H.-T. 2009, *ApJ*, 694, 664
Masada, Y., Takiwaki, T., & Kotake, K. 2011, *ApJ*, submitted
Müller, B., Janka, H.-T., & Marek, A. 2012, arXiv:1202.0815
Nordhaus, J., Burrows, A., Almgren, A., & Bell, J. 2010, *ApJ*, 720, 694
Ohnishi, N., Kotake, K., & Yamada, S. 2006, *ApJ*, 641, 1018
Shibata, M., Kiuchi, K., Sekiguchi, Y., & Suwa, Y. 2011, *Progress of Theoretical Physics*, 125, 1255
Sumiyoshi, K., Yamada, S., Suzuki, H., Shen, H., Chiba, S., & Toki, H. 2005, *ApJ*, 629, 922
Suwa, Y., Kotake, K., Takiwaki, T., Whitehouse, S. C., Liebendörfer, M., & Sato, K. 2010, *PASJ* (Letters), 62, L49
Takiwaki, T. & Kotake, K. 2011, *ApJ*, 743, 30
Takiwaki, T., Kotake, K., & Suwa, Y. 2012, *ApJ*, 749, 98
Thorne, K. S. 1981, *MNRAS*, 194, 439
Wang, L. & Wheeler, J. C. 2008, *ARAA*, 46, 433
Woosley, S. E., Heger, A., & Weaver, T. A. 2002, *Reviews of Modern Physics*, 74, 1015

Discussion

KEIICHI NISHIKAWA: Could you explain about GR model? Do you include MHD?

KEI KOTAKE: The metric evolution is solved by the BSSN formalism with the HRSC scheme to evolve GR hydrodynamics. The code is already written in MHD, but we have only run non-magnetized models at present. As you would imply, MHD simulations in GR presumably in the context of collapsar should be very interesting.

SERGEY MOISEENKO: MHD mechanism of SN explosion gives explosion energy 10^{51} erg which is enough for explaining the observed kinetic energy.

KEI KOTAKE: That's right, but *only if* the precollapse core has strong B-fields with rapid rotation. But I agree with you that the MHD mechanism should be revisited by our GR (or more importantly) 3D models.

SEAN COUCH: Have you tried any rotating models?

KEI KOTAKE: Not yet, but this is what we are currently undertaking.

Death of Massive Stars: Supernovae and Gamma-Ray Bursts
Proceedings IAU Symposium No. 279, 2012
P. Roming, N. Kawai & E. Pian, eds.

doi:10.1017/S1743921312012811

A shallow water analogue of asymmetric core-collapse, and neutron star kick/spin

Thierry Foglizzo[1], **Frédéric Masset**[2,1], **Jérôme Guilet**[3,1] **and Gilles Durand**[1]

[1]Lab. AIM Paris-Saclay, CEA/Irfu Univ. Paris-Diderot CNRS/INSU, 91191, France
[2]Instituto de Ciencias Fisicas, UNAM, Av. Universidad s/n, 62210 Cuernavaca, Mor., Mexico
[3]DAMTP, University of Cambridge, Centre for Math. Sciences, Cambridge CB3 0WA, UK

Abstract. Massive stars end their life with the gravitational collapse of their core and the formation of a neutron star. Their explosion as a supernova depends on the revival of a spherical accretion shock, located in the inner 200km and stalled during a few hundred milliseconds. Numerical simulations suggest that the large scale asymmetry of the neutrino-driven explosion is induced by a hydrodynamical instability named SASI. Its non radial character is able to influence the kick and the spin of the resulting neutron star. The SWASI experiment is a simple shallow water analog of SASI, where the role of acoustic waves and shocks is played by surface waves and hydraulic jumps. Distances in the experiment are scaled down by a factor one million, and time is slower by a factor one hundred. This experiment is designed to illustrate the asymmetric nature of core-collapse supernova.

Keywords. supernovae, neutron stars, instabilities, numerical simulations, experiments

1. Introduction

The stellar core-collapse and explosion signing the death of a massive star and the birth of a neutron star is a complex problem involving nuclear densities, neutrino interactions and transport, general relativistic corrections and multi-dimensional magnetohydrodynamics. The favored explosion scenario for moderately rotating cores relies on the delayed neutrino driven mechanism proposed by Bethe & Wilson (1985). A spherical accretion shock stalls during a few hundred milliseconds immediately after the formation of a neutron star, while neutrinos diffuse out of the neutrinosphere and eventually revive the shock. Transverse motions are generated by hydrodynamical instabilities developing after the shock bounce: entropy gradients induced by neutrino heating generate convection on the scale of the gain region (e.g. Janka & Müller 1996), while the Standing Accretion Shock Instability (SASI) is responsible for large scale oscillations (Blondin *et al.* 2003).

Some potential consequences of SASI. Large scale transverse motions contribute to the explosion by lengthening the exposure time of large coherent regions of the post shock flow to the neutrino flux, leading to an asymmetric explosion (Marek & Janka 2009). The $l = 1$ character of SASI allows for a significant pulsar kick, as demonstrated by numerical simulations in 2D and 3D (Scheck *et al.* 2004, 2006, Nordhaus *et al.* 2010, Wongwathanarat *et al.* 2010). Whether the spiral mode of SASI could significantly influence the spin of the neutron star is more debated (Blondin & Mezzacappa 2007, Yamasaki & Foglizzo 2008, Iwakami *et al.* 2009, Fernandez 2010, Rantsiou *et al.* 2011).

The hydrodynamics of core-collapse. Interpreting the outcome of the most realistic simulations benefits from understanding much simpler models where neutrino transport and interactions are replaced by ad-hoc cooling functions, and the infalling gas is approximated as time-independent. Blondin & Mezzacappa (2007) simply considered an ideal adiabatic gas for their first 3D simulation of SASI. They discovered a dominant spiral mode and its potential consequences on the spin the neutron star. The experimental approach described below and in more details in Foglizzo *et al.* (2012, hereafter FMGD) is built upon a similar degree of simplification, in the 2D equatorial plane of the stellar core.

2. SWASI, a Shallow Water Analogue of a Shock Instability

Inviscid shallow water equations. In a gravitational field g, surface gravity waves with a wavelength longer than the depth H of the fluid propagate with a velocity $c = (gH)^{\frac{1}{2}}$. The classical hydraulic jump seen in a kitchen sink marks the sudden transition between a thin layer of fast fluid ($v > c$) and a thicker layer of slower fluid ($v < c$). The variations of depth in the liquid are analogous to the variations of density in a compressible gas. The Froude number $\mathrm{Fr} \equiv v/c$ is equivalent to the Mach number comparing the gas velocity to the velocity of sound waves. A hydraulic jump in shallow water is analogous to a shock in a gas. Averaged over the depth of the fluid, the inviscid shallow water equations describing the flow of water along a surface $z = H_{\mathrm{grav}}(r)$ are identical to the equations describing an isentropic gas with an adiabatic index $\gamma = 2$ in a potential $\Phi(r) = gH_{\mathrm{grav}}(r)$:

$$\frac{\partial H}{\partial t} + \nabla \cdot (Hv) = 0, \tag{2.1}$$

$$\frac{\partial v}{\partial t} + (\nabla \times v) \times v + \nabla \left(\frac{v^2}{2} + c^2 + \Phi \right) = 0. \tag{2.2}$$

Choosing a surface shape $H_{\mathrm{grav}}(r) \equiv H_{\mathrm{grav}}^{\mathrm{jp}} \times r_{\mathrm{jp}}/r$ such that $\Phi(r)$ mimics the gravitational potential GM_{NS}/r of the central neutron star, the free fall timescale $t_{\mathrm{ff}}^{\mathrm{jp}}$ in the 2D shallow water model is related to the free fall time scale $t_{\mathrm{ff}}^{\mathrm{sh}}$ in its astrophysical analogue as follows:

$$\frac{t_{\mathrm{ff}}^{\mathrm{sh}}}{t_{\mathrm{ff}}^{\mathrm{jp}}} \equiv \left(\frac{r_{\mathrm{sh}}}{r_{\mathrm{jp}}} \right) \left(\frac{r_{\mathrm{sh}} g H_{\mathrm{grav}}^{\mathrm{jp}}}{GM_{\mathrm{NS}}} \right)^{\frac{1}{2}}. \tag{2.3}$$

Remembering that the mechanism responsible for SASI in a gas is based on the interaction of acoustic waves and advected perturbations (Foglizzo *et al.* 2007, Scheck *et al.* 2008, Fernandez & Thompson 2009, Foglizzo 2009, Guilet & Foglizzo 2012), we expect a similar instability in shallow water based on the interaction of surface gravity waves and vorticity perturbations.

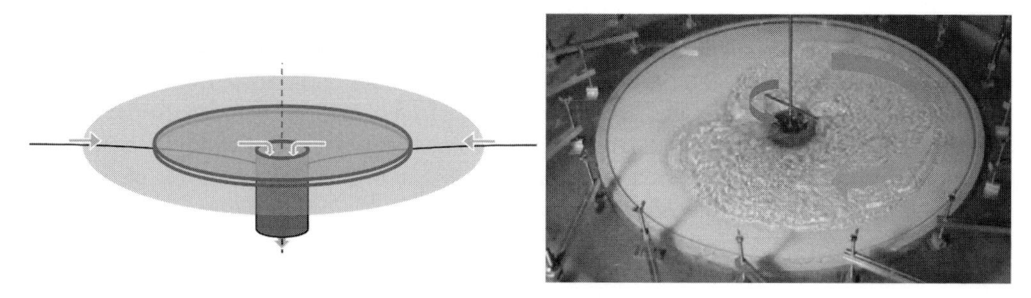

Figure 1. In the SWASI experiment, water flows radially inward from an annular reservoir down a potential well, and spills over the edge of an inner cylinder. On the right picture, the dynamical evolution of the hydraulic jump is dominated by a spiral mode. A light horizontal bar is used to visualize the budget of angular momentum.

Inner boundary. Like in the adiabatic simulations of Blondin & Mezzacappa (2007), the flow is extracted at the inner boundary rather than settling down onto the neutron star surface after intense neutrino cooling. In the experiment, the height of the upper edge of the inner cylinder defines a pressure threshold over which the fluid is evacuated by spilling over.

Viscous drag in the experiment. The viscous drag has been measured in the experiment in the stationary flow and modeled as a drag force $\bar{\nu}v/H^2$ in Eq. 2.2, with an effective viscosity coefficient $\bar{\nu} = 0.03\text{cm}^2/\text{s}$. The viscous drag is negligible after the hydraulic jump. The injection radius is 32cm, the radius of the inner cylinder is 4cm, the radius of the hydraulic jump is typically 20cm, and the flow rate is typically 1L/s.

Perturbative analysis. The perturbative analysis of the 2D model of the experiment revealed a dominant $m = 1$ instability in most of the parameter space (FMGD). For a given radius of the hydraulic jump, the variation of the injection slit and injection velocity in the stationary flow correspond to variations of the pre-jump Froude number Fr_{jp} and flow velocity v_{jp} , which we measure in units of the local inviscid free fall velocity v_{ff}. The flow is linearly unstable over a domain approximately described by $v_{\text{jp}}/v_{\text{ff}} > 0.5$ and $\text{Fr}_{\text{jp}} > 3$ (Fig. 4 in FMGD), which is barely affected by the viscous drag.

Experimental and numerical results in the linear regime. The experiment shows a robust $m = 1$ instability which grows from random noise. The oscillations of the hydraulic jump can either be sloshing of rotating, as expected from the linear stability analysis. Oscillation periods in the range 2-4 seconds have been measured in the experiment for a range of initial jump radii from 15 cm to 25 cm. Comparing these measurements to the prediction of the perturbative analysis and to the linear phase of 2D simulations showed very good agreement (Fig. 3 of FMGD), thus validating the 2D model of the experiment.

Experimental and numerical results in the non linear regime. When the instability is vigorous enough to reach non linear amplitudes, a systematic transition to a spiral mode has been observed, both in the experiment and in the 2D simulations. This non linear symmetry breaking had also been observed in the 3D numerical simulations of Blondin & Mezzacappa (2007).

Angular momentum budget. A freely rotating horizontal bar has been used in the experiment to visualize the angular momentum in the inner regions of the flow (Fig. 1). As the radius of the rotating hydraulic jump increases, the rotation of this bar is opposite to the rotation of the hydraulic jump. Both the experiment and the 2D simulations illustrate the partition of angular momentum between a rotating wave and the vorticity advected towards the accretor. The SWASI experiment illustrates the possible impact of SASI on the pulsar spin (Blondin & Mezzacappa 2007, Fernandez 2010).

3. Perspectives

Theory of core-collapse. The experimental limitations being different from the numerical ones, the SWASI experiment is a complementary tool to address some hydrodynamical questions in the theory of core-collapse. The non linear evolution of the 2D inviscid shallow water model presented here is remarkably similar to the 3D evolution of SASI in an adiabatic gas, despite the different value of the adiabatic index and the absence of entropy perturbations. A further comparison between these dynamical systems can help us characterize the role of the inner boundary, test the saturation mechanism of SASI (e.g. Guilet *et al.* 2010), and identify the physical mechanism responsible for the non linear symmetry breaking. This can help us understand the degree of asymmetry of the explosion, the strength of the pulsar kick and the possible effect of SASI on the pulsar spin. A new experiment is under construction at CEA-Saclay with improved accuracy.

Allowing for a global rotation, it should be able to describe the 2D evolution of SASI during the collapse of a rotating stellar core.

Public outreach. The SWASI experiment is simple and robust. It demonstrates some dynamical aspects of supernova theory using human timescales and sizes with a low construction cost. This experiment could contribute to scientific outreach towards students, researchers and the general public. A simplified version, currently designed at CEA-Saclay, will be proposed to universities and science museums.

References

Bethe, H. A. & Wilson, J. R. 1985, *ApJ*, 295, 14
Blondin, J. M., Mezzacappa, A., & DeMarino, C. 2003, *ApJ*, 584, 971
Blondin, J. M. & Mezzacappa, A. 2007, *Nature*, 445, 58
Fernandez, R. 2010, *ApJ*, 725, 1563
Fernandez, R. & Thompson, C. 2009, *ApJ*, 697, 1827
Foglizzo, T. 2009, *ApJ*, 694, 820
Foglizzo, T., Galletti, P., Scheck, L., & Janka, H.-Th. 2007, *ApJ* 654, 1006
Foglizzo, T., Masset, F., Guilet, J., & Durand, G. 2012, *Phys. Rev. Lett.* 108, 051103 (FMGD)
Guilet, J. & Foglizzo, T. 2012, *MNRAS* 421, 546
Guilet, J., Sato, J., & Foglizzo, T. 2010, *ApJ* 713, 1350
Iwakami, W., Kotake, K., Ohnishi, N., Yamada, S., & Sawada, K. 2009, *ApJ*, 700, 232
Janka, H.-T. & Müller, E. 1996, *A&A*, 306, 167
Marek, A. & Janka, H.-Th. 2009, *ApJ*, 694, 664
Nordhaus, J. & Brandt, T., Burrows, A., Livne, E., Ott, C. 2010, *Phys. Rev. D* 82, 103016.
Rantsiou, E., Burrows, A., Nordhaus, J., & Almgren, A. 2011, *ApJ*, 732, 57.
Scheck, L., Plewa, T., Janka, H.-Th., Kifonidis, K., & Müller, E. 2004, *Phys. Rev. Lett.*, 92, 011103
Scheck, L., Kifonidis, K., Janka, H. T., & Müller, E. 2006, *A&A*, 457, 963
Scheck, L., Janka, H.-Th., Foglizzo, T., & Kifonidis, K. 2008, *A&A*, 477, 931
Wongwathanarat, A., Janka, H.-T., & Müller, E. 2010, *ApJ* 725, L106.
Yamasaki, T. & Foglizzo, T. 2008, *ApJ*, 679, 607

Discussion

SUWA: Do you have any plan to extend the experiment in 3D ?

FOGLIZZO: Unfortunately not. The shallow water analogy is fundamentally restricted to a 2D approximation of the astrophysical flow.

COUCH: I would have thought viscous forces would be important in the experiment, in contrast to the core-collapse context.

FOGLIZZO: The viscous drag appears to be significant only in the shallowest regions of the flow, ahead of the hydraulic jump. The unstable eigenmodes are barely affected by viscosity, but their non linear saturation could be.

Death of Massive Stars: Supernovae and Gamma-Ray Bursts
Proceedings IAU Symposium No. 279, 2012
P. Roming, N. Kawai & E. Pian, eds.

doi:10.1017/S1743921312012823

Spectropolarimetry of Type Ibc Supernovae

Masaomi Tanaka[1], Koji S. Kawabata[2], Takashi Hattori[3], Paolo A. Mazzali[4,5], Kentaro Aoki[3], Masanori Iye[1], Keiichi Maeda[6], Ken'ichi Nomoto[6], Elena Pian[7], Toshiyuki Sasaki[3], and Masayuki Yamanaka[2,8]

[1]National Astronomical Observatory, Mitaka, Tokyo, Japan
email: masaomi.tanaka@nao.ac.jp

[2]Hiroshima Astrophysical Science Center, Hiroshima University, Higashi-Hiroshima, Hiroshima, Japan

[3]Subaru Telescope, National Astronomical Observatory of Japan, Hilo, HI

[4]Max-Planck Institut für Astrophysik, Karl-Schwarzschild-Strasse 2 D-85748 Garching bei München, Germany

[5]Istituto Naz. di Astrofisica-Oss. Astron., vicolo dell'Osservatorio, 5, 35122 Padova, Italy

[6]Institute for the Physics and Mathematics of the Universe, University of Tokyo, Kashiwa, Japan

[7]Istituto Naz. di Astrofisica-Oss. Astron., Via Tiepolo, 11, 34131 Trieste, Italy

[8]Department of Physical Science, Hiroshima University, Higashi-Hiroshima, Hiroshima, Japan

Abstract. Studying a multi-dimensional structure of supernovae (SNe) gives important constraints on the mechanism of the SN explosion. Polarization measurement is one of the most powerful methods to study the explosion geometry of extragalactic SNe. Especially, Type Ib/c SNe are the ideal targets because the core of the explosion is bare. We have performed spectropolarimetric observations of Type Ib/c SNe with the Subaru telescope. We detect a rotation of the polarization angle across the line, which is seen as a loop in the $Q-U$ plane. This indicates that axisymmetry is broken in the SN ejecta. Adding our new data to the sample of stripped-envelope SNe with high-quality spectropolarimetric data, five SNe out of six show a loop in the $Q-U$ plane. This implies that the SN explosion commonly has a non-axisymmetric, three-dimensional geometry.

Keywords. supernovae: general, techniques: polarimetric

---

1. Introduction

The explosion mechanism of supernovae (SNe) has been unclear for a long time after the first concept by Burbidge *et al.* (1957) and the first numerical simulation by Colgate & White (1966). According to modern numerical simulations, a successful explosion cannot be obtained in one-dimensional simulations. And it is suggested that multi-dimensional effects, such as convection (e.g. Herant *et al.* (1994)) and Standing Accretion Shock Instability (SASI, e.g. Blondin *et al.* (2003)) are important.

Given these circumstances, it is important to obtain observational constraints on the multi-dimensional geometry of SN explosions. Spectropolarimetry is one of the most powerful methods to study the multi-dimensional geometry of extragalactic SNe (see Wang & Wheeler (2008) for a review). To study the explosion geometry with spectropolarimetry, Type Ib/c SNe are the ideal targets because the core of the explosion is bare. We have performed spectropolarimetric observations of Type Ib/c SNe with the Subaru telescope. In these proceedings, we summarize spectropolarimetric properties of Type Ibc SNe.

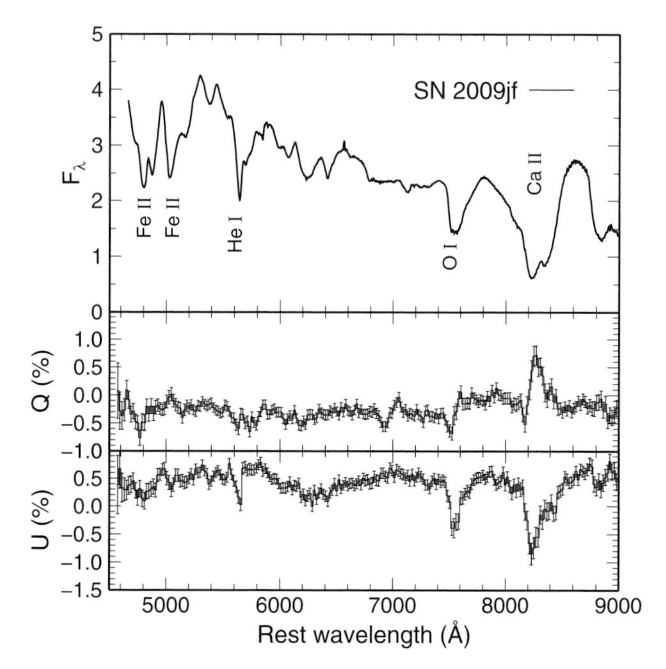

Figure 1. Total flux spectrum and polarization spectrum of SN 2009jf at 9.3 days after maximum. The total flux is shown in unit of 10^{-15} erg s^{-1} cm^{-2} Å^{-1}. For the polarization spectrum, the contribution of the interstellar polarization is *not* corrected for.

2. Subaru Spectropolarimetry of SNe 2009jf and 2009mi

We have performed spectropolarimetric observations of Type Ib SN 2009jf and Type Ic SN 2009mi with the Subaru telescope equipped with the Faint Object Camera and Spectrograph (FOCAS, Kashikawa *et al.* (2002)). The data for SNe 2009jf and 2009mi were obtained on UT 2009 October 24.3 (MJD = 55128.3) and 2010 January 8.3 (MJD=55204.3), respectively. These epochs correspond to 9.3 and 26.5 days after the B band maximum (MJD = 55118.96 for SN 2009jf according to Sahu *et al.* (2011), and MJD = 55177.8 for SN 2009mi, based on our observations). More details of the observations are given in Tanaka *et al.* (2012).

Figure 1 shows the observed total flux and polarization spectrum of SN 2009jf. We clearly detect the changes in the polarization at the strong absorption lines, such as He I, O I, and Ca II. The data around the strong lines are shown in the $Q-U$ plane in Figure 2. It is clear that the data at the Ca II and O I lines occupy different regions in the $Q-U$ plane, indicating different spatial distributions between Ca II and O I. Such a difference is also clearly seen in other SNe e.g. Type Ib SN 2008D (Maund *et al.* (2009)) and Type IIb SN 2008ax (Chornock *et al.* (2011)).

A more interesting feature is the shape of the polarization data in the $Q-U$ plane. The Ca II and O I lines in SN 2009jf show a loop at these lines. This means that the polarization angle varies with Doppler velocity. As suggested by e.g. Kasen *et al.* (2003), Maund *et al.* (2007a), Maund *et al.* (2007b), this loop indicates that axisymmetry is broken in the SN ejecta. We also detect a similar (although less significant) loop in SN 2009mi. Loops in the $Q-U$ plane have also been observed in Type Ia SNe (Wang *et al.* (2003a), Kasen *et al.* (2003), Chornock & Filippenko (2008), Patat *et al.* (2009), Tanaka *et al.* (2010)).

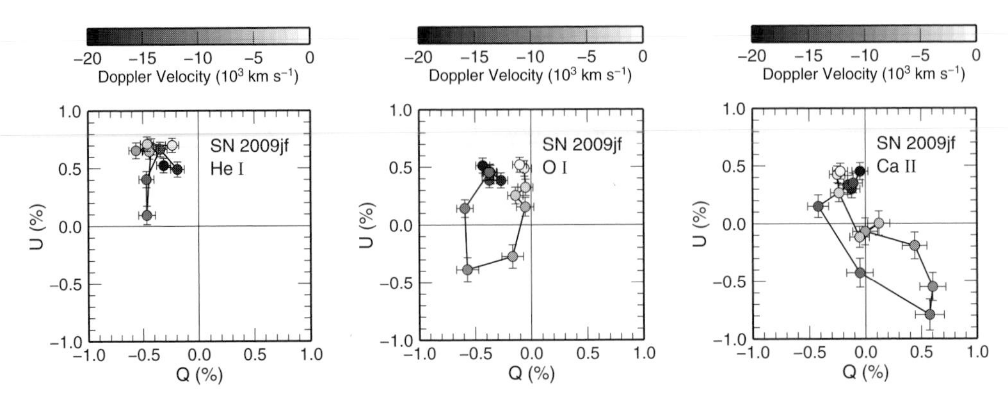

Figure 2. Polarization data of SN 2009jf in the $Q-U$ plane. Data around the He I (left), O I (center) and Ca II (right) lines are shown. Different data points show the polarization at different Doppler velocities as shown in the bar above the plots.

Table 1. Summary of Spectropolarimetric Observations of Type Ibc Supernovae

Object	Type	Loop?	Reference
SN 2005bf	Ib	yes	1,2
SN 2008D	Ib	yes	3
SN 2009jf	Ib	yes	4
SN 2002ap	Ic broad	yes	5,6,7
SN 2007gr	Ic	no	8
SN 2009mi	Ic	yes	4

References: [1]Maund *et al.* (2007a), [2]Tanaka *et al.* (2009a), [3]Maund *et al.* (2009), [4]Tanaka *et al.* (2012) [5]Kawabata *et al.* (2002), [6]Wang *et al.* (2003b), [7]Leonard *et al.* (2002), [8]Tanaka *et al.* (2008)

3. Spectropolarimetric Properties of Type Ibc Supernovae

In Table 1, we list up Type Ibc SNe with high-quality spectropolarimetric data. As also noted by Wang & Wheeler (2008), all SNe show non-zero polarization, which means that stripped-envelope SNe generally have asymmetric explosion geometry. In addition, five SNe out of six show loops in $Q-U$ plane. This implies that a non-axisymmetric, 3D geometry is common in stripped-envelope SNe.

This is somewhat surprising since the line profiles in the nebular spectra, which are another probe of the explosion geometry, are nicely modelled by a bipolar geometry (e.g. Mazzali *et al.* (2001), Maeda *et al.* (2002), Mazzali *et al.* (2005), Maeda *et al.* (2008), Modjaz *et al.* (2008), Tanaka *et al.* (2009b), Taubenberger *et al.* (2009), Maurer *et al.* (2010), but see also Milisavljevic *et al.* (2010)). It must be noted, however, that spectropolarimetry is sensitive to the outer ejecta while nebular line profile is sensitive to the inner ejecta. Thus, for example, 3D perturbation onto a bipolar structure can also be consistent with the observations.

It must be interesting to look for a possible relation between spectropolarimetric properties and nebular line profiles. In fact, we did such an attempt, but could not find any significant correlation because of the small sample size (Tanaka *et al.* 2012). It is important to increase spectropolarimetric samples to further study fully multi-dimensional structure of SNe.

References

Blondin, J. M., Mezzacappa, A., & DeMarino, C. 2003, *ApJ*, 584, 971
Burbidge, E. M., Burbidge, G. R., Fowler, W. A., & Hoyle, F. 1957, *Reviews of Modern Physics*, 29, 547
Chornock, R. & Filippenko, A. V. 2008, *AJ*, 136, 2227
Chornock, R., *et al.* 2011, *ApJ*, 739, 41
Colgate, S. A. & White, R. H. 1966, *ApJ*, 143, 626
Herant, M., Benz, W., Hix, W. R., Fryer, C. L., & Colgate, S. A. 1994, *ApJ*, 435, 339
Kasen, D., *et al.* 2003, *ApJ*, 593, 788
Kashikawa, N., *et al.* 2002, *PASJ*, 54, 819
Kawabata, K. S., *et al.* 2002, *ApJ* (Letters), 580, L39
Leonard, D. C., Filippenko, A. V., Chornock, R., & Foley, R. J. 2002, *PASP*, 114, 1333
Maeda, K., Nakamura, T., Nomoto, K., Mazzali, P. A., Patat, F., & Hachisu, I. 2002, *ApJ*, 565, 405
Maeda, K., *et al.* 2008, *Science*, 319, 1220
Maund, J. R., Wheeler, J. C., Patat, F., Baade, D., Wang, L., & Höflich, P. 2007a, *MNRAS*, 381, 201
Maund, J. R., Wheeler, J. C., Patat, F., Wang, L., Baade, D., & Höflich, P. A. 2007b, *ApJ*, 671, 1944
Maund, J. R., Wheeler, J. C., Baade, D., Patat, F., Höflich, P., Wang, L., & Clocchiatti, A. 2009, *ApJ*, 705, 1139
Maurer, J. I., *et al.* 2010, *MNRAS*, 402, 161
Mazzali, P. A., *et al.* 2005, *Science*, 308, 1284
Mazzali, P. A., Nomoto, K., Patat, F., & Maeda, K. 2001, *ApJ*, 559, 1047
Milisavljevic, D., Fesen, R. A., Gerardy, C. L., Kirshner, R. P., & Challis, P. 2010, *ApJ*, 709, 1343
Modjaz, M., Kirshner, R. P., Blondin, S., Challis, P., & Matheson, T. 2008, *ApJ* (Letters), 687, L9
Patat, F., Baade, D., Höflich, P., Maund, J. R., Wang, L., & Wheeler, J. C. 2009, *A&A*, 508, 229
Sahu, D. K., Gurugubelli, U. K., Anupama, G. C., & Nomoto, K. 2011, *MNRAS*, 383
Tanaka, M., Kawabata, K. S., Maeda, K., Hattori, T., & Nomoto, K. 2008, *ApJ*, 689, 1191
Tanaka, M., *et al.* 2009a, *ApJ*, 699, 1119
Tanaka, M., *et al.* 2009b, *ApJ*, 700, 1680
Tanaka, M., *et al.* 2010, *ApJ*, 714, 1209
Tanaka, M., *et al.* 2012, *ApJ*, submitted
Taubenberger, S., *et al.* 2009, *MNRAS*, 397, 677
Wang, L., *et al.* 2003a, *ApJ*, 591, 1110
Wang, L., Baade, D., Höflich, P., & Wheeler, J. C. 2003b, *ApJ*, 592, 457
Wang, L. & Wheeler, J. C. 2008, *ARA&A*, 46, 433

Discussion

COUCH: (Comment) SASI is not necessary to break axisymmetry. Bipolar magnetorotational explosions will break axisymmetry in 3D simulations.

CHORNOCK: I agree that $Q-U$ loops are likely a signature of large-scale structures in the ejecta, but how do you explain the different behaviors of different lines? Do we expect oxygen and calcium to have such different spatial distributions?

TANAKA: Oxygen is synthesized at the pre-SN stage while calcium can be synthesized by the explosion. So, qualitatively, these two elements can have different distributions in the SN ejecta. In fact, helium, which is also synthesized at the pre-SN stage, occupies a similar region to oxygen in the $Q-U$ plane in SN 2009jf.

Death of Massive Stars: Supernovae and Gamma-Ray Bursts
Proceedings IAU Symposium No. 279, 2012
P. Roming, N. Kawai & E. Pian, eds.

doi:10.1017/S1743921312012835

Gravitational waves and gamma-ray bursts

Alessandra Corsi[1]
for the LIGO Scientific Collaboration and Virgo collaboration

[1]LIGO laboratory, California Institute of Technology
MS 100-36, Pasadena, CA 91125 (USA)
email: corsi@caltech.edu

Abstract. Gamma-Ray Bursts are likely associated with a catastrophic energy release in stellar mass objects. Electromagnetic observations provide important, but indirect information on the progenitor. On the other hand, gravitational waves emitted from the central source, carry direct information on its nature. In this context, I give an overview of the multi-messenger study of gamma-ray bursts that can be carried out by using electromagnetic and gravitational wave observations. I also underline the importance of joint electromagnetic and gravitational wave searches, in the absence of a gamma-ray trigger. Finally, I discuss how multi-messenger observations may probe alternative gamma-ray burst progenitor models, such as the magnetar scenario.

Keywords. gamma rays: bursts, gravitational waves, stars: neutron, supernovae: general

---

1. Introduction

During the last 15 years, thanks to satellite missions like *BeppoSAX* (Boella *et al.* 1997), *Swift* (Gehrels *et al.* 2004), and *Fermi* (Atwood *et al.* 2009; Meegan *et al.* 2009), our progress in the understanding of gamma-ray bursts (GRBs) has been quite spectacular. We now know that GRBs are cosmological events related to a catastrophic energy release in stellar mass objects. Energy dissipation within a highly relativistic "fireball", presumably emitted in the form of a jet (e.g., Rhoads 1999; Sari *et al.* 1999; Kumar & Panaitescu 2000; Frail *et al.* 2001), is believed to power the observed γ-ray flash (prompt emission) and the subsequent "afterglow" (e.g., Blandford & McKee 1976; Rees & Mészáros 1992; Mészáros & Rees 1993a,b; Piran 2004; Mészáros 2006).

Traditionally, GRBs have been divided in two main categories, long and short ones (e.g. Kouveliotou *et al.* 1993), depending on the duration of their prompt γ-ray emission ($\lesssim$ 2 s for the short bursts, $\gtrsim$ 2 s for the long ones). These two populations of bursts are thought to be related to two different progenitor models: "collapsars" for the long-soft bursts (e.g., Woosley 1993; MacFadyen & Woosley 1999; Piran 2004; Mészáros 2006 ; Woosley & Bloom 2006, and references therein), and the merger of binary systems of compact objects such as neutron star (NS)-NS or black-hole (BH)-NS systems, for the short-hard ones (e.g., Eichler *et al.* 1989; Narayan *et al.* 1992; Janka *et al.* 1999; Belczynski *et al.* 2002; Rosswog 2005; Belczynski *et al.* 2006; Faber *et al.* 2006).

Despite the recent issues raised in the classification of short and long GRBs based solely on their prompt emission properties (e.g, Zhang *et al.* 2009), the general picture of two classes of bursts related to two main progenitor models, still holds. The collapsar scenario is observationally supported by the fact that long GRBs occur in galaxies with high specific star formation rate (e.g., Christensen *et al.* 2004; Castro Cerón *et al.* 2006; Fruchter *et al.* 2006; Levesque *et al.* 2010), and that at least some long GRBs have been observed to be associated with core-collapse supernovae (SNe) of rare type (Galama *et al.* 1998; Berger *et al.* 2003; Hjorth *et al.* 2003; Malesani *et al.* 2004; Campana *et al.* 2006;

Pian *et al.* 2006). Indeed, the question of what makes some stars die as SNe and some other as relativistic GRBs, is not solved yet (e.g., Woosley & Bloom 2006, and references therein). Indications that progenitors of short bursts belong, on average, to old stellar populations with a typical lifetime of several Gyr (Barthelmy *et al.* 2005; Berger *et al.* 2005; Gehrels *et al.* 2005; Villasenor *et al.* 2005; Bloom *et al.* 2006; Gal-Yam *et al.* 2008; O'Shaughnessy *et al.* 2008), provide support to the binary merger scenario.

Collapsars and binary mergers leading to the formation of a BH plus an accretion disk (e.g., Woosley 1993; Fryer *et al.* 1996; MacFadyen & Woosley 1999; Rosswog *et al.* 1999; Ruffert & Janka 1999; Narayan *et al.* 2001), have the potential to power the GRB fireball via the energy released from the accretion of the disk onto the newly formed BH. While the formation of a BH plus disk system is common to both progenitor models, the more compact scale of the NS-NS (or BH-NS systems), and the less massive debris left over after merger, are invoked to explain the shorter duration and the smaller isotropic energies of these bursts with respect to long ones.

Within the standard fireball model, once the fireball is launched from the central engine, the observed radiation is explained as synchrotron and/or inverse Compton emission from electrons accelerated in internal and external shocks (e.g., Sari 1997; Kobayashi *et al.* 1997; Sari *et al.* 1998; Granot *et al.* 1999; Dermer *et al.* 2000; Sari & Esin 2001), taking place at distances $\gtrsim 10^{13}$ cm from the central source. High-energy (GeV) tails observed in some GRBs have challenged the internal-external shock fireball model in its simplest formulation (Hurley *et al.* 1994; Baring & Harding 1997; Abdo *et al.* 2009; Kumar & Barniol Duran 2009; Ackermann *et al.* 2010; Corsi *et al.* 2010; De Pasquale *et al.* 2010; Ghirlanda *et al.* 2010; Giuliani *et al.* 2010; Abdo *et al.* 2011; Asano & Mészáros 2011; Mészáros & Rees 2011; Toma *et al.* 2011; Zhang *et al.* 2011). However, it remains true that the electromagnetic emission from GRBs, being produced at large distances from the central engine, provides indirect information on the progenitor. On the other hand, gravitational waves (GWs) emitted from the progenitor could directly probe its nature.

2. GRB-triggered searches for GWs

Being related to catastrophic events involving stellar-mass objects, GRBs are good candidates for the detection of GWs (Kochanek & Piran 1993; Finn *et al.* 1999; van Putten 2001, 2002; Kobayashi & Mészáros 2003). Coalescing binaries, thought to be associated with short bursts, are one of the most promising GW sources (e.g., Phinney 1991; Cutler *et al.* 1993; Zhuge *et al.* 1994; Flanagan & Hughes 1998; Abadie *et al.* 2010a; Shibata & Taniguchi 2011, and references therein) for detectors like the Laser Interferometer Gravitational-Wave Observatory (LIGO; Abbott *et al.* 2009b) and Virgo (Acernese *et al.* 2008a; Accadia *et al.* 2011). For such systems, a chirp signal should be emitted in GWs during the in-spiral, followed by a burst-type signal associated with the merger, and subsequently a signal from the ring-down phase of the newly formed BH (e.g., ?Kobayashi & Mészáros 2003; Berti *et al.* 2009, and references therein). The last, initially deformed, is expected to radiate GWs until reaching a Kerr geometry (Kobayashi & Mészáros 2003).

In the collapsar scenario, relevant for long GRBs, the high rotation required to form the centrifugally supported disk that powers the GRB, should produce GWs via bar (e.g., Houser *et al.* 1994; New *et al.* 2000; Baiotti *et al.* 2007; Dimmelmeier *et al.* 2008) or fragmentation instabilities that might develop in the collapsing core (see e.g., Fryer & New 2003; Ott 2009, for recent reviews) and/or in the disk (Davies *et al.* 2002; Fryer *et al.* 2002; Kobayashi & Mészáros 2003; Piro & Pfahl 2007). Moreover, asymmetrically infalling matter is expected to perturb the final BH geometry, leading to a ring-down phase (?Kobayashi & Mészáros 2003).

LIGO and Virgo have been carrying out electromagnetically triggered searches for GWs (Abbott *et al.* 2005, 2007; Acernese *et al.* 2007; Abbott *et al.* 2008b,a; Acernese *et al.* 2008b; Abbott *et al.* 2009a; Abadie *et al.* 2010b; Abbott *et al.* 2010; Abadie *et al.* 2011, 2012) over the past decade (for bar detectors electromagnetically triggered searches, see e.g. Astone *et al.* 1999, 2002, 2005; Baggio *et al.* 2005). The LIGO Scientific Collaboration operates two LIGO observatories in the U.S. along with the GEO600 detector (Grote & LIGO Scientific Collaboration 2010) in Germany. Together with Virgo, located in Italy, they form a detector network capable of detecting GW signals arriving from all directions.

GRB-triggered searches for GWs by LIGO and Virgo have targeted both the chirp signal expected in the case of short GRBs during the NS-NS or BH-NS in-spiral, and short unmodeled pulses of GWs that may be expected during the merger/collapse, and ring-down phases of short/long GRBs (Abbott *et al.* 2005; Acernese *et al.* 2007; Abbott *et al.* 2008b,a; Acernese *et al.* 2008b; Abadie *et al.* 2010b; Abbott *et al.* 2010; Abadie *et al.* 2012). These searches have adopted on-source time windows of few minutes (long GRBs) or few seconds (short GRBs) around the GRB trigger time. In fact, for long GRBs, the time delay between the GW signal and γ-ray trigger is thought to be dominated by the time necessary for the fireball to push through the stellar envelope of the progenitor (10-100 s; Zhang & Mészáros 2004). On the other hand, for short GRBs, the NS-NS/BH-NS merger is believed to occur quickly, and be over within a few seconds (naturally accounting for the short nature of these bursts). It is estimated that triggered searches for GWs in few minutes time-windows yield a factor of ≈ 2 improvement in sensitivity with respect to untriggered ones (Kochanek & Piran 1993).

The most exciting results from LIGO GRB-triggered searches of GWs are probably represented by the cases of the short GRBs 070201 (Abbott *et al.* 2008a; Ofek *et al.* 2008) and 051103 (Abadie *et al.* 2012; Hurley *et al.* 2010), whose error boxes overlap with nearby galaxies (M31 for GRB 070201; M81 for GRB 051103). A NS-NS binary merger scenario occurring in such hosts was excluded by LIGO with rather high confidence (Abbott *et al.* 2008a; Abadie *et al.* 2012). However, the possibility that GRB 070201 and GRB 051103 are related to (extra-galactic) soft gamma-ray repeaters (SGR) giant flares (for a recent review, see ?, and references therein), could not be ruled out. Indeed, LIGO upper-limits for short unmodeled pulses of GWs from GRB 070201 and GRB 051103, are above the maximum GW energy emissible in SGR giant flares (Ioka 2001; Corsi & Owen 2011).

3. GW-triggered searches for GRBs

A very appealing prospect is represented by the possibility of using GWs to trigger electromagnetic (radio, optical, X-ray) follow-ups of GW sources (e.g., ?Bloom *et al.* 2009; Metzger & Berger 2012). The discovery of off-axis optical or radio afterglows (Mészáros *et al.* 1998; Granot *et al.* 2002; Janka *et al.* 2006; van Eerten & MacFadyen 2011) triggered via the (non-beamed) GW emission from the GRB progenitors, would yield a dramatic confirmation of the "jet model", map out the beaming distribution, and provide fundamental inputs to models of relativistic outflows. Radio follow-ups, in particular, are an effective tool to identify relativistic and mildly relativistic outflows (e.g., Kulkarni *et al.* 1998; Soderberg *et al.* 2010; Nakar & Piran 2011) in the absence of a γ-ray trigger.

Finding electromagnetic counterparts to GW triggers is technically challenging due to imperfect localization of the GW signal and uncertainty regarding the relative timing of the GW and electromagnetic emissions. The localization of LIGO-Virgo triple-coincidence GW triggers can yield error-areas of $\sim 100\,\mathrm{deg}^2$, possibly spread over different patches of the sky (see e.g. Fig. 3 in Abadie *et al.* 2012). The problem of following-up with optical (or radio, or X-ray) telescopes in such a large error-area can be partially mitigated by: (i) restricting the search for electromagnetic counterparts to transients in nearby galaxies (within the LIGO-Virgo horizon distance); (ii) by noticing that the most promising electromagnetic counterparts of GW events detectable by LIGO and Virgo are expected to be "exotic" (rare) ones (e.g., the orphan afterglow of a GRB, and/or the "kilonova" from a binary merger - see Kulkarni 2005; Metzger *et al.* 2010).

In 2009-2010, LIGO and Virgo, together with partner electromagnetic observatories, performed their first "LOOC-UP" - Locating and Observing Optical Counterparts to Unmodeled Pulses of gravitational waves - experiment (Kanner *et al.* 2008; Abadie *et al.* 2012, and references therein). At the time, there were two operating LIGO interferometers (Abbott *et al.* 2009b), each with 4-km arms (one near Hanford, Washington, the other in Livingston Parish, Louisiana). The Virgo 3-km arms detector (Acernese *et al.* 2008a; Accadia *et al.* 2011) located near Cascina (Italy), was also operating. The LOOC-UP search has established a baseline for low-latency analyses with the next-generation GW detectors (Advanced LIGO and Advanced Virgo; Acernese *et al.* 2009; Harry & LIGO Scientific Collaboration 2010). The collaboration between GW and electromagnetic observatories is likely to continue to develop over the next few years, as the scientific community gets ready for a global network of advanced GW detectors.

4. GRBs and magnetars: prospects for multi-messenger studies

The forthcoming years may see the development of new GW searches in coincidence with GRBs, aimed at answering some of the questions opened by recent observations. In particular, a compelling result from *Swift* has been the discovery that the "normal" power-law behavior of long GRB X-ray light curves is often preceded at early times by a steep decay, followed by a shallower-than-normal decay (e.g. Nousek *et al.* 2006; Zhang *et al.* 2006). The steep-to-shallow and shallow-to-normal decay transitions are separated by two break times, $100\ \mathrm{s} \lesssim T_{break,1} \lesssim 500\ \mathrm{s}$ and $10^3\ \mathrm{s} \lesssim T_{break,2} \lesssim 10^4\ \mathrm{s}$. It has been suggested that the shallow phase may be attributed to a continuous energy injection by a long-lived central engine, with progressively reduced activity (Zhang *et al.* 2006).

Newborn magnetars, besides being candidate GRB progenitors (e.g., Usov 1992; Thompson 1994; Bucciantini *et al.* 2007; Metzger *et al.* 2007), have also been proposed to account for shallow decays or plateaus observed in GRB light curves (Dai & Lu 1998; Zhang & Mészáros 2001; Fan & Xu 2006; Yu & Huang 2007; Metzger *et al.* 2008; Xu *et al.* 2009; Rowlinson *et al.* 2010). Independent support for the magnetar scenario comes from the observation of SN 2006aj, associated with the nearby sub-energetic GRB 060218, suggesting that the SN-GRB connection may extend to a much broader range of stellar masses than previously thought, possibly involving two different mechanisms: a "collapsar" for the more massive stars collapsing to a BH, and a newborn (highly-magnetized) NS for the less massive ones (Mazzali *et al.* 2006).

Several studies have shown how magnetars dipole losses may indeed explain the flattening observed in GRB afterglows (Dai & Lu 1998; Zhang & Mészáros 2001; Fan & Xu 2006; Yu & Huang 2007; Dall'Osso *et al.* 2011; Bernardini *et al.* 2012). Corsi & Mészáros (2009) have explored a scenario in which the newly born magnetar left over after the GRB explosion undergoes a secular bar-mode instability (Lai & Shapiro 1995), thus producing a bar-like GW signal associated to the electromagnetic plateau, potentially detectable by the advanced ground-based interferometers like LIGO and Virgo (up to distances of $\sim 100\,\mathrm{Mpc}$). Compared to current analyses that GW detectors are carrying out (see Section 2), this scenario (Corsi & Mészáros 2009) involves a new class of GW signals, with a longer duration ($10^3 - 10^4$ s) and a different frequency evolution. Data analysis techniques for the search of longer duration GW signals possibly applicable to GRB searches, are being developed (e.g., Thrane *et al.* 2011).

5. Prospects and conclusions

The LIGO interferometers are being upgraded to the next-generation Advanced detectors (Harry & LIGO Scientific Collaboration 2010), that are expected to become operational around 2015. Virgo will also be upgraded to become Advanced Virgo (Acernese *et al.* 2009). Additionally, the new LCGT detector (Kuroda & LCGT Collaboration 2010) is being built in Japan. These advanced detectors are expected to detect compact binary coalescences, possibly at a rate of dozens per year after reaching design sensitivity (Abadie *et al.* 2010a), so that the short-GRB progenitor scenario may finally be probed directly. Long-standing open questions (e.g., is the jet model for GRBs correct? Why do some massive stars die as SNe and others as relativistic GRBs?), or other issues raised by more recent observations (such as the difficulties in the long-short GRB classification; the role of magnetars as GRB progenitors; the link between long GRBs and SGRs, etc.), will greatly benefit from joint GW studies. The advanced GW detectors will provide a totally new view of the transient sky (Márka *et al.* 2010, 2011). The prospects for this new era of astronomy are exciting, and promise a return of big scientific impact.

Acknowledgments

LIGO was constructed by the California Institute of Technology and Massachusetts Institute of Technology with funding from the National Science Foundation and operates under cooperative agreement PHY-0757058. This paper has LIGO Document Number LIGO-P1200042.

References

Abadie, J., *et al.* 2011, *ApJ* (Letters), 734, L35
Abadie, J., *et al.* 2010a, *Classical and Quantum Gravity*, 27, 173001
Abadie, J., *et al.* 2010b, *ApJ*, 715, 1453
Abadie, J., *et al.* 2012, ArXiv: 1201.4413
Abbott, B., *et al.* 2005, *Phys. Rev. D*, 72, 042002
Abbott, B., *et al.* 2007, *Phys. Rev. D.*, 76, 062003
Abbott, B., *et al.* 2008a, *ApJ*, 681, 1419
Abbott, B., *et al.* 2008b, *Phys. Rev. D*, 77, 062004
Abbott, B. P., *et al.* 2010, *ApJ*, 715, 1438
Abbott, B. P., *et al.* 2009a, *Reports on Progress in Physics*, 72, 076901
Abbott, B. P., *et al.* 2009b, *ApJ* (Letters), 701, L68
Abdo, A. A., *et al.* 2011, *ApJ* (Letters), 734, L27
Abdo, A. A., *et al.* 2009,*Science*, 323, 1688
Accadia, T., *et al.* 2011, *Classical and Quantum Gravity*, 28, 114002
Acernese, F., *et al.* 2009, Note VIR-027A09
Acernese, F., *et al.* 2008a, *Classical and Quantum Gravity*, 25, 184001
Acernese, F., *et al.* 2008b, *Classical and Quantum Gravity*, 25, 225001
Acernese, F., *et al.* 2007, *Classical and Quantum Gravity*, 24, 671
Ackermann, M., *et al.* 2010, *ApJ*, 716, 1178
Asano, K. & Mészáros, P. 2011, *ApJ*, 739, 103
Astone, P., *et al.* 2005, *Phys. Rev. D*, 71, 042001
Astone, P., *et al.* 1999, *Astroparticle Physics*, 10, 83
Astone, P., *et al.* 2002, *Phys. Rev. D*, 66, 102002

Atwood, W. B., *et al.* 2009, *ApJ*, 697, 1071
Baggio, L., *et al.* 2005, *Physical Review Letters*, 95, 081103
Baiotti, L., de Pietri, R., Manca, G. M., & Rezzolla, L. 2007, *Phys. Rev. D*, 75, 044023
Baring, M. G. & Harding, A. K. 1997, *ApJ*, 491, 663
Barthelmy, S. D., *et al.* 2005, *Nature*, 438, 994
Belczynski, K., Bulik, T., & Rudak, B. 2002, *ApJ*, 571, 394
Belczynski, K., Perna, R., Bulik, T., Kalogera, V., Ivanova, N., & Lamb, D. Q. 2006, *ApJ*, 648, 1110
Berger, E., *et al.* 2003, *Nature*, 426, 154
Berger, E., *et al.* 2005, *Nature*, 438, 988
Bernardini, M. G., Margutti, R., Mao, J., Zaninoni, E., & Chincarini, G. 2012, *A&A*, 539, A3
Berti, E., Cardoso, V., & Starinets, A. O. 2009, *Classical and Quantum Gravity*, 26, 163001
Blandford, R. D. & McKee, C. F. 1976, *Physics of Fluids*, 19, 1130
Bloom, J. S., *et al.* 2009, ArXiv: 0902.1527
Bloom, J. S., *et al.* 2006, *ApJ*, 638, 354
Boella, G., Butler, R. C., Perola, G. C., Piro, L., Scarsi, L., & Bleeker, J. A. M. 1997, *A&AS*, 122, 299
Bouhou, B. & for the ANTARES Collaboration, the LIGO Scientific Collaboration, the Virgo Collaboration. 2012, ArXiv: 1201.2840
Bucciantini, N., Quataert, E., Arons, J., Metzger, B. D., & Thompson, T. A. 2007, *MNRAS*, 380, 1541
Campana, S., *et al.* 2006, *Nature*, 442, 1008
Castro Cerón, J. M., Michałowski, M. J., Hjorth, J., Watson, D., Fynbo, J. P. U., & Gorosabel, J. 2006, *ApJ* (Letters), 653, L85
Christensen, L., Hjorth, J., & Gorosabel, J. 2004, *A&A*, 425, 913
Corsi, A., Guetta, D., & Piro, L. 2010, *ApJ*, 720, 1008
Corsi, A. & Mészáros, P. 2009, *ApJ*, 702, 1171
Corsi, A. & Owen, B. J. 2011, *Phys. Rev. D*, 83, 104014
Cutler, C., *et al.* 1993, *Physical Review Letters*, 70, 2984
Dai, Z. G. & Lu, T. 1998, *A&A*, 333, L87
Dall'Osso, S., Stratta, G., Guetta, D., Covino, S., de Cesare, G., & Stella, L. 2011, *A&A*, 526, A121
Davies, M. B., King, A., Rosswog, S., & Wynn, G. 2002, *ApJ* (Letters), 579, L63
De Pasquale, M., *et al.* 2010, *ApJ* (Letters), 709, L146
Dermer, C. D., Chiang, J., & Mitman, K. E. 2000, *ApJ*, 537, 785
Dimmelmeier, H., Ott, C. D., Marek, A., & Janka, H.-T. 2008, *Phys. Rev. D*, 78, 064056
Echeverria, F. 1989, *Phys. Rev. D*, 40, 3194
Eichler, D., Livio, M., Piran, T., & Schramm, D. N. 1989, *Nature*, 340, 126
Faber, J. A., Baumgarte, T. W., Shapiro, S. L., & Taniguchi, K. 2006, *ApJ* (Letters), 641, L93
Fan, Y. & Xu, D. 2006, *MNRAS*, 372, L19
Finn, L. S., Mohanty, S. D., & Romano, J. D. 1999, *Phys. Rev. D*, 60, 121101
Flanagan, É. É. & Hughes, S. A. 1998, *Phys. Rev. D*, 57, 4535
Frail, D. A., *et al.* 2001, *ApJ* (Letters), 562, L55
Fruchter, A. S., *et al.* 2006, *Nature*, 441, 463
Fryer, C. L., Benz, W., & Herant, M. 1996, *ApJ*, 460, 801
Fryer, C. L., Holz, D. E., & Hughes, S. A. 2002, *ApJ*, 565, 430
Fryer, C. L. & New, K. C. 2003, *Living Reviews in Relativity*, 6
Gal-Yam, A., *et al.* 2008, *ApJ*, 686, 408
Galama, T. J., *et al.* 1998, *Nature*, 395, 670
Gehrels, N., *et al.* 2004, *ApJ*, 611, 1005
Gehrels, N., *et al.* 2005, *Nature*, 437, 851
Ghirlanda, G., Ghisellini, G., & Nava, L. 2010, *A&A*, 510, L7
Giuliani, A., *et al.* 2010, *ApJ* (Letters), 708, L84
Granot, J., Panaitescu, A., Kumar, P., & Woosley, S. E. 2002, *ApJ* (Letters), 570, L61

Granot, J., Piran, T., & Sari, R. 1999, *ApJ*, 527, 236
Grote, H., & LIGO Scientific Collaboration. 2010, *Classical and Quantum Gravity*, 27, 084003
Harry, G. M., & LIGO Scientific Collaboration. 2010, *Classical and Quantum Gravity*, 27, 084006
Hjorth, J., *et al.* 2003, *Nature*, 423, 847
Houser, J. L., Centrella, J. M., & Smith, S. C. 1994, *Physical Review Letters*, 72, 1314
Hurley, K., *et al.* 1994, *Nature*, 372, 652
Hurley, K., *et al.* 2010, *MNRAS*, 403, 342
Ioka, K. 2001, *MNRAS*, 327, 639
Janka, H.-T., Aloy, M.-A., Mazzali, P. A., & Pian, E. 2006, *ApJ*, 645, 1305
Janka, H.-T., Eberl, T., Ruffert, M., & Fryer, C. L. 1999, *ApJ* (Letters), 527, L39
Kanner, J., Huard, T. L., Márka, S., Murphy, D. C., Piscionere, J., Reed, M., & Shawhan, P. 2008, *Classical and Quantum Gravity*, 25, 184034
Kobayashi, S. & Mészáros, P. 2003, *ApJ*, 589, 861
Kobayashi, S., Piran, T., & Sari, R. 1997, *Apj*, 490, 92
Kochanek, C. S. & Piran, T. 1993, *ApJ* (Letters), 417, L17
Kouveliotou, C., Meegan, C. A., Fishman, G. J., Bhat, N. P., Briggs, M. S., Koshut, T. M., Paciesas, W. S., & Pendleton, G. N. 1993, *ApJ* (Letters), 413, L101
Kulkarni, S. R., *et al.* 1998, *Nature*, 395, 663
Kulkarni, S. R. 2005, ArXiv: astro-ph/0510256
Kumar, P. & Barniol Duran, R. 2009, *MNRAS*, 400, L75
Kumar, P. & Panaitescu, A. 2000, *ApJ* (Letters), 541, L9
Kuroda, K., & LCGT Collaboration. 2010, *Classical and Quantum Gravity*, 27, 084004
Lai, D. & Shapiro, S. L. 1995, *ApJ*, 442, 259
Levesque, E. M., Berger, E., Kewley, L. J., & Bagley, M. M. 2010, *AJ*, 139, 694
MacFadyen, A. I. & Woosley, S. E. 1999, *ApJ*, 524, 262
Malesani, D., *et al.* 2004, *ApJ* (Letters), 609, L5
Márka, S. & for LIGO Scientific Collaboration, Virgo Collaboration. 2011, *Classical and Quantum Gravity*, 28, 114013
Márka, S. & LIGO Scientific Collaboration, Virgo Collaboration. 2010, *Journal of Physics Conference Series*, 243, 012001
Mazzali, P. A., *et al.* 2006, *Nature*, 442, 1018
Meegan, C., *et al.* 2009, *ApJ*, 702, 791
Mereghetti, S. 2008, *A&ARv*, 15, 225
Mészáros, P. 2006, *Reports on Progress in Physics*, 69, 2259
Mészáros, P. & Rees, M. J. 1993a, *ApJ* (Letters), 418, L59
Mészáros, P. & Rees, M. J. 1993b, *ApJ*, 405, 278
Mészáros, P. & Rees, M. J. 2011, *ApJ* (Letters), 733, L40
Mészáros, P., Rees, M. J., & Wijers, R. A. M. J. 1998, *ApJ*, 499, 301
Metzger, B. D. & Berger, E. 2012, *ApJ*, 746, 48
Metzger, B. D., *et al.* 2010, *MNRAS*, 406, 2650
Metzger, B. D., Quataert, E., & Thompson, T. A. 2008, *MNRAS*, 385, 1455
Metzger, B. D., Thompson, T. A., & Quataert, E. 2007, *ApJ*, 659, 561
Nakar, E. & Piran, T. 2011, *Nature*, 478, 82
Narayan, R., Paczynski, B., & Piran, T. 1992, *ApJ* (Letters), 395, L83
Narayan, R., Piran, T., & Kumar, P. 2001, *ApJ*, 557, 949
New, K. C. B., Centrella, J. M., & Tohline, J. E. 2000, *Phys. Rev. D*, 62, 064019
Nousek, J. A., *et al.* 2006, *ApJ*, 642, 389
Ofek, E. O., *et al.* 2008, *ApJ*, 681, 1464
O'Shaughnessy, R., Belczynski, K., & Kalogera, V. 2008, *ApJ*, 675, 566
Ott, C. D. 2009, *Classical and Quantum Gravity*, 26, 063001
Phinney, E. S. 1991, *ApJ* (Letters), 380, L17
Pian, E., *et al.* 2006, *Nature*, 442, 1011
Piran, T. 2004, *Reviews of Modern Physics*, 76, 1143
Piro, A. L. & Pfahl, E. 2007, *ApJ*, 658, 1173

Rau, A. *et al.* 2009, *PASP*, 121, 1334
Rees, M. J. & Mészáros, P. 1992, *MNRAS*, 258, 41P
Rhoads, J. E. 1999, *ApJ*, 525, 737
Rosswog, S. 2005, *ApJ*, 634, 1202
Rosswog, S., Liebendörfer, M., Thielemann, F.-K., Davies, M. B., Benz, W., & Piran, T. 1999, *A&A*, 341, 499
Rowlinson, A., *et al.* 2010, *MNRAS*, 409, 531
Ruffert, M. & Janka, H.-T. 1999, *A&A*, 344, 573
Sari, R. 1997, *ApJ* (Letters), 489, L37
Sari, R. & Esin, A. A. 2001, *ApJ*, 548, 787
Sari, R., Piran, T., & Halpern, J. P. 1999, *ApJ* (Letters), 519, L17
Sari, R., Piran, T., & Narayan, R. 1998, *ApJ* (Letters), 497, L17
Shibata, M. & Taniguchi, K. 2011, *Living Reviews in Relativity*, 14, 6
Soderberg, A. M., et. al. 2010, *Nature*, 463, 513
Sylvestre, J. 2003, *ApJ*, 591, 1152
Thompson, C. 1994, *MNRAS*, 270, 480
Thrane, E., *et al.* 2011, *Phys. Rev. D*, 83, 083004
Toma, K., Wu, X.-F., & Mészáros, P. 2011, *MNRAS*, 415, 1663
Usov, V. V. 1992, *Nature*, 357, 472
van Eerten, H. J. & MacFadyen, A. I. 2011, *ApJ* (Letters), 733, L37
van Putten, M. H. P. M. 2001, *ApJ* (Letters), 562, L51
van Putten, M. H. P. M. 2002, *ApJ* (Letters), 575, L71
Villasenor, J. S., *et al.* 2005, *Nature*, 437, 855
Woosley, S. E. 1993, *ApJ*, 405, 273
Woosley, S. E. & Bloom, J. S. 2006, *ARA&A*, 44, 507
Xu, M., Huang, Y., & Lu, T. 2009, *Research in Astronomy and Astrophysics*, 9, 1317
Yu, Y. & Huang, Y. 2007, *Chinese Journal of Astronomy and Astrophysics*, 7, 669
Zhang, B., Fan, Y. Z., Dyks, J., Kobayashi, S., Mészáros, P., Burrows, D. N., Nousek, J. A., & Gehrels, N. 2006, *ApJ*, 642, 354
Zhang, B. & Mészáros, P. 2001, *ApJ* (Letters), 552, L35
Zhang, B. & Mészáros, P. 2004, *International Journal of Modern Physics A*, 19, 2385
Zhang, B., *et al.* 2009, *ApJ*, 703, 1696
Zhang, B.-B., *et al.* 2011, *ApJ*, 730, 141
Zhuge, X., Centrella, J. M., & McMillan, S. L. W. 1994, *Phys. Rev. D*, 50, 6247

Discussion

ASTRAATMADJA: Is the angular resolution of the GW detectors good enough to search for an electromagnetic counterpart? Do you also intend to look for neutrino signals?

CORSI: The error-area for triple coincidence GW events from the LIGO-Virgo network is ~ 100 deg^2, much bigger than e.g. the $\approx 2''$ FWHM of a telescope like the Palomar 48-inch (Rau *et al.* 2009). While a large number of optical transients is expected to be found in the GW error-area, the problem can be mitigated by selecting only the most promising for a GW detection (in nearby galaxies and likely of "exotic", rare type). Joint searches for GWs and high energy neutrinos (though, currently, not specifically within the LOOC-UP experiment) are indeed being performed (see e.g., Bouhou *et al.* 2012, and references therein).

METZGER: In the magnetar scenario proposed for explaining GRB plateaus, can sufficiently rapid rotation be maintained in the presence of enhanced early spin-down by neutrino emission?

CORSI: Sufficiently high rotation should be maintained to explain the observed plateaus: typically, a $(1-5)$ ms magnetar with $B \sim (1-10)\times 10^{14}$ G is required from modeling of the X-ray light curves with plateaus (e.g., Zhang & Mészáros 2001; Yu & Huang 2007; Xu *et al.* 2009; Dall'Osso *et al.* 2011). As you have shown (Metzger *et al.* 2007), enhanced spin-down by neutrino emission at earlier timescales may be an issue, but likely only for the shortest periods and highest magnetic fields in these ranges.

Death of Massive Stars: Supernovae and Gamma-Ray Bursts
Proceedings IAU Symposium No. 279, 2012
P. Roming, N. Kawai & E. Pian, eds.

doi:10.1017/S1743921312012847

3D Core-Collapse Supernova Simulations: Neutron Star Kicks and Nickel Distribution

Annop Wongwathanarat, Hans-Thomas Janka, and Ewald Müller

Max-Planck Institut für Astrophysik, Karl-Schwarzschild-Straße 1, D-85748 Garching, Germany
email: annop@mpa-garching.mpg.de

Abstract. We perform a set of neutrino-driven core-collapse supernova (CCSN) simulations studying the hydrodynamical neutron star kick mechanism in three-dimensions. Our simulations produce neutron star (NS) kick velocities in a range between ∼100-600 km/s resulting mainly from the anisotropic gravitational tug by the asymmetric mass distribution behind the supernova shock. This stochastic kick mechanism suggests that a NS kick velocity of more than 1000 km/s may as well be possible. An enhanced production of heavy elements in the direction roughly opposite to the NS recoil direction is also observed as a result of the asymmetric explosion. This large scale asymmetry might be detectable and can be used to constrain the NS kick mechanism.

Keywords. (stars:) supernovae: general, (stars:) pulsars: general

1. Introduction

Observations of young pulsars show that they possess space velocities approximately 250-400 km/s on average (Hobbs *et al.* 2005; Faucher-Giguère & Kaspi 2006). Even velocities beyond 1000 km/s, which are much larger than the velocity of SN progenitor stars, are observed (Chatterjee *et al.* 2005). Therefore, two possibilities exist for the mechanism to produce these high NS velocities. Either the NS acquires its velocity at the time of its birth during the SN explosion or it gradually builds up its velocity at a later time (see, *e.g.*, Lai *et al.* 2001 for review). We aim to investigate one of the proposed mechanisms, namely the hydrodynamically driven kick mechanism, to answer the question whether or not this mechanism can produce large NS recoil velocities. While simulations in two spatial dimensions have been performed by Scheck *et al.* (2004,2006) and Nordhaus *et al.* (2010,2011), Wongwathanarat *et al.* (2010b) performed a small set of 3D simulations for the first time. Here, we extend our previous study by considering a larger set of parameters.

2. Numerical methods

All simulations are performed with the explicit finite-volume Eulerian multi-fluid hydrodynamics code PROMETHEUS (Fryxell *et al.* 1991; Müller *et al.* 1991a,b). The multi-dimensional hydrodynamic equations are solved in spherical polar coordinates using the dimensional splitting method of Strang (1968), the piecewise parabolic reconstruction scheme (Colella & Woodward 1984), and a Riemann solver for real gases (Colella & Glaz 1985). The AUSM+ fluxes of Liou (1996) are used to replace the fluxes computed by the Riemann solver inside grid cells with strong grid-aligned shocks to prevent odd-even decoupling (Quirk 1994). Advection of nuclear species is treated by the Consistent Multi-fluid Advection (CMA) scheme as described in Plewa & Müller (1999).

2.1. *Initial models and additional physics*

We considered three different non-rotating SN progenitor models: a $15\,M_{\odot}$ progenitor, s15s7b2 of Woosley & Weaver (1995), a $15\,M_{\odot}$ star evolved by Limongi *et al.* (2000), and a $20\,M_{\odot}$ progenitor star for SN1987A developed by Shigeyama & Nomoto (1990). In the following, these models are called W15, L15, and N20, respectively. These progenitors are evolved through collapse and core bounce in one dimension using the PROMETHEUS-VERTEX code, which includes an energy-dependent neutrino transport solver (A. Marek and R. Buras, private communication). Our simulations then started from the 1D output data at approximately 15 ms after the core bounce. To break the spherical symmetry, random seed perturbations of 0.1% amplitude are imposed on the radial velocity field since our code otherwise preserves exact spherical symmetry of the initial state.

The SN explosion is initiated by imposing a suitable neutrino luminosity at the inner grid boundary. The volume inside the inner grid boundary contains the innermost $\sim 1.1\,M_{\odot}$ of the PNS, which are excised and replaced by a point mass situated at the coordinate origin. We apply a simplified grey neutrino transport using the "ray-by-ray" approach following Scheck *et al.* (2006). Newtonian self-gravity of the stellar fluid in 3D is taken into account by solving Poisson's equation in its integral form using the spherical harmonics expansion technique of Müller & Steinmetz (1995). The monopole term is corrected for general relativistic effects as described in Scheck *et al.* (2006) and Arcones *et al.* (2007). We use the subnuclear equation of state of Janka & Müller (1996), and include a small alpha-reaction nuclear network consisting of 13 alpha group nuclei (^{4}He to ^{56}Ni) and a tracer nucleus tracing neutron-rich material as described in Kifonidis *et al.* (2003).

2.2. *Domain discretization and grid setup*

It is well known that discretizing a sphere using the usual spherical polar (or latitude-longitude) grid leads to a severe timestep restriction due to the CFL condition in the polar regions. Therefore, to ease this problem, we employed the "Yin-Yang" overlapping grid technique developed by Kageyama & Sato (2004) for simulations in geophysical science. It consists of two geometrically identical grid patches which contain only the low-latitude part of a spherical polar grid. Thus, when compared to the spherical polar grid, a much larger timestep can be chosen (up to a factor of 40 larger for an angular resolution of 2°). Details regarding the numerical implementation of the Yin-Yang grid for astrophysical self-gravitating systems are given in Wongwathanarat *et al.* (2010a).

Our standard grid configuration consists of $400(r) \times 47(\theta) \times 137(\phi) \times 2$ grid cells corresponding to an angular resolution of 2°. Three simulations denoted with the suffix "lr" are conducted with a lower angular grid resolution of 5° ($400(r) \times 18(\theta) \times 54(\phi) \times 2$ grid cells). The radial grid resolution is kept constant at $\Delta r = 0.3$ km from the inner grid boundary out to a radius of approximately 100 km (depending on the model). Outside of this radius the radial grid is logarithmically spaced. The outer grid boundary R_{ob} is placed sufficiently far out (approximately at 18000 km) to prevent the SN shock wave from leaving the computational grid during the simulated epoch. Hydrostatic equilibrium is assumed at the inner, retracting radial grid boundary R_{ib}, while a free outflow boundary condition is employed at the outer one.

3. Results

We have calculated 15 models in 3D considering three different non-rotating progenitor stars. We varied the neutrino luminosities imposed at the inner grid boundary to obtain explosions with different energies and onset times. Because the development of

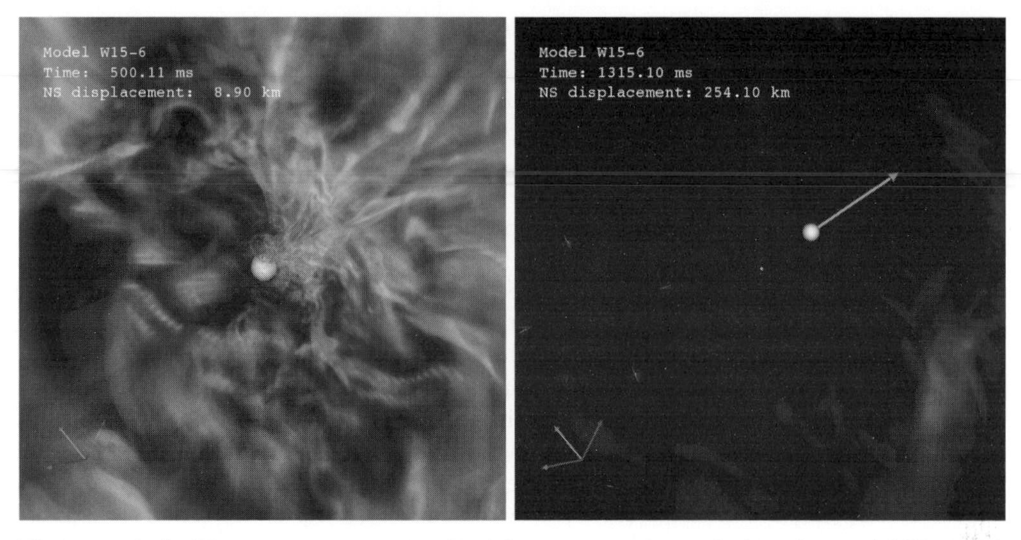

Figure 1. Left: Ejecta asymmetry visualized by a ray-casting technique for model W15-6. A large-scale ($\ell = 1$) asymmetry of the ejecta mass distribution resulting in a large NS kick velocity can clearly be seen shortly after the onset of the explosion. It exhibits the strongest amplitude roughly at $t = 500$ ms after bounce. As shown in the figure, denser clumps (red color) remain close to the NS (white ball) exerting a large gravitational pull towards the upper right direction. Right: The NS and its velocity vector (orange arrow) pointing to the upper right direction shown at the end of the simulation, $t = 1.3$ s after bounce. The displacement of the NS from the coordinate origin (white dot at the center of the figure) is obtained by performing a time-integration of the NS recoil velocity.

hydrodynamic instabilities in our simulations proceeds chaotically, changing only the initial seed perturbation pattern allows us to obtain explosions with morphologically different ejecta asymmetries. Tables summarizing explosion and NS properties obtained from these models can be found in Wongwathanarat *et al.* (2010b) and Wongwathanarat *et al.* (2012).

The NS recoil velocity, $\vec{v}_{ns}$, as a function of time t is calculated by assuming conservation of linear momentum, *i.e.*, $\vec{v}_{ns}(t) = -\vec{P}_{gas}(t)/M_{ns}(t)$, where $\vec{P}_{gas}$ and M_{ns} are the total linear momentum of the ejecta and the NS mass, respectively. We define the NS radius, R_{ns}, to be the radius where the gas density $\rho = 10^{11}$ g/cm^3. From our set of 15 models, we obtain a maximum NS kick velocity of 437 km/s (model W15-6) at 1.3 s after bounce, when the NS is still accelerating with 222 km/s^2. We expect the NS recoil velocity of this model to grow well beyond 600 km/s. This claim is supported by model W15-2 which has a slightly lower velocity at 1.3 s after bounce than model W15-6. Model W15-2 leads to a NS recoil velocity of 575 km/s at 3.3 s after bounce. Following the analysis of Scheck *et al.* (2006), we integrate various hydrodynamical forces acting on the NS to evaluate the contribution by each force to the NS acceleration and find that the gravitational tug is the dominant force. The tug force by asymmetrically distributed dense clumps lagging behind in the ejecta will act for a timescale of a few seconds. Figure 1 shows the large-scale ($\ell = 1$) asymmetry of the mass distribution behind the SN shock for model W15-6. This kick mechanism is different from that where the NS kick velocity originates from many small thrusts of downflow material pushing the NS in random directions (*e.g.*, Spruit & Phinney 1998).

The asymmetric explosion also leads to an asymmetry in the production of heavy elements such as ^{56}Ni. We observe that the SN shock's strength is largest in the direction roughly opposite to the recoil direction of the NS for models which exhibit strong NS recoils. The SN shock compresses the ejecta to higher temperature in this direction, leading to higher abundances of explosively produced heavy elements. We calculated the total mass of each element contained in each hemisphere using the NS recoil direction as the north pole. It turned out that a clear relative difference between the two hemispheres in the abundances of elements heavier than ^{28}Si can be expected. This relative difference can amount up to 50% in mass. This is different from the result obtained by Fryer & Kusenko (2006) who investigated NS kicks resulting from asymmetric neutrino emission, the so-called neutrino-driven kick. In that case, more heavy elements are predicted in the direction of the NS recoil. Such an asymmetry is likely to be preserved during the later evolution of the explosion, and if observed, it might provide a strong constraint to the NS kick mechanism.

References

Arcones, A., Janka, H., & Scheck, L. 2007, *A&A*, 467, 1227

Chatterjee, S., Vlemmings, W. H. T., Brisken, W. F., Lazio, T. J. W., Cordes, J. M., Goss, W. M., Thorsett, S. E., Fomalont, E. B., Lyne, A. G., & Kramer, M. 2005, *ApJ*, 630, L61

Colella, P. & Glaz, H. M. 1985, *J. Comput. Phys.*, 59, 264

Colella, P. & Woodward, P. R. 1984, *J. Comput. Phys.*, 54, 174

Faucher-Giguère, C. & Kaspi, V. M. 2006 *ApJ*, 643, 332

Fryer, C. L. & Kusenko, A. 2006, *ApJS*, 163, 335

Fryxell, B., Müller, E., & Arnett, D. 1991, *ApJ*, 367, 619

Hobbs, G., Lorimer, D. R., Lyne, A. G., & Kramer, M. 2005, *MNRAS*, 360, 974

Janka, H. & Müller, E. 1996, *A&A*, 306, 167

Kageyama, A. & Sato, T. 2004, *Geochemistry Geophysics Geosystems*, 5

Kifonidis, K., Plewa, T., Janka, H., & Müller, E. 2003, *A&A*, 408, 621

Lai, D., Chernoff, D. F., & Cordes, J. M. 2001, *ApJ*, 549, 1111

Limongi, M., Straniero, O., & Chieffi, A. 2000, *ApJS*, 129, 625

Liou, M.-S. 1996, *J. Comput. Phys.*, 129, 364

Müller, E., Fryxell, B., & Arnett, D. 1991a, *A&A*, 251, 505

Müller, E., Fryxell, B., & Arnett, D. 1991b, in *European Southern Observatory Conference and Workshop Proceedings*, edited by I. J. Danziger & K. Kjaer, vol. 37, 99

Müller, E. & Steinmetz, M. 1995, *Comput. Phys. Commun.*, 89, 45

Nordhaus, J., Brandt, T. D., Burrows, A., Livne, E., & Ott, C. D. 2010, *Phys. Rev. D*, 82, 103016

Nordhaus, J., Brandt, T. D., Burrows, A., & Almgren, A. 2011, arxiv:1112.3342

Plewa, T. & Müller, E. 1999, *A&A*, 342, 179

Quirk, J. J. 1994, *Int. J. Num. Meth. Fluids*, 18, 555

Scheck, L., Kifonidis, K., Janka, H., & Müller, E. 2006, *A&A*, 457, 963

Scheck, L., Plewa, T., Janka, H.-T., Kifonidis, K., & Müller, E. 2004, *Phys. Rev. Lett.*, 92, 011103

Shigeyama, T. & Nomoto, K. 1990, *ApJ*, 360, 242

Spruit, H. & Phinney, E. S. 1998, *Nature*, 393, 139

Strang, G. 1968, *SIAM J. Numer. Anal.*, 5, 506

Wongwathanarat, A., Hammer, N. J., & Müller, E. 2010a, *A&A*, 514, A48

Wongwathanarat, A., Janka, H., & Müller, E. 2010b, *ApJL*, 725, L106

Wongwathanarat, A., Janka, H., & Müller, E. 2012, to be submitted to *A&A*

Woosley, S. E. & Weaver, T. A. 1995, *ApJS*, 101, 181

Death of Massive Stars: Supernovae and Gamma-Ray Bursts
Proceedings IAU Symposium No. 279, 2012
P. Roming, N. Kawai & E. Pian, eds.

doi:10.1017/S1743921312012859

Core Collapse in Rotating Massive Stars and LGRBs

Aldo Batta

Instituto de Astronomía, Universidad Nacional Autónoma de México, Apdo. postal 70-264
Ciudad Universitaria, D.F., México
email: abatta@astro.unam.mx

Abstract. The collapse of massive rotating stellar cores and the associated accretion is thought to power long GRBs. The physical scale and dynamics of the accretion disk are initially set by the angular momentum distribution in the progenitor, and the physical conditions make neutrino emission the main cooling agent in the flow. We have carried out an initial set of calculations of the collapse of rotating polytropic cores in three dimensions, making use of a pseudo-relativistic potential and a simplified cooling prescription. We focus on the effects of self gravity and cooling on the overall morphology and evolution of the flow for a given rotation rate in the context of the collapsar model. For the typical cooling times expected in such a scenario we observe the appearance of strong instabilities on a time scale, t_{cool}, following disk formation. Such instabilities and their gravitational interaction with the black hole (BH) produce significant variability in the energy loss and accretion rates, which would translate into neutrino cooling variations when a more realistic neutrino cooling scheme is implemented in future work.

Keywords. Accretion disks, instabilities, hydrodynamics

1. Introduction

From GRB afterglow observations of the optical and X-ray band, it has been possible to locate the origin of these sources at cosmological distances. They also show that long GRBs can sometimes be associated with a core collapse SN (with no H lines, i.e. type Ib or Ic) taking place at the same time and at the same place. A review by Woosley & Bloom (2006) shows the existing evidence for the link of long GRB at low redshift with type Ic SNe, and the progenitor mechanisms currently explored.

In this work we will study the *collapsar model* proposed by Woosley (1993), to explain the formation of a GRB from a pre-supernova (PreSN) star whose shutdown of nuclear reactions at the central region induces the collapse of the core, and later on of the whole star. Eventually, the core will collapse into a BH that would be able to accrete the remaining infalling material from the star. With the high temperatures and densities the material reaches at the disk, neutrino emmision becomes the main cooling mechanism. These neutrinos may provide the energy to power the GRB, with the obtained neutrino luminosities (L_{ν}) from the collapsar, contributing to the production of a relativistic jet through neutrino annihilation.

Important works have been carried out in 2D and 3D (e.g. MacFadyen & Woosley 1999, and Taylor *et al.* 2010, respectively) for this collapsar model, but there is not a complete study on the importance of structure formation in the accretion flow and or heating/cooling mechanisms. This work focuses on the study of the effects of self gravity and cooling on 3D simulations of the collapse of a rotating polytropic envelope onto a BH in the context of the collapsar model. More specific details on the initial conditions and the input physics are given in the next section.

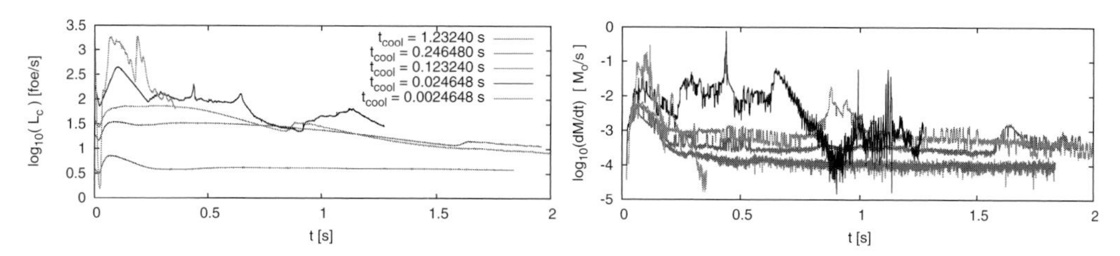

Figure 1. Logarithm of the energy loss rate L_c (left panel) in foes for models with $t_{cool} = 1.23$ s (red line), $t_{cool} = 0.24$ s (blue line), $t_{cool} = 0.12$ s (pink line), $t_{cool} = 0.02$ s (orange line) and $t_{cool} = 0.002$ s (gray line) and logarithm of the BH accretion mass rate $\dot{M}$ in solar masses per second (right panel).

2. Overview

For the densities and temperatures expected near the BH, one can estimate the neutrino cooling time scale in the gas. By considering an approximate EOS (internal energy u) composed by relativistic non-degenerated electron-positron pairs and an ideal gas, and a neutrino cooling prescription q_ν (Narayan *et al.* 2001), one can estimate the neutrino cooling time scale as follows:

$$t_\nu = u/q_\nu \qquad \text{where } u = \frac{3}{2}\frac{kT\rho}{\mu m_p} + \frac{11}{4}aT^4,\ q_\nu \simeq 5 \times 10^{33} T_{11}^9 + 9.0 \times 10^{23} \rho T_{11}^6 \tag{2.1}$$

This turns into a neutrino cooling time scale ranging from a few seconds to $\sim 10^{-4}$ s for $\rho = 10^{11}$ to 10^{12} g cm^{-3} and $T = 10^{10}$ to 10^{11} K.

With this t_ν timescale in mind, we study the importance of the neutrino cooling efficiency by means of a simplified cooling prescription with a characteristic cooling time (t_{cool}). Each fluid element is cooled according to

$$\frac{du}{dt} = -\frac{u}{t_{cool}} \tag{2.2}$$

where $t_{cool} = 1.23, 0.24, 0.12, 0.024,$ and 0.002 s are the cooling times explored in our simulations. For a fixed rigid body rotation rate (just below breakup) we studied the differences obtained by using each of these cooling times on the collapse and accretion of $2.5M_\odot$ polytropic envelopes onto a $2M_\odot$ BH. These envelopes where constructed by solving the Lane-Emdem hydrostatic equilibrium equation for a $\gamma = 5/3$ and $4.5M_\odot$ polytropic star with a radius of $R_s \simeq 1700$ km. All simulations were made using a personal version of the code GADGET-2 (Springel *et al.* 2001, Springel 2005).

Mass Accretion and Energy Loss Rates. By looking at the accretion and energy loss rates ($\dot{M}$ and L_c respectively) showed in Fig. 1 we can see that increasing cooling efficiency translates into more profuse and more intense variations in both L_c and $\dot{M}$. These strong variations seem to occur at the same time on both quantities, which implies they must be produced by the same phenomena. Given that by construction, strong variations on L_c are only possible if there is an increase in the internal energy u of the gas, there must be a contraction or shock at the disk material producing such increase in order to obtain intense variations in L_c. By looking after structure formation at the disk, we will be able to see if it is responsible for the intense variations on both L_c and $\dot{M}$.

Cooling Efficiency & Structure Formation. In order to obtain information about the formation of non-axisymmetric instabilities, we performed a Fourier transform of the

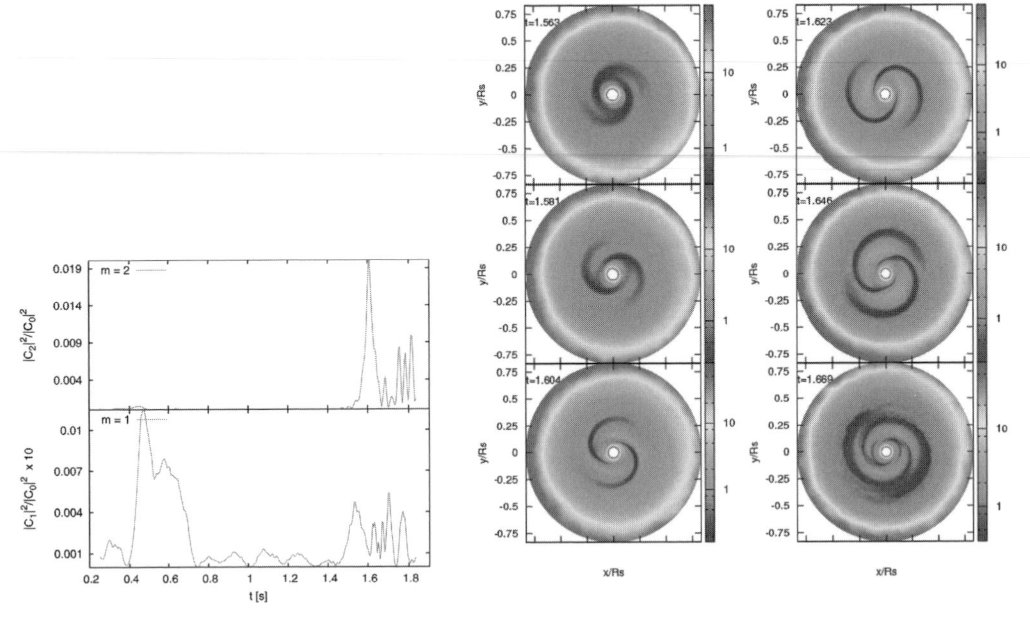

Figure 2. Relative power $|c_m|^2 = |C_m|^2/|C_0|^2$ ($m = 1, 2, 3, 4$) of the azimuthal distribution of mass (Φ_M) Fourier transform of the model with $t_{cool} = 0.24$ s (left panel) and evolution of the Toomre parameter Q at times $\simeq 1.6$ s (right panel). Deep blue regions have $Q < 1$. There is noticeable spiral structure forming at the same time strong variations on L_c and $\dot{M}$ appear.

azimuthal distribution of mass $\Phi_M = \int[\int \rho(\phi, r, z)dz]r\ dr$ (as in Zurek & Benz 1986) defining the amplitude of the mode m by

$$C_m = \frac{1}{2\pi}\int_0^{2\pi} e^{im\phi}\Phi_M\, d\phi \tag{2.3}$$

The relative power $|c_m|^2 = |C_m|^2/|C_0|^2$ indicates the intensity of m spiral arms compared to the disk integrated mass. If such spiral structures are present in the disk, they should be visible in density or internal energy maps; moreover, they should also be visible as unstable regions by plotting the Toomre parameter $Q_T = \kappa c_s/(\pi G\Sigma)$, determined by the superficial density Σ, the local sound speed c_s and the epicyclic frequency κ.

Figure 2 shows the relative power $|c_m|^2$ as a function of time (left panel) and the Toomre parameter evolution near $t = 1.6$ s (right panel) for the model with $t_{cool} = 0.24$ s. As shown in Fig 1, the strong variation on L_c and $\dot{M}$ is caused by the formation of two spiral arms shown as an intense increase in $|c_2|^2$ at $t \simeq 1.6$ s. Further increase of the cooling efficiency induces formation of gas clumps due to the collapse of highly unstable regions. These can be seen on Fig. 3, showing the relative power $|c_m|^2$ as a function of time (left panel) and the Toomre parameter evolution near $t = 0.4$ s (right panel) for the model with $t_{cool} = 0.024$ s. Intense structure formation episodes shown in the left panel, coincide in time with intense variations shown in both L_c and $\dot{M}$. Some of these variations are considerably narrower and are due to the close encounter of a gas clump with the BH, which produces an intense spiral arm of material falling onto the BH.

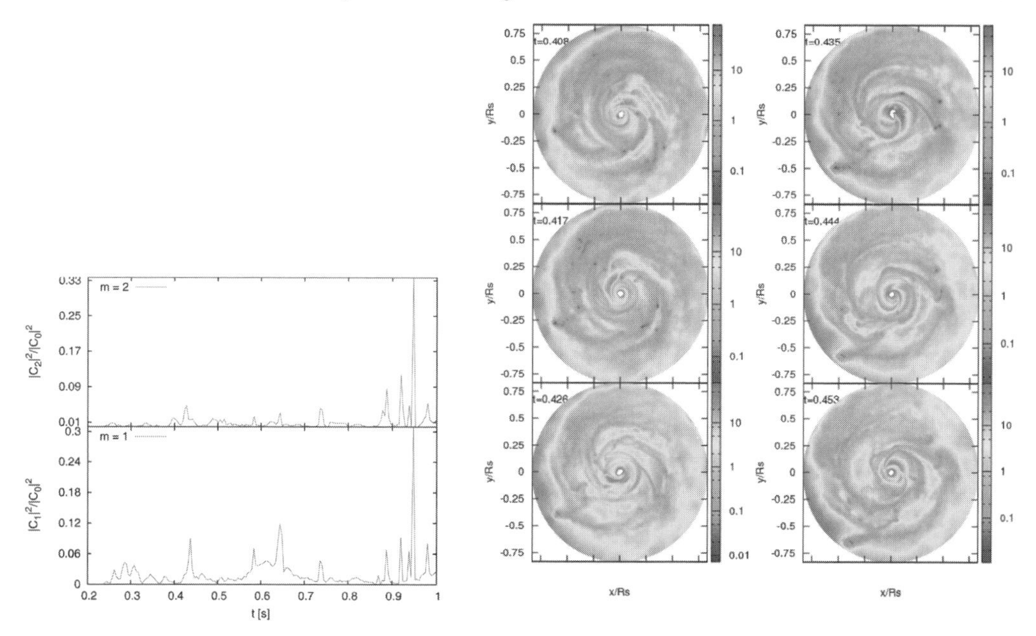

Figure 3. Relative power $|c_m|^2 = |C_m|^2/|C_0|^2$ $(m = 1, 2, 3, 4)$ for the azimuthal's distribution of mass (Φ_M) Fourier transform of the model with $t_{cool} = 0.024$ s (left panel) and evolution of the Toomre parameter Q at times $\simeq 0.4$ s (right panel). Deep blue regions have $Q < 0.1$. There is clump formation since early times and noticeable spiral structure forming at the same time strong variations on L_c and $\dot{M}$ appear.

3. Implications

As shown by our simulations, if neutrino cooling is efficient enough to induce the formation of structure at the accretion disk, the energy loss and accretion rates would have intense variations whose duration depends on the time those structures remain. Such intense and localized variations could change significantly the neutrino energy deposition on the infalling layers that have not reached the disk yet. Neutrino cooling efficiency (and therefore, structure formation) will depend on the initial rotation rate of the progenitor (Taylor *et al.* 2010). In the case of really collapsed and massive structures (such as gas clumps and intense spiral arms), the intrinsic symmetry break in the gravitational interaction with the BH could be intense enough to considerably move the BH from its original position. Allowing the BH to move could change its gravitational interaction with intense structures, and therefore, the accretion and energy loss rates obtained from the disk. Further work will explore such scenario.

References

MacFadyen, A. I. & Woosley, S. E. 1999, *ApJ* 524, 262
Narayan, R., Piran, T., & Kumar, P. 2001, *ApJ*, 557, 949
Springel, V., Yoshida, N., & White, S. D. M. 2001, *New Astronomy*, 6, 51
Springel, V. 2005, *MNRAS*, 364, 1105
Taylor, P. A., Miller, J. C., & Podsiadlowski, P. 2010, arxiv:1006.4624v1

Discussion

METZGER: What about the ultimate fate of the clumps? Do you expect to have any runaway cooling?

BATTA: Clumps that are not accreted by the BH or stay in orbits around the BH get scattered to outer regions, even though they may not get to far away given that they have to interact with the rest of the gas at the disk and the infalling material. Ultimately if the clump becomes very dense, cooling will be less efficient due to increased optical depth to neutrinos.

Death of Massive Stars: Supernovae and Gamma-Ray Bursts
Proceedings IAU Symposium No. 279, 2012
P. Roming, N. Kawai & E. Pian, eds.

doi:10.1017/S1743921312012860

SN 2010jp (PTF10aaxi): A Jet-driven Type II Supernova

Nathan Smith[1]†, S. Bradley Cenko[2], Nat Butler[2], Joshua S. Bloom[2], Mansi M. Kasliwal[3], Assaf Horesh[3], Shrinivas R. Kulkarni[3], Nicholas M. Law[4], Peter E. Nugent[2,5], Eran O. Ofek[3], Dovi Poznanski[2,5], Robert M. Quimby[3], Branimir Sesar[3], Sagi Ben-Ami[6], Iair Arcavi[6], Avishay Gal-Yam[6], David Polishook[6], Dong Xu[6], Ofer Yaron[6], Dale A. Frail[7], & Mark Sullivan[8]

[1]Steward Observatory, University of Arizona, 933 North Cherry Ave., Tucson, AZ 85721, USA

[2]Department of Astronomy, University of California, Berkeley, CA 94720-3411, USA

[3]Cahill Center for Astrophysics, California Institute of Technology, Pasadena, CA, 91125, USA

[4]Dunlap Institute for Astronomy and Astrophysics, University of Toronto, 50 St. George Street, Toronto M5S 3H4, Ontario, Canada

[5]Computational Cosmology Center, Lawrence Berkeley National Laboratory, 1 Cyclotron Road, Berkeley, CA 94720, USA

[6]The Weizmann Institute of Science, Rehovot 76100, Israel

[7]National Radio Astronomy Observatory, P.O. Box O, Socorro, NM 87801, USA

[8]Department of Physics, University of Oxford, Keble Road, Oxford, OX13RH, UK

Abstract. We present photometry and spectroscopy of the peculiar Type II supernova SN 2010jp, also named PTF10aaxi. The light curve exhibits a linear decline with a relatively low peak absolute magnitude of only −15.9 (unfiltered), and a low radioactive decay luminosity at late times that suggests a low synthesized nickel mass of about 0.003 $M_{\odot}$ or less. Spectra of SN 2010jp display an unprecedented *triple-peaked* Hα line profile, showing: (1) a narrow central component that suggests shock interaction with a dense circumstellar medium (CSM); (2) high-velocity blue and red emission features centered at −12,600 and +15,400 km s^{-1}; and (3) very broad wings extending from −22,000 to +25,000 km s^{-1}. We propose that this line profile indicates a bipolar jet-driven explosion, with the central component produced by normal SN ejecta and CSM interaction at mid and low latitudes, while the high-velocity bumps and broad line wings arise in a nonrelativistic bipolar jet. Jet-driven SNe II are predicted for collapsars resulting from a wide range of initial masses above 25 $M_{\odot}$, especially at the sub-solar metallicity consistent with the SN host environment. It also seems consistent with the apparently low ^{56}Ni mass that may accompany black hole formation. We speculate that the jet survives to produce observable signatures because the star's H envelope was very low mass, having been mostly stripped away by the previous eruptive mass loss.

Keywords. ISM: jets and outflows, supernovae: general

---

1. Introduction

Many theoretical studies of core-collapse suggest that breaking spherical symmetry may be an essential ingredient for producing a successful SN explosion (Blondin *et al.* 2003; Buras *et al.* 2006a, 2006b; Burrows *et al.* 2006, 2007). An extreme case of breaking spherical symmetry involves jet-driven explosions (Khokhlov *et al.* 1999; Höflich *et al.* 2001; Maeda & Nomoto 2003; Wheeler *et al.* 2000; Couch *et al.* 2009). Strongly collimated

† Email: nathans@as.arizona.edu

jets that expel the surrounding stellar envelopes may arise from accretion onto newly-formed black holes as in the "collapsar" model (MacFadyen & Woosley 1999; MacFadyen *et al.* 2001), or by magnetohydrodynamic (MHD) mechanisms in the collapse and spin-down of highly magnetized and rapidly rotating neutron stars, or magnetars (LeBlanc & Wilson 1970; Bodenheimer & Ostriker 1974; Wheeler *et al.* 2000; Thompson *et al.* 2004; Bucciantini *et al.* 2006; Burrows *et al.* 2007; Komissarov & Barkov 2007; Dessart *et al.* 2008; Metzger *et al.* 2010; Piro & Ott 2011). A jet-driven explosion may be important for producing successful SNe from high-mass stars above $\sim$25 $M_\odot$ (e.g., MacFadyen *et al.* 2001; Heger *et al.* 2003), which might otherwise collapse quietly to a black hole.

While there is strong evidence of collimated jets in broad-lined SNe Ic from the relativistic jets that produce observable gamma-ray bursts (GRBs) (Woosley & Bloom 2006; Galama *et al.* 1998; Matheson *et al.* 2003; Mazzali *et al.* 2005; see also Soderberg *et al.* 2009), there is to-date no substantial evidence for collimated jets in Type II supernovae. Jets are expected to not survive passage through the larger and more massive H envelope of a typical RSG (MacFadyen *et al.* 2001; Höflich *et al.* 2001), and this may help account for some of the time-dependent asymmetry inferred from polarization studies of SNe II-P (Leonard *et al.* 2001, 2006).

In this paper we discuss the recent peculiar Type II SN 2010jp (also called PTF10aaxi), which shows evidence for a fast (although non-relativistic) collimated jet that survived passage through its H envelope, perhaps because of significant previous mass loss. This

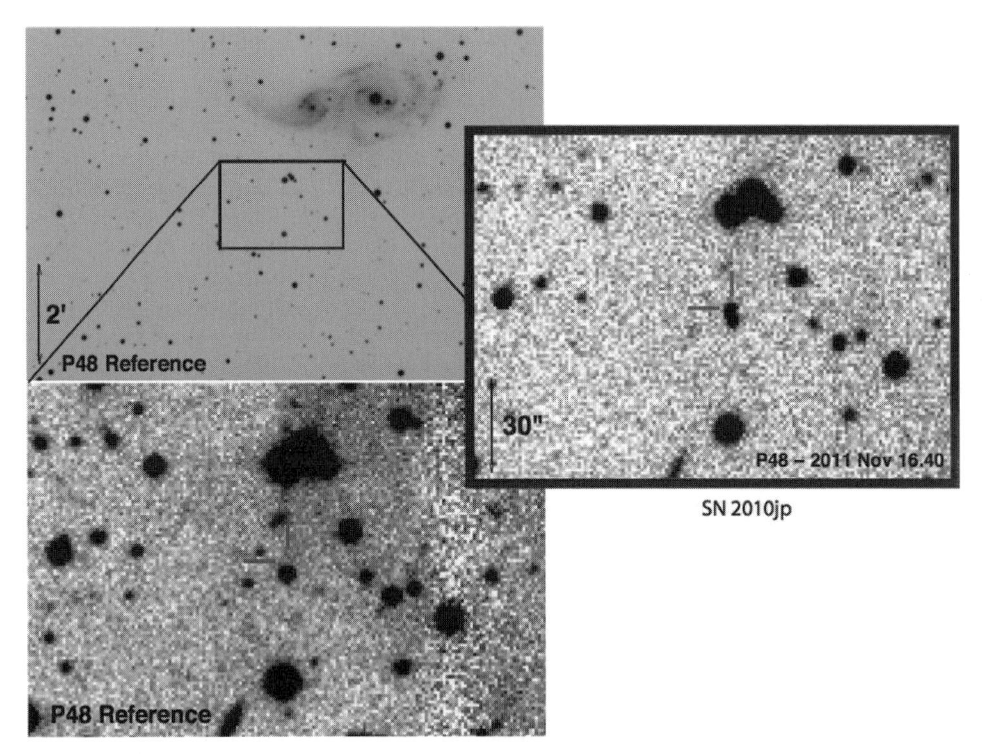

Figure 1. Finder chart for SN 2010jp (PTF10aaxi). *Top Left:* Wide-field P48 image of the SN field, including pre-discovery frames taken from 2009 November to 2010 March. The interacting galaxy pair IC 2163 / NGC 2207 is clearly visible about 2 arcminutes to the north. *Bottom Left:* Zoomed-in portion of the same P48 reference image of the location of SN 2010jp. *Right:* P48 image of SN2010jp. The SN position, indicated with the red tick marks, is a few arcseconds north of a foreground star, which appears blended with the P48 angular resolution.

talk gives only an overview of work that is now published in Smith *et al.* (2012); more detail about the analysis and additional references can be found there.

2. Observation Summary

Imaging/Host Galaxy. Images of the site of SN 2010jp obtained with the Palomar 48-inch (P48) telescope (R-band) are shown in Figure 1. This provides our first important clue about the nature of SN 2010jp: it occurs in a remote environment. There is nothing visible at the position of SN 2010jp in pre-explosion images, and upper limits there suggest a very faint dwarf galaxy at least 40 times fainter than the SMC. Alternatively, SN 2010jp may be in the far outer regions of the pair of interacting galaxies NGC 2207/IC 2163 (top panel of Fig. 1). NGC 2207 has a redshift of z=0.0091, which is the same as SN 2010jp, so if SN 2010jp is associated with these host galaxies, it is located 33 kpc out from their center. In either case, SN 2010jp probably occurred in a relatively low-metallicity environment that is comparable to the SMC or even more metal-poor.

Photometry. We analyzed photometry obtained with the Palomar 48-inch (P48) telescope (R-band), the Palomar 60-inch (P60) telescope (g, r, and i bands), and UV/optical photometry obtained with *Swift* (uvw2, uvm2, uvw1, u, b, and v bands). The P48 R-band light curve is shown in Fig. 2 (additional photometry is presented in Smith *et al.* 2012). Basically, the light curve of SN 2010jp looks like an underluminous light curve of a Type II-L or IIn supernova. Its peak absolute magnitude is about −15.9, reached about 20 days after discovery, from which it fades slowly until 110 days after discovery. After that point, it appears to fade faster (Fig. 2), although this late phase fading or a possible radioactive decay tail are not well constrained. Figure 2 shows a representative upper limit to the radioactive decay luminosity, corresponding to a ^{56}Ni mass of only 0.003 $M_{\odot}$. This indicates that SN 2010jp seems to have produced very little ^{56}Ni compared to normal core-collapse SNe.

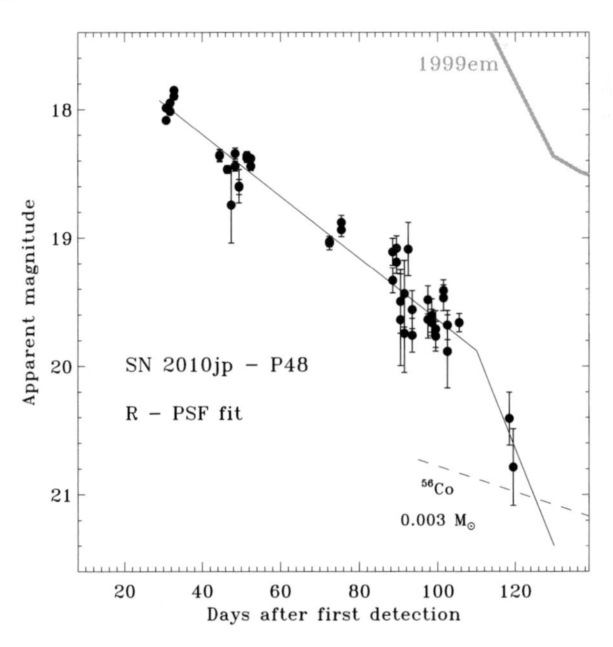

Figure 2. Light curve of SN 2010jp, plotted as days after the first detection. We show R-band magnitudes from the Palomar 48-in telescope, reduced using PSF-fitting photometry (see Smith *et al.* 2012).

Spectroscopy. We obtained a series of visual-wavelength spectra with the *MMT* (bluechannel spectrograph), the Palomar 5m (Double-beam spectrograph), and the Keck telescope (LRIS spectrograph). Our series of spectra are shown in Figure 3. The most striking aspect of the spectra in Figure 3 is the strong Hα line with an extremely unusual triple-peaked line profile, simultaneously showing extremely broad wings ($\pm$25,000 km s^{-1}) and a narrow core (few 10^2 km s^{-1}). A more detailed view of the Hα line is shown in Figure 4. This triple-peaked profile is the central aspect of the observations that point toward the presence of a collimated jet in SN 2010jp. Also noteworthy is the marked absence of any lines in the spectrum other than the H Balmer lines (i.e., no Ca II or Fe II lines that are usually seen in Type II SNe), which supports the conjecture that SN 2010jp arose in a low-metallicity environment.

Radio/X-ray. The site of SN 2010jp was observed at radio wavelengths with the EVLA and at X-rays using *Swift*. Details of the observations are presented in Smith *et al.* (2012). Both sets of observations yielded only non-detections at a significant level, which provide important arguments that the multiple peaks in the Hα profile in SN 2010jp were not associated with accretion onto a massive black hole.

3. Interpretation

More detail about the interpretation of SN 2010jp is given in Smith *et al.* (2012), but here we briefly list the essential components of the observations:

1. SN 2010jp occurred in a remote, low-metallicity environment.

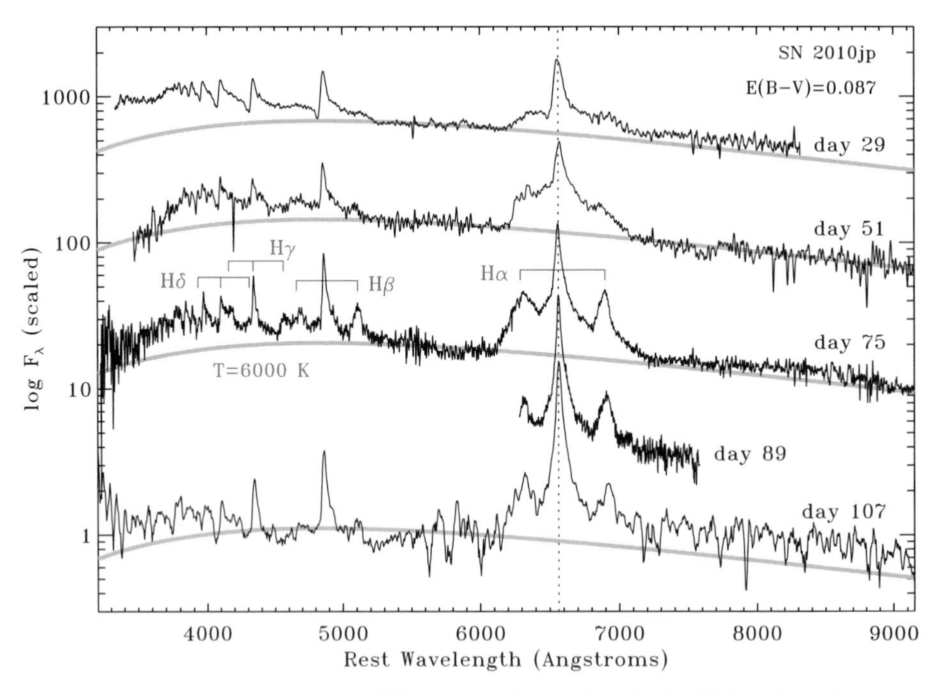

Figure 3. Visual-wavelength spectra of SN 2010jp obtained with the MMT, the Palomar 200-in telescope, and Keck, listed as days after discovery. The day 89 spectrum plotted in red is a high-resolution spectrum from the MMT. The blue and red high-velocity components of Hα, Hβ, Hγ, and Hδ are marked in orange brackets on the day 75 spectrum taken with Keck/LRIS.

2. A strange triple-peaked Hα profile (the triple peak is also seen in other lines of the Balmer series). On either side of the central narrow peak (see below), we see distinct blue and redshifted peaks at roughly ±15,000 km s^{-1}.

3. The central narrow peak has a width of about 800 km s^{-1}, resolved in our day 89 MMT spectrum. This probably corresponds to the speed of the CSM of the progenitor star, where the central part of the line profile arises from CSM interaction.

4. From the light curve, SN 2010jp was not particularly luminous for core-collapse SNe, with a peak absolute R magnitude of about −15.9 mag. This is on the faint tail of the luminosity function for SNe II-L and IIn (Li *et al.* 2011), which tend to have similarly shaped light curves (see additional figures in Smith *et al.* 2012). CSM interaction therefore does not provide a substantial luminosity boost for SN 2010jp. The progenitor star's mass-loss rate was probably not more than about 10^{-3} $M_\odot$ yr^{-1} (Smith *et al.* 2009).

5. From the late time luminosity, we find that SN 2010jp had a ^{56}Ni mass less than 0.003 $M_\odot$, which is ∼10 times less than a prototypical SN II-P like SN 1999em. This low ^{56}Ni mass could indicate either that the progenitor star had a relatively low initial mass near 8 $M_\odot$, or that it was underluminous because it lost much of the radioactive material into a black hole.

6. The non-detection of radio or X-ray emission argues against an interpretation where the blue and red peaks in the Hα line arise from accretion around a massive black hole, as

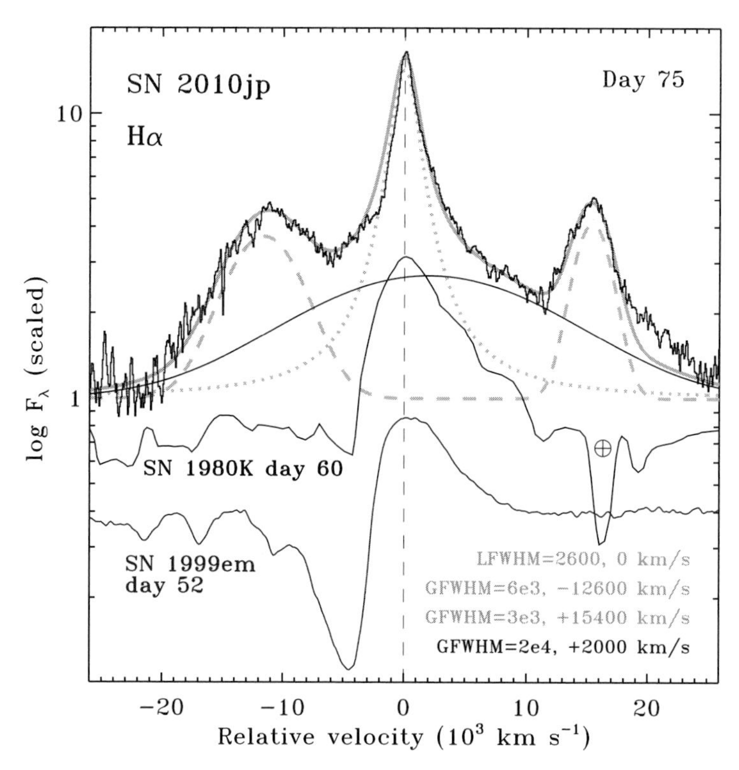

Figure 4. The Hα profile of SN 2010jp on day 75, decomposed into multiple contributing features (thin black and grey curves) with the sum of all individual components shown in orange. GFWHM and LFWHM denote Gaussian or Lorentzian FWHM values and centroid velocities. For comparison, we also show the Hα profile of the normal SN II-P 1999em from Leonard *et al.* (2002), plotted in red, as well as the Hα line in the Type II-L SN 1980K observed on day ∼60 (Barbieri *et al.* 1982) in blue.

in the class of double-peaked emitters among AGN (Halpern & Filippenko 1988). There are other arguments against this interpretation based on the time evolution of velocity, the requirements on black hole mass, and the remote environment (see Smith *et al.* 2012).

We speculate that the observed properties of SN 2010jp arises from a superposition of two physical scenarios. The first is a relatively traditional Type IIn explosion, where the rapidly expanding low-mass H envelope of the star collides with dense pre-existing CSM ejected recently by the progenitor. This can produce the blue continuum, the narrow emission cores of the Balmer lines, and some of the underlying broad emission profiles (see Smith *et al.* 2008). Most of the emitting volume corresponds to this component (regions 1, 2, and 3 in Figure 5). The second component is that SN 2010jp also produces a fast bipolar jet, tilted out of the plane of the sky, which gives rise to the fast blue and red emission features in Hα, and some emission in the very broad line wings. The combination of these two scenarios is depicted schematically in Figure 5. The two isolated red and blue emitting components must arise in a collimated geometry, as explained further in Smith *et al.* (2012).

The detection of a collimated jet in a Type II explosion is unprecedented. Models of jet-powered SNe have been published for fully stripped-envelope progenitors, which yield

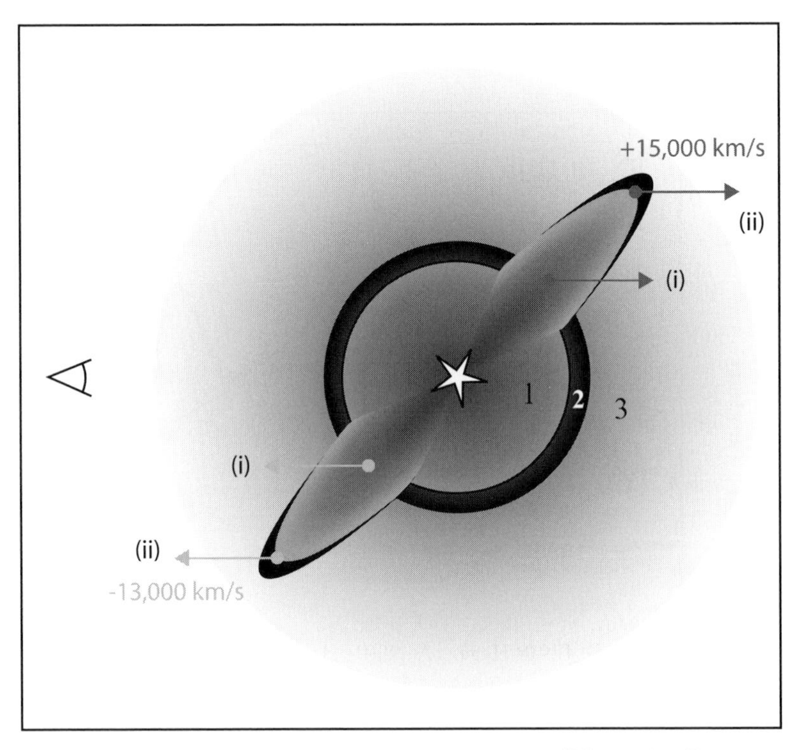

Figure 5. Cartoon of the possible jet-powered geometry in SN 2010jp. Regions 1, 2, and 3 correspond to the unshocked SN ejecta (inner gradient), the CSM interaction region (dark), and the pre-shock CSM (outer gradient), respectively. The upper-left and lower-right quadrants in this cartoon are the same as for a conventional SN IIn with CSM interaction, corresponding to low/mid-latitudes in the explosion. The lower-left and upper-right quadrants depict a tilted fast bipolar jet breaking through the otherwise spherical CSM interaction shell. An observer located to the left would see a combination of the spectrum from a conventional SN IIn, plus blue and red emission peaks in emission lines arising in either the unshocked jet material (i) or at the reverse shock in the jet (ii). We suggest that a bipolar jet such as this causes the blue and red bumps at $-13{,}000$ and $+15{,}000$ km s^{-1} observed in the Hα profile of SN 2010jp.

SNe Ibc and GRBs. Theoretical models also predict jet-driven SNe II for a wide range of initial masses exceeding 25 $M_\odot$ (e.g., Heger *et al.* 2003), from collapsars that yield black holes. These are expected to be more common at sub-solar metallicity (Heger *et al.* 2003), due to the expectation of weaker metallicity-dependent mass loss.

We consider two possible specific mechanisms that may power the emission in the fast blue and red peaks of the H Balmer series: (i) A jet-powered explosion might mix significant quantities of ^{56}Ni to high velocities in the polar regions of a thin H envelope. It is interesting that the Hα profile we observe bears a striking resemblance to the distribution of ^{56}Ni velocities in simulations of jet-powered SNe (Couch *et al.* 2009; see their Figure 12). (ii) Alternatively, the fast blue and red Hα bumps may be excited by CSM interaction. Fast H-bearing SN ejecta that cross the reverse shock of the jet will also yield two very fast but localized velocity components (Figure 5). Further study of additional examples of similar types of SNe, possibly including spectropolarimetry, might help solve this ambiguity.

References

Blondin, J. J. M., Mezzacappa, A., & DeMarino, C. 2003, *ApJ*, 584, 971

Bodenheimer, P. & Ostriker, J. P. 1974, *ApJ*, 191, 465

Bucciantini, N., Thompson, T. A., Aarons, J., Quataert, E., & Del Zanna, L. 2006, *MNRAS*, 368, 1717

Buras, R., Rampp, M., Janka, H. T., & Kifonidis, K. 2006a, *A&A*, 447, 1049

Buras, R., Janka, H. T., Rampp, M., & Kifonidis, K. 2006b, *A&A*, 457, 281

Burrows, A., Livne, E., Dessart, L., Ott, C. D., & Murphy, J. 2006, *ApJ*, 640, 878

Burrows, A., Dessart, L., Livne, E., Ott, C. D., & Murphy, J. 2007, *ApJ*, 664, 416

Couch, S. M., Wheeler, J. C., & Milosavljević, M. 2009, *ApJ*, 696, 953

Dessart, L., Burrows, A., Livne, E., & Ott, C. D. 2008, *ApJ*, 673, L43

Galama, T. J., *et al.* 1998, *Nature*, 395, 670

Halpern, J. P. & Filippenko, A. V. 1988, *Nature*, 331, 46

Heger, A., Fryer, C. L., Woosley, S. E., Langer, N., & Hartmann, D. H. 2003, *ApJ*, 591, 288

Höflich, P. A., Kholkov, A., & Wang, L. 2001, in *20th Texas Symp. on Relativistic Astroph.*, eds. J.C. Wheeler, & H. Martel (New York: AIP), 459

Khokhlov, .M., Höflich, P. A., Oran, E. S., Wheeler, J. C., Wang, L., & Chtchelkanova, A. Y. 1999, *ApJ*, 524, L107

Komissarov, S. S. & Barkov, M. V. 2007, *MNRAS*, 382, 1029

LeBlanc, J. M. & Wilson, J. R. 1970, ApJ 161, 541

Leonard, D. C., Filippenko, A. V., Ardila, D. R., & Brotherton, M. S. 2001, *ApJ*, 553, 861

Leonard, D. C., *et al.* 2006, *Nature*, 440, 505

Li, W., *et al.* 2011, *MNRAS*, 412, 1441

MacFadyen, A. I. & Woosley, S. E. 1999, *ApJ*, 524, 262

MacFadyen, A. I., Woosley, S. E., & Heger, A. 2001, *ApJ*, 550, 410

Maeda, K. & Nomoto, K. 2003, *ApJ*, 598, 1163

Matheson, T., *et al.* 2003, *ApJ*, 599, 394

Mazzali, P. A., *et al.* 2005, *Science*, 308, 1284

Metzger, B. D., Giannios, D., Thompson, T. A., Bucciantini, N., & Quataert, E. 2010, arXiv:1012.0001

Piro, A. L. & Ott, C. D. 2011, *ApJ*, 736, 108

Smith, N., Chornock, R., Li, W., Ganeshalingam, M., Silverman, J. S., Foley, R., Filippenko, A. V., & Barth, A. J. 2008a, *ApJ*, 686, 467

Smith, N., Hinkle, K. H., & Ryde, N. 2009, *AJ*, 137, 3558

Smith, N., *et al.* 2012, *MNRAS*, 420, 1135

Soderberg, A., *et al.* 2009, *Nature*, 463, 513

Thompson, T. A., Chang, P., & Quataert, E. 2004, *ApJ*, 611, 380

Wheeler, J. C., Yi, I., Höflich, P., & Wang, L. 2000, *ApJ*, 537, 810
Woosley, S. E. & Bloom, J. S. 2006, *ARAA*, 44, 507

Discussion

KUNCARAYAKTI: Are there any plans to study the host galaxy in more detail? (If it is an invisible dwarf galaxy.)

SMITH: Yes, we would like to do this, but it will be difficult and will require HST due to the relatively bright field star located only an arcsecond away from the SN position.

CROWTHER: What is a massive star doing so far from its likely host? Does the redshift match that of the spiral galaxy pair?

SMITH: Good question. Runaway? Tidal stream? eh. Anyway, the redshifts do match.

FOLATELLI: Have you considered a merger for your supernova SN 2010jp? That could explain the distance to any star forming region.

SMITH: Sure, a late merger may also be a possibility, giving angular momentum, etc., but we don't have any observational constraints that can test this hypothesis. Strong asymmetry in the CSM might be an interesting avenue to pursue, though. [Author's note: There was an interesting paper by R. Chevalier posted to the arXiv after this meeting, which is relevant here.]

MIRABEL: Comment: The massive progenitor of SN 2010jp may have been formed in the low metallicity debris of the colliding galaxies that form faint "tidal dwarf galaxies".

SMITH: I suppose this is possible; the red images of the SN 2010jp environment show some faint diffuse emission connecting back to the pair of interacting galaxies.

MAEDA: Do you have an idea about the rates of these kinds of objects? If you look at the phases earlier than the late phase with the triple-peaked H-alpha, are there any SNe IIn/II-L similar to this one.

SMITH: We have not seen anything similar to this before, so they are rare. I'm hesitant to quote a rate based on one relatively low-luminosity object in a rare/remote environment not sampled by the KAIT survey, but I'd guess substantially less than 1% of all core collapse SNe.

MAZZALI: Regarding a comment made by another audience member in the discussion (which was not recorded above) that the Hα profile could be produced by motion of the SN ejecta through an inhomogeneous shell: this comment does not apply here; if screening by a shell were the source of the Hα profile, the ejecta would need to be moving at 50,000 km s^{-1} or more.

Death of Massive Stars: Supernovae and Gamma-Ray Bursts
Proceedings IAU Symposium No. 279, 2012
P. Roming, N. Kawai & E. Pian, eds.

doi:10.1017/S1743921312012872

Host Galaxies of Gamma-Ray Bursts

Emily M. Levesque[1,2]

[1]CASA, Department of Astrophysical and Planetary Sciences, University of Colorado
389-UCB, Boulder, CO 80309, USA [2]Einstein Fellow
email: Emily.Levesque@colorado.edu

Abstract. Host galaxies are an excellent means of probing the natal environments that generate gamma-ray bursts (GRBs). Recent work on the host galaxies of short-duration GRBs has offered new insights into the parent stellar populations and ages of their enigmatic progenitors. Similarly, surveys of long-duration GRB (LGRB) host environments and their ISM properties have produced intriguing new results with important implications for long GRB progenitor models. These host studies are also critical in evaluating the utility of LGRBs as potential tracers of star formation and metallicity at high redshifts. I will summarize the latest research on LGRB host galaxies, and discuss the resulting impact on our understanding of these events' progenitors, energetics, and cosmological applications.

Keywords. gamma rays: bursts, galaxies: abundances, galaxies: starburst

1. Introduction

Long-duration gamma-ray bursts (LGRBs), associated with the core-collapse deaths of massive stars, are among the most energetic events observed in our universe. As a result, they are widely cited as powerful and potentially unbiased tracers of the star formation and metallicity history of the universe out to $z \sim 8$ (e.g. Bloom *et al.* 2002, Fynbo *et al.* 2007, Chary *et al.* 2007, Savaglio *et al.* 2009). However, in recent years potential biases in the star-forming galaxy population sampled by LGRBs have become a matter of debate. Recent work on a small number of nearby LGRBs suggested a connection between LGRBs and low-metallicity environments (e.g. Fruchter *et al.* 2006, Wainwright *et al.* 2007). Nearby host galaxies appeared to fall below the luminosity-metallicity and mass-metallicity relations for star-forming galaxies out to $z \sim 1$ (e.g. Modjaz *et al.* 2008, Kocevski *et al.* 2009, Levesque *et al.* 2010a,b). These results could potentially introduce key biases that would impact the use of LGRBs as cosmic probes.

A metallicity bias, or some correlation between metallicity and LGRB host or explosive properties, is indeed expected under the most commonly-cited progenitor model for LGRBs, the collapsar model (Woosley 1993). Under the classical assumptions of stellar evolutionary theory, the progenitor is a single rapidly-rotating massive star which maintains a high enough angular momentum over its lifetime to generate an LGRB from core-collapse to an accreting black hole. In addition, LGRBs have been observationally associated with broad-lined Type Ic supernovae (e.g. Galama *et al.* 1998, Stanek *et al.* 2003, Malesani *et al.* 2004, Modjaz *et al.* 2006, Starling *et al.* 2011), requiring the progenitors to have shed mass, and therefore angular momentum, as a means of stripping away their outer H and He shells. Mass loss rates for these evolved massive stars are dependent on stellar winds (Vink & de Koter 2005), which in turn are dependent on the stars' metallicity (Kudritzki 2002, Vink *et al.* 2001). For young massive stars, the metallicities of their natal environments can be adopted as the metallicities of the stars themselves.

It therefore stands to reason that the wind-driven mass loss rates in high-metallicity environments would rob the stars of too much angular momentum, preventing them from rotating rapidly enough to produce a LGRB and suggesting that LGRBs should either be restricted to low-metallicity environments (e.g. Hirschi *et al.* 2005, Yoon *et al.* 2006, Woosley & Heger 2006), or produce weaker explosions at higher metallicities (MacFadyen & Woosley 1999).

The work presented here originally aimed to observationally confirm and quantify the predicted role of metallicity in LGRB production and progenitor evolution. However, the results illustrate that the effects of metallicity on LGRBs are complex and do not agree with these expectations, suggesting that the predictions of stellar evolutionary theory and progenitor models may require further development.

2. The Mass-Metallicity Relation for LGRBs

In Levesque *et al.* (2010a,b) we conducted a uniform rest-frame optical spectroscopic survey of $z < 1$ LGRB host galaxies, using the Keck telescopes at Mauna Kea Observatory and the Magellan telescopes at Las Campanas Observatory. The sample was restricted to confirmed long-duration bursts with well-associated and observable host galaxies. From these spectra we were able to determine a number of key parameters for the star-forming LGRB host galaxies, including metallicity, ionization parameter, young stellar population age, SFR, and stellar mass. The primary metallicity diagnostic used in this work was the ([OIII] λ5007 + [OIII] λ4959 + [OII] λ3727)/Hβ (R_{23}) diagnostic (Kewley & Dopita 2002, Kobulnicky & Kewley 2004); for our full sample we found an average R_{23} metallicity of log(O/H) + 12 = 8.4 $\pm$ 0.3. Our stellar mass estimates were determined using the *Le Phare* code (Ilbert *et al.* 2009), fitting multi-band photometry for the host galaxies (Savaglio *et al.* 2009) to stellar population synthesis models adopting a Chabrier IMF, the Bruzual & Charlot synthetic stellar templates, and the Calzetti extinction law (Bruzual & Charlot 2003, Chabrier 2003, Calzetti *et al.* 2000). The fitting yielded a stellar mass probability distribution for each host galaxy, with the median of the distribution serving as our estimate of the final stellar mass. For our sample, we found a mean stellar mass of $\log(M_*/M_\odot) = 9.25^{+0.19}_{-0.23}$.

These metallicities and stellar masses were used to construct a mass-metallicity relation for LGRB host galaxies, which we plot in Figure 1. For comparison, we also compare our results to two samples of star-forming galaxies with comparable redshifts. The nearby ($z < 0.3$) LGRB hosts are compared to $\sim 53,000$ star-forming SDSS galaxies, while the intermediate-redshift ($0.3 < z < 1$) hosts are compared to 1,330 galaxies from the Deep Extragalactic Evolutionary Probe 2 (DEEP2) survey (Tremonti *et al.* 2004, Zahid *et al.* 2011). Surprisingly, we found a strong and statistically significant correlation between stellar mass and metallicity for LGRB hosts out to $z < 1$ (Pearson's $r = 0.80$, $p = 0.001$), with the relation showing no evidence for a clear metallicity cut-off above which LGRBs cannot be formed - instead, the overall LGRB mass-metallicity relation is offset from the mass-metallicity relation for star-forming galaxies by an average of -0.42 ± 0.18 dex in metallicity. The phenomenological explanation for this offset is unclear.

3. Energetics and Host Metallicity in LGRBs

Lacking observational evidence for a pure cut-off metallicity for LGRB formation, we instead consider the possibility that LGRBs at high metallicity may simply produce less energetic explosions; that is, explosions with a lower isotropic ($E_{\gamma,iso}$) or beaming-corrected ($E_\gamma = E_{\gamma,iso} \times 1 - cos(\theta_j)$ energy release in the gamma-ray regime, where

θ_j is the GRB jet opening angle) energy release (Frail *et al.* 2001). Previous studies of several local LGRBs suggested a strong correlation between these parameters, with high-metallicity LGRBs producing markedly-lower $E_{\gamma,iso}$ (Stanek *et al.* 2006). By combining energetic parameters available in the literature with our previously-determined host galaxy metallicities, we were able to reproduce this comparison (Levesque *et al.* 2010c; Figure 2). A comparison with redshift was considered as well, to highlight any

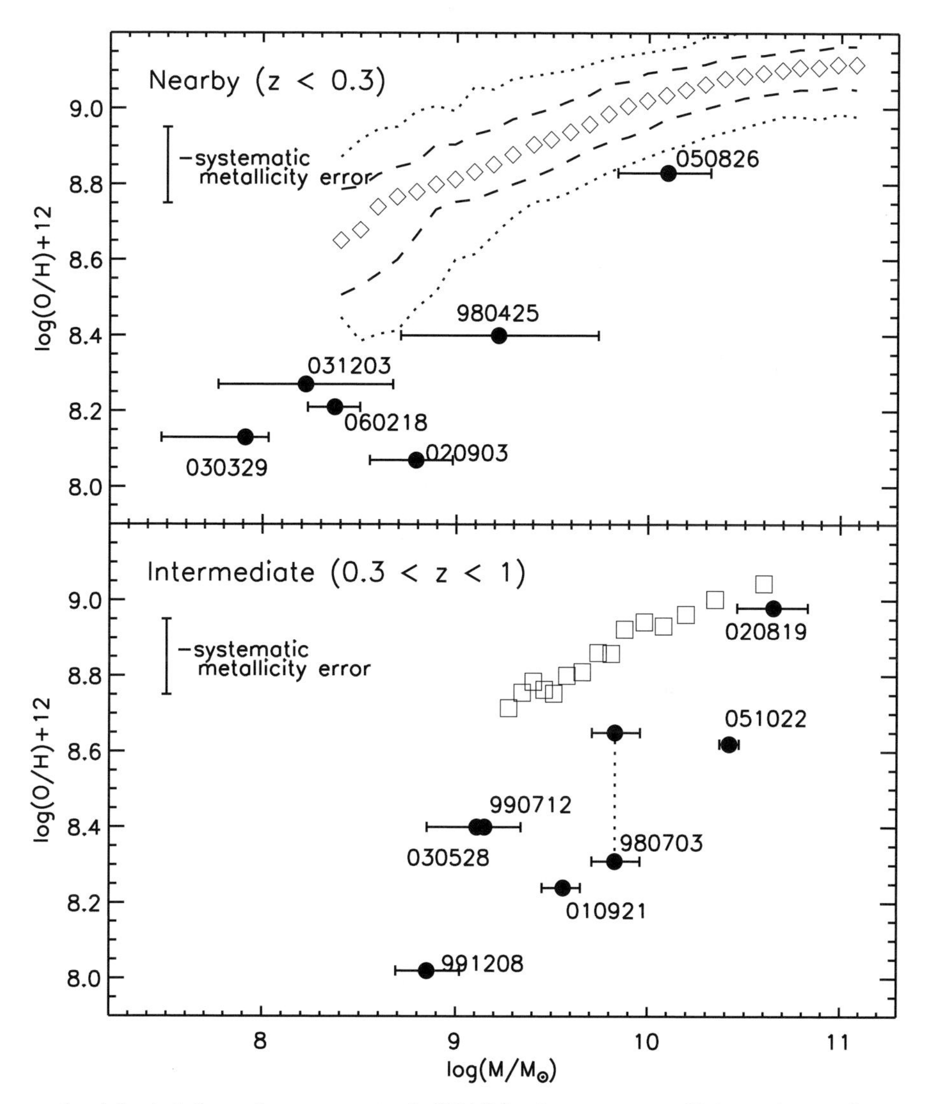

Figure 1. Adapted from Levesque *et al.* (2010b); the mass-metallicity relation for nearby ($z < 0.3$, top) and intermediate-redshift ($0.3 < z < 1$, bottom) LGRB host galaxies (filled circles). The nearby LGRB hosts are compared to binned mass-metallicity data for a sample of ~53,000 SDSS star-forming galaxies, where the open diamonds represent the median of each bin and the dashed/dotted lines show the contours that include 68%/95% of the data (Tremonti *et al.* 2004). For the intermediate-redshift hosts we plot binned mass-metallicity data for a sample of 1330 emission line galaxies from the DEEP2 survey (open squares; Zahid *et al.* 2011). For the $z = 0.966$ host galaxy of GRB 980703, where we cannot distinguish between the upper and lower metallicities given by the R_{23} diagnostic, we plot both metallicities and connect the data point with a dotted line to indicate their common origin from the same host spectrum.

potential correlation that may appear as an artifact of metallicity evolution with redshift. However, we found that there is no statistically significant correlation between metallicity and redshift, *or* between metallicity and $E_{\gamma,iso}$ or E_{γ}. This result is at odds with the previously-predicted and tentatively-observed inverse correlation, and appears to demonstrate that metallicity has no clear impact on the final explosive properties and gamma-ray energy release of an LGRB progenitor.

4. Spatially-Resolved Host Studies of LGRBs

It is important to note one strong limitation of current LGRB studies: their reliance on global metallicities. For the majority of LGRB hosts at $z \geqslant 0.3$, pinpointing the LGRB explosion site and acquiring site-specific spectra within the small, faint host galaxies is a difficult proposition. However, these site-specific studies *are* possible for a key sample of seven nearby spatially-resolved LGRB host galaxies. For this subset of hosts we can determine metallicities and star-formation rates directly at the LGRB host site as well as in the surrounding star-forming regions of the galaxy. This allows us to pinpoint the precise environments that produce LGRBs and place these sites in context with their global host galaxy properties. Despite the enormous value of such observations, only a small handful of spatially-resolved LGRB hosts have been previously studied. Christensen *et al.* (2008) obtained integral field unit spectroscopy of the $z = 0.008$ host galaxy of GRB 980425, determining metallicities at 23 different sites across the host. Thöne *et al.*

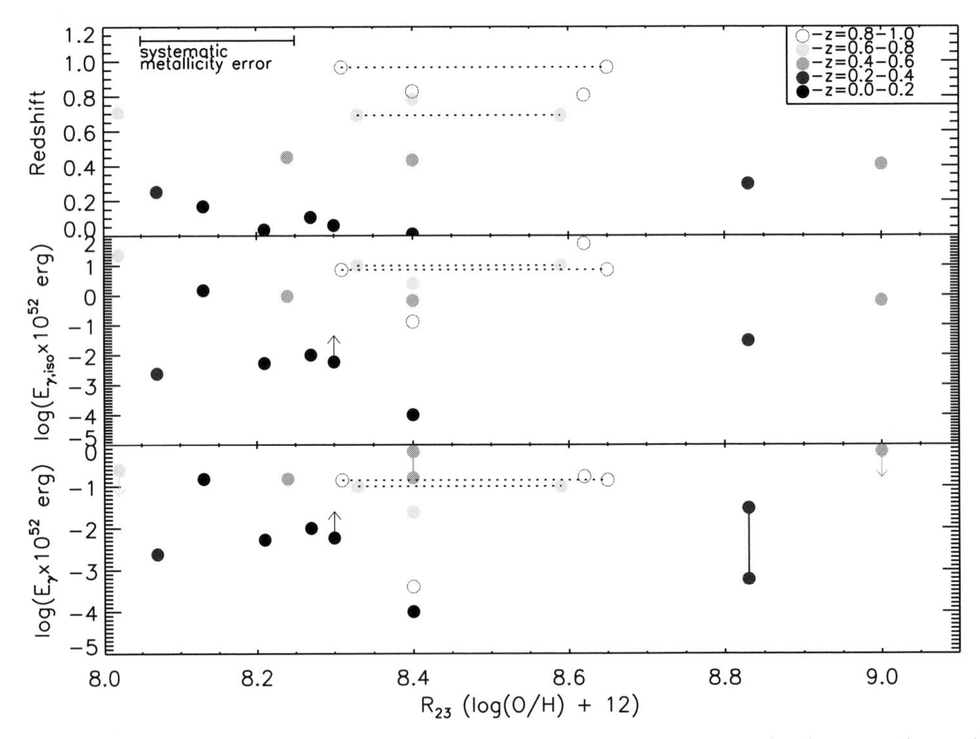

Figure 2. Adapted from Levesque *et al.* (2010c); metallicity vs. redshift (top), $E_{\gamma,iso}$ (center), and E_{γ} (bottom). The hosts are separated into redshift bins in order to better illustrate redshift effects. Two hosts with both lower- and upper-branchR_{23} metallicities (the hosts of GRB 020405 at $z = 0.691$ and GRB 980703 at $z = 0.966$) are shown as lower and upper data points connected by dotted lines. Upper and lower limits are indicated by arrows. Hosts with both upper *and* lower limits on their E_{γ} values are shown as data points connected by solid lines.

(2008) examined spatially-resolved ISM properties in the $z = 0.089$ host galaxy of GRB 060505. Levesque *et al.* (2010d) measured high metallicities at both the nucleus and GRB explosion site within the massive $z = 0.410$ host galaxy of GRB 020819B.

Most recently, Levesque *et al.* (2011) presented a detailed analysis of the GRB 100316D host environment at $z = 0.059$. By obtaining longslit spectra of the host complex at two different position angles using LDSS3 on Magellan, we were able to extract spatially-resolved profiles for a number of key diagnostic emission features, thus constructing metallicity and star formation rate profiles across the host that focused on both the specific LGRB explosion site and the diffuse emission of the host complex. Based on this analysis, we determined that GRB 100316D happened near the lowest-metallicity and most strong star-forming region of the host complex. However, this work also revealed only a very weak metallicity gradient within the host complex. Combined with the previous studies of nearby LGRB hosts, we found that, within this small sample, "host" or "global" metallicities were comparable to metallicities at the GRB explosion sites (Figure 3), suggesting that global metallicities may indeed be valid proxies for explosion site metallicities in higher-redshift LGRB host galaxy studies. Expanding this work to the remaining resolved LGRB hosts (the hosts of GRBs 020903, 030329, and 060218) would allow us to further explore this interesting result.

5. What's Next?

Based on the work described above, the role of metallicity in LGRB production and progenitor evolution remains a mystery. LGRBs occur preferentially in low-metallicity environments but do not show any evidence of a cut-off metallicity above which LGRB production is suppressed. There is also no statistically significant correlation between

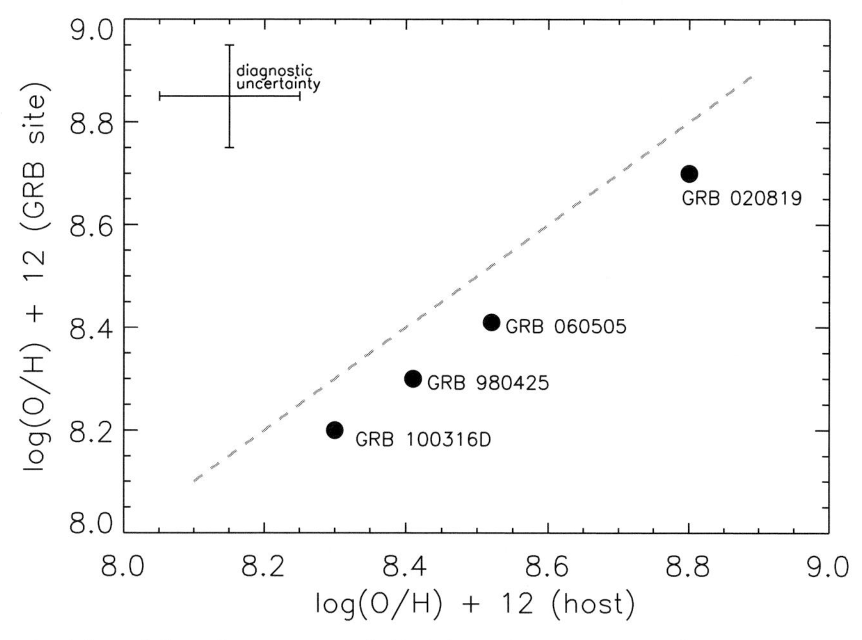

Figure 3. Adapted from Levesque *et al.* (2011); explosion site metallicities vs. average host metalicities for the current sample of previously-studied nearby LGRB host galaxies. All four explosion sites fall ~0.1 dex below the theoretical relation where explosion site metallicity and host metallicity are identical, plotted here as a gray dashed line, though this is within the uncertainty of the metallicity diagnostics.

the gamma-ray energy release of LGRBs and the metallicity of their host environments. Finally, it appears that these results cannot be attributed to effects of local metallicities within a globally-sampled host, given that several nearby LGRB hosts show evidence of minimal metallicity gradients and explosion site metallicities that are representative of the global environment. In light of these results, it is worth considering alternative models of LGRBs progenitors and stellar evolution, as well as new analytical means of examining metallicity effects in LGRBs. For example, it is possible that additional comparisons with other explosive properties of LGRBs, such as X-ray fluence or blastwave velocity, could still reveal a correlation with host metallicity.

Alternative progenitor scenarios, such as magnetars or binary channels, could also potentially agree with these observed results. Binary progenitor scenarios in particular are an intriguing possibility. One of the most common binary progenitor scenarios for LGRBs invokes a terminal common envelope phase where the outer envelope is ejected and the stellar cores coalesce. This manner of binary is predicted to occur at a higher rate - but *not* exclusively - in low-metallicity environments due to stellar wind effects, with weaker stellar winds permitting the evolution of binaries at closer proximities (Podsiadlowski *et al.* 2010). A second common progenitor model considers an interim common envelope phase, where the outer envelope is ejected, followed by a contact binary phase. This is also predicted to occur at a higher rate in low-Z environments, due to a widening range of Roche lobe radii that can permit a binary to enter and survive an interim common envelope phase while still maintaining Roche lobe overflow (Linden *et al.* 2010).

Finally, in addition to new progenitor scenarios, new treatments of stellar evolution with rotation are also compelling. Detailed treatments of differential rotation in massive stars have profound effects on the properties and populations of massive stars (Ekstrom *et al.* 2012, Levesque *et al.* 2012), and at low metallicities these effects are expected to be further enhanced (Leitherer 2008). Georgy *et al.* (2012) recently examined the effects of rotation on the production of evolved massive stars, supernovae, and LGRBs, using the new stellar rotation models of Ekstrom *et al.* (2012) at solar metallicity, and were able to produce favorable conditions for LGRB formation in 40-60$M_\odot$ stars at solar metallicity. Indeed, the latest stellar rotation models actually overproduce the predicted rate of LGRBs, although the introduction of additional parameters in the stellar interiors, such as strong coupling of the core to the stellar surface due to interior magnetic fields, could decrease this rate and bring predictions of the models into very good agreement with observations.

Collaborators on this work included Megan Bagley, Edo Berger, Ryan Chornock, Andrew Fruchter, John Graham, Lisa Kewley, and H. Jabran Zahid. The author is supported by NASA through Einstein Postdoctoral Fellowship grant number PF0-110075 awarded by the Chandra X-ray Center, which is operated by the Smithsonian Astrophysical Observatory for NASA under contract NAS8-03060.

References

Bloom, J. S., Kulkarni, S. R., & Djorgovski, S. G. 2002, *ApJ*, 121, 1111

Bruzual, G. & Charlot, S. 2003, *MNRAS*, 344, 1000

Calzetti, D., Armus, L., Bohlin, R. C., Kinney, A. L., Koornneef, J., & Storchi-Bergmann, T. 2000, *ApJ*, 533, 682

Chabrier, G. 2003, *PASP*, 115, 763

Chary, R., Berger, E., & Cowie, L. 2007, *ApJ*, 671, 272

Christensen, L., Vreeswijk, P. M., Sollerman, J., Thöne, C. C., Le Floc'h, E., & Wiersema, K. 2008, *A&A*, 490, 45

Ekström, S., *et al.* 2012, *A&A*, 537, 146
Frail, D. A., *et al.* 2001, *ApJ*, 562, 55
Fruchter, A. S. *et al.* 2006, *Nature*, 441, 463
Fynbo, J. P. U., Hjorth, J., Malesani, D., Sollerman, J., Watson, D., Jakobsson, P., Gorosabel, J., & Jaunsen, A. O. 2007, *arXiv:astro-ph/0703458v2*
Galama, T. J. *et al.* 1998, *Nature*, 395, 670
Georgy, C., Ekström, S., Meynet, G., Massey, P., Levesque, E. M., Hirshi, R., Eggenberger, P., & Maeder, A. 2012, *A&A*, in press (arXiv:1203.5243)
Hirschi, R., Meynet, G., & Maeder, A. 2005, *A&A*, 443, 581
Ilbert, O., *et al.* 2009, *ApJ*, 690, 1236
Kewley, L. J. & Dopita, M. A. 2002, *ApJS*, 142, 35
Kobulnicky, H. A. & Kewley, L. J. 2004, *ApJ*, 617, 24
Kocevski, D., West, A. A., & Modjaz, M. 2009, *ApJ*, 702, 377
Kudritzki, R. P. 2002, *ApJ*, 557, 389
Leitherer, C. 2008, *IAU Symp. 255, Low-Metallicity Star Formation: From the First Stars to Dwarf Galaxies*, ed. L. K. Hunt, S. Madden, & R. Schneider (Cambridge: CUP), 305
Levesque, E. M., Berger, E., Kewley, L. J., & Bagley, M. M. 2010a, *AJ*, 139, 694
Levesque, E. M., Kewley, L. J., Berger, E., & Jabran Zahid, H. 2010b, *AJ*, 140, 1557
Levesque, E. M., Soderberg, A. M., Kewley, L. J., & Berger, E. 2010e, *ApJ*, 725, 1337
Levesque, E. M., Kewley, L. J., Graham, J. F., & Fruchter, A. S. 2010c, *ApJ*, 712, L26
Levesque, E. M., Berger, E., Soderberg, A. M., & Chornock, R. 2011, *ApJ*, 739, 23
Levesque, E. M., Leitherer, C., Ekström, S., Meynet, G., & Schaerer, D. 2012, *ApJ*, in press (arXiv:1203.5109)
Linden, T., Kalogera, V., Sepinsky, J. F., Prestrwich, A., Zezas, A., & Gallagher, J. S. 2010, *ApJ*, 725, 1984
MacFadyen, A. I. & Woosley, S. E. 1999, *ApJ*, 524, 262
Malesani, D., *et al.* 2004, *ApJ*, 609, L5
Modjaz, M. *et al.* 2006, *ApJ*, 645, L21
Modjaz, M., *et al.* 2008, *AJ*, 135, 1136
Podsiadlowski, P., Ivanova, N., Justham, S., & Rappaport, S. 2010, *MNRAS*, 406, 840
Savaglio, S., Glazebrook, K., & Le Borgne, D. 2009, *ApJ*, 1091, 182
Stanek, K. Z. *et al.* 2003, *ApJ*, 591, L17
Stanek, K. Z., *et al.* 2006, *Acta Astron.*, 56, 333
Starling, R. L. C., *et al.* 2011, *MNRAS*, 411, 2792
Thöne, C. C., *et al.* 2008, *ApJ*, 676, 1151
Tremonti, C. A., *et al.* 2004, *ApJ*, 613, 898
Vink, J. S., de Koter, A., & Lamers, H. J. G. L. M. 2001, *A&A*, 369, 574
Vink, J. S. & de Koter, A. 2005, *A&A*, 442, 587
Wainwright, C., Berger, E., & Penprase, B. E. 2007, *ApJ*, 657, 367
Woosley, S. E. 1993, *ApJ*, 405, 273
Woosley, S. E. & Bloom, J. S. 2006, *ARA&A*, 44, 507
Yoon, S.-C., Langer, N., & Norman, C. 2006, *A&A*, 460, 199
Zahid, H. J., Kewley, L. J., & Bresolin, F. 2011, *ApJ*, 730, 137

Discussion

NOBUYUKI: The GRB host studies have a selection bias against optically dark GRBs. How does this bias affect your conclusion on GRB progenitor models?

LEVESQUE: It is true that GRB host studies are biased against optically dark GRBs, since with a handful of exceptions it is difficult to confirm their host associations. However, the implications of this work for the future of GRB progenitor modeling remain the same even without this sample. If anything, including this sample of dark GRBs further encourages the pursuit of alternative progenitor scenarios, since some studies suggest that dark GRBs are caused by the production of GRBs in dusty - and potentially higher-metallicity - environments.

KATZ: If restricted to high energies, can a maximum metallicity be ruled out?

LEVESQUE: Unfortunately, no. The two high-metallicity LGRB hosts in our sample are GRB 020819B and GRB 050826, which have energies on the order of 10^{50}-10^{52} erg and are consistent with the energies of other "cosmological" bursts, so there is no apparent maximum metallicity even if we only consider these higher-energy LGRBs.

VINK: Regarding alternatives for quasi-homogeneous evolution models: we have recently found a subset of rotating Galactic Wolf-Rayet stars from linear polarimetry (Vink *et al.* 2011, A&A Letters). Their surface velocities are only of order $\sim$100 km s^{-1} but their cores may rotate more rapidly if coupling due to B-fields is not so efficient.

LEVESQUE: This is a very interesting result, and highlights the importance of modeling stellar rotation and careful treatments of stellar interiors. A very interesting possibility is that LGRB progenitor atmospheres are *not* well-coupled to their cores. This would allow high mass loss and angular momentum rates at the stellar surface without removing angular momentum from the core, where a high rotation rate is critical for LGRB production. If LGRB progenitors are sufficiently decoupled in this manner, it would be a possible single-star mechanism for producing LGRBs at high metallicity. Georgy *et al.* (2012) examines this in more detail.

ZHANG: Without looking at the prompt emission properties, what fraction of short GRB hosts can be immediately identified based on the host galaxy information alone?

LEVESQUE: The nature of host galaxies can, *in some cases*, be used to identify whether a burst is short or long. If a GRB is observed in an elliptical galaxy with no star formation, we can safely conclude that it is a "short" GRB, or a GRB with a compact object progenitor, since LGRBs are restricted to actively star-forming galaxies with young massive star populations. However, the inverse is not true - if a GRB is observed in a star-forming galaxy, we cannot therefore conclude that it is a LGRB, since short GRBs have also been observed in star-forming host galaxies.

Death of Massive Stars: Supernovae and Gamma-Ray Bursts
Proceedings IAU Symposium No. 279, 2012
P. Roming, N. Kawai & E. Pian, eds.

doi:10.1017/S1743921312012884

The soft X-ray landscape of gamma-ray bursts: thermal components

Rhaana Starling[1*], Kim Page[1] and Martin Sparre[2,1]

[1]Department of Physics and Astronomy, University of Leicester,
University Road, Leicester LE1 7RH, UK
email: `rlcs1@le.ac.uk; klp5@le.ac.uk`

[2]Dark Cosmology Centre, Niels Bohr Institute, University of Copenhagen,
Juliane Maries Vej 30, 2100 Copenhagen Ø, Denmark
email: `sparre@dark-cosmology.dk`

[*]Royal Society Dorothy Hodgkin Fellow

Abstract. The repository of GRB (gamma-ray burst) observations made by the *Swift* X-ray Telescope, now consisting of over 650 bursts, is a valuable and unique resource for the study of GRB X-ray emission. The observed soft X-ray spectrum typically arises from an underlying power law continuum, absorbed by gas along the line-of-sight. However, particularly at early times in a burst's evolution the continuum emission is not always understood and may comprise multiple components including thermal emission unexpected in the standard model. A thermal X-ray component has been discovered in two very unusual GRBs, perhaps suggesting an association only with this subset of events. However, evidence exists for thermal emission from more typical examples and here we present a new discovery of one such case and describe a systematic search for thermal components among all early GRB X-ray spectra.

Keywords. gamma rays: bursts, X-rays: bursts

1. Introduction

The landmark discovery of thermal X-ray emission in the first few thousand seconds of the evolution of *Swift* GRB 060218 (Campana *et al.* 2006), followed four years later by a similar discovery in the spectra of GRB 100316D (Starling *et al.* 2011), were important steps in building a picture of the soft X-ray landscape of GRBs. However, both the origin and prevalence of such spectral components still evades understanding. GRBs 060218 and 100316D were classed as X-ray Flashes (XRFs), being softer and less energetic ($E_{\rm iso} \sim$ few $\times 10^{49}$ erg) when compared with their classical counterparts. They also had very long, $>$ 1000 s, burst durations with more slowly evolving X-ray light curves than is typical (see Fig. 1), and are among the closest known GRBs at $z = 0.033$ and 0.059 respectively. Their blackbody temperatures appeared to cool over time, while the emitting radii increased (Campana *et al.* 2006; Olivares *et al.* 2012 and this volume). It is not clear whether this unusual pair defines a seperate class of transient objects or forms a subenergetic tail to the classical long GRB population. Possibly most importantly, these two objects are firmly associated with supernovae (SN), evident in optical spectroscopy (SN2006aj, Campana *et al.* 2006; Pian *et al.* 2006; Mazzali *et al.* 2006, and SN2010bh, Starling *et al.* 2011; Bufano *et al.* 2012), begging the question of whether the thermal X-ray components could be a part of the GRB-SN connection. If this tenth-of-a-keV blackbody-like emission were to come from the emerging supernova itself, as the shock front breaks out of the star (e.g. Nakar, this volume; Nakar & Sari 2012), then GRB X-ray emission could provide the earliest glimpses of these distant explosions.

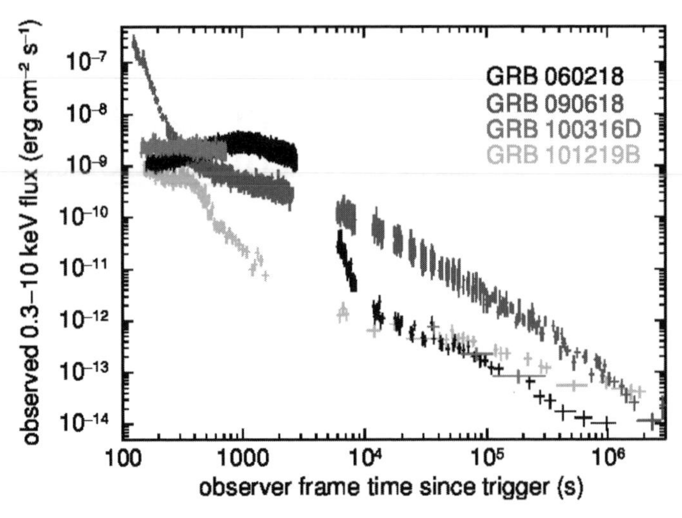

Figure 1. *Swift* X-ray light curves of the four GRBs reported to require thermal X-ray components in their early spectra: GRBs 060218 and 100316D show an initial flat evolution, GRB 090618 is initially the most X-ray bright and GRB 101219B (reported here) is initially the least X-ray bright and shows the slowest late-time decay.

X-ray emission from supernova shock breakouts had been predicted prior to the explosion of GRB 060218/SN2010bh, and was in fact observed serendipitously for SN2008D (Soderberg *et al.* 2008; Chevalier & Fransson 2008). Bromberg *et al.* (this volume) propose that these low luminosity events are in fact failed GRBs in which the jet could not break out of the star (Bromberg, Nakar & Piran 2011; Bromberg *et al.* 2012). Alternative explanations may lie in the central engine or cocoon emission.

The next development in this story came through a thorough analysis of GRB 090618. The early soft X-ray emission in this typical $z = 0.54$ GRB showed a more complex spectrum than could be fit with an absorbed power law. A thermal component was required, which had similar properties to those seen in the two earlier examples whilst being somewhat hotter (0.9 keV [restframe] cooling to 0.3 keV) and requiring larger radius and higher luminosity (see Page *et al.* 2011 for full details). This discovery suggested a link between the low luminosity events and classical long GRBs, although one should be careful in making this association before understanding of the origins of the thermal emission which could differ per GRB. Interestingly, GRB 090618 also has an associated SN, identified photometrically by Cano *et al.* (2011).

To investigate the extent of this phenomenon, we study the X-ray emission of SN-associated GRBs in detail, and have performed a systematic search for thermal X-ray components in all suitable GRBs from the large *Swift* X-ray Telescope (XRT) GRB Repository† (Evans *et al.* 2009).

2. GRB X-ray spectra

The basic shape The underlying GRB afterglow continuum in the 0.3–10 keV X-ray band takes the form of a power law or broken power law. This is expected in the fireball model (Sari 1997), where the interaction of the GRB jet with the surrounding medium results in a shock at which particles are accelerated producing synchrotron emission characterised by a set of power laws and breaks that move with time. The power law

† `www.swift.ac.uk/xrt_products/`

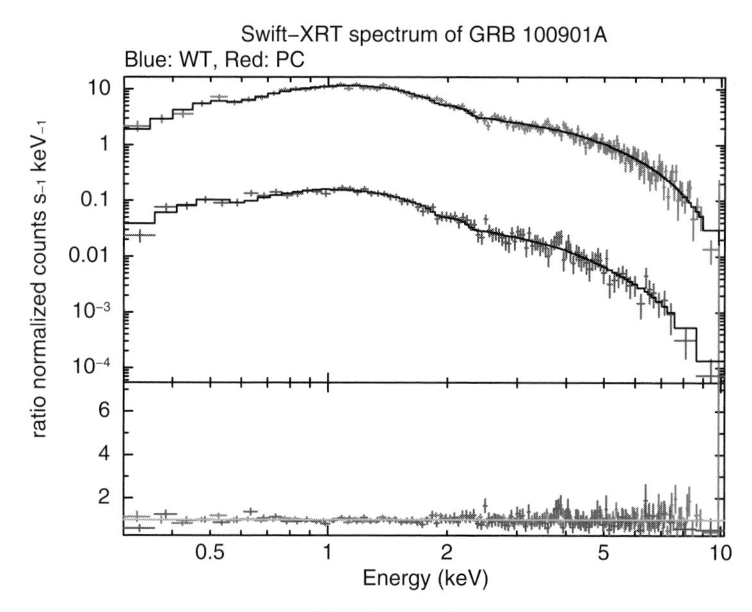

Figure 2. Example spectra from the *Swift* XRT GRB Repository. The early, Windowed Timing mode spectrum is shown in blue while the later, Photon Counting mode spectrum is shown in red. The lower panel shows the data to model ratio for an absorbed power law which is, in this case (GRB 100901A), an excellent fit. The curved profile of this spectrum is caused by a combination of instrumental response and absorption at soft X-ray energies by line-of-sight material.

is attenuated at soft X-ray energies (< 2 keV) by both Galactic and host galaxy absorbing gas, and potentially also by discrete intervening systems and/or the intergalactic medium. The total observed soft X-ray emission, 0.3–2 keV in *Swift* XRT data, could be a complex mix of evolving emission components and absorption components at varying redshifts, but in practice most GRB spectra can be adequately modelled with a single power law, Γ, absorbed by a fixed column of gas in our own galaxy, $N_{\rm H,Gal}$ as measured by Kalberla *et al.* (2005), plus an intrinsic column, $N_{\rm H,intrinsic}$, at the GRB redshift (Fig. 2).

The Swift XRT X-ray afterglow sample The GRB-dedicated *Swift* satellite (Gehrels *et al.* 2004) has been in operation since November 2004. The XRT (Burrows *et al.* 2005) takes data in two main operational modes, beginning in Windowed Timing (WT) mode for high count rates and automatically switching to Photon Counting (PC) mode as the GRB afterglow fades (Fig. 2). X-ray positions, light curves and spectra are among the automated science-grade products that are promptly available for each *Swift* XRT-detected GRB at the UK Swift Science Data Centre†, comprising a sample of over 650 GRBs (as of March 2012) suitable for statistical studies. Searching this database for excess soft X-ray emission above an absorbed power law which could be best fitted with a blackbody, we began a systematic search for further examples of thermal X-ray components in GRBs with known redshift (Sparre *et al.* in preparation).

† `www.swift.ac.uk`

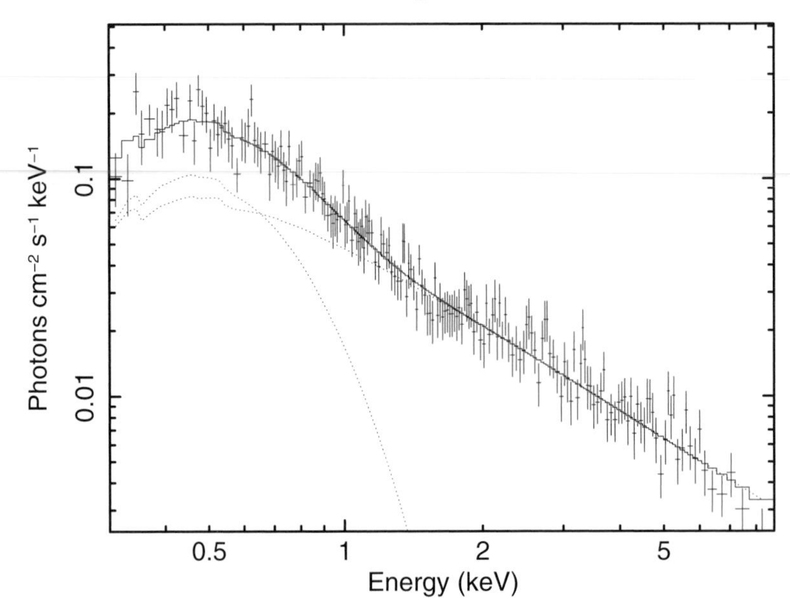

Figure 3. Average (160–540 s) *Swift* XRT WT mode spectrum of GRB 101219B. This spectrum is unfolded, meaning the instrumental response has been removed. The observed curvature here is largely due to absorption. The solid line shows the best fit absorbed BB+PL model, while the dashed lines show the conrtributions of the BB and PL respectively.

3. GRB 101219B: a new example of thermal X-ray emission

Our search revealed that a thermal X-ray component is required in GRB 101219B. We measure a temperature of 0.2 keV and corresponding luminosity 10^{47} erg s^{-1} for this blackbody; the unfolded X-ray spectrum is shown in Fig. 3. These properties are in the same ball-park as those measured for GRBs 060218, 100316D and 090618, whilst contributing the lowest fraction of the unabsorbed 0.3–10 keV flux (11% compared with 20% to $\geqslant$50%). This source lies at $z = 0.559$, comparable to GRB 090618, and has a spectroscopically identified SN (SN2010ma, Sparre *et al.* 2011). Its prompt emission properties cross the boundary between classical and low luminosity GRBs (e.g. $E_{\rm iso} \sim$ few $\times 10^{51}$ erg, see also Sakamoto, this volume), and the X-ray light curve can be compared with GRBs 060218, 100316D and 090618 in Fig. 1. This interesting discovery will be reported in detail in Starling *et al.* (in preparation).

4. Complicating factors and realistic limits on the recovery of similar blackbody components

In our search for further thermal X-ray components we take care to account for complexities which may affect spectral modelling, particularly at the earliest epochs and for the low luminosity events. These include tracking of spectral peak energy and flaring activity, and it is important to obtain as accurate as possible a measurement of absorbing column density and underlying continuum spectral shape. We fit the excess curvature of spectra with a single temperature perfect blackbody: this seems an simplistic way to represent this component but until we know its true origin we can only make this first approximation, which thus far fits the X-ray data well.

We have also performed monte carlo simulations to assess under which conditions a blackbody of the type discovered in GRB 101219B can be clearly recovered. Thus

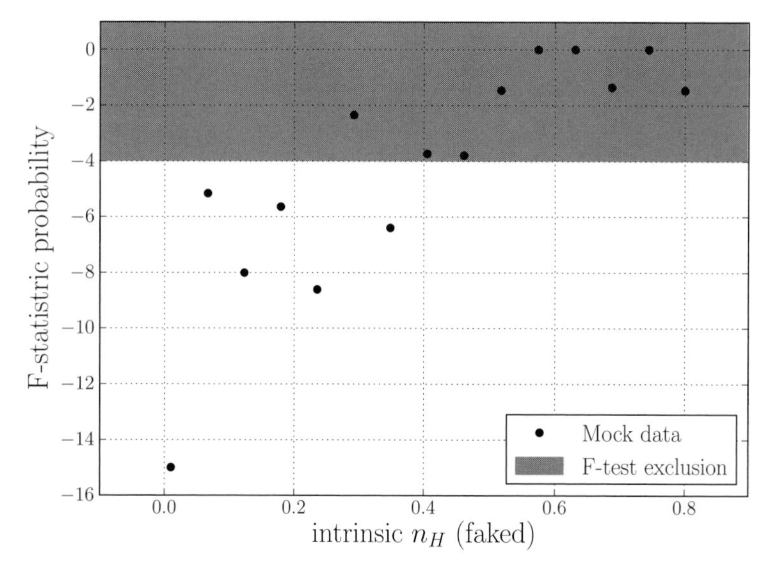

Figure 4. Simulated spectra reveal the conditions under which we might expect to recover a thermal component in *Swift* XRT data similar in nature to that found in GRB 101219B. In particular, a high intrinsic column density, $N_{\rm H,intrinsic} \geqslant 4 \times 10^{21}$ cm^{-2}, in GRB 101219B would have rendered us unable to unequivocally identify such a feature.

far the simulations are showing that we would be unable to identify such a component if the intrinsic column density were higher than around 4×10^{21} cm^{-2} (Fig. 4), while within a reasonable range the power law slope does not play a significant role in thermal component detectability. These results will appear in Sparre *et al.* (in preparation).

References

Bromberg, O., Nakar, E., & Piran, T, 2011, *ApJ* (Letters), 739, L55

Bromberg, O., Nakar, E., Piran, T., & Sari, R. 2012, *ApJ*, 749, 110

Bufano, F., Pian, E., Sollerman, J., Benetti, S., Pignata, G., Valenti, S., Covino, S., D'Avanzo, P., Malesani, D., Cappellaro, E., Della Valle, M., Fynbo, J., Hjorth, J., Mazzali, P. A., Reichart, D. E., Starling, R. L. C., Turatto, M., Vergani, S. D., Wiersema, K., Amati, L., Bersier, D., Campana, S., Cano, Z., Castro-Tirado, A. J., Chincarini, G., D'Elia, V., de Ugarte Postigo, A., Deng, J., Ferrero, P., Filippenko, A. V., Goldoni, P., Gorosabel, J., Greiner, J., Hammer, F., Jakobsson, P., Kaper, L., Kawabata, K. S., Klose, S., Levan, A. J., Maeda, K., Masetti, N., Milvang-Jensen, B., Mirabel, F. I., Moller, P., Nomoto, K., Palazzi, E., Piranomonte, S., Salvaterra, R., Stratta, G., Tagliaferri, G., Tanaka, M., & Wijers, R. A. M. J. 2012, *ApJ*, in press, arXiv:1111.4527

Burrows, D. N., Hill, J. E., Nousek, J. A., Kennea, J. A., Wells, A., Osborne, J. P., Abbey, A. F., Beardmore, A., Mukerjee, K., Short, A. D. T., Chincarini, G., Campana, S., Citterio, O., Moretti, A., Pagani, C., Tagliaferri, G., Giommi, P., Capalbi, M., Tamburelli, F., Angelini, L., Cusumano, G., Bräuninger, H. W., Burkert, W., & Hartner, G. D. 2005, *Space Sci. Revs*, 120, 165

Campana, S., Mangano, V., Blustin, A. J., Brown, P., Burrows, D. N., Chincarini, G., Cummings, J. R., Cusumano, G., Della Valle, M., Malesani, D., Mészáros, P., Nousek, J. A., Page, M., Sakamoto, T., Waxman, E., Zhang, B., Dai, Z. G., Gehrels, N., Immler, S., Marshall, F. E., Mason, K. O., Moretti, A., O'Brien, P. T., Osborne, J. P., Page, K. L., Romano, P., Roming, P. W. A., Tagliaferri, G., Cominsky, L. R., Giommi, P., Godet, O., Kennea, J. A., Krimm, H., Angelini, L., Barthelmy, S. D., Boyd, P. T., Palmer, D. M., Wells, A. A., & White, N. E. 2006, *Nature*, 442, 1008

Cano, Z., Bersier, D., Guidorzi, C., Margutti, R., Svensson, K. M., Kobayashi, S., Melandri, A., Wiersema, K., Pozanenko, A., van der Horst, A. J., Pooley, G. G., Fernandez-Soto, A., Castro-Tirado, A. J., Postigo, A. De Ugarte, Im, M., Kamble, A. P., Sahu, D., Alonso-Lorite, J., Anupama, G., Bibby, J. L., Burgdorf, M. J., Clay, N., Curran, P. A., Fatkhullin, T. A., Fruchter, A. S., Garnavich, P., Gomboc, A., Gorosabel, J., Graham, J. F., Gurugubelli, U., Haislip, J., Huang, K., Huxor, A., Ibrahimov, M., Jeon, Y., Jeon, Y.-B., Ivarsen, K., Kasen, D., Klunko, E., Kouveliotou, C., Lacluyze, A., Levan, A. J., Loznikov, V., Mazzali, P. A., Moskvitin, A. S., Mottram, C., Mundell, C. G., Nugent, P. E., Nysewander, M., O'Brien, P. T., Park, W.-K., Peris, V., Pian, E., Reichart, D., Rhoads, J. E., Rol, E., Rumyantsev, V., Scowcroft, V., Shakhovskoy, D., Small, E., Smith, R. J., Sokolov, V. V., Starling, R. L. C., Steele, I., Strom, R. G., Tanvir, N. R., Tsapras, Y., Urata, Y., Vaduvescu, O., Volnova, A., Volvach, A., Wijers, R. A. M. J., Woosley, S. E., & Young, D. R. 2011, *MNRAS*, 413, 669

Chevalier, R. A. & Fransson, C. 2008, *ApJ*, 683, L135

Evans, P. A., Beardmore, A. P., Page, K. L., Osborne, J. P., O'Brien, P. T., Willingale, R., Starling, R. L. C., Burrows, D. N., Godet, O., Vetere, L., Racusin, J., Goad, M. R., Wiersema, K., Angelini, L., Capalbi, M., Chincarini, G., Gehrels, N., Kennea, J. A., Margutti, R., Morris, D. C., Mountford, C. J., Pagani, C., Perri, M., Romano, P., & Tanvir, N. 2009, *MNRAS*, 397, 1177

Gehrels, N., Chincarini, G., Giommi, P., Mason, K. O., Nousek, J. A., Wells, A. A., White, N. E., Barthelmy, S. D., Burrows, D. N., Cominsky, L. R., Hurley, K. C., Marshall, F. E., Mszros, P., Roming, P. W. A., Angelini, L., Barbier, L. M., Belloni, T., Campana, S., Caraveo, P. A., Chester, M. M., Citterio, O., Cline, T. L., Cropper, M. S., Cummings, J. R., Dean, A. J., Feigelson, E. D., Fenimore, E. E., Frail, D. A., Fruchter, A. S., Garmire, G. P., Gendreau, K., Ghisellini, G., Greiner, J., Hill, J. E., Hunsberger, S. D., Krimm, H. A., Kulkarni, S. R., Kumar, P., Lebrun, F., Lloyd-Ronning, N. M., Markwardt, C. B., Mattson, B. J., Mushotzky, R. F., Norris, J. P., Osborne, J., Paczynski, B., Palmer, D. M., Park, H.-S., Parsons, A. M., Paul, J., Rees, M. J., Reynolds, C. S., Rhoads, J. E., Sasseen, T. P., Schaefer, B. E., Short, A. T., Smale, A. P., Smith, I. A., Stella, L., Tagliaferri, G., Takahashi, T., Tashiro, M., Townsley, L. K., Tueller, J., Turner, M. J. L., Vietri, M., Voges, W., Ward, M. J., Willingale, R., Zerbi, F. M., & Zhang, W. W. 2004, *ApJ*, 611, 1005

Kalberla, P. M. W., Burton, W. B., Hartmann, Dap, Arnal, E. M., Bajaja, E., Morras, R., & Pöppel, W. G. L. 2005, *A&A*, 440, 775

Mazzali, P. A., Deng, J., Nomoto, K., Sauer, D. N., Pian, E., Tominaga, N., Tanaka, M., Maeda, K., & Filippenko, A. V. 2006, *Nature*, 442, 1018

Nakar, E. & Sari, R. 2012, *ApJ*, 747, 88

Olivares E., F., Greiner, J., Schady, P., Rau, A., Klose, S., Krühler, T., Afonso, P. M. J., Updike, A. C., Nardini, M., Filgas, R., Nicuesa Guelbenzu, A., Clemens, C., Elliott, J., Kann, D. A., Rossi, A., & Sudilovsky, V. 2012, *A&A*, in press, arXiv:1110.4109

Page, K. L., Starling, R. L. C., Fitzpatrick, G., Pandey, S. B., Osborne, J. P., Schady, P., McBreen, S., Campana, S., Ukwatta, T. N., Pagani, C., Beardmore, A. P., & Evans, P. A. 2011, *MNRAS*, 416, 2078

Pian, E., Mazzali, P. A., Masetti, N., Ferrero, P., Klose, S., Palazzi, E., Ramirez-Ruiz, E., Woosley, S. E., Kouveliotou, C., Deng, J., Filippenko, A. V., Foley, R. J., Fynbo, J. P. U., Kann, D. A., Li, W., Hjorth, J., Nomoto, K., Patat, F., Sauer, D. N., Sollerman, J., Vreeswijk, P. M., Guenther, E. W., Levan, A., O'Brien, P., Tanvir, N. R., Wijers, R. A. M. J., Dumas, C., Hainaut, O., Wong, D. S., Baade, D., Wang, L., Amati, L., Cappellaro, E., Castro-Tirado, A. J., Ellison, S., Frontera, F., Fruchter, A. S., Greiner, J., Kawabata, K., Ledoux, C., Maeda, K., Møller, P., Nicastro, L., Rol, E., & Starling, R. 2006, *Nature*, 442, 1011

Sari, R. 1997, *ApJ* (Letters), 489, L37

Soderberg, A. M., Berger, E., Page, K. L., Schady, P., Parrent, J., Pooley, D., Wang, X.-Y., Ofek, E. O., Cucchiara, A., Rau, A., Waxman, E., Simon, J. D., Bock, D. C.-J., Milne, P. A., Page, M. J., Barentine, J. C., Barthelmy, S. D., Beardmore, A. P., Bietenholz, M. F., Brown, P., Burrows, A., Burrows, D. N., Byrngelson, G., Cenko, S. B., Chandra, P., Cummings, J. R., Fox, D. B., Gal-Yam, A., Gehrels, N., Immler, S., Kasliwal, M., Kong, A. K. H., Krimm, H. A., Kulkarni, S. R., Maccarone, T. J., Mészáros, P., Nakar, E., O'Brien,

P. T., Overzier, R. A., de Pasquale, M., Racusin, J., Rea, N., & York, D. G. 2008, *Nature*, 453, 469

Sparre, M., Sollerman, J., Fynbo, J. P. U., Malesani, D., Goldoni, P., de Ugarte Postigo, A., Covino, S., D'Elia, V., Flores, H., Hammer, F., Hjorth, J., Jakobsson, P., Kaper, L., Leloudas, G., Levan, A. J., Milvang-Jensen, B., Schulze, S., Tagliaferri, G., Tanvir, N. R., Watson, D. J., Wiersema, K., & Wijers, R. A. M. J. 2011, *ApJ* (Letters), 735, 24

Starling, R. L. C., Wiersema, K., Levan, A. J., Sakamoto, T., Bersier, D., Goldoni, P., Oates, S. R., Rowlinson, A., Campana, S., Sollerman, J., Tanvir, N. R., Malesani, D., Fynbo, J. P. U., Covino, S., D'Avanzo, P., O'Brien, P. T., Page, K. L., Osborne, J. P., Vergani, S. D., Barthelmy, S., Burrows, D. N., Cano, Z., Curran, P. A., de Pasquale, M., D'Elia, V., Evans, P. A., Flores, H., Fruchter, A. S., Garnavich, P., Gehrels, N., Gorosabel, J., Hjorth, J., Holland, S. T., van der Horst, A. J., Hurkett, C. P., Jakobsson, P., Kamble, A. P., Kouveliotou, C., Kuin, N. P. M., Kaper, L., Mazzali, P. A., Nugent, P. E., Pian, E., Stamatikos, M., Thöne, C. C., & Woosley, S. E. 2011, *MNRAS*, 411, 2792

Discussion

KATZ: Is a 0.14 keV thermal component robustly measured in the spectrum?

STARLING: Yes, I believe it can be. Of course you only see the high energy tail of it, and that itself is absorbed, but for example in GRB 060218 that 0.1–0.2 keV component completely dominated the soft X-ray flux for a time, and we get a consistent picture from the broadband data, seeing the thermal component cooling from $\sim$1–0.3 keV in 090618 and down into the UV/optical in 100316D and 060218. Even for the least prominent example the uncertainties on the BB temperature for example cannot be accounted for solely by say instrument calibration uncertainties.

CHORNOCK: In addition to the fitting uncertainties, I am concerned about the physical interpretation. For example, 090618 has a BB radius of 6×10^{12} cm. What is your physical interpretation? This is larger than the expected radii of the progenitors.

STARLING: I don't have a physical interpretation! Tomorrows' session will include discussion of shock breakout, proposed by some for 060218. The radius measured for that source ($\sim$ few $\times$ 10^{11} cm) by Campana *et al.* was, they said, consistent with the radius of a BSG or possibly a WR star with a thick wind. I agree you'd be having to stretch this for 090618 - it is the most extreme in its BB properties and will be interesting to see if/how it fits together with the low luminsoity GRBs.

KAWAI: The spectrum modelled with an additional soft X-ray thermal model may also be modelled with a partial absorption model with a single power law emission component and patchy absorber.

STARLING: We haven't tried complex absorption models - I suspect they would pose extra free parameters that would be difficult to cope with in typical GRB X-ray spectra. It is probable that absorption in more complicated than the Galactic+intrinsic model generally adopted, but in what way is difficult to determine until the advent of such missions as Athena.

PERLEY: Should we be concerned, when using $N_{\rm H}$ as a diagnostic of redshift or environment, that these estimates might be affected by an underlying thermal component? Can you rule out thermal emission for other GRBs?

STARLING: No, for the vast majority of GRBs you don't need to worry. The afterglow will usually far outshine any thermal component. With our next set of simulations we will be able to set limits on the presence of thermal soft X-ray emission for any given burst, but we haven't done those yet.

Death of Massive Stars: Supernovae and Gamma-Ray Bursts
Proceedings IAU Symposium No. 279, 2012
P. Roming, N. Kawai & E. Pian, eds.

doi:10.1017/S1743921312012896

The Local Environments of Core-Collapse SNe within Host Galaxies

Joseph P Anderson[1], Stacey M Habergham[2], Phil A James[2] & M Hamuy[1]

[1]Departamento de Astronomía, Universidad de Chile, Casilla 36-D, Santiago, Chile
[2]Astrophysics Research Institute, Liverpool John Moores University, Twelve Quays House, Egerton Wharf, Birkenhead, CH41 1LD
email: anderson@das.uchile.cl

Abstract. We present constraints on core-collapse supernova progenitors through observations of their environments within host galaxies. This is achieved through 2 routes. Firstly, we investigate the spatial correlation of supernovae with host galaxy star formation using pixel statistics. We find that the main supernova types form a sequence of increasing association to star formation. The most logical interpretation is that this implies an increasing progenitor mass sequence going from the supernova type Ia arising from the lowest mass, through the type II, type Ib, and the supernova type Ic arising from the highest mass progenitors. We find the surprising result that the supernova type IIn show a lower association to star formation than type IIPs, implying lower mass progenitors. Secondly, we use host HII region spectroscopy to investigate differences in environment metallicity between different core-collapse types. We find that supernovae of types Ibc arise in slightly higher metallicity environments than type II events. However, this difference is not significant, implying that progenitor metallicity does not play a dominant role in deciding supernova type.

Keywords. (stars:) supernovae: general, (ISM:) HII regions

1. Introduction

Mapping the links between progenitors and observed transients has become a key goal of supernova (SN) science, helping us to understand stellar evolution processes while also strengthening our confidence in the use of SNe to understand the Universe. In a number of cases the progenitor stars of core-collapse (CC) SNe have been directly identified on pre-explosion imaging (e.g. Smartt 2009). This allows one to gain 'direct' knowledge of progenitor masses when observed luminosities are compared to stellar models. However, the rarity of such detections limits the statistical results which can be gained. One can also relate overall host galaxy properties to the relative rates of SNe to infer progenitor properties (e.g. Prieto, Stanek & Beacom 2008, Boissier & Prantzos 2009). While this can attain significant statistics, the presence of multiple stellar populations within galaxies complicates the implications which can be drawn. Our method is intermediate to these; we investigate differences in CC SN progenitor characteristics using observations of the exact environments at the discovery positions of their explosions. Other recent examples using similar techniques include: Leloudas *et al.* (2010), Modjaz *et al.* (2011) and Kelly & Kirshner (2011).

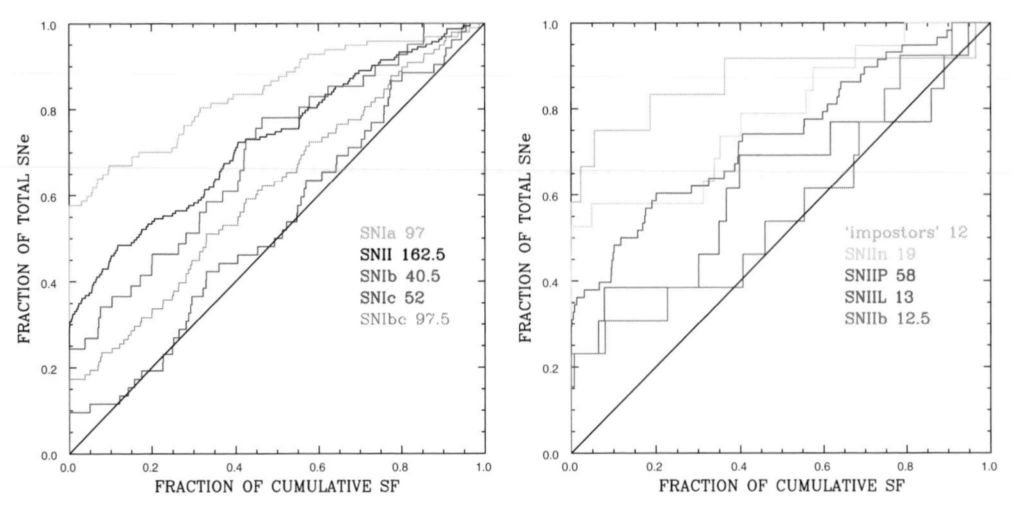

Figure 1. Left a): cumulative pixel statistics of the main SN types. SNIa are in green, SNII black, SNIb red and SNIc blue (combined SNIbc magenta). The black diagonal line illustrates a flat distribution that accurately traces the SF. As a distribution moves away to the left from this diagonal it displays a lower association with SF. **Right b):** Same as a) but for SNII sub-types. 'Impostors' are shown in green, SNIIn cyan, SNIIP blue, SNIIL red, and SNIIb magenta.

2. Spatial correlations with star formation

We have obtained host galaxy Hα imaging for 260 CC SNe. The sample can be separated into 58 SNIIP, 13 IIL, 12.5 IIb (one object has indistinct type), 19 IIn, 12 'impostors' plus 48 SNII (no sub-type), and 40.5 SNIb, 52 SNIc plus 5 SNIb/c. To analyse the association of each SN with the SF of its host galaxy we use our 'NCR' (Normalised Cumulative Rank pixel function) pixel statistic presented in James & Anderson (2006). To produce an NCR value for each SN we proceed with the following steps. First the sky-subtracted Hα image pixels are ordered into a rank of increasing pixel count. Alongside this we form the cumulative distribution of this ranked increasing pixel count. Negative cumulative values are then set to zero and all other values are normalised to the total flux summed over all pixels. It follows that each pixel has a value between 0 and 1, where values of 0 correspond to zero flux values while a value of 1 means the pixel has the highest count within the image. We measure an NCR value for each SN within our sample and proceed to build distributions for each CC SN type. The resulting cumulative distributions are displayed in Figure 1. In both of these figures (a and b), as a distribution moves closer to the black diagonal line from the top left hand side of the plots, it is showing a higher correlation with bright HII regions. In Fig. 1a) we see a clear sequence of increasing association to SF going from the SNIa, through the SNII, the SNIb and finally the SNIc. The most logical way to interpret differences in associations to SF is that more massive stars will be more highly correlated with bright HII regions. This is because these stars are a) more likely to produce sufficient ionizing flux to produce bright HII regions, and b) they will have shorter lifetimes and therefore have less time to move away from host HII regions. The implication is then that we observe a sequence of increasing progenitor mass. This starts with the SNIa arising from the lowest mass, followed by the SNII, then the SNIb, and finally the SNIc arising from the most massive progenitors. These trends were first seen in a smaller sample published in Anderson & James (2008). However, the current data set is the first to clearly separate the SNIb from the SNIc, with the latter now being seen to be significantly more associated with SF, and hence arising from higher

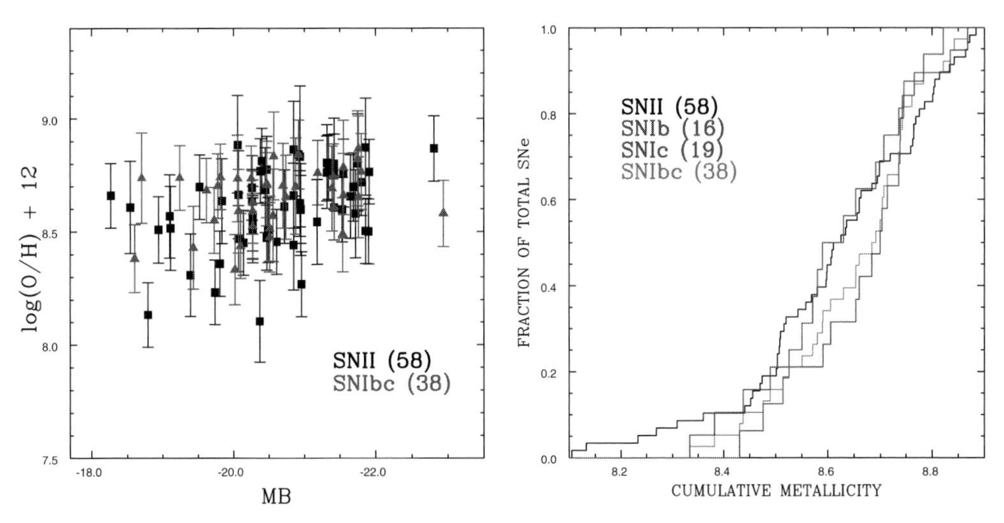

Figure 2. Left a): Host galaxy absolute B-band magnitudes against derived environment metallicities. SNII are shown in black squares while SNIbc are shown in red triangles. **Right b):** Cumulative metallicity distributions. The SNII distribution is plotted in black and the SNIbc distribution is shown in magenta. We also show the individual SNIb (red) and SNIc (blue) distributions.

mass progenitors (these results, plus discussion will be presented in Anderson *et al.* in preparation). On the sub-types plotted in Fig. 1b) we find the interesting result that both the 'impostors' and SNIIn show a lower degree of association with SF than the SNIIP. While there is likely to be a strong selection effect with respect to the 'impostors', for the SNIIn this implies that these events primarily arise from relatively low mass progenitors, contrary to general consensus. Finally, we find some suggestive constraints that both the SNIIL and SNIIb arise from higher mass progenitors than SIIP.

3. Host HII region spectroscopy

For 96 CC SNe (58 SNII and 38 SNIbc) we have also obtained host HII region optical spectroscopy of their immediate environments which we use to determine region metallicities. To achieve this we use the diagnostics of Pettini & Pagel (2004) which give gas-phase oxygen metallicities from emission line ratios. In Figure 2 we show the resulting distributions of each CC SN sample. Fig. 2a) plots the absolute B-band magnitude of host galaxies against derived environment metallicities. While the mean metallicity of SNIbc is higher than that of the SNII, this difference is not significant, and there is no clear offset between the distributions. In Fig. 2b) we display the cumulative metallicity distributions of the SNII, SNIb, SNIc and the combined SNIbc. While we see a metallicity trend going from the SNII in the lowest, through the SNIb to the SNIc within the highest metallicity regions, the difference between these distributions is statistically insignificant. Overall we see that SNIbc and SNII arise from similar metallicity environments, implying similar progenitor metallicities. While this is the only study thus far to derive 'direct' environment metallicities to probe differences between SNII and SNIbc events (a smaller sample was presented in Anderson *et al.* 2010), other recent results have been published on differences between SNIb, SNIc and LGRBs, finding different results (Leloudas *et al.* 2011, Modjaz *et al.* 2011, and talks at this symposium by the same authors).

4. Conclusions

Through observations of the environments of CC SNe we have presented progenitor constraints. We find that the main SN types can be ordered into a sequence of increasing progenitor mass: SNIa-SNII-SNIb-SNIc. With respect to progenitor metallicity we find no significant difference between the SNII and SNIbc populations. This argues that mass is the dominant (over metallicity) progenitor characteristic that influences resulting SN type.

Acknowledgments J.A. acknowledges support from FONDECYT grant 3110142, and grant ICM P10-064-F (Millennium Center for Supernova Science), with input from 'Fondo de Innovacion para la Competitividad, del Ministerio de Economia, Fomento y Turismo de Chile'.

References

Anderson, J. P. & James, P. A., 2008 *MNRAS*, 390, 1527
Anderson, J. P. *et al.*, 2010 *MNRAS*, 407, 2660
Boissier, S. & Prantzos, N., 2009, *A&A*, 503, 137
James, P. A. & Anderson, J. P., 2006 *A&A*, 453, 57
Kelly, P. L. & Kirshner, R. P., 2011 *arXiv*, 1110.1377
Leloudas, G., *et al.*, 2010 *A&A*, 518, 29
Leloudas, G., *et al.*, 2011 *A&A*, 530, 95
Modjaz, M., *et al.* 2011 *ApJ*, 731, 4
Pettini, M. & Pagel, B. E. J., 2004 *MNRAS*, 348, 59
Prieto, J. L., Stanek, K. Z., & Beacom, J. F. 2008, *ApJ*, 673, 999
Smartt, S. J., 2009, *ARAA*, 47, 63

Discussion

CROWTHER: Regarding the SNIIn results; if these are linked to Luminous Blue Variable (LBV) eruptions, locally LBVs span a wide range of mass with most shying away from HII regions (except notably Eta Carinae!)

ANDERSON: SNIIn show a lower association to HII regions than SNIIP. The (to-date) progenitors of SNIIP are 8-16 $M_{\odot}$. Given that our results suggest that SNIIn arise from lower masses than SNIIP, LBV progenitors seem inconsistent with this picture.

CROWTHER: How would your statistics differ if you limited your sample volume, given that many SNIIP are faint and so inherently trace the lowest mass progenitors?

ANDERSON: We have split our sample by host recession velocity. If there were a significant effect with distance then we would expect NCR values to differ with distance. We do not observe this.

PERLEY: A significant fraction of Ics are probably dust-obscured. How do you think dust extinction -either impacting the images (dust-lanes) or the non-detection of some events- might affect your results?

ANDERSON: SNIc are strongly associated with SF. If either Hα emission were being obscured, or SNIc were missed due to dust, this would only strengthen our result; more SNIc than we observe would be instrinsically associated with HII regions.

Death of Massive Stars: Supernovae and Gamma-Ray Bursts
Proceedings IAU Symposium No. 279, 2012
P. Roming, N. Kawai & E. Pian, eds.

doi:10.1017/S1743921312012902

The Optically Unbiased GRB Host (TOUGH) Survey

Páll Jakobsson[1], Jens Hjorth[2], Daniele Malesani[2], Johan P. U. Fynbo[2], Thomas Krühler[2], Bo Milvang-Jensen[2] and Nial R. Tanvir[3]

[1]Centre for Astrophysics and Cosmology, Science Institute, University of Iceland, Dunhagi 5, 107 Reykjavík, Iceland

[2]Dark Cosmology Centre, Niels Bohr Institute, University of Copenhagen, Juliane Maries Vej 30, 2100 Copenhagen Ø, Denmark

[3]Department of Physics and Astronomy, University of Leicester, University Road, Leicester, LE1 7RH, UK

Abstract. We present the results from our *Swift*/VLT legacy survey, a VLT Large Programme aimed at characterizing the host galaxies of a homogeneously selected sample of *Swift* gamma-ray bursts (GRBs). The immediate goals are to determine the host luminosity function, study the effects of reddening, determine the fraction of Lyα emitters in the hosts, and obtain redshifts for targets without a reported one. We have carefully selected a sample, obeying strict and well-defined criteria: 69 targets in total. Among the results is a large optical detection rate, the lack of extremely red objects (only one possible case in the sample), and 15 new GRB redshifts with the mean redshift of the host sample assessed to be $\langle z \rangle \gtrsim 2.2$.

Keywords. dust, extinction, galaxies: fundamental parameters, gamma rays: bursts

1. Introduction

Determining the statistical properties of gamma-ray bursts (GRBs) has long been compromised by inhomogeneous selection and a bias against optically dark bursts. With *Swift* (Gehrels *et al.* 2004) it has become possible to construct much more uniform samples, and to target the host galaxies even of optically faint bursts via X-Ray Telescope (XRT) localizations, for which redshifts could not be determined from the afterglows.

We have been securing GRB host galaxy information for a homogeneous sample of 69 *Swift* GRBs with a large programme at the Very Large Telescope (VLT). We aim at selecting a representative, unbiased sample of GRB host galaxies. To optimize the survey, we focus on the systems with the best observability, which also have the best available information. We thus select *Swift* bursts with the following properties:

- A long-duration burst detected automatically by the Burst Alert Telescope.
- The Galactic extinction in the direction to the burst has to be $A_V \leqslant 0.5$ mag.
- The Sun-to-field distance, at the time of the GRB, has to be $\theta_{\rm Sun} \geqslant 55°$.
- Good observability from the VLT ($-70° < \delta < +27°$).
- Prompt XRT localization distributed within 12 hours.
- Happened between March 2005 and August 2007.
- A small XRT error circle (radius $\leqslant 2''$).

Imposing these restrictions does not bias the sample towards optically bright afterglows; instead each GRB in the sample has favourable observing conditions, i.e. useful follow-up observations are likely to be secured. The bursts fulfilling all the criteria, 69 targets in total, make up The Optically Unbiased GRB Host (TOUGH) sample. More than 75% of the sample have an optical/near-IR afterglow (OA) (52/69) and 55% already

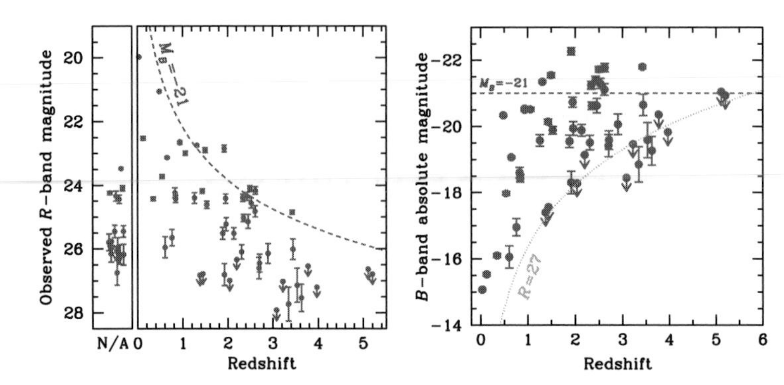

Figure 1. Left: The R-band host magnitude as a function of redshift for all the bursts in the TOUGH sample. Upper limits are shown with arrows. Hosts without a reported redshift are plotted on the left side of the diagram. The dashed curve shows a galaxy with an absolute B-band magnitude of -21 assuming $F_\nu \propto \nu^{-0.5}$. **Right**: The absolute B-band magnitude ($F_\nu \propto \nu^{-0.5}$) as a function of redshift for all the hosts in the left panel with a reported redshift. The dotted curve shows a galaxy with an observed magnitude of $R = 27$.

have a reported reliable redshift (38/69) in the range $0.033 < z < 6.3$. This should be compared to the full *Swift* sample that only has around 30% redshift completion.

2. Host magnitudes and colours

The first step in the TOUGH program is to obtain sufficiently deep images of the targets in the R- (typically restframe UV) and K-band (typically restframe visual). The characteristic magnitude limits of the survey are $R \sim 27$ and $K \sim 21.5$ (Vega photometric system). Around 80% (54/69) of the hosts are detected in the R band (Fig. 1), being mostly subluminous (between 1% and roughly 100% of L^*) as previous findings have indicated (e.g. Le Floc'h *et al.* 2003; Fruchter *et al.* 2006). *HST* observations have recently been carried out to search for some of the missing hosts in the sample.

Could a large fraction of faint hosts go undetected in our survey? The probability for a galaxy to be detected within an afterglow error circle by chance depends on the magnitude of the galaxy. The number of galaxies per arcmin^2 has been well determined to deep limits in the various Hubble deep fields. To limits of $R = 24$, 26 and 28 there are about 2, 6 and 13 galaxies arcmin^{-2} (e.g. Fynbo *et al.* 2000). Hence, the probability to find an $R = 24$ galaxy by chance in an error circle with a $0\farcs5$ radius is about 0.4%. For a $R = 28$ galaxy the probability is about 8%. If the error circle is defined only by the X-ray afterglow with a $2''$ radius we expect a random $R = 24$ and $R = 28$ galaxy in 6% and 72% of the error boxes. There are 17 bursts with X-ray positions only in the TOUGH sample (75% have an OA). Of these the faintest host has $R \sim 26$. Hence, chance projection should hence not be a serious concern and we are confident that the vast majority of the host candidates are real.

Only 35% of the hosts are detected in the K band. The corresponding $R - K$ color is shown in Fig. 2 with the hosts predominantly lying in the range $2 < R - K < 5$. There is only a single possible extremely red object (ERO) with $R - K = 5.6$; this GRB had no reported OA to a limit of around $R > 24$ at 1 hour after the burst (Fynbo *et al.* 2009). Although this might indicate the presence of dust, a chance association cannot be excluded between the GRB and this galaxy, given that the XRT error circle is among the largest in the sample. Figure 2 also confirms earlier findings (e.g. Le Floc'h *et al.* 2003) that GRB hosts mostly have blue colors (even considering bursts with no reported OA).

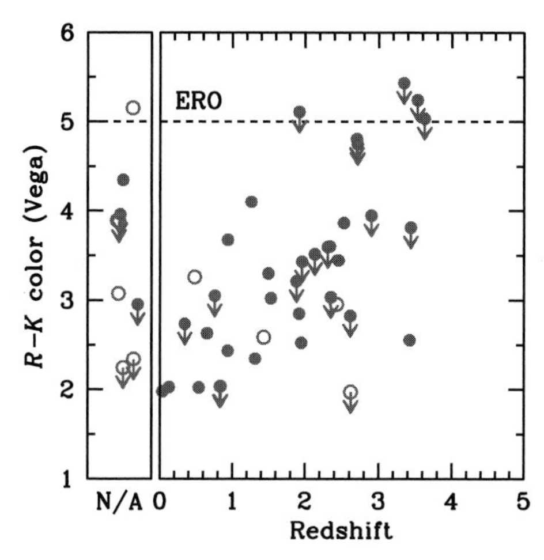

Figure 2. The $R-K$ color of the TOUGH hosts as a function of redshift. The left side represents objects without, as yet, a measured redshift. The filled (open) circles represent host galaxies of GRBs with (without) a detected OA. Only one ERO ($R-K > 5$) is visible in the sample.

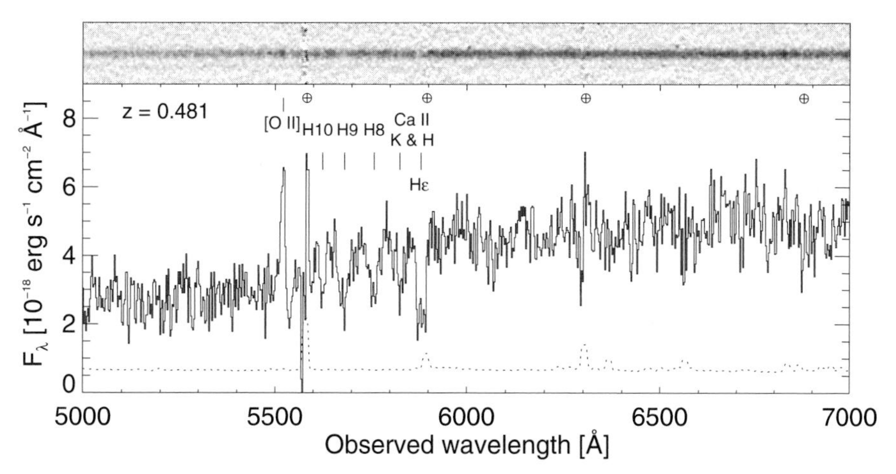

Figure 3. One- and two-dimensional spectra of a host galaxy in the TOUGH sample. Line features are marked with vertical lines, whereas telluric features and skyline residuals are marked with ⊕. The error spectrum is plotted as a dotted line.

3. Spectroscopic observations

We attempted spectroscopic observations of all hosts with $R \lesssim 25.5$ that did not have a reported reliable redshift. We have determined the redshift of 15 GRB host galaxies (an example is shown in Fig. 3) whose average redshift is $\langle z \rangle = 1.8$, significantly lower than the overall *Swift* GRB mean redshift. The low value is most likely the result of targeting the brightest galaxies in the sample. Furthermore, we have estimated redshift limits for an additional five hosts and inferred that three burst redshifts reported in the literature are erroneous (GRBs 060306, 060814 and 060908). Figure 4 shows how the redshift distribution has changed due to our measurements. For more details about the spectroscopic results we refer to Jakobsson *et al.* (2012), Krühler *et al.* (2012) and Milvang-Jensen *et al.* (2012).

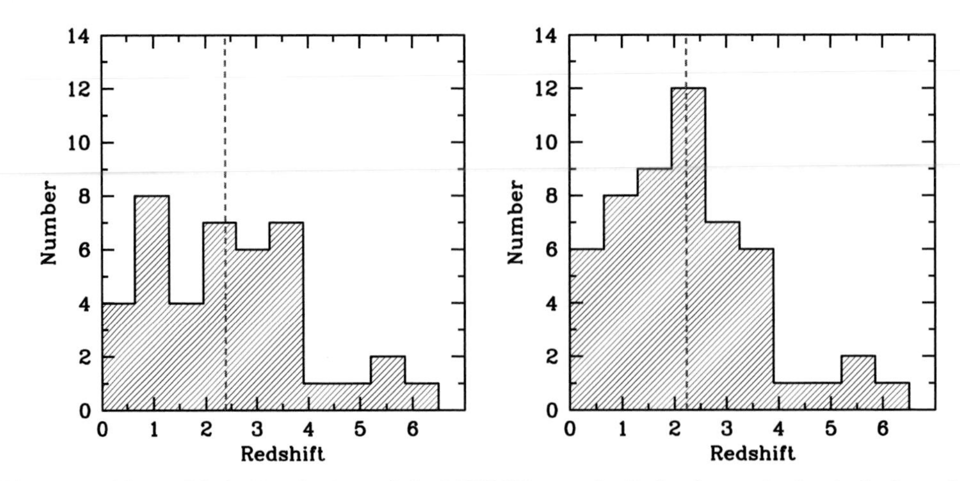

Figure 4. The redshift distribution of the TOUGH sample. In both panels the dashed vertical line is the mean redshift. **Left**: Before we started our observations, 38 redshifts were considered secure with $\langle z \rangle = 2.39$. **Right**: Our TOUGH spectroscopic observations added 15 new redshifts and demonstrated that three redshift reported in the literature were erroneous. Here $\langle z \rangle = 2.23$.

4. Implications and prospects

We have now reached a point in GRB research where a single burst rarely elucidates and illuminates our general understanding of the field. It is important to focus on well-defined samples and population studies, where systematics and biases can be minimized. *Swift* has made it possible to build such a sample and thanks to new available instrumentation, such as the VLT/X-shooter (Vernet *et al.* 2011), we can continue to follow this track into the future.

References

Gehrels, N., *et al.* 2004, *ApJ*, 611, 1005
Le Floc'h, E., *et al.* 2003, *A&A*, 400, 499
Fruchter, A. S., *et al.* 2006, *Nature*, 441, 463
Fynbo, J. P. U., Freudling, W., & Møller, P. 2000, *A&A*, 355, 37
J. P. U. Fynbo, *et al.* 2009, *ApJS*, 185, 526
Jakobsson, P., *et al.* 2012, *ApJ*, in press
Krühler, T., *et al.* 2012, *ApJ*, submitted
Milvang-Jensen, B., *et al.* 2012, *ApJ*, submitted
Vernet, J., *et al.* 2011, *A&A*, 536, 105

Discussion

ASTRAATMADJA: Based on the redshift distribution, what can we tell about the star-formation history of the Universe?

JAKOBSSON: It is possible that star formation at high redshifts has been significantly underestimated. Even at $z \sim 2$ it appears that the galaxy luminosity function has a substantially steeper faint-end slope than locally. Alternatively, it could be that GRB production is substantially enhanced in the conditions of early star formation, beyond the metallicity-dependent rate correction already applied. In the long run, large complete samples of GRB redshifts should shed light on whether the GRB rate is proportional to SFR or whether other effects play an important role.

Death of Massive Stars: Supernovae and Gamma-Ray Bursts
Proceedings IAU Symposium No. 279, 2012
P. Roming, N. Kawai & E. Pian, eds.

doi:10.1017/S1743921312012914

The locations of SNe Ib/c and their comparison to those of WR stars and GRBs

Giorgos Leloudas[1,2]

[1]The Oskar Klein Centre, Department of Physics, Stockholm University, 106 91 Stockholm, Sweden
email: giorgos.leloudas@fysik.su.se

[2]Dark Cosmology Centre, Niels Bohr Institute, University of Copenhagen, 2100 Copenhagen, Denmark

Abstract.
The locations of long GRBs and stripped supernovae are compared to those of their favored progenitors, WR stars, and their sub-classes. Compared to Leloudas *et al.* (2010), we have doubled the number of galaxies with suitable WR data. In the combined sample, WC stars are found, on average, in brighter locations than WN stars. The WN distribution is fully consistent with the one of SNe Ib, while it is inconsistent with those of SNe II, Ic and GRBs. The WC distribution is both consistent with SNe Ib and Ic. It is inconsistent with SNe II, and marginally consistent with GRBs. Furthermore, we present a spectroscopic study of the locations of SNe Ib/c. The average metallicity in the environments of SNe Ic is found to be a little higher than for SNe Ib, but the difference is small and not significant within our sample. Under the assumption that the SN regions were formed in an instantaneous burst of star formation, we find that a fraction of them appear older than what is allowed in order to host SNe Ib/c from single massive stars. Within this framework, these SNe must come from lower mass binaries.

Keywords. supernovae: general, gamma rays: bursts, stars: Wolf-Rayet, galaxies: abundances

---

1. Introduction: stripped supernovae and open questions

Type Ib/c supernovae (SNe) are among the few astronomical objects that do not show H in their spectra. It is therefore believed that they are the explosions of stars that have shed their outer H envelope (SNe Ib), or even (most of) their He envelope (SNe Ic). For this reason, they are known as *stripped* SNe. To date, there has been no direct progenitor detection of a SN Ib/c in pre-explosion images (Smartt 2009). Nevertheless, it is widely believed that stripped SNe are the explosions of Wolf-Rayet (WR) stars, exactly because these stars appear to be H-free. Depending on their spectroscopic appearance, WR stars can be broadly divided into nitrogen-rich (WN) and carbon-rich (WC) stars. A comprehensive review of WR stars, their sub-classes, and their properties is given by Crowther (2007). WN stars are expected to die as SNe Ib, while WC stars as SNe Ic, although variations to this general picture might exist, depending on the progenitor metallicity and initial mass (e.g. Georgy *et al.* 2009). Even if the connection of SNe Ib/c to WR stars seems like the only plausible scenario, it is important to stress that it awaits confirmation, especially in its details.

Another fundamental question concerning stripped SNe and their progenitors is the *way* that they lose their H envelopes. Massive stars lose mass through stellar winds and this is one of the main theoretical channels leading to a SN Ib/c. For single stars to reach the WR phase it is estimated that they need to start their lives with an initial mass $>$ 25 $M_\odot$ (Crowther 2007). Another way for a star to lose its H envelope, is through evolution in a binary system (e.g. Podsiadlowski *et al.* 1992). In this scenario, the outer layers of the primary star are expelled through interaction with its binary companion (Roche-lobe overflow or common envelope evolution). It is important to note that through the binary channel, it is possible to obtain stripped stars from initial masses substantially lower than from the single progenitor channel.

Direct searches for SN progenitors have not yielded any results yet and, even in the best case, are not expected to give more than a handful of detections during the next decade. Studies of the SN properties themselves (light-curves, spectra, etc) are informative, but they suffer from small numbers and the difficulty of acquiring suitable homogeneous datasets. Another powerful means of probing the nature of these explosions is by studying their environments. Although indirect, 'environmental' methods have statistical power (as large samples can be constructed) and can be used in complementary manner to direct methods in order to attack the same problem from different angles. Here, we concentrate on some of these methods as they have been applied to stripped SNe: in Section 2 we compare the locations of SNe Ib/c (and long GRBs) within their host galaxy with those of WR stars, while in Section 3 we describe direct spectroscopic observations of the birthplaces of SNe Ib/c, with the purpose of studying their metallicity and stellar population ages.

2. Comparison with locations of WR stars

The main driver motivating this comparison is the following: *if WR stars are the progenitors of stripped SNe and long GRBs, they must also be found in similar locations within their hosts.* The distribution of long GRBs with respect to their host galaxy light was studied by Fruchter *et al.* (2006), while those of different SN sub-types by Kelly *et al.* (2008). The statistical diagnostic used is the *fractional flux*, which describes the fraction of light in host locations fainter than the explosion locations, over the integrated galaxy light. Simply put, a fractional flux value of 0 means that the explosion took place in a location without flux, while a value of 1 means that the explosion took place in the brightest location of the host. A similar statistic was used by Anderson & James (2008), while other authors (e.g. Larsson *et al.* 2007) have used this notion in theoretical studies. Fruchter *et al.* (2006) showed that GRBs showed a strong preference for occurring in the brightest locations of their host. Kelly *et al.* (2008), studying a more nearby sample of SN hosts drawn from SDSS, showed that SNe II, Ib, and Ic occur in progressively brighter locations, with the distribution of SNe Ic being broadly consistent with the one of GRBs.

In Leloudas *et al.* (2010), hereafter L10, we extended this analysis to compare with the locations of WR stars within their host galaxies. For this purpose, we used the work of Crowther and collaborators, who in the recent years have mapped the population of WR stars in a number of nearby galaxies. What makes these papers very attractive for this kind of study is that a detailed estimate of the *completion level* of the surveys is given. We have excluded from our study many galaxies whose information on the WR population is known, but suffer from some kind of incompleteness. At the time of publication of L10, this analysis was possible for two galaxies: M 83 (Hadfield *et al.* 2005) and NGC 1313 (Hadfield & Crowther 2007). At present, suitable data are available for two more galaxies: NGC 7793 (Bibby & Crowther 2010) and NGC 5068 (Bibby & Crowther 2012). In this proceeding, we extend our analysis to include these new data. It should thus be regarded as a continuation and complementary to the original paper (L10).

A conceptual difference between our approach and the one of Fruchter *et al.* (2006) and Kelly *et al.* (2008), is that these studies examine many host galaxies with one explosion per galaxy, while we only examine a few galaxies with many potential explosion progenitors. It is therefore important to take into account the general properties of these individual galaxies during the comparison, to make sure that they are not radically different. Metallicity is a particularly important factor, as both the total number of WR stars and the ratio of WC to WN stars strongly depend on it. In L10, we showed that M 83 (a galaxy with super-solar metallicity) is a very typical representative of the SN Ib/c host sample, while NGC 1313 is both fainter and more metal poor. The two new galaxies have somewhat intermediate properties between these two. As discussed by Bibby & Crowther (2010, 2012), they both demonstrate important metallicity gradients encompassing both solar and LMC values. Another important difference is that these galaxies are more nearby. To enable the comparison, it is necessary to resample them and make them appear as if they were at the median redshift of the SN host sample (L10).

Figure 1 contains a comparison of the WN and the WC distributions of these 4 galaxies (panels A and B). The WN distributions are not identical: in the case of NGC 1313, it almost follows the host galaxy light, while the others move progressively toward brighter locations. As a matter of fact, this appears to be happening as a function of host metallicity. At this stage, we just point this out as an interesting finding, although it might disappear with a larger sample, and it is not straightforward to explain. Our favorite explanation is that at the low metal content of NGC 1313, it is more difficult for WR stars to lose all their H, and that WN stars might retain enough H to appear spectroscopically as a SN II (Georgy *et al.* 2009). In the case of the WC distributions, the one that stands out is NGC 5068. This is the only galaxy in this sample, for which the WC distribution is not skewed to brighter pixels than the WN, as also pointed out by Bibby & Crowther (2012). One of the main conclusions of L10 was that WC stars are found on average in brighter locations (and therefore closer to star-forming regions) of their hosts than WN stars. This statement is therefore now true for 3 out of 4 galaxies. It will be interesting to see how this will evolve in the future with a larger sample.

It is possible to combine the distributions from the different galaxies, although this approach has weaknesses because all distributions have particularities (both physical and related to the observation methods). Especially M 83 stands out from the rest of the sample: (i) it has a super-solar metallicity that is very different from the average metallicities of the other 3 galaxies (that ignoring metallicity gradients are not that different); (ii) it has a WR population far greater than all the others added together (and therefore any combined sample would be heavily weighted towards M 83). For this reason, it was decided to only combine the distributions from the other 3 galaxies. These are presented in panel C of Fig. 1, while M 83 is shown separately in panel D. In addition, for M 83, comparisons are made after the removal of the bulge light (Kelly *et al.* 2008), since the WR survey is incomplete in this region of the galaxy (Hadfield *et al.* 2005).

The results for M 83 were discussed in L10. Briefly, WC stars are found (on average) in brighter locations than WN stars and both distributions are statistically compatible with SNe Ib/c. They are incompatible with SNe II, while WC stars are marginally compatible with GRBs. In the combined sample, which now contains over 100 WN and WC stars including the most secure photometric candidates, we observe again that the average WC distribution appears in brighter pixels than the average WN distribution. A KS test between the two yields $p = 4.2\%$. The WN distribution is highly compatible with SNe Ib ($p = 98.4\%$) while it is highly incompatible with SNe II, SNe Ic and GRBs (p-values of 0.1%, 0.7% and 0% respectively). The WC distribution is again mostly compatible with SNe Ib ($p = 37.0\%$), but also with SNe Ic ($p = 12.3\%$). It is marginally compatible with GRBs ($p = 1.9\%$), while the association with SNe II is excluded ($p = 0\%$). We believe that these results are in nice agreement with the theoretical expectations and with the idea that WC stars originate from more massive progenitors than WN stars.

WN stars sometimes evolve to become WC stars, depending on the initial mass and metallicity. In a Monte Carlo simulation we allowed a fraction of WN stars to be included to the WC fractional flux distribution. For the combined sample, we estimate that this fraction is of the order of 20% (see L10 for details). The simulation showed that this effect does not alter our general conclusions, although the specific p-values change. Similar simulations were executed by L10 for M 83 as well as for the effect of photometric candidates and the WR number errors.

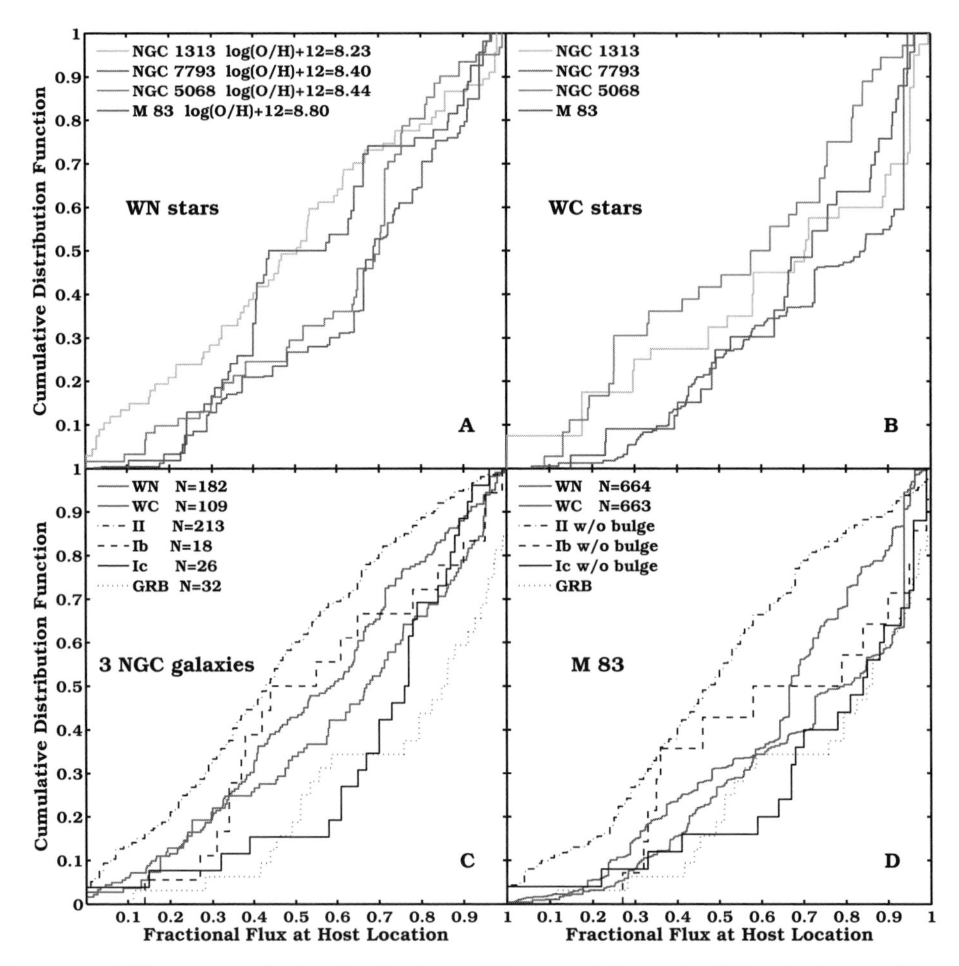

Figure 1. Different cumulative distributions of fractional fluxes for WR sub-classes, SNe and GRBs. Panel A shows the distributions of WN stars in 4 galaxies. The average metallicities of these galaxies are also indicated, showing that there is perhaps a metallicity effect. Panel B shows the same but for WC stars. With the exception of NGC 5068, all WC distributions are pushed towards brighter locations in their hosts than WN stars. In panel C, the WN and WC distributions of NGC 1313, NGC 7793 and NGC 5068 have been combined and are compared to the SN and GRB distributions of Kelly *et al.* (2008) and Fruchter *et al.* (2006). WC stars are on average found in brighter locations than WN stars. M 83 that is quite different from the other 3 galaxies in terms of metallicity and number of WR stars, is shown separately in panel D. In this panel, the distributions are shown after the removal of the bulge light (Kelly *et al.* 2008).

3. Direct spectroscopic observations

There are a number of predictions that are related to the single progenitor scenario, and that could be used to test it. First, mass stripping must be the result of stellar winds. Stellar winds are driven by metals and their strength increases with metallicity (e.g. Vink & de Koter 2005). It is therefore anticipated that SNe Ic will be found, on average, in more metal-rich environments than SNe Ib, if they result from massive stars. Second, massive stars have short lifetimes. Therefore, if SNe Ib/c come from single massive stars, they must be found close to areas of recent star formation.

Metallicities. The role of metallicity at the locations of SNe, has been examined by many authors. Until recently, however, most studies were either restricted to special events (Sollerman *et al.* 2005, Modjaz *et al.* 2008), or were based on abundances measured at the host galaxy nucleus (e.g. Prieto *et al.* 2008) or on proxies of metallicity (Boissier & Prantzos 2009, Anderson & James 2009). Direct metallicity measurements at the locations of normal stripped SNe were presented by Anderson *et al.* (2010), Modjaz *et al.* (2011) and Leloudas *et al.* (2011).

Within our sample, we find an average metallicity of 8.52±0.05 dex for SNe Ib and 8.60±0.08 dex for SNe Ic (Leloudas *et al.* 2011). These numbers are given in the N2 scale of Pettini & Pagel (2004) and include the systematic uncertainty related to the calibration of this relation. The contamination from stellar continuum was removed by fitting Bruzual & Charlot (2003) models, but this procedure did not significantly affect our results. We conclude that there might be hints that SNe Ic are found in more metal-rich environments than SNe Ib, but this difference is far from being statistically significant. A KS test between the two distributions gives the non-decisive p-value of 17%.

The other two studies did not yield consistent results: Anderson *et al.* (2010) find almost equal metallicities between the two sub-classes, but Modjaz *et al.* (2011) report on a difference of 0.20 dex and a KS test p-value of 1%. This might be partly due to the different biases affecting the different studies. Our sample, unlike the one by Anderson *et al.* (2010) and similar to the one of Modjaz *et al.* (2011), comprises SNe in both targeted and non-targeted surveys (in almost equal ratios) and there is evidence that the latter are found in lower metallicity than the former (Leloudas *et al.* 2011).

Even in the last two studies, however, it is true that this bias is not properly quantified. In addition, the difference between SNe Ib and Ic, if it exists, appears to be small and comparable to the systematic errors affecting the metallicity calibrators. For this reason, if a significant difference is to be sought, a large and well-controlled sample will be needed (see also talk by M. Modjaz; this volume). In any case, we think that it is unlikely that this difference can yield conclusive answers on the issue of binarity of stripped SNe.

Stellar ages. An age estimator sensitive to the most recent star formation is the equivalent width (EW) of Hα (e.g. Leitherer *et al.* 1999). It is thus possible, using our direct spectroscopic observations (Leloudas *et al.* 2011), to constrain the ages of the local stellar populations in the vicinity of the SN explosions. It is not important for our purposes that this method is only sensitive to the youngest, ionizing, stellar populations probed by our slit, exactly because what we want is to place a *lower limit* on the ages of the stars at the SN locations. The question we want to address is whether there are any SN regions where all stars are older than what is allowed by evolutionary models for single massive stars. In that case, by exclusion, the progenitors should be binary systems.

In order to estimate the ages of the young stellar populations, we used the measured Hα EW and compared with the predictions of Starburst99 (Leitherer *et al.* 1999) *for instantaneous star formation* (see discussion concerning this limitation below). Our metallicity measurements were used to select the appropriate table. The ranges we derived (Fig. 2) are conservative and include the measurement uncertainty, and the range of corresponding possible ages and IMFs given by Leitherer *et al.* (1999).

These lower limits were then compared with the lifetimes of massive stars, as predicted by the Geneva evolutionary models (Meynet & Maeder 2003, 2005, Georgy *et al.* 2009). According to these authors, at low metallicity ($Z = 0.008$), the lower mass limit above which single stars explode as SNe Ib/c is 30 $M_{\odot}$, while at solar values ($Z = 0.02$) it becomes 25 $M_{\odot}$. The corresponding lifetimes for these limiting stars are ~7.6 and 8.7 Myr, respectively. These ages represent *upper limits* because more massive stars will explode even sooner. Consequently, single massive progenitors of SNe Ib/c can only be found in the gray shaded area of Fig. 2. The constraints for He-poor SNe Ic are even stricter, as they should result from even more massive stars (> 39 $M_{\odot}$; Georgy *et al.* 2009) and have even shorter lifetimes. As a matter of fact, in Fig. 2 there is no allowed

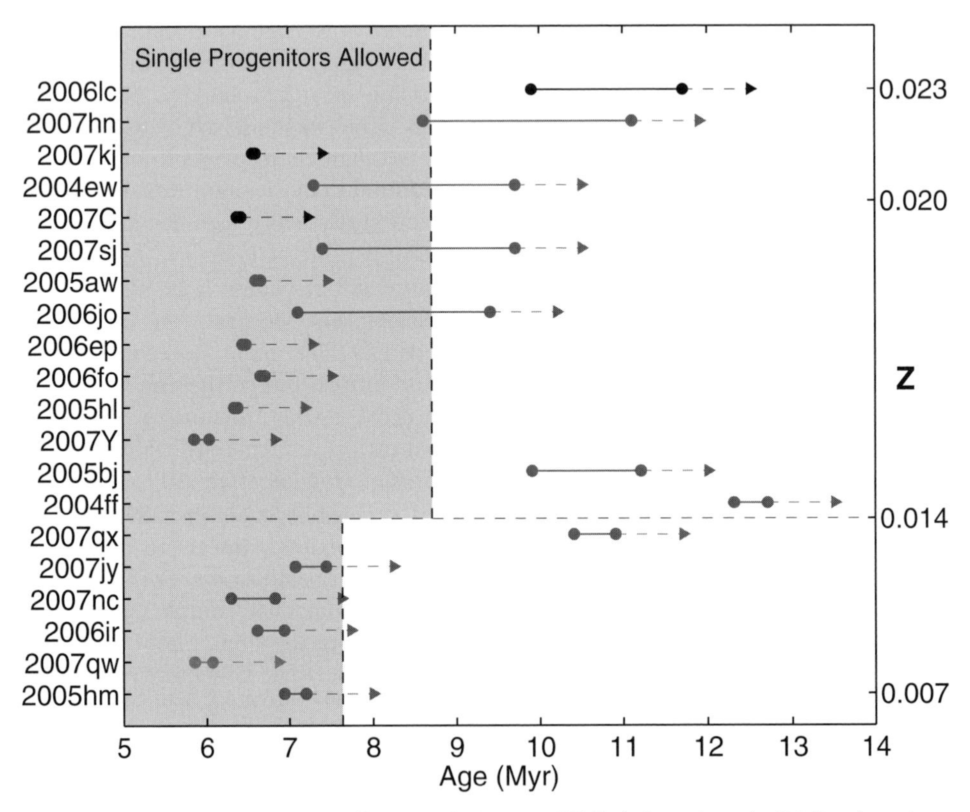

Figure 2. Ages of the youngest stellar populations at SN Ib/c locations (solid lines), estimated through comparing the measured Hα EW with the predictions of Starburst99. These set a lower limit (dashed arrow) to the age of the SN birthplace. The SN environments were ordered by ascending metallicity (some key values are indicated on the righthand axis). The SNe are color coded according to their type: SNe Ib in blue, SNe Ic in red and intermediate SNe Ib/c in black (an interloper SN Ia is colored in magenta). The horizontal dashed line separates the host regions into two metallicity groups, depending on whether a comparison with $Z = 0.008$ or $Z = 0.02$ models is more appropriate. The vertical dashed lines denote the predictions of the Geneva evolutionary codes for the lifetimes of single massive stars of 25 $M_{\odot}$ at $Z = 0.02$ and 30 $M_{\odot}$ at $Z = 0.008$. The allowed ages for single progenitors of SNe Ib/c are within the gray-shaded area. The more massive SNe Ic, however, have lifetimes < 5 Myr (i.e. outside this graph).

area for SNe Ic at all. Therefore, at least four SNe in Fig. 2 seem to have age constraints that are incompatible with the single progenitor channel, and the number is doubled if we consider the last argument concerning SNe Ic. By exclusion, an important fraction (20-35%) of the SNe examined should have lower-mass binary progenitors (without this possibility being excluded for the other SNe).

There is one main caveat in our considerations and this is that we have only considered the case of instantaneous star formation. If the SN birthplace is still forming stars, no such strict limits can be placed through the Hα emission. We have no means of assessing the validity of such an assumption for the individual cases and, for this reason, it is not possible to unambiguously rule out single progenitors for these 'discrepant' SNe. Nevertheless, we can propose them as good candidates for binarity and suggest that this possibility is also investigated by other complementary means. In addition, it will be interesting to study if the fraction of these discrepancies persists in even larger samples. To our knowledge, this is the first (and remains the only) attempt to statistically constrain the nature of SN Ib/c progenitors with this method.

Acknowledgments: I am grateful to Joanne Bibby for providing me with pre-processed images and files for NGC 7793 and NGC 5068, and to Max Stritzinger for providing comments on the manuscript. This work is supported by the Swedish Research Council through grant No. 623-2011-7117.

References

Anderson, J. P. & James, P. A. 2008, *MNRAS*, 390, 1527
Anderson, J. P. & James, P. A. 2009, *MNRAS*, 399, 559
Anderson, J. P., Covarrubias, R. A., James, P. A., Hamuy, M., & Habergham, S. M. 2010, *MNRAS*, 1175
Bibby, J. L. & Crowther, P. A. 2010, *MNRAS*, 405, 2737
Bibby, J. L. & Crowther, P. A. 2012, *MNRAS*, 420, 3091
Boissier, S. & Prantzos, N. 2009, *A&A*, 503, 137
Bruzual, G. & Charlot, S. 2003, *MNRAS*, 344, 1000
Crowther, P. A. 2007, *ARAA*, 45, 177
Fruchter, A. S., Levan, A. J., Strolger, L., *et al.* 2006, *Nature*, 441, 463
Georgy, C., Meynet, G., Walder, R., Folini, D., & Maeder, A. 2009, *A&A*, 502, 611
Hadfield, L. J., Crowther, P. A., Schild, H., & Schmutz, W. 2005, *A&A*, 439, 265
Hadfield, L. J. & Crowther, P. A. 2007, *MNRAS*, 381, 418
Kelly, P. L., Kirshner, R. P., & Pahre, M. 2008, *ApJ*, 687, 1201
Larsson, J., Levan, A. J., Davies, M. B., & Fruchter, A. S. 2007, *MNRAS*, 376, 1285
Leitherer, C., Schaerer, D., Goldader, J. D., *et al.* 1999, *ApJS*, 123, 3
Leloudas, G., Sollerman, J., Levan, A. J., *et al.* 2010, *A&A*, 518, 29
Leloudas, G., Gallazzi, A., Sollerman, J., *et al.* 2011, *A&A*, 530, 95
Meynet, G. & Maeder, A. 2003, *A&A*, 404, 975
Meynet, G. & Maeder, A. 2005, *A&A*, 429, 581
Modjaz, M., Kewley, L., Kirshner, R. P., *et al.* 2008, *AJ*, 135, 1136
Modjaz, M., Kewley, L., Bloom, J. S., *et al.* 2011, *ApJL*, 731, 4
Pettini, M. & Pagel, B. E. J. 2004, *MNRAS*, 348, 59
Podsiadlowski, P., Joss, P. C., & Hsu, J. J. L. 1992, *ApJ*, 391, 246
Prieto, J. L., Stanek, K. Z., & Beacom, J. F. 2008, *ApJ*, 673, 999
Smartt, S. J. 2009, *ARAA*, 47, 63
Sollerman, J., Östlin, G., Fynbo, J. P. U., *et al.* 2005, *New Astron.*, 11, 103
Vink, J. S. & de Koter, A. 2005, *A&A*, 442, 587

Discussion

PAUL CROWTHER: I have two comments: (1) Extragalactic giant HII regions have an age spread of 10+ Myr and are strongly biased towards the youngest (most massive) episode of star formation. (2) Our extragalactic WR surveys are only sensitive to high mass WR stars while the probable progenitors of many/most SNe Ib/c (He star primary and bright BSG secondary) cannot be detected since the WR wind is swamped by its companion. Not even in the Milky Way.

GIORGOS LELOUDAS: Indeed, it is true that studying unresolved regions can be problematic. That is why in our paper we did not claim that we can accurately pinpoint the age of the SN progenitor, but only placed a lower limit, based exactly on the signatures of the youngest stars in the region. But this is ok, since we wanted to compare with upper limits of single stars lifetimes. It is true that we can probably not do better than this, but, even like this, we discovered some potential discrepancies. Concerning the incompleteness of the surveys, I think this is of course important to take under consideration. But we have to try to do as good as we can with the information we have.

NATHAN SMITH: You showed that WN stars and Type II SNe have roughly the same distribution in their host galaxies. This is interesting, because we have strong evidence that most SNe IIP come from lower masses (8-17 $M_{\odot}$) while WN-type Wolf-Rayet stars come from much higher initial masses. This argues strongly against the interpretation that the different locations of different SN types within their host galaxies are due to different initial masses. While you did not make this claim in your talk, it is widely assumed to be the case so it is important to point out that this interpretation is probably wrong. It appears that some systematic effect other than the initial mass determines the different distributions of SN types within their host galaxies.

GIORGOS LELOUDAS: One thing that is important to note, is that this observation is only true for galaxies of lower metallicity, and only statistically significant for NGC 1313. As I said, perhaps this is not that surprising, because there have been arguments (by the Geneva group) that at low metallicity WN stars might actually explode as SNe of Type II, because they retain enough hydrogen to be observed as such. In addition, the 'red supergiant problem' (the missing higher mass SN II progenitors) indicates exactly that we are missing some part of the picture concerning initial masses. I think that we should take the observational evidence and try to explain it, rather than simply using it to dismiss some assumptions that seem logical. Indeed, I did not make this claim (initial mass follows light distribution) directly in my talk, as this was not important for my purposes. I restricted myself in comparing locations of explosions and potential progenitors and investigating their compatibility. However, I have no doubt that there is some kind of relation between the two. And even if it is not a strict correlation in the case of broad-band blue light, it is certainly more informative in the case of Hα (see, for example, Joe Anderson's talk).

Death of Massive Stars: Supernovae and Gamma-Ray Bursts
Proceedings IAU Symposium No. 279, 2012
P. Roming, N. Kawai & E. Pian, eds.

doi:10.1017/S1743921312012926

Dust and metal column densities in GRB host galaxies

P. Schady[1],* T. Dwelly[1] M. J. Page[2] J. Greiner[1] T. Krühler[3] S. Savaglio[1] S. Oates[2] A. Rau[1] and the GROND and UVOT teams

[1]Max-Planck-Institut für Extraterrestrische Physik, Giessenbackstraße 1, 85748 Garching, Germany
*email: pschady@mpe.mpg.de

[2]The UCL Mullard Space Science Laboratory, Holmbury St Mary, Surrey, RH5 6NT, UK

[3]Dark Cosmology Centre, Niels Bohr Institute, University of Copenhagen, Juliane Maries Vej 30, 2100 Copenhagen, Denmark

Abstract. The immensely bright and intrinsically simple afterglow spectra of gamma-ray bursts (GRBs) have proven to be highly effective probes of the interstellar dust and gas in distant, star-forming galaxies. Despite significant progress, many aspects of the host galaxy attenuating material are still poorly understood. There is considerable discrepancy between the amount of X-ray and optical afterglow absorption, with the former typically an order of magnitude higher than what would be expected from the optical line absorption of neutral element species. Similar inconsistencies exist between the abundance of interstellar dust derived from spectroscopic and photometric data, and the relation between the line-of-sight and integrated host galaxy interstellar medium (ISM) remains unclear. In these proceedings we present our analysis on both spectroscopic and photometric multi-wavelength GRB afterglow data, and summarise some of the more recent results on the attenuation properties of the ISM within GRB host galaxies.

Keywords. gamma rays: bursts, galaxies: ISM, ISM: dust, extinction, ISM: abundances

1. Introduction

A causal connection between long gamma-ray bursts (GRBs) and massive star formation is now well established, and their immensely bright afterglows therefore pinpoint regions of distant star formation independent of galaxy luminosity. Moreover, due to their featureless, broadband afterglow, the imprint from attenuating material within the host galaxy can be studied in detail, providing a truly unique view of the interstellar medium (ISM) within distant star forming galaxies. Their incredibly bright and multi-wavelength afterglow not only illuminates the gas and dust within the star forming regions of the host galaxy, but also that of the interstellar material in the disk and halo of the galaxy, and intervening intergalactic medium along the GRB line-of-sight.

In these proceedings we summarise our work in the context of others on the various components of attenuating gas, metals and dust within the host galaxies of GRBs. Our analysis makes use of X-ray through to near-infrared (nIR) spectroscopic and photometric data, comprising one of the most complete studies on the attenuation properties of the ISM within GRB host galaxies. In particular, we shall address the nature of the excess X-ray absorption and discuss the location of this "missing" absorbing gas, as well as discuss the observed variation in the dust extinction properties of GRB host galaxies.

These proceedings are divided into two main sections, where in section 2 we focus on afterglow absorption from intervening gas and metals, and in section 3 we discuss properties of dust extinction along the GRB line-of-sight, and its relation with the global host galaxy characteristics. A summary and future prospects is provided in section 4.

2. Afterglow absorption from intervening gas and metals

From the increasing sample of GRB spectroscopic observations that cover the neutral hydrogen Lyman-α absorption feature at ultraviolet (UV) wavelengths (rest-frame 1215Å), it is becoming clear that GRB host galaxies have high column densities of cold neutral gas ($T \leqslant 10^3$ K). Any corresponding ionised hydrogen would not be detected, and neutral hydrogen measurements thus require an ionisation correction to determine the total column density of atomic hydrogen, $N_{\rm H}$. A large fraction of GRB hosts have such large $N_{\rm H}$ values, however, that the ionisation correction is negligible and the neutral absorbing gas component is known as a damped-Lyman-α (DLA) system (log $(N_{\rm HI}/{\rm cm}^{-2}) > 20.3$). The survival of certain species, such as Mg I, and time varying Fe II and Ni II fine-structure lines, place this neutral gas component at a few hundred parsecs from the GRB (e.g. Vreeswijk *et al.* 2007), within the ISM of the host galaxy. The neutral ISM is also traced by low ionisation species detected in the UV, whereas highly-ionised species (e.g. O IV, C IV, Si IV and N V) possibly probe the hot gas (T$\sim 10^4$ K) within the circumburst environment of the GRB (Prochaska *et al.* 2008), as well as a contribution from gas in the rest of the galaxy (Fox *et al.* 2008).

In contrast to the specific regions of gas that can be identified from UV spectra, X-ray spectroscopic observations provide measurements of the total column density of gas along the line-of-sight, probing both the cold neutral gas as well as the warm ionised regions. Soft X-rays with energies < 0.8 keV are absorbed by medium-weight metals along the line-of-sight, which is predominantly in the form of oxygen and to a lesser extent carbon and nitrogen. Furthermore, the cross-section of oxygen remains relatively unchanged regardless of ionisation state (Verner & Yakovlev 1995), making it a good proxy for the host galaxy total oxygen column density. In addition to this, soft X-rays can also be absorbed by metals locked in dust grains, although optical afterglow dust extinction measurements indicate that most GRB lines-of-sight are relatively dust-poor.

In comparing X-ray and optical afterglow measurements, Watson *et al.* 2007 noted considerable discrepancy between the two. They found the equivalent neutral hydrogen column density derived from soft X-ray absorption (assuming solar abundances), $N_{\rm H,X}$, to be typically an order of magnitude higher than the optically derived $N_{\rm H}$. This led the authors to conclude that the former probed a significant column of ionised gas that is transparent to UV photons.

In order to explore the properties of this additional absorbing component, in Schady *et al.* (2011) we combined the results from the analysis of optical and X-ray spectra, and also of broadband afterglow SEDs, with the aim of providing greater insight on the ionisation state of the host galaxy medium along the line-of-sight to GRBs. Instead of neutral hydrogen, we used absorption lines of singly-ionised metals to trace the neutral gas within GRB host galaxies that could then be compared directly with the soft X-ray absorption, which measures primarily the oxygen column density (neutral and ionised), $N_{\rm O,X}$. This removes the need for hydrogen column density and host galaxy metallicity measurements, which are frequently uncertain or unavailable, thus increasing our sample size. We took a sample of 26 GRBs with detected optical spectroscopic absorption lines in at least one of Zn II, Si II, S II or Fe II with which to probe the host galaxy neutral gas, as well as with detected soft X-ray absorption or upper limit.

2.1. *Total versus neutral gas*

To trace the neutral gas within a GRB host galaxy, we preferentially used the ion Zn II or S II, which if not available was substituted by Si II or otherwise Fe II. This is due to the inherent uncertainties present in the dust depletion of the refractory elements Si II, and

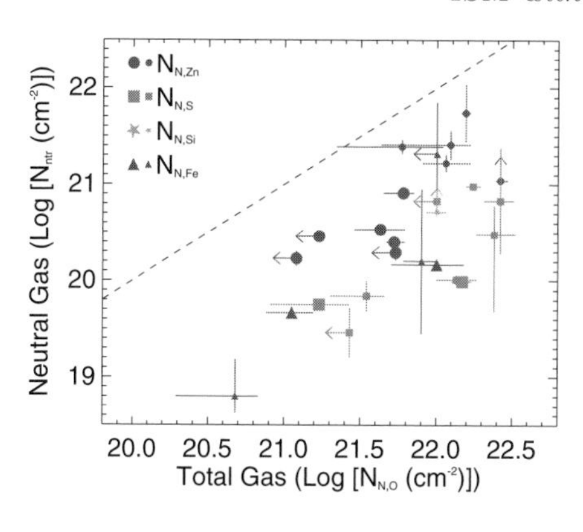

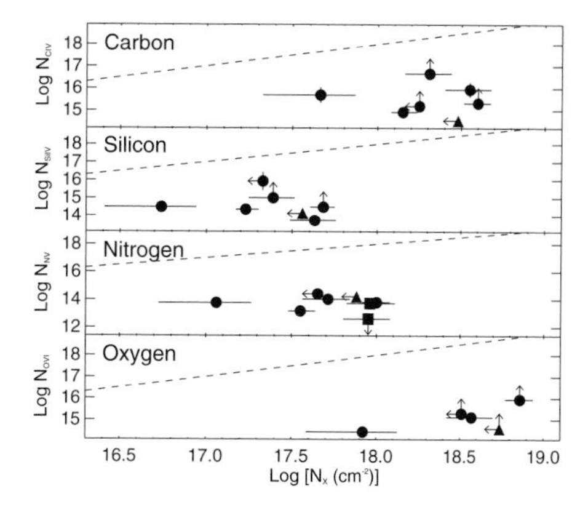

Figure 1. ***Left:*** Host galaxy neutral gas column density, $N_{\rm ntr}$, against host galaxy total gas column density, $N_{\rm N,O}$, along the line-of-sight to a sample of 26 GRBs. For each GRB, $N_{\rm ntr}$ is derived from either the column density of Zn II (red circles), S II (orange squares), Si II (green stars) or Fe II (blue triangles), where a correction for dust depletion is applied to $N_{\rm N,Si}$ and $N_{\rm N,Fe}$. Smaller data points correspond to those taken from low- or mid-resolution spectra, and larger data points are taken from high-resolution spectra (R $>$ 10,000). The dashed line corresponds to where $N_{\rm ntr}$ is equal to $N_{\rm N,O}$. ***Right:*** Logarithmic host galaxy column density of the highly ionised atoms C IV, Si IV, N V and O VI against the total logarithmic column densities of C, Si, N and O in the top, second, third and bottom panels, respectively. The total column densities are derived from X-ray observations, denoted by N_X. The C IV, Si IV, N V and O VI measurements plotted as circles are all taken from Fox *et al.* (2008), the N V data plotted as squares are from Prochaska *et al.* (2008), and data taken from D'Elia *et al.* (2010) are plotted as triangles. In all four panels, the dashed line corresponds to where the normalised soft X-ray column density is equal to the column density of the corresponding highly ionised atom.

especially Fe II. When column densities for both Zn II and S II were available, the one with the smaller error bars was used. When only Si II or Fe II absorption measurements were available, a dust correction was applied to the measured column density. We then corrected for cosmic abundance variances by normalising all our measurements to each other using the abundances from Asplund *et al.* (2009). We denote the normalised column densities of oxygen, zinc, silicon, sulphur and iron by $N_{N,O}$, $N_{N,Zn}$, $N_{N,Si}$, $N_{N,S}$ and $N_{N,Fe}$, respectively.

In the left panel of Fig. 1 we show the neutral gas column density, $N_{\rm ntr}$, represented by either $N_{\rm N,Zn}$ (red circles), $N_{\rm N,S}$ (orange squares), $N_{\rm N,Si}$ (green stars) or $N_{\rm N,Fe}$ (blue triangles), against $N_{\rm N,O}$, which represents the total column of gas (neutral and partially ionised), where both $N_{\rm N,Si}$ and $N_{\rm N,Fe}$ have been corrected for dust depletion. The dashed line in Fig. 1 (left panel) corresponds to where the column density of neutral gas is equal to the total column density of gas as probed by our X-ray absorption measurements. From this figure it is clear that for all GRBs, $N_{\rm N,O}$ is larger than the column density of neutral gas, $N_{\rm ntr}$, by around an order of magnitude, implying that over 90% of the gas along the line-of-sight is ionised.

There is an indication that the difference between the soft X-ray and neutral metal absorption increases for smaller column densities. This would suggest that the fraction of ionised gas is not dependent on the global galaxy properties, but on local conditions, such as would be the case if the GRB itself were the dominant source of ionising photons. Nevertheless, this trend is not significant, and we can thus only speculative at this stage.

2.2. *Highly ionised gas*

To explore further the ionisation state of the X-ray absorbing gas, we looked into the contribution from highly ionised gas, such as C IV, Si IV, N V and O VI, that unlike low-ionised gas, can survive within the circumburst environment of the GRB (Fox *et al.* 2008;

Prochaska *et al.* 2008). We used the high-ionised gas absorption measurements for nine GRBs reported in Fox *et al.* (2008), Prochaska *et al.* (2008), and D'Elia *et al.* (2010). To compare the amount of soft X-ray absorbing gas with the fraction of highly ionised gas, we then normalised the soft X-ray absorption measurements to the cosmic abundance of the element in question being compared (i.e. either Si, C, N or O). Our results are shown in Fig. 1 (right panel), which compares the logarithmic column density of the highly ionised atoms N_{CIV}, N_{SIV}, N_{NV} and N_{OVI} against the soft X-ray absorption column densities normalised to C, Si, N and O cosmic abundances in the top, second, third and bottom panel, respectively. The normalised soft X-ray absorption column density is denoted by N_X to indicate the column density of C, Si, N or O as determined from our soft X-ray absorption measurements. In all four panels the dashed line represents where the normalised soft X-ray column density measurements are equal to the column density of the highly ionised atom represented in each corresponding panel.

It is clear from this figure that all data points lie several orders of magnitude below the dashed lines, indicating that the highly ionised gas within GRB hosts only makes up a small fraction ($< 0.01\%$) of the soft X-ray absorption. The left-hand panel in Fig. 1, on the other hand, indicates that $\sim 90\%$ of the gas probed by the soft X-rays is ionised. If the large fraction of this gas is in a lower ionised state, then there should be a signature of this in the detection of strong absorption lines from intermediate ionisation lines with ionising potentials (IPs) between ~ 20 eV and ~ 50 eV (i.e. Si II and Si IV). This is, however, not the case. For example, Al III, which has an IP of 28 eV, is frequently observed to be weaker than Al II, which has an IP of 19 eV. This would, therefore, suggest that most of the ionised gas probed by the soft X-rays is in an ultra-ionised state, with IPs larger than ~ 200 eV. The signature left by such an ultra-ionised gas would lie predominantly in the soft X-ray energy range, which is already heavily absorbed by oxygen and carbon (both neutral and ionised). Resolving the absorption from an ultra-ionised gas is, therefore, beyond the spectral capabilities of current fast-response X-ray telescopes and will require a future era of rapid-response, high-resolution X-ray telescopes.

Other possibilities that have been raised to explain the origins of this soft X-ray excess absorption have been an underestimate of the absorption from metals within the Milky Way, absorption from intervening systems (Campana *et al.* 2012), or from a diffuse and local 'warm-hot' intergalactic medium, or WHIM (Behar *et al.* 2011).

3. Afterglow dust extinction

Another source of attenuation is in the absorption and scattering of light, or *extinction*, caused by intervening dust. The amount of dust extinction along a single line-of-sight is, for the most part, an inverse function of wavelength, and this dependence on wavelength is known as the *dust extinction curve*. The extinction curve is a function of both the grain size distribution and composition of the extinguishing dust, and the effect that differences in environmental conditions have on the observed extinction curve can be seen within the local Universe, in the average extinction curve observed within the Milky Way, and the Large and Small Magellanic Clouds (LMC and SMC respectively) (see Fig. 2). The dominant differences between these extinction curves is in the prominence of an extinction feature centred at ~ 2175Å (most pronounced along Milky Way sight lines, negligible within the SMC), and in the slope of the extinction curve at UV and optical wavelengths, which is flattest within the Milky Way, and gets progressively steeper from LMC to SMC sight lines.

The single or broken power-law GRB afterglow spectrum allows the broad, smooth shape cut-out by dust-extinction to be measured with relative ease, making GRBs ideal

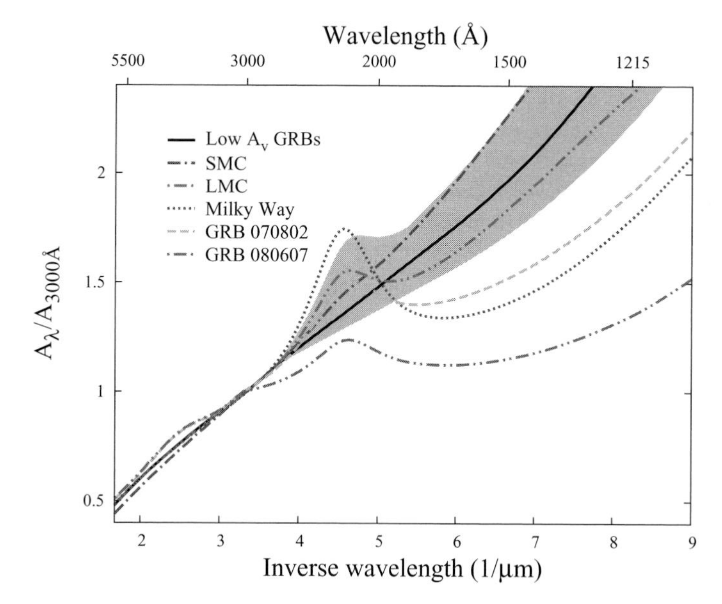

Figure 2. Best-fit mean GRB host extinction law as derived from simultaneous SED fits normalised at $A_{3000\text{Å}}$ (black) and corresponding 90% confidence region (grey) (Schady *et al.* 2012). Also shown for comparison are the mean SMC (red, dot-dash), LMC (pink, dot-dot-dash) and Milky Way (blue, dotted) extinction laws (Pei 1992), as well as the best-fit extinction curves to the two heavily extinguished GRBs, GRB 070802 (green, dashed; Elíasdóttir *et al.* 2009) and GRB 080607 (orange, dot-dot-dashed; Perley *et al.* 2011).

probes to the extinction properties of dust beyond the local Universe. Typically it is found that for GRBs with well-observed afterglow spectral energy distributions (SEDs), the amount of host galaxy extinction is small (total V-band or *visual* extinction $\langle A_V \rangle < 0.3$ (e.g Kann *et al.* 2006, Schady *et al.* 2010, Schady *et al.* 2012), and has a nIR to UV wavelength dependence similar to that along SMC lines-of-sight. However, recent dedicated GRB rapid-response programmes with nIR wavelength coverage are now revealing a sample of GRBs ($\sim 10\%$) that have been significantly extinguished ($A_V > 1$) by host galaxy dust (Greiner *et al.* 2011), many of which show evidence for much flatter, Milky Way-like host extinction curves than observed in the hosts of only moderately extinguished GRBs. There are now four GRBs with spectroscopically confirmed detections of the 2175Å absorption bump at the redshift of the GRB host galaxy (Zafar *et al.* 2011), as well as several GRBs with a 2175Å extinction feature detected in the GRB SED. The causes for this range in the total host galaxy dust extinction, and in the shape of the extinction curves of GRB hosts are still poorly understood.

To investigate the origin of these observed variations further, in Schady *et al.*(2012) we used predominantly *Swift* data to quantify the range observed in GRB host extinction curves and investigate the possible environmental factors contributing to the measured variations. The results from this study were consistent with a dustier host galaxy having a flatter GRB host extinction curve, and more prominent 2175Å bump (see Fig. 2). Nevertheless, this analysis also emphasised the need for accurate, near-IR photometry to be able to increase the sample to include more dust-extinguished GRB afterglows, as well as to better study the relation between A_V and the extinction curve properties.

3.1. *Heavily dust-extinguished GRB sightlines*

Around 40% of GRBs have their optical afterglow attenuated by either neutral gas within the intergalactic medium, or more commonly, from dust within the host galaxy (Perley *et al.* 2009, Greiner *et al.* 2011). For these so-called 'dark' GRBs, the arcsecond position provided by the optical afterglow is therefore not available, introducing a clear bias against follow-up observations of their host galaxies. The improved positional accuracy of GRBs provided by the rapid response of *Swift* and (semi-)robotic ground-based telecopes, in particular those equipped with nIR instruments, has significantly improved our capabilities over the last half decade to study highly dust-extinguished GRBs and their host galaxies. A leading instrument in acquiring broad-band observations that extend into the near-IR wavelength range has been the GRB optical and nIR detector (GROND), commissioned in July 2007. GROND is unique in providing simultaneous $g'r'i'z'$ and near-IR JHK observations of the early-time afterglow, and has proved highly effective at detecting and measuring host galaxy dust extinction for a large sample of GRBs (Greiner *et al.* 2011), and increasing the host galaxy visual extinction distribution out to $A_V > 1$. Other nIR instruments such as PAIRITEL have also contributed.

This reduction in the selection effects against dusty lines-of-sight has also seen an increase in the variety of GRB host galaxy properties. The majority of optically-selected host galaxies, making up the majority of the population, are young, star-forming, low-mass galaxies (Le Floc'h *et al.* 2003; Savaglio *et al.* 2009), with sub-solar metallicities. Recent follow-up observations of the host galaxies of heavily extinguished GRBs have nevertheless indicated the environmental conditions traced by GRBs to be much more diverse than indicated by previous, optically biased observations.

3.1.1. *Relation of extinction curves with A_V and global galaxy properties*

There is empirical evidence that the prominence of the 2175Å extinction bump and the overall flatness of the extinction curve is related to the total visual extinction, A_V, along the line-of-sight. The few GRB host extinction curves with spectroscopically detected 2175Å extinction bump, also correspond to lines-of-sight with the largest visual extinction. There is some evidence that the host galaxies of highly dust-extinguished GRBs are typically more massive, luminous, and chemically evolved than the typical host galaxies of relatively unextinguished GRBs (e.g. Krühler *et al.* 2011). Nevertheless, despite such observed trends, the relation between the GRB line-of-sight, and the global host galaxy properties remains far from unclear.

One of the most dust-extinction GRBs (GRB 080607; Prochaska *et al.* 2009) also had one of the flattest extinction curves and most prominent 2175Å bumps detected along a GRB line-of-sight. Its host galaxy properties were also atypical when compared to optically-selected samples (Chen *et al.* 2011), with a stellar mass $M_* \sim 8\times10^9\ M_\odot$, almost an order of magnitude larger than the mean stellar mass of optically selected samples ($\langle M_* \rangle \sim 10^9\ M_\odot$; Savaglio *et al.* 2009), as well as being highly reddened ($R - K > 5$; Chen *et al.* 2011). This is also the only GRB to have a robust detection of molecular hydrogen absorption in its afterglow spectrum (Prochaska *et al.* 2009).

On the other hand, GRB 070306 and GRB 100621A were two other heavily extinguished GRBs ($A_V \sim 5.5$ and $A_V \sim 3.8$ respectively), but both with very blue host galaxies, with $R-K$ colours comparable to the host galaxies of relatively unextinguished GRBs (Krühler *et al.* 2011), indicative of a very clumpy distribution of dust. Furthermore, although GRB 070306 had one of the largest stellar masses measured for a long GRB host galaxy ($2 \times 10^{10}\ M_\odot$; Krühler *et al.* 2011), the host stellar mass for GRB 100621 ($10^9\ M_\odot$) was comparable to that of optically bright afterglow host galaxies.

3.1.2. *Dust-to-metals ratio*

The visual extinction-to-metals column density ratio, $A_V/N_{\rm H,X}$, within GRB host galaxies along the GRB line-of-sight has been investigated in a number of papers, and these ratios have been typically found to be much higher than the ones observed in the Local Group (e.g. Galama & Wijers *et al.* 2001, Stratta *et al.* 2004, Starling *et al.* 2007, Schady *et al.* 2007,2010). The large majority of these studies have, nevertheless, been biased towards unextinguished lines-of-sight, and recent samples of GRBs with $A_V > 4$ have shown metals-to-dust ratios significantly below what is typically measured for GRB afterglows, and more in line with measurements from the Local Group (Krühler *et al.* 2011). Krühler *et al.*(2011) also found a strong anti-correlation between the metals-to-dust ratio and A_V along the GRB sight-line, which they speculate could be evidence of two physically independent absorbers: a dust-free, ionized plasma in the GRB circum-burst environment, and another dusty and more distant, thus less ionised cloud.

4. Summary and future prospects

The move towards combining both imaging and spectroscopic data from across the spectrum, and ongoing programmes dedicated at pushing down observational biases in GRB afterglow and host galaxy samples are providing us which a much more rounded view on the local and global environments of GRBs, and on the effect that the GRB explosion has on the surrounding circumburst and interstellar material.

The large UV radiation field implied by a situation in which $\sim 90\%$ of the gas within GRB host galaxies is in an ultra-ionised state has consequences for the effect of massive star formation on the surrounding environment. Direct detection of an ultra-ionised gas component would require early-time, high spectral-resolution X-ray afterglow data, which is currently unattainable. Nevertheless, a useful verification would be to model the UV radiation field and circum- and interstellar properties required to satisfy observations.

A significant contribution to the absorption of low-energy X-rays from a WHIM component, on the other hand, would have important implications to our understanding and investigation of 'the missing baryons problem'. The contribution to soft X-ray absorption from intervening gas external to the host galaxy could be further investigated through analysis on the dependence of the X-ray-to-optical afterglow absorption on redshift.

Finally, the tantalising evidence that GRBs may reside in more varied environmental conditions than previously speculated (i.e. with no metallicity cap), would suggest that GRBs make better tracers of the cosmic star formation rate density than had been thought, and certainly places GRBs as truly unique probes of the dust properties and ionisation state of the ISM within high-z, star-forming galaxies. How the attenuation along the GRB sight line relates to the global host galaxy properties is not yet clear, and this requires further investigation through dedicated, broadband (including nIR) afterglow follow-up campaigns to acquire a larger sample of *unbiased*, and thus more representative GRB afterglow and host galaxy observations.

References

Asplund, M., *et al.* 2009, *AEA&A*, 47, 481
Behar, E., *et al.* 2011, *ApJ*, 734, 26
Campana, S., *et al.* 2012, *MNRAS*, 421, 169
Chen, H.-W., *et al.* 2011, *ApJ*, 727, L53
D'Elia, V., *et al.* 2010, *A&A*, 523, 36
Elíasdóttir, Á., *et al.* 2009, *ApJ*, 697, 1725
Fox *et al.* 2008, *A&A*, 491, 189
Galama, T. J. & Wijers, R. A. M. J. 2001, *ApJ*, 549, L209
Greiner, J., *et al.* 2011, *A& A*, 526, A30
Kann, D. A., Klose, S., & Zeh, A. 2006, *ApJ*, 641, 993

Krühler, T., *et al.* 2011, *A&A*, 534, 108
Pei, Y. C. 1992, *ApJ*, 395, 130
Perley, D. A., *et al.* 2009, *AJ*, 138, 1690
Perley, D. A., *et al.* 2011, *AJ*, 141, 36
Prochaska, J. X., *et al.* 2008, *ApJ*, 685, 344
Prochaska, J. X., *et al.* 2009, *ApJ*, 691, L27
Savaglio, S., Glazebrook, K., & Le Borgne, D. 2009, *ApJ*, 691, 182
Schady, P., *et al.* 2007, *MNRAS*, 377, 273
Schady, P., *et al.* 2010, *MNRAS*, 401, 2773
Schady, P., *et al.* 2011, *A&A*, 525, 113
Schady, P., *et al.* 2012, *A&A*, 537, 15
Starling, R. L. C., *et al.* 2007, *ApJ*, 661, 787
Stratta, G., Fiore, F., Antonelli, L. A., Piro, L., & De Pasquale, M. 2004, *ApJ*, 608, 846
Vreeswijk *et al.* 2007, *A&A*, 468, 83
Verner, D. A., & Yakovlev, D. G. 1995, *A&AS*, 109, 124
Watson, D., *et al.* 2006, *ApJ*, 652, 1011
Watson, D., *et al.* 2007, *ApJ*, 660, 101
Zafar, D., *et al.* 2011, *A&A*, 532, 143

Discussion

DAVID BURROWS: X-ray absorption traces the total column density of metals, not just the gas phase. Also, it is difficult to get substantial X-ray absorption from highly (ultra-) ionised gas, because the temperature of that gas is very high, and the density is very low unless it is transient.

SCHADY: I agree on both those points. On the first point, I do not, however, believe that the contribution to the X-ray absorption from metals locked up in dust is large. The particular sample of GRBs that I used in my investigation into the soft X-ray excess had, by selection, high signal-to-noise optical afterglow spectra, and thus little dust extinction. As to the latter point, I would certainly argue that if the "missing gas" is indeed in a ultra-ionised state, then it must be the GRB itself that produced the high UV-radiation field required to form such a gas.

RHAANA STARLING: Do your results suggest a solar abundance pattern is correct for GRB hosts?

SCHADY: There is currently no evidence to believe otherwise. However, a different host galaxy solar abundance pattern to solar, such as an over-abundance of α-elements, cannot account for the large soft X-ray absorption excess observed.

DANIEL PERLEY: You mentioned that one explanation for the discrepancy between X-ray inferred gas abundances and abundances inferred from optical spectroscopy could be the presence of gas in a very highly ionised state. Of course, the GRB itself is a prolific source of ionising radiation. Can you address the possibility that ionisation from the GRB could be the explanation for this discrepancy?

SCHADY: As already highlighted by David, such a highly ionised gas cannot be in equilibrium, and must thus be in a transient state. In this case, ionisation by the GRB itself could produce the short-lived, highly ionised gas. Furthermore, the most plausible location for this highly ionised plasma would be within a dense region very near-in to the GRB, which again, would favour the GRB as the source of the ionising photons. Further modelling would, nevertheless, be necessary to test the validity of such a scenario.

Death of Massive Stars: Supernovae and Gamma-Ray Bursts
Proceedings IAU Symposium No. 279, 2012
P. Roming, N. Kawai & E. Pian, eds.

doi:10.1017/S1743921312012938

Type Ib/c Supernovae with and without Gamma-Ray Bursts

Maryam Modjaz

CCPP, New York University, 4 Washington Place
New York City, NY, 10003, USA
email: mmodjaz@nyu.edu

Abstract. While the connection between Long Gamma-Ray Bursts (GRBs) and Type Ib/c Supernovae (SNe Ib/c) from stripped stars has been well-established, one key outstanding question is what conditions and factors lead to each kind of explosion in massive stripped stars. One promising line of attack is to investigate what sets apart SNe Ib/c **with** GRBs from those **without** GRBs. Here, I briefly present two observational studies that probe the SN properties and the environmental metallicities of SNe Ib/c (specifically broad-lined SNe Ic) with and without GRBs. I present an analysis of expansion velocities based on published spectra and on the homogeneous spectroscopic CfA data set of over 70 SNe of Types IIb, Ib, Ic and Ic-bl, which triples the world supply of well-observed Stripped SNe. Moreover, I demonstrate that a meta-analysis of the three published SN 1b/c metallicity data sets when including only values at the SN positions to probe natal oxygen abundances, indicates at very high significance that indeed SNe Ic erupt from more metal-rich environments than SNe Ib, while SNe Ic-bl with GRBs still prefer, on average, more metal-poor sites than those without GRBs.

Keywords. (stars:) supernovae: general, gamma rays: bursts

1. Introduction

Stripped supernovae (SNe) and long-duration Gamma-Ray Bursts (long GRBs) are nature's most powerful explosions from massive stars. They energize and enrich the interstellar medium, and, like beacons, they are visible over large cosmological distances. However, the mass and metallicity range of their progenitors is not known, nor the detailed physics of the explosion (see reviews by Woosley & Bloom 2006; Smartt 2009). Stripped-envelope SNe (i.e, SNe of Types IIb, Ib, and Ic, e.g., Filippenko 1997) are core-collapse events whose massive progenitors have been stripped of progressively larger amounts of their outermost H and He envelopes (Fig. 1). In particular, broad-lined SNe Ic (SNe Ic-bl) are SNe Ic whose line widths approach 20,000−30,000 km s^{-1} around maximum light (see below) and whose optical spectra show no trace of H and He.

For the last 15 years, the exciting connection between long GRBs and SNe Ic-bl, the only type of SNe observed accompanying long GRBs (for reviews, see Woosley & Bloom 2006; Hjorth & Bloom 2011; Modjaz 2011), and the existence of many more SNe Ic-bl **without** GRBs raises the question of what distinguishes SN-GRB progenitors from those of ordinary SNe Ic-bl without GRBs. Viewing angle effects are probably not the reason why SNe Ic-bl do not show an accompanied GRB (Soderberg *et al.* 2006). Based on radio upper-limits, only $\sim 1\%$ of SNe Ib/c appear to be accompanied by GRBs (Soderberg *et al.* 2010). One promising line of attack is to investigate what sets apart SNe Ib/c **with** GRBs from those **without** GRBs to elucidate the conditions and progenitors of these two types of explosions. While of course there are numerous possible avenues (for a recent review see e.g., Modjaz 2011), I will here adopt a two-pronged approach, given the short amount of

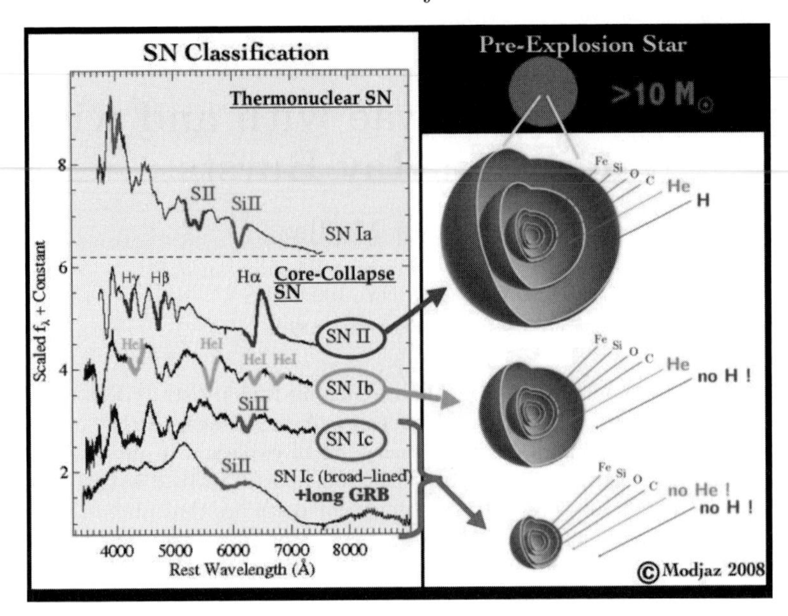

Figure 1. Mapping between different types of core-collapse SNe (*left*) and their corresponding progenitor stars (*right*). *Left*: Representative observed spectra of different types of SNe. Broad--lined SN Ic are the only type of SNe seen in conjunction with GRBs. Not shown are some of the other H-rich SN members (SNe IIn and very luminous SNe). *Right*: Schematic drawing of massive ($\geqslant$ 8–10 M$_\odot$) stars before explosion, with different amounts of intact outer layers, showing the "onion-structure" of different layers of elements that result from successive stages of nuclear fusion during the massive stars' lifetimes (except for H). This figure is found at http://cosmo.nyu.edu/ mmodjaz/research.html.

time. First, I focus on comparing the optical spectra of SNe Ib/c with and without GRBs, since early-time optical spectra are used for identifying the spectral features of different explosions and probe the bulk of the ejected stellar material, in particular the outermost layers. Secondly, I present a meta-analysis of published measured metallicities at the explosion site of SNe Ib/c with and without GRBs. Metallicity is expected to strongly impact the lives and deaths of stars due to the metallicity dependence of mass loss (e.g., Vink & de Koter 2005) and its subsequent link to rotation and angular momentum content of the stellar core. The main thrust of my talk is that now a number of different groups, including ours, have contributed to gathering large data-sets, whose analysis can to lead to robust statistical conclusions and interesting insights into different populations of SNe with and without GRBs.

2. Optical Spectra and Expansion Velocities of SNe Ib/c with and without GRBs

While the observational hallmark of a SNe Ic-bl is, by definition, its high expansion velocities (which, when modeled in combination with light curves, yields high energies, sometimes above 10^{52} erg, i.e. 10 times more than the canonical CCSN, and thus motivated some to call them "HyperNovae"), there are debates within the community whether such SNe Ic-bl can be robustly distinguished from "normal" SNe Ic and whether there are systematic differences between SNe Ic-bl with and without GRBs. Prior work involving synthetic models based on Monte Carlo radiative transfer codes (Mazzali *et al.* 2009 and reference therein), while important, has included only a few normal SNe Ic and a few SN Ic-bl without GRBs, thus not yet providing a large sample.

Here we are using the spectra from the CfA sample of Stripped SNe (Modjaz *et al.*, in prep), as well as those from the literature (see references in Modjaz *et al.*, in prep) to compare the absorption velocities as traced by Fe II λ5169 of different kinds of SNe Ic. Spectral synthesis studies have shown that this and other Fe lines are good tracers of the photospheric velocity, since they do not saturate (Branch *et al.* 2002). With the largest sample of spectra to date, we find that SN Ic-bl **with** GRBs have the **highest** absorption velocities (25,000−35,000 km s^{-1} at maximum V-light), while SNe Ic-bl **without** GRBs have **lower** velocities (between 15,000−25,000 km s^{-1} at maximum V-light), and normal SN Ic have the lowest absorption velocities (8,000−15,000 km s^{-1}). We caution that because of severe blending, specifically in SNe Ic-bl, the Fe II λ5169 line could be blended with other nearby lines such that it may compromise the velocity measurements. However other, more isolated, lines (e.g. Si II) also indicate high velocities for SN Ic-bl.

3. Measured Metallicities at the Explosion Sites of SNe Ib/c with and without GRBs

Since direct SN Ib/c progenitor detection attempts via deep pre-explosion images have not been successful (Smartt 2009) and are impossible for GRBs, we employ a complimentary approach: we study the host galaxy environments in order to discern any systematic trends as a function of explosion type that may characterize their stellar progenitors. Specifically, massive stars at different metallicities are expected to live and die differently, due to the metallicity dependence of mass loss and its subsequent link to rotation and angular momentum content of the stellar core (e.g., Crowther 2007). Since the early work on GRB host metallicities and their comparison with SDSS galaxies as well as with SN galaxies, the field of environmental metallicity studies has experienced a tremendous growth (see discussions in e.g., Levesque *et al.* 2010b; Modjaz 2011; and references therein, as well as contributions in this volume).

Here, we outline the recipe for state-of-the-art metallicity analysis, specifically of the oxygen abundance from nebular HII region emission lines, first formalized in Modjaz *et al.* (2008): 1) In order to probe the natal oxygen abundance, obtain spectra at the position of the SN or GRB (because of metallicity gradients in spiral galaxies); 2) Include only SNe with secure SN ID (i.e., ideally from multi-epoch SN spectra to monitor for any potential classification changes) and also (only) from untargetted surveys in order to mitigate any selection effects; 3) Employ spectrographs with a large wavelength range in order to observe emission lines ([OII] λ3727 to Hα and [NII] λ6584) and to compute abundances with different diagnostics, since there are systematic differences between different diagnostics (Kewley & Ellison 2008); 4) Remove stellar absorption in spectra when necessary, and 5) Obtain a good handle on uncertainty budget and propagate line flux and reddening uncertainties into abundance measurement errors via Monte Carlo simulations.

While different groups have recently arrived at different conclusions about whether there is a statistically significant trend of metallicity with Stripped SN subtype (Anderson *et al.* 2010; Modjaz *et al.* 2011; Leloudas *et al.* 2011; see also proceedings by Anderson, Leloudas in this volume), not all measured metallicities reported in the Anderson and Leloudas samples are at the position of the SNe. Thus, we conducted a meta-analysis of all samples (Modjaz *et al.* 2008; Anderson *et al.* 2010; Modjaz *et al.* 2011; Leloudas *et al.* 2011) with the best-possible quality-control and following the above state-of-the-art

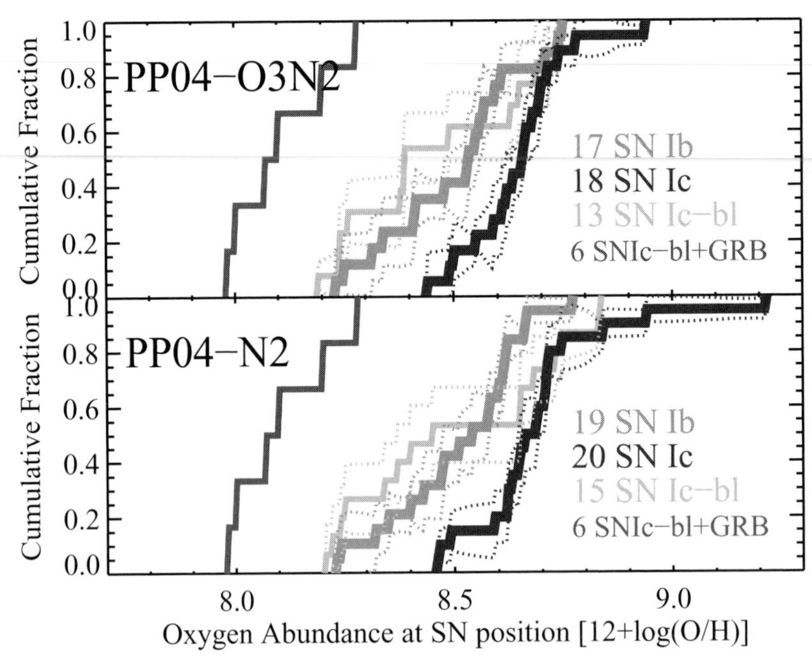

Figure 2. Cumulative fraction (solid lines) of measured oxygen abundances at the SN position of different types of SNe Ib/c with and without GRBs and their confidence bands (dotted lines), based on the meta-analysis of the SN samples in the literature (Modjaz *et al.* 2008; Anderson *et al.* 2010; Modjaz *et al.* 2011; Leloudas *et al.* 2011; and references therein) and in the two scales of Pettini & Pagel (2004) that all samples had in common. The ordinate indicates the fraction of the SN population with metallicities less than the abscissa value. The confidence bands are shown around each cumulative trend, which we computed via bootstrap with 10,000 realizations based on the metallicity measurements and their associated reported uncertainties. SNe Ic (the demise of the most heavily stripped stars that lost much, if not all, of both their H and He layers) are systematically in more metal-rich environments than SNe Ib (SNe arising from less stripped stars that retained their He layer). SN Ic-bl with GRBs are still at consistently lower oxygen abundances than SN Ic-bl without GRBs. The radio-loud SN Ic-bl 2009bb (Soderberg *et al.* 2010) is not included in this plot, since no GRB was detected, but would add one data point at $12+\log(O/H)_{PP04-O3N2}=8.9$ (Levesque *et al.* 2010b).

recipe. Now that we have larger samples to draw from, we only included oxygen abundance measurements at the exact SN explosion sites (within the slit) of SNe with solid IDs and also from untargeted surveys, to have the best handle possible on the natal metallicity estimates of SNe with well-determined SN types over a large metallicity baseline, the ultimate goal of the study. Figure 2 shows the result of our metal-analysis, namely the cumulative distributions of local metallicities for different types of stripped CCSNe (SNe Ib, Ic, Ic-bl without GRBs, SNe Ic-bl with GRBs) from the combined samples. We find that with a combined and large sample size, the sites of SNe Ic do indeed have higher oxygen abundances than SNe Ib, and with a higher statistical significance than in the individual samples. There is only a 0.1% (2%) probability in the PP04-N2 (PP04-O3N2) scale that the oxygen abundances of the 19 (17) SNe Ib and of 20 (18) SNe Ic are drawn from the same parent population, which are different on average by ~ 0.2 dex. Here we have taken advantage of the power of statistics by combining the hard work of three different groups.

In addition, SN Ic-bl with GRBs still prefer, on average, more metal-poor environments than those without GRBs (see Fig. 2 in Modjaz *et al.* 2011), with the GRB metallicity-luminosity relation offset to lower metallicities, but without a cut-off metallicity above which GRB production would be suppressed (Levesque *et al.* 2010a). Since the host galaxies of both samples span similar ranges in galaxy luminosity (i.e., even to luminous GRB host galaxies of $M_B = -21\,\mathrm{mag}$), dust effects are most likely not the reason for the offset to low metallicity. However, while these results are intriguing, the next step is to conduct a thorough and extensive host galaxy study with a large single-survey, untargeted, spectroscopically classified, and homogeneous collection of stripped SNe, something we are currently undertaking with the Palomar Transient Factory (PTF).

M.M. acknowledges current support from the NYU ADVANCE Women-in-Science Travel Grant funded by the NSF ADVANCE-PAID award Number HRD-0820202 and prior support from Hubble Fellowship grant HST-HF-51277.01-A.

References

Anderson, J. P. *et al.* 2010, *MNRAS*, 407, 2660
Branch, D., *et al.* 2002, *ApJ*, 566, 1005
Crowther, P. A. 2007, *ARAA*, 45, 177
Filippenko, A. V. 1997, *ARAA*, 35, 309
Hjorth, J. & Bloom, J. S. 2011, Chapter 9 in "Gamma-Ray Bursts" (arXiv:1104.2274)
Kewley, L. J. & Ellison, S. L. 2008, *ApJ*, 681, 1183
Leloudas, G., *et al.* 2011, *A&A*, 530, A95
Levesque, E. M., Berger, E., Kewley, L. J., & Bagley, M. M. 2010a, *AJ*, 139, 694
Levesque, E. M., *et al.* 2010b, *ApJ*, 709, L26
Mazzali, P. A., Deng, J., Hamuy, M., & Nomoto, K. 2009, *ApJ*, 703, 1624
Modjaz, M., *et al.* 2008, *AJ*, 135, 1136
Modjaz, M., *et al.* 2011, *ApJ*, 731, L4
Modjaz, M. 2011, *AN*, 332, 434
Pettini, M. & Pagel, B. E. J. 2004, *MNRAS*, 348, L59
Smartt, S. J. 2009, *ARAA*, 47, 63
Soderberg, A. M., *et al.* 2006, *ApJ*, 638, 930
Soderberg, A. M., *et al.* 2010, *Nature*, 463, 513
Vink, J. S. & de Koter, A. 2005, *A&A*, 442, 587
Woosley, S. E. & Bloom, J. S. 2006, *ARAA*, 44, 507

Discussion

ANDERSON (Q): Is there a bias in PTF for studying SNe in dim hosts? As a community we have to be careful not to over-emphasize them.

MODJAZ (A): As we had discussed in Sydney, PTF has the current strategy of making sure to obtain spectroscopic classification of SNe in low-luminosity hosts, so that we have a complete view of what kinds of SNe erupt in the neglected bin of low-luminosity galaxies. However, this selection should not skew the results of demographic studies – the SN IDs are obtained in those low-luminosity galaxies independent of the SN type. But I agree that this strategy would not lend itself well for computing SN rates if there is no well-defined selection function.

RYAN CHORNOCK (Q): Do you see signs of Wolf-Rayet stars in the spectra of the HII regions at the explosion sites?

MODJAZ (A): For some of them, yes, and a current student of mine, David Fierroz, is analyzing the data – the published spectra, as well as other ones we obtained from Keck.

Death of Massive Stars: Supernovae and Gamma-Ray Bursts
Proceedings IAU Symposium No. 279, 2012
P. Roming, N. Kawai & E. Pian, eds.

doi:10.1017/S174392131201294X

Unveiling the fundamental properties of Gamma-Ray Burst host galaxies

Sandra Savaglio

Max Planck Institute for extraterrestrial Physics,
Giessenbachstr., 85741, Garching, Germany
email: savaglio@mpe.mpg.de

Abstract. The galaxies hosting the most energetic explosions in the universe, the gamma-ray bursts (GRBs), are generally found to be low-mass, metal-poor, blue and star forming. However, the majority of the targets investigated so far (less than 100) are at relatively low redshift, $z < 2$. We know that at low redshift, the cosmic star formation is predominantly in small galaxies. Therefore, at low redshift, long-duration GRBs, which are associated with massive stars, are expected to be in small galaxies. Preliminary investigations of the stellar mass function of $z < 1.5$ GRB hosts does not indicate that these galaxies are different from the general population of nearby star-forming galaxies. At high-z, it is still unclear whether GRB hosts are different. Recent results indicate that a fraction of them might be in dusty regions of massive galaxies. Remarkable is the a super-solar metallicity measured in the interstellar medium of a $z = 3.57$ GRB host.

Keywords. Gamma rays: bursts, observations, ISM: abundances, galaxies: evolution

1. Introduction

GRB host galaxies are traditionally thought to be associated with small, metal- and dust-poor, star forming galaxies. However, most of these galaxies studied in detail are at redshift smaller than $z = 1.5$. We still do not know whether the situation at higher redshift is different. We know with high confidence that the progenitor of most GRBs, the long-duration ones, is a massive and short-lived star ($M > 30$ $M_\odot$; Heger *et al.* 2003). Therefore, their hosts are likely star forming galaxies. The star-forming galaxy population experienced major changes in the global history of the universe, for instance, for the star-formation rate (SFR), the galaxy stellar mass, and the merger rate.

The cosmic star-formation rate dropped by a factor of 50 from $z \sim 1.8$ (Hayes *et al.* 2010), and it transited from large galaxies in the past to small galaxies today (Juneau *et al.* 2005). The evolution of the fraction of major mergers of massive galaxies in the same time interval dropped by a factor of ~ 25 (Bluck *et al.* 2012). The assembly growth of galaxy stellar mass shows that the mass density increased by only a factor of $\sim 50\%$ since $z = 1$ (Sobral *et al.* 2012). The mass-metallicity relation in the local universe, and its redshift evolution, show that large galaxies reached high metallicities early on, while small galaxies are chemically more slow (Savaglio *et al.* 2005). These relations are affected by the galaxy SFR, such that, for a given stellar mass, metallicities tend to be lower for higher SFRs (Mannucci *et al.* 2010).

Therefore, if from one hand it is not surprising that, in the local universe, most GRBs occur in small star-forming and metal-poor galaxies, at high redshift, more massive galaxies, active and metal-rich galaxies might have hosted a large fraction of the events. Numerous new and deeper observations start to suggest that the canonical view might be affected by a combination of the difficulty of detecting distant targets (Krühler *et al.*

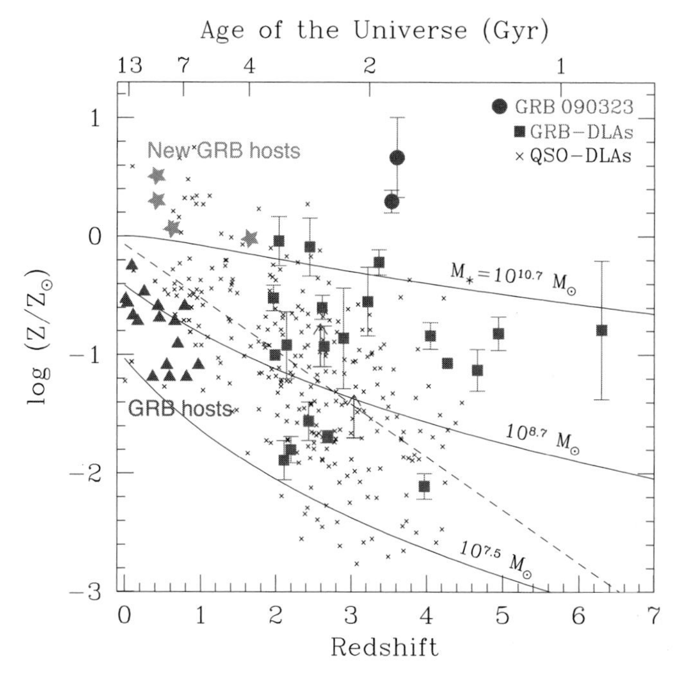

Figure 1. Metallicity as a function of redshift in galaxies. Metallicities in the two absorbers detected in GRB 090323 (blue dots) are from Savaglio *et al.* (2012). Red squares are metallicities of other GRB-DLAs. Blue triangles are GRB host metallicities measured from emission lines at low redshift (Savaglio *et al.* 2009). More recent metal determinations show high metallicities (Levesque *et al.* 2010; Perley *et al.* 2011; Krühler *et al.* 2012; Niino *et al.* 2012). Metallicities in other high-z absorbers detected in QSO spectra (QSO-DLAs) are black crosses. The dashed line is the linear correlation for QSO-DLA points. Solid curves are average metallicities expected for star-forming galaxies with different stellar masses (Savaglio *et al.* 2005).

2011) and the redshift evolution of galaxy fundamental parameters. The peculiar strong double absorption system recently studied in the spectrum of the distant GRB 090323 ($z = 3.57$), and the measured super-solar metallicity (Fig. 1) supports the same idea (Savaglio *et al.* 2012). One of the next major goals is to study the stellar-mass function of galaxies hosting GRBs.

2. The stellar mass of GRB hosts

The investigation of the mass function (MF) of GRB hosts is vaguely possible, if at all, at low redshift only, due to the small number statistics. The hosting galaxy is studied in only half of all GRBs with known redshift (more than 240), mainly at $z < 2.3$ (87 galaxies, 71% of the total). The galaxy MF is a fundamental mean through which the cosmic change of galaxies can be identified. Widely investigated in the local universe (e.g., Baldry *et al.* 2008), and at high redshift (Santini *et al.* 2011), it shows that small galaxies are the most common ones at $z = 0$, and even more so at $z > 2$.

The MF was never derived for galaxies hosting GRBs, due to the small number statistics and the difficulty in defining the sample completeness. Nevertheless, the identification of GRB hosts, which is, to a first-order approximation, independent of the galaxy brightness, makes the investigation of the $z > 0$ MF in the low stellar-mass regime ($M_* < 10^{10}$ M$_\odot$) a possible task. That will help to establish whether these galaxies belong to a unique population, or naturally fill the low-mass end of the galaxy MF. The 45 GRB hosts studied by Savaglio *et al.* (2009) did not show evidence for deviation from normal galaxies. In

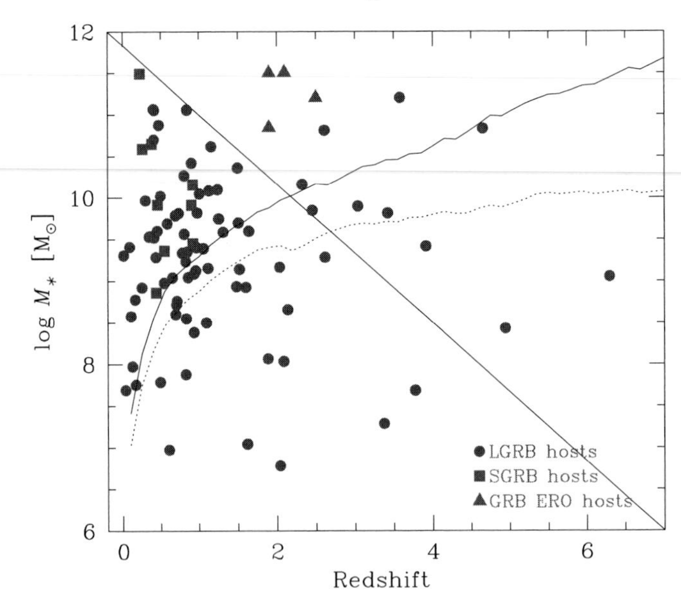

Figure 2. Stellar mass of GRB hosts as a function of redshift. Circles and squares are galaxies associated with long and short GRBs with spectroscopic redshift, respectively (Savaglio *et al.* in prep.). Triangles are galaxies with photometric redshifts associated with extremely red objects (see Hunt *et al.* 2011 for details). The solid and dashed lines represent the stellar mass as a function of redshift of a galaxy with AB K-band magnitude $m_K = 24.3$, and old stellar population or constant SFR, respectively.

our ongoing work (Savaglio *et al.*, in prep.), the GRB host sample is more than 2.5 times larger. Fig. 2 displays the preliminary stellar mass as a function of redshift for 88 hosts with spectroscopic redshift, and 4 hosts associated with extremely red objects (EROs) with photometric redshift. A large fraction of GRB hosts are massive, in particular those associated with EROs (see Hunt *et al.* 2011, and details therein).

To investigate the GRB host MF, one needs to select the sample according to the volume limited criteria. This is the most critical part, relatively under control for a large sample. In any magnitude-limited galaxy sample, the mass distribution is dominated by massive galaxies. When one corrects for volume, then low-mass galaxies dominate. The sample selection for GRB hosts is very different. If 100% of detected GRBs were followed up and galaxy masses measured, then part of the relevant volume calculation would be the flux limit of the GRB-detecting telescopes (e.g., *Swift*) as a function of redshift.

In our preliminary study, we just investigated the shape of the MF, by normalizing it to that of field galaxies. The MF was estimated for 31 GRB hosts at $z < 1.5$ and stellar mass $M_* > 10^{9.25}$ M$_\odot$. Field galaxies in the local universe (Baldry *et al.* 2008) and star-forming galaxies at $z \sim 1$ (Pozzetti *et al.* 2010; Gilbank *et al.* 2011) indicate that the general shape of the MF did not change remarkably. Today we have ~ 3 times more galaxies with $M_* > 3\times10^8$ M$_\odot$ than back then. The GRB-host MF shows a similar shape. GRB hosts have a reputation of being small star-forming galaxies, but this is not apparent from their relatively flat MF.

The most important issue to solve now is the primary selection function of *Swift* in a luminosity vs. redshift plot. We will have to determine whether the actual luminosity distribution of GRBs (in γ-rays) is a gaussian, a power law, or a Schechter luminosity function. Once this is under control, we can apply the volume-limited calculation.

3. Conclusion

The importance of investigating galaxies hosting GRBs has become evident because of their connection with the most active, observationally hostile, and remote regions of the universe. Recent works have shown that some (long) GRB hosts are metal-rich, massive, and dusty, with high star-formation rates, in contrast with the low-metallicity, low-mass hosts commonly found at $z < 1.5$. The impact on cosmology is still limited by the small number statistics. One important step forward is the multi-wavelength (from X-ray to radio) approach, which can statistically quantify the importance of red galaxies associated with dark GRBs. Exploitation of the long wavelength regime is now possible thanks to the capabilities of new ground-based telescopes (SCUBA-2, APEX, ALMA, ATCA) and space missions (Spitzer, Herschel, WISE). This will ultimately establish whether some high-z GRB hosts are connected to dusty sub-millimeter galaxies.

Acknowledgement

The author is particularly indebted to S. Basa, J. Greiner, K. Glazebrook, L. Hunt, T. Krühler, D. Le Borgne, M. Michałowski, E. Palazzi, A. Rau, A. Rossi, and P. Schady.

References

Baldry, I. K., Glazebrook, K., & Driver, S. P., 2008, *MNRAS*, 388, 945
Bluck, A. F. L., Conselice, C. J., Buitrago, F., *et al.* 2012, *ApJ*, 747, 34
Gilbank, D. G., Bower, R. G., Glazebrook, K., *et al.*, 2011, *MNRAS* , 414, 304
Hayes, M., Schaerer, D., & Östlin, G. 2010 *A&A*, 509, L5
Heger, A., Fryer, C. L., Woosley, S. E., Langer, N., & Hartmann, D. H. 2003 *ApJ*, 591, 288
Hunt, L., Palazzi, E., Rossi, A., *et al.* 2011, *ApJ*, 736, L36
Juneau, S., Glazebrook, K., Crampton, D., *et al.* 2005, *ApJ*, 619, L135
Krühler, T., Fynbo, J. P. U., Geier, S., *et al.*, 2012, *A&A*, submitted, arXiv:1203.1919
Krühler, T., Greiner, J., Schady, P., *et al.* 2011, *A&A*, 534, A108
Levesque, E. M., Kewley, L. J., Graham, J. F., & Fruchter, A. S. 2010, *ApJ*, 712, L26
Mannucci, F., Cresci, G., Maiolino, R., Marconi, A., & Gnerucci, A., 2010, *MNRAS*, 408, 2115
Niino, Y., Hashimoto, T., Aoki, K., *et al.* 2012, *PASJ*, submitted, arXiv:1204.0583
Perley, D. A., Modjaz, M., Morgan, A. N., *et al.* 2011, *ApJ*, submitted, arXiv:1112.3963
Pozzetti, L., Bolzonella, M., Zucca, E., *et al.*, 2010, *A&A*, 523, A13
Santini, P., Fontana, A., Grazian, A., *et al.*, 2012, *A&A*, 538, A33
Savaglio, S., Glazebrook, K., Le Borgne, D., *et al.*, 2005, *ApJ*, 635, 260
Savaglio, S., Glazebrook, K., & Le Borgne, D., 2009, *ApJ*, 691, 182
Savaglio, S., Rau, A., Greiner, J., *et al.* 2012, *MNRAS*, 420, 627
Sobral, D., Smail, I., Best, P. N., *et al.*, 2012, *MNRAS*, submitted, arXiv:1202.3436

Discussion

PERLEY: Regarding the point about long-wavelength observations being essential at $z > 1.5$, I certainly agree. In the past year we've been getting ground-based sub-millimeter/radio data on several dark GRB hosts and so far seeing very few bright ULIRG-like systems. Since it was mentioned in your abstract, I was wondering if you had any Herschel observations that might shed light on this question as well.

SAVAGLIO: We are in the process of receiving Herschel data for a sample of 13 GRB hosts. They are selected for being associated with dark GRBs. A few are already detected. We will study in detail the spectral energy distributions (SED) spanning roughly a factor of 1,000 in wavelength. The SED fitting will give stellar mass, bolometric luminosity, and star-formation rate at redshifts $z \leqslant 2$.

Death of Massive Stars: Supernovae and Gamma-Ray Bursts
Proceedings IAU Symposium No. 279, 2012
P. Roming, N. Kawai & E. Pian, eds.

doi:10.1017/S1743921312012951

Star Formation in the Early Universe

Kazuyuki Omukai

Department of Physics, Kyoto University, Kyoto, Japan
email: omukai@tap.scphys.kyoto-u.ac.jp

Abstract. In low-metallicity environments, massive stars are more easily formed than in the solar neighborhood. In this article, we see the following examples of low-mass star formation. We first describe the first star formation in the universe and argue that they are typically ordinary-sized massive stars, rather than very massive ($> 100 M_\odot$) ones. Next, we see how the metal-enrichment changes the thermal evolution of gas, thereby causing the shift of characteristic stellar mass towards lower mass. Finally, we discuss the possibility of forming supermassive stars in some special conditions in young galaxies.

Keywords. Star Formation, Pop II Stars, Pop III Stars

---

1. Metallicity and Massive Star Formation

Stars are formed as a result of gravitational collapse of dense molecular cores and subsequent accretion of ambient matter. In the standard picture of low-mass star formation, the accretion rate is approximately given by dividing the Jeans mass, $M_{\rm Jeans}$, by the free-fall time $t_{\rm ff}$: $\dot{M}_* \sim M_{\rm Jeans}/t_{\rm ff} \simeq c_{\rm s}^3/G \simeq 2 \times 10^{-6} M_\odot/{\rm yr} \left(\frac{T}{10{\rm K}}\right)^{3/2}$, where $c_{\rm s}$ is the sound speed in the dense core.

In trying to apply the same formation scenario to massive stars ($>$ a few 10s $M_\odot$) just with prolonged accretion, however, we meet the following difficulties. First, with accretion rates as low as $10^{-6} M_\odot/{\rm yr}$, the time needed to form a massive star (with mass M_*) $t_{\rm acc} = M_*/\dot{M}_*$ exceeds the stellar lifetime $t_{\rm OB} \simeq 3 \times 10^6$yr. Second, the radiation pressure onto the dust grains in the accretion flow becomes stronger than the ram pressure of the flow, thereby halting the accretion (e.g., Wolfire & Cassinelli 1987).

In low-metallicity (and thus low dust-depletion) gas, due to inefficient cooling, clouds are warmer than the present-day counterparts, which results in the higher accretion rate. In addition, the reduced dust opacity means lower radiation pressure. Those effects mitigate the above mentioned obstacles in massive star formation. We thus naturally expect that massive stars are more abundant in low-metallicity environments.

In the following, we see specific examples of low-metallicity star formation. In Sec. 2, we describe the first star formation in the universe and argue that they are typically ordinary-sized massive stars, rather than very massive ($> 100 M_\odot$) ones as previously considered. Next, in Sec. 3, we see how the metal-enrichment changes the thermal evolution of gas, thereby causing a shift of the characteristic stellar mass towards lower mass. Finally, in Sec. 4, we discuss the possibility of forming supermassive stars (SMSs) in some special conditions in young galaxies.

2. First Stars: Massive but Not Very Massive

According to our current understanding of star formation, the dense core of a gas cloud gravitationally contracts in a non-homologous run-away fashion, whereby the densest parts become denser faster than the rest of the cloud. In a primordial gas cloud one or a few embryo protostars are formed near the center. The initial mass of these embryo protostars is only $\simeq 0.01\ M_{\odot}$; the bulk of the dense core material remains in the surrounding envelope and is subsequently drawn toward the protostar(s) through gravity. With the typical angular momentum of dense cores, the centrifugal barrier allows only a small amount of infalling gas to accrete directly onto the star. Instead, a circumstellar disk is formed and the gas is accreted onto the central star through the disk. The final mass of these first stars is fixed, when the mass accretion terminates.

Here, we report the calculation by Hosokawa *et al.* (2011). We follow the radiation hydrodynamic evolution in the vicinity of an accreting protostar, incorporating thermal and chemical processes in the primordial gas in a direct manner. We also follow the evolution of the central protostar self-consistently by solving the detailed structure of the stellar interior with zero metallicity as well as the accretion flow near the stellar surface. We configure the initial conditions by using the results of a three-dimensional cosmological simulation, which followed the entire history from primeval density fluctuations to the birth of a small "seed" protostar at the cosmological redshift 14. Specifically, when the maximum particle number density reaches 10^6 cm^{-3} in the cosmological simulation, we consider a gravitationally bound sphere of radius 0.3 pc around the density peak, which encloses a total gas mass of $\simeq 300\ M_{\odot}$. We reduce the three-dimensional data to an axisymmetric structure by averaging over azimuthal angles.

At its birth a very small protostar of $\sim 0.01\ M_{\odot}$ is surrounded by a molecular gas envelope of $\sim 1\ M_{\odot}$, which is quickly accreted onto the protostar. Atomic gas further out initially has too much angular momentum to be accreted directly and a circumstellar disk forms. The infalling atomic gas first hits the disk plane roughly vertically at supersonic velocities. A shock front forms; behind the shock the gas cools, settles onto the disk, and its hydrogen is converted to the molecular form via rapid gas phase three-body reactions. The molecular disk extends out to $\simeq 400$ AU from the protostar, when the stellar mass is $10\ M_{\odot}$. Accretion onto the protostar proceeds through this molecular disk as angular momentum is transported outward. The accretion rate onto the protostar is $\simeq 1.6 \times 10^{-3}\ M_{\odot}\ \mathrm{yr}^{-1}$ at this moment.

Fig. 1 depicts the highly dynamical behavior of the circumstellar material in the late KH contraction phase. When the stellar mass is $20\ M_{\odot}$, an ionized region rapidly expands in a bipolar shape perpendicular to the disk, where gas is cleared away (Fig. 1-a). At this moment the disk extends out to $\simeq 600$ AU. The disk is self-shielded against the stellar H_2-dissociating ($11.2\ \mathrm{eV} \leqslant h\nu \leqslant 13.6\ \mathrm{eV}$) as well as the ionizing radiation. The HII region continues to grow and finally breaks out of the accreting envelope. At $M_* \simeq 25\ M_{\odot}$, the size of the bipolar HII region exceeds 0.1 pc (Fig. 1-b). Because of the high pressure of the heated ionized gas, the opening angle of the ionized region also increases as the star grows (Fig. 1-c). Shocks propagate into the envelope preceding the expansion of the ionized region. The shocked gas is accelerated outward at a velocity of several kilometers per second. The shock even reaches regions shielded against direct stellar UV irradiation. The outflowing gas stops the infall of material from the envelope onto the disk. Without the replenishment of disk material from the envelope the accretion rate onto the protostar decreases (Fig. 2). In addition, the absence of accreting material onto the circumstellar disk means that the disk is exposed to the intense ionizing radiation from the star. The resulting photoevaporation of disk gas also reduces the accretion rate onto the protostar. The photoevaporated gas escapes toward the polar direction within the ionized region. The typical velocity of the flow is several tens of kilometers per second, comparable to the sound speed of the ionized gas, which is high enough for the evaporating flow to escape from the gravitational potential well of the dark matter halo.

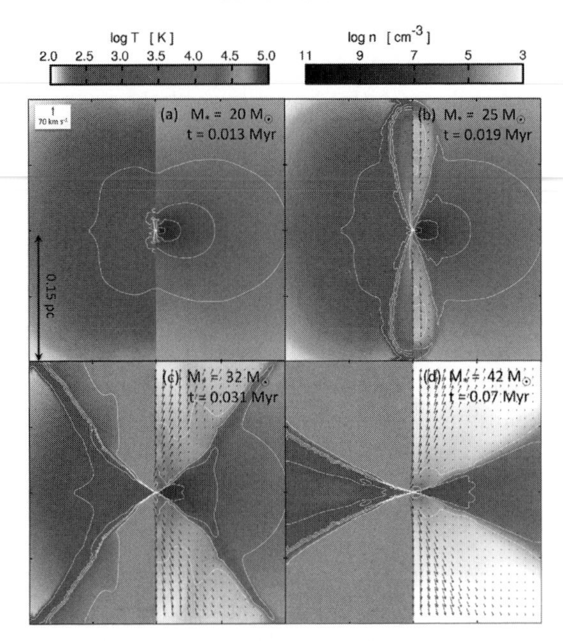

Figure 1. UV radiative feedback from the primordial protostar. The spatial distributions of gas temperature (*left*), number density (*right*), and velocity (*right, arrows*) are presented in each panel for the central regions of the computational domain. The four panels show snapshots at times, when the stellar mass is $M_* = 20\ M_\odot$ (*panel a*), 25 $M_\odot$ (*b*), 35 $M_\odot$ (*c*), and 42 $M_\odot$ (*d*). The elapsed time since the birth of the primordial protostar is labeled in each panel. Figure from Hosokawa *et al.* (2011)

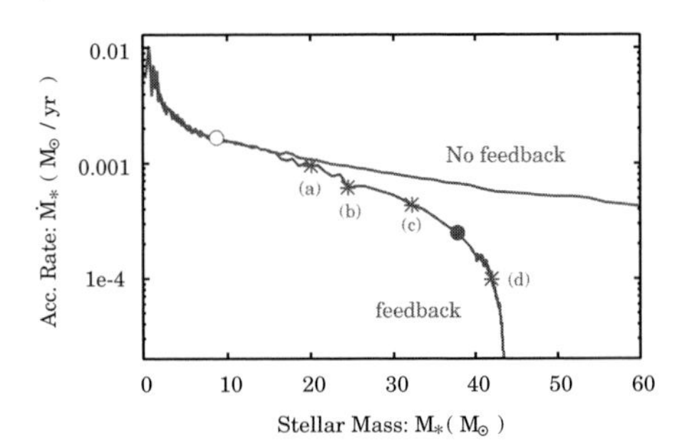

Figure 2. Evolution of the accretion rate onto the primordial protostar. The blue line represents the evolution, which includes the effect of UV radiative feedback from the protostar. The red line depicts a numerical experiment with no UV feedback. The open and filled circles denote the characteristic epochs of the protostellar evolution, beginning of the KH contraction and the protostar's arrival to the ZAMS. The panels (a) - (d) in Figure 1 show the snapshots at the moments marked with the asterisks. Figure from Hosokawa *et al.* (2011)

When central nuclear hydrogen burning first commences at a stellar mass of 35 $M_\odot$, it is via the pp-chain normally associated with low mass stars. The primordial material does not have the nuclear catalysts necessary for CNO-cycle hydrogen burning. Because the pp-chain alone cannot produce nuclear energy at the rate necessary to cover the radiative energy loss from the stellar surface, the star continues to contract until central temperatures and densities attain values that enable the 3-α process (helium burning). The product of helium burning is carbon and once the relative mass abundance of carbon reaches $\sim 10^{-12}$, CNO-cycle hydrogen burning takes over as the principal source of nuclear energy production, albeit at much higher central densities and temperatures than

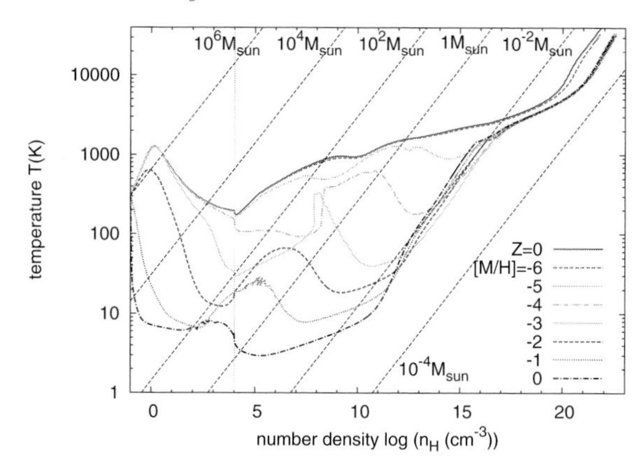

Figure 3. Temperature evolution at the center of pre-stellar clouds with different metallicities. Those with metallicities [M/H]=-∞ (Z=0), -6, -5, -4, -3, -2, -1 and 0 are shown. No external radiation is imposed. The lines for constant Jeans mass are indicated by thin dotted lines. The thermal evolution is computed by a one-zone model up to the number density of 10^4cm^{-3} (dashed horizontal line), and by a hydrodynamical model for higher density. Figure from Omukai *et al.* (2010)

stars with solar abundances. These first generation ZAMS (zero age main sequence) stars are thus more compact and hotter than their present day counterparts of equal mass. By the time the star attains 40 $M_{\odot}$, the entire region above and below the disk (Fig. 1-d) is ionized. Mass accretion is terminated when the stellar mass is 45 $M_{\odot}$ (Fig. 2).

Our calculation shows that the first stars regulate their growth by their own radiation. The first stars are not extremely massive, but rather similar in mass to the O-type stars in our Galaxy. This resolves a long-standing enigma regarding the elemental abundance patterns of the Galactic oldest metal-poor stars, which contain atmospheric signatures from the earliest generation of stars. If a significant number of first stars had masses in excess of 100 $M_{\odot}$, they would end their lives through pair-instability supernovae, expelling heavy elements that would imprint a characteristic nucleosynthetic signature to the elemental abundances in metal-poor stars. However, no such signatures have been detected in the metal-poor stars in the Galactic halo. For example, metals released by pair-instability supernovae are characterized by a small relative abundance of Zn to Fe, in contrast to observations. Detailed spectroscopic studies of extremely metal-deficient stars rather suggest that the metal-poor stars were born in an interstellar medium that had been metal-enriched by supernovae with ordinary massive stellar progenitors. If first stars have masses in the range of several tens $M_{\odot}$, as predicted by our calculations, and ultimately explode as core-collapse supernovae, the associated nucleosynthetic signatures are consistent with those observed in hyper-metal-poor stars.

3. Effect of Metal-Enrichment: Pop III/II Transition

The first stars in the universe were massive, typically $\sim$ a few 10s $M_{\odot}$. Stars in the solar neighborhood, on the other hand, are typically low-mass objects: the initial mass function exhibits a peak around $0.1 - 1M_{\odot}$. The origin of these characteristic masses can be attributed to the thermal evolution of pre-stellar clouds (e.g., Larson 2005).

In Figure 3, we show the temperature evolution of star-forming cores with different metallicities (Omukai *et al.* 2010). This is calculated by using a radiation hydrodynamical calculation assuming spherical symmetry. The initial density of the hydrodynamical calculation is taken at 10^4cm^{-3}, until which, we followed the evolution by a one-zone model. Here, the dust to metal ratio is assumed the same as the local ISM value. The dotted lines indicate the constant Jeans mass.

The temperature of metal-free gas has a minimum around $10^4 \mathrm{cm}^{-3}$, where the rotational levels of H_2 reach LTE and its cooling rate saturates. Up to this point, the rapid cooling allows the temperature to decrease with increasing density, i.e., the effective ratio of specific heat $\gamma = d\log p/d\log\rho < 1$. In such a condition, clouds in shapes other than the sphere, e.g., filaments, which eventually fragment into smaller pieces, can also collapse gravitationally. However, once the temperature begins to increase with density, filamentary clouds are no longer able to collapse due to increasing pressure support, and only spherical clouds can collapse thereafter. As a consequence, spherical clouds are formed at this moment. Since spherical clouds are hard to fragment, the typical epoch of fragmentation is around this temperature minimum, and the Jeans mass there is imprinted as the characteristic fragmentation mass scale. This has been confirmed by numerical simulations adopting turbulent initial conditions (Li *et al.* 2003; Jappsen *et al.* 2005). In the case of zero-metallicity gas, this characteristic fragmentation mass is given by the Jeans mass around $10^4 \mathrm{cm}^{-3}$ and is about $1000 M_{\odot}$.

On the other hand, the solar-metallicity gas has a temperature minimum at $\sim 10^5 \mathrm{cm}^{-3}$. This is due to the cooling by infrared emission of the dust, which makes the temperature decreases until the thermal coupling of the gas and dust is reached. The Jeans mass at this temperature minimum is $\sim M_{\odot}$, which agrees with the observed characteristic mass in the solar neighborhood.

How about the slightly metal-enriched clouds? Below the metallicity $[M/H] = -6$, the temperature evolution is the same as that of zero-metallicity. With metallicity of $[M/H] \sim -(4-5)$, the evolutionary track of the temperature has two minima: the first one at lower density is due to line-emission cooling, i.e., H_2 and HD cooling at low ($[M/H] <\sim -3.5$) metallicity, and to C and O fine-structure line cooling at higher metallicity. The second temperature minimum at higher density is due to dust cooling. The existence of the two temperature minima corresponds to two fragmentation epochs. Note that the fragmentation mass scales, which are given by the Jeans masses at the temperature minima, depend only moderately on the metallicity, being very large, 100-$1000 M_{\odot}$ when associated to line-cooling, and low, 0.1-$1 M_{\odot}$, when associated to dust-cooling. It is only dust-cooling that enables low-mass fragmentation, which implies that dust is indispensable for the formation of low-mass stars in low-metallicity gas.

Then, how much metallicity, or correctly speaking, the amount of dust, is needed for causing fragmentation and thus producing low-mass cores? This threshold value is often called *the critical metallicity* in literature (e.g., Schneider *et al.* 2002). To pin down this value, we have studied the evolution of pre-stellar cores in the dust-cooling phase by SPH simulations (Tsuribe & Omukai 2006). We assumed that the cores are initially moderately elongated and have some random velocity perturbations. We found that (i) important quantities for telling whether the cores fragment are the elongation of cores defined by (axis ratio) - 1, and the elongation factor, that is, how many times the elongation grows from the initial value at the onset of dust-cooling phase; (ii) if the elongation is order of unity at the onset of dust-cooling phase and the elongation factor calculated by the linear theory exceeds $\simeq 3$, the core becomes very elongated with the actual elongation factor ~ 30 and finally fragments into many pieces during the dust-cooling phase. By applying this criterion to the thermal evolution presented in Figure 3, we find that as long as a core is moderately elongated at the beginning of the dust cooling, it will fragment for $[M/H] > -5$.

Thus, as a rule of thumb, we may say that the critical metallicity is $[M/H]_{\mathrm{cr}} \simeq -5$, although the exact value depends on the dust model we assume and may change by as much as an order of magnitude (e.g., Schneider *et al.* 2006). In any case, we conclude that only with slight metal enrichment, low-mass fragments can be produced as long as a sizable fraction of metals have condensed in the dust.

4. Possibility of Supermassive Star Formation

In this section, we consider the possibility of SMS formation in a rare environment in first galaxies. If formed, these can be the seeds of supermassive black holes that exist already when the age of the universe is $\lesssim 1$ Gyr (e.g., Mortlock *et al.* 2011).

The possibility of massive seed BH formation by direct collapse of SMS ($\gtrsim 10^5$ M$_\odot$) has been considered by some authors. Specifically, SMS formation in massive halos ($T_{\rm vir} \gtrsim 10^4$ K) irradiated with strong far ultraviolet (FUV) radiation has often been studied (e.g., Shang, Bryan & Haiman 2010). Since the H_2 molecule, the main coolant in the primordial gas, is photo-dissociated with strong FUV radiation in the Lyman and Werner bands, clouds under such an environment collapse isothermally at $\sim$ 8000 K by Lyα cooling without fragmentation, if they are massive enough ($\gtrsim 10^5$ M$_\odot$). As an outcome of such collapse, SMSs are expected to form. A massive seed BH as a remnant of SMS collapse reduces the growth time to 10^9 M$_\odot$ within 0.46 Gyr and mitigates the growth-time problem by a big margin. This scenario, however, has a serious drawback: for this mechanism of SMS formation to work, extremely strong FUV radiation $J_{21}^{\rm LW} \gtrsim 10^2 - 10^3$ (in units of 10^{-21} erg s^{-1} cm^{-2} Hz^{-1} sr^{-1}) is required (Omukai 2001; Shang, Bryan & Haiman 2010), while the fraction of halos irradiated with such intense FUV fields with $J_{21}^{\rm LW} \gtrsim 10^3$ is estimated to be $\lesssim 10^{-6}$ at $z \sim 10$ (Dijkstra *et al.* 2008), i.e., only extremely rare halos satisfy the condition for SMS formation. Although the above scenario might be still viable considering the rarity of high-z SMBHs, it is worthwhile to explore another possibility.

Here, we propose a new scenario for SMS formation in post-shock gas of cold accretion flows in forming first galaxies. Recent numerical simulations of galaxy formation have revealed that, in halos with virial temperature $T_{\rm vir} \gtrsim 10^4$ K, the accreting cold gas penetrates deep to the center through dense filamentary flows (e.g., Dekel & Birnboim 2006). The supersonic flows collide with each other and the resultant shock develops a hot and dense ($\sim 10^4$ K and $\sim 10^3$ cm^{-3}) gas near the center. By studying thermal evolution of the shocked gas, we find that, if the post-shock density is high enough for the H_2 vibrational levels to reach the local thermodynamic equilibrium (LTE), the efficient collisional dissociation suppresses H_2 cooling, and the gas cannot cool below several thousand K (Figure 4). Massive clouds with $\gtrsim 10^5$ M$_\odot$ formed by fragmentation of the post-shock layer subsequently collapse isothermally at $\sim$ 8000 K by the Lyα cooling. Without further fragmentation, monolithic collapse of the clouds results in the SMS formation. Note that, unlike the previous SMS formation mechanism, strong FUV radiation is not required in our scenario.

The mass of protostars formed via the isothermal collapse is initially only $\sim 10^{-2}$ $M_\odot$ but quickly increases via mass accretion. The expected accretion rates are $0.1-1$ $M_\odot$ yr^{-1}, more than 100 times higher than the standard value $\simeq 10^{-3}$ $M_\odot$ yr^{-1} expected for Pop III star formation. The stellar mass could reach 10^{5-6} $M_\odot$ in $\sim$ 1 Myr with this very rapid mass accretion. General relativity predicts that such a supermassive star (SMS) becomes unstable and collapses to form a BH, which subsequently swallows most of the surrounding stellar material.

We present here the evolution of protostars under such extreme conditions of rapid mass accretion (Fig. 4). Our results show that rapid accretion with $\dot{M}_* \gtrsim 10^{-2}$ $M_\odot$ yr^{-1} causes the star to bloat up like a red giant. The stellar radius increases monotonically

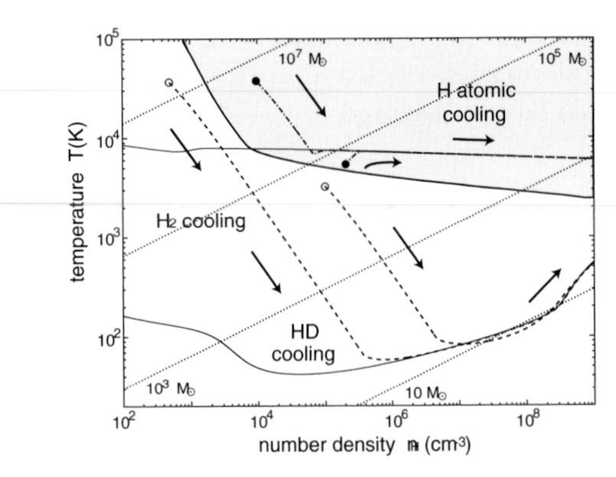

Figure 4. The temperature evolution of primordial gas after heating by the cold accretion shock for the initial ionization degree $x_{e,0} = 10^{-2}$ and H_2 fraction $x_{H_2,0} = 10^{-6}$. The evolutionary tracks are shown with dashed and dash-dotted lines for four combinations of initial temperature and density, each indicated by two open and filled circles. From low-density or low-temperature initial conditions (dashed lines starting from the open circles), the temperature decreases below $\sim$ 100 K owing to the H_2 and HD cooling. On the other hand, from dense and hot initial conditions (dash-dotted lines from the filled circles), the clouds do not cool below $\sim$ 8000 K and subsequently collapse almost isothermally by the H atomic cooling. The triangle symbol on each track indicates the epoch when the post-shock layer fragments by the gravitational instability. The thick solid line, the domain above which is hatched, divides the initial conditions leading to these two different ways of thermal evolution. The two thin solid lines show the temperature evolution by the H atomic cooling (upper) and the H_2 and HD cooling (lower), respectively. The diagonal dotted lines (lower-left to upper-right) indicate the constant Jeans masses, whose values are denoted by numbers in the Figure. Figure from Inayoshi & Omukai (2012)

with stellar mass and reaches $\simeq 7000\ R_\odot (\simeq 30$ AU) at a mass of $10^3\ M_\odot$. Unlike the cases with lower accretion rates previously studied, the mass-radius relation in this phase is almost independent of the assumed accretion rate. In addition, the effective temperature of those bloated stars is almost constant at $T_{\rm eff} \simeq 5000$ K due to the strong temperature-dependence of H^- absorption opacity.

We have calculated the early evolution until the stellar mass reaches $10^3\ M_\odot$. The subsequent evolution remains unexplored because of convergence difficulties with the current numerical codes. If the star continues to expand following the same mass-radius relation also for $M_* > 10^3\ M_\odot$, the stellar radius at $10^5\ M_\odot$ would be $\simeq$ 400 AU. Since the stellar effective temperature remains $\simeq$ 5000 K, the star hardly emits ionizing photons during accretion. Therefore, it is unlikely that stellar growth is limited by the radiative feedback via formation of an HII region as discussed in Section 2. Such massive "super-giant" protostars could be the progenitors that eventually evolve to the observed SMBHs in the early universe.

References

Dekel, A. & Birnboim, Y. 2006, *MNRAS*, 368, 2
Dijkstra, M., Haiman, Z., Mesinger, A., & Wyithe, J. S. B. 2008, *MNRAS*, 391, 1961
Hosokawa, T., Omukai, K., Yoshida, N., & Yorke, H. W. 2011, *Science*, 334, 1250
Hosokawa, T., Omukai, K., & Yorke, H. W. 2012, ArXiv:1203.2613
Inayoshi, K. & Omukai, K. 2012, *MNRAS*, in press (arXiv:1202.5380)
Jappsen, A.-K., Klessen, R. S., Larson, R. B., Li, Y., & Mac Low, M.-M. 2005, *A&A*, 435, 611
Larson, R. B. 2005, *MNRAS*, 359, 21
Li, Y., Klessen, R. S., & Mc Low, M.-M. 2003, *ApJ*, 592, 975
Mortlock, D. J. *et al.* 2011, *Nature*, 474, 616

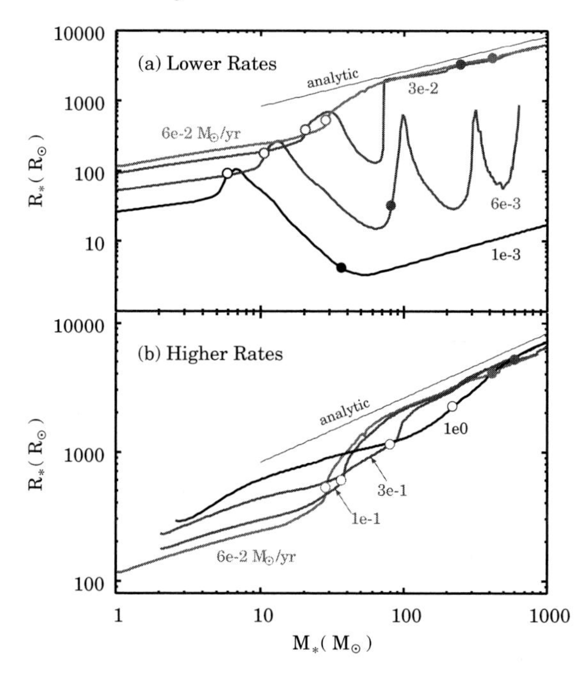

Figure 5. Evolution of the protostellar radius for various accretion rates. *Upper panel:* the different curves represent the cases with $\dot{M}_* = 10^{-3}\ M_\odot\ \mathrm{yr}^{-1}$, $6 \times 10^{-3}\ M_\odot\ \mathrm{yr}^{-1}$, $3 \times 10^{-2}\ M_\odot\ \mathrm{yr}^{-1}$, and $6 \times 10^{-2}\ M_\odot\ \mathrm{yr}^{-1}$. The open and filled circles on each curve denote the epoch when $t_{\mathrm{KH}} = t_{\mathrm{acc}}$ and when the central hydrogen burning begins, respectively. *Lower panel:* same as the upper panel but for higher accretion rates of $1 \times 10^{-2}\ M_\odot\ \mathrm{yr}^{-1}$, $0.1\ M_\odot\ \mathrm{yr}^{-1}$, $0.3\ M_\odot\ \mathrm{yr}^{-1}$, and $1\ M_\odot\ \mathrm{yr}^{-1}$. The case with $6 \times 10^{-2}\ M_\odot\ \mathrm{yr}^{-1}$ is illustrated in both panels as a reference. For the cases with $0.3\ M_\odot\ \mathrm{yr}^{-1}$ and $1\ M_\odot\ \mathrm{yr}^{-1}$ hydrogen fusion has not ignited by the time the stellar mass reaches $10^3\ M_\odot$. In the both panels the thin green line represents the mass-radius relation given by an analytic expression. Figure from Hosokawa, Omukai, & Yorke (2012)

Omukai, K. 2001, *ApJ*, 546, 635
Omukai, K., Hosokawa, T., & Yoshida, N. 2010, *ApJ*, 722, 1793
Schneider, R., Ferrara, A., Natarajan, P., & Omukai, K. 2002, *ApJ* 571, 30
Schneider, R., Omukai, K., Inoue, A. K., & Ferrara, A. 2006, *MNRAS*, 369, 1437
Shang, C., Bryan, G. L., & Haiman, Z. 2010, *MNRAS*, 402, 1249
Tsuribe, T. & Omukai, K. 2006, *ApJ*, 642, L61
Wolfire M. G. & Cassinelli, J. P. 1987, *ApJ*, 319, 850

Discussion

MIRABEL: In your simulations with what frequency are multiple systems form?

OMUKAI: In the simulation in my talk, we limited the calculation to the axisymmetric case and so we cannot discuss the multiplicity. In other calculations with Machida *et al.* (2008), we found that with some rotation, sufficiently flat disks are formed during the prestellar collapse, which eventually fragment into binary or multiple systems. Due to the difference in temperature evolution, with the same amount of initial angular momentum, binary formation appears easier in the primordial gas.

Death of Massive Stars: Supernovae and Gamma-Ray Bursts
Proceedings IAU Symposium No. 279, 2012
P. Roming, N. Kawai & E. Pian, eds.

doi:10.1017/S1743921312012963

Luminosities, Masses and Star Formation Rates of Galaxies at High Redshift

Andrew J. Bunker

Department of Physics, University of Oxford
Denys Wilkinson Building, Keble Road, Oxford OX1 3RH, U.K.
email: a.bunker1@physics.ox.ac.uk

Abstract. There has been great progress in recent years in discovering star forming galaxies at high redshifts ($z > 5$), close to the epoch of reionization of the intergalactic medium (IGM). The WFC3 and ACS cameras on the Hubble Space Telescope have enabled Lyman break galaxies to be robustly identified, but the UV luminosity function and star formation rate density of this population at $z = 6 - 8$ seems to be much lower than at $z = 2 - 4$. High escape fractions and a large contribution from faint galaxies below our current detection limits would be required for star-forming galaxies to reionize the Universe. We have also found that these galaxies have blue rest-frame UV colours, which might indicate lower dust extinction at $z > 5$. There has been some spectroscopic confirmation of these Lyman break galaxies through Lyman-α emission, but the fraction of galaxies where we see this line drops at $z > 7$, perhaps due to the onset of the Gunn-Peterson effect (where the IGM is opaque to Lyman-α).

Keywords. galaxies: evolution – galaxies: formation – galaxies: starburst – galaxies: high-redshift – ultraviolet: galaxies

---

1. Introduction

In the past decade, the quest to observe the most distant galaxies in the Universe has rapidly expanded to the point where the discovery of $z \simeq 6$ star-forming galaxies has now become routine. Deep imaging surveys with the *Hubble Space Telescope (HST)* and large ground based telescopes have revealed hundreds of galaxies at $z \simeq 6$ (Bunker *et al.* 2004, Bouwens *et al.* 2006, 2007). These searches typically rely on the Lyman break galaxy (LBG) technique pioneered by Steidel and collaborators to identify star-forming galaxies at $z \approx 3 - 4$ (Steidel *et al.* 1996, 1999). Narrow-band searches have also proved successful in isolating the Lyman-α emission line in redshift slices between $z = 3$ and $z = 7$ (e.g., Ouchi *et al.* 2008; Ota *et al.* 2008). Rapid follow-up of Gamma Ray Bursts has also detected objects at $z \sim 6$, and most recently one spectroscopically confirmed at $z = 8.2$ (Tanvir *et al.* 2009).

Parallel to these developments in identifying high redshift objects has been the discovery of the onset of the Gunn-Peterson (1965) effect. This is the near-total absorption of the continuum flux shortward of Lyman-α in sources at $z > 6.3$ due to the intergalactic medium (IGM) having a much larger neutral fraction at high redshift. The Gunn-Peterson trough was first discovered in the spectra of SDSS quasars (Becker *et al.* 2001, Fan *et al.* 2001, 2006). This defines the end of the reionization epoch, when the Universe transitioned from a neutral IGM. Latest results from WMAP indicate the mid-point of reionization may have occurred at $z \approx 11$ (Dunkley *et al.* 2009). The source of necessary ionizing photons remains an open question: the number density of high redshift quasars is insufficient at $z > 6$ to achieve this (Fan *et al.* 2001, Dijkstra *et al.* 2004). Star-forming galaxies at high redshift are another potential driver of reionization, but we must first determine their rest-frame UV luminosity density to assess whether they are plausible sources; the escape fraction of ionizing photons from these galaxies, along with the slope of their UV spectra, are other important and poorly-constrained factors in determining whether star formation is responsible for the ionization of the IGM at high redshift.

2. The Star Formation Rate at High Redshift

Early results on the star formation rate density at $z \approx 6$ were conflicting, with some groups claiming little to no evolution to $z \sim 3$ (Giavalisco *et al.* 2004, Bouwens *et al.* 2003) while other work suggested that the star formation rate density at $z \approx 6$ was significantly lower than in the well-studied LBGs at $z = 3-4$ (Stanway *et al.* 2003). The consensus which has now emerged from later studies is that the abundance of *luminous* galaxies is substantially *less* at $z \approx 6$ than at $z \approx 3$ (Stanway *et al.* 2003, Bunker *et al.* 2004, Bouwens *et al.* 2006, Yoshida *et al.* 2006, McLure *et al.* 2009). If this trend continues to fainter systems and higher redshifts, then it may prove challenging for star-forming galaxies to provide the UV flux needed to fulfill reionisation of the intergalactic medium at (e.g. Bunker *et al.* 2004). Importantly, analysis of the faint-end of the luminosity function at high redshift has revealed that feeble galaxies contribute an increased fraction of the total UV luminosity, suggesting that the bulk of star formation (and hence reionizing photons) likely come from lower luminosity galaxies not yet adequately probed even in deep surveys.

Until recently, extending this work to the $z \approx 7$ universe has been stunted by small survey areas (from space) and low sensitivity (from the ground). There is evidence of old stellar populations in $z \sim 4-6$ galaxies from measurements of the Balmer break in Spitzer/IRAC imaging (Eyles *et al.* 2005,2007; Stark *et al.* 2007, 2009), which must have formed at higher redshift (see Fig. 3). However, the age-dating and mass-determination of these stellar populations has many uncertainties, so searching directly for star formation at redshifts $z \geqslant 7$ is critical to measure the evolution of the star formation rate density, and address the role of galaxies in reionizing the universe.

The large field of view and enhanced sensitivity of the recently-installed Wide Field Camera 3 (WFC3) on *HST* has made great progress in identifying larger samples of $z > 6$ Lyman break galaxies, as it covers an area 6.5 times that of the previous-generation NICMOS NIC3 camera, and has better spatial sampling, better sensitivity and a filter set better tuned to identifying high-redshift candidates through their colours. In Bunker *et al.* (2010) we presented an analysis of the recently-obtained WFC3 near-infrared images of the *Hubble* Ultra Deep Field (UDF). We have previously used the i'-band and z'-band ACS images to identify LBGs at $z \approx 6$ through their large $i' - z'$ colours (the i'-drops, Bunker *et al.* 2004). Using this ACS z'-band image in conjunction with the new WFC3 Y-band we could search for galaxies at $z \approx 7$, with a spectral break between these two filters (the z'-drops). We also anaylsed the J and H-band WFC3 images to eliminate low-redshift contaminants of the z'-drop selection through their near-infrared colours (Fig. 1), and also to look for Lyman-break galaxies at higher redshifts (the Y- and J-drops at $z \approx 8$ & $z \approx 10$).

3. The Role of Star Forming Galaxies in Reionization

We can use the observed Y-band magnitudes of objects in the z'-drop sample to estimate their star formation rate from the rest-frame UV luminosity density. In the absence of dust obscuration, the relation between the flux density in the rest-UV around $\approx 1500\,\text{Å}$ and the star formation rate (SFR in $M_{\odot}\,\text{yr}^{-1}$) is given by $L_{UV} = 8 \times 10^{27}\ SFR\ ergs\,s^{-1}\,Hz^{-1}$ from Madau, Pozzetti & Dickinson (1998) for a Salpeter (1955) stellar initial mass function (IMF) with $0.1\,M_\odot < M^* < 125\,M_\odot$. This is comparable to the relation derived from the models of Leitherer & Heckman (1995) and Kennicutt (1998). However, if a Scalo (1986) IMF is used, the inferred star formation rates will be a factor of ≈ 2.5 higher for a similar mass range. The redshift range is not surveyed with uniform sensitivity to UV luminosity (and hence star formation rate). We have a strong luminosity bias towards the lower end of the redshift range, due to the effects of

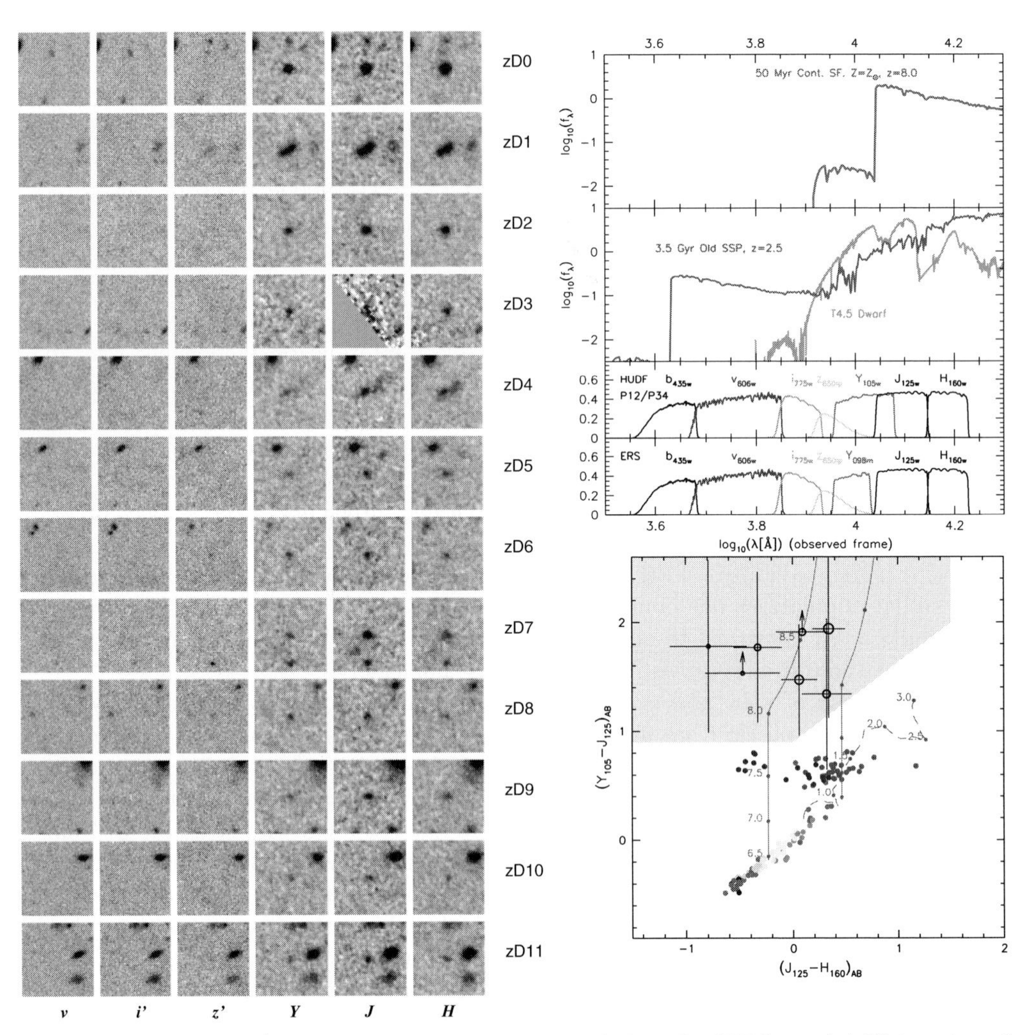

Figure 1. Left: The z'-band dropouts at $z \sim 7$ identified in the WFC 3 and ACS imaging of the *Hubble* Ultra Deep Field (Bunker *et al.* 2010). The top candidate, zD0, is actually an intermediate-redshift supernova (rather than a high redshift Lyman break galaxy) which occurred during the more recent WFC 3 imaging from 2009 but was absent in the 2003 ACS images.
Top Right: The WFC 3 and ACS filter set on *HST* used, along with the spectral template for a $z \sim 7$ star-forming galaxy, and two potential interloper populations: a low mass star in our own Galaxy, and an evolved galaxy at $z \approx 2.5$ (from Wilkins *et al.* 2011a).
Bottom Right: The use of a two-colour diagram to reject the low-redshift interlopers and select candidate $z \sim 8$ Y-drop galaxies (Lorenzoni *et al.* 2011). The dotted line is the locus of evolved galaxy colours as a function of redshift; the lower points are stellar colours at zero redshift, and the near-vertical lines denote star-forming galaxies with spectral slopes $\beta = -3$ (left) and $\beta = 0$ (right). The points with error bars are candidate high redshift galaxies in the HUDF09 field from Lorenzoni *et al.* (2011).

increasing luminosity distance with redshift and also the Lyman-α break obscuring an increasing fraction of the filter bandwidth. Using an "effective volume" approach (e.g., Steidel *et al.* 1999) accounts for this luminosity bias, we infer a total star formation rate density at $z \approx 7$ of $0.0034\, M_{\odot}\, yr^{-1}\, Mpc^{-3}$, integrating down to $0.2\, L^{*}_{UV}$ of the L^{*} value at $z = 3$ (i.e. $M_{UV} = -19.35$) and assuming a UV spectral slope of $\beta = -2.0$. If instead we integrate down to $0.1\, L^{*}_{z=3}$ $(0.2\, L^{*}_{z=6})$ and the total star formation rate density is $0.004\, M_{\odot}\, yr^{-1}\, \mathrm{Mpc}^{-3}$. These star formation rate densities are a factor of ~ 10 *lower* than at $z \sim 3-4$, and even a factor of $1.5 - 3$ below that at $z \approx 6$ (Bunker *et al.* 2004; Bouwens *et al.* 2006).

We can compare our measured UV luminosity density at $z \approx 7$ (quoted above as a corresponding star formation rate) with that required to ionize the Universe at this redshift. Madau, Haardt & Rees (1999) give the density of star formation required for reionization (assuming the same Salpeter IMF as used in this paper):

$$\dot{\rho}_{SFR} \approx \frac{0.005\, M_{\odot}\, \mathrm{yr}^{-1}\, \mathrm{Mpc}^{-3}}{f_{esc}} \left(\frac{1+z}{8}\right)^{3} \left(\frac{\Omega_b\, h_{70}^2}{0.0457}\right)^{2} \left(\frac{C}{5}\right)$$

We have updated equation 27 of Madau, Haardt & Rees (1999) for a more recent concordance cosmology estimate of the baryon density from Spergel *et al.* (2003). C is the clumping factor of neutral hydrogen, $C = \left\langle \rho^2_{\mathrm{H}I} \right\rangle \left\langle \rho_{\mathrm{H}I} \right\rangle^{-2}$. Early simulations suggested $C \approx 30$ (Gnedin & Ostriker 1997), but more recent work including the effects of reheating implies a lower concentration factor of $C \approx 5$ (Pawlik *et al.* 2009). The escape fraction of ionizing photons (f_{esc}) for high-redshift galaxies is highly uncertain (e.g., Steidel, Pettini & Adelberger 2001, Shapley *et al.* 2006), but even if we take $f_{esc} = 1$ (no absorption by H I) and a very low clumping factor, this estimate of the star formation density required is a factor of 1.5–2 higher than our measured star formation density at $z \approx 7$ from z'-drop galaxies in the UDF. For faint end slopes of α $-1.8 \rightarrow -1.6$ galaxies

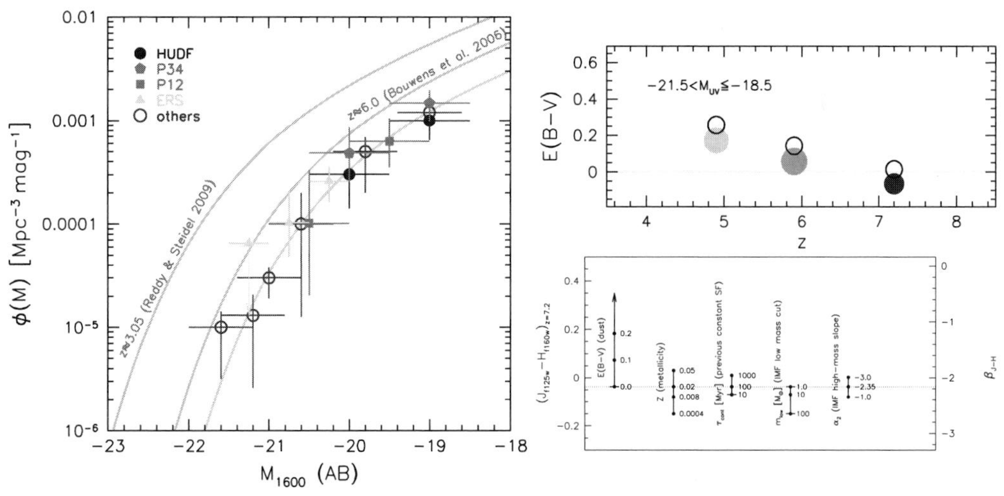

Figure 2. Left: the rest-frame UV luminosity function of star-forming galaxies at $z \approx 7$ is dramatically lower than at $z \approx 2$, and even than $z \approx 6$ (from Wilkins *et al.* 2011a).
Top Right: The evolution in the rest-UV colours of bright Lyman break galaxies; at higher redshifts, the galaxies are much more blue (from Wilkins *et al.* 2011b).
Bottom Right: Various factors can affect the rest-UV spectral slopes of star-forming galaxies, such as IMF, metallicity, star formation history and dust (from Wilkins *et al.* 2011b). The blue colours at high redshift might imply these early galaxies were much less dust reddened than the corresponding lower-redshift populations.

with $L > 0.2\,L^*$ account for $24-44\%$ of the total luminosity (if there is no low-luminosity cut-off for the Schechter function), so even with a steep faint-end slope at $z \approx 7$ we still fall short of the required density of Lyman continuum photons required to reionize the Universe, unless the escape fraction is implausibly high ($f_{esc} > 0.5$) and/or the faint end slope is $\alpha < -1.9$ (much steeper than observed at $z = 0-6$). At $z \approx 8$, the situation is even more extreme (Lorenzoni *et al.* 2011), as the luminosity function is even lower (Fig. 4).

However, the assumption of a solar metallicity Salpeter IMF may be flawed: the colours of $z \sim 6$ i'-band drop-outs are very blue (Stanway, McMahon & Bunker 2005), with $\beta < -2$, and the new WFC3 J- and H-band images show that the $z \approx 7$ z'-drops also have blue colours on average (Figure 2). A slope of $\beta < -2$ is bluer that for continuous star formation with a Salpeter IMF, even if there is no dust reddening. Such blue slopes could be explained through low metallicity, or a top-heavy IMF, which can produce between 3 and 10 times as many ionizing photons for the same 1500 Å UV luminosity (Schaerer 2003 – see also Stiavelli, Fall & Panagia 2004). Alternatively, we may be seeing galaxies at the onset of star formation, or with a rising star formation rate (Verma *et al.* 2007), which would also lead us to underestimate the true star formation rate from the rest-UV luminosity. We explore the implications of the blue UV spectral slopes in $z \geqslant 6$ galaxies in Wilkins *et al.* (2011b).

We have also targeted many of our high redshift Lyman break galaxies with spectroscopy from the ground, using Keck/DEIMOS (Bunker *et al.* 2003), Gemini/GMOS (Stanway *et al.* 2004) and most recently VLT/XSHOOTER (Bunker *et al.* 2012, Fig. 5). While we have seen Lyman-α emission in several $z \approx 6$ galaxies, the frequency seems to drop at $z > 7$, perhaps due to absorption of this line in the increasingly-neutral IGM.

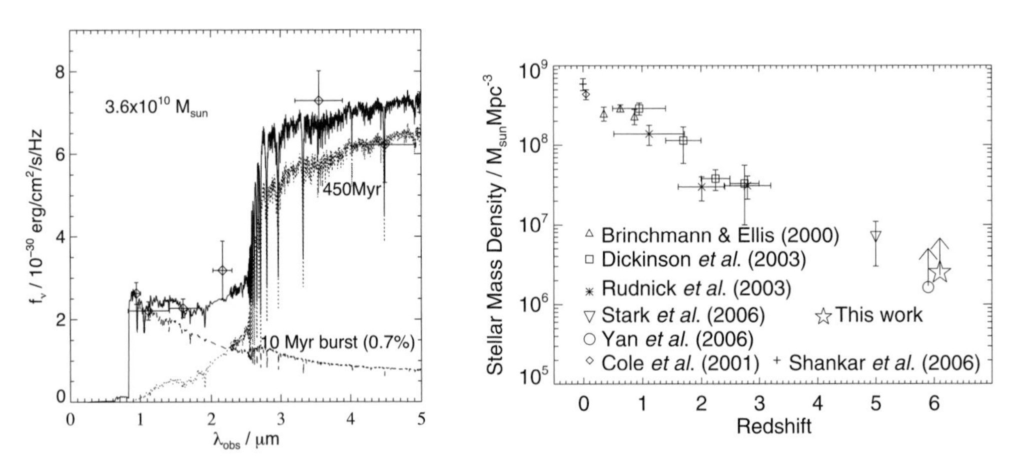

Figure 3. Left: the spectral energy distribution from *Spitzer/IRAC* and *HST* data for a $z \sim 6$ star-forming galaxy (Eyles *et al.* 2005). If we fit two stellar populations (ongoing star formation and an older burst), we find that there are stars as old as 450Myr at $z \approx 6$ (within 900Myr of the Big Bang). This pushes the formation epoch back to $z' sim 10$, around the time to reionization. **Right:** The assembly of stellar mass at high redshift, derived from *Spitzer* measurements of Lyman break galaxies (Eyles *et al.* 2007). Our high redshift points are lower limits because the Lyman break selection only picks out actively star-forming galaxies (and not post-burst systems).

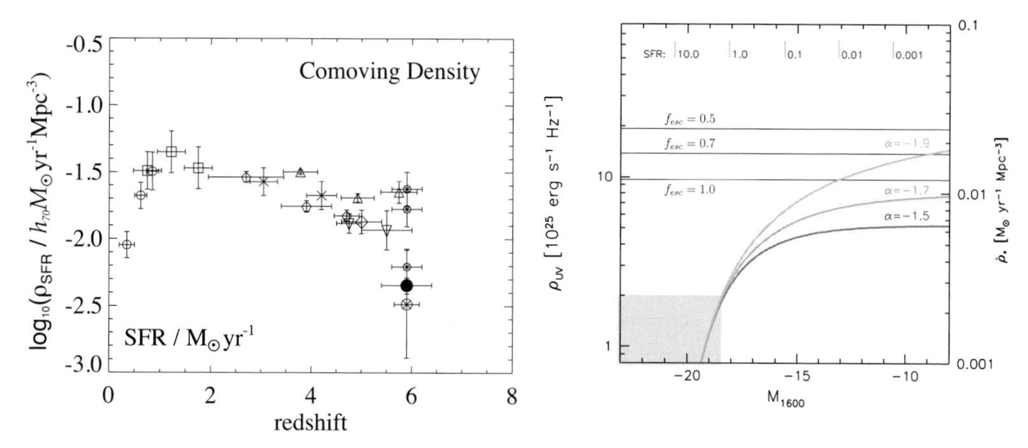

Figure 4. Left: the evolution of the star formation rate density (the "Madau–Lilly" diagram) based on Lyman break galaxies. Our work at $z \sim 6$ (Bunker *et al.* 2004) indicates a much lower level of activity than at $z \approx 2 - 4$.
Right: the ionising flux produced (y-axis) by the observed luminosity function of star-forming galaxies at $z \approx 8$, integrated down to a faint limiting magnitude (plotted on the x-axis). Where the integrated Schechter function curves cross the horizontal lines, there is sufficient ionising photons to keep the Universe ionised. As can be seen, a high escape fraction is required, and most of the photons would have to come from faint galaxies well below our observational limit (the shaded grey box in the lower left). Figure from Lorenzoni *et al.* (2011).

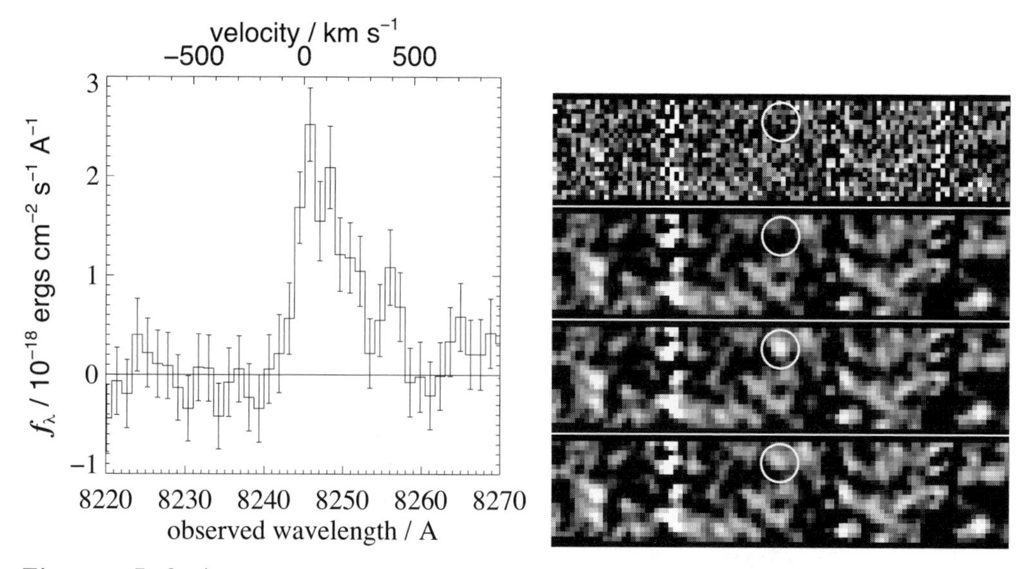

Figure 5. Left: A growing subset of the $z \approx 6$ Lyman break population have been spectroscopically confirmed through Lyman-α emission (figure from Bunker *et al.* 2003). At higher redshifts, $z > 7$, the spectroscopy has been less successful, perhaps because the resonant Lyman-α line is absorbed by optically-thick neutral hydrogen in the Inter-Galactic Medium during the Gun-n-Peterson era.
Right: we have been unable to confirm the claimed spectroscopic detection of Lyman-α at $z = 8.55$ by Lehnert *et al.* (2010) with SINFONI in the HUDF Y-band drop-out $YD3$ from Bunker *et al.* (2010). Our 5-hour XSHOOTER spectrum is shown at the top, and Gaussian-smoothed below. The expected position of the line emission is denoted by a circle. The bottom two panels have fake sources of the same flux as the Lehnert *et al.* emission line added in, with two different velocity widths. We should have detected the line at $\approx 4\,\sigma$ level if it was real (from Bunker *et al.* 2012).

4. Conclusions

The availability of WFC3 and ACS on the *Hubble* Space Telescope has made the discovery of Lyman break galaxies at $z > 5$ possible. From these galaxies, we have shown that the star formation rate density declines at $z > 5$. Based on their rest-frame UV luminosity function, it appears that a high escape fraction for Lyman continuum photons and a huge contribution from faint sources well below current detection limits would be needed for star forming galaxies to keep the Universe ionised at these redshifts. We have been able to measure stellar masses and ages from Spitzer in some instances. We also have discovered that the high redshift Lyman break galaxies have much bluer rest-frame UV colours than the $z \approx 4$ population, perhaps indicative of less dust obscuration at higher redshifts. Spectroscopic follow-up is ongoing, but we have confirmed Lyman-α emission from several $z \approx 6$ galaxies, although the success rate drops at $z > 7$, perhaps due to the onset of Gunn-Peterson absorption.

Acknowledgments

I gratefully acknowledge the involvement of my collaborators on these projects – Stephen Wilkins, Elizabeth Stanway, Richard Ellis, Silvio Lorenzoni, Joseph Caruana, Matt Jarvis, Samantha Hickey, Daniel Stark, Mark Lacy, Kuenley Chiu, Richard McMahon and Laurence Eyles. I thank the IAU and the conference organisers for an excellent Symposium.

References

Becker R. H. *et al.*, 2001, *AJ*, 122, 2850

Bouwens, R. J., *et al.* 2003, *ApJ*, 595, 589

Bouwens R. J., Illingworth G. D., Blakeslee J. P., & Franx M., 2006, *ApJ*, 653, 53

Bouwens R. J., Illingworth G. D., Franx M., & Ford H., 2007, *ApJ*, 670, 928

Bunker A. J., Stanway E. R., Ellis R. S., & McMahon R. G., 2004, *MNRAS*, 355, 374

Bunker, A. J., Wilkins, S., Ellis, R. S., Stark, D. P., Lorenzoni, S., Chiu, K., Lacy, M., Jarvis, M. J., & Hickey, S. 2010, *MNRAS*, 409, 855

Dijkstra, M., Haiman, Z., & Loeb, A. 2004, *ApJ* 613, 646

Dunkley, J., *et al.* 2009, *ApJS*, 180, 306

Eyles, L. P., Bunker, A. J., Stanway, E. R., Lacy, M., Ellis, R. S., & Doherty, M. 2005, *MNRAS*, 364, 443

Eyles, L. P., Bunker, A. J., Ellis, R. S., Lacy, M., Stanway, E. R., Stark, D. P., & Chiu, K. 2007, *MNRAS*, 374, 910

Fan X., *et al.*, 2001, *AJ*, 122, 2833

Fan, X., *et al.* 2006, *AJ*, 132, 117

Giavalisco, M., *et al.* 2004, *ApJ*, 600, L103

Gnedin, N. Y. & Ostriker, J. P. 1997, *ApJ*, 486, 581

Gunn, J. E. & Peterson, B. A. 1965, *ApJ*, 142, 1633

Hickey, S., Bunker, A. J., Jarvis, M., Chiu, K., & Bonfield, D., 2009, *MNRAS*, *in press* arXiv:0909.4205

Iye M. *et al.*, 2006, *Nature*, 443, 186

Kennicutt, R. C., 1998, *ARA&A*, 36, 189

Leitherer, C. & Heckman, T. M. 1995, *ApJS*, 96, 9

Lorenzoni, S., Bunker, A. J., Wilkins, S., Stanway, E., Jarvis, M. J., & Caruana, J. 2011, *MNRAS*, 414, 1455

Madau P., Ferguson H. C., Dickinson M. E., Giavalisco M., Steidel C. C., Fruchter A., 1996, *MNRAS*, 283, 1388

Madau P., Pozzetti L., & Dickinson M., 1998, *ApJ*, 498, 106

Madau P., Haardt F., & Rees M., 1999, *ApJ*, 514, 648
McLure, R. J., Cirasuolo, M., Dunlop, J. S., Foucaud, S., & Almaini, O. 2009, *MNRAS*, 395, 2196
Ouchi, M., *et al.* 2008, *ApJS*, 176, 301
Ota, K., *et al.* 2008, *ApJ*, 677, 12
Pawlik, A. H., Schaye, J., & van Scherpenzeel, E. 2009, *MNRAS*, 394, 1812
Reddy, N. A. & Steidel, C. C. 2009, *ApJ*, 692, 778
Salpeter E. E., 1955, *ApJ*, 121, 161
Scalo, J. M. 1986, *Fundamentals of Cosmic Physics*, 11, 1
Shapley, A. E., Steidel, C. C., Pettini, M., Adelberger, K. L., & Erb, D. K. 2006, *ApJ*, 651, 688
Spergel, D. N., *et al.* 2003, *ApJS*, 148, 175
Stanway E. R., Bunker A. J., & McMahon R. G., 2003, *MNRAS*, 342, 439
Stanway E. R., Glazebrook K., Bunker A. J., *et al.* 2004, *ApJ*, 604, L13
Stanway E. R., McMahon R. G., & Bunker A. J., 2005, *MNRAS*, 359, 1184
Stark, D. P., Ellis, R. S., Bunker, A., Bundy, K., Targett, T., Benson, A., & Lacy, M. 2009, *ApJ*, 697, 1493
Stark, D. P., Bunker, A. J., Ellis, R. S., Eyles, L. P., & Lacy, M. 2007, *ApJ*, 659, 84
Steidel C. C., Giavalisco M., Pettini M., Dickinson M., & Adelberger K. L., 1996, *ApJ*, 462, 17
Steidel C. C., Adelberger K. L., Giavalisco M., Dickinson M., & Pettini M., 1999, *ApJ*, 519, 1
Stiavelli, M., Fall, S. M., & Panagia, N. 2004, *ApJ*, 610, L1
Tanvir, N. R., *et al.* 2009, *Nature*, 461, 1254
Wilkins, S. M., Bunker, A. J., Lorenzoni, S., & Caruana, J. 2011, *MNRAS*, 411, 23
Wilkins, S. M., Bunker, A. J., Stanway, E. R., Lorenzoni, S., & Caruana, J. 2011b, *MNRAS*, 417, 717
Yoshida M., *et al.*, 2006, *ApJ*, 653, 988
Verma, A., Lehnert, M. D., Förster Schreiber, N. M., Bremer, M. N., & Douglas, L. 2007, *MNRAS*, 377, 1024

Death of Massive Stars: Supernovae and Gamma-Ray Bursts
Proceedings IAU Symposium No. 279, 2012
P. Roming, N. Kawai & E. Pian, eds.

doi:10.1017/S1743921312012975

Star Formation and the Metallicity Aversion of Long-Duration Gamma-Ray Bursts

John F. Graham[1,2] **and Andrew S. Fruchter**[1]

[1]Space Telescope Science Institute,
3700 San Martin Dr, Baltimore MD 21218, USA

[2]Dept. of Physics and Astronomy, Johns Hopkins University,
3400 N. Charles St, Baltimore MD 21218, USA

Abstract. It has been suggested that the apparent bias of long-duration GRBs (LGRBs) to low metallicity environments might be a result of the fact that star-formation is anti-correlated with metallicity. However, if this were the cause, one would expect other indicators of star formation, such as Type II and Type Ic SNe to demonstrate a similar bias. Here we show that local Type Ic and Type II SNe track the star-formation weighted metallicity distribution of the SDSS galaxies. In contrast LGRBs are typically found at far-lower metallicities than would be expected based on the distribution of star-formation. This is true even when one takes into account so-called "dark bursts". Indeed, while we will present data that show that some LGRBs form at very high metallicities, these objects enter the sample because of the large effective search volume produced by their bright hosts. The bias of LGRBs to low metallicity is real and must be related to a mechanism which is crucial in their formation.

Keywords. Gamma Rays: Bursts, Galaxy: Abundances, Stars: Formation.

While the existence of an apparent bias of long-duration GRBs (LGRBs) occurring in low metallicity environments is now generally accepted, Mannucci *et al.* (2011) proposes that this may be the byproduct of their "fundamental relation" that star-formation is anti-correlated with metallicity and not an intrinsic environmental preference in LGRB formation. Here we analyze across, LGRB & SNe hosts as well as the general star forming galaxy population to determine whether low metallicity environments is indeed the intrinsic selection preference for LGRB formation. This work grew out of ongoing work looking at high metallicity LGRB host galaxies with a larger collaboration including Emily Levesque, Lisa Kewley, Jarle Brinchmann, Andrew Levan, Nial Tanvir, Sandy Patel, Greg Aldering, and Saul Perlmutter.

Beginning with updated populations in Modjaz *et al.* (2008) we make the following additions in populations and methodologies: Considering all LGRBs regardless of an observed associated SN event to increase the LGRB population to 14 bursts with 3 bursts being at high metallicity. Constraining the SDSS general star-forming galaxies and supernovae to a redshift of less then 0.04 to effectively transform our population from a magnitude to a volume limited sample. Introducing Type II SNe within this redshift range and selected from galaxies in the SDSS to provide a more numerous and consistently measured population of supernovae than the broad-lined type Ic population (see figure 2). Adopting a star-formation rate weighted sample of the general star-forming galaxies to graphically show the distribution of star-formation within our volume limited SDSS sample (again see figure 2). The latter was obtained by collecting all star-formation, selecting with a random number generator a specific bit of that star-formation and then selecting the specific galaxy that contributed it and repeating this process many times to collect a sample of galaxies based on their star-formation. Where previously it appeared that supernovae were biased toward bright host galaxies comparison, it is now visually apparent that both supernovae populations have the same star-formation distribution as the SDSS.

To provide a more analytical comparison we devise a method for directly comparing the populations on a normalized cumulative distribution plot (figure 4). For the star-forming SDSS galaxies within the redshift range mandated in the preceding paragraph we use their normalized cumulative star-formation. Assuming that LGRB and supernovae track the star-formation in which they occur this can be directly compared to the sorted fraction of LGRB and supernovae hosts vs. their increasing metallicities. (Note: this does not require any estimates of the SFR for the LGRB or supernovae hosts). For both non LGRB broad-lined type Ic supernovae and Type II SNe within this redshift range of the SDSS these host populations track the distribution of star-forming SDSS galaxies quite well suggesting that there is no additional metallicity constraint on the supernovae formation. The LGRBs however display a profound preference for lower metallicities than the other populations. Excluding the three high metallicity LGRB hosts the remainder of the population occur at metallicities containing only 5 to 10 % of the star-formation. The exceedingly good match of the supernovae populations to the available star-formation excludes argument that the formation of LGRB progenitors is anti-correlated with metallicity. Thus the metallicity disparity between LGRB and broad-lined type Ic supernovae, as originally shown in Modjaz *et al.* (2008), is critical to the LGRB formation process.

While the preceding methodology looks at the distribution in metallicity, the star-formation of the LGRB or supernovae hosts is assumed to be typical. To check this assumption we proceed to directly compare the distribution in star-formation rates of the LGRB and Type II SNe hosts to the general star-forming SDSS galaxy population. This comparison is performed by taking the star-formation rate of each LGRB and SN host and asking what fraction of the total star-formation in the general SDSS galaxy population occurs in galaxies with less star-formation than the target to generate a fractional value of total star-formation for each LGRB and SN host. These fractional values are then sorted and plotted against the fraction of each object type. Should this distribution of star-formation within each object type follow the general star-forming SDSS galaxy population then an LGRB host with its star-formation at, e.g., the top of the bottom quartile would be expected to be about consistent with 25 % of the star-formation in the general SDSS galaxy population and so on, such that the population tracks a diagonal line. Naturally this requires that for the Type II supernovae where the redshift is constrained to be low the general SDSS galaxy population is also so constrained and conversely for the LGRBs where no redshift constraint is added that the general SDSS galaxy population span the full range of the SDSS (the actual difference in the redshift ranges, while a source of error, is small since the bias towards the star-formation being grossly dominated by the brightest objects is well established in both redshift ranges). As seen in figure 5, both the LGRBs and our larger general Type II SNe population track the diagonal well, indicating a good correspondence between the star-formation rates of the two host populations and the general SDSS galaxy population, and suggesting that star-formation rate is correlated with LGRB and SNe formation. Curiously an attempt to generate a Type II SNe population from untargeted surveys failed this test quite badly and we believe it to be an issue with the Supernovae Factory to detect SNe in high surface brightness backgrounds.

From these analyzes we conclude that the low host metallicities observed for LGRBs are the direct consequence of an intrinsic preference for low metal content in the progenitor system during typical LGRB formation. This is consistent with the myriad of theoretical models requiring high progenitor rotation to which metal accelerated mass loss would likely be detrimental. The small subset of LGRBs found to occur in high metallicity hosts

Fig. 1 to 3.— Metallicity vs. B band absolute galaxy luminosity of LGRB (squares) and broad-lined Type Ic SNe hosts (circles). In the background SDSS (small points forming a cloud) and TKRS (small diamonds) galaxies are shown. SNe found through targeted galaxy surveys are not filled.

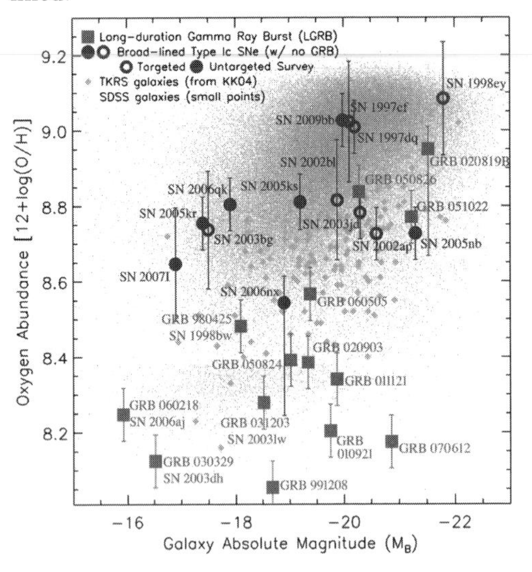

Figure 1. Site metallicities of LGRBs & SNe. While using site metallicities, initially done in Modjaz *et al.* (2008), is ideal for comparing the progenitor metallicities of LGRB and SNe events it introduces a bias when comparing with general galaxy populations such as the SDSS & TKRS populations thus we switch to using central metallicity in subsequent figures and analysis.

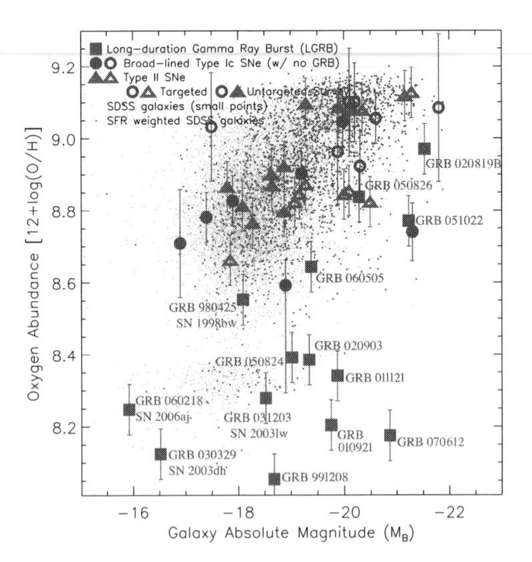

Figure 2. Central metallicities in a 0.0209 < z < 0.04 redshift cut to convert the SDSS into a volume limited sample. Since cutting the broad-lined Ic's would unacceptably degrade the sample size and they are already at low redshift we retain them and introduce a type II SNe population (purple triangles) strictly within the redshift range. LGRBs are shown purely for reference. We also select a subset of the redshift cut SDSS population in a star-formation weighted random manner to provide a graphical representation of the star-formation distribution of the SDSS. Note that this matches both SNe populations quite well as plotted in figure 4.

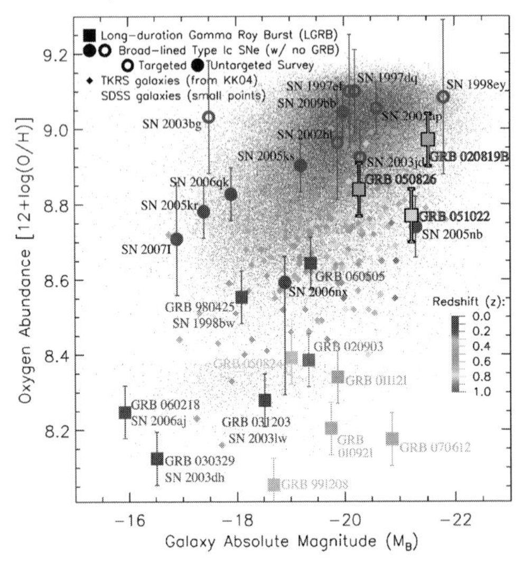

Figure 3. Central metallicity of all objects with color now showing redshift. This provides a graphical representation of the effect of redshift on the luminosity metallicity relation. Note that the three high metallicity LGRBs (highlighted) are at typical metallicity for galaxies of their luminosity and redshift. This is not consistent with the otherwise observed LGRB metal aversion. Note: The color version of this figure (where color indicates redshift) is available only in the online proceedings.

and presumably with similarly high metallicity progenitors objects enter the sample due to the large effective search volume produced by their bright hosts and the order of magnitude larger amount of star-formation available at these higher metallicities. Since the non LGRB broad-lined type Ic supernovae population is consistent with the comparably low redshift subset of the general star-forming galaxy population, the LGRBs consistency with the general star-forming galaxy population by proxy excludes IMF differences or, as argued by Mannucci *et al.* (2011), star-formation being anti-correlated with metallicity as an explanation of the LGRB populations low metallicity. Thus the bias of LGRBs to low metallicity is real and must be related to a mechanism which is crucial in their formation.

References

Mannucci, F., Salvaterra, R., & Campisi, M. A. 2011, *MNRAS*, 414, 1263

Modjaz, M., Kewley, L., Kirshner, R. P., *et al.* 2008, *AJ*, 135, 1136

Prieto, J. L., Stanek, K. Z., & Beacom, J. F. 2008, *ApJ*, 673, 999

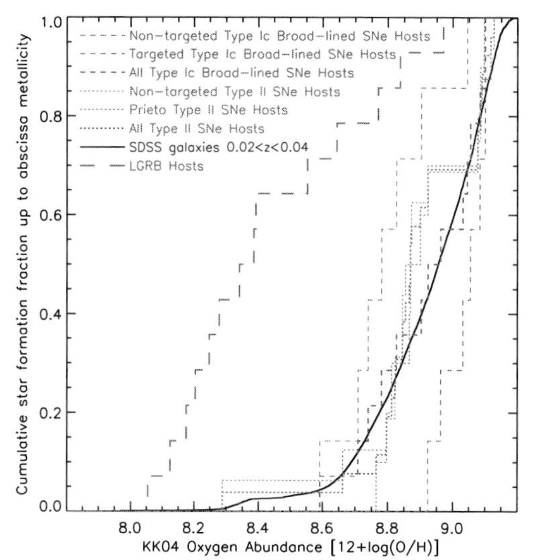

Figure 4. Star-formation cumulative summation fraction plots vs. core metallicity for the redshift range of 0.0209 to 0.04 with LGRB and SNe populations overplotted. LGRB host galaxies are shown in red, broad-lined Type Ic SNe hosts in blue, and Type II SNe hosts in purple. For the later two populations, the non targeted population is shown with a dashed line, the targeted (or potentially targeted population with a dotted line and the union of both with a solid line. The SDSS star-formation galaxy population is shown in back. Only the SDSS galaxies, including both the general star-forming galaxy and the type II SNe host populations are limited to the redshift cut range. No such cuts are applied to the LGRB or broad-lined type Ic SNe populations. While omitted for clarity using the site metallicity values for the broad-lined Type Ic supernovae populations does not shift any of those consistent with the LGRBs. This strongly suggests that LGRB hosts prefer host galaxies of metallicity considerably lower than would be obtained simply by following star-formation while both types of supernova are consistent with the star-formation distribution.

Figure 5. Normalized cumulative distribution plot of star-formation fraction values for various object classes. The abscissa values are obtained by determining for each LGRB and SN the fraction of total star-formation occurring in galaxies with lower star-formation rates than the objects host galaxy. For the type II SNe the star-formation of a volume limited SDSS sample was used spanning the same redshift range as the SNe sample whereas for the LGRBs the entire SDSS sample was used to match the selection biases in the LGRB host observations. The Prieto *et al.* (2008) type II SNe sample (blue line - "All" in the legend refers to this sample including SNe found through both targeted and non-targeted surveys and does not include the other dedicated non-targeted sample) tracks the diagonal reference line with astonishingly good accuracy. The LGRB sample (red line) also tracks the diagonal quite well indicating that the LGRB host galaxy sample has a typical star-formation distribution. Thus only metallicity itself remains to explain the discordant results shown in figure 4. Unfortunately however a similar comparison of the non targeted supernovae sample (green line) shows a clear off diagonal disagreement due to what we suspect is a failure of the Supernovae Factory to detect SNe in high surface brightness backgrounds.

Discussion

PERLEY: I think it's now becoming fairly convincing that GRBs can occur even at high metallicity, but also that they do so at a much lower rate than at lower metallicity. So the next step is to try to quantify this trend. Do you have an estimate of what the high metallicity (say $> 0.5\ Z_{\odot}$) rate might be relative to the lower metallicity rate?
GRAHAM: Correct on the lower rate & next step. Also such rate estimate should be relative to the available star formation in the metallicity range. The small number of high metallicity LGRBs makes me unwilling to give a solid number but it looks like high metallicity LGRB formation is suppressed by at least a factor of five and probably more than one and less than two orders of magnitude compared with low metallicity LGRB formation (relative to the underlying available star formation at those metallicities).

FYNBO: Is the completeness of the SDSS uniform across the luminosity range you use and if not does this affect your analysis.
GRAHAM: The $0.02 > z \geqslant 0.04$ redshift cut SDSS sample provides reasonable luminosity completeness for galaxies brighter than M_B -18 mag. The fraction of LGRBs, SNe & SDSS star formation in fainter galaxies is roughly consistent and small - thus unlikely to be significantly biasing.

Death of Massive Stars: Supernovae and Gamma-Ray Bursts
Proceedings IAU Symposium No. 279, 2012
P. Roming, N. Kawai & E. Pian, eds.

doi:10.1017/S1743921312012987

Nucleosynthesis in neutrino-driven, aspherical Population III supernovae

Shin-ichiro Fujimoto[1], Masa-aki Hashimoto[2], Masaomi Ono[3], and Kei Kotake[4],

[1]Kumamoto National College of Technology, 2659-2 Suya, Goshi, Kumamoto 861-1102, Japan
email: fujimoto@ec.knct.ac.jp
[2]Kyushu University, Fukuoka 810-8560, Japan
[3]Kyoto University, Kyoto, 606-8502, Japan
[4]National Astronomical Observatory Japan, Tokyo, 181-8588, Japan

Abstract. We investigate explosive nucleosynthesis during neutrino-driven, aspherical supernova (SN) explosion aided by standing accretion shock instability (SASI), based on two-dimensional hydrodynamic simulations of the explosion of 11, 15, 20, 25, 30 and $40M_{\odot}$ stars with zero metallicity. The magnitude and asymmetry of the explosion energy are estimated with simulations, for a given set of neutrino luminosities and temperatures, not as in the previous study in which the explosion is manually and spherically initiated by means of a thermal bomb or a piston and also some artificial mixing procedures are applied for the estimate of abundances of the SN ejecta.

By post-processing calculations with a large nuclear reaction network, we have evaluated abundances and masses of ejecta from the aspherical SNe. We find that matter mixing induced via SASI is important for the abundant production of nuclei with atomic number $\geqslant 21$, in particular Sc, which is underproduced in the spherical models without artificial mixing. We also find that the IMF-averaged abundances are similar to those observed in extremely metal poor stars. However, observed [K/Fe] cannot be reproduced with our aspherical SN models.

Keywords. Supernovae: general, nuclear reactions, nucleosynthesis, abundances.

---

1. Introduction

Recent high resolution spectroscopy of metal poor stars (MPSs) showed that the scatters of abundance ratio of the stars around averaged abundance-ratios are small (e.g. Cayrel *et al.* (2004)). The observed ratios are well reproduced by IMF-averaged ejecta of supernovae (SNe) of Population (Pop) III stars, as in Tominaga *et al.* (2007) and Heger & Woosley (2010). The spherical models of Tominaga *et al.* (2007) and Heger & Woosley (2010) however require artificial matter mixing of SN ejecta to reproduce the observed abundances. Abundances of nuclei with $Z \geqslant 21$ in SN ejecta with aspherical model of Joggerst *et al.* (2009) and Joggerst *et al.* (2010), in which the explosion is spherically triggered but matter mixing through Rayleigh-Taylor (RT) instabilities is taken into account during later explosion phase, are underproduced compared with the observed abundances. Standing accretion shock instability (SASI) is a reliable candidate to initiate bipolar oscillations of a stalled shock. Successful explosion could occur via efficient neutrino heating induced via the bipolar oscillations (e.g. Marek & Janka(2009) and Suwa *et al.*(2010)). In addition to the importance of SASI to the explosion, material mixing due to SASI may change abundances and masses of the SN ejecta.

In the present work, we examine explosive nucleosynthesis in neutrino-driven, aspherical SNe of Pop III stars aided by SASI, based on two-dimensional (2D) hydrodynamic simulations of the SN explosion of the stars. In §2 we briefly describe a numerical code for hydrodynamic calculation of SN, initial setup, and properties of the explosion. In §3, we present abundances of the SN ejecta. Finally we will summarize our results in §4.

2. Hydrodynamic simulations of supernovae

For 2D hydrodynamic simulations of neutrino-driven SNe, we employ a numerical code, which is based on the ZEUS-2D code, as in Ohnishi *et al.* (2006) and Fujimoto *et al.* (2011). We assume that the fluid is axisymmetric and that neutrinos are isotropically emitted from the neutrino spheres with given luminosities and with the Fermi-Dirac distribution of given temperatures. Initial conditions for the hydrodynamic simulations are similarly arranged as in Fujimoto *et al.* (2011). The central region inside 50 km in radius is excised to follow a long-term postbounce evolution. We impose velocity perturbations to the unperturbed radial velocity in a dipolar manner, and follow the postbounce evolution.

We have performed the simulations for six non-rotating progenitors, whose main-sequence masses, $M_{\rm ms}$, are 11, 15, 20, 25, 30 and 40 $M_\odot$ with zero metallicity in Heger & Woosley (2010), for 1 – 2s after the core bounce, when a shock front has reached to a layer with $r = 10,000\,$km in almost all directions. We consider models with neutrino temperatures, T_{ν_e}, $T_{\bar{\nu}_e}$, and T_{ν_x} as 4, 5, and 10 MeV, respectively, as in Ohnishi *et al.* (2006) and Fujimoto *et al.* (2011). We take the input neutrino luminosities from $L_{\nu_e,\rm min}$ to $L_{\nu_e,\rm max}$ for a given $M_{\rm ms}$, because the revival of the stalled bounce shock occurs only for models with $L_{\nu_e} \geqslant L_{\nu_e,\rm min}$ and also because for models with $L_{\nu_e} > L_{\nu_e,\rm max}$, the star explodes too early for the SASI to grow. Here $L_{\nu_e,\rm min}$ are 1.2, 2.0, 4.3, 7.0, 6.0, and 21.0 in units of $10^{52}\,{\rm erg\,s^{-1}}$ for $M_{\rm ms}/M_\odot$ = 11, 15, 20, 25, 30, and 40, respectively, $L_{\nu_e,\rm max}$ are 6.0, 8.0, 8.0, 20.0, 10.0, and 28.0 in units of $10^{52}\,{\rm erg\,s^{-1}}$, respectively.

We note that abundances of SN ejecta are shown to depend not directly on L_ν and T_ν, but on the explosion energy and the mass of the proto-neutron star, as shown in Fujimoto *et al.* (2011). It should be emphasized that magnitude and asymmetry of the explosion energy are evaluated for the simulations. Moreover, we stress that estimated masses of the central remnant are less than $2.5M_\odot$, which is lighter than the maximum, baryonic mass of a neutron star ($2.61M_\odot$ for Shen EOS in O'Connor & Ott (2011)), for all models, and the onset of the explosion is earlier than the time for black hole formation estimated by O'Connor & Ott (2011), in particular for models with explosion energies $>10^{52}\,$erg.

3. Abundances of supernova ejecta

Next we calculate abundances and masses of the SN ejecta in a same manner as in Fujimoto *et al.* (2011). Ejecta that is located on the central part of the star ($\leqslant 10,000\,$km) before the core collapse has high enough maximum temperatures for elements heavier than C to burn explosively. We therefore follow abundance evolution of the ejecta from the central region using a nuclear reaction network, which includes 463 nuclides from neutron, proton to Kr, while the abundances of ejecta from outside the region are set to be those before the core collapse in Heger & Woosley (2010).

Figure 1 shows masses of Fe, M(Fe) as a function of $E_{\rm exp}$ in the SN ejecta for all models (left panel) and [Mg/Fe] as a function of $E_{\rm exp}$ for all models (right panel), where [A/B] is defined as $\log[(X({\rm A})/X({\rm B}))\,/\,(X({\rm A})_\odot/X({\rm B})_\odot)]$ with mass fractions of X(A) and X(B) for nuclei A and B. We find that M(Fe) roughly correlates with $E_{\rm exp}$. For $M_{\rm ms} = 11M_\odot$ ($15M_\odot$), M(Fe) ranges from $5\times10^{-5}M_\odot$ ($2\times10^{-3}M_\odot$) to $3\times10^{-2}M_\odot$ ($1\times10^{-1}M_\odot$), while for a given $M_{\rm ms}$ larger than $20M_\odot$, M(Fe) are comparable because

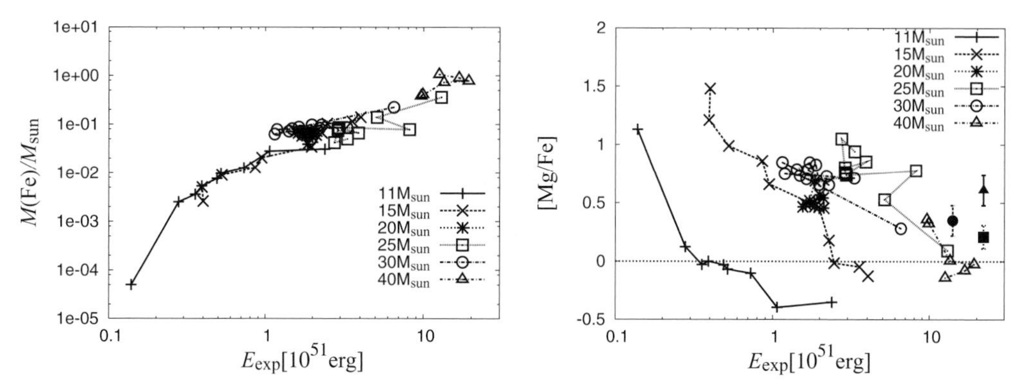

Figure 1. left: masses of Fe as a function of E_{exp} for all models. **right**: [Mg/Fe] as a function of E_{exp} for all models. Square, triangle, and circle with a vertical error bar describe observed [Mg/Fe] in Cayrel *et al.* (2004), in Cayrel *et al.* (2004) with corrections of Andrievsky *et al.* (2008), in which NLTE effects are taken into account, and in Preston *et al.* (2006), respectively.

of comparable E_{exp}. Moreover, [Mg/Fe] are found to roughly anti-correlate with E_{exp} (right panel in Fig. 1). The observed [Mg/Fe] of Cayrel *et al.* (2004) with the NLTE corrections of Andrievsky *et al.* (2008) (triangle with a vertical error bar) is comparable to [Mg/Fe] estimated for $M_{\mathrm{ms}} = 20, 25$ and $30M_\odot$ and for models with $M_{\mathrm{ms}} = 15M_\odot$ and $E_{\mathrm{exp}} \sim 0.5 \times 10^{51}$erg, in which ejected masses of ^{56}Ni are $0.02 - 0.03M_\odot$. [Mg/Fe] for models with $M_{\mathrm{ms}} = 11M_\odot$ and $M(\mathrm{Fe}) > 0.005M_\odot$ are lower than the observed [Mg/Fe]. Pop III SN of $M_{\mathrm{ms}} = 11$ and $15M_\odot$ could be a faint SN. Also for models with $E_{\mathrm{exp}} > 10^{52}$erg [Mg/Fe] are lower than the observed ones.

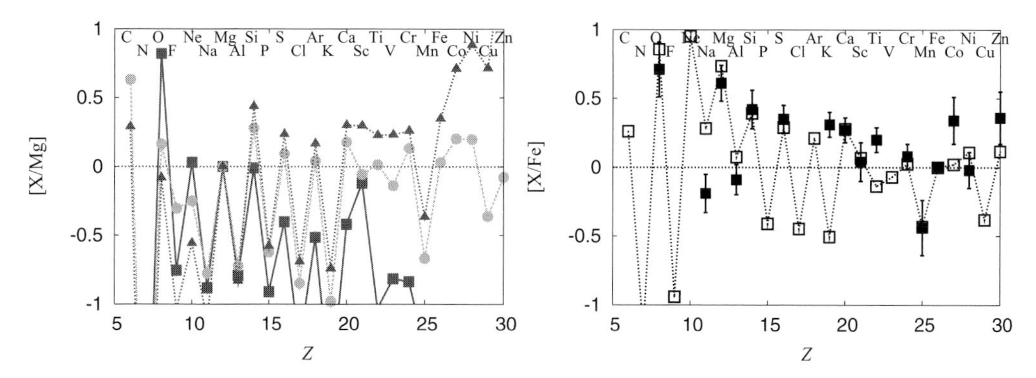

Figure 2. left: [X/Mg] as a function of Z of the SN ejecta for $M_{\mathrm{ms}} = 11M_\odot$ and $E_{\mathrm{exp}} = 0.18 \times 10^{51}$erg (line with filled triangles), 0.19×10^{51}erg (line with filled circles) and 0.91×10^{51}erg (line with filled squares). **right**: IMF-averaged [X/Fe] as a function of Z (dotted line with squares) and averaged [X/Fe] observed in MPSs (filled squares). Here IMF-averaging is performed over models with $M_{\mathrm{ms}} = 11, 15, 20, 25, 30$, and $40\ M_\odot$ whose E_{exp} are 0.91, 1.00, 1.06, 1.24, 1.09, and 7.05 in units of 10^{51}erg, respectively.

Figure 2 shows [X/Mg] as a function of the atomic number Z of the SN ejecta for $M_{\mathrm{ms}} = 11M_\odot$ (left panel), and [X/Fe] averaged with the Salpeter IMF as a function of Z as well as the average of observed abundance ratios in Cayrel *et al.* (2004) with the NLTE corrections of Andrievsky *et al.* (2008) (right panel).

We find that nuclei with $Z \geqslant 21$, which are underproduced in the aspherical model of Joggerst *et al.* (2010), are appreciably produced even for cases with lower explosion energies $E_{\mathrm{exp}} < 1 \times 10^{51}$erg (left panel in Fig. 2). Material mixing induced via SASI is therefore important for the enhancement of nuclei with $Z \geqslant 21$ in the SN ejecta, since the mixing via RT instabilities during the later explosion phase is taken into account in the aspherical model of Joggerst *et al.* (2010). Moreover, we find that the IMF-averaged [X/Fe] well reproduce the observed average ratio (solid line with filled squares in right panel), other than K, as in Tominaga *et al.* (2007) and Heger & Woosley (2010). Ti and Co are slightly underproduced, while [Na/Fe] is overproduced compared with the observed ratio because of large [Na/Fe] for $M_{\rm ms} \geqslant 15 M_\odot$.

4. Summary

We have investigated explosive nucleosynthesis during neutrino-driven, aspherical SN explosion aided by SASI, based on 2D hydrodynamic simulations of the explosion of six Pop III stars, whose masses are 11, 15, 20, 25, 30, and $40 M_\odot$. The magnitude and asymmetry of the explosion energy are estimated with the simulations, for a given set of neutrino luminosities and temperatures. By post-processing calculations with the large nuclear reaction network, we have evaluated abundances and masses of ejecta from the aspherical SNe.

We find that masses of Fe roughly correlate with the explosion energies, while [Mg/Fe] anti-correlate. Comparison between estimated and observed [Mg/Fe] suggests that Pop III SNe of a progenitor with lower mass ($\leqslant 15 M_\odot$) may be faint SNe. Nuclei with $Z \geqslant 21$, which are underproduced in the aspherical model of Joggerst *et al.* (2010), are appreciably produced. Material mixing induced via SASI is therefore important for the enhancement of nuclei with $Z \geqslant 21$ in the SN ejecta. Moreover, we evaluate IMF-averaged abundances using the abundances and masses of the ejecta in our 2D models. We find that the evaluated abundance ratios are similar to the averaged observed ratios in MPSs (Cayrel *et al.* (2004)), as shown in the spherical models by Tominaga *et al.* (2007) and Heger & Woosley (2010), although in their models, the explosion is manually and spherically initiated by means of a thermal bomb or a piston, and also artificial mixing procedures are required for the reproduction of the observed ratios. However, observed [K/Fe] cannot be reproduced with our 2D SN models.

References

Cayrel, R., Depagne, E., Spite, M., *et al.* 2004, *A&A*, 416, 1117
Tominaga, N., Umeda, H., & Nomoto, K. 2007, *ApJ*, 660, 516
Heger, A. & Woosley, S. E. 2010, *ApJ*, 724, 341
Joggerst, C. C., Woosley, S. E., & Heger, A. 2009, *ApJ*, 693, 1780
Joggerst, C. C., Almgren, A., Bell, J., *et al.* 2010, *ApJ*, 709, 11
Marek, A., & Janka, H.-T.2009, ApJ, 694, 664
Suwa, Y., Kotake, K., Takiwaki, T., Whitehouse, S., & Liebendörfer, M. 2010, *PASJ*, 62, L49.
Ohnishi, N., Kotake, K., & Yamada, S. 2006, *ApJ*, 641, 1018
Fujimoto, S., Hashimoto, M., Ono, M., & Kotake, K. 2011, *ApJ*, 738, 61
O'Connor, E. & Ott, C. D. 2011, *ApJ*, 730, 70
Andrievsky, S. M., Spite, M., Korotin, S. A., *et al.* 2008, *A&A*, 481, 481
Preston, G. W., Sneden, C., Thompson, I. B., *et al.* 2006, *AJ*, 132, 85

Discussion

K. Nomoto: Comment: Nucleosynthesis in hypernovae models is sensitive to the degree of fallback of material onto a black hole. Especially, the fallback of ^{56}Ni results in faint SNe and large [Mg/Fe] in your models.

Death of Massive Stars: Supernovae and Gamma-Ray Bursts
Proceedings IAU Symposium No. 279, 2012
P. Roming, N. Kawai & E. Pian, eds.

doi:10.1017/S1743921312012999

Gamma-Ray Bursts as Cosmological Probes

Tomonori Totani

Department of Astronomy, Kyoto University,
Sakyo-ku, Kyoto 606-8502, Japan
email: `totani@kusastro.kyoto-u.ac.jp`

Abstract. The status and prospects for gamma-ray bursts (GRBs) as cosmological probes are reviewed. Long duration GRBs can potentially be used as an indicator of star formation rate (SFR), though GRB rate might be systematically different from SFR, by the effect of e.g. metallicity. There are several papers claiming that the cosmic GRB rate history is different from that of SFR in the sense that GRB rate is relatively higher than SFR at higher redshifts, which may be explained by the metallicity effect. However, considering the large uncertainties about the efficiency of GRB afterglow detection and redshift determination, it would be conservative to state that the observed GRB rate is roughly consistent with the star formation history. GRBs can also be used as a unique and powerful tool to reveal the reionization history. However, there is practically no progress in this direction since the first GRB-based useful constraint on reionization in 2005 (GRB 050904). The bottleneck now is the insufficient sensitivity of near-infrared spectroscopy, even with 8m class telescopes. The planned 30m class telescopes will bring the next breakthruough. Finally, GRBs can potentially be used as a standard candle to study cosmology by a geometrical test. However, there are still many steps for GRBs to overcome before it produces a result that has strong impact on the cosmology community in the precision cosmology era.

Keywords. cosmology: miscellaneous, gamma rays: bursts

---

1. Introduction

Although more than 30 years have passed since the discovery of gamma-ray bursts (GRBs) [see, e.g., Mészáros (2002), Piran (2004) for recent reviews], it was rather recent that GRBs became widely recognized as a unique tool of cosmological studies and exploring the early universe. It was just before GRBs were proved to have the cosmological origin in 1997 that the first attempt to study the cosmic star formation history by using GRBs was made (Totani 1997, Wijers *et al.* 1998). The potential use of GRBs as a probe of the reionization history of intergalactic medium (IGM) was also pointed out (Miralda-Escudé 1998) soon after GRBs were proven to be cosmological sources. At that time, it was thought that GRBs can probe the universe at most up to modest redshifts of $z \sim 3$ by instruments available at that time or in the near future. Such notion was, however, soon discarded by the discoveries of extremely luminous GRBs like GRB 971214 and 990123, which could be detected even at redshifts beyond 10. Then it did not take long time before astrophysicists started to discuss GRBs as a promising lighthouse to study the extremely high redshift universe (e.g., Lamb & Reichart 2000), potentially giving cosmologically important information including the population III star formation and the reionization history.

During 2000–2004, satellites such as the BeppoSAX and the HETE-2 continued to discover more and more GRBs, and there was important progress including the establishment of the firm connection between long duration GRBs and energetic supernovae. However, the distance of GRBs did not extend to very high redshift, with the highest record of $z = 4.5$ (Andersen *et al.* 2000). The launch of the *Swift* satellite in 2004 then allowed the GRB community to search for fainter and more distant GRBs with improved detection rate. The first breakthrough was brought by GRB 050904 at $z = 6.3$ (Kawai *et al.* 2006; Totani *et al.* 2006), and it kept its record more than two years until broken by GRB 080913 at $z = 6.7$ (Greiner *et al.* 2009). Furthermore, in 2009 GRB 090423 was discovered at $z = 8.3$ (Salvaterra *et al.* 2009; Tanvir *et al.* 2009), placing GRBs at the top rank in the list of the most distant astronomical source populations, far beyond the most distant galaxy at $z = 6.96$.

Now fifteen years have already passed since gamma-ray bursts were revealed to be objects at cosmological distances. Since then, various possibilities of using GRBs as cosmological probes have been discussed, and applied to derive useful cosmological information. (Here, I use the term "cosmology" in a relatively wide context including galaxy formation, not only more cosmological issues such as cosmological parameters, dark energy, etc.)

Here I review the status and prospects of the three important applications of GRBs as cosmological probes: GRBs as an indicator of star formation rate and cosmic star formation history, GRBs as a probe of cosmic reionization, and GRBs as a standard candle to study cosmic expansion by a geometrical test.

2. GRBs as a Star Formation Indicator

The idea of using GRBs as a SFR indicator is simple; since GRBs are related to the death of massive stars, the time from star formation to GRB events are much shorter than the cosmological time scale, and hence GRB rate is expected to trace SFR. Compared with various SF indicators of galaxies, GRBs have advantages of (1) being free from extinction by dust, (2) reaching to very high redshifts, and (3) free from the detection limit about host galaxy luminosity. Note that this is applicable only for long duration GRBs, for which the association with massive supernovae has been established. Short duration GRBs are known to occur also in galaxies having old stellar populations, and hence it cannot be a direct SFR indicator (see below).

There are a number of papers that tried to reconstruct GRB rate history from the observed GRB redshift distribution, and compare it with the cosmic star formation history based on SFR estimates of galaxies at various redshifts. Many papers found and claimed that GRB rate evolution is systematically different from that of SFR, in the sense that GRB rate becomes relatively higher than SFR with increasing redshifts (see Campisi *et al.* 2010; Qin *et al.* 2010; Wanderman *et al.* 2010 for recent results and references therein for earlier ones). This may indicate some physical effects about the condition for GRB events to occur. The most popular scenario is the metallicity effect; theoretically GRBs are expected to preferentially occur in low metallicity environment, because low metallicity is useful to suppress stellar mass loss and keep large total stellar mass and angular momentum. If the low metallicity preference of GRBs is correct, indeed we expect higher GRB rate than SFR with increasing redshift.

However, this result might be a result of selection effects. The difficulty to reconstruct a correct GRB rate history from observation is the treatment of GRB detection efficiency. This efficiency includes not only that for detecting the prompt gamma-ray emission, but also those for afterglow detection and redshift determination. There are many observational steps to determine redshift for a GRB, and the selection effects must be complicated.

In fact, recent studies using latest, relatively homogeneous samples indicate that the major origin of "dark" GRBs is large extinction of optical afterglow. This means that the past GRB sample might have been biased to less dusty, low mass and low metallicity host galaxies. Using the latest sample taking into account this effect, Elliott *et al.* (2012)

found that GRB rate history is consistent with the star formation history. [But see also Salvaterra *et al.* (2012) who found an opposite result using another complete sample taking into account detection efficiency.] This demonstrates that the importance and difficulty of correctly incorporating selection effects and detection efficiency in the rate studies of GRBs.

As for short-duration GRBs, currently the sample size of GRBs with known redshifts is still small, making it difficult to derive strong constraint on the rate evolution over the cosmic time. However, more detailed studies will become possible in the near future. What is important for rate studies for short GRBs is the delay time distribution (DTD), since we know that short GRBs occur with a large delay time from star formation. A hint is coming from the rate studies for type Ia supernovae (SNe Ia). Recent studies indicate that DTD of SNe Ia is described as $\propto t^{-1}$ (Totani *et al.* 2008; Maoz & Mannucci 2012). It is interesting to note that this form of DTD is a general prediction of souces whose delay time is determined by gravitational wave radiation in binaries (Totani 1997; Totani *et al.* 2008). Therefore the t^{-1} DTD of SNe Ia can easily be explained if the progenitor of SNe Ia is double-degenerate (two white dwarfs) binaries, though it does not simply exclude another popular progenitor scenario of SNe Ia (single-degenerate binaries). The double neutron star binaries or NS-BH binaries, which are discussed as a popular scenario of short GRBs, also predict this form of DTD, and future short GRB rate studies should examine the consistency between short GRB rate evolution and SFR evolution under this type of DTD.

3. GRBs as a Probe of Reionization

The famous Gunn-Peterson (GP) test tells us that the IGM is highly ionized at $z \lesssim 5$, while the observations of the cosmic microwave background radiation (CMB) indicates that the universe became neutral at the recombination epoch of $z \sim 1100$. The reionization of the IGM is believed to have occurred during $z \sim 6$–20 by the first stars and/or quasars, and the precise epoch and nature of the reionization is one of the central topics in the modern cosmology (see e.g. Loeb & Barkana 2001 for recent reviews). The dramatic increase of the optical depth of the Lyα forest with increasing redshift at $z \gtrsim 5.2$ and the subsequent discovery of broad and black troughs of Lyα absorption (the GP troughs) in the spectra of $z \gtrsim 6$ quasars indicate that we are beginning to probe the epoch of reionization. On the other hand, the polarization observation of the CMB by the Wilkinson Microwave Anisotropy Probe (WMAP) indicates a much higher redshift of reionization, $z \sim 10$. Some theorists have argued that the hydrogen in the IGM could have been reionized twice.

Because the cross section of the Lyα resonance absorption is so large, the light blueward of the Lyα wavelength at the source is completely attenuated if the IGM neutral fraction $x_{\rm HI} \equiv n_{\rm HI}/n_{\rm H}$ is larger than $\sim 10^{-3}$, and hence the Lyα trough of $z \sim 6$ quasars gives a constraint of only $x_{\rm HI} \gtrsim 10^{-3}$. The cross section becomes much smaller for longer wavelength photons than the Lyα resonance, and the spectral shape of the red damping wing of the GP trough can potentially be used to measure $x_{\rm HI}$ more precisely (Miralda-Escudé 1998). However, applying this method to quasars is problematic because of the uncertainties in the original unabsorbed quasar spectra and the proximity effect, i.e., the ionization of surrounding IGM by strong ionizing flux from quasars. Therefore, though some authors suggested $x_{\rm HI} \gtrsim 0.1$ using $z \sim 6$ quasar spectra, these estimates are generally model dependent.

The Lyα line emission is seriously attenuated if it is embedded in the neutral IGM, and hence the Lyα line emissivity of galaxies at $z \gtrsim 6$ is another probe of the reionization.

Therefore the detection of Lyα emission from many galaxies at $z \gtrsim 6$ may indicate that the universe had already been largely ionized at that time. However it should not be naively interpreted as implying small $x_{\rm HI}$, since these Lyα emitters (LAEs) are selected by strong Lyα emission and hence they may be biased to those in ionized bubbles created by themselves or clusters of undetected sources. On the other hand, the Lyman-break galaxies (LBGs) selected by broad-band colors are free from the selection bias about Lyα emission, but the statistical nature of Lyα emission from LBGs at $z \gtrsim 5$ is not yet well understood, compared with those at $z \sim 3$. Furthermore, the Lyα line emission from lower-redshift starburst galaxies is often redshifted with respect to the systemic velocity of galaxies, and such a relative redshift will lead to a higher detectability of LAEs at $z \gtrsim 6$, indicating a possible systematic uncertainty in the reionization study by Lyα emission.

GRBs have a few advantages as a probe of the cosmic reionization, compared with quasars or LAEs/LBGs. GRB afterglows are much brighter than LAEs/LBGs and comparable to or even brighter than quasars if they are observed quickly enough after the explosion. The Lyα or ultraviolet luminosity of host galaxies is irrelevant to the detectability of GRBs, and hence GRBs can probe less biased regions in the early universe, while quasars and bright LAEs/LBGs are likely biased to regions of rapid structure formation with strong clustering. In most cases it is expected that the IGM ionization state around GRB host galaxies had not yet been altered by strong ionizing flux from quasars. Finally, the spectrum of GRB afterglows has a much simpler power-law shape than complicated lines and continuum of quasars and LAEs/LBGs, and hence model uncertainty can greatly be reduced. Especially, a detailed fitting analysis of the damping wing of the GP trough in an afterglow spectrum may lead to a precise determination of $x_{\rm HI}$.

In the following of this section, I summarize the knowledge we obtained from GRB 050904 (the only useful constraint on reionization by GRBs), and discuss the present status and future prospects.

3.1. *Lessons from GRB 050904*

The GRB 050904 was discovered by the *Swift* on 2005 September 4 at 01:51:44 UT, and follow-up photometric observations of the afterglow found a strong spectral break between optical and near-infrared (NIR) bands, indicating a very high redshift of $z \sim 6$. This suggestion was confirmed by the subsequent spectroscopic observation by the Subaru Telescope, which found metal absorption lines at $z = 6.295$ and the corresponding Lyman break and red damping wing (Kawai *et al.* 2006). This opened a new era of GRB observations at redshifts that are close to the cosmic reionization and comparable to those of the most distant galaxies and quasars.

The detection of red damping wing by neutral hydrogen could have a significant impact on reionization if it is caused by neutral hydrogen in intergalactic medium. However, it could also be caused by neutral hydrogen in the host galaxy of the GRB. In fact, many damped Lyα systems (DLAs) have been observed in GRB afterglow spectra, like those in quasar spectra. The key issue is how to discriminate between the two possibilities.

Since the damping wing by intergalactic hydrogen is a superposition of hydrogen at many different redshifts, the shape and location of the wing are slightly different from those of DLA in the host galaxy. Although the shape difference is difficult to discriminate with the quality of GRB 050904 spectrum, the location of IGM wing is shifted to blueward compared with DLA wings and they can be discriminated if the host redshift is precisely known. Because of the high quality of the afterglow spectrum, several absorption lines were detected, giving strong constraints on the exact redshift of the host galaxy. There are several physical effects that could produce systematic shift of observed absorption lines from the true systemic redshift of the host, as discussed in Totani *et al.* (2006), and these must be carefully taken into account when one derives constraint on reionization. Another feature useful to discriminate between IGM and DLA is Lyβ feature, because it also gives a clear constraint on the host redshift.

By using the absorption line information and Lyβ feature, Totani *et al.* (2006) showed that the damping wing is caused by DLA, and absence of evidence for IGM wing gives an upper limit on neutral hydrogen fraction at $z = 6.3$. They derived $x_{\rm HI} < 0.17$ and 0.60 at 68 and 95% C.L., respectively. This is complementary to the lower bounds on $x_{\rm HI}$ from quasar GP tests, proving that GRBs are a powerful and unique tool to study reionization.

3.2. *GRB 080913 and GRB 090423*

In contrast to GRB 050904, no useful constraints on reionization was derived from afterglow spectra of GRB 080913 and 090423. The main reason is that the signal-to-noise is much lower than that of GRB 050904, even though these two bursts were observed at much shorter time after the burst compared with the case of GRB 050904 (3 days). This is simply because the absolute luminosity of GRB 050904 afterglow was exceptionally bright. The absolute luminosities of GRB afterglows are widely distributed over more than two orders of magnitude. Compared with other past afterglows after scaled to the same redshift and band filter, GRB 050904 optical afterglow was one of the brightest among all the afterglows observed in the past. On the other hand, GRB 080913 and 090423 had afterglow luminosities that are close to the average. The situation is even more severe for GRB 090423, because its spectrum at $z = 8.3$ is only accessible with near-infrared spectrographs, whose sensitivity is much worse than optical spectrographs used to marginally detect Lyman breaks of GRB 050904 and 080913 spectra. Both the spectra of GRBs 080913 and 090423 were taken by VLT, and this indicates that even the state-of-art 8m-class telescopes are not sufficient to study reionization by using GRBs of normal luminosities.

3.3. *Future Prospects*

Since the currently available large telescopes do not have enough power, we need more powerful instruments in the future to probe reionization by GRBs. An obvious solution is planned 30m-class telescopes and the James Webb Space Telescope. These facilities would have more than 100 times better spectroscopic sensitivities against point sources for Lyman break at $z \gtrsim 7$ than current 8m-class telescopes, and it would be possible to perform detailed studies of afterglow spectra including absorption lines, even for normal-luminosity GRB afterglows. However, the ultimate limitation for the reionization study by GRBs would come from their event rate. The statistics of very high-z GRBs searched by *Swift* (only 3 detected GRBs at $z > 6$ in five years) indicates that the fraction of very high redshift GRBs is not very high. Therefore, it is important to launch GRB satellites during the era of 30m telescopes and JWST, which have high sensitivity to trigger faint gamma-ray flux, and a wide field of view to increase the GRB detection rate. There are several proposals for such missions, e.g., *EXIST*. It should also be noted that, even if the expected event rate is not very high, only a few events of very high-z GRBs would have strong impact on reionization study, because GRBs are the only one source population that can be used to precisely measure $x_{\rm HI}$ by damping wing.

4. GRBs as a Standard Candle

There are several types of correlations between GRB absolute luminosity (or total emitted gamma-ray energy) and spectral features, e.g., spectral peak energy ($E_{\rm peak}$) versus isotropic energy ($E_{\rm iso}$)(Amati *et al.* 2002), or $E_{\rm peak}$ versus isotropic luminosity ($L_{\rm iso}$) (Yonetoku *et al.* 2004). These correlation can be used to correct and calibrate GRB luminosities to a standard candle, in a similar way to that of the period-luminosity relation of Cepheid variables or the peak luminosity versus stretch relation for type Ia supernovae. A number of papers already tried to use gamma-ray bursts as a standard candle and derive some constraints on cosmological parameters. Advantages of GRBs compared with the other cosmological standard candle, SNe Ia, are (1) being free from the extinction in host galaxies and (2) reach to much higher redshifts.

However, unfortunately, results on cosmological parameters by GRBs have not yet had strong impact on the cosmological community. The main reason for this is the still very large systematic uncertainty of GRBs as a standard candle. The dispersion around the mean relations (e.g., a factor of several for $E_{\rm peak}$ - $E_{\rm iso}$) is much larger than that of SNe Ia (~10–20%). In the cosmology community, many people think that even SNe Ia is now limited by systematic uncertainties, and there is a huge effort to standardize SNe Ia with higher accuracy to derive reliable constraints on the dark energy parameter such as $w \equiv p/\rho$ and its time derivative.

An obvious concern about the correlations of GRB quantities used to standardize GRBs is the selection effect. Most of the proposed relations tell us that the total energy or luminosity of GRBs increases with increasing characteristic photon energy of GRB spectra. It should be noted that this is something expected by a selection effect about a fixed photon energy range of GRB detectors. When GRBs are preferentially detected at a fixed observed photon energy range, the rest-frame spectral peak energy would increase with increasing redshifts, and at the same time, the absolute luminosity will also increase by the detection flux limit of the detector.

It is probably impossible to explain all the observed GRB correlations simply by the selection effect, and the astrophysical origin of the observed correlation is a quite interesting topic. However, such a selection effect is certainly significant when GRBs are applied to precision cosmology. To conclude, although GRBs may potentially be a standard candle that is useful for cosmological tests, there are still many steps for GRBs to overcome before producing a cosmological result having a strong impact on the cosmological community.

References

Amati, L., Frontera, F., Tavani, M., *et al.* 2002, *A&A*, 390, 81
Andersen, M. I., *et al.* 2000, *A&A*, 364, L54
Campisi, M. A., Li, L.-X., & Jakobsson, P. 2010, *MNRAS*, 407, 1972
Elliott, J., Greiner, J., Khochfar, S., *et al.* 2012, *A&A*, 539, A113
Greiner, J., *et al.* 2009, *ApJ*, 693, 1610
Kawai, N., *et al.* 2006, *Nature*440, 184
Lamb, D. Q. & Reichart, D. E. 2000, *ApJ*, 536, 1
Loeb, A. & Barkana, R. 2001, *ARA&A*, 39, 19
Maoz, D. & Mannucci, F. 2011, arXiv:1111.4492
Mészáros, P. 2002, *ARA&A*, 40, 137
Miralda-Escudé, J. 1998, *ApJ*, 501, 15
Piran, T. 2004, *Rev. Mod. Phys.* 76, 1143
Qin, S.-F., Liang, E.-W., Lu, R.-J., Wei, J.-Y., & Zhang, S.-N. 2010, *MNRAS*, 406, 558

Salvaterra, R. *et al.* 2009, *Nature*, 461, 1258
Salvaterra, R. *et al.* 2012, *ApJ*, 749, 68
Tanvir, N. R., *et al.* 2009, *Nature*, 461, 1254
Totani, T. 1997, *ApJ*, 486, L71
Totani, T., *et al.* 2006, PASJ, 58, 485
Totani, T., Morokuma, T., Oda, T., Doi, M., & Yasuda, N. 2008, *PASJ*, 60, 1327
Wanderman, D. & Piran, T. 2010, *MNRAS*, 406, 1944
Wijers, R. A. M. J., Bloom, J. S., Bagla, J. S., & Natarajan, P. 1998, *MNRAS*, 294, L13
Yonetoku, D., Murakami, T., Nakamura, T., *et al.* 2004, *ApJ*, 609, 935

Discussion

AMATI: Concerning the use of GRBs for cosmological parameters, there is evidence that the $E_{\rm peak}$-$E_{\rm iso}$ ($L_{\rm iso}$) correlations are not significantly affected by selection effects.

TOTANI: It depends on the required accuracy as a standard candle whether the selection effect is significant or not. Currently, the scatter along the mean relation of GRBs is still much larger than SNe Ia, and only weak constraints on Ω_M and Ω_Λ have been derived. In such analysis, the selection effect may not be significant. But I guess it should be significant if GRBs are used to derive much more precise cosmological constraints, such as w.

Death of Massive Stars: Supernovae and Gamma-Ray Bursts
Proceedings IAU Symposium No. 279, 2012
P. Roming, N. Kawai & E. Pian, eds.

doi:10.1017/S1743921312013002

On the intrinsic nature of the updated luminosity time correlation in the X-ray afterglows of GRBs

Maria G. Dainotti[1,2], Vahe' Petrosian[2], and Jack Singal[2]

[1]Astronomical Observatory, Jagellonian University
ul. Orla 171, 31-501 Cracow, Poland
email: `dainotti@oa.uj.edu.pl`, `mariagiovannadainotti@yahoo.it`

[2]Department of Physics & Astronomy, University of Stanford, Via Pueblo Mall,
email: `vahep@stanford.edu`, `jacks@slac.stanford.edu`

Abstract. Gamma-ray bursts (GRBs) observed up to redshifts $z > 9.3$ are fascinating objects to study due to their still unexplained relativistic outburst mechanisms and a possible use to test cosmological models. Our analysis of all GRB afterglows with known redshifts and definite plateau (100 GRBs) reveals not only that the luminosity $L^*_X(T_a)$ - break time T^*_a correlation, called hereafter LT, (Dainotti *et al.* 2010a) is confirmed with higher value of the Spearman correlation coefficient for the new updated sample, but also reveals its intrinsic nature throughout the analysis of the Efron & Petrosian (1992) test. The above mentioned test is performed to check if there is redshift evolution in both the luminosity and time. This test shows that the correlation still holds probing that its nature is intrinsic and it is not due to selection biases. The novelty of this approach is that the Efron & Petrosian method has been applied for the first time for a two parameter correlation that involves not only luminosities, but also time. Notwithstanding the intrinsic nature of the correlation, the correction of the observables for the effect of redshift evolution does not lead to a significantly tighter correlation and thus to a better redshift estimator. Therefore, the usage of the L^*_a correlation is limited, at least with the present data analysis, to constrain physical models of plateau emission. With an enlarged data sample in the future the aim will be to make the luminosity time correlation a useful redshift estimator.

Keywords. cosmological parameters - gamma-rays bursts: general, -radiation mechanisms: non-thermal

1. Introduction

GRBs are the farthest, up to z = 9.46, and the most powerful, up to 10^{54} ergs/s, objects ever observed in the Universe. Finding out universal properties which could be revealed by looking for strict relations among their observables plays a crucial role in understanding processes responsible for GRBs. But, GRBs seem to be everything but standard candles, with their energetics spanning over 8 orders of magnitude. Notwithstanding the variety of their different peculiarities, some common features have been identified thanks to the observation of GRBs by the *Swift* satellite which provides a rapid follow-up of the afterglows in several wavelengths with better coverage than previous missions. *Swift* revealed a more complex behavior of the lightcurves than the broken power-law assumed in the past (O' Brien *et al.* 2006). In the lightcurves observed by Swift one can identify two, three and even more segments in the afterglows. The second segment, when it is flat, is called plateau emission. A significant step forward in determining common features in the afterglow lightcurves has been made by the analysis of the X-ray afterglow lightcurves of the full sample of *Swift* GRBs showing that they may be fitted by the same analytical expression (Willingale *et al.* 2007). This provides the unprecedented opportunity to look for universal features that would allow us to recognize if GRBs are standard candles. Therefore, studies of correlations between GRB observables, E_{iso} - E_{peak} (Lloyd & Petrosian 1999, Amati *et al.* 2009), E_{γ} - E_{peak} (Ghirlanda *et al.* 2004, Ghirlanda *et al.* 2006), L - E_{peak} (Schaefer 2003, Yonekotu 2004), L - V (Fenimore & Ramirez - Ruiz 2000, Riechart *et al.* 2001, Norris *et al.* 2000) are the attempts pursued in this direction. However, the problem of large data scatter in the considered luminosity correlations (Butler *et al.* 2009, Yu *et al.* 2009) and a possible impact of detector thresholds on cosmological standard candles (Shahmoradi & Nemiroff 2009, Petrosian 1998, Petrosian *et al.* 1999, Petrosian 2002) are also debated issues (Cabrera *et al.* 2007) and should be taken into account.

Within the framework of correlations a new phenomenological one for long GRBs (Dainotti *et al.* 2010a, Dainotti *et al.* 2011, Dainotti *et al.* 2008) between the luminosity at the end of the plateau phase, L^*_X and its duration, T^*_a has been discovered (We denote with $*$ the rest frame quantities). In particular, the established behaviour is $\log L^*_X = \log a + b \log T^*_a$, where a (the normalization parameter) and b (the slope) are constants obtained by the fitting procedure. The above anticorrelation has already been confirmed in the literature (Ghisellini *et al.* 2008, Yamazaki 2009) and it is also a useful test for theoretical models (Cannizzo & Gehrels 2009, Cannizzo *et al.* 2011, Dall'Osso *et al.* 2011, Bernardini *et al.* 2011). Here, we study the LT correlation in order to test its intrinsic nature and what is its intrinsic slope, because this is the first step to cast light on the nature of the plateau emission and provide further constrains on the theoretical models. We have found the true power law of the LT correlation corrected for possible data truncation due to the instrumental threshold. This step is necessary to assess the possibile usage of the LT correlation as a distance estimator.

2. Data Analysis

We have extented the analysis of the LT correlation using a sample of 100 afterglows of all GRBs, long, short with extended emission and X-ray Flashes with known redshifts (from 0.08 to 9.4) detected by Swift from 2005 January up to 2011 May, for which the light curves include early X-ray Telescope (XRT) data and can be fitted by the Willingale's *et al.* phenomenological model 1. The present data analysis presents a modification compared to previous papers (Dainotti *et al.* 2008, Dainotti *et al.* 2010), in which the Swift BAT+XRT lightcurves of GRBs were fitted assuming that the time rise of the afterglow, t_a started at the same time as the beginning of the decay phase of the prompt emission, T_p, namely $t_a = T_p$. In such a way the fitting of the afterglow was related throughout this parameter to the prompt emission. In the present analysis we aim to have an independent measure of the afterglow parameters, therefore we have left t_a free of vary.

For further details of the computation of the source rest -frame luminosity and the spectral fitting procedure see Dainotti *et al.* (2010), Evans *et al.* (2009).

The best fit for the slope of the new observed $L^*_X(T_a)$-T^*_a correlation with 100 GRBs is -1.6 while in the previous sample with 77 GRBs the slope was roughly -1.1. We investigated several hypotheses to explain the change of the slope in the correlation: a) If the method of the fitting changed the result, it would mean that the correlation is model dependent and it would not be intrinsic; b) the changes are caused by the redshift evolution of both luminosity and time redshift evolution which is different in the two samples due to the difference in the redshift distribution. The two samples have 53 GRBs in common.

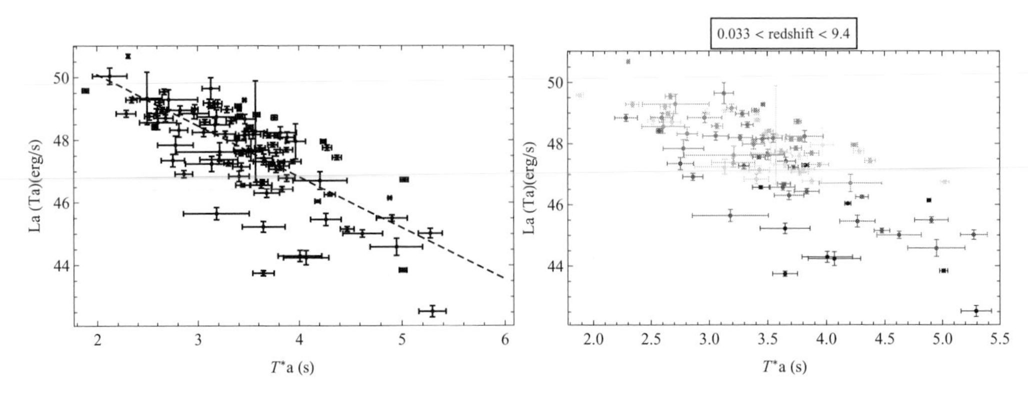

Figure 1. Left panel: L_X^* vs T_a^* distribution for the sample of 100 afterglows of long GRB and short GRB with extended emission, with the fitted correlation line in black. Right panel: L_X^* vs T_a^* distribution divided in 8 redshift bin, black $z \leqslant 0.69$, brown $0.69 < z \leqslant 1.0$, blue $1.0 < z \leqslant 1.49$, cyan $1.49 < z \leqslant 2.15$, yellow $2.15 < z \leqslant 2.67$, green $2.67 < z \leqslant 3.2$, pink $3.2 < z \leqslant 4.1$ and red for $z > 4.1$

3. Methodology and results

To answer the previous questions it is imperative to first determine the true correlations among the variables, not those introduced by the observational selection effects, before obtaining the individual distributions of the variables themselves and the correlations among them. With the term selection effect or bias we refer to the distortion of statistical analysis, resulting from the method of collecting samples, namely gathering data from a certain satellite with a certain flux limit would prevent us from seeing a truly representative sample of events. Therefore, we present here in left panel, see Fig. 1 the observed $L_X^*(T_a)$-T_a^* distribution divided in 8 equipopulated redshift bins. We have checked that the correlation coefficient of every subsample and the slope are still compatible in 1 σ, but for the redshift bin between $3.2 < z < 4.1$ (pink points) for which the correlation coefficient is negligible. So, the test applied in Dainotti *et al.* (2011a) alone is not sufficient to guarantee the lack of evolution in redshift of the LT correlation. Therefore, we have applied the Efron & Petrosian (1992) technique (EP) that corrects for instrumental threshold selection effect and redshift evolution and has been already successfully applied to GRBs (Lloyd & Petrosian 1999). In general, the first step required for this kind of investigation is the determination of whether the variables of the distributions, L_X^* and T_a^*, are correlated with redshift or are statistically independent. For example, the correlation between L_X^* and the redshift, z, is what we call luminosity evolution, and independence of these variables would imply absence of such evolution.

The EP method prescribed how to remove the correlation by defining new and independent variables. Therefore, following the approach used for quasars and blazars (Singal *et al.* 2011a, 2011b), the new variables will not evolve with redshift, namely they will not be affected by redshift evolution, and the correction will be the following $L'_X \equiv L_X^*/g(z)$, where the function $g_z = (1+z)^{k_L}$ describes the luminosity evolution, and $T'_a \equiv T_a^*/f(z)$ where $f(z) = (1+z)^{k_T}$ describes the time evolution. We denote with $'$ the not-evolved quantities. The EP method deals with data subsets that can be constructed to be independent of the truncation limit suffered by the entire sample. This is done by creating 'associated sets', which include all objects that could have been observed given a certain limiting luminosity. A specialized version of the Kendall-rank correlation coefficient τ, a statistic tool used to measure the association between two measured quantities, takes into account the associated sets and not the whole sample and produces a single parameter whose value directly rejects or accepts the hypothesis of independence. The values of k_L and k_T for which $\tau_{L,z} = 0$ and $\tau_{T,z} = 0$ are the ones that best fit the luminosity and time evolution, respectively. With these values of k_L and k_T we can determine the not-evolved T'_a and L'_X. In the space of T'_a and L'_X we have applied again the method of the associated sets to derive the best estimate of the intrinsic slope of the correlation. We have tested this procedure using Monte Carlo simulations designed to resemble the observations, but with known distributions of uncorrelated and correlated luminosities, L'_X, and time, T'_a, and subjected to a truncation similar to the actual data. The simulations confirm the results obtained from the EP technique with the observed data. We can conclude that the LT correlation exists indeed with an intrinsic correlation slope ranging from −1.6 to −2.0. There is no relevant redshift evolution both in time and luminosity, so the correlation is preserved. Therefore, the LT correlation can be used to discriminate among theoretical models and to cast light on the nature of the plateau emission.

References

Amati, L., Frontera, F., & Guidorzi, C. *A&A*, 508, 173.

Bernardini, M. G. *et al.* 2011, accepted on *A&A* arXiv 1112.1058B

Butler N. R., Kocevski D., & Bloom J. S., 2009, *ApJL*, 694, 76.

Cabrera, J. I., Firmani, C., Avila-Reese, V., *et al.* 2007, *MNRAS*, 382, 342

Cannizzo, J. K. & Gehrels, N., 2009, *ApJ*, 700, 1047

Cannizzo, J. K., Troja, E., & Gehrels, N., 2011, *ApJ*, 734, 35C

Dall'Osso, S. *et al.* 2011, *A&A*, 526A, 121D

Dainotti, M. G., Cardone, V. F. & Capozziello, S. 2008, *MNRAS* 391L, 79D

Dainotti, M. G., *et al.* 2010, *ApJL*, 722, L215

Dainotti, M. G., *et al.* 2011, *ApJ*, 730, 135D

Dainotti, M. G., M. Ostrowski & Willingale, R., 2010, *MNRAS*, 418, 2202D

Evans, P., *et al. MNRAS*, 2009, 397, 1177

Efron, B. & Petrosian, V., 1992, *ApJ*, 399, 345

Fenimore, E. E. & Ramirez-Ruiz, E. 2000, *ApJ*, 539, 712v

Ghirlanda, G., Ghisellini, G., & Lazzati, D. 2004, *ApJ*, 616, 331

Ghirlanda G., Ghisellini G. & Firmani C., 2006, *New Journal of Physics*, 8, 123.

Ghisellini G., *et al.* 2008, *A&A*, 496, 3, 2009.

Lloyd, N., & Petrosian, V. *ApJ*, 511, 550, 1999

Norris, J. P., Marani, G. F., & Bonnell, J. T., 2000, *ApJ*, 534, 248

Yonetoku, D., *et al.* 2004, *ApJ*, 609, 935Y

O'Brien, P. T., Willingale, R., Osborne, J., *et al.* 2006, *ApJ*, 647, 1213

Petrosian, V. 1998, AAS, 193, 8702P

Petrosian, V., *et al.* 1999 *ASPC*, 190, 235P

Petrosian, V. 2002, *ASPC*, 284, 389P

Riechart, D. E., Lamb, D. Q., Fenimore, E. E., Ramirez-Ruiz, E., & Cline, T. L., 2001, *ApJ*, 552, 57

Sakamoto, T., Hill, J., Yamazaki, R., *et al.* 2007, *ApJ*, 669, 1115.

Shahmoradi, A. & Nemiroff R. J. 2009, *AIPC*. 1133, 425S

Schaefer, B. E., 2003, *ApJ*, 583, L67

Singal, J., *et al.*, 2011, *ApJ*, 743, 104S

Singal, J., *et al.*, 2011, *ApJ* accepted arXiv1106.3111S

Yamazaki, R. 2009, *ApJ*, 690, L118

Yu, B., Qi, S., & Lu, T. 2009, *ApJL*, 705, L15

Willingale, R. W., *et al.*, *ApJ*, 2007, 662, 1093

Discussion

AMATI LORENZO: Did you perform simulations in order to evaluate selection effects introduced by the cuts of low accurate variables, Lx-Ta?

DAINOTTI MARIA G.: Indeed, we have performed simulations that confirms that the correlation is intrinsic and the more accurate the measurement of the variables the tigher the correlation is implying the existence of a subclass of GRBs with well definite properties.

Death of Massive Stars: Supernovae and Gamma-Ray Bursts
Proceedings IAU Symposium No. 279, 2012
P. Roming, N. Kawai & E. Pian, eds.

doi:10.1017/S1743921312013014

Pair-Instability Explosions: observational evidence

Avishay Gal-Yam

Department of Particle Physics and Astrophysics,
Weizmann Institute of Science, 76100 Rehovot, Israel
email: avishay.gal-yam@weizmann.ac.il

Abstract. It has been theoretically predicted many decades ago that extremely massive stars that develop large oxygen cores will become dynamically unstable, due to electron-positron pair production. The collapse of such oxygen cores leads to powerful thermonuclear explosions that unbind the star and can produce, in some cases, many solar masses of radioactive ^{56}Ni. For many years, no examples of this process were observed in nature. Here, I briefly review recent observations of luminous supernovae that likely result from pair-instability explosions, in the nearby and distant Universe.

---

1. Introduction

The pair-instability explosion mechanism (e.g., Rakavy & Shaviv 1967; Barkat, Rakavy & Sack 1967; Bond *et al.* 1984; Heger & Woosley 2002; Scannapieco *et al.* 2005; Waldman 2008) was predicted to occur during the evolution of very massive stars that develop oxygen cores above a critical mass threshold ($\sim 50\,\mathrm{M}_\odot$). These cores achieve high temperatures at relatively low densities (e.g., Fig. 1 of Waldman 2008). Significant amounts of electron-positron pairs are created prior to oxygen ignition; loss of pressure support, rapid contraction, and explosive oxygen ignition follow, leading to a powerful explosion that disrupts the star. Extensive theoretical work indicates this result is unavoidable for massive oxygen cores; when the core mass in question is large enough ($\sim 100\,\mathrm{M}_\odot$; e.g., Heger & Woosley 2002; Waldman 2008) many solar masses of radioactive nickel are naturally produced. Such Ni-rich events will be extremely luminous and therefore easy to observe. On the other hand, oxygen cores that are massive enough to become pair unstable were predicted to evolve, according to most stellar-evolution models, only in stars of exceedingly large initial masses (many hundreds times $\mathrm{M}_\odot$; e.g., Yoshida & Umeda 2010; though see Langer *et al.* 2007), unless the stars are assumed to have very low initial metallicity. For this reason it was often assumed that pair-instability supernovae (PISN) only occurred among population III stars at very high redshifts. Recently, we have shown that luminous events that match the predictions of PISN models very well (SLSN-R; Gal-Yam 2012) do occur in dwarf galaxies in the local Universe (e.g., SN 2007bi Gal-Yam *et al.* 2009).

2. Early candidates

SN 1999as (Knop *et al.* 1999) was one of the first genuine Superluminous Supernovae (SLSN; Gal-Yam 2012) discovered. It was initially analyzed by Hatano *et al.* (2001). As shown in Gal-Yam *et al.* (2009; Fig. 1) this object was similar to the likely PISN event SN 2007bi during its photospheric phase (reaching -21.4 mag absolute at peak). The analysis of Hatano *et al.* (2001) suggests physical attributes (^{56}Ni mass, kinetic energy, and ejected mass) that are close to, but somewhat lower than, those of SN 2007bi. Unfortunately,

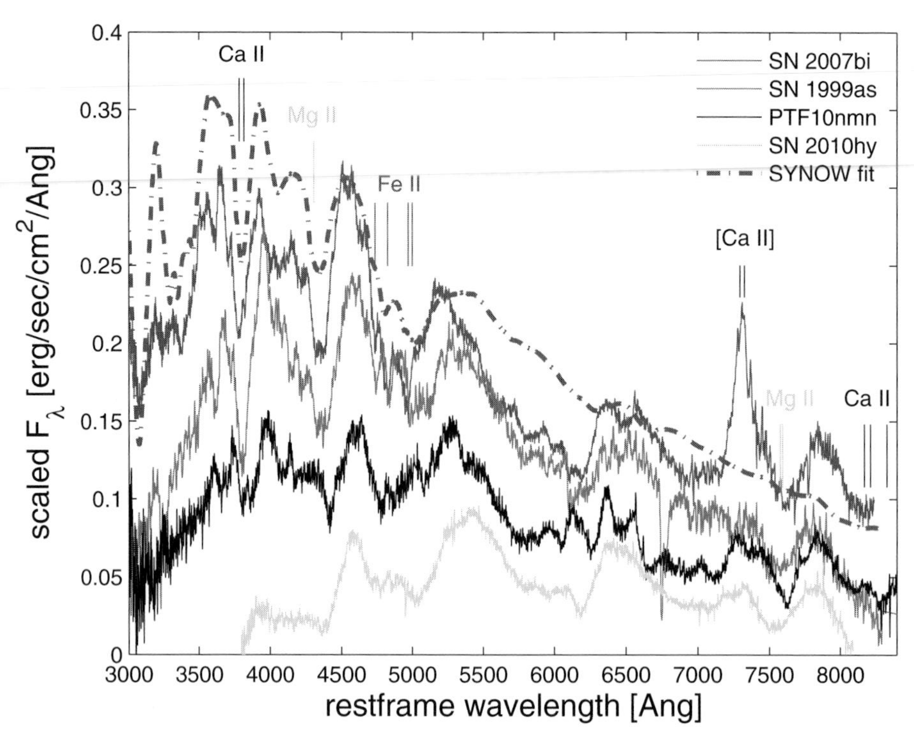

Figure 1. Photospheric spectra of SLSN-R events SN 2007bi (blue, from Gal-Yam *et al.* 2009), SN 1999as (magenta, from Gal-Yam *et al.* 2009; Nugent *et al.* 2012, in preparation), PTF10nmn (black, from Yaron *et al.* 2012, in preparation) and SN 2010hy (cyan; S. B. Cenko, private communication); all spectra were obtained close to peak. Identification of prominent spectral features as well as a synthetic SYNOW fit (red, from Gal-Yam *et al.* 2009) are also shown. Figure taken from Gal-Yam 2012.

no late-time data have been published for this object, so it is impossible to conduct the same analysis carried for SN 2007bi, but the similarities suggest this may also have been a PISN. Interestingly, late-time photometry presented for the first time by K. Nomoto during a presentation in this symposium may argue against this possibility.

3. SN 2007bi: the first likely detection of a pair-instability event

The first well-observed example of a likely PISN was SN 2007bi, discovered by the PTF "dry run" experiment. An extensive investigation of this object and its physical nature is presented in Gal-Yam *et al.* (2009). The most prominent physical characteristic of this object (the prototype of the SLSN-R group), the large ^{56}Ni mass, is well-measured in this case using both the peak luminosity ($R = -21.3$ mag) and the cobalt decay tail, followed for >500 days. Estimates derived from the observations, as well as via comparison to other well-studied events (SN 1987A and SN 1998bw) converge on a value of $M_{^{56}Ni} \approx 5\,M_\odot$. The large amount of radioactive material powers a long-lasting phase of nebular emission, during which the optically thin ejecta are energized by the decaying radio nucleides. Analysis of late-time spectra obtained during this phase (Gal-Yam *et al.* 2009) provides independent confirmation of the large initial ^{56}Ni mass via detection of strong nebular emission from the large mass of resulting ^{56}Fe, as well as the integrated emission from all elements, powered by the remaining ^{56}Co.

Estimation of other physical parameters of the event, in particular the total ejected mass (which provides a lower limit on the progenitor star mass), its composition, and the kinetic energy it carries, is more complicated. There are no observed signatures of hydrogen in this event (either in the ejecta or traces of CSM interaction) so the ejecta mass directly constrains the mass of the exploding helium core, which is likely dominated by oxygen and heavier elements. Gal-Yam *et al.* (2009) use scaling relations based on the work of Arnett (1982), as well as comparison of the data to custom light-curve models, and derive an ejecta mass of $M \approx 100\,M_\odot$. Analysis of the nebular spectra provides an independent lower limit on the mass of $M > 50\,M_\odot$, with a composition similar to that expected from theoretical models of massive cores exploding via the pair-instability process. Moriya *et al.* (2010) postulate a lower ejecta mass ($M = 43\,M_\odot$); this difference becomes crucial to the controversy about the explosion mechanism of these giant cores (see below). In any case there is no doubt this explosion was produced by an extremely massive star, with the most massive exploding heavy-element core we know. The same scaling relations used by Gal-Yam *et al.* (2009) also indicate extreme values of ejecta kinetic energy (approaching $E_k = 10^{53}$ erg). Finally, the integrated radiated energy of this event over its very long lifetime is high ($>10^{51}$ erg).

4. Additional events

Recently, the Lick Observatory Supernova Survey (LOSS; Filippenko *et al.* 2001) using the 0.75m Katzman Automatic Imaging Telescope (KAIT) discovered the luminous Type Ic SN 2010hy (Kodros *et al.* 2010; Vinko *et al.* 2010); Following the discovery by KAIT this event was also recovered in PTF data (and designated PTF10vwg). It is interesting to note that while the LOSS survey is operating in a targeted mode looking at a list of known galaxies, by performing image subtraction on the entire KAIT field of view it is effectively running in parallel also an untargeted survey of the background galaxy population (as noted by Gal-Yam *et al.* 2008 and Li *et al.* 2011). It is during this parallel survey that KAIT detected this interesting rare SN, residing in an anonymous dwarf host. While final photometry is not yet available for this event, preliminary KAIT and PTF data indicate a peak magnitude of -21 mag or brighter, and it is spectroscopically similar to other SLSN-R (Fig. 1) suggesting it is also likely a member of this class.

PTF has discovered another likely PISN, PTF10nmn (Yaron *et al.* 2012, in preparation; Fig. 2). The object is similar to SN 2007bi both in terms of its light curve (Fig. 2) and spectra (Fig. 1). Objects of this sub-class are exceedingly rare (this is observationally the rarest class among the SLSN classes; Gal-Yam 2012), and thus additional examples are scarce. In addition to this single event from PTF, the Pan-Starrs 1 survey may have discovered another similar object at a higher redshift (Kotak *et al.* 2012, in preparation), while Cooke *et al.* (2012) may have recovered events at even higher redshifts (up to $z \sim 4$) in archival SNLS data. Assembling a reasonable sample of such events may thus be a time-consuming process.

5. The physical properties of SLSN-R and their PISN nature

Of all classes of super-luminous SNe, this seems to be the best understood. SLSN-R events are powered by large amounts (several $M_\odot$) of radioactive ^{56}Ni (hence the suffix "R"), produced during the explosion of a very massive star. The radioactive decay chain $^{56}\mathrm{Ni} \rightarrow {}^{56}\mathrm{Co} \rightarrow {}^{56}\mathrm{Fe}$ deposits energy via γ-ray and positron emission, that is thermalized and converted to optical radiation by the expanding massive ejecta. The luminosity of the peak is broadly proportional to the amount of radioactive ^{56}Ni, while the late-time decay

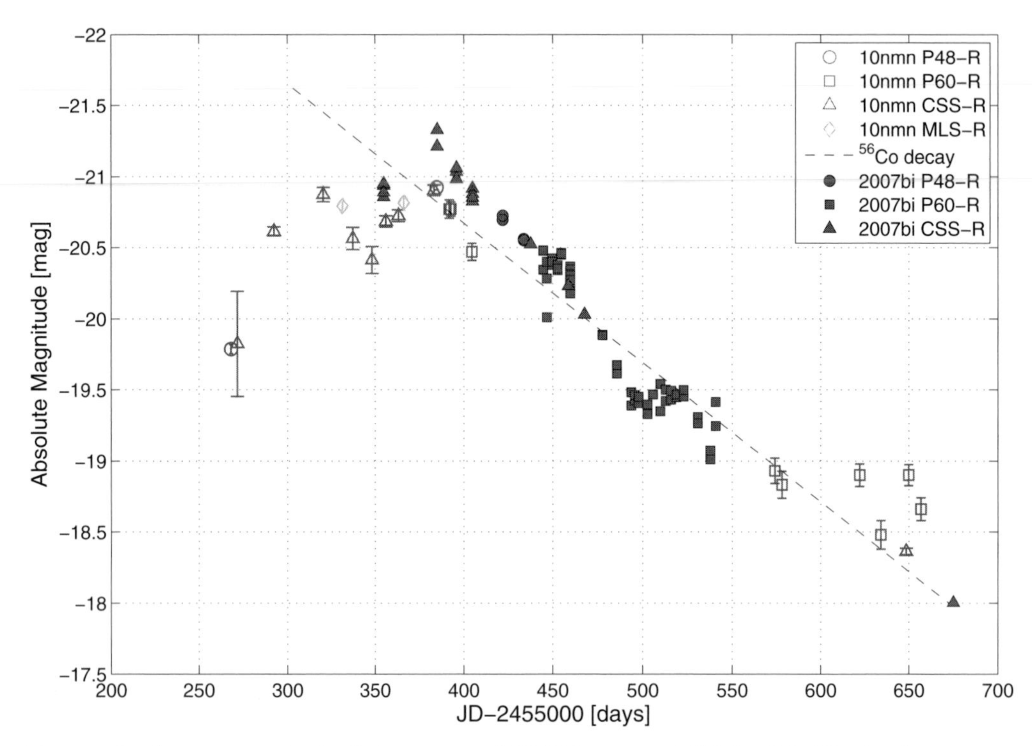

Figure 2. Comparison of the light curves of PTF10nmn and SN 2007bi. The luminous peak and slow decline are similar, indicating a large mass of ^{56}Ni is powering these explosions, mixed into a large total mass of ejecta. From Yaron *et al.* 2012 (in preparation).

(which in the most luminous cases begins immediately after the optical peak) follows the theoretical ^{56}Co decay rate (0.0098 mag day^{-1}). The luminosity of this "cobalt radioactive tail" can be used to infer an independent estimate of the initial ^{56}Ni mass.

Considering all available data, it seems there is agreement about the observational properties of this class and their basic interpretation: very massive star explosions that produce large quantities of radioactive ^{56}Ni. A controversy still exists about the underlying explosion mechanism that leads to this result, either very massive oxygen cores ($M > 50\,M_{\odot}$) become unstable to electron-positron pair production and collapse (Gal-Yam *et al.* 2009), or else slightly less massive cores ($M < 45\,M_{\odot}$) evolve all the way till the common iron-core-collapse process occurs (Moriya *et al.* 2010).

Umeda & Nomoto (2008) and Moriya *et al.* (2010) show that if one considers a carbon-oxygen core with a mass of $\sim 43\,M_{\odot}$ (just below the pair-instability threshold), which explodes with an ad-hoc large explosion energy ($>10^{52}$ erg), one can produce the required large amounts of nickel (Umeda & Nomoto 2008), as well as recover the light curve shape of the SLSN-R prototype, SN 2007bi (Moriya *et al.* 2010). Since both the pair-instability model and the massive core-collapse model fit the light curve shape of SN 2007bi equally well; and progenitors of pair-instability explosions have larger cores and thus larger initial stellar masses, which are, assuming a declining initial mass function, intrinsically more rare, Yoshida & Umeda (2011) favor the core-collapse model.

The two models (pair instability vs. core-collapse) agree about the nickel mass, but strongly differ in their predictions about the *total* ejected mass. Total heavy-element masses above the $50\,M_{\odot}$ threshold would indicate a core that is bound to become pair unstable, and will rule out the core-collapse model. Gal-Yam *et al.* (2009) estimated the

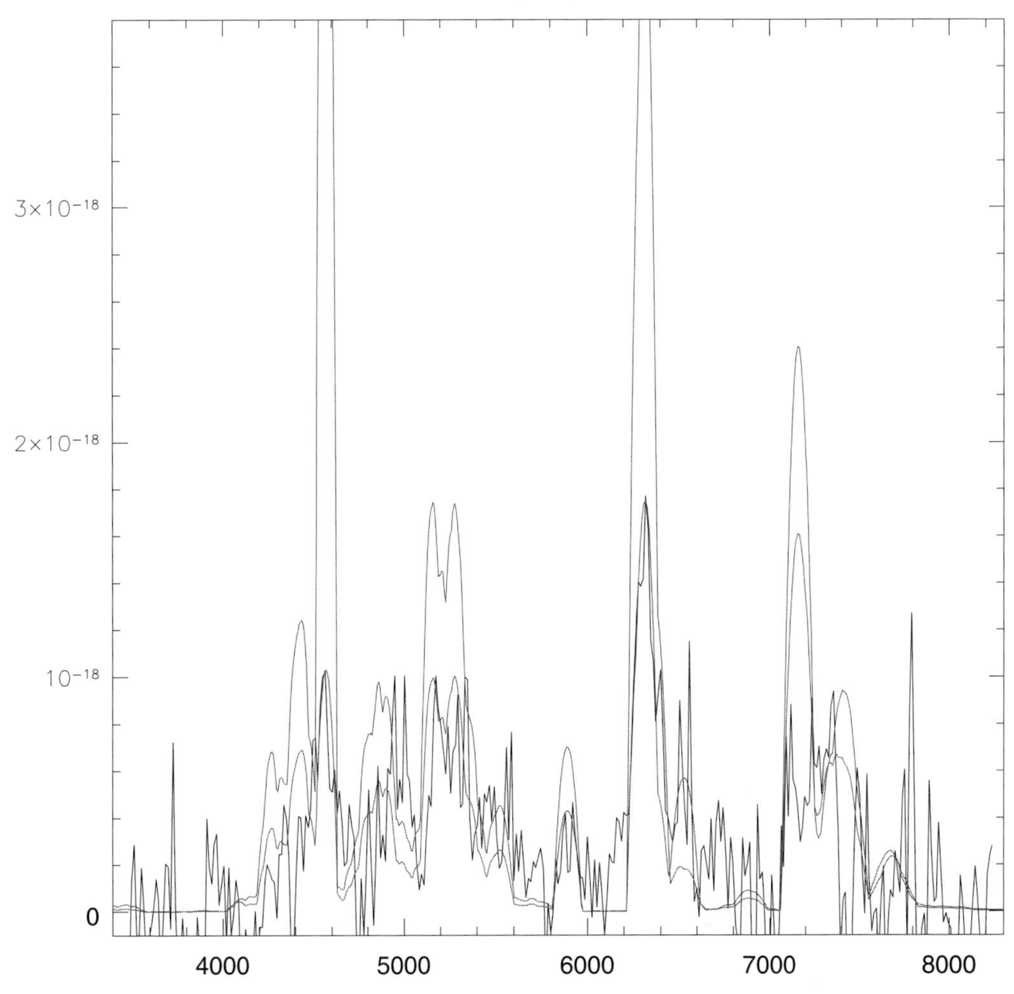

Figure 3. A model of the nebular emission expected from ejecta with the composition given by the massive core-collapse model of Moriya *et al.* (2010, blue), compared to the nebular model from Gal-Yam *et al.* 2009 (red) based on pair-instability event with a composition expected from the models of Heger & Woosley (2002). The massive core-collapse model (blue) has a similar amount of radioactive nickel mixed into a smaller total ejecta mass, significantly over-predicting the observed line strengths in the nebular spectrum of SN 2007bi (black).

total ejected heavy-element mass of SN 2007bi in several ways, including modelling of the nebular spectra of this event. The core-collapse model of Moriya *et al.* 2010 does not fit these data (Fig. 3); this model assumes a similar amount of radioactive ^{56}Ni and lower total ejected mass (to avoid the pair-instability) leading to very strong nebular emission lines that are not consistent with the data. Thus this model is not viable for this prototypical SLSN-R object, supporting instead a pair instability explosion as originally claimed.

It remains to be seen whether the massive core-collapse model does manifest in nature (the prediction would be for SLSNe showing large amounts of radioactive nickel but relatively small amounts of total ejecta). As a final note, it should be stressed that while the stellar evolution models considered by Yoshida & Umeda (2011) require stars with exceedingly large initial masses ($> 310\,M_{\odot}$) to form pair-unstable cores at the moderate

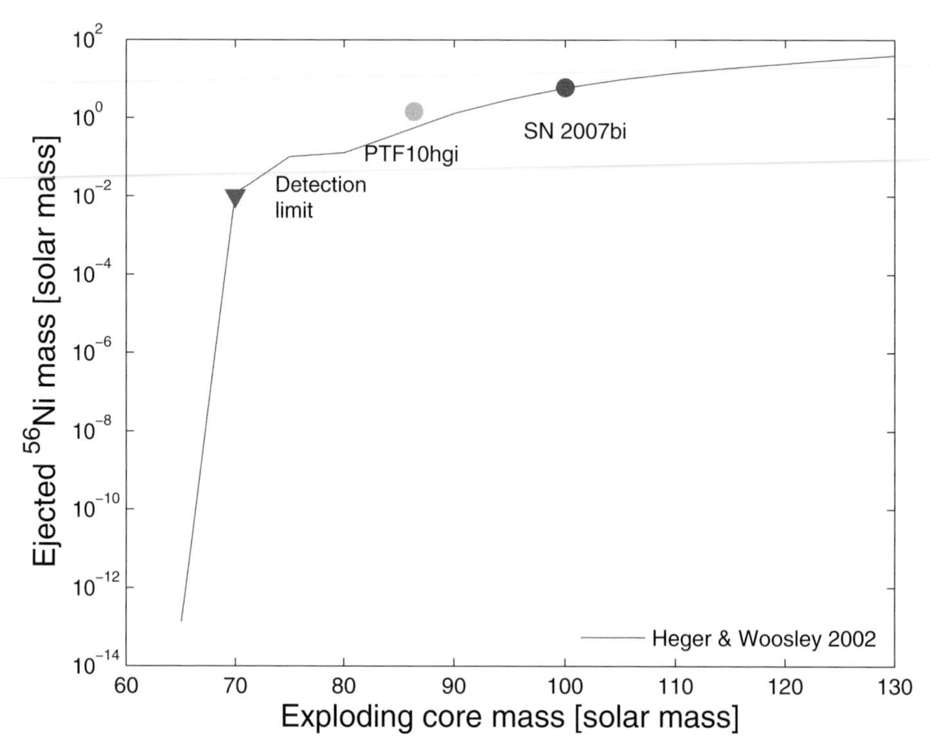

Figure 4. The relation between the synthesized ^{56}Ni mass and the He-core mass (which should roughly equal the total ejected mass for Type I events) based on the (non-rotating and non–magnetic) models of Heger & Woosley (2002). Superposed are the data points for SN 2007bi (Gal-Yam *et al.* 2009) and PTF10hgi (previously unpublished). We also mark the approximate upper limit on object detectability ($M_{^{56}\mathrm{Ni}} = 0.01\,M_{\odot}$) below which these events are expected to be too faint to be discovered anywhere except for the nearest galaxies, where the expected rate of these events is probably prohibitively low.

metallicity indicated for SN 2007bi (Young *et al.* 2010), alternative models (Langer *et al.* 2007) predict that stars with much lower initial masses ($150 - 250\,M_{\odot}$) explode as pair-instability SNe at SMC- or LMC-like metallicities, though they may have to be tweaked to explain the lack of hydrogen in observed SLSN-R spectra.

Assuming, for the sake of the current discussion, that observed SLSN-R explosions do arise from the pair instability, a clear prediction of the relevant theoretical models (e.g., Heger & Woosley 2002, Waldman 2008) is that for each luminous, ^{56}Ni-rich explosion (from a core around $100\,M_{\odot}$) there would be numerous less luminous events with smaller ^{56}Ni masses but large ejecta masses ($M > 50\,M_{\odot}$). These should manifest as events with very slow light curves (long rise and decay times) and yet moderate or even low peak luminosities (Fig. 4).

6. Host galaxies

Young *et al.* (2010) present a detailed study of the host galaxy of SN 2007bi. They find the host is a dwarf galaxy (with luminosity similar to that of the SMC), with relatively low metallicity ($Z \approx Z_{\odot}/3$) - somewhere between those of the LMC and SMC. So, while the progenitor star of this explosion probably had sub-solar metal content, there is no evidence that it had very low metallicity. The host galaxy of SN 1999as is more luminous (and thus likely more metal-rich) than that of SN 2007bi, but still fainter than typical giant galaxies like the Milky Way (Neill *et al.* 2011), while the host of PTF10nmn seems to be as faint or fainter than that of SN 2007bi (Yaron *et al.* 2012, in preparation). It thus seems this class of objects typically explode in dwarf galaxies. This is likely yet another aspect of the difference between the population of massive star explosions in observed in dwarf galaxies compared to giant hosts (Arcavi *et al.* 2010).

7. Rates

An estimate of the rate of SLSN-R can be derive from a rough estimate of the rate of SLSN-I provided by Quimby *et al.* (2011) based on statistics of events detected by the Texas Supernova Survey (TSS), which, normalizing the rate of SLSN-I at $z \approx 0.3$ relative to that of SNe Ia, yields a SLSN-I rate of $\sim 10^{-8}\,\mathrm{Mpc}^{-3}\,\mathrm{y}^{-1}$. Both the reported discovery statistics as well as unpublished PTF counts suggest that SLSN-R are rarer by about a factor of five, correcting for their slightly lower peak luminosities. This rate is substantially lower than the rates of core-collapse SNe ($\sim 10^{-4}\,\mathrm{Mpc}^{-3}\,\mathrm{y}^{-1}$), and is also well below those of rare sub-classes like broad-line SNe Ic ("hypernovae"; $\sim 10^{-5}\,\mathrm{Mpc}^{-3}\,\mathrm{y}^{-1}$) or long Gamma-Ray Bursts ($> 10^{-7}\,\mathrm{Mpc}^{-3}\,\mathrm{y}^{-1}$; Podsiadlowski *et al.* 2004; Guetta & Della Valle 2007). Interestingly, in is comparable to the rate recently predicted by Pan, Loeb & Kasen (2012). I believe this suggests that SLSN-R are indeed the rarest type of explosions studied so far, and quite possibly arise from stars that are at the very top of the IMF.

References

Arcavi, I., Gal-Yam, A., Kasliwal, M. M., *et al.* 2010, *ApJ*, 721, 777

Arnett, W. D. 1982, *ApJ*, 253, 785

Barkat, Z., Rakavy, G., & Sack, N. 1967, *Physical Review Letters*, 18, 379

Bond, J. R., Arnett, W. D., & Carr, B. J. 1984, *ApJ*, 280, 825

Cooke, J., *et al.* 2012, *Nature*, submitted

Filippenko, A. V., Li, W. D., Treffers, R. R., & Modjaz, M. 2001, *IAU Colloq. 183: Small Telescope Astronomy on Global Scales*, 246, 121

Gal-Yam, A., Maoz, D., Guhathakurta, P., & Filippenko, A. V. 2008, *ApJ*, 680, 550

Gal-Yam, A., Mazzali, P., Ofek, E. O., *et al.* 2009, *Nature*, 462, 624

Gal-Yam, A. 2012, *Science*, in press

Guetta, D. & Della Valle, M. 2007, *ApJ*, 657, L73

Hatano, K., Branch, D., Nomoto, K., *et al.* 2001, *Bulletin of the American Astronomical Society*, 33, 838

Heger, A. & Woosley, S. E. 2002, *ApJ*, 567, 532

Knop, R., Aldering, G., Deustua, S., *et al.* 1999, *IAUC*, 7128, 1

Kodros, J., Cenko, S. B., Li, W., *et al.* 2010, *Central Bureau Electronic Telegrams*, 2461, 1

Langer, N., Norman, C. A., de Koter, A., *et al.* 2007, *A&A*, 475, L19

Li, W., Leaman, J., Chornock, R., *et al.* 2011, *MNRAS*, 412, 1441

Moriya, T., Tominaga, N., Tanaka, M., Maeda, K., & Nomoto, K. 2010, *ApJ*, 717, L83

Neill, J. D., Sullivan, M., Gal-Yam, A., *et al.* 2011, *ApJ*, 727, 15

Pan, T., Loeb, A., & Kasen, D. 2012, *MNRAS*, 2979

Podsiadlowski, P., Mazzali, P. A., Nomoto, K., Lazzati, D., & Cappellaro, E. 2004, *ApJ*, 607, L17

Quimby, R. M., Kulkarni, S. R., Kasliwal, M. M., *et al.* 2011, *Nature*, 474, 487

Rakavy, G. & Shaviv, G. 1967, *ApJ*, 148, 803

Umeda, H. & Nomoto, K. 2008, *ApJ*, 673, 1014

Vinko, J., Wheeler, J. C., Chatzopoulos, E., Marion, G. H., & Caldwell, J. 2010, *Central Bureau Electronic Telegrams*, 2476, 1

Waldman, R. 2008, *ApJ*, 685, 1103

Yoshida, T. & Umeda, H. 2011, *MNRAS*, 412, L78

Young, D. R., Smartt, S. J., Valenti, S., *et al.* 2010, *A&A*, 512, A70

Death of Massive Stars: Supernovae and Gamma-Ray Bursts
Proceedings IAU Symposium No. 279, 2012
P. Roming, N. Kawai & E. Pian, eds.

doi:10.1017/S1743921312013026

Asymmetry in Supernovae

Keiichi Maeda

Kavli Institute for the Physics and Mathematics of the Universe, University of Tokyo,
5-1-5 Kashiwanoha, Kashiwa, Chiba 277-8583, Japan
email: keiichi.maeda@ipmu.jp

Abstract. Asymmetry in the innermost part of the supernova (SN) ejecta is a key to understanding their explosion mechanisms. Late-time spectroscopy is a powerful tool to investigate the issue. We show what kind of geometry is inferred for different types of SNe – core-collapse SNe Ib/c, those associated with Gamma-Ray Bursts (GRBs), and thermonuclear SNe Ia –, and discuss implications for the explosion mechanisms, observational diversities, and cosmological applications. For SNe Ib/c, the data show the clear deviation from spherical symmetry, and they are most consistent with the bipolar-type explosion as the characteristic geometry. Detailed modeling of optical emissions from SN 1998bw associated with GRB980425 indicates that this SN was in the extreme end of the bipolar explosion, suggesting that the explosion mechanisms of canonical SNe Ib/c and GRB-SNe are different. The situation is different for SNe Ia. Late-time spectra indicate the deviation from spherical symmetry, but for SNe Ia the explosion is asymmetric between two hemispheres, i.e., one-sided explosions. The diversities arising from different viewing directions can nicely explain (a part of) observational diversities of SNe Ia, and correcting this effect may improve the standard-candle calibration of SNe Ia for cosmology.

Keywords. supernovae: general – gamma rays: bursts

1. Introduction

Supernovae (SNe) are classified into several types by characteristics in their spectra around the maximum light (e.g., Filippenko 1997), and the spectral typing has been linked to the progenitor scenarios through the amount of the stellar envelope present at the time of explosion (e.g., Nomoto *et al.* 1995). Stars that retain their H envelope produce SNe with H-rich spectra, classified as SNe II. Stars that have lost at least a large fraction, if not all, of the H envelope produce He-rich SNe IIb, H-deficient SNe Ib, and both H and He-deficient SNe Ic, in order of increasing degree of the envelope stripping. They are believed to be an explosion of a massive star, i.e., core-collapse (CC) SNe. SNe Ia show neither H nor He, and they are further characterized by a strong Si absorption. The spectral characteristics of SNe Ia matches well the standard scenario that they are a result of a thermonuclear explosion of a white dwarf, reaching (nearly) the Chandrasekhar limiting mass (e.g., Nomoto *et al.* 1984; Woosley & Weaver 1986).

Some SNe Ic ('broad-line SNe Ic') show broad absorption features, indicating a higher expansion velocity and a larger energy than other SNe ($E_{\rm K} \sim 10^{51}$ erg; e.g., Iwamoto *et al.* 1998). The most energetic broad-line SNe Ic reach $E_{\rm K} \gtrsim 10^{52}$ erg, and some of them are associated with long-soft Gamma-Ray Bursts (hereafter GRBs for the long-soft class) as exemplified by GRB-SNe 1998bw and 2003dh (Galama *et al.* 1998; Hjorth *et al.* 2003). The association has been regarded as solid evidence for the popular scenario that GRBs are the outcome of the collapse of a stripped envelope C+O star (e.g., Woosley 1993). A majority of SNe Ic are not associated with GRBs: clarifying the origin of the difference will shed light on the still-unresolved explosion mechanism(s) of both of them.

SNe Ia are extremely useful for cosmology (Permutter *et al.* 1999; Riess *et al.* 1998).

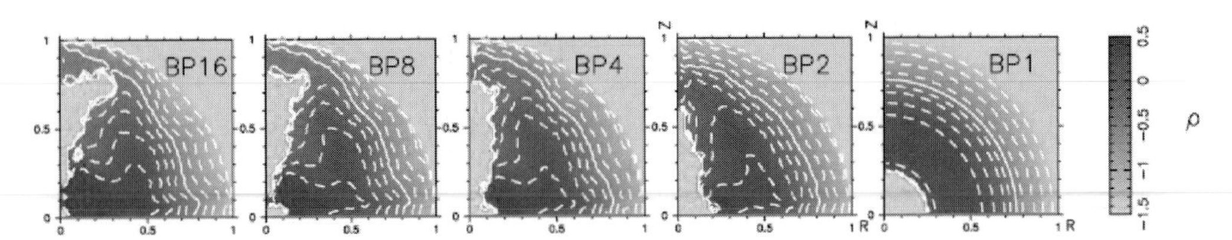

Figure 1. Examples of bipolar SN Ib/c explosion models. The degree of asphericity is expressed by a parameter BP in this model sequence, which is larger for more extremely bipolar explosions (Maeda *et al.* 2006a).

They are mature standardized candles, since their luminosity can be estimated through a phenomenological relation between the peak luminosity and the light curve decline rate (Phillips *et al.* 1999). Their maximum-light spectra however show diversities, highlighted by different absorption velocities and the speed of the velocity decrease (Benetti *et al.* 2005). This is likely related to details of the explosion process.

For both CC SNe (plus GRBs) and SNe Ia, the explosion mechanisms have not been clarified in detail. Recent simulations suggest that a key to understanding their explosion mechanisms is the geometry of the explosion. Different scenarios predict different types of geometry in the explosion, thus *observationally* deriving the geometry of the innermost part of SN ejecta is of highest importance. In this review, we summarize what has been learned through late-time spectroscopy of SNe about their explosion mechanisms.

2. Possible explosion geometry and expected features

The explosion mechanisms of both CC SNe and SNe Ia have not been clarified in detail. For CC SNe, recent theoretical studies have shown that the standard delayed neutrino heating mechanism requires significant deviation from spherical symmetry so that the process works effectively. The initial results in 2D simulations show the development of the bipolar-type explosion through the standing accretion shock instability (SASI; e.g., Blondin *et al.* 2003). It is not clear yet if this is an artifact of the imposed axisymmetry – so far the agreement seems controversial (e.g., Blondin & Mezzacappa 2007; Iwakami *et al.* 2009; Takiwaki *et al.* 2012). In addition to the SASI scenario, there are other possible mechanisms that can/should generate characteristic geometry in the explosion – for example, if the magnetic field plays an important role (Takiwaki *et al.* 2009), as has been suggested for GRB-SNe and some CC SNe, it likely generates the bipolar-type structure. In CC SNe, the materials in the direction of the stronger shock suffer from the more energetic explosive nucleosynthesis, while in the other directions the pre-explosion composition will be preserved . We thus expect that the distribution of Fe (created as ^{56}Ni at the strong shock) follows the geometry of the explosion. Distribution of O (i.e., pre-SN composition) will reflect the explosion geometry as well, where the 'hole' in the O distribution indicates the strong shock wave in that direction. As an example, the bipolar explosion predicts that Fe is distributed along the axis of symmetry, while O is distributed as a 'torus' (Fig. 1: from Maeda *et al.* 2006a).

The possible importance of bulk asymmetry in SN Ia explosions has been recognized recently (e.g., Kasen *et al.* 2009; Maeda *et al.* 2010a). Within the standard delayed detonation scenario (Khokhlov 1991), where the explosion is first triggered by ignition of deflagration sparks near the center then energized further by the transition to the detonation wave, an issue is how the initial sparks are ignited. Perturbations within the progenitor, e.g., by convection, could naturally result in the off-set ignition (Kuhlen *et al.* 2006). If this is true, it is predicted that stable Fe-peak elements are produced by the

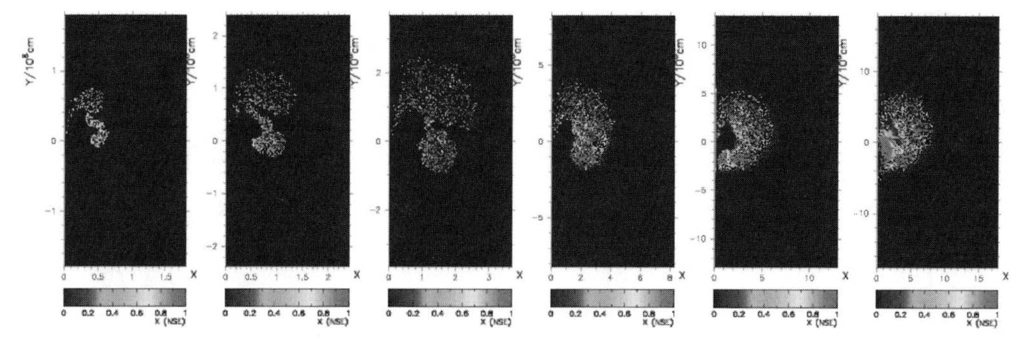

Figure 2. Temporal evolution of distribution of Fe-peaks (at 0.7, 0.9, 1.1, 1.3, 1.5, 1.7 seconds after the ignition, from left to right), in an offset delayed detonation model (Maeda *et al.* 2010b; see also Kasen *et al.* 2009). Shown here is the mass fraction of the Fe-peaks, from 0 (blue/dark) to 1 (red/bright). The deflagration phase (producing stable Fe-peaks) is highly asymmetric, while the detonation wave (producing ^{56}Ni) expands nearly isotropically.

deflagration (including stable Ni) at an off-set from the explosion center, while radioactive ^{56}Ni (which decays into Co then into Fe) is produced by the detonation more or less in a spherically symmetric manner (e.g., Maeda *et al.* 2010b; Fig. 2). Other explosion models also predict large deviation from spherical symmetry; for example, the prompt detonation following the merger of two white dwarfs will be highly asymmetric (Röpke *et al.* 2012; Sim *et al.* 2012).

3. Late-time spectroscopy to study explosion geometry

As the SN ejecta expand, the ejecta become transparent at about 100 or 200 days after the explosion. Afterwards, the effect of the radiation transfer is minimal to alter the line profile. At this late epoch, the SN spectrum is nebular, showing emission lines mostly of forbidden transitions. Thanks to the homologous nature of the SN expansion, the Doppler shift indicates where the photon was emitted. A photon emitted from the near/far side of the ejecta is blueshifted/redshifted. Thus, the line profile can be used as a direct tracer of the distribution of the emitting materials within the SN ejecta.

The general behavior of the line profile for different distributions of material is model independent. If the distribution is spherical, then the center of the line should be at the rest wavelength of the transition, and the line profile should be symmetric between the blue and red. If the ejecta is axisymmetric and also symmetric with respect to the equator (including the bipolar distribution), the line profile should be symmetric with respect to the rest wavelength, but the peak position is generally different from the rest wavelength (e.g., double peaks). If the distribution is still axisymmetric but there is an imbalance between two hemispheres (i.e., a one-sided distribution), then the line neither has to peak at the rest wavelength, nor has to be symmetric.

4. Asymmetry in SNe Ib/c and connection to GRBs

Detailed study of the geometric nature of the explosion of SNe Ib/c has been initially developed for GRB-SN Ic 1998bw. SN Ic 1998bw was discovered in association with GRB 980425 (Galama *et al.* 1998). Modeling the early phase observations (through two or three months after the explosion) showed that it is a highly energetic explosion (Iwamoto *et al.* 1998). Its late-time spectra showed the velocity of Fe higher than O (Mazzali *et al.* 2001), and a highly bipolar explosion was suggested (Maeda *et al.* 2002). Meanwhile, its explosion geometry was intensively studied based on multi-D radiation

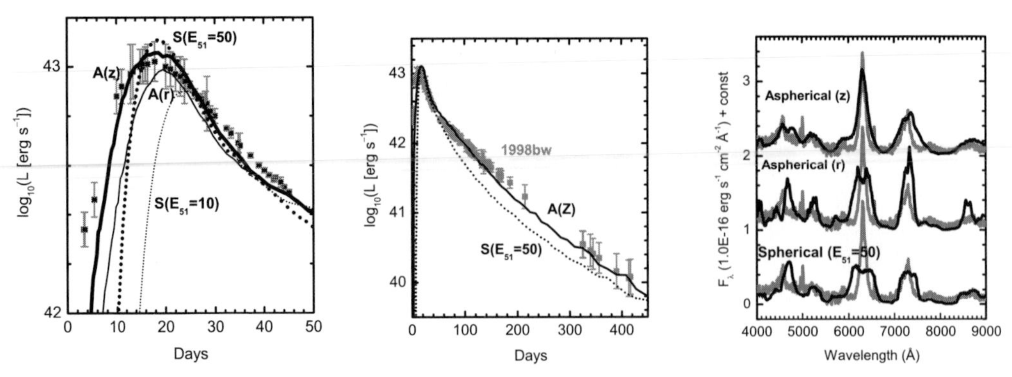

Figure 3. Radiation transfer models for GRB-SN 1998bw based on the bipolar model (Fig. 1). The left and middle panels show the bolometric light curve of SN 1998bw, as compared with the bipolar model BP8 (denoted by 'A') with $E_{\rm K} = 2 \times 10^{52}$ erg as seen on-axis (z; thick solid) and off-axis (r; thin solid), and the spherical model BP1 (denoted by 'S'; dotted lines). The right panel shows the late-time spectrum of SN 1998bw as compared with the model BP8 from two different directions and with the spherical model BP1.

transfer calculations (Fig. 3; Maeda *et al.* 2006a; Maeda *et al.* 2006b; Tanaka *et al.* 2007) applied to the observations of SN 1998bw. They concluded that no spherical models can consistently reproduce the observations from early through late phases, although a bipolar explosion model can reproduce these observations consistently. Within the framework of the bipolar explosion models, they showed that the parameters for the explosion can be constrained into a small parameter space: the main-sequence mass $M_{\rm ms} \sim 40 M_{\odot}$, the explosion energy $E_{\rm K} \sim 10 - 20 \times 10^{51}$ erg (smaller than the spherically symmetric estimate by a factor of a few), the degree of bipolarity BP $\sim 8 - 16$ (Fig. 1), and the viewing direction from the pole $\theta \sim 0 - 30°$.

In SNe Ib/c, O is the dominant element, and [OI] $\lambda\lambda$6300, 6363 is the strongest in their late-time spectra. Luckily, this wavelength range is free from contamination from other lines. Maeda *et al.* (2002) predicted that a bipolar explosion should produce double peaks in the [OI] if viewed from the equatorial direction as O is distributed in a torus – thus, if other SNe share a similar bipolar geometry with SN 1998bw, then there must be SNe showing the double peaks in the [OI] profile. Indeed, such a double peaked profile was discovered after the prediction, in the late-time spectrum of SN 2003jd – a broad-lined SN Ic similar to SN 1998bw, but without an associated GRB (Mazzali *et al.* 2005). This led us to suggest that SN 2003jd was similar to SN 1998bw but viewed off-axis.

We coordinated late-time spectroscopy of SNe Ib/c with *Subaru* and *VLT* telescopes, and investigated the profile of the [OI]. Given that the line profile is only a function of the geometry and the viewing direction, the number of SNe Ib/c that show the doubly-peaked [OI] line to the total number of SNe Ib/c sample should provide a measure for the typical degree of asphericity. Figure 4 shows the [OI] line profiles of 18 SNe IIb/Ib/c (Maeda *et al.* 2008). We found that the doubly-peaked profile is not rare, suggesting that the asphericity is a generic feature in SNe IIb/Ib/Ic (see also Modjaz *et al.* 2008; Taubenberger *et al.* 2009). As shown in Figure 4, the characteristic profile is either a single peak (S) or double peaks (D), and in both cases the line profile is rather symmetric with respect to the rest wavelength if the observation is performed at $\gtrsim 200$ days (see, however, Taubenberger *et al.* 2009). These features require the deviation from spherical symmetry, but the axisymmetry and the symmetry with respect to the equator should be roughly preserved – the straightforward interpretation is a bipolar-type geometry. Quantitatively, the fraction of SNe showing the doubly-peaked [OI] is $39 \pm 11\%$. This indicates that the

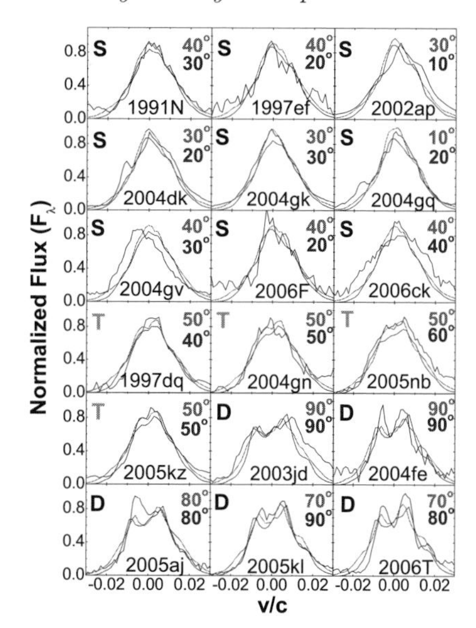

Figure 4. The line profile of [OI] in late-time spectra of 18 SNe IIb/Ib/Ic (black). Also shown are the synthetic line profiles for models BP8 and BP2 (see Fig. 1).

typical degree of the asphericity is moderate, i.e., BP $= 2$ (Fig. 1). This is smaller than the degree of asphericity derived for SN 1998bw (BP $= 8 - 16$).

5. Asymmetry in SNe Ia and cosmological application

The offset ignition model (Fig. 2) predicts that an emission from the deflagration ash (showing an offset) should show a variation in the wavelength for different viewing directions, while that from the detonation ash (spherically distributed) should show little variation (Maeda *et al.* 2010b). We have investigated emission processes within the SN Ia ejecta in late-phases (Maeda *et al.* 2010c; see also Mazzali *et al.* 2007), and pointed out that the [Fe III] blend at 4,700Å is emitted from the detonation ash while [Fe II]λ7155, [Ni II]λ7378, and some NIR [Fe II] lines are emitted from the deflagration ash.

This analysis allowed us to investigate the explosion geometry though the existing late-time spectra. Figure 5 shows the [Fe III] blend at 4,700Å and [Ni II]λ7378 for 12 SNe Ia. The [Fe III] blend does not show significant shift for all SNe Ia. On the other hand, we discovered that [Ni II]λ7378 does show variations in its central wavelength for different SNe Ia; some SNe show redshift and others blueshift. Also, the wavelength of [Ni II]λ7378 does not show significant temporal evolution when late-time spectra at multiple epochs are available. These observational behaviors reject the possibility that the shift is produced by non-complete transparency or by contaminations from other lines.

Although the above observations indicate that an asymmetric, offset ignition is a generic feature of SNe Ia, the argument is based on a few lines only. This can be further tested for individual SNe for which late-time spectra are obtained with wide wavelength coverage, using various lines. This test has been performed for SN 2003hv. We have found that the lines in its late-time optical through mid-IR spectrum can indeed be divided into two groups, one showing a blueshift (with similar degree) and the other showing virtually no shift, and that grouping into these two categories is fully consistent with the above expectation. A model introducing the offset velocity of $\sim 3,500$ km s^{-1} reproduces all the emission lines, without other parameters invoked to explain these different behaviors for different lines (Maeda *et al.* 2010c).

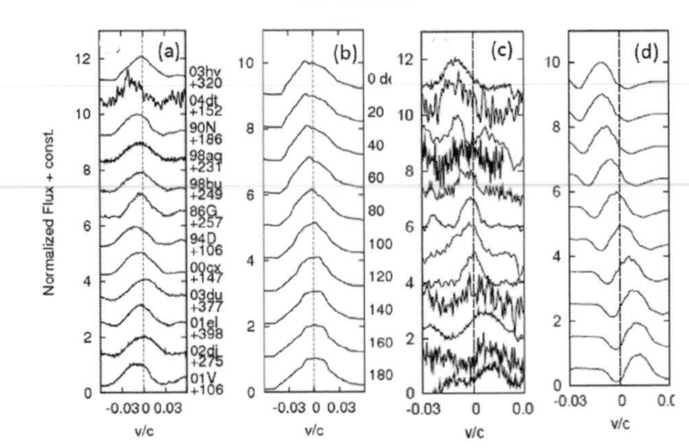

Figure 5. Late-time spectra of SNe Ia (Maeda *et al.* 2010c). (a) Observed and (b) model line profiles of [Fe III] blend at 4700Å and (c) observed and (d) model line profiles of [Ni II] λ7378. The model profiles are shown for various viewing directions.

The finding has implications for explaining observational diversity and the cosmological applications. Spectral properties of SNe Ia are not uniform despite the standardized candle nature in the light curve (Benetti *et al.* 2005). In the early photospheric phase, materials moving toward the observer produce blueshifted absorption lines. The absorption velocity decreases, following the recession of the photosphere. The velocity gradient ($\dot{v}_{\rm Si}$) is defined as a rate of the decrease in the Si II λ6355 absorption velocity. SNe that show $\dot{v}_{\rm Si} > 70$ km s^{-1} day^{-1} are classified as the high velocity gradient (HVG) group, while those showing smaller $\dot{v}_{\rm Si}$ are called the low velocity gradient (LVG) SNe Ia. It has been shown that $\dot{v}_{\rm Si}$ is *not* correlated with the decline rate. In addition, SNe Ia show a variation in their peak colors (e.g., Folatelli *et al.* 2010). The origin of these diversities beyond the 'one-parameter' description (by the decline rate) has not yet been clarified.

If the explosion is asymmetric, it raises an interesting question – whether the observed diversities could be related to the diversities arising from the explosion asymmetry and different viewing directions. Figure 6 provides a comparison between the velocity gradient ($\dot{v}_{\rm Si}$) and the late-time emission-line velocity shift ($v_{\rm neb}$; measured by the Doppler shift of [Fe II]λ7155 and [Ni II]λ7378) as investigated by Maeda *et al.* 2010a). These two quantities turned out to be related, with HVG SNe showing redshift in late-time spectra. The correlation indicates that there is no intrinsic difference in HVG and LVG SNe, but the different appearance is merely a consequence of different viewing directions.

SNe Ia show a variation in their peak $B - V$ color. This is another major issue in SN cosmology since the estimate of extinction (thus distance) relies on the intrinsic color of SNe Ia. By comparing the 'viewing direction' and the color of SNe Ia, we have shown that that the intrinsic color variation is at the level of $B - V \sim 0.2$ mag, and the variation within this level can be attributed to the difference arising from different viewing directions (Maeda *et al.* 2011).

6. Conclusions

In this review, we discussed how the geometry in the SN explosions has been obtained through the late-time spectroscopy. Our findings are summarized as follows.

• **SNe Ib/c:** The late-time [OI] profiles show the deviation from spherically symmetric explosions. A straightforward interpretation is that the characteristic geometry is (moderately) bipolar-like.

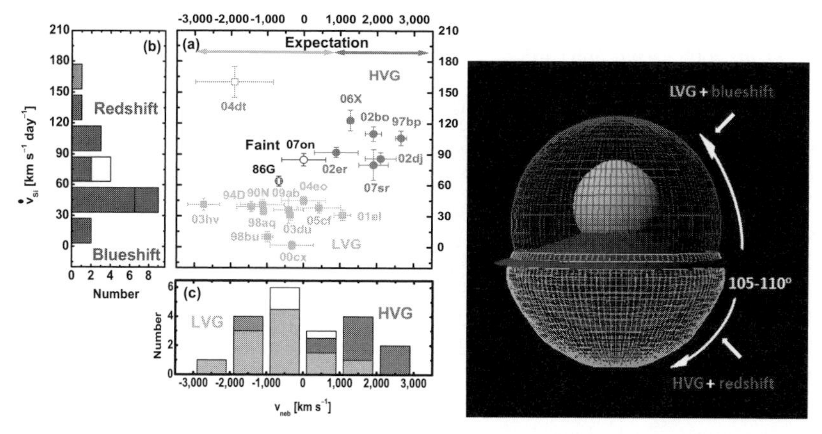

Figure 6. The left panel shows a comparison between the velocity gradient ($\dot{v}_{\rm Si}$) and the late-time emission line velocity shift ($v_{\rm neb}$) (Maeda *et al.* 2010a). The right panel shows a schematic picture of SN Ia ejecta derived from the late-time spectra and its relation to the velocity gradients. Depending on the viewing direction an SN appears either as HVG or LVG in this scenario.

- **GRB-SN 1998bw:** Detailed radiation transfer modeling coupled with the late-time spectrum analysis indicates a bipolar explosion with a more extreme degree of asymmetry than in other SNe Ib/c. This supports the idea that the explosion mechanism of GRB-associated broad-line SNe Ic is different from canonical SNe Ib/c. Also, the energy derived for the bipolar explosion model is a factor of a few smaller than the value obtained assuming a spherically symmetric explosion, and this effect must be taken into account in considering the explosion mechanism and energetics.
- **SNe Ia:** Late-time spectrum analysis suggests that the explosion has a bulk offset in its mechanism. Distribution of different isotopes are consistent with the expectation from the delayed detonation model, but further study is necessary to investigate if different explosion mechanisms can explain the behavior as well. In any case, correlations between the late-time emission line velocity shift and other observables in the early phases (i.e., velocity gradient, $B-V$ color) suggest that the combination of the asymmetric explosion and various viewing directions is a source of at least a part of observational diversities of SNe Ia. Understanding this effect is important for future use of SNe in cosmology.

Acknowledgements

This research is supported by World Premier International Research Center Initiative (WPI Initiative), MEXT, Japan and by Grant-in-Aid for Scientific Research (23740141).

References

Benetti, S., *et al.* 2005, *ApJ*, 623, 1011
Blondin, J. M., Mezzacappa, A., & DeMarino, C. *ApJ*, 584, 971
Blondin, J. M. & Mezzacappa, A. 2007, *Nature*, 445, 58
Filippenko, A. V. 1997, *ARAA*, 35, 309
Folatelli, G., *et al.* 2010, *AJ*, 139, 120
Galama, T. J., *et al.* 1998, *Nature*, 395, 670
Hjorth, J., *et al.* 2003, *Nature*, 423, 847
Iwakami, W., *et al.* 2009, *ApJ*, 700, 232
Iwamoto, K., *et al.* 1998, *Nature*, 395, 672
Kasen, D., Röpke, F., & Woosley, S. E. 2009, *Nature*, 460, 869
Khokhlov, A. M. 1991, *A&A*, 245, 114

Kuhlen, M., Woosley, S. E., & Glatznaier, G. A. 2006, *ApJ*, 640, 407
Maeda, K., *et al.* 2002, *ApJ*, 565, 405
Maeda, K., *et al.* 2006a, *ApJ*, 640, 854
Maeda, K., Mazzali, P. A., & Nomoto, K. 2006b, *ApJ*, 645, 1331
Maeda, K., *et al.* 2008, *Science*, 319, 1220
Maeda, K., *et al.* 2010a, *Nature*, 466, 82
Maeda, K., *et al.* 2010b, *ApJ*, 712, 624
Maeda, K., *et al.* 2010c, *ApJ*, 708, 1703
Maeda, K., *et al.* 2011, *MNRAS*, 413, 3075
Mazzali, P. A., *et al.* 2001, *ApJ*, 559, 1047
Mazzali, P. A., *et al.* 2005, *Science*, 308, 1284
Mazzali, P. A., *et al.* 2007, *Science*, 315, 825
Modjaz, M., *et al.* 2008, *ApJ*, 687, L9
Nomoto, K., Thielemann, F.-K., & Yokoi, K. 1984, *ApJ*, 286, 644
Nomoto, K., Iwamoto, K., & Suzuki, T. 1995, *Phys. Rep.*, 256, 173
Permutter, S., *et al.* 1999, *ApJ*, 517, 565
Phillips, M. M., *et al.* 1999, *AJ*, 118, 1766
Riess, A., *et al.* 1998, *AJ*, 116, 1009
Röpke, F. K., *et al.* 2012, *ApJ*, 750, 19
Sim, S., *et al.* 2012, *MNRAS*, 420, 3003
Takiwaki, T., Kotake, K., & Sato, K. 2009, *ApJ*, 691, 1360
Takiwaki, T., Kotake, K., & Suwa, Y. 2012, *ApJ*, 749, 98
Tanaka, M., *et al.* 2007, *ApJ*, 668, L19
Taubenberger, S., *et al.* 2009, *MNRAS*, 397, 677
Woosley, S. E. & Weaver, T. A. 1986, *ARAA*, 24, 205
Woosley, S. E. 1993, *ApJ*, 405, 273

Discussion

MOISEENKO: Can strictly spherically symmetrical supernovae exist in nature?

MAEDA: There is probably no 'strictly' spherical explosions in nature. SNe of different types show different degrees of deviation from spherical symmetry, for example this is more extreme in SNe Ib/c than SNe Ia as we see in the data, and there may be a class of SNe that are less asymmetric than those we have analyzed. Among the proposed explosion mechanisms, pair-instability SNe do not require the asymmetry in its mechanism, and they seem to be a candidate of a more or less spherically symmetric explosion.

MODJAZ: For SNe Ib/c, there is a worry for [OI]$\lambda\lambda$6300, 6364 that its doublet nature potentially produces the double peaks. Are you planning to include the radiative transfer models for other lines, e.g., OIλ7774, Ca II?

MAEDA: In my personal view point, the argument for explaining the [OI] double peaks by the doublet nature has not been well justified. The center of the observed double peaks is close to 6300Å in the late phases or even bluer in the earlier phases, while the doublet center should be redder than 6300Å. Thus the interpretation requires one to introduce artificially very high optical depths. However, such an effect is not seen in SNe II where the doublet is clearly detected separately, despite the expectation that SNe II are more dense than SNe Ib/c and such an effect, if it exists, should be clearly seen. In any case, it is very important and interesting to look into other lines as an independent check for the geometrical interpretation. Yes, we are planning to provide more detailed radiation transfer models to make the full analysis of the spectra, not only for the geometry study but also for studying other properties of SNe as well.

Death of Massive Stars: Supernovae and Gamma-Ray Bursts
Proceedings IAU Symposium No. 279, 2012
P. Roming, N. Kawai & E. Pian, eds.

doi:10.1017/S1743921312013038

Detecting the First Supernovae in the Universe with JWST

Daniel J. Whalen[1,2]

[1]McWilliams Fellow, Department of Physics, Carnegie Mellon University, Pittsburgh, PA 15213 email: dwhalen@lanl.gov

[2]Theoretical Division (T-2), Los Alamos National Laboratory, Los Alamos, NM 87545

Abstract. Massive Population III stars die as pair-instability supernovae (PI SNe), the most energetic thermonuclear explosions in the universe with energies up to 100 times those of Type Ia or Type II SNe. Their extreme luminosities may allow them to be observed from the earliest epochs, revealing the nature of Pop III stars and the primitive galaxies in which they reside. We present numerical simulations of Pop III PI SNe done with the radiation hydrodynamics code RAGE and calculations of their light curves and spectra performed with the SPECTRUM code. We find that 150 - 250 $M_\odot$ PI SNe will be visible to the James Webb Space Telescope (JWST) out to $z \sim 30$ and to $z \sim 15 - 20$ in all-sky NIR surveys by the Wide Field Infrared Survey Telescope (WFIRST).

Keywords. early universe – stars:early type – galaxies:high-redshift – supernovae:general – methods:numerical

---

1. Introduction

The first stars in the universe are thought to form at $z \sim$ 20 - 30 in 10^5 - 10^6 $M_\odot$ cosmological halos (Bromm *et al.* 2001, Nakamura & Umemura 2001, Abel *et al.* 2002). Numerical models suggest that they are 20 - 500 $M_\odot$ (O'Shea & Norman 2007) and form one per halo, in binaries (Turk *et al.* 2009), or in small multiples of up to a dozen (Clark *et al.* 2011, Stacy *et al.* 2010, Greif *et al.* 2011). Their extreme ionizing luminosities (Schaerer 2002) create large H II regions and expel most of the gas from the halo (Whalen *et al.* 2004, Kitayama *et al.* 2004, Alvarez *et al.* 2006, Abel *et al.* 2007). Pop III stars are central to understanding the nature of primeval galaxies, early cosmological reionization and chemical enrichment, and the origin of supermassive black holes. Unfortunately, although they are very bright individual Pop III stars lie beyond the realm of direct detection by current and upcoming surveys.

The final fates of Pop III stars primarily depend on their main sequence masses at birth. If their masses are 140 - 260 $M_\odot$, they die in spectacular thermonuclear explosions known as pair-instability supernovae, with energies up to 100 times greater than those of Type Ia or Type II SNe (Heger & Woosley 2002). Such explosions completely unbind the star, leaving no compact remnant. Because they can be hundreds of thousands of times brighter than their progenitors, Pop III PI SNe are prime candidates for detection by upcoming missions such as JWST and WFIRST that would directly probe the Pop III IMF for the first time. However, unlike Type Ia SNe now being used to constrain cosmic acceleration, photons from Pop III SNe must traverse the Lyman alpha forest, the vast tracts of intervening neutral hydrogen clouds and filaments that absorb or scatter most of them out of our line of sight. Numerical predictions of their detection therefore require absorption by the Lyman alpha forest in addition to cosmological reddening and accurate

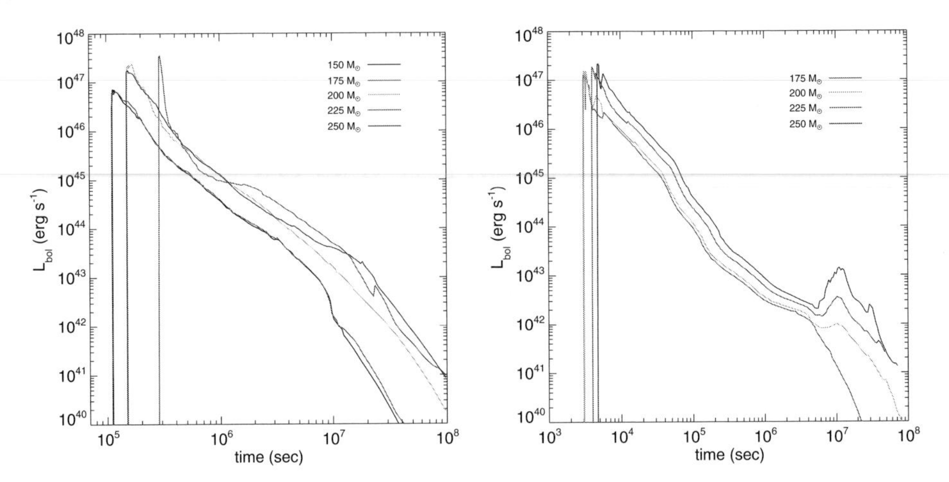

Figure 1. Rest frame bolometric luminosities for all 9 PI SNe out to 3 yr. Left panel: u-series. Right panel: z-series.

light curves (see Scannapieco *et al.* 2005, Joggerst & Whalen 2011, Kasen *et al.* 2011, and Pan *et al.* 2011 for earlier studies of PI SNe at low redshift).

We present numerical simulations of Pop III PI SNe and their light curves and spectra with the Los Alamos National Laboratory (LANL) RAGE and SPECTRUM codes. We convolve our spectra with cosmological reddening, absorption by the Lyman alpha forest according to the prescription of Madau 1995 and Su *et al.* 2011, and instrument filters to find detection thresholds in redshift for JWST and WFIRST.

2. RAGE/SPECTRUM

RAGE (Radiation Adaptive Grid Eulerian) is a multidimensional adaptive mesh refinement (AMR) radiation hydrodynamics code that couples second-order conservative Godunov hydrodynamics to grey or multigroup flux-limited diffusion to simulate strongly radiating flows (Gittings *et al.* 2008). RAGE utilizes the extensive LANL OPLIB database of atomic opacities† (Magee *et al.* 1995). We describe the physics implemented in our RAGE runs and why it is needed to capture the features of our light curves in Fryer *et al.* 2010 and Frey *et al.* 2012: multispecies advection, grey FLD radiation transport with 2T physics, energy deposition from the radioactive decay of ^{56}Ni, and no self-gravity. We include mass fractions for 15 elements, the even numbered elements predominantly synthesized in PI SNe. We post-process RAGE profiles with the SPECTRUM code to obtain spectra with 14900 energies. SPECTRUM accounts for Doppler shifts and time dilation due to the relativistic expansion of the ejecta and calculates intensities of emission lines and the attenuation of flux along the line of sight with monochromatic OPLIB opacities. Our spectra thus capture both limb darkening and the absorption lines imprinted on the flux by intervening material in the ejecta and wind.

3. Pop III PI SN Light Curves

We show bolometric luminosities for 150, 175, 200, 225 and 250 M$_\odot$ Pop III PI SNe in Figure 1. U-series progenitors are red hypergiants and z-series progenitors are compact blue giants. Total luminosities at shock breakout typically exceed 10^{47} erg/s and are primarily x-rays in the z-series and both x-rays and hard UV in the u-series. Radiation

† http://aphysics2/www.t4.lanl.gov/cgi-bin/opacity/tops.pl

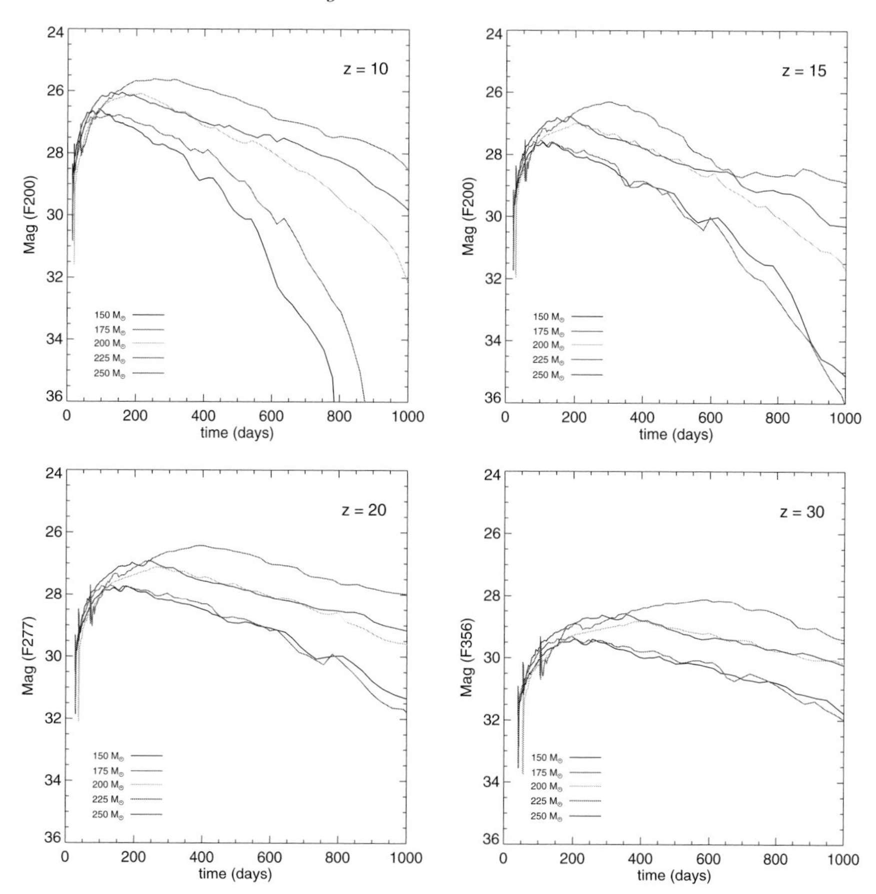

Figure 2. Pop III u-series PI SN light curves for the *JWST* NIRCam. The optimum filter for each redshift is noted on the y-axis labels and the times on the x-axes are for the observer frame.

temperatures at breakout are lower in the u-series because the shock must do more PdV work on its surroundings prior to exiting the star than in the z-series. U-series PI SNe exhibit a more extended plateau in luminosity than z-series explosions that is reminiscent of Type II-p SNe, whose progenitors are also though to be red giants. In general, z-series explosions are somewhat dimmer than u-series explosions of the same mass because they manufacture significantly less ^{56}Ni, whose decay powers the light curves at intermediate to late times. The resurgence in luminosity at $\sim 10^7$ s occurs when the photosphere descends into the hot ^{56}Ni layer in the frame of the ejecta. This feature is present in both series but is less evident in the u-series because luminosities before and after the bump are greater.

Our Pop III PI SN bolometric luminosities are ~100 times those of Type Ia and II SNe because of their much greater explosion energies and because they synthesize up to 50 $M_\odot$ of ^{56}Ni, in contrast to Type Ia SNe that create $\sim 1.5\ M_\odot$. Also, PI SNe remain bright for up to 3 yr, rather than 3 - 5 months. This is primarily due to the longer radiation diffusion timescales in the much more massive ejecta:

$$t_d \sim \kappa^{\frac{1}{2}} {M_{ej}}^{\frac{3}{4}} E^{-\frac{3}{4}}, \tag{3.1}$$

where κ is the average opacity of the ejecta, M_{ej} is the mass of the ejecta, and E is the explosion energy.

4. JWST/WFIRST Detection Thresholds

We show JWST NIRCam light curves at 2.0, 2.0, 2.77 and 3.56 μm at z = 10, 15, 20 and 30, respectively, for our five u-series SNe in Figure 2. Given that $z \sim 20$ light curves would be stretched to nearly 60 yr in duration in the earth frame, one might think that PI SNe at that epoch would not appear as transients today. However, the expansion and cooling of the fireball in the rest frame creates an NIR flux in the earth frame that varies over 1000 days, the typical duration of a high-redshift protogalactic survey. NIRCam photometry detection thresholds for deep surveys are mag 31 - 32, so it is clear that JWST will detect PI SNe out to $z \sim 30$ and even perform spectroscopy on them.

While JWST might be sensitive enough to detect $z \sim 30$ SNe, are its search fields too narrow to encounter one in its operational lifetime? New calculations indicate that a few Pop III PI SNe will be present NIRCam surveys at any given time (Hummel *et al.* 2011). Furthermore, as can be seen in the z = 15 and 20 panels above, WFIRST, with its proposed sensitivity of mag 26.5 at 2.2 μm, will detect Pop III PI SNe at z = 15 - 20 in much greater numbers in all-sky surveys. The detection of the first supernovae in the universe will be one of the most spectacular results in extragalactic astronomy in the coming decade, unveiling for the first time the stars that ended the cosmic Dark Ages.

References

Abel, T., Bryan, G. L., & Norman, M. L. 2002, *Science*, 295, 93
Abel, T., Wise, J. H., & Bryan, G. L. 2007, *ApJL*, 659, L87
Alvarez, M. A., Bromm, V., & Shapiro, P. R. 2006,
Bromm, V., Ferrara, A., Coppi, P. S., & Larson, R. B. 2001, *MNRAS*, 328, 969
Clark, P. C., Glover, S. C. O., Smith, R. J., Greif, T. H., Klessen, R. S., & Bromm, V. 2011, *Science*, 331, 1040
Frey, L. H., Even, W., Whalen, D. J., Fryer, C. L., Hungerford, A. L., Fontes, C. J., & Colgan, J. 2012, ArXiv e-prints
Fryer, C. L., Whalen, D. J., & Frey, L. 2010, in *American Institute of Physics Conference Series*, Vol. 1294, American Institute of Physics Conference Series, ed. D. J. Whalen, V. Bromm, & N. Yoshida, 70–75
Gittings, M. *et al.* 2008, *Computational Science and Discovery*, 1, 015005
Greif, T. H., Springel, V., White, S. D. M., Glover, S. C. O., Clark, P. C., Smith, R. J., Klessen, R. S., & Bromm, V. 2011, *ApJ*, 737, 75
Heger, A. & Woosley, S. E. 2002, *ApJ*, 567, 532
Hummel, J., Pawlik, A., Milosavljevic, M., & Bromm, V. 2011, ArXiv e-prints
Joggerst, C. C. & Whalen, D. J. 2011, *ApJ*, 728, 129
Kasen, D., Woosley, S. E., & Heger, A. 2011, *ApJ*, 734, 102
Kitayama, T., Yoshida, N., Susa, H., & Umemura, M. 2004, *ApJ*, 613, 631
Madau, P. 1995, *ApJ*, 441, 18
Magee, N. H., Abdallah *et al.* 1995, in *Astronomical Society of the Pacific Conference Series*, Vol. 78, Astrophysical Applications of Powerful New Databases, ed. S. J. Adelman & W. L. Wiese, 51
Nakamura, F. & Umemura, M. 2001, *ApJ*, 548, 19
O'Shea, B. W. & Norman, M. L. 2007, *ApJ*, 654, 66
Pan, T., Kasen, D., & Loeb, A. 2011, ArXiv e-prints
Scannapieco, E., Madau, P., Woosley, S., Heger, A., & Ferrara, A. 2005, *ApJ*, 633, 1031
Schaerer, D. 2002, *A&A*, 382, 28
Stacy, A., Greif, T. H., & Bromm, V. 2010, *MNRAS*, 403, 45
Su, J. *et al.* 2011, *ApJ*, 738, 123
Turk, M. J., Abel, T., & O'Shea, B. 2009, *Science*, 325, 601
Whalen, D., Abel, T., & Norman, M. L. 2004, *ApJ*, 610, 14
Whalen, D., van Veelen, B., O'Shea, B. W., & Norman, M. L. 2008, *ApJ*, 682, 49

Discussion

NOMOTO: If the core-collapse supernova is a hypernova with an energy of 3×10^{52} erg, up to what redshift can such an explosion be observed with JWST?

WHALEN: Preliminary simulations suggest that 15 - 40 $M_\odot$ Pop III core-collapse supernovae will be visible in the NIR out to $z \sim 7$ with JWST. My guess is that hypernovae could be seen out to $z \sim 10$ - 15. They would be an exciting discovery because this is the era of protogalaxy formation, and such SNe would reveal the positions of primeval galaxies on the sky.

IOKA: How do you find Pop III SNe? The FOV of JWST is not large.

WHALEN: Recent calculations by Hummel *et al.* (2011) suggest that Pop III PI SN rates at $z \sim 20$ will ensure that a few are present in any given JWST search field over typical survey times. We also find that the Wide-Field Infrared Survey Telescope (WFIRST), with its proposed sensitivity of 26.5 mag at 2.2 μm, will be able to see PI SNe out to $z \sim 15$ - 20 and would detect much larger numbers of such events. In fact, this redshift range may be optimal for detecting PI SNe because of the rise of Lyman-Werner UV H_2 photodissociating backgrounds from the first generations of Pop III stars. The LW background is thought to suppress Pop III star formation in minhalos until they have grown to larger masses that likely result in more massive stars at slightly lower redshifts (see O'Shea & Norman 2008).

COUCH: Do the shock breakout light curves account for light travel time effects across the surface of the star?

WHALEN: They do to an extent because the radiation flow is calculated in 1D spherical geometry with flux-limited diffusion. A higher-order transport method such as implicit Monte Carlo would do a better job of this because it better captures the angular dependence of the radiation field. However, radiation – matter coupling effects prevent all the photons from exiting the shock at once when it breaks through the surface of the star and broaden the light curve by several light-crossing times, effectively erasing light-crossing time effects.

Death of Massive Stars: Supernovae and Gamma-Ray Bursts
Proceedings IAU Symposium No. 279, 2012
P. Roming, N. Kawai & E. Pian, eds.

doi:10.1017/S174392131201304X

X-rays, γ-rays and neutrinos from collisionless shocks in supernova wind breakouts

Boaz Katz[1], Nir Sapir[2] and Eli Waxman[2]

[1]Institute for Advanced Study, Princeton, NJ 08540, USA
email: boazka@ias.edu

[2]Dept. of Particle Phys. & Astrophys., Weizmann Institute of Science, Rehovot 76100, Israel
email: nir.sapir@weizmann.ac.il
email: eli.waxman@weizmann.ac.il

Abstract. Some of the observed bursts of X-rays/Gamma-rays associated with supernovae (SNe) as well as very luminous SNe may result from the breakout of the SN shock from an optically thick wind surrounding the progenitor. We show that in such scenarios a collisionless shock necessarily forms during the shock breakout. An intense non-thermal flash of $\lesssim 1$ MeV gamma rays, hard X-rays and multi-TeV neutrinos is produced simultaneously with and following the typical soft X-ray breakout emission, carrying similar or larger energy than the soft emission. The non-thermal flash is detectable by current X-ray telescopes and may be detectable out to 10's of Mpc by km-scale neutrino telescopes.

Keywords. (stars:) supernovae: general, shock waves, radiation mechanisms: nonthermal

---

1. Introduction

One of the most exciting, relatively recent, developments in the study of SNe, is the association with type Ib/c SNe of bursts of hard X-rays/γ-rays, lasting for tens to thousands of seconds (e.g. Galama *et al.* 1998, Hjorth *et al.* 2003, Malesani *et al.* 2004, Campana *et al.* 2006, Soderberg *et al.* 2008, Vergani *et al.* 2010).

What are these bursts? A list of the confirmed associations, including some of the main observable properties of the bursts, is given in table 1. Perhaps the most important feature of these bursts is that they are different from one another by several orders of magnitude in energy, time scale and rate. In addition, most of these bursts must be very rare compared to SNe. To allow a rough estimation of the typical rates, the volume in which each of these bursts was detected is provided in the 4th column of the table. With the exception of SN1998bw and SN2008D, 100-10000 type Ibc SNe have occurred in each of these volumes within the typical 10 yr lifetime of the relevant γ-ray detectors (assuming a conservative low rate of SNe Ibc of $10^{-5}\mathrm{Mpc}^{-3}\mathrm{yr}^{-1}$) with only ~ 1 X-ray/gamma-ray event. The extremely bright radio afterglow of SN1998bw, implies that this too is a rare event (e.g. Soderberg *et al.* 2010). The only exception left is the burst associated with 2008D, which could be common among Ib SNe (e.g. Mazzali *et al.* 2008).

The huge differences in energy, time scale and rate of these events implies that the physical processes involved in different bursts may be very different. In particular, the exciting fact that the burst properties of SN2003dh are similar to cosmological gamma-ray bursts (GRBs) does not imply that the other (much weaker and common) events are related to cosmological GRBs.

It has long been suggested that an intense burst of soft X-ray radiation is expected to be emitted at the initial phases of every SN explosion, once the radiation mediated shock

Table 1. 'GRB/XRF'-SNe

SNe	d [Mpc]	Detector's FOV [deg^2]	$V_{obs} = FOV \times d^3/3$[Mpc^3]	$E_{\gamma,iso}$[erg]	T[s]
1998bw	40	800[2]	3×10^3	10^{48}	30
2003dh	700	5000[3]	3×10^8	3×10^{52}	30
2003lw	500	80[4]	10^6	10^{50}	30
2006aj	140	4000[5]	10^6	5×10^{49}	2000
2008D	30	0.16[6]	0.3[a]	10^{46}	200
2010bh	250	4000[5]	10^7	10^{50}	2000

Notes:
[1] Data taken from Fan *et al.* (2011), except for SNe 2008D which is taken from Soderberg *et al.* (2008)
Burst detected by [2]BeppoSax, [3]HETE-II, [4]Integral, [5]Swift-BAT, [6]Swift-XRT
[a] This small volume does not represent the rate since there are $\ll 1$ galaxies on average within this volume, while the detector is focused on galaxies most of the time

reaches the edge of the star (e.g. Colgate 1974, Falk *et al.* 1978). It is natural to study the possibility that some of these events are related to such breakouts and several such claims have been made based on the bursts properties (e.g. Kulkarni *et al.* 1998, Campana *et al.* 2006, Soderberg *et al.* 2008). If detected, the properties of the shock breakout burst allow a robust determination of important progenitor properties including its radius.

The main challenge to the association of these events with the predicted properties of shock breakouts is that their spectrum is much harder than expected. The spectrum of typical shock breakouts is expected to be soft, sharply declining beyond a few hundreds of eV's (e.g. Falk *et al.* 1978, Ensman & Burrows 1991, Matzner & Mckee 1999), in contrast to the observed hard spectrum extending to multi-keV energies in all of the observed bursts.

This concern is relaxed by the results of radio and X-ray afterglow observations on scales of days-months which showed the presence of fast material $\Gamma\beta \gtrsim 0.2 - 1$, carrying energies similar to the burst energies. At these inferred fast shock velocities, departure from equilibrium may imply very high electron temperatures, reaching tens or hundreds of keV (Weaver 1976, Katz *et al.* 2010) which naturally account for the hard spectra. Note that the large amount of energy carried by very fast ($\Gamma\beta \sim 1$) ejecta (with the exception of SN2008D) poses a challenge in itself to SNe explosion models. SN2008D, which does not suffer the latter problem and could be a relatively common phenomenon among SNe, is a particularly appealing shock breakout candidate.

A second challenge is that the progenitor radius inferred from the breakout interpretation, $R \approx 10^{12}(E_\gamma/10^{47}\text{ erg})^{1/2}\beta^{-1/2}$ cm (e.g. Matzner & McKee 1999, Katz *et al.* 2012), is larger than the expected radius $R \lesssim 10^{11}$ cm of WR progenitors believed to produce Ibc SNe, which agrees with inferred radii from followup UV and optical observations (e.g. Campana *et al.* 2006, Soderberg *et al.* 2008, Rabinak & Waxman 2011).

If the star is surrounded by a sufficiently optically thick shell of circumstellar matter (CSM), e.g. a high density wind, the breakout occurs within the shell at radii much larger than the progenitor and may solve this problem. Several observed γ-ray/X-ray flashes associated with SNe have been suggested to be such cases (Tan *et al.* 2001, Campana *et al.* 2006, Soderberg *et al.* 2008).

Breakout outbursts of slower shocks, $v_{\rm sh} \sim 0.03c$, have been suggested to account for strong optical/UV transients (Ofek *et al.*2010) and very luminous SNe (e.g. Quimby *et al.* 2007, Smith *et al.* 2007, Miller *et al.* 2009). In order to explain the high energy (reaching 10^{51} erg) emitted in these SNe, CSM parameters were suggested such that the diffusion time scale through the CSM is comparable to the dynamical time scale, $R/v_{\rm sh}$ (e.g. Quimby *et al.* 2007, Smith *et al.* 2007b, Moriya *et al.* 2011, Chevalier & Irwin 2011, contribution to this proceedings by T. Moriya). If true, the observed emission is, by construction, the breakout outburst from the CSM.

Following breakout, the radiation mediated shock is expected to become a collisionless shock, leading to the emission of gamma-rays and neutrinos (e.g. Waxman & Loeb 2001). In the absence of a (significant) wind, the small mass of the shell shocked by the collisionless shock implies that only a small fraction, $\lesssim 10^{-2}$, of the breakout energy is converted to such high energy radiation. Moreover, the formation of a collisionless (or collisional) shock is controversial (e.g. Klein & Chevalier 1978, Epstein 1981, Sapir *et al.* 2011), since the light shell may be accelerated to sufficiently high velocity by the escaping radiation.

In this letter we show that if the progenitor is surrounded by an optically thick CSM, e.g. a dense wind, a collisionless shock is necessarily created during the breakout, and that an energy comparable to or greater than the breakout energy is emitted by quasi-thermal particles in high energy ($\gtrsim 50$ keV) photons, and by accelerated protons in high energy ($\gtrsim 1$ TeV) neutrinos. The latter is an extension of the study of high energy emission from the interaction of the ejecta with a dense optically thin CSM (Murase *et al.* 2010).

2. Formation of a collisionless shock

Consider for simplicity a piston moving with a constant velocity $v = 10^9\ \mathrm{cm\,sec^{-1}}\, v_9$ through an optically thick fully ionized hydrogen wind with a density profile

$$\rho(r) = \frac{c}{v}\frac{m_p}{\sigma_T R_{\rm br}}(r/R_{\rm br})^{-2}, \tag{2.1}$$

where $R_{\rm br} = 10^{14} R_{14}$ cm is a normalization parameter with dimensions of length. A shock propagates ahead of the piston with velocity $v_{\rm sh} \sim v$. As long as the optical depth across the shock transition region, $\Delta_{\tau,\rm sh} \sim c/v$, is much smaller than the optical depth of the system, $\Delta_\tau = (c/v)R_{\rm br}/r$, the post-shock radiation is confined. Once the shock reaches $r \sim R_{\rm br}$, the width of the shock becomes comparable to the size of the system and a significant fraction of the post-shock energy can be emitted during one dynamical time scale, $R_{\rm br}/v$. This emission is the breakout outburst discussed above.

The material lying ahead of the piston must be accelerated to velocities approaching v by some process. At large optical depth, where the radiation mediated shock is sustained, the radiation accelerates the material by Compton scattering off the electrons. The maximal velocity to which a fluid shell can be accelerated by this process is given by

$$v_{\rm max} = \frac{E_\gamma/c}{4\pi r^2}\frac{\sigma_T}{m_p}, \tag{2.2}$$

where $E_\gamma = \int L_\gamma dt$ is the radiation energy emitted through the fluid shell and r is its initial position. This maximum velocity is achieved if all of the flowing photons move radially and the shell does not expand considerably during the passage of the radiation. In this case, a fraction $\sigma_T/(4\pi r^2)$ of the momentum E_γ/c carried by the radiation is transferred on average to each proton.

E_γ is limited by the thermal energy accumulated in the post shock region which, in turn, is limited by $0.5M(r)v^2$, where $M(r) = 4\pi(c/\kappa v)R_{\rm br}r$ is the wind mass inward of r. Using equation (2.2) an upper limit for $v_{\rm max}$, $v_{\rm max} < 0.5(R_{\rm br}/r)v$, is obtained. This implies that beyond a radius of $0.5R_{\rm br}$ the shock can no longer be mediated by radiation and must be transformed into a collisional or a collisionless shock. Since the ion plasma frequency, $\omega_p = (4\pi\rho e^2/m_p^2)^{1/2} \sim 10^9 v_9^{-1/2} R_{14}^{-1/2}\ \mathrm{sec}^{-1}$, is many orders of magnitudes larger than the ion Coulomb collision rate per particle, $\nu_C = \rho\sigma_C v/m_p \sim 2\times 10^{-2} R_{14}^{-1} v_9^{-4}\ \mathrm{sec}^{-1}$ (e.g. Waxman & Loeb 2001), the shock will be collisionless, i.e. mediated by collective plasma instabilities.

3. Emission from thermal electrons.

The collisionless shock heats the protons on a time scale of ω_p^{-1} to a temperature roughly given by

$$T_p \sim \frac{3}{16} m_p v^2 \sim 0.4 v_9^2 m_e c^2 \sim 0.2 v_9^2\ \mathrm{MeV}. \tag{3.1}$$

The electron temperature depends on the unknown amount of collisionless heating. A lower limit for the electron temperature can be obtained by assuming that there is no collisionless heating.

The collisional heating rate of the electrons due to Coulomb collisions with the protons is given by

$$-\frac{dT_p}{dt} = \frac{dT_e}{dt} \approx \lambda_{ep}\sqrt{\frac{2}{\pi}\frac{m_e}{m_p}} T_p \left(\frac{T_e}{m_e c^2}\right)^{-3/2} n_d \sigma_T c, \tag{3.2}$$

where $\lambda_{ep} = 30\lambda_{ep,1.5}$ is the Coulomb logarithm and it was assumed that $T_e/(m_e c^2) \gg T_p/(m_p c^2)$. The fastest possible cooling source for thermal electrons is Inverse Compton (IC) scattering of the local radiation field, which carries a significant fraction $\epsilon_\gamma \lesssim 1$ of the post shock energy and is given by

$$\frac{dT_e}{dt} = -\frac{2}{3}\frac{4T_e}{m_e c^2}\sigma_T c U_\gamma, \tag{3.3}$$

where $U_\gamma = \epsilon_\gamma n_d T_d$ is the photon energy density and n_d is the shocked material proton density. Assuming $U_\gamma \lesssim n_d T_p$ (equivalently, $\epsilon_\gamma \lesssim 1$) we find

$$\frac{T_e}{m_e c^2} \gtrsim 0.6\left(\frac{m_e}{m_p}\lambda_e/\epsilon_\gamma\right)^{2/5} \Rightarrow T_e \gtrsim 60\ \mathrm{keV}\lambda_{ep,1.5}^{2/5}. \tag{3.4}$$

The time it takes the protons to lose a significant fraction of their energy (which is of the order of the total available energy) is

$$\begin{aligned} t_\mathrm{p} &= T_p\left(\frac{dT_p}{dt}\right)^{-1} \lesssim 0.6\left(\lambda_{ep}\frac{m_e}{m_p}\right)^{-2/5} \epsilon_\gamma^{-3/5}(n_d\sigma_T c)^{-1} \\ &\sim 3\lambda_{ep,1.5}^{-2.5}\epsilon_\gamma^{-3/5}(n_d\sigma_T c)^{-1}. \end{aligned} \tag{3.5}$$

The proton cooling time is thus much shorter than the dynamical time $R_\mathrm{br}/v = (c/v)^2/(n\sigma_T c)$, where $n = \rho/m_p$ is the proton number density in the pre-shocked region and is smaller than n_d by the compression factor. This is not surprising. While the shock is radiation mediated, radiation energy equal to the mechanical energy is generated on each shock crossing time scale. At breakout, the shock crossing time scale equals the dynamical scale and radiation with energy density comparable to the total energy density must be generated during the dynamical time scale. In fact, since the electron temperature is higher than that expected in a corresponding radiation mediated shock, the emission efficiency is even higher.

The shock is strongly radiative and the energy is efficiently converted to radiation. The typical photon energies are expected to be of the same order of magnitude as the electron energies, i.e. $\gtrsim 60$ keV. The calculation of the emitted spectrum is beyond the scope of this paper. We note that since the initial photon energies are much lower (~ 1 eV assuming equilibrium) we expect that the spectrum hardens continuously with time and that on the breakout time scale, significant emission is likely emitted at all intermediate energies.

We conclude that gamma-rays/hard X-rays will be emitted with total energy comparable to that of the breakout energy

$$E_\gamma = \frac{4\pi R_{\rm br}^2}{\sigma_T} m_p c v \sim 10^{49} v_9 R_{14}^2 \text{ erg} \tag{3.6}$$

on a time scale similar to the breakout time $t \sim R/v \sim 1 R_{14} v_9^{-1}$d with typical luminosity

$$L_\gamma \sim 10^{45} R_{14} v_9^2 \text{ erg sec}^{-1}. \tag{3.7}$$

When the shock expands, it will remain radiative beyond $R_{\rm br}$ and the total emitted energy, integrated over longer time scales, may be significantly larger than that of the breakout energy.

4. Accelerated protons: Non-thermal emission energy.

Relativistic particles (CRs) accelerated in the collisionless shock that forms due to the collision of the SN ejecta with dense interstellar material may emit high energy gamma rays and neutrinos due to the interaction with the dense material (Murase *et al.* 2010). The collisionless shock that was shown above to be produced during breakout from a dense wind is a constrained example of such interaction and may be a source of detectable high energy neutrinos and gamma rays. Here we focus on the emission from accelerated protons and their products. In what follows it is assumed that the accelerated protons carry a fraction $\epsilon_{CR} = 0.1\epsilon_{CR,-1}$ of the post shock energy and have a flat power law energy distribution, $\varepsilon^2 dn/d_\varepsilon \propto \varepsilon^0$.

The cooling time of a relativistic accelerated proton due to inelastic pp collisions is roughly given by

$$\begin{aligned} t_{pp} &= (0.2(\rho/m_p)\sigma_{pp}c)^{-1} = 5\frac{\sigma_T}{\sigma_{pp}}\left(\frac{v}{c}\right)^2 \frac{R_{\rm br}}{v} \\ &\sim 0.1 v_9^2 \frac{R_{\rm br}}{v}. \end{aligned} \tag{4.1}$$

Hence, for slow enough shock velocities, $v/c \lesssim 0.1$, protons accelerated at breakout efficiently convert their energy to neutrinos, gamma-rays and pairs by pion production and decay (and muon decay). In this section we restrict the discussion to $v/c \lesssim 0.1$. For such shock velocities, the amount of energy emitted by relativistic protons during breakout is expected to be roughly a fraction ϵ_{CR} of the energy emitted by the thermal particles. Using Eq. (3.6) we have $E_{\rm Non-thermal} \sim 10^{48} \epsilon_{CR,-1} R_{14}^2 v_9$ erg.

At later stages, $t_{pp} v/r$ grows linearly with r, and as long as it is smaller than unity, the energy converted into pions increases linearly with the accumulated mass. The radius r_{pp} at which the proton energy loss time is equal to the dynamical time, $t_{pp} v/r_{pp} = 1$, is

$$r_{pp} \sim 10 v_9^{-2} R_{\rm br}. \tag{4.2}$$

Beyond this radius, the fraction of energy converted to pions drops like $1/r$ ($t_{pp} \propto \rho^{-1} \propto r^2$ while the available energy increases linearly with r) implying a logarithmic increase in the total emitted energy. Given that in reality, $v(r)$ is slowly declining, the total contribution to the non thermal fluence from $r > r_{pp}$ is of order unity compared to fluence produce up to this radius. The total emitted energy is therefore given by

$$E_{\rm Non-thermal} \sim 10^{49} \epsilon_{{\rm CR},-1} R_{14}^2 v_9^{-1} \ {\rm erg}. \tag{4.3}$$

4.1. *Accelerated protons: Maximal proton energy.*

The maximal proton energy is limited by the time available for acceleration which is the shorter of the dynamical time and energy loss time. The acceleration time depends on the unknown magnetic field value and the loss time depends on the unknown target photon energy distribution. Nevertheless, we next demonstrate that protons are very likely to be accelerated to at least multi-TeV energies.

Assuming Bohm diffusion, the acceleration time to energy ε is given by

$$t_{\rm acc} = \frac{\varepsilon}{(v/c)^2 eBc} \sim 2 \times 10^{-7} \frac{\varepsilon_{\rm TeV}}{\epsilon_B^{1/2} v_9^{3/2} R_{14}^{1/2}} \frac{R_{\rm br}}{v}, \tag{4.4}$$

where B is the post shock magnetic field and $\epsilon_B = B^2/(8\pi\rho v^2)$ is roughly the fraction of postshock energy carried by it. For TeV CRs, the acceleration time is thus much shorter than the dynamical time and the pp energy loss time. For protons in the range $10 - 1000$ TeV the strongest possible cooling mechanism is photo-production of pions, with cooling time

$$t_{p\gamma} = (0.2 n_\gamma \sigma_{p\gamma} c)^{-1} \gtrsim 5 \frac{\sigma_T}{\sigma_{p\gamma}} \frac{h\nu_\gamma}{m_p c^2} \frac{R_{\rm br}}{v}, \tag{4.5}$$

where $n_\gamma(h\nu_\gamma)$ is the target photon number density (typical energy) and we conservatively assumed that $n_\gamma = \rho v^2/(h\nu_\gamma)$. Photo-production of pions occurs if the proton energy is higher than the threshold, $\sim m_\pi m_p c^4/(h\nu_\gamma) \sim 0.13(h\nu_\gamma/\ {\rm MeV})^{-1}$ TeV. The possible presence of many ~ 1 MeV photons implies that photo-production may be important for 1 TeV protons. Photo production is not important if the target photons have $\sim 1\,$eV energies, as assumed in (Murase *et al.* 2011). Given the constraint $n_\gamma \lesssim \rho v^2/(h\nu_\gamma)$, the strongest losses for protons of energy $\varepsilon = \varepsilon_{\rm TeV}$ TeV occurs for target photons having a typical energy of $h\nu_\gamma \sim m_\pi m_p c^4/\varepsilon \sim 0.13\ {\rm MeV} \varepsilon_{\rm TeV}^{-1}$. Using this in Eq. (4.5) we obtain

$$t_{p\gamma} \gtrsim 1 \varepsilon_{\rm TeV}^{-1} \frac{R_{\rm br}}{v}. \tag{4.6}$$

Comparing Eq. (4.6) to Eq. (4.4) we conclude that acceleration to multi TeV energies is possible for $\epsilon_B \gtrsim 10^{-13} v_9^{-3} R_{14}^{-1} \varepsilon_{\rm TeV}^4$, implying that reaching energies well above 1 *TeV* is very likely.

We verified that proton CRs do not suffer significant losses due to Inverse Compton and Synchrotron emission during the acceleration time, and that the resulting pions and muons do not suffer significant energy losses due to these processes before decaying. Finally, note that the maximal proton energy is increasing with radius since the ratio of proton acceleration time to dynamical time is independent of radius ($t_{\rm acc} \propto B^{-1} \propto \rho^{-1/2} \propto r$) while the ratio of all the loss times to the dynamical time decreases with radius.

4.2. *Accelerated protons: Multi TeV neutrinos.*

Roughly a third of the non thermal energy Eq. (4.3) will be emitted in muon neutrinos (and anti-neutrinos) and a significant fraction of this energy may be emitted beyond TeV energies. In the neutrino energy range of one to hundred TeV, the effective area for muon neutrinos of a Cherenkov neutrino detector like IceCube is increasing linearly with energy, approximately as $10^{-6}\varepsilon_{\rm TeV}A$, where $A = 10^{10}A_{10}$ cm^2 is the geometrical cross-section of the detector. The number of muons induced by one to hundred TeV neutrinos is therefore independent of the neutrino spectrum in this range and is given by

$$N_\mu \sim 5\frac{E_{\nu_\mu,1-100\rm TeV}/10^{51}\ \rm erg}{(d/100\ \rm Mpc)^2} \sim 1\frac{\epsilon_{p,-1}R_{15}^2}{v_9(d/100\ \rm Mpc)^2} \tag{4.7}$$

where d is the distance to the SN and where we optimistically assumed that 1/3 of the non thermal emission, Eq. (4.3), is in multi TeV neutrinos.

4.3. *Accelerated protons: Gamma-rays.*

High energy gamma-rays and pairs with energies reaching multi TeV energy will be generated with a comparable rate to that of the neutrinos. The pairs will emit further high energy gamma-rays by Inverse Compton interactions with the radiation field. Emission below ~ 1 MeV will be mixed with the emission from the thermal electrons. Emission at a photon energy $h\nu \gtrsim$ MeV may be suppressed by the large optical depth for pair creation, which depends on the density of photons with energies above the pair production threshold $h\nu_T \gtrsim m_ec^2/(h\nu)$.

An upper limit to the optical depth for pair creation at a given photon energy, $h\nu$, can be obtained by using the fact that the total energy density of photons of any frequency is smaller than ρv^2. Assuming that the energy of photons per unit logarithmic frequency does not exceed $\epsilon_{0.1} \times 10\%$ of ρv^2, and focusing on the radius $10R_{\rm br}$ at which the protons are still efficiently cooled (for $v \sim 10^9$ cm sec^{-1}, see Eq. (4.2)), we find

$$\tau_{\gamma\gamma} \lesssim \epsilon\frac{v}{c}\frac{m_p}{m_e}\frac{h\nu}{m_ec^2} \sim \epsilon_{0.1}0.6v_9\frac{h\nu}{m_ec^2}. \tag{4.8}$$

The emitted spectrum is suppressed by at most $\sim \tau^{-1}$. In this 'worst case scenario', such bursts will be too faint to be observable by high energy ($h\nu \sim 1$ GeV) gamma-ray detectors such as Fermi.

5. Discussion

We have shown that shock breakouts in optically thick winds will necessarily be accompanied by high energy radiation from a collisionless shock that inevitably forms on the time scale of the breakout outburst.

Low luminosity GRBs associated with SNe have been suggested to be the outbursts associated with fast shocks $v \gtrsim 0.1c$ breaking out of dense optically thick winds. As we have shown here, a significant fraction of the observed radiation, or even most of it, may be generated by the collisionless shock that will form during the breakout.

If the slow, $v/c \sim 0.03$, breakout interpretation of events such as PTF09u (Ofek *et al.* 2010) is correct, a significant amount of energy, $E \sim 10^{51}$ erg, is expected to be emitted in hard X/γ-rays reaching energies $h\nu \gtrsim 50$ keV, Eq. (3.4), and multi-TeV neutrinos (see also Murase *et al.* 2010). X-rays at lower energies are likely to be emitted with similar efficiency and would be easily detected by instruments like the X-ray telescope (XRT; Burrows *et al.* 2005) on board *Swift* (Gehrels *et al.* 2004) or the Chandra X-ray observatory. We note that more detailed followup analysis suggests that the hard X-ray signal may be partly absorbed (Chevalier & Irwin 2012) or significantly delayed (Svirski *et al.* 2012, contribution to this proceedings by Ehud Nakar) due to the interaction of the radiation with the thick wind. TeV neutrinos may be detectable by experiments like IceCube, see Eq. (4.7), provided such events are sufficiently common and a similar event occurs at a distance $d \lesssim 100$ Mpc (compared to ~ 300 Mpc for PTF09uj).

Finally, we note that if the CSM breakout explanation of very luminous SNe (Quimby *et al.* 2007) is correct, our analysis implies that these events should be accompanied by strong high energy X-ray emission. Lack off (Miller *et al.* 2009), or very weak (Smith *et al.* 2007), X-ray emission from some of these events challenges this interpretation.

B.K. is supported by NASA through Einstein Postdoctoral Fellowship awarded by the Chandra X-ray Center, which is operated by the Smithsonian Astrophysical Observatory for NASA under contract NAS8-03060. The research of E.W and N.S is partially supported by ISF, Minerva and PBC grants.

References

Balberg, S. & Loeb, A. 2011, *MNRAS*, 478
Burrows, D. N., *et al.* 2005, *Space Sc. Revs*, 121, 165
Campana, S., *et al.* 2006, *Nature*, 442, 1008
Chevalier, R. A. & Irwin, C. M. 2011, *ApJ*, 729, L6
Chevalier, R. A. & Irwin, C. M. 2012, *ApJ* (Letters), 747, L17
Colgate, S. A. 1974, *ApJ*, 187, 321
Ensman, L. & Burrows, A. 1992, *ApJ*, 393, 742
Epstein, R. I. 1981, *ApJ*, 244, L89
Falk, S. W. 1978, *ApJ*, 225, L133
Fan, Y.-Z., Zhang, B.-B., Xu, D., Liang, E.-W., & Zhang, B. 2011, *ApJ*, 726, 32
Galama, T. J., Vreeswijk, P. M., van Paradijs, J., *et al.* 1998, *Nature*, 395, 670
Gehrels, N., *et al.* 2004, *ApJ*, 611, 1005
Hjorth, J., Sollerman, J., Møller, P., *et al.* 2003, *Nature*, 423, 847
Katz, B., Budnik, R., & Waxman, E. 2010, *ApJ*, 716, 781
Katz, B., Sapir, N., & Waxman, E. 2012, *ApJ*, 747, 147
Klein, R. I. & Chevalier, R. A. 1978, *ApJ*, 223, L109
Kulkarni, S. R., Frail, D. A., Wieringa, M. H., *et al.* 1998, *Nature*, 395, 663
Lasher, G. J. & Chan, K. L. 1979, *ApJ*, 230, 742
Malesani, D., Tagliaferri, G., Chincarini, G., *et al.* 2004, *ApJ* (Letters), 609, L5
Miller, A. A., *et al.* 2009, *ApJ*, 690, 1303
Moriya, T., Tominaga, N., Blinnikov, S. I., Baklanov, P. V., & Sorokina, E. I. 2011, *MNRAS*, 415, 199
Murase, K., Thompson, T. A., Lacki, B. C., & Beacom, J. F. 2011, *Phys. Rev. D*, 84, 043003
Nakar, E. & Sari, R. 2010, *ApJ*, 725, 904
Ofek, E. O., *et al.* 2010, *ApJ*, 724, 1396
Quimby, R. M., Aldering, G., Wheeler, J. C., Höflich, P., Akerlof, C. W., & Rykoff, E. S. 2007, *ApJ* (Letters), 668, L99
Rabinak, I. & Waxman, E. 2011, *ApJ*, 728, 63
Sapir, N., Katz, B., & Waxman, E. 2011, *ApJ*, 742, 36
Smith, N., *et al.* 2007, *ApJ*, 666, 1116
Smith, N. & McCray, R. 2007, *ApJ* (Letters), 671, L17
Soderberg, A. M., Nakar, E., Berger, E., & Kulkarni, S. R. 2006, *ApJ*, 638, 930
Soderberg, A. M., *et al.* 2008, *Nature*, 453, 469
Soderberg, A. M., Chakraborti, S., Pignata, G., *et al.* 2010, *Nature*, 463, 513
Svirski, G., Nakar, E., & Sari, R. 2012, arXiv:1202.3437
Tan, J. C., Matzner, C. D., & McKee, C. F. 2001, *20th Texas Symposium on relativistic astrophysics*, 586, 638
Vergani, S. D., D'Avanzo, P., Levan, A. J., *et al.* 2010, *GRB Coordinates Network*, 10512, 1
Wang, X.-Y., Li, Z., Waxman, E., & Mészáros, P. 2007, *ApJ*, 664, 1026
Waxman, E. & Loeb, A. 2001, *Phys. Rev. Lett.*, 87, 071101
Waxman, E., Mészáros, P., & Campana, S. 2007, *ApJ*, 667, 351

Death of Massive Stars: Supernovae and Gamma-Ray Bursts
Proceedings IAU Symposium No. 279, 2012
P. Roming, N. Kawai & E. Pian, eds.

doi:10.1017/S1743921312013051

Relativistic and Newtonian Shock Breakouts

Ehud Nakar

Raymond and Beverly Sackler School of Physics & Astronomy, Tel Aviv University,
Tel Aviv 69978, Israel

Abstract. Observations of the first light from a stellar explosion can open a window to a wealth of information on the progenitor system and the explosion itself. Here I briefly discuss the theoretical expectation of that emission, comparing Newtonian and relativistic breakouts. The former takes place in regular core-collapse supernovae (SNe) while the latter is expected in SNe that are associated with gamma-ray bursts (GRBs), extremely energetic SNe (e.g., SN2007bi) and white dwarf explosions (e.g., type Ia and .Ia SNe, accretion induced collapse). I present the characteristic observable signatures of both types of breakouts, when spherical. Finally, I discuss Newtonian shock breakouts through wind, which produce a very luminous signal, with an X-ray component that is weak around the breakout, and becomes brighter afterwards.

Any observable stellar explosion must break out of the stellar surface. The breakout emission is the first light seen from an explosion, and it probes the outer layers of the star. The signal from a SN breakout was first calculated four decades ago. It was immediately realized that detection of the breakout emission, or of the emission that comes shortly after it, are excellent probes of progenitor properties such as its radius and mass, and the wind it throws just prior to its explosion. It is also a probe of the explosion energy and geometry. These are all important properties, which may, for example, shed light on the evolution of massive stars. It is therefore important to provide theoretical predictions of the observed signature of shock breakouts in different scenarios. Here I briefly summarize the main properties of shock breakout pulses in the cases of Newtonian and relativistic breakouts from the surface of a star and of a Newtonian breakout from a thick stellar wind. I consider only spherically symmetric cases. All the results described here, including references and comparison to previous studies, are discussed in length in Nakar & Sari (2010), Nakar & Sari (2012) and Svirski *et al.* (2012).

Newtonian Breakout

When a shock is generated within a stellar envelope it accelerates in the decreasing density as it approaches the stellar surface. The shock breaks out once $\tau = c/v_s$ where τ is the optical depth towards the observer, c is the speed of light and v_s is the shock speed. The total energy in the breakout pulse is therefore determined by the stellar radius R_* and by the breakout velocity v_{bo}. It depends strongly on R_*, reaching $\sim 10^{48}$ erg in a typical red-supergiant (RSG) SNe compared to $\sim 10^{45}$ erg in a typical Wolf-Rayet (WR) SNe. The duration of the breakout pulse is dominated by photon travel time and it is therefore (in spherical symmetry) simply R_*/c. The breakout luminosity, therefore, varies weakly between different progenitor radii and it is typically $\sim 10^{44} - 10^{45}$ erg/s. The breakout temperature depends strongly on v_{bo}. Slow breakouts, $v_{bo} < 15,000$ km/s results in a UV flash with temperature $T_{bo} = 10 - 100$ eV, but faster shocks results in a keV or even harder breakouts. The reason is that the radiation behind fast shocks falls out of thermal equilibrium due to photon starvation. Thus, the small number of post shock photons shares the entire post shock energy, achieving a higher energy per photon.

Relativistic Breakout

A key process in relativistic shock breakouts is the formation of pairs, which play two roles. First, pair formation regulates photon production in the shock, preventing the post shock temperature to climb above ~ 200 keV. As a result and in contrast to Newtonian shocks, there is no strong dependence of the temperature immediately behind the shock on its velocity or Lorentz factor. Second, pairs significantly increase the optical depth of the plasma. As a result, the temperature at which the emission is released to the observer is set by pair annihilation to be ~ 50 keV in the outflow rest frame.

The physical setup after the shock end crossing the star is similar in some aspects to that of a pair-radiation, baryon loaded, fireball (as often discussed in the context of gamma-ray bursts). The pair loaded plasma accelerates as long as it is still opaque without significant loses of internal energy to bulk motion energy. In fact, as a result of the post shock energy density profile, the fastest moving parts at the front of the flow gain energy at the expense of slower moving parts that push from behind.

As discussed above the photons are released from the breakout layer once its rest frame temperature is ~ 50 keV. Therefore the observed breakout temperature is $T_{bo} \approx 50\gamma_{bo}$ keV, where γ_{bo} is the Lorentz factor of the breakout layer when it becomes transparent. the duration of the breakout pulse is set by photon travel time, $t_{bo} \approx R_{bo}/c\gamma_{bo}^2$. Calculating the energy in the breakout pulse, E_{bo} is slightly more complicated, but it also depends only on R_{bo} and γ_{bo} (see Nakar & Sari 2012). Therefore, there are three generic observables (E_{bo}, t_{bo} and T_{bo}) that depend on only two parameters R_{bo} and γ_{bo}. Thus, the observed energy, temperature and time scales of the breakout overconstrain R_{bo} and γ_{bo}, where any two observables out of these three are enough to determine both R_{bo} and γ_{bo}. In case all three observables are measured, they must satisfy

$$t_{bo} \sim 20 \text{ s} \left(\frac{E_{bo}}{10^{46} \text{ erg}}\right)^{1/2} \left(\frac{T_{bo}}{50 \text{ keV}}\right)^{-\frac{9+\sqrt{3}}{4}}, \tag{0.1}$$

if the source of the flare is a quasi-spherical relativistic shock breakout. This relation can be used to test whether any observed gamma-ray flare is consistent with a relativistic breakout origin or not.

Newtonian Breakout through a thick wind

When an exploding star is surrounded by a wind that is thick enough, the radiation dominated shock remains confined to the wind after it crosses the star and the breakout takes place at $R_{bo} \gg R_*$. Here I shortly describe cases where the wind is so thick that the breakout takes place only days after the explosion. The breakout luminosity here is almost similar to that of the Newtonian breakout through a stellar surface, i.e., $\sim 10^{44}$ erg/s. But, given the long duration of the breakout pulse, the released energy is significantly larger. In the most extreme cases, when the breakout time is ~ 60 days, the entire explosion energy can be released in the breakout pulse. Thus, shock breakouts through a thick wind are promising candidates for some of the observed super-luminous SNe.

An interesting property of wind breakouts is that the shock makes a transition from radiation mediated to collisionless following the breakout. As a result the emission contains two spectral components - soft (optical/UV) and hard (X-rays and possibly soft gamma-rays). During the breakout, the soft component temperature can vary significantly from one event to another ($10^4 - 10^6$ K), where events with longer breakout time are generally softer. The hard component is always a small fraction, $\sim 10^{-4}$, of the breakout emission, and its fraction of the total luminosity rises quickly afterwards, becoming dominant at $\sim 10 - 50\ t_{bo}$. Therefore, in addition for being extremely luminous, wind shock breakouts are expected to show very weak X-ray emission around the time that the optical-UV light curve peaks, but it should increase with time afterwards.

Acknowledgement
This research was partially supported by an ERC starting grant (GRB-SN 279369)

References

Nakar, E. & Sari, R. 2010, *ApJ*, 725, 904
Nakar, E. & Sari, R. 2012, *ApJ*, 747, 88
Svirski, G., Nakar, E., & Sari, R. 2012, arXiv:1202.3437

Death of Massive Stars: Supernovae and Gamma-Ray Bursts
Proceedings IAU Symposium No. 279, 2012
P. Roming, N. Kawai & E. Pian, eds.

doi:10.1017/S1743921312013063

Probing explosion geometry of core-collapse supernovae with light curves of the shock breakout

Akihiro Suzuki[1] and Toshikazu Shigeyama[2]

[1]Center for Computational Astrophysics, National Astronomical Observatory of Japan, Japan
email: asuzuki@cfca.jp

[2]Research Center for the Early Universe, School of Science, University of Tokyo, Japan

Abstract. We consider supernova shock breakout in aspherical core-collapse supernovae. We perform hydrodynamical calculations to investigate the propagation of a strong shock wave in a compact star and the subsequent emergence from the surface. Using the results combined with a simple emission model based on blackbody radiation, we clarify how aspherical energy depositions affect shock breakout light curves.

Keywords. supernovae: general, shock waves, X-rays: bursts

---

1. Introduction

The final evolutionary state of a massive star is a violent explosion of the star as a core-collapse supernova. The gravitational collapse of the iron core triggered by photodisintegration process deposits the gravitational energy into the central region of the star, which generates a strong shock wave propagating in the stellar interior. The emergence of the shock wave from the surface is accompanied by a UV/X-ray flash, which is known as a supernova shock breakout.

We observe supernovae by using electromagnetic radiation only after the shock breakout phase, which means that detection of the shock breakout emission is very difficult. However, recent developments in observational techniques have gradually allowed us to detect such phenomena. On the theoretical side, there are lots of earlier studies on shock breakout emissions: pioneering works by Grassberg *et al.* (1971), Arnett & Falk (1976), Chevalier (1976), Falk (1978), and Klein & Chevalier (1978); semi-analytical considerations by e.g., Matzner & McKee (1999); and one-dimensional radiation-hydrodynamical calculations by Shigeyama *et al.* (1988), Ensman & Burrows (1992), and so on. Recently, several effects on temporal and spectral features of shock breakout emissions have been extensively considered. For example, the importance of the deviation of the shocked matter from thermal equilibrium, which is crucial for predicting spectra of shock breakout emission, has been investigated by, e.g., Katz *et al.* (2010), Nakar & Sari (2010). The so-called bulk comptonization process may cause the deviation in the spectra from the Planck function (Wang *et al.* 2007; Suzuki & Shigeyama 2010). Moriya *et al.* (2011) and Chevalier & Irwin (2011) considered shock breakout in a dense wind. In addition, results from 1D radiation-hydrodynamical calculations have been compared in detail with the observed shock breakout emission from a type II supernova, SNLS-04D2dc (Tominaga *et al.* 2009).

However, most works have assumed spherical symmetry. In fact, deviation from spherical symmetry is also important in predicting shock breakout light curves. The importance is pointed out by Couch *et al.* (2009), Suzuki & Shigeyama (2010), and Couch *et al.*

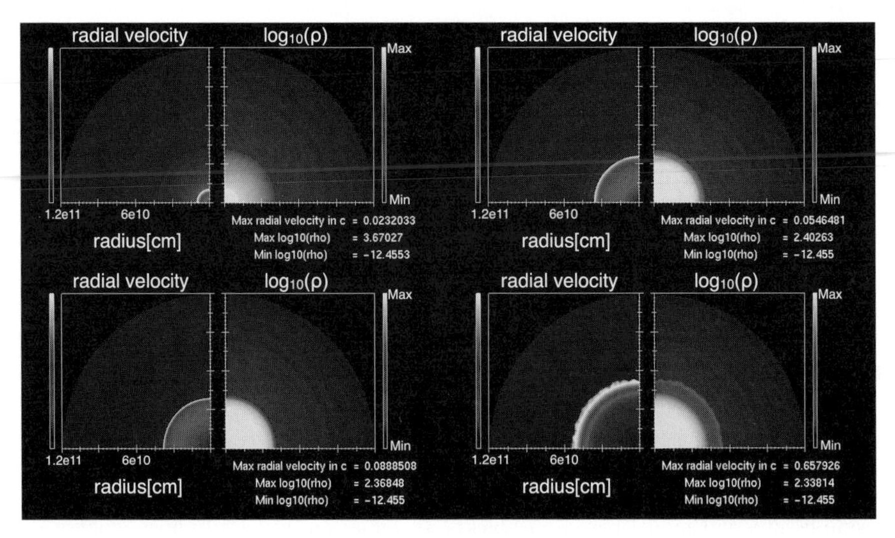

Figure 1. Snapshots of radial velocity and density maps at $t = 1$ (top left), 36 (top right), 37 (bottom left), and 38 (bottom right) sec in the spherical case ($\alpha = 0$).

(2011). In this letter, we show results from hydrodynamical calculations of aspherical shock breakout on the surface of a compact progenitor, which is expected to explode as a type Ic supernova.

2. Hydrodynamical calculations

Hydrodynamical calculations are performed by using a two-dimensional special relativistic hydrodynamics code in spherical coordinates (r, θ) developed by one of the authors. The radial coordinate r ranges from $r_{\rm in} = 3 \times 10^8$ cm to $r_{\rm out} = 1.2 \times 10^{11}$ cm and the angular coordinate θ from 0 to $\pi/2$. The computational domain is covered by 1024×128 meshes. The progenitor model is a $14{\rm M}_\odot$ CO core with a radius of 4×10^{10} cm, which is taken from Woosley & Heger (2006). For the equation of state, an ideal gas with an adiabatic index of $\gamma = 4/3$ is assumed. We inject an energy of $E_{\rm total} = 10^{51}$ ergs from the inner boundary as the kinetic energy. Aspherical explosions are realized by assuming the following condition on the kinetic energy flux at the inner boundary,

$$\frac{\rho_{\rm in} v_r^3}{2} = \frac{E_{\rm total}}{4\pi r_{\rm in}^2 \tau_{\rm in}} [1 + \alpha \cos(2\theta)] \quad \text{for } t < \tau_{\rm in}. \tag{2.1}$$

Here $\rho_{\rm in}$ is the density at the inner boundary and $\tau_{\rm in}$ is the duration of the energy injection, which is fixed to be 0.1 sec in this study. The introduced parameter α ($0 \leqslant \alpha \leqslant 1.0$) controls the asphericity of the explosion. For $\alpha > 0$, a jet-like explosion is realized.

In this paper, we show results of our calculations with $\alpha = 0$ and 0.8 (referred to as spherical and aspherical cases hereafter). Figure 1 represents snapshots of radial velocity and density maps in the spherical case. In this case, the shock breakout occurs at all points on the stellar surface in the same manner. In Figure 2, snapshots at similar epochs in the aspherical case are shown. One can see clear deviations from the spherical case arising from the aspherical energy deposition at the core.

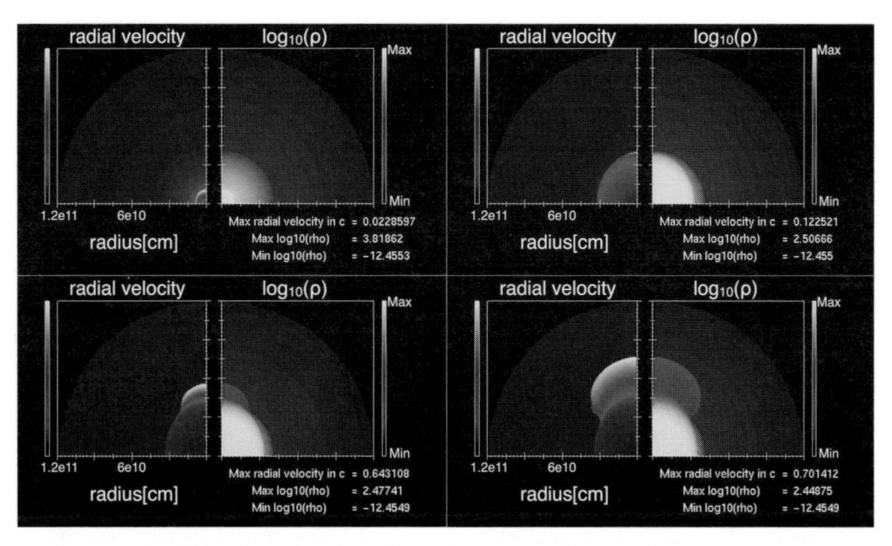

Figure 2. Snapshots of radial velocity and density maps at $t = 1$ (top left), 35 (top right), 36 (bottom left), and 37 (bottom right) sec in the aspherical case ($\alpha = 0.8$).

3. Calculation of light curves

We calculate light curves of the shock breakout emission as seen by distant observers in the following simple way. We consider two observers located 100 stellar radii away from the star with inclination angles $\Theta = 0°$, and $90°$. At first, results of our 2D calculations are mapped into 3D space. Next, we identify the photosphere, where the Thomson optical depth measured from each observer is unity. Assuming that black body radiation is emitted from the photospheres, we calculate light curves of the emission. Here, the radiation temperature $T_{\rm ph}$ is derived by assuming that the pressure p at the photosphere is dominated by the radiation pressure, $p = a_{\rm r}T_{\rm ph}^4/3$, where $a_{\rm r}$ is the radiation constant.

Figure 3 represents the resultant light curves. As shown in the left panel of the figure, in the case with the viewing angle of $\Theta = 0°$, the shock breakout emission is less luminous than the spherical case and the luminosity rapidly decreases at $t \sim 173$ s. On the other hand, in the case with the viewing angle of $\Theta = 90°$ (the right panel), the light curve shows a plateau-like feature after the rising and then the luminosity gradually approaches that of the spherical case.

4. Discussions and conclusions

In the previous sections, we have seen that the aspherical energy deposition at the core of a compact star affects its shock breakout emission. Then, the question is what makes the differences between the light curves. In fact, the differences arise from a simple geometrical effect as explained in the following.

At first, we consider the aspherical case with the viewing angle of $\Theta = 0°$. As seen in Figure 2, the shock breakout occurs in the region around the jet axis at first. In this case, the area of the emitting region is smaller than that in the spherical case, where the shock breakout simultaneously occurs at all points on the surface. This makes the luminosity lower than that in the spherical case. After the emergence of the shock wave from the surface, the ejected matter rapidly expands and then covers the star, which prevents the observer from seeing the shock breakout emission from the region near the equatorial plane. Therefore, the observer sees the emission from the cooling ejecta instead of the

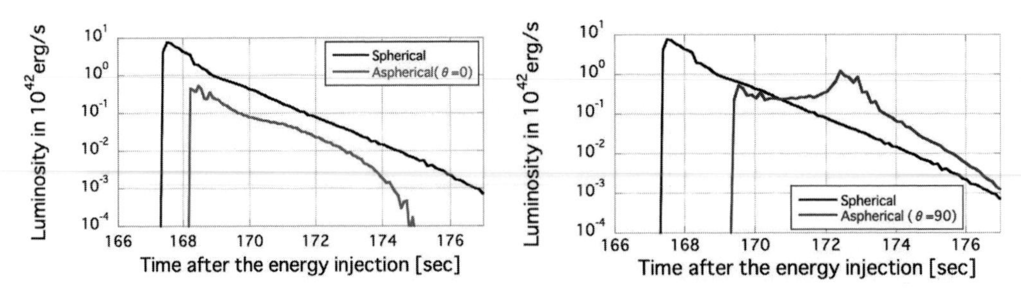

Figure 3. Luminosities of the shock breakout emissions as a function of time measured from the energy injection. In the left panel, light curves of the spherical case and the aspherical case with a viewing angle $\Theta = 0°$ are compared. In the right panel, same as the left panel but with a viewing angle $\Theta = 90°$.

shock breakout emission. This is the reason why the luminosity drops at $t \sim 173$ s as seen in the left panel of Figure 3.

On the other hand, in the case of the viewing angle of $\Theta = 90°$, the observer can see the shock breakout occuring near the equatorial plane. Furthermore, the duration of the shock breakout emission becomes longer than that of the spherical case because the emergence of the shock wave from the surface near the equatorial plane is delayed from that around the jet axis. This delay corresponds to the duration of the plateau-like phase seen in the light curve of the right panel of Figure 3.

Finally, we address the caveat of how the calculations of the light curves are made. As pointed out by several authors, deviation from thermal equilibrium may be important in the shocked envelope. In other words, the shocked matter may outshine in a way different from the black body radiation assumed here. Thus, we should note that predicting spectra of shock breakout emission in the simplified manner as was done in this study has large uncertainties. However, the behavior of light curves revealed here should be seen even when we include such effects, because they merely arise from geometrical effects.

References

Arnett, W. D. & Falk, S. W. 1976, *ApJ*, 210, 733

Chevalier, R. A. 1976, *ApJ*, 207, 872

Chevalier, R. A. & Irwin, C. M. 2011, *ApJL*, 729, L6

Couch, S. M., Wheeler, J. C., & Milosavljević, M. 2009, *ApJ*, 696, 953

Couch, S. M., Pooley, D., Wheeler, J. C., & Milosavljević, M. 2011, *ApJ*, 727, 104

Ensman, L. & Burrows, A. 1992, *ApJ*, 393, 742

Falk, S. W. 1978, *ApJL*, 225, L133

Grassberg, E. K., Imshennik, V. S., & Nadyozhin, D. K. 1971, *ApSS*, 10, 28

Katz, B., Budnik, R., & Waxman, E. 2010, *ApJ*, 716, 781

Klein, R. I. & Chevalier, R. A. 1978, *ApJL*, 223, L109

Matzner, C. D. & McKee, C. F. 1999, *ApJ*, 510, 379

Moriya, T., Tominaga, N., Blinnikov, S. I., Baklanov, P. V., & Sorokina, E. I. 2011, *MNRAS*, 415, 199

Nakar, E. & Sari, R. 2010, *ApJ*, 725, 904

Shigeyama, T., Nomoto, K., & Hashimoto, M. 1988, *A&Ap*, 196, 141

Suzuki, A. & Shigeyama, T. 2010, *ApJL*, 717, L154

Suzuki, A. & Shigeyama, T. 2010, *ApJ*, 719, 881

Tominaga, N., Blinnikov, S., Baklanov, P., *et al.* 2009, *ApJL*, 705, L10

Wang, X.-Y., Li, Z., Waxman, E., & Mészáros, P. 2007, *ApJ*, 664, 1026

Woosley, S. E. & Heger, A. 2006, *ApJ*, 637, 914

Death of Massive Stars: Supernovae and Gamma-Ray Bursts
Proceedings IAU Symposium No. 279, 2012
P. Roming, N. Kawai & E. Pian, eds.

doi:10.1017/S1743921312013075

Magnetars and Gamma Ray Bursts

Niccolò Bucciantini

INAF, Osservatorio Astrofisico di Arcetri, L.go Fermi 5, 50125, Firenze, Italia
email: niccolo@arcetri.astro.it

Abstract. In the last few years, evidences for a long-lived and sustained engine in Gamma Ray Bursts (GRBs) have increased the attention to the so called *millisecond-magnetar* model, as a competitive alternative to the standard collapsar scenario. I will review here the key aspects of the *millisecond magnetar* model for Long Duration Gamma Ray Bursts (LGRBs). I will briefly describe what constraints present observations put on any engine model, both in terms of energetics, outflow properties, and the relation with the associated Supernova (SN). For each of these I will show how the millisecond magnetar model satisfies the requirements, what are the limits of the model, how can it be further tested, and what observations might be used to discriminate against it. I will also discuss numerical results that show the importance of the confinement by the progenitor star in explaining the formation of a collimated outflow, how a detailed model for the evolution of the central engine can be built, and show that a wide variety of explosive events can be explained by different magnetar parameters. I will conclude with a suggestion that magnetars might be at the origin of the Extended Emission (EE) observed in a significant fraction of Short GRBs.

Keywords. (magnetohydrodynamics:) MHD, stars: neutron, stars: magnetic fields, stars: winds, outflows, (stars:) supernovae: general, gamma rays: bursts

---

1. Introduction

The key idea behind the so called *magnetar* model for LGRBs (Usov 1992,Thompson 1994,Wheeler *et al.* 2000,Thompson *et al.* 2004) assumes that the collapse of the core of a massive star at the end of its life leads to a rapidly rotating proto-neutron star (NS; period $\sim$ 1 ms), with a strong surface magnetic field (B $\geqslant 10^{15}$ G). There are reasons to believe that these two properties might be related by dynamo processes (Thompson & Duncan 1995, Thompson & Duncan 1996), even if this might be debated and requires progenitors with rapidly rotating cores. The maximum energy that can be stored in a rotating NS is $\sim 2 \times 10^{52}$erg, and the typical timescale over which this energy can be extracted and delivered to the surrounding medium is $\sim$ 100s for a magnetic field $\sim 3 \times 10^{15}$G. These energies and timescales are compatible with almost every LGRB observed. Moreover, the formation of a proto-NS is fundamental for a successful supernova explosion (NSs are known to produce relativistic outflows; Bucciantini 2008), and there are evidences that massive stars might not necessarily end their life forming Black Holes (Muno *et al.* 2006, Belczynski & Taam 2008 , Gaensler *et al.* 2005, Wachter *et al.* 2008, DeLaney *et al.* 2006 , Vink 2008, Morton *et al.* 2007). Indeed the rotational energy available in a millisecond proto-NS is more than sufficient to unbind the envelope of even a 40-60 $M_{\odot}$ star (Metzger *et al.* 2011), and, for a strong magnetic field mass losses can outpace accretion (Dessart *et al.* 2008) and the wind might halt the fallback of marginally bound ejecta (Bucciantini *et al.* 2009)

2. Energetics and timescales for proto-NS winds

Once a proto-NS is formed it will cool via neutrino emission in a typical Kelvin-Helmholtz timescale $t_{KH} \sim 10 - 100$ s (Pons *et al.* 1999). Neutrinos deposit heat in a neutrinosphere and at about 500-800 ms after core bounce the density surrounding the proto-NS can drop to the point where a neutrino driven wind develops (Arcones *et al.* 2007, Thompson *et al.* 2001). For proto-NS with pulsar like magnetic fields, this wind carries negligible energy (Thompson *et al.* 2001). However, for magnetar like magnetic fields, the wind is magnetocentrifugally accelerated and far more energetic (Thompson *et al.* 2004). As the neutrino luminosity decreases, and the mass loss rate drops, the wind becomes progressively more magnetized eventually reaching relativistic speeds (Thompson *et al.* 2004, Bucciantini *et al.* 2006).

In a recent paper, Metzger *et al.* (2011) have developed a full model for the spin-down evolution of a proto-NS, taking into account the neutrino cooling, the magnetic torque by the wind, and the possible inclination of the magnetic axis with respect to the spin axis. The model can be used to estimate the energy flux in the wind and its magnetization parameter σ which is defined as the ratio of Poynting flux over kinetic energy flux, and can be thought of as a proxy for the maximum achievable Lorentz factor.

Four phases can be distinguished in the overall evolution (Metzger *et al.* 2011):

• An early non relativistic phase, that lasts ~ 1 s after bounce, when the proto-NS is still hot and contracting, the wind is magnetized, but with typical terminal speeds $\sim 0.1c$.

• An intermediate mildly relativistic phase that lasts ~ 10 s after bounce. The proto-NS relaxes to a radius ~ 20 km and begins to spin-down, the wind mass loss rate drops, and σ increases from values of ~ 1 to ~ 10. The wind is now relativistic but still confined inside the progenitor.

• The GRB phase, after the wind breaks out of the progenitor and starts to accelerate in the circumstellar medium. Due to the decreased neutrino luminosity and related mass loss rate, the magnetization rapidly increases to values of $\sigma \sim 10^2 - 10^3$, while energy losses are still high ($\sim 10^{49}$ erg s^{-1}). This phase lasts for about 100 s, when the neutrino luminosity drops sharply below the threshold to drive a baryon loaded wind.

• A late activity phase that begins once the baryon loaded wind ceases, to be replaced by a leptonic wind once the density in the magnetosphere drops below the threshold for pair production. The wind luminosity is smaller ($\sim 10^{48}$ erg s^{-1}) but the typical spin-down time is now longer (~ 100 s), as expected to explain some aspects of the late activity observed in some LGRBs.

We want to emphasize here that energy losses are due to a magnetized stellar wind, analogous to the Solar wind and other stellar winds, and are not due to magnetic dipole radiation. Interestingly for $\sigma \gg 1$ the two have the same value, but in the early and mildly relativistic phases the energy losses in the wind can exceed even by a factor of 10 what can be estimated based on dipole radiation alone.

3. Collimation and acceleration of the outflow

As opposed to accretion disks around black holes that are known to power collimated outflow in the form of relativistic jets, magnetars are supposed to produce essentially spherical outflows. Relativistic outflows cannot self collimate (Lyubarsky & Eichler 2001), and at large distances from the Light Cylinder the wind structure should closely follow the split monopole solution (Bucciantini *et al.* 2006) where most of the energy flux is in the equatorial region. Moreover in monopolar relativistic outflows, the terminal Lorentz factor can only be as high as $\sigma^{1/3}$ (Arons 2002). This is called the σ problem. It is evident that the GRB outflow that is observed cannot coincide with the the steady state spherical wind emerging directly from the proto-NS magnetosphere.

We know that as the outflow emerges from the interior of the host star it collimates into a jet that punches through the stellar envelope, creating a channel where material coming from the central engine can flow (e.g., Matzner 2003). Afterglow observations (*jet breaks*; Rhoads 1999) and GRB energetics (comparison of the total energy derived from late radio afterglow observations with respect to the prompt emission) confirm that a collimated flow is present.

Simple energy considerations demonstrate that the surrounding stellar envelope provides an efficient confining medium even for a very energetic proto-magnetar wind. It is indeed the interaction with the progenitor star that provides the collimating agent to channel the spherical magnetar wind into a polar jet. By analogy to bipolar wind bubbles (Königl & Granot 2002, Begelman & Li 1992), the interaction of the wind from the spinning-down magnetar with the surrounding star could facilitate collimation. It has been shown, under different assumptions, that once the interaction with the surrounding progenitor is taken into account, this can in fact occur (Bucciantini *et al.* 2007, Bucciantini *et al.* 2008, Komissarov & Barkov 2007, Bucciantini *et al.* 2009).

The physical picture is based on an analogy to the case of pulsar wind nebulae (PWNe; Komissarov & Lyubarsky 2004, Del Zanna *et al.* 2004): the magnetar wind is confined by the surrounding (exploding) stellar envelope, as a result a *magnetar wind nebula* (MWN) forms where the wind magnetic field is compressed. If the toroidal magnetic field in the bubble is sufficiently strong, due to the *tube of toothpaste* effect the bubble expands primarily in the polar direction while a negligible amount of energy is transferred to the SN envelope.

The issue of collimation is strictly related to the problem of the acceleration because deviations from the strict monopole geometry can substantially enhance the terminal Lorentz factor, as well as time dependent effects. In the millisecond magnetar, the collimation of the outflow and the formation of a wind nebula becomes a key features if one wants to build a GRB engine. A key assumption here is that magnetic energy can be efficiently dissipated/converted into kinetic energy (Arons 2002, Lyubarsky & Kirk 2001 , Kirk & Skjæraasen 2003). This can happen in various ways. For an oblique rotator, the striped magnetic field might reconnect and dissipate at the wind termination shock of the nebula as it slows down (Lyubarsky 2005). It is not unlikely that instabilities and dissipation might be at work inside the nebula itself or in the jet (Begelman & Li 1992, Moll *et al.* 2008). Modulation of the outflow by the confining walls of the channel formed inside the progenitor might enhance magnetic to kinetic energy conversion (Granot *et al.* 2011).

Results show that the opening angle of the jet, as it punches trough the star and later emerges into the circumstellar medium, is of order $5-10^{\circ}$ and appears to be independent of the dissipative properties of the magnetar wind, as long as the magnetization inside the MWN reaches equipartition. It is also shown that the outflow can accelerate rapidly as soon as it emerges from the progenitor star.

4. SN association and late activity

It is now well established that long-duration GRBs are associated with core-collapse SNe, in particular with the subclass of SNe Ic-BL (BL = broad line; Woosley & Bloom 2006, Della Valle 2006, Zhang 2007). Interestingly, the converse is not true (Soderberg 2006, Woosley & Bloom 2006), and the search for orphan afterglows shows that within a high confidence level the hypothesis that every broad-lined SN harbors a GRB can be ruled out. Moreover, even if the SNe associated with long-duration GRBs tend to be more luminous than the average sample, they are not particularly unusual among the class of BL SN in terms of their energies, photospheric velocities, and Ni masses.

It can be debated if this class of core-collapse SNe that are unusually energetic, asymmetric (as revealed by spectra-polarimetry), and produce significant amounts of Ni are powered by a diverse central engine (a failed GRB), or if the GRB engine is only a possible outcome of the conditions leading to those supernovae: are hypernovae due to a GRB-like engine, or vice-versa? Higher energies, axisymmetry, and Nickel production are the three aspects that must be considered in the GRB-SN association.

In the magnetar model nearly all of the spindown energy of the neutron star escapes in the polar channel (Komissarov & Barkov 2007, Bucciantini *et al.* 2009). There is very little coupling between the exploding star and the GRB engine. This seems to apply both to the low (dissipative) and high (non dissipative) σ limits. The interesting implication of this result is that a proto-magnetar powering a GRB is unlikely to contribute significantly to energizing the SN shock as a whole (although it clearly does so in the polar region), at least on timescales $\gtrsim$1 sec after core bounce. This is a key property of magnetized outflows and in principle it is not specific to a particular central engine. Specifically, we suspect that the same results will apply also to winds from accretion disks (Proga *et al.* 2003) that will likely escape via the polar channel rather than transferring energy to the SN shock as has been previously hypothesized (Arons 2003).

Given the relatively *on-axis* viewing angle of observed GRBs, high velocity ejecta might be observed; high velocity O and Ne can also be produced by the jet blowing out stellar material that had been processed during stellar evolution (Mazzali *et al.* 2006). A jet might lead to unique observable signatures in the ejecta at late times (as may be the case for Cas A; Wheeler *et al.* 2008).

A separate issue relates *magnetar* engines and the production of excess ^{56}Ni that are observed in hypernovae. It has been shown that the temperature at which explosive nucleosynthesis of ^{56}Ni happens ($\gtrsim 5\times 10^9$ K; Woosley *et al.* 2002) is not attained even at relatively early times. This happens because, by the time the jet-plume emerges outside the SN shock, the density of the progenitor into which it propagates is $\sim 10^{4-5}$ g cm^{-3}. At these densities Ni production requires a shock moving at nearly the speed of light, significantly faster than what can be achieved at these early times (Komissarov & Barkov 2007, Bucciantini *et al.* 2009, Metzger *et al.* 2007). However, $\sim 10^{-2} M_\odot$ of high speed ($v \simeq 0.1-0.2\,c$) Ne and O can be created because these have lower threshold temperatures for successful explosive nucleosynthesis.

As a side point, the specific angular momentum required for a millisecond magnetar engine is $J \simeq 3\times 10^{15} R_{10}^2 P_1^{-1}$ cm^{2}s^{-1} (Thompson *et al.* 2004), which is about a factor of five smaller than what is required for the formation of an accretion disk for the black-hole accretion-disk model (MacFadyen & Woosley 1999). This implies that if the core has enough angular momentum to power a *collapsar* engine, then it has enough to create a millisecond magnetar.

The millisecond magnetar model for LGRBs is particularly interesting in view of the so called *late activity*. Late activity manifests itself in the afterglow up to 10^{4-5} seconds after the prompt emission (Campana *et al.* 2005, Vaughan *et al.* 2006, Cusumano *et al.* 2006 , Nousek *et al.* 2006, O'Brien *et al.* 2006, Willingale *et al.* 2007), either as a shallow decay or plateau of the light curve, or with the presence of flares (Burrows *et al.* 2005, Falcone & The Swift Xrt Team 2006 , Chincarini *et al.* 2007). Late activity requires a persistent engine at times much longer than the typical duration of the prompt emission to provide continuous injection of energy.

In the millisecond magnetar model, late time injection can take the form of a leptonic wind, not dissimilar to standard pulsars. The amount of energy that can be released in this scenario, even if smaller than the prompt emission, can produce the shallow decay phase that is observed (Yu & Dai 2007, Metzger *et al.* 2011).

More intriguing is the presence of flares that can carry a substantial fraction of energy compared to the prompt emission. A possibility is that these flares are the signature of magnetic readjustments, within the proto-magnetar that give rise to bursting activity not dissimilar to what is observed during giant bursts in SGRs (Thompson & Duncan 1995, Thompson & Duncan 1996, Woods 2004, Mereghetti 2008). The magnetic energy stored in canonical magnetars is smaller than the rotational energy required for a GRB engine, however the internal magnetic field might be much higher than the surface value (Braithwaite & Spruit 2006, Braithwaite 2008).

5. Validating the model for LGRBs

As shown before, the magnetar model can reproduce many aspects of the observed phenomenology in LGRBs. It is interesting to evaluate if and how one can distinguish a magnetar from a different engine (i.e. a Black Hole). Unfortunately, the dynamical properties of magnetized outflows, once the value of Lorentz factor and of σ are set, are largely independent on the conditions at injection. A more promising discriminant might be the composition: in particular within the magnetar model one expects a transition from a baryon loaded wind to a leptonic dominated outflow at ~ 100 sec after bounce.

Perhaps the bigger discriminant is the available energy. A magnetar can store at most a few times 10^{52} erg of energy. The detection of a GRB with higher total energy, could rule out a magnetar as its engine. Determining with accuracy the total energy of a GRB is non trivial. The prompt emission must be corrected for beaming (Cenko *et al.* 2010) and off-axis effects (van Eerten *et al.* 2010), while the late radio emission is often assumed to originate from a Sedov phase to be converted into a kinetic energy (Shivvers & Berger 2011). There is a small set of very energetic GRBs (Cenko *et al.* 2010) that are marginally compatible with a magnetar engine. However for the vast majority, and for those for which we have good data, the inferred energies are always a few 10^{51} ergs (Shivvers & Berger 2011).

There is also a set of GRBs with a long prompt emission characterized by several events lasting ~ 100 sec and separated by quiescence periods of about $200-400$ sec. Unlike for a BH scenario where one might invoke bursty mass accretion, the magnetar spin-down is smooth. However, the gamma-ray luminosity might not be a good tracer of the energy injection, depending on the efficiency of particle acceleration in the outflow. In the recent paper by Metzger *et al.* (2011) it was shown that several expected correlations, like the Amati relation, can be recovered in the magnetar model assuming magnetic dissipation to be at the origin of the radiation mechanism.

6. Short GRBs with extended emission

The standard LGRB/SGRB dichotomy has recently been challenged by several 'hybrid' events that conform to neither class (e.g. Zhang 2007; Bloom *et al.* 2008). All together $\sim 1/4$ of *Swift* SGRBs are accompanied by extended X-ray emission lasting for $\sim 10-100$ s with a fluence approximately greater than that of the GRB itself (see Norris & Bonnell 2006 and Perley *et al.* 2009 for a compilation of events). The hybrid nature and common properties of these events ('Short GRB' + ~ 100 s X-ray tail) have motivated the introduction of a new subclass: Short GRBs with Extended Emission (SGRBEEs).

It was moreover recently discovered that some SGRBs are followed by an X-ray 'plateau' ending in a very sharp break (GRB 980515; Rowlinson *et al.* 2010; Troja *et al.* 2008; Lyons *et al.* 2010), which is difficult to explain by circumstellar interaction alone. Although the connection of this event to SGRBEEs is unclear, it nevertheless provides additional evidence that the central engine is active at late times. The long duration and high fluence of the extended emission of SGRBEEs poses a serious challenge to the NS merger scenario, because in this model both the prompt and extended emission are necessarily powered by black hole accretion. It is in particular difficult to understand how such a high accretion rate is maintained at very late times. Metzger *et al.* 2008 recently proposed that SGRBEEs result from the birth of a rapidly spinning proto-magnetar, created by a NS-NS merger or the AIC of a WD. In this model the short GRB is powered by the accretion of the initial torus (similar to standard NS merger models), but the EE is powered by a relativistic wind from the proto-magnetar at later times, after the disk is disrupted. Although a NS remnant is guaranteed in the case of AIC, the merger of a double NS binary could also leave a stable NS remnant. The interaction of the relativistic proto-magnetar wind with the expanding ejecta was investigated by Bucciantini *et al.* (2012), with a focus on the confining role of the ejecta and its dependence on the wind power and on the ejecta mass and density profile. The model thus predicts a class of events for which the EE is observable with no associated short GRB. These may appear as long-duration GRBs or X-Ray Flashes unaccompanied by a bright supernova and not solely associated with massive star formation, which may be detected by future all-sky X-ray survey missions.

References

Arcones, A., Janka, H.-T., & Scheck, L. 2007, *A&A*, 467, 1227

Arons, J. 2002, *Neutron Stars in Supernova Remnants*, ASPC, 271, 71

Arons, J. 2003, *ApJ*, 589, 871

Belczynski, K. & Taam, R. E. 2008, *ApJ*, 685, 400

Begelman, M. C. & Li, Z. 1992, *ApJ*, 397, 187

Bloom, J. S., Butler, N. R., & Perley, D. A. 2008, *American Institute of Physics Conference Series*, 1000, 11

Braithwaite, J. & Spruit, H. C. 2006, *A&A*, 450, 1097

Braithwaite, J. 2008, *MNRAS*, 386, 1947

Bucciantini, N., Thompson, T. A., Arons, J., Quataert, E., & Del Zanna, L. 2006, *MNRAS*, 368, 1717

Bucciantini, N., Quataert, E., Arons, J., Metzger, B. D., & Thompson, T. A. 2007, *MNRAS*, 380, 1541

Bucciantini, N., Quataert, E., Arons, J., Metzger, B. D., & Thompson, T. A. 2008, *MNRAS*, 383, L25

Bucciantini, N. 2008, 40 *Years of Pulsars: Millisecond Pulsars, Magnetars and More*, 983, 186

Bucciantini, N., Quataert, E., Metzger, B. D., Thompson, T. A., Arons, J., & Del Zanna, L. 2009, *MNRAS*, 396, 2038

Bucciantini, N., Metzger, B. D., Thompson, T. A., & Quataert, E. 2012, *MNRAS*, 419, 1537

Burrows, D. N., *et al.* 2005, *Science*, 309, 1833

Campana, S., *et al.* 2005, *ApJL*, 625, L23

Cenko, S. B., Frail, D. A., Harrison, F. A., *et al.* 2010, *ApJ*, 711, 641

Chincarini, G., *et al.* 2007, *ApJ*, 671, 1903

Cusumano, G., *et al.* 2006, *ApJ*, 639, 316

DeLaney, T., Gaensler, B. M., Arons, J., & Pivovaroff, M. J. 2006, *ApJ*, 640, 929

Della Valle, M. 2006, *Chinese Journal of Astronomy and Astrophysics Supplement*, 6, 010000

Del Zanna, L., Amato, E., & Bucciantini, N. 2004, *A&A*, 421, 1063

Dessart, L., Burrows, A., Livne, E., & Ott, C. D. 2008, *ApJL*, 673, L43
Falcone, A. D., *The Swift Xrt Team 2006, AIP, Gamma-Ray Bursts in the Swift Era*, 836, 386
Gaensler, B. M., McClure-Griffiths, N. M., Oey, M. S., Haverkorn, M., Dickey, J. M., & Green, A. J. 2005, *ApJL*, 620, L95
Granot, J., Komissarov, S. S., & Spitkovsky, A. 2011, *MNRAS*, 411, 1323
Kirk, J. G. & Skjæraasen, O. 2003, *ApJ*, 591, 366
Komissarov, S. S. & Lyubarsky, Y. E. 2004, *MNRAS*, 349, 779
Komissarov, S. S. & Barkov, M. V. 2007, *MNRAS*, 382, 1029
Königl, A. & Granot, J. 2002, *ApJ*, 574, 134
Lyons, N., O'Brien, P. T., Zhang, B., *et al.* 2010, *MNRAS*, 402, 705
Lyubarsky, Y. & Kirk, J. G. 2001, *ApJ*, 547, 437
Lyubarsky, Y. E. & Eichler, D. 2001, *ApJ*, 562, 494
Lyubarsky, Y. 2005, *Advances in Space Research*, 35, 1112
MacFadyen, A. I. & Woosley, S. E. 1999, *ApJ*, 524, 262
Matzner, C. D. 2003, *MNRAS*, 345, 575
Mazzali, P. A., *et al.* 2006, *ApJ*, 645, 1323
Mereghetti, S. 2008, *The Astronomy and Astrophysics Review*, 15, 225
Metzger, B. D., Thompson, T. A., & Quataert, E. 2007, *ApJ*, 659, 561
Metzger, B. D., Quataert, E., & Thompson, T. A. 2008, *MNRAS*, 385, 1455
Metzger, B. D., Giannios, D., Thompson, T. A., Bucciantini, N., & Quataert, E. 2011, *MNRAS*, 413, 2031
Moll, R., Spruit, H. C., & Obergaulinger, M. 2008, *A&A*, 492, 621
Morton, T. D., Slane, P., Borkowski, K. J., Reynolds, S. P., Helfand, D. J., Gaensler, B. M., & Hughes, J. P. 2007, *ApJ*, 667, 219
Muno, M. P., *et al.* 2006, *ApJL*, 636, L41
Norris, J. P. & Bonnell, J. T. 2006, *ApJ*, 643, 266
Nousek, J. A., *et al.* 2006, *ApJ*, 642, 389
O'Brien, P. T., *et al.* 2006, *ApJ*, 647, 1213
Panov, I. V. & Janka, H.-T. 2009, *A&A*, 494, 829
Perley, D. A., Metzger, B. D., Granot, J., *et al.* 2009, *ApJ*, 696, 1871
Piran, T. 1999, *Physics Reports*, 314, 575
Pons, J. A., Reddy, S., Prakash, M., Lattimer, J. M., & Miralles, J. A. 1999, *ApJ*, 513, 780
Proga, D., MacFadyen, A. I., Armitage, P. J., & Begelman, M. C. 2003, *ApJL*, 599, L5
Rhoads, J. E. 1999, *ApJ*, 525, 737
Rowlinson, A., O'Brien, P. T., Tanvir, N. R., *et al.* 2010, *MNRAS*, 409, 531
Shivvers, I. & Berger, E. 2011, *ApJ*, 734, 58
Soderberg, A. M. 2006, *Gamma-Ray Bursts in the Swift Era, AIP*, 836, 380
Thompson, C. 1994, *MNRAS*, 270, 480
Thompson, C. & Duncan, R. C. 1995, *MNRAS*, 275, 255
Thompson, C. & Duncan, R. C. 1996, *ApJ*, 473, 322
Thompson, T. A., Burrows, A., & Meyer, B. S. 2001, *ApJ*, 562, 887
Thompson, T. A. 2003, *ApJL*, 585, L33
Thompson, T. A., Chang, P., & Quataert, E. 2004, *ApJ*, 611, 380
Troja, E., King, A. R., O'Brien, P. T., Lyons, N., & Cusumano, G. 2008, *MNRAS*, 385, L10
Usov, V. V. 1992, *Nature*, 357, 472
van Eerten, H., Zhang, W., & MacFadyen, A. 2010, *ApJ*, 722, 235
Vaughan, S., *et al.* 2006, *ApJ*, 638, 920
Vink, J. 2008, *Advances in Space Research*, 41, 503
Wachter, S., Ramirez-Ruiz, E., Dwarkadas, V. V., Kouveliotou, C., Granot, J., Patel, S. K., & Figer, D. 2008, *Nature*, 453, 626
Willingale, R., *et al.* 2007, *ApJ*, 662, 1093
Wheeler, J. C., Yi, I., Höflich, P., & Wang, L. 2000, *ApJ*, 537, 810
Wheeler, J. C., Maund, J. R., & Couch, S. M. 2008, *ApJ*, 677, 1091
Woods, P. M. 2004, *Advances in Space Research*, 33, 630

Woosley, S. E., Heger, A., & Weaver, T. A. 2002, *Reviews of Modern Physics*, 74, 1015
Woosley, S. E. & Bloom, J. S. 2006, *ARA&A*, 44, 507
Yu, Y. W. & Dai, Z. G. 2007, *A&A*, 470, 119
Zhang, B. 2007, *Chinese Journal of Astronomy and Astrophysics*, 7, 1

Discussion

B. ZHANG: So you don't believe the magnetar model for Superluminous SNe

N. BUCCIANTINI: A millisecond magnetar might have enough energy to power a Superluminous SN, but this energy, extrtacted by a magnetized wind, is only weakly coupled with the SN shock. I do not think that a magnetar can energize a SN shock, early enough to drive the nucleosynthesys of $\sim 0.5M_{\odot}$ of ^{56}Ni.

B. ZHANG: For NS-NS making a magnetar have you considered how a supermassive magnetar form? What kind of NS equation of state is needed.

N. BUCCIANTINI: For NS-NS merger resulting in a long lived magnetar, one needs peculiar conditions: two low-mass NSs must be involved; a few tenths $M_{\odot}$ must be lost either by strong neutrino driven winds or during the merger itself; the EoS must be particularly stiff. The existence of a $2M_{\odot}$ NS suggests that the EoS might allow for massive magnetars.

S. MOISEENKO: In our simulations of magneto-rotational supernovae explosion we found that magnetic field can reach the values $10^{14} - 10^{15}$ G, but it is chaotic magnetic field which can be reduced in a short time due to reconnection. How to make a neutron star with so strong magnetic field?

N. BUCCIANTINI: It is true that MRI, and other instabilities migh enhance strongly the magnetic field, but this happens at small scales, and the resulting field is mostly chaotic and tends to dissipate rapidly. One of the ideas behind the origin of the strong magnetar magnetic field, is that dynamo processes are at work. The key idea here is that a mean field dynamo operates. Investigating the possibility of mean field dynamo requires a full 3D geometry, with enough resolution to properly sample the parameter space in term of viscosity and resistivity. 2D simulation will all be subject to Cowling antidynamo theorem, so they can never lead to large scale fields. To my knowledge investigation of MHD Supernovae in the full 3D regime is very demanding and quite limited. We do observe magnetar, so nature must find a way to produce them.

Death of Massive Stars: Supernovae and Gamma-Ray Bursts
Proceedings IAU Symposium No. 279, 2012
P. Roming, N. Kawai, E. Pian, eds.

doi:10.1017/S1743921312013087

Are short GRBs powered by magnetars?

Paul T. O'Brien[1] **and Antonia Rowlinson**[2]

[1]Department of Physics & Astronomy, University of Leicester
University Road, Leicester, LE1 7RH, United Kingdom
email: paul.obrien@leicester.ac.uk

[2]Astronomical Institute "Anton Pannekoek", University of Amsterdam, Postbus 94249, 1090
GE Amsterdam, The Netherlands
email: b.a.rowlinson@uva.nl

Abstract. The standard model for a short duration Gamma-Ray Burst (GRB) involves the merger of a neutron star binary system, resulting in a black hole which accretes for a brief period of time. However, some of the short-duration GRBs observed by the Swift satellite show features in their light curves which are difficult to explain in this model. As an alternative, we examine the light curves of the Swift short GRB sample to see if they can be explained by the presence of a highly magnetised, rapidly rotating pulsar, or magnetar. We find that magnetars may be present in a large fraction of short bursts, and discuss briefly how this model can be tested using the next generation of gravity-wave observatories.

Keywords. gamma rays: bursts, (stars:) pulsars: general, black hole physics

1. Introduction

The *Swift* satellite (Gehrels *et al.* 2004) has observed a number of short gamma-ray bursts (SGRBs) X-ray afterglows (e.g. Gehrels *et al.* 2005). The properties of the light curves and their inferred host galaxies have provided support for the most popular compact binary merger progenitor theory, i.e. the coalescence of two neutron stars (e.g. Lattimer & Schramm 1976). However, without the coincident observation of gravitational waves by observatories like LIGO (Laser Interferometry Gravitational-wave Observatory) we are missing the "smoking gun" observation for this progenitor theory. It is interesting that, as with the long-duration GRBs (LGRBs), many of the SGRBs show features in their X-ray light curves which suggest a long-lived central engine, for example late time flares and plateaus (e.g., Nousek *et al.* 2006; O'Brien *et al.* 2006).

The presence of long-lived features is particularly problematic for SGRB progenitor theories as accretion is expected to end within a few seconds and only a small fraction of the merger mass is available ($0.01 - 0.1 M_{\odot}$), although this is dependant on the NS equation of state. An alternative energy source, if one can be found, is an attractive solution. One such source is that in some GRBs rather than a black hole (BH), a highly magnetised, rapidly rotating pulsar, or magnetar, may be formed with enough rotational energy to prevent gravitational collapse (e.g., Usov 1992; Duncan & Thompson 1992). The rotational energy is then released as gravitational waves and electromagnetic radiation, causing the magnetar to spin down. If the magnetar is sufficiently massive it may reach a critical point at which differential rotation is no longer able to support it, resulting in collapse to a BH. Assuming constant radiative efficiency, the energy injection from the magnetar would produce a plateau in the X-ray light curve Zhang & Mészáros (2001) and would be followed by a steep decay if the magnetar collapses to a BH or a more gentle decline if it does not collapse.

We have recently proposed candidates for such a multi-stage progenitor system among both LGRBs (Troja *et al.* 2007; [Lyons *et al.* 2010]lyons2009) and SGRBs (Rowlinson *et al.* 2010). The likelihood of producing a magnetar is dependent on the equation of state of neutron stars. Morrison, Baumgarte, & Shapiro (2004) showed that the rotation of the NS could increase the maximum mass by $\sim 50\%$ and hence NS mergers could often result in a NS. Ozel *et al.* (2010) further showed that in such a merger the collapse to a BH can be delayed or not occur at all. Thus, it seems reasonable to assume that many NS binary mergers could result in a magnetar in a SGRB.

Here we consider all *Swift* detected SGRBs with $T_{90} \leqslant 2$ s, observed until March 2012 with an X-ray afterglow or which were promptly slewed to and observed by the X-ray Telescope (XRT), and identify those with a plateau phase in their light-curves suggesting ongoing central engine activity. For the 28 SGRBs with sufficient data, we fit the 0.3–10 Kev X-ray light curves to search for the signature of a magnetar (with or without collapse to a BH). This work is described in detail in Rowlinson *et al.* (2012, submitted to MNRAS).

2. Magnetar model fits

The model used here is as described in Zhang & Mészáros (2001) and used by Troja *et al.* (2007), Lyons *et al.* (2010), Rowlinson *et al.* (2010). This model is consistent with the late-time residual spin-down phase driving a relativistic magnetar wind as described in Metzger *et al.* (2011). We fit the equations below with an additional underlying power-law component whose decay rate is governed by the curvature effect (Kumar & Panaitescu 2000). We use the fitted values of the magnetic field and initial spin period to explain the luminosity of the X-ray plateau and its duration.

$$B^2_{p,15} = 4.2025 I^2_{45} R_6^{-6} L^{-1}_{0,49} T^{-2}_{em,3}$$
$$P^2_{0,-3} = 2.05 I_{45} L^{-1}_{0,49} T^{-1}_{em,3}$$

where $T_{em,3}$ is the plateau duration in 10^3 s, $L_{0,49}$ is the initial plateau luminosity in 10^{49} erg s^{-1}, I_{45} is the moment of inertia in units of 10^{45}g cm^2, $B_{p,15}$ is the magnetic field strength at the poles in units of $10^{15}G$, R_6 is the radius of the neutron star in 10^6cm and $P_{0,-3}$ is the initial period of the compact object in milliseconds. These equations apply to the electromagnetic dominated spin down regime. We have assumed that the

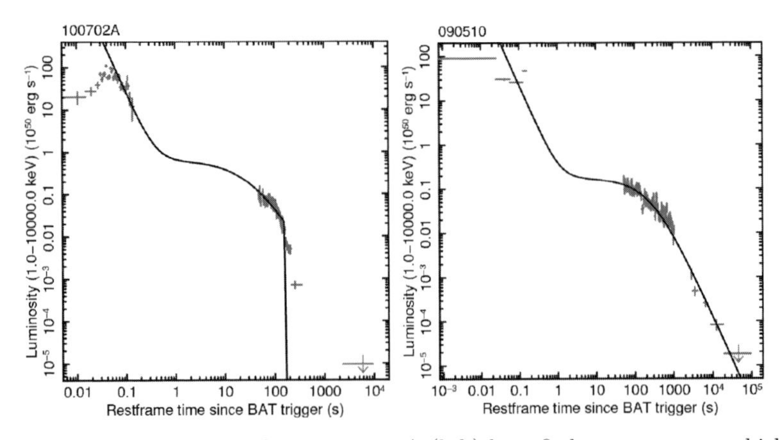

Figure 1. Model fit to two short GRBs: 100702A (left) best-fit by a magnetar which collapses to a black hole, and 090510 (right) which survives as a magnetar.

emission is 100% efficient and isotropic. The equations of vacuum dipole spin-down given above neglect the enhanced angular momentum losses due to neutrino-driven mass loss, which are important at early times after the magnetar forms Metzger *et al.* (2011). Nevertheless, these expressions reasonably approximate the spin-down of very highly magnetised neutron stars of most relevance in this paper. Isotropic emission is also a reasonable assumption for relatively powerful magnetar winds, since (unlike following the collapse of a massive star) the magnetar outflow cannot be confined efficiently by the relatively small quantity of surrounding material expected following a binary merger.

3. Results

Of the 28 SGRBs fitted, 21 provide a good or possible fit to the magnetar model (75%) while the other 8 provide poor fits, although in some cases this may simply be due to insufficient data. Example fits are shown in Fig. 1 for GRB 100702A and GRB090510. GRB 100702A shows a sharp drop which is consistent with collapse of the magnetar to a BH (after 167s in this case). For GRB 090510, the fit does not require collapse to a BH so we class that as an object where the magnetar survives. Among the 21 SGRBs with good/possible fits, 8 collapse to a BH (38%) within the first few hundred seconds.

The derived magnetic field strengths and initial spin periods are shown in Fig. 2 for the 28 SGRBs fitted. All objects are to the right of the shortest allowed spin period. All of the magnetar candidates are in the $10^{15}\mathrm{G} \leqslant B \leqslant 10^{17}\mathrm{G}$ region, but some are rotating relatively slowly, particularly where they do not require a collapse to a BH.

4. Gravitational wave signals

If the magnetar model we propose here is correct, gravitational wave signals may be detectable from all three stages the systems can be in: inspiral to form magnetar, spin down and collapse to BH. In Table 1 we give a comparison of the distances to

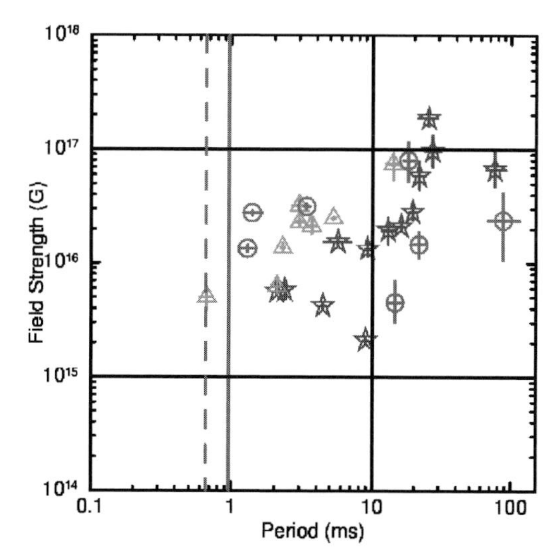

Figure 2. The derived magnetic fields and spin periods. The solid (dashed) vertical lines near 1 ms represent the spin break up periods for a 1.4 $M_\odot$ and 2.1 $M_\odot$ NS respectively (Lattimer & Prakash 2004). The allowed region for an unstable magnetar is assumed such that the initial rotation period needs to be $\leqslant$ 10 ms (Usov 1992) and the magnetic field $10^{15}\mathrm{G} \leqslant B \leqslant 10^{17}\mathrm{G}$ (Thompson 2007). Stars: good fit with a stable magnetar; circles: good fit with an unstable magnetar which collapses to a BH; and triangles: poor fit to the model.

Phase	A-LIGO limit (Mpc)	ET Limit (Mpc)
Inspiral (Abadie *et al.* (2010))	445	5900
Magnetar Spindown (Corsi & Mészáros (2009))	<85	<570
Collapse to BH (Novak (1998))	100	1300

Table 1. Gravitational wave luminosity distance limits for A-LIGO and ET for the different regimes in this magnetar model (based on the predicted amplitudes given in the listed references).

which these signals could be detected by Advanced LIGO (A-LIGO) and the proposed Einstein Telescope (ET) based on the gravitational wave amplitudes discussed in the cited references and assuming a sensitivity of $h \sim 4 \times 10^{-24}$ for A-LIGO and $h \sim 3 \times 10^{-25}$ (Hild *et al.* 2011) for ET.

The chances of a simultaneous electromagnetic and gravitational wave detection are modest for A-LIGO but high for ET. The detection of such multiple gravitational wave signals corresponding to distinct light curve features would be a "smoking gun" test of the magnetar model.

References

Abadie J., *et al.*, 2010, *CQGra*, 27, 173001

Corsi A. & Mészáros P., 2009, *ApJ*, 702, 1171

Duncan R. C. & Thompson C., 1992, *ApJ*, 392, L9

Gehrels N., *et al.*, 2004, *ApJ*, 611, 1005

Gehrels, N., *et al.* 2005, *Nature*, 437, 851

Hild S., *et al.*, 2011, *CQGra*, 28, 094013

Kumar P. & Panaitescu A., 2000, ApJ, 541, L51

Lattimer, J. M. & Schramm, D. N., 1976, *ApJ*, 210, 549

Lattimer J. M. & Prakash M., 2004, *Science*, 304, 536

Lyons N., O'Brien P. T., Zhang B., Willingale R., Troja E., & Starling R. L. C., 2010, *MNRAS*, 402, 705

Metzger B. D., Giannios D., Thompson T. A., Bucciantini N., & Quataert E., 2011, *MNRAS*, 413, 2031

Morrison I. A., Baumgarte T. W., & Shapiro S. L., 2004, *ApJ*, 610, 941

Nousek J. A., *et al.*, 2006, *ApJ*, 642, 389

Novak J., 1998, *Phys. Rev. D*, 57, 4789

O'Brien, P. T., *et al.*, 2006, *ApJ*, 647, 1213

Ozel F., Psaltis D., Ransom S., Demorest P., & Alford M., 2010, *ApJ*, 724, L199

Rowlinson, A., *et al.* 2010, *MNRAS*, 409, 531

Thompson T. A., 2007, *Rev. Mexicana AyA*, 27, 80

Troja E., *et al.*, 2007, *ApJ*, 665, 599

Usov V. V., 1992, *Nature*, 357, 472

Zhang B. & Mészáros P., 2001, *ApJ*, 552, L35

Discussion

CORSI: In the unstable magnetar scenario, where is the afterglow emission and is there any spectral difference during the plateau phase between stable and unstable magnetar cases?

O'BRIEN: We include an underlying power-law component in the fits, which may be the afterglow or the off-axis (curvature) emission. Within the uncertainties, there are no spectral difference

Death of Massive Stars: Supernovae and Gamma-Ray Bursts
Proceedings IAU Symposium No. 279, 2012
P. Roming, N. Kawai & E. Pian, eds.

doi:10.1017/S1743921312013099

Population III Gamma-Ray Burst

Kunihito Ioka, Yudai Suwa, Hiroki Nagakura, Rafael S. de Souza, and Naoki Yoshida

KEK Theory Center, 1-1 Oho, Tsukuba 305-0801, Japan
email: kunihito.ioka@kek.jp

Abstract. Gamma-ray bursts (GRBs) are unique probes of the first generation (Pop III) stars. We show that a relativistic gamma-ray burst (GRB) jet can potentially pierce the envelope of a very massive Pop III star even if the Pop III star has a supergiant hydrogen envelope without mass loss, thanks to the long-lived powerful accretion of the envelope itself. While the Pop III GRB is estimated to be energetic ($E_{\gamma,iso} \sim 10^{55}$ erg), the supergiant envelope hides the initial bright phase in the cocoon component, leading to a GRB with a long duration $\sim$1000$(1+z)$ s and an ordinary isotropic luminosity $\sim 10^{52}$ erg s^{-1} ($\sim 10^{-9}$ erg cm^{-2} s^{-1} at redshift $z \sim 20$), although these quantities are found to be sensitive to the core and envelope mass. We also show that Pop III.2 GRBs (which are primordial but affected by radiation from other stars) occur >100 times more frequently than Pop III.1 GRBs, and thus should be suitable targets for future X-ray and radio missions. The radio transient surveys are already constraining the Pop III GRB rate and promising in the future.

Keywords. gamma rays: bursts, stars: Population III

My talk is mainly based on Suwa & Ioka (2011), Nagakura, Suwa & Ioka (2011), de Souza, Yoshida & Ioka (2011), Ioka & Mészáros (2005), Ioka (2003).

1. Introduction

The ancient era of the first generation stars (Population III; Pop III) – the end of the dark age – is still an unexplored frontier in the modern cosmology. Gamma-Ray Bursts (GRBs) are potentially powerful probes of the Pop III era. The highest redshift of GRB is increasing rapidly, almost approaching $z = 10$ (GRB 050904 at $z = 6.3$, GRB 080913 at $z = 6.7$, GRB 090423 at $z = 8.2$, GRB 090429B at $z = 9.4$). The GRBs are presumed to manifest the gravitational collapse of a massive star – a collapsar – to a black hole with an accretion disk, launching a collimated outflow (jet) with a relativistic speed. The massive stars quickly die within the Pop III era. The GRBs, the most luminous objects in the Universe, are detectable in principle out to redshifts $z \sim 100$, while their afterglows are observable up to $z \sim 30$ (Ioka & Mészáros 2005, Ioka 2003).

The first stars are predicted to be predominantly very massive $\gtrsim 100 M_{\odot}$ (Abel, Bryan & Norman 2002, Bromm, Coppi, & Larson 2002). The central part collapses first to a tiny ($\sim 0.01 M_{\odot}$) protostar, followed by the rapid accretion of the surrounding matter to form a massive first star (Omukai & Palla 2003, Yoshida, Omukai, & Hernquist 2008). The stars with 140–260$M_{\odot}$ are expected to undergo the pair-instability supernovae without leaving any compact remnant behind, while those above $\sim 260 M_{\odot}$ would collapse to a massive ($\sim 100 M_{\odot}$) black hole with an accretion disk, potentially leading to scaled-up collapsar GRBs (Komissarov & Barkov 2010, Mészáros & Rees 2010, Suwa & Ioka 2011, Nagakura, Suwa & Ioka 2011). The Pop III GRB rate would be rare ~ 0.1–10 yr^{-1} but within reach (de Souza, Yoshida & Ioka 2011). These GRBs also mark the formation of the first black holes, which may grow to supermassive black holes (BHs) via merger or accretion.

However, the zero-metal stars could have little mass loss by the line driven wind, and thereby have a large ($R_* \sim 10^{13}$ cm) hydrogen envelope at the end of life (red supergiant (RSG) phase). Especially for Pop III stars, the mass accretion continues during the main sequence phase, so that the chemically homogeneous evolution induced by rapid rotation might not work (Ohkubo *et al.* 2009). Their extended envelopes may suppress the emergence of relativistic jets out of their surface even if such jets were produced (Matzner 2003). The observed burst duration $T \sim 100$ s, providing an estimate for the lifetime of the central engine, suggests that the jet can only travel a distance of $\sim cT \sim 10^{12}$ cm before being slowed down to a nonrelativistic speed. This picture is also supported by the nondetections of GRBs associated with type II supernovae. Nevertheless, this may not apply to the Pop III GRBs because the massive stellar accretion could enhance the jet luminosity and duration and therefore enable the jet to break out the first stars.

We discuss the jet propagation in the first stars using the Pop III stellar structure to estimate the jet luminosity via accretion and to predict the observational main characters of the Pop III GRBs, such as energy and duration. The stellar structure also determines the jet head velocity inside the star. We treat both the jet luminosity and its penetrability with the same stellar structure consistently for the first time.

2. Progenitor structure and jet propagation

We employ three representative progenitors in Fig. 1. Red line shows the density profile of Pop III star with 915 $M_\odot$ (model Y-1 of Ohkubo *et al.* 2009). Blue indicates the GRB progenitor with 16 $M_\odot$. Green line represents the progenitor of ordinary core-collapse supernovae with 15 $M_\odot$. The density profiles are roughly divided into two parts: core and envelope. The GRB progenitor (WR star) does not have hydrogen envelope, while Pop III and RSG keep their envelope so that these stars experience the envelope expansion triggered by core shrinkage after the main sequence.

We can calculate the accretion rate, $\dot{M}$, using these density profiles. The accretion timescale of matter at a radius r to fall to the center of the star is roughly equal to the free-fall timescale, $t_{ff} \approx \sqrt{r^3/GM_r}$. Then we can evaluate the accretion rate at the

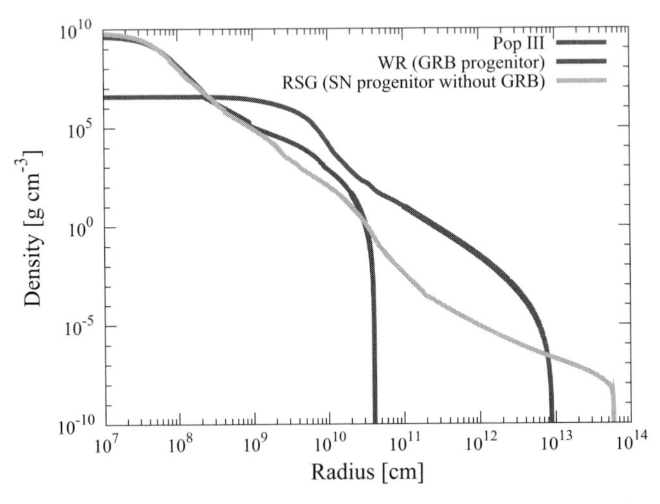

Figure 1. Density profiles of investigated models. Red, blue, and green lines correspond to Pop III star ($M = 915M_\odot$), Walf-Rayet star (WR; GRB progenitor, $M = 16M_\odot$), and red supergiant (RSG; SN progenitor without GRB, $M = 15M_\odot$), respectively. Pop III and RSG have a hydrogen envelope, which expands to a large radius, while WR has only a core.

center as $\dot{M} = dM_r/dt_{ff}$. The rotation law inside the star as well as the jet production mechanism are unknown so that we introduce an efficiency parameter to connect the (free-fall) mass accretion rate and jet luminosity, $\eta = L_j/\dot{M}c^2$, which will be normalized by the observed GRBs.

The jet head velocity $c\beta_h$ inside the star is determined by the pressure balance between the jet ($\sim L_j/\pi\theta_j^2 r^2 c$) and the envelope ($\sim \rho c^2 \beta_h^2$). The shocked jet and shocked envelope go sideways and become the cocoon component (Matzner 2003, Ioka *et al.* 2011).

Figure 2 shows our results. For the Pop III star, the core accretion ends at a few seconds, which is hidden by the stellar envelope. However the envelope accretion still continues and the jet finally breaks out the stellar envelope. Therefore we conclude that the GRB jet can break out the Pop III first star. The key is the envelope accretion that can enable the jet breakout from the envelope itself. Note that we confirm that the jet cannot break out of a RSG, which is consistent with the nondetections of GRBs associated with type II supernovae.

3. Characteristic of Pop III GRB

From Fig. 2, we can find that the duration of Pop III GRB at redshift z is

$$T_{\rm GRB} = T_{90}(1+z) \approx 30000 \text{ s} \left(\frac{1+z}{20}\right), \tag{3.1}$$

which is much longer than the canonical duration of GRBs, ~ 20 s. The total isotropic-equivalent energy of Pop III GRB is

$$E_{\gamma,\rm iso} = \varepsilon_\gamma E_{\rm iso} \approx 1.2 \times 10^{55} \left(\frac{\varepsilon_\gamma}{0.1}\right) \text{ erg}, \tag{3.2}$$

where ε_γ is the conversion efficiency from the jet kinetic energy to gamma rays. It should be noted that this value is comparable to the largest $E_{\gamma,\rm iso}$ ever observed, $\approx 9 \times 10^{54}$ erg

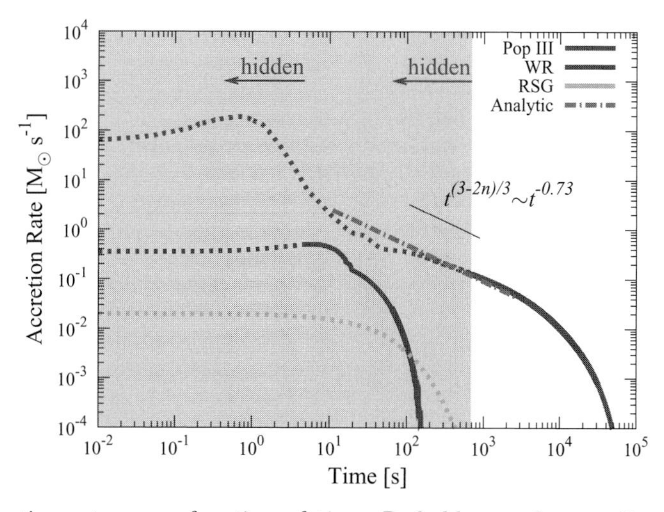

Figure 2. Accretion rates as a function of time. Red, blue, and green lines show Pop III, WR, and RSG, respectively. Dotted regions represent the jet propagating inside the star, while the solid regions correspond to the time after the jet breakout. Solid lines give information of observables (e.g., duration and energetics of GRB). On the other hand, dotted regions show the hidden energy inside the star that goes into the nonrelativistic cocoon component. The gray dot-dashed line represents the analytic model.

for GRB 080916C. Since the large isotropic energy is stretched over the long duration, the expected flux just after the breakout is not so bright,

$$F = \frac{\varepsilon_\gamma L_{\rm iso}}{4\pi r_L^2} \sim 10^{-9}\ {\rm erg\ cm^{-2}\ s^{-1}}, \tag{3.3}$$

which is smaller than the *Swift* Burst Array Telescope (BAT) sensitivity, $\sim 10^{-8}$ erg cm^{-2} s^{-1}. However, there must be a large variety of the luminosity as ordinary GRBs so that more luminous but rare events might be observable by BAT.

The above discussions strongly depend on the envelope mass because the stellar radius is highly sensitive to the envelope mass. We derive analytical dependences on the model parameters in Suwa & Ioka (2011), Nagakura, Suwa & Ioka (2011). The matter entrainment from the envelope is also crucial for the fireball dynamics and the GRB spectra (Ioka 2010). Since the envelope of the Pop III star is different from that of present-day stars, the GRB appearance is also likely distinct from the observed ones (Ioka *et al.* 2011). These are interesting future problems.

The mass of Pop III star could be much smaller than $\sim 1000 M_\odot$ down to $\sim 40 M_\odot$ if we consider the photoevaporation of the accretion disk, the disk fragmentation, or the Pop III star formation from once-ionized gas (i.e., Pop III.2 star). The jet breakout is also possible for these light Pop III stars as shown by Nagakura, Suwa & Ioka (2011).

We also calculate the Pop III GRB rate, and find that Pop III.2 GRBs (which are primordial but affected by radiation from other stars) occur > 100 times more frequently than Pop III.1 GRBs, and thus should be suitable targets for future GRB missions. The radio afterglows of Pop III GRBs are observable up to $z \sim 30$ (Ioka & Mészáros 2005, Ioka 2003). Interestingly, the radio transient searches are already constraining the Pop III GRB rate. Future surveys by EVLA, LOFAR and SKA are promising.

References

Abel, T., Bryan, G. L., & Norman, M. L. 2002 *Science*, 295, 93
Bromm, V., Coppi, P. S., & Larson, R. B. 2002 *ApJ*, 564, 23
de Souza, R. S., Yoshida, N., & Ioka, K. 2011 *A&A*, 533, A32
Ioka, K. 2003, *ApJ*, 598, L79
Ioka, K. & Mészáros, P. 2005 *ApJ*, 619, 684
Ioka, K. 2010 *Prog. Theor. Phys.*, 124, 667
Ioka, K., Ohira, Y., Kawakana, N., & Mizuta, A. 2011 *Prog. Theor. Phys.*, 126, 555
Komissarov, S. S., & Barkov, M. V. 2010 *MNRAS*, 402, L25
Matzner, C. D. 2003 *MNRAS*, 345, 575
Mészáros, P. & Rees, M. J. 2010 *Apj*, 715, 967
Nagakura, H., Suwa, Y., & Ioka, K. 2011 *arXiv*:1104.5691
Ohkubo, T., Nomoto, K., Umeda, H., Yoshida, N., & Tsuruta, S. 2009 *Apj*, 706, 1184
Omukai, K., & Palla, F. 2003 *Apj*, 589, 677
Suwa, Y., & Ioka, K. 2011 *ApJ*, 726, 107
Yoshida, N., Omukai, K., & Hernquist, L. 2008 *Science*, 321, 669

Discussion

QUESTION: How about a binary Pop III GRB?

KUNIHITO IOKA: We study a GRB from a single Pop III star. However our analytical criteria for the jet breakout are also useful for the binary case.

QUESTION: What is the detection rate of Pop III GRBs by X-ray?

KUNIHITO IOKA: ~ 1/yr at most. Please see de Souza, Yoshida & Ioka (2011) for details.

Death of Massive Stars: Supernovae and Gamma-Ray Bursts
Proceedings IAU Symposium No. 279, 2012
P. Roming, N. Kawai & E. Pian, eds.

doi:10.1017/S1743921312013105

Formation and evolution of black hole and accretion disk in collapse of massive stellar cores

Yuichiro Sekiguchi

Yukawa Institute for Theoretical Physics, Kyoto University, Kyoto 606-8502, Japan
email: sekig@yukawa.kyoto-u.ac.jp

Abstract. We describe the results of our numerical simulations of the collapse of a massive stellar core to a BH, performed in the framework of full general relativity incorporating finite-temperature equation of state and neutrino cooling. We adopt a 100 $M_{\odot}$ presupernova model calculated by Umeda & Nomoto (2008), which has a massive core with a high value of entropy per baryon. Changing the degree of rotation for the initial condition, we clarify the dependence of the outcome on this. When the rotation is rapid enough, the shock wave formed at the core bounce is deformed to be a torus-like shape. Then, the infalling matter is accumulated in the central region due to the oblique shock at the torus surface, hitting the hypermassive neutron star (HMNS) and dissipating the kinetic energy there. As a result, outflows can be launched. The HMNS eventually collapses to a BH and an accretion torus is formed around it. We also found that the evolution of the BH and torus depends strongly on the rotation initially given.

Keywords. Stellar core collapse, black hole, accretion disk, gamma-ray burst

1. Introduction

The observational associations (for a review, see Woosley & Bloom (2006)) between LGRBs and supernovae has provided strong support to the so-called collapsar model (Woosley (1993); MacFadyen & Woosley (1999)). In the collapsar model, a central core of a massive star is required to be rotating rapidly enough that a massive accretion disk can be formed around a BH. The observational association of LGRBs with Type Ic (and b) SNe and the requirement that the relativistic jets have to reach the stellar surface (Zhang & Woosley (2004)) raise the serious problem that according to stellar evolution calculations, it is very difficult to produce pre-collapse cores which satisfy both the requirement of the collapsar model and the association of Type Ib/c SNe, if magnetic torques and standard mass-loss rates are taken into account (Woosley & Heger (2006)).

To resolve the above dilemma, several models have been proposed (see Fryer *et al.* (2007) for a review). All of the proposed progenitor models of LGRBs are anomalous in the sense that they are different from the progenitors of ordinary SNe (see Sekiguchi & Shibata (2011) for a discussion). Qualitatively speaking, LGRB progenitor cores may be modeled by a rapidly rotating, higher-entropy core, regardless of their formation processes. Based on this assumption, we performed simulations of a massive stellar core with higher values of entropy collapsing to a BH.

2. Setting

As a representative model of a high entropy core, we adopt a presupernova core of $100M_{\odot}$ model calculated by Umeda & Nomoto (2008) (hereafter denoted by UN100). The model has an iron core of large mass $M_{\rm core} \approx 3.2M_{\odot}$ and the central value of entropy per baryon is $s \approx 4k_B$, which is much larger than that of an ordinary presupernova

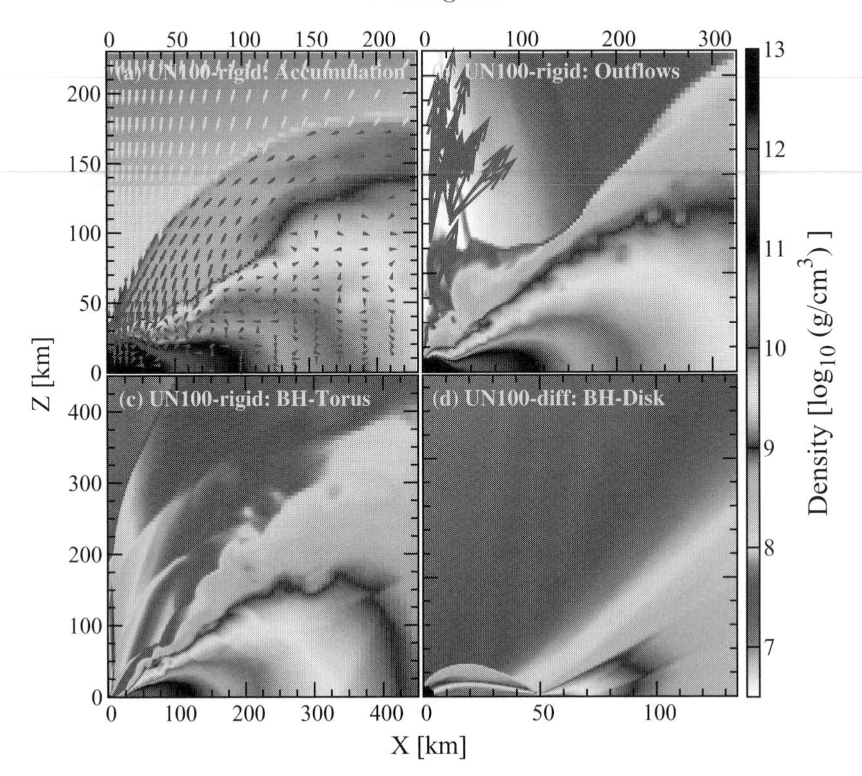

Figure 1. Contours of the rest-mass density in the x-z plane for UN100-rigid at (a) $t = 645$ ms (the mass accumulation onto the HMNS), (b) 1256 ms (the launch of outflows), and (c) 2225 ms (time-variable BH-Torus) , and (d) for UN100-diff at $t = 1150$ ms (quiet BH-Disk).

core for which $s \lesssim 1k_B$. Because the model UN100 is non-rotating, we add rotational profiles according to, $\Omega(\varpi) = \Omega_0 \frac{R_0^2}{R_0^2+\varpi^2}\mathcal{F}_{\rm cut}$, where $\varpi = \sqrt{x^2+y^2}$. We fix the central angular velocity as $\Omega_0 = 1.2$ rad/s and consider two values of R_0; a rigid rotation model ($R_0 = \infty$, referred to as UN100-rigid) and a differential rotation ($R_0 = R_{\rm core}$, referred to as UN100-diff) model†. The cut-off factor $\mathcal{F}_{\rm cut}$ is introduced so that the matter in the outer region would not escape from the computational domain.

The simulations are performed using a full GR code recently developed, Sekiguchi (2011). We assume axial and equatorial symmetries of the spacetime and the so-called Cartoon method (e.g., Shibata (2003)) is adopted. In numerical simulations, we adopt a nonuniform grid, in which the grid spacing is increased according to the rule $dx_{j+1} = (1+\delta)dx_j$, where δ is a constant. In addition, a regridding technique (e.g., Sekiguchi & Shibata (2005)) is adopted. We set an infalling boundary condition at the outer boundary.

3. Results

The dynamics of the collapse in UN100-rigid and UN100-diff is similar to that in the the ordinary SN simulations, until the shock wave formed in the core bounce stalls. However, the dynamics of UN100-diff and UN100-rigid (and the ordinary SN) in the later phase is qualitatively different as follows. Figure 1 shows contour plots of the rest-mass density in the x-z plane at selected time slices for UN100-diff (d) and UN100-rigid (a)–(c). (i) Due to the faster rotation of the outer region in UN100-rigid, the shock wave is deformed to be a torus-like configuration (see Fig. 1(a)), which is one of the

† We note that Ω_0 is much smaller than the Kepler value $(M_{\rm core}/R_{\rm core}^3)^{1/2} \approx 5.2$ rad/s.

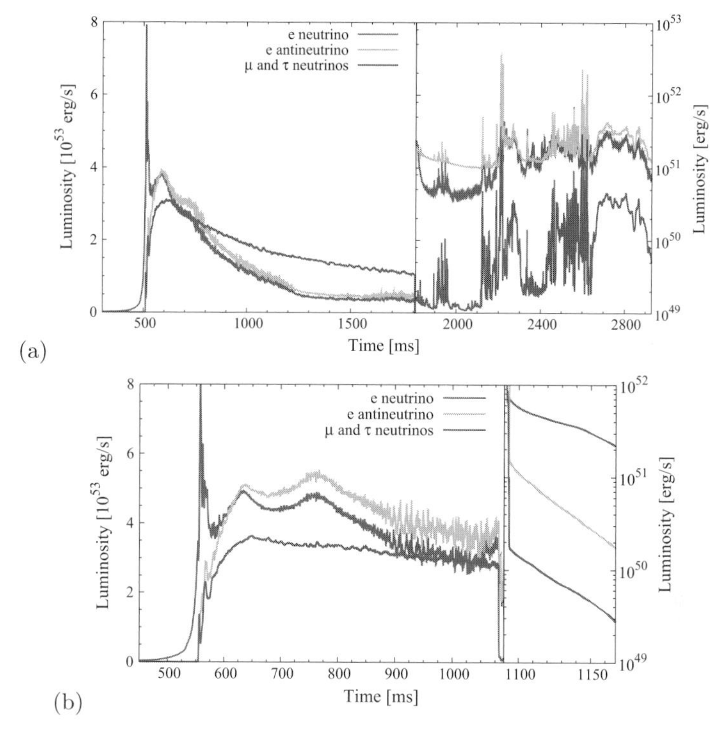

Figure 2. Time evolution of neutrino luminosities for (a) UN100-rigid and (b) UN100-diff.

characteristic features in UN100-rigid. (ii) At the shock, the kinetic energy associated with the motion perpendicular to the shock surface is dissipated but that associated with the parallel component is preserved. This implies that the infalling material is eventually accumulated in the central region and its kinetic energy is dissipated at the surface of the hypermassive neutron star (HMNS). During this process, oscillations of the HMNS are excited as the infalling matter hits it and the shock waves gain the thermal energy via PdV work and propagate outward. (iii) Due to the accumulation of the matter onto the HMNS and the resulting shock heating, outflows are launched from the polar surface of the HMNS, forming shocks (see Figs. 1(b)). Note that in the present code, neutrino heating is not taken into account and exploring the fate of the thermally driven outflows in the presence of the neutrino heating is an interesting subject. We plan to pursue this issue in the near future. (iv) We found, as another novel feature of dynamics, that the BH-torus system shows time variability (see Figs. 1(c)). Such a time variability has not been seen in UN100-diff (see Figs. 1(d)).

Figures 2(a) and (b) show the time evolution of neutrino luminosities. The neutrino luminosities shows a precipitation when the BH is formed. The total neutrino luminosity emitted from the HMNS and from the torus around the BH amounts to $L_{\nu,\mathrm{tot}} \sim 10^{53}$ and $\sim 10^{51}$ ergs/s, respectively. By contrast with the case of UN100-diff, the neutrino luminosities show a violent time variability which is maintained for $\gtrsim 1$ s. Such a long-term high luminosity and a time variability may be associated with the time variability that LGRBs show. *It is remarkable that the above qualitative differences in dynamics between UN100-diff and UN100-rigid stem from a small difference in the initial angular velocity profile in the outer region* (see Fig. 3). Taking into account the dependence of the neutrino pair annihilation rate on the geometry of the torus $\dot{E}_{\nu\bar{\nu}}$ would be given by Beloborodov

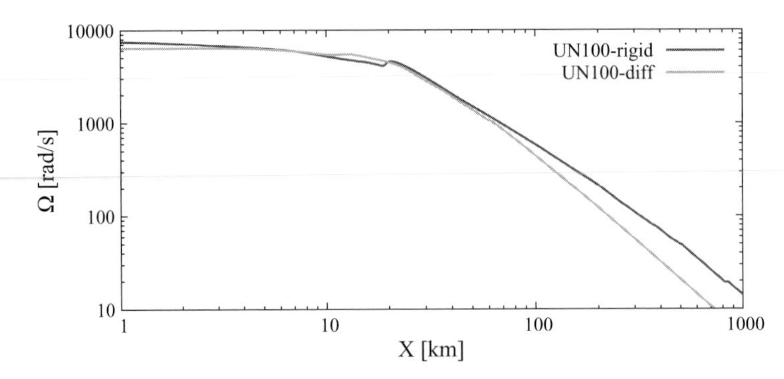

Figure 3. Profiles of the rotational angular velocity along the equator just before the BH formation for UM100-diff (green curve) and UM100-rigid (red curve).

(2008), $\dot{E}_{\nu\bar{\nu}} \sim 10^{48}$ ergs/s $\left(\frac{100\,\mathrm{km}}{R_{\mathrm{fun}}}\right)\left(\frac{0.1}{\theta_{\mathrm{fun}}}\right)^2\left(\frac{E_\nu + E_{\bar{\nu}}}{10\,\mathrm{MeV}}\right)\left(\frac{L_\nu}{10^{51}\,\mathrm{ergs/s}}\right)\left(\frac{L_{\bar{\nu}}}{10^{51}\,\mathrm{ergs/s}}\right)\sin^2\Theta$, where R_{fun} and θ_{fun} are the characteristic radius and the opening angle of the funnel region. Θ denotes the collision angle of the neutrino pair. Thus a low-luminosity LGRB could be explained. In the HMNS phase, by contrast, the neutrino luminosity is huge as $L_\nu \gtrsim 10^{53}$ ergs/s and hence, the deposition rate would be very large as $\dot{E}_{\nu\bar{\nu}} \sim 3 \times 10^{52}$. If the outflows launched due to the mass accumulation mechanism can penetrate the stellar envelope, a system composed of a long-lived HMNS and a geometrically thick torus may be a promising candidate of the central engine of GRBs of relatively short duration.

Future prospects: Recently, we have developed a formulation of general relativistic radiation transfer (Shibata *et al.* (2011)). We have already performed general relativistic radiation magnetohydrodynamics (GRRMHD) simulations for the evolution of a system composed of a BH and a surrounding torus with a simplified treatment of microphysics (Shibata & Sekiguchi (2012)). Furthermore, we have succeeded in implementing a code which can solve the neutrino transfer with a detailed microphysics (in preparation). Using this code, we plan to perform simulations of the stellar core collapse to explore a SN explosion mechanism and the formation of a BH in full general relativity.

YS thanks H. Umeda for providing us the presupernova model (UN100) adopted in this work. Numerical simulations were performed on SR16000 at YITP of Kyoto University, on SX9 and XT4 at CfCA of NAOJ, and on the NEC SX-8 at RCNP in Osaka University. This work was supported by Grant-in-Aid for Scientific Research (21018008, 21105511, 23740160) and HPCI Strategic Program of Japanese MEXT.

References

Beloborodov, A. M. 2008, *AIP Conf. Proc.* 1054, 51
Fryer, C. L., *et al.* 2007, *PASP*, 119, 1211
Sekiguchi, Y. & Shibata, M. 2005, *Phys. Rev. D*, 71, 084013
Sekiguchi, Y. & Shibata, M. 2011, *ApJ*, 737, 6
Shibata, M. 2003, *Phys. Rev. D*, 67, 024033
Shibata, M. & Sekiguchi, Y. 2012, *Prog. Theor. Phys.*, 127, 535
Shibata, M., *et al.* 2011, *Prog. Theor. Phys.*, 125, 1255
Umeda, H. & Nomoto, K. 2008, *ApJ*, 673, 1014
Woosley, S. E. 1993, *ApJ*, 405, 273
Woosley, S. E. & Bloom, J. S. 2006, *ARAA*, 44, 507
Woosley, S. E. & Heger, A. 2006, *ApJ*, 637, 914
Zhang, W. & Woosley, S. E. 2004, *ApJ*, 608, 365

Death of Massive Stars: Supernovae and Gamma-Ray Bursts
Proceedings IAU Symposium No. 279, 2012
P. Roming, N. Kawai & E. Pian, eds.

doi:10.1017/S1743921312013117

Concluding Remarks

Elena Pian[1], Nobuyuki Kawai[2], and Peter W. A. Roming[3]

[1]Scuola Normale Superiore di Pisa, Piazza dei Cavalieri 7, I-56126 Pisa, Italy
email: elena.pian@sns.it

[2]Department of Physics, Tokyo Institute of Technology, 2-12-1 Ookayama, Meguro-ku, Tokyo 152-8551, Japan

[3]Southwest Research Institute, Space Science & Engineering Division, PO Drawer 28510, San Antonio, TX 78228-0510, USA

Abstract. At the end of IAU Symposium 279, Shri Kulkarni delivered the concluding remarks. This paper presents a summary of his comments as interpreted by the Chairs of the Science Organizing Committee.

Keywords. supernovae: general, gamma rays: bursts

This symposium confirmed the existing general consensus that stellar deaths do not represent an incomplete chapter, but rather an unfinished book. Not only are we not clear about stellar paths that lead to certain classes of progenitors, but the ingredients that take part in this evolution - such as mass loss, angular momentum, magnetic fields - are also obscure, debated and controversial. The core-collapse progenitor "spectrum" defines a nice sequence that goes from red/blue supergiants and Wolf-Rayet stars, all presumably ending their lives as neutron stars (NS); to Type IIn SNe and massive stripped stars, probably producing black holes (BH); to the most luminous known explosions, superluminous supernovae, likely proceeding from massive stars with large radii; and finally pair-instability supernovae, thermonuclear explosions originating from the most massive stars in the Universe (more than $100\,M_\odot$). While the boundary between white dwarfs and neutron stars is rather well determined, the gap between neutron stars and black holes is still debated, and may hide a variety of compact objects - described by a range of equations of state - that may account for the diversity we see in the electromagnetic display of stripped envelope supernovae (SNe) and gamma-ray bursts (GRBs).

During this conference, we saw the clear emergence of numerous questions and controversies such as:

- What (rare) type of massive stars end their lives as GRBs?
- What is the mass spectrum of Population III stars ($40\,M_\odot$ versus $400\,M_\odot$)
- What factors produce NS versus BH?
- What factors genreate NS versus magnetars?
- Are stellar collisions important for certain outcomes?
- Are there some long GRBs without SNe?
- Is there a fundamental difference between GRB 980425 - the close-by (35 Mpc) prototype of GRB-SN association - and classical GRBs, i.e. those at "cosmological" redshifts?

In parallel, other questions and puzzles were raised with answers that are very likely to be complex in nature:

- What factors determine mass loss rates?

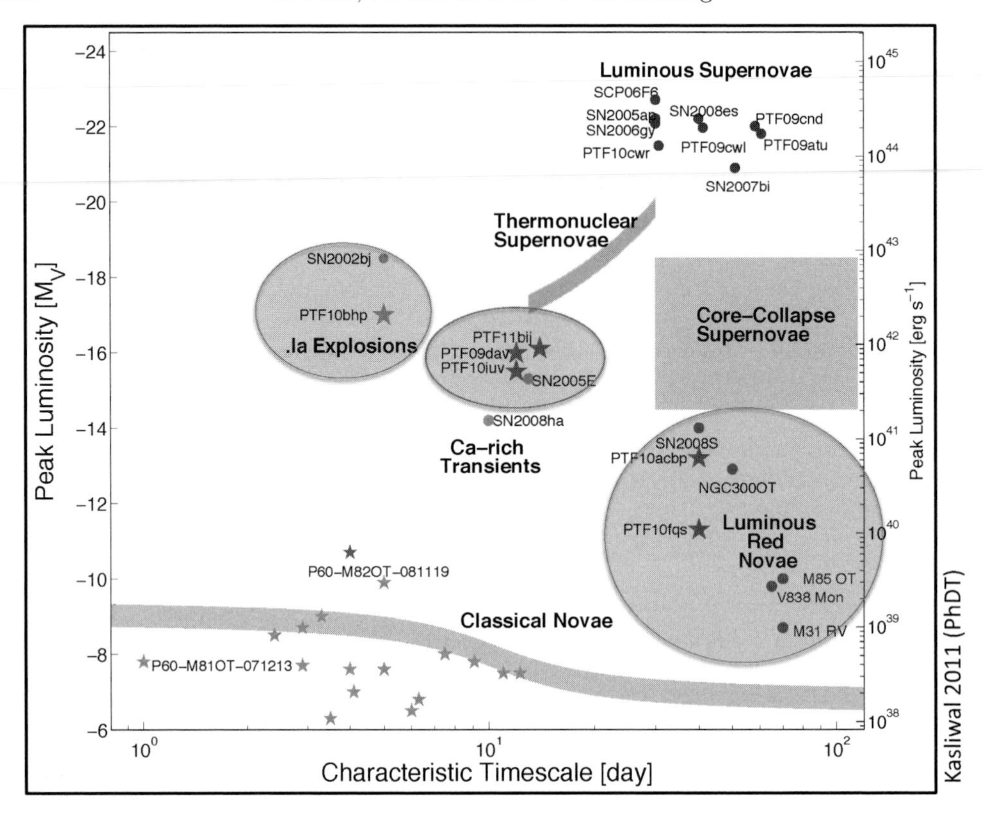

Figure 1. The transient zoo from Kasliwal (2011). Reprinted with permission.

- How does the environment shape the IMF?
- What is the role of metallicity?
- What determines the retaining (or radiating) of angular momentum?
- Do (slowly rotating) BH outcomes produce detectable SNe?
- The explosion mechanism(s) issue for core-collapse SNe is rapidly escaping a general approach, because numerous peculiarities arise. The behavior of SN 2010jp was a case in point. Are there many cases which require bipolar explosion mechanism?
- The "Christmas burst", GRB 101225A, characterized by an early prominent thermal component and possibly related to the explosion of a helium star in a binary system, is a clear reminder that we are wandering in a heterogeneous zoo, still without solid diagnostic tools at hand to recognize and interpret diversity and variety (see Figure 1).
- How is energy carried in long GRBs?
- Are relativistic jets Poynting dominated?
- What determines the jet opening angles?
- Are ultra-high energy cosmic rays produced by stellar deaths?
- What re-ionized the early Universe?

Recently, the importance of the standing accretion shock instability (SASI) has been recognized as a fundamental ingredient (although probably not necessary, but certainly sufficient) to create asymmetry in the explosion. This was nicely simplified and represented during the conference in a shallow water analogue that explains the basic properties of the phenomenon by keeping it close to our "everyday life" experience and common sense.

Zooming out from explosion sites, we saw many attempts at investigating stellar deaths through the analysis of their environments: star forming regions and host galaxies. Cumulative distributions of light in galaxies provide hints toward identifying similarities and differences in GRB and SNe progenitors, and clearly GRB hosts are akin to those of Type Ic SNe, the most stripped type of core-collapse supernova, i.e. associated with the death of a massive star core that has lost both its hydrogen and helium envelopes. While this is not completely unexpected (because whenever a supernova is identified in association with a GRB or XRF, this is usually of Type Ic), it suggests that Type Ic SNe may be the true parent population of long GRBs. A number of them are likely the most energetic and perhaps more will be identified that are able to produce "engines", i.e. jets. How we can distinguish the signature of these engines (which are expected to have non-thermal spectra, but may occasionally have a thermal component leading to a cocoon) from that of shock breakout emission (the classical blast wave following collapse stalling and rebouncing) is one of the current challenges of exploration in this field.

A frontier of the observational approach is represented by polarimetry at high energies (the detection of gamma-ray polarimetry in GRB 100826A by the Japanese experiment IKAROS/GAP is pioneering in this sense) and non-electromagnetic messengers (neutrinos, cosmic rays, gravitational waves).

Finally, particular attention deserves the issue of metallicity: massive stars exist locally (examples abound in our own Galaxy and in the Magellanic Clouds, in isolated clusters or in large star-forming complexes), and we know beyond a doubt that their final fate depends on metallicity. How exactly this parameter governs and determines the explosion and how prominently it shapes the small and large scale environment of the GRB is a matter of hot debate, because the observational information is still too limited to allow a satisfactory quantitative analysis that can provide objective conclusions.

Reference

Kasliwal, M. M. 2011, Bridging the Gap: Elusive Explosions in the Local Universe, PhD Dissertation, California Institute of Technology (http://resolver.caltech.edu/CaltechTHESIS:05162011-09434522)

Death of Massive Stars: Supernovae and Gamma-Ray Bursts
Proceedings IAU Symposium No. 279, 2012
P. Roming, N. Kawai & E. Pian, eds.

doi:10.1017/S1743921312013130

Searching for X-ray counterparts of Fermi Gamma-ray pulsars in Suzaku observations

Yu Aoki[1], Takahiro Enomoto[1], Yoichi Yatsu[1], Nobuyuki Kawai[1], Takeshi Nakamori[2], Jun Kataoka[2] and P. Saz Parkinson[3]

[1]Dept. of Physics, Tokyo Institute of Technology,
B2-12-1 Ookayama Meguro Tokyo, 152-8551, Japan
email: aoki@hp.phys.titech.ac.jp

[2]Research Institute for Science and Engineering, Waseda University, Japan

[3]Santa Cruz Institute for Particle Physics, Dept. of Physics and Dept. of Astronomy and Astrophysics, University of California at Santa Cruz, Santa Cruz, CA 95064, USA

Abstract. We report the Suzaku follow-up observations of the Gamma-ray pulsars, 1FGL J0614,13328, J1044.55737, J1741.82101, and J1813.31246, which were discovered by the Fermi Gamma-ray observatory. Analysing Suzaku/XIS data, we detected X-ray counterparts of these pulsars in the Fermi error circle and interpreted their spectra with absorbed power-law functions. These results indicate that the origin of these X-ray sources is non-thermal emission from the pulsars or from Pulsar Wind Nebulae (PWNe) surrounding them. Moreover we found that J1741.82101 exhibits a peculiar profile: spin-down luminosity vs flux ratio between X- and gamma-rays is unusually large compared to usual radio pulsars.

Keywords. Pulsar, Pulsar wind nebulae, Gamma-ray pulsars. Fermi.

---

1. Introduction

The Fermi gamma-ray observatory has discovered more than three thousands of gamma-ray sources. About 10 percent of the newly found objects were categorized into pulsars and pulsar wind nebulae. We focused on the newly found gamma-ray pulsars to study their nature. For this purpose we observed four bright gamma-ray pulsars with the Suzaku X-ray observatory. Here we present a summary of the observations.

2. Observation and analysis

Suzaku is the fifth Japanese X-ray observatory with 4 X-ray telescopes. Thanks to the low-earth orbit the particle background is low compared with the other large X-ray observatories. So it is suited for searching diffuse emissions in the universe. We conducted observations on 2010. Each exposure time was 20 ks.

First we performed astrometry. From the obtained X-ray images we discovered point sources clearly. These locations are consistent with the Fermi's error circles. The position of these point sources are also consistent with the timing position determined by radio follow-up observations. These positional coincidences strongly support that the discovered X-ray sources are the X-ray counterparts of gamma-ray pulsars.

Next we studied the spectroscopy. In order to accumulate source photons we chose circle regions with a radius of 1 arcmin centered on the pulsars. The obtained X-ray spectra were well modelled by absorbed power-law functions. The results of spectral fitting are summarized in Table. 1. 1FGL J1813,3-1246 shows a very flat spectrum with a photon index of 0.8 that cannot be explained by the standard acceleration model. While the other pulsars show relatively flat spectra.

Table 1. Fit results with power-law model or interstellar absorbed power-law model for the spectra of four pulsars

name	n_H [$\times 10^{22}$cm^2]	Index	Flux(0.5-10.0keV) [erg/s/cm^2]	χ^2/D.O.F
J0614.1-3328	–	$2.63^{+0.30}_{-0.27}$	$5.73 \pm 0.67 \times 10^{-14}$	11.95/14
J1044.5-5737	–	$2.00^{+0.36}_{-0.34}$	$7.50^{+1.22}_{-1.12} \times 10^{-14}$	5.56/8
J1741.8-2101	$0.28^{+0.11}_{-0.09}$	$2.81^{+0.23}_{-0.20}$	$6.56^{+1.35}_{-0.97} \times 10^{-13}$	7.44/24
J1813.3-1246	$1.36^{+0.31}_{-0.27}$	0.79 ± 0.16	$1.27 \pm 0.06 \times 10^{-12}$	9.25/20

3. Discussion

We estimate the X-ray luminosity from the X-ray flux for each pulsar using equation (1) which is the empirical formula between X-ray luminosity and spindown luminosity (Kanai D-thesis).

$$\frac{L_X}{10^{32}\text{erg/s}} = (2.99 \pm 1.08) \times \left(\frac{L_{sd}}{10^{36}\text{erg/s}}\right)^{1.10\pm0.14} \tag{3.1}$$

Then we calculated the distance and Gamma-ray luminosity of the four pulsars assuming $L_{X,\Gamma} = 4\pi d^2 f F_{X,\Gamma}$ from obtained X-ray flux, X-ray luminosity, and Gamma-ray flux.

Next, we compared the four pulsars to other Fermi pulsars used in Kanai D-thesis. Fig. 1 and 2 show the spin-down luminosity vs. Gamma-ray luminosity and spin-down luminosity vs. the X/Gamma flux ratio, respectively.

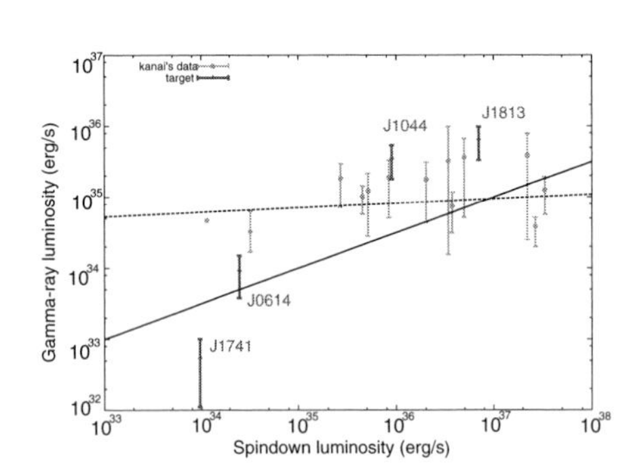

Figure 1. Relation between spindown luminosity and Gamma-ray luminosity for four pulsars and other Fermi pulsars

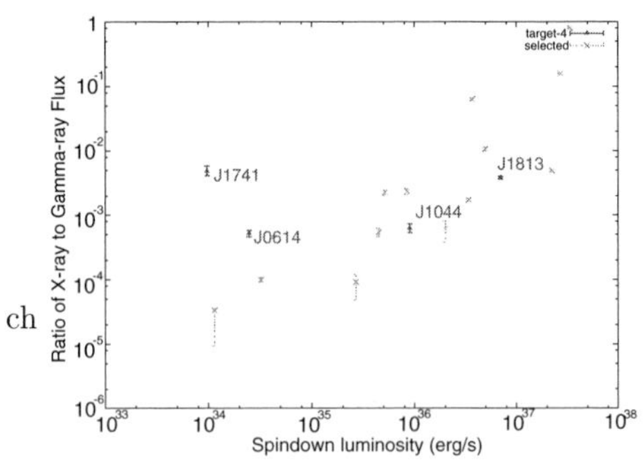

Figure 2. Relation between spindown luminosity and the ratio of X-ray and Gamma-ray flux for four pulsars and other Fermi pulsars

References

Amari, S., Hoppe, P., Zinner, E., & Lewis R. S. 1995, *Meteoritics*, 30, 490

Anders, E. & Zinner, E. 1993, *Meteoritics*, 28, 490

Bernatowicz, T. J., Messenger, S., Pravdivtseva, O., Swan, P., & Walker, R. M. 2003, *Geochim. Cosmochim. Acta*, 67, 4679

Busso, M., Gallino, R., & Wasserburg, G. J. 1999, *ARAA*, 37, 239

Croat, T. K., Stadermann, F. J., & Bernatowicz, T. J. 2005, *ApJ*, 631, 976

Draine, B. T. 2003, *ARAA*, 41, 241

Hoppe, P. & Zinner, E. 2000, *J. Geophys. Res.*, A105, 10371

Hoppe, P., Ott, U., & Lugmair, G. W. 2004, *New Astron. Revs*, 48, 171

Kanai.Y., 2010, Ph.D. thesis, Tokyo Institute of Technology

Lodders, K. & Fegley, B. 1998, *Meteorit. Planet. Sci.*, 33, 871

Meyer, B. S., Clayton, D. D., & The, L.-S. 2000, *ApJ* (Letters), 540, L49

Death of Massive Stars: Supernovae and Gamma-Ray Bursts
Proceedings IAU Symposium No. 279, 2012
P. Roming, N. Kawai & E. Pian, eds.

doi:10.1017/S1743921312013142

Temporal Evolution of GRB Spectra: Leptonic and Hadronic

Katsuaki Asano[1] **and Peter Mészáros**[2]

[1]Interactive Research Center of Science, Tokyo Institute of Technology, 2-12-1 Ookayama, Meguro-ku, Tokyo 152-8550, Japan
email: asano@phys.titech.ac.jp

[2]Department of Astronomy & Astrophysics; Department of Physics; Center for Particle Astrophysics; Pennsylvania State University, University Park, PA 16802
email: nnp@astro.psu.edu

Abstract. As the Fermi observatory has revealed, the GRB light curves show variant behaviours in different energy bands. Especially, the onset of GeV emission tend to lag that at lower energy. Various models to explain the GeV-delay, including early afterglow models or hadronic models, have been proposed. We have developed a time-dependent code for emission processes with one-zone approximation. The temporal evolution of GRB spectra is discussed based on leptonic inverse Compton and hadronic cascade models. This offers important predictions for future observations such as CTA.

Keywords. gamma rays: bursts, radiation mechanisms: nonthermal, cosmic rays

---

1. Introduction

Most of the GRB spectra peak around the MeV range, and are well fitted by the usual Band function. In the standard picture, this component is explained by synchrotron emission from accelerated electrons, while alternative models have been proposed such as photospheric emission. The recent detection of GeV photons with *Fermi*-LAT has opened up the possibility of constraining such models. In some objects *Fermi* has also found in the GeV energy range additional spectral components (Abdo *et al.* 2009a,b, Ackermann *et al.* 2011, 2011). Moreover, the onset of the GeV emission tends to be delayed relative to the onset of the main MeV emission (Abdo *et al.* 2009c).

The photon statistics above the GeV range provided by *Fermi* are not sufficient to distinguish between the internal or external origin (Ghisellini *et al.* 2010, Kumar & Barniol Duran 2010) of the high-energy emission. However, it is expected that future multi-GeV observations with atmospheric Cherenkov telescope arrays such as CTA will drastically improve the data quality, owing to their large effective area. Lightcurves and spectral evolution measurements expected from such telescopes should provide critical information on the GRB physics. To discriminate between the emission models, detailed temporal-spectral evolution studies for various situations (e.g. Pe'er 2008, Vurm & Poutanen 2009, Daigne *et al.* 2011) are needed.

2. Leptonic and Hadronic Lightcurves

In our recent studies (Asano & Mészáros 2011, 2012), the temporal-spectral evolution of the prompt emission of GRBs is simulated numerically for both leptonic and hadronic models (e.g. Böttcher & Dermer 1998, Gupta & Zhang 2007, Asano *et al.* 2009a,b, 2010). We consider internal dissipation regions in possible models for explaining the delayed onset of the GeV emission. The numerical code can follow the evolution of the particle energy distributions in a relativistically expanding shell. For sufficiently weak magnetic fields, leptonic inverse Compton (IC) models can reproduce the few seconds delay of the onset of GeV photon emission observed by *Fermi*-LAT, due to the slow growth of the target photon field for IC scattering. However, even for stronger magnetic fields, the GeV delay can be explained with hadronic models, due to the long acceleration timescale of protons and the continuous photo-pion production after the end of the particle injection. While the FWHMs of the MeV and GeV lightcurves are almost the same in one-zone leptonic models, the FWHM of the 1-30 GeV lightcurves in hadronic models are significantly wider than those of the 0.1-1 MeV lightcurves. The amount of the GeV delay depends on the importance of the Klein-Nishina effect in both the leptonic and hadronic models. The amounts of escaped neutrons in our simulations are within the acceptable range for acting as UHECR sources. Since we have adopted large Γ and R values in order to simulate GRBs where GeV photons can escape from the source, the resultant neutrino spectra are hard enough to avoid the current flux limit constraints from IceCube (Abbasi *et al.* 2012). Our hadronic model also predicts a delayed onset of the neutrino emission, which is more pronounced than the corresponding GeV photon delay. If neutrinos are eventually observed, this point can also be directly tested. The quantitative differences in the lightcurves for various models may be further tested with future atmospheric Cherenkov telescopes such as CTA.

References

Abbasi, R. *et al.* 2012, *Nature*, 484, 351
Abdo, A. A. *et al.* 2009a, *Nature*, 462, 331
Abdo, A. A. *et al.* 2009b, *ApJ*, 706 L138
Abdo, A. A. *et al.*, 2009c, *Science*, 323, 1688
Ackermann, M. *et al.*, 2010, *ApJ*, 716, 1178
Ackermann, M. *et al.*, 2011, *ApJ*, 729, 114
Asano, K., Guiriec, S., & Mészáros, P. 2009a, *ApJ*, 705 L191
Asano, K., Inoue, S., & Mészáros, P. 2009b, *ApJ*, 699, 953
Asano, K., Inoue, S., & Mészáros, P. 2010, *ApJ*, 725, L121
Asano, K. & Mészáros, P. 2011, *ApJ*, 739, 103
Asano, K. & Mészáros, P. 2012, in preparation
Böttcher, M. & Dermer, C. D. 1998, *ApJ*, 499, L131
Daigne, F., Bošnjak, Ž., & Dubus, G. 2011, *A&A*, 526, 110
Ghisellini, G. *et al.* 2010, *MNRAS*, 403, 926
Gupta, N. & Zhang, B., 2007, *MNRAS*, 380, 78
Kumar, P. & Barniol Duran, R. 2010, *MNRAS*, 409, 226
Pe'er, A. 2008, *ApJ*, 682, 463
Vurm, I. & Poutanen, J. 2009, *ApJ*, 698, 293

Death of Massive Stars: Supernovae and Gamma-Ray Bursts
Proceedings IAU Symposium No. 279, 2012
P. Roming, N. Kawai & E. Pian, eds.

doi:10.1017/S1743921312013154

Detecting TeV γ-rays from GRBs with km^3 neutrino telescopes

Tri L. Astraatmadja[1,2]

[1]Nikhef – National Institute for Subatomic Physics,
Science Park 105 1098 XG Amsterdam, The Netherlands

[2]LION – Leiden Institute of Physics, Leiden University,
PO Box 9504 2300 RA Leiden, The Netherlands
email: t.astraatmadja@nikhef.nl

Abstract. Observing TeV photons from GRBs can greatly enhance our understanding of their emission mechanisms. Under-sea/ice neutrino telescopes—such as ANTARES in the Mediterranean Sea or IceCube at the South Pole—can also operate as a γ-ray observatory by detecting downgoing muons from the electromagnetic cascade induced by the interaction of the photons with the Earth's atmosphere. Theoretical calculations of the number of detectable muons from single GRB events, located at different redshifts and zenith distances, have been performed. The attenuation by pair production of TeV photons with cosmic infrared background photons has also been included.

Keywords. astroparticle physics, gamma-ray burst: general, elementary particles, methods: analytical

---

The ANTARES neutrino telescope is currently operating in the Mediterranean Sea, 40 km offshore Toulon (France) at a depth of 2475 m. With an instrumented volume of ~ 0.01 km^3, it is the largest neutrino telescope in the Northern Hemisphere. Although optimized to detect upgoing neutrino-induced muons, it can also detect downgoing photon-induced muons and thus operate as a gamma-ray telescope. Because of its large collecting area, wide field of view, and high duty cycle, there is a potential then for ANTARES to detect TeV photons emitted from gamma-ray burst (GRBs).

TeV photons from GRBs are produced from electron Inverse Compton emission or π^0 decay from $p\gamma$ interactions (Asano & Inoue, 2007). Searching for these photons could help us not only in understanding the mechanisms of GRB emission but also in identifying the possible source of Ultra High-Energy Cosmic Rays (UHECR).

Along their path from the source to the Earth, TeV photons interact with ambient IR photons. They annihilate themselves, creating pairs of electron–positron in the process. The transparency of the universe to TeV photons depends on the photons's energy and the distance to the source. Once the surviving TeV photons reach the top of the Earth's atmosphere, they will interact with atmospheric particles and initiate particle showers. Muons are produced from these showers mainly (Halzen *et al.* 2009) through 1) photoproduction, in which TeV photons interact with atmospheric nuclei and produce pions, followed by the decay of the pion into a positive muon and a muon antineutrino; and 2) direct muon-pair production, where muons are created directly via the channel $\gamma + N \rightarrow N + \mu^+ + \mu^-$, where N is a nucleus of the atmosphere.

Muon production through the first channel dies away with increasing energy, but the cross section of muon-pair production increase with photon energy. At TeV regime it is thus the dominant muon-producing channel. As the muons travels downward toward the detector at the bottom of the sea, they lose their energy through ionization and radiative

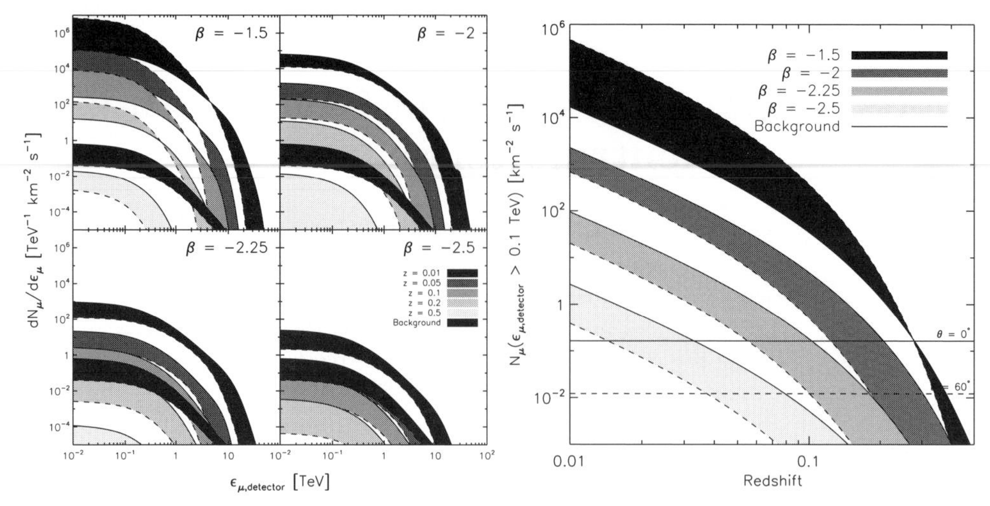

Figure 1. Left: The muon flux at detector depth (2475 m) for single GRBs emitted from different redshifts as indicated by the color coding on the legend. The black horizontal lines show the background flux from cosmic ray-induced muons parametrized by Gaisser (1991), assuming a search cone with an opening angle of 1°. For each color, the muon flux drawn by the dashed-line is the flux from zenith distance $\theta = 60°$ while that drawn by the solid line is the flux straight from the zenith (i.e. $\theta = 0$). The filled-area then defines all the possible flux from all zenith distances between $\theta = 0$ and $\theta = 60°$. Attenuation is determined by using a model by Finke *et al.* (2010). Each square is a plot for different high-energy photon spectral index β of the GRB. **Right:** The number of muons with energies $\epsilon_{\mu,\mathrm{detector}} > 0.1$ TeV per km^2 per seconds for GRB sources with various high-energy photon spectral index β. The muon counts are plotted as a function of the redshift of the GRB. The two horizontal lines are the background levels at $\theta = 0$ and $\theta = 60°$.

processes (Barrett *et al.* 1952). There is thus a minimum energy threshold for the muons to survive its journey to the bottom of the sea: For a vertical depth of 2475 m, the surface energy of the muons must be larger than ~ 1 TeV.

Consequently, only the very nearby GRBs can be detected significantly. To obtain at least 3σ detection significance, a GRB has to be located at redshift $z \lesssim 0.07$ if the detector's muon effective area is $A^{\mu}_{\mathrm{eff}} \sim 0.01$ km^2, or $z \lesssim 0.15$ if the muon effective area is $A^{\mu}_{\mathrm{eff}} \sim 1$ km^2 (Astraatmadja, 2011). The annual probability that such an event will occur (Wang & Dai, 2011) is very small ($P \sim 2 \times 10^{-4}$) but nevertheless it has occured in the past (Butler *et al.* 2007, 2010).

References

Asano, K. & Inoue, S. 2007, *ApJ* 671, 645

Astraatmadja, T. L. 2011, *MNRAS* 418, 1774

Barrett, P. H., Bollinger, L. M., Cocconi, G., Eisenberg, Y., & Greisen, K. 1952, *Rev. Mod. Phys.* 24, 133

Butler, N. R., Kocevski, D., Bloom, J. S., & Curtis, J. L. 2007, *ApJ* 671, 656

Butler, N. R., Bloom, J. S., & Poznanski, D. 2010, *ApJ* 711, 495

Finke, J. D., Razzaque, S., & Dermer, C. D. 2010, *ApJ* 712, 238

Gaisser, T. K. 1991, *Cosmic Rays and Particle Physics*, Cambridge, UK: Cambridge University Press

Halzen, F., Kappes, A., & Ó Murchadha, A. 2009, *Phys. Rev. D* 80, 083009

Wang, F. Y. & Dai, Z. G. 2011, *ApJ* 727, L34

Death of Massive Stars: Supernovae and Gamma-Ray Bursts
Proceedings IAU Symposium No. 279, 2012
P. Roming, N. Kawai & E. Pian, eds.

doi:10.1017/S1743921312013166

Neutrinos from GRBs and their detection with ANTARES

Tri L. Astraatmadja[1,2] **on behalf of the ANTARES Collaboration**

[1]Nikhef – National Institute for Subatomic Physics,
Science Park 105 1098 XG Amsterdam, The Netherlands

[2]LION – Leiden Institute of Physics, Leiden University,
PO Box 9504 2300 RA Leiden, The Netherlands
email: t.astraatmadja@nikhef.nl

Abstract. The detection principle of ANTARES and its sensitivity to GRB neutrinos will be discussed. Latest analysis of ANTARES data in coincidence with GRB direction and time of occurence will also be presented, as well as the prospects of neutrino detection with KM3NeT, the km^3 neutrino telescope that will succeed ANTARES.

Keywords. astroparticle physics, gamma-ray burst: general, elementary particles

---

Most of our knowledge on the universe come from photons, and it is rightly so: they are abundantly produced, stable, and electrically neutral. However, photons interact electromagnetically with matter and thus could be quickly absorbed in matter. Other astrophysical carriers of information are cosmic rays, which are primarily protons accelerated to up to 10^{20} eV. Protons however are charged particles and are deflected by magnetic fields to a random direction, making it difficult to pinpoint their source. Neutrinos are electrically neutral and only interact with matter through the weak nuclear force. Thus they can travel through great distance and point back through the source.

High-energy neutrinos are produced from the interactions of protons with photons through the Δ^+ resonance, which produce neutral and charged mesons. The neutral mesons will then decay into gamma-rays while the charged mesons will decay into neutrinos. Calculating the neutrino energy spectrum based on the observed properties of the gamma-ray spectrum is commonly done using the formulation derived by Waxman & Bahcall (1997).

The ANTARES neutrino telescope consists of a three-dimensional array of 885 light-sensitive photomultipliers (PMT). The PMTs are grouped into triplets, with 25 triplets forming one detector line. ANTARES has 12 such lines. The total instrumented volume of the detector is approximately 0.01 km^3. It is located at a depth of 2475 m at the bottom of the Mediterranean Sea, 40 km offshore Toulon in the south of France.

When a neutrino passes through the Earth it will interact via the charged current interaction to produce a muon that would come out of the other side of the Earth. As it traverses the sea, the muon will produce Cherenkov light because its velocity is greater than the velocity of light in seawater. The PMTs that comprised ANTARES could detect these Cherenkov photons by recording the position of the PMT that got hit by a photon and the time of occurence. If we look for hits that are causally connected in space and time, we could accurately reconstruct the muon track.

Data coinciding with 40 GRBs in 2007 have been analysed. No neutrino event has been observed, but upper limit with 90% confidence level has been set. KM3NeT, the much larger successor of ANTARES, is currently being planned and is expected to have an instrumented volume of 1 km^3. The first scientific data is anticipated in 2014.

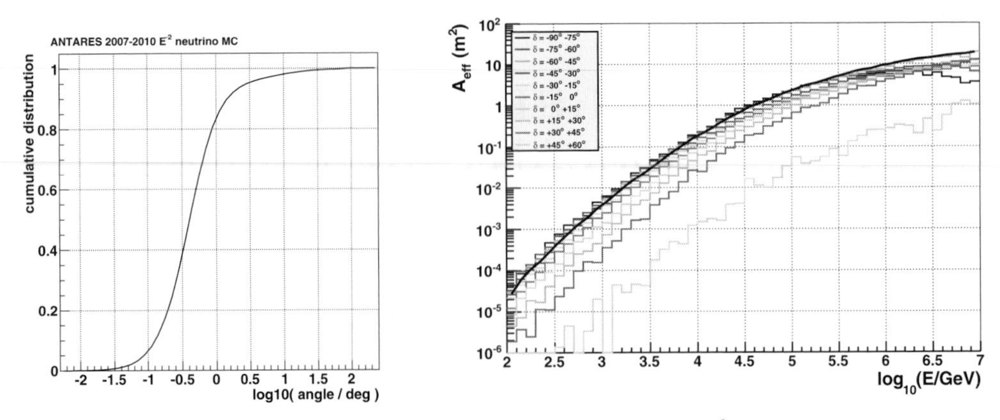

Figure 1. Left: Cumulative distribution of the angular error for ϵ_ν^{-2} neutrino events that pass the quality cuts. The median angular error is 0.46°. **Right:** The neutrino effective area in 10 declination bands of 15 degrees as a function of neutrino energy, visibility included. The black line is zenith angle-averaged effective area.

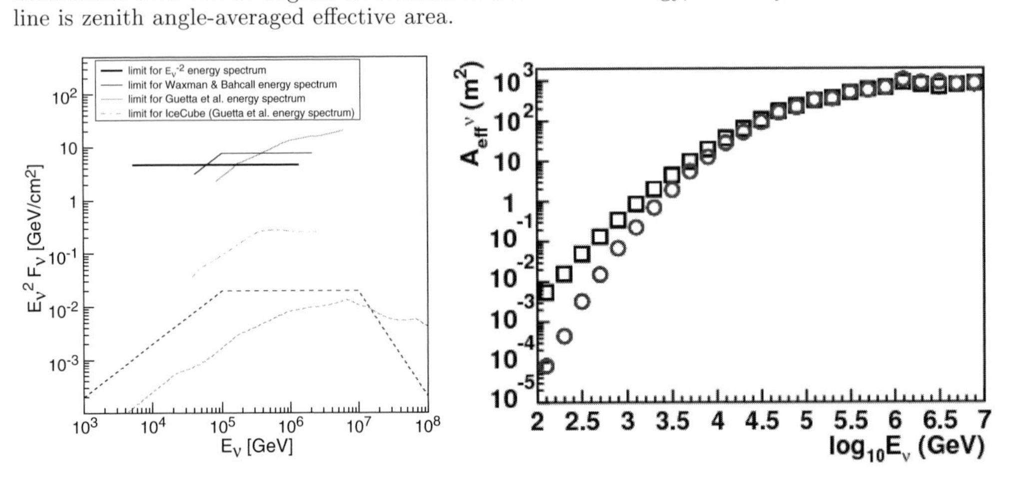

Figure 2. Left: Upper limits as a function of neutrino energy on the muon-(anti)neutrino fluence F_ν at 90% confidence level (CL), for 40 selected GRBs and for three different neutrino energy spectra. The total prompt emission duration of the 40 selected GRBs is 2114 s. Four different limits are shown: limit for the ϵ_ν^{-2} energy spectrum, the Waxman & Bahcall (1999) energy spectrum, the Guetta *et al.* (2004) energy spectrum, and the IceCube 90% CL upper limit from Abbasi *et al.* (2011). **Right:** Neutrino effective areas for the future KM3NeT detector. The plot is for upgoing neutrinos. The circles are after quality cuts optimised for searches for point-like neutrino sources and the squares are for looser quality cuts ensuring reasonable angular resolution.

References

Abbasi, R., *et al.* 2009, *ApJ* 701, L47

Abbasi, R., *et al.* 2011, *Phys. Rev. Lett.* 106, 141101

Aharonian, F., Buckley, J., Kifune, T., & Sinnis, G. 2008, *Rep. Prog. Phys.* 71, 096901

Guetta, D., Hooper, D., Alvarez-Muñiz, J., Halzen, F., & Reuveni, E. 2004, *APh* 20, 429

Waxman, E. & Bahcall, J. 1997, *Phys. Rev. Lett.* 78, 2292

Waxman, E. & Bahcall, J. 1999, *Phys. Rev. D* 59, 023002

Death of Massive Stars: Supernovae and Gamma-Ray Bursts
Proceedings IAU Symposium No. 279, 2012
P. Roming, N. Kawai & E. Pian, eds.

doi:10.1017/S1743921312013178

Spectropolarimetry of Type IIn SN2010jl: Peering Into the Heart of a Monster

Franz E. Bauer[1,2,3], Paula Zelaya[1,3], Alejandro Clocchiatti[1,3], and Justyn Maund[4]

[1]Pontificia Universidad Católica de Chile, Departamento de Astronomía y Astrofísica, Casilla 306, Santiago 22, Chile [2]Space Science Institute, 4750 Walnut Street, Suite 205, Boulder, Colorado 80301 [3]Millennium Center for Supernova Science (MCSS), [4]Astrophysics Research Centre, School of Mathematics and Physics, Queen's University Belfast, Belfast BT7 1NN, UK

Abstract. We report results for two epochs of spectropolarimetry on the luminous type IIn SN2010jl, taken at ≈36 and 85 days post-explosion with VLT FORS2-PMOS. The high signal-to-noise data demonstrate distinct evolution in the continuum and the broad lines point to a complex origin for the various emission components and to a potentially common polarization signal for the type IIn class even over 1-2 orders of magnitude in luminosity output.

Keywords. (stars:) supernovae: individual (SN2010jl), techniques: spectroscopic,polarimetric

1. Introduction

SN2010jl was discovered on Nov 3.52 UT, 2010 with a broad visual magnitude of 12.9 (CBET 2532) and was later isolated to have an explosion date prior to Oct 9.60 UT, 2010 (Stoll *et al.* 2011). It was spectroscopically classified as a prompt strongly-interacting type IIn (CBET 2536), and is ultraviolet- and X-ray-bright, as expected for its class (ATEL 3012). The SN resides in the metal-poor UGC 5189A at 48.9 Mpc (ATEL 3010), giving it a peak uncorrected $M_V = -20.55$, roughly 1–3 mags more luminous than typical IIn ($M_V = -17$ to -19.3; Richardson *et al.* 2002; Kiewe *et al.* 2012). Type IIn SNe are characterized by intense, composite-profile emission lines, a lack of the broad P-Cygni absorption troughs, and X-ray/radio emission. The properties are explained as arising from a strong interaction between the fast-moving SN ejecta and a dense, slow-moving CSM, created by significant and likely rapid mass loss of the progenitor star (e.g., Chugai & Danziger 1994). Yet many questions remain regarding the mass loss history of these progenitors, the geometry of both the explosion and CSM, and the origin of the broad emission lines (e.g., Chugai 2001). The geometrical insight gained from spectropolarimetry may allow a critical approach to these open issues, especially for rare luminous type IIn's like SN2010jl, which could provide extreme tests of models. The only other type IIn SNe to be observed in polarized light were SN1998S (Leonard *et al.* 2000; Wang *et al.* 2001) and SN1997eg (Hoffman *et al.* 2008) to provide points of comparison.

2. Data

We obtained 1hr spectropolarimetric integrations of SN2010jl on Nov. 14.3 UT, 2010 (E1) and Jan. 2.2 UT, 2011 (E2) with FORS2 PMOS on the 8m VLT-UT1. These correspond to 35.7 and 84.6 days after earliest detection. SN2010jl was also observed via spectropolarimetry by Patat *et al.* 2011 (hereafter P11) at lower S/N (2.7hr) and spectral resolution on Nov. 18.2 UT, 2010 with the Calar Alto 2m telescope. A 3rd epoch observed on days ≈530-540 is still being analyzed. Spectral reductions, calibrations, and error estimates were performed following the prescriptions described in Patat & Romaniello (2006). Following P11, we assume negligible (<0.3%) interstellar polarization (ISP); this is consistent with the nearly complete depolarization seen in narrow Hα at both epochs, despite a 0.5% change in continuum polarization.

3. Results and Discussion

Our high S/N spectra provide several useful insights into the properties of SN2010jl. The E1 spectrum provides a significant improvement upon the Nov 18, 2010 spectrum of P11, and confirms many of their findings, while the E2 spectrum allows us to understand the evolution of the various polarized components. Here are our main findings:

• The narrow Hα lines presumably stem from the slow-moving and possibly clumpy circumstellar wind material outside of the SN shock and polarized regions. This recombination region provides a large electron scattering shell to naturally explain the various observed polarization signals and narrow-line depolarization.

• The continuum is strongly polarized, and is stronger toward the blue. Values of 1.7–2.5% on E1 and 1.2–2.0% on E2 imply axial ratios of $\lesssim$0.7 and $\lesssim$0.8 for the continuum-generating region, respectively. On average, $Q_{\rm cont} \approx 0.0\%$ with no apparent evolution, while $U_{\rm cont} \approx 2.0\%$ on E1, decreasing to $\sim 1.5\%$ by E2, implying a clear axis to the polarized signal and presumed continuum asymmetry, which is slowly becoming more isotropic. The polarized continuum fades faster than the actual continuum, particularly in the blue, while the strength of total and polarized broad Hα does not appear to evolve, indicating that these components originate from distinct regions.

• No obvious line features are seen in polarization percentage, although broad Hα and Hβ lines are clearly distinguished in polarized flux. Other line complexes are presumably strongly polarized as well, since they are individually distinct in Stokes Q, U, and polarization angle plots. Together, this implies strong depolarization is present in most lines at both epochs, presumably due to continuum dilution. Continuum-subtracted polarization percentages (and hence asymmetries) for the lines are generally higher than the nominal continuum values reported above. The variations in the evolving polarization strength and angle for line emission of broad H, He, and Fe implies that the various elements likely arise in somewhat distinct asymmetric regions: H and He from shocked CSM possibly associated with swept up material from the SN shock, and Fe from fast-moving shocked ejecta.

4. Acknowledgments

We gratefully thank the Programa de Financiamiento Basal, Iniciativa Cientifica Milenio grant P10-064-F (MCSS), with input from "Fondo de Innovación para la Competitividad, del Ministerio de Economía, Fomento y Turismo de Chile", FONDECYT Regular #1101024 and Beca de Doctorado, FONDAP-CATA 15010003, and the Royal Society.

References

Chugai, N. N. 2001, *MNRAS*, 326, 1448
Chugai, N. N. & Danziger, I. J. 1994, *MNRAS*, 268, 173
Hoffman, J. L., *et al.* 2008, *ApJ*, 688, 1186
Kiewe, M., *et al.* 2012, *ApJ*, 744, 10
Leonard, D. C., Filippenko, A. V., Barth, A. J., & Matheson, T. 2000, *ApJ*, 536, 239
Patat, F. & Romaniello, M. 2006, *PASP*, 118, 146
Patat, F., *et al.* 2011, *A&A*, 527, L6 (P11)
Richardson, D., *et al.* 2002, *AJ*, 123, 745
Stoll, R., *et al.* 2011, *ApJ*, 730, 34
Wang, L., Howell, D. A., Höflich, P., & Wheeler, J. C. 2001, *ApJ*, 550, 1030

Death of Massive Stars: Supernovae and Gamma-Ray Bursts
Proceedings IAU Symposium No. 279, 2012
P. Roming, N. Kawai & E. Pian, eds.

doi:10.1017/S174392131201318X

Super iron-rich gas towards GRB 080310: SN yields or dust destruction?

A. De Cia[1], C. Ledoux[2], P. Vreeswijk[1], A. Fox[2,3], A. Smette[2], P. Petitjean[4], G. Björnsson[1], J. Fynbo[5], J. Hjorth[5] and P. Jakobsson[1]

[1]Centre for Astrophysics and Cosmology, Reykjavík, Iceland - email: annalisa@raunvis.hi.is
[2]European Southern Observatory, Santiago, Chile - [3]STScI, Baltimore, MD 21218, USA
[4]Institut d'Astrophysique de Paris, France - [5]Dark Cosmology Centre, Copenhagen, Denmark

Abstract. We performed Rapid-Response Mode (RRM) VLT/UVES high-resolution UV/ optical spectroscopy of the GRB 080310 afterglow, starting 13 min after the burst trigger, in order to investigate the ISM in the GRB host galaxy. The four spectra show remarkable features at z_{GRB}, including a low $\log N(\text{H I}) = 18.7$ and time-variable absorption from ground-state and excited levels of Fe II and Fe III, the latter being observed for the first time in a GRB afterglow. These observations indicate i) ongoing photo-ionization of the surrounding gas due to the GRB radiation and ii) Fe and Cr overabundances in the host galaxy ISM. We derive ionic column densities through a four-component Voigt-profile fit of the absorption lines and investigate the pre-burst ionization level of the gas with CLOUDY photo-ionization modelling. The resulting intrinsic [Si/Fe] $= -1.4$ ([C/H] $= -1.3$, [O/H] < -0.8, [Si/H] $= -1.2$, [Cr/H] $= +0.7$ and [Fe/H] $= +0.2$) for the whole line profile - and even more extreme for one of the absorption components - cannot be explained with current models of SN yields. Dust destruction may contribute to the marked iron overabundance, possibly induced by the burst. The overall high iron enhancement along the line-of-sight also suggests little recent star formation in the host galaxy.

Keywords. gamma rays: bursts, galaxies: ISM, galaxies: abundances

---

1. Introduction

Absorption-line spectroscopy of GRB afterglows can probe the gaseous and metal content of the host galaxy ISM out to high redshift (e.g., Prochaska *et al.*, 2007). In particular, high-resolution spectroscopy provides reliable estimates of the metallicity, relative abundances, ionization states and kinematics of the gas along the line-of-sight in the host galaxy (e.g., Ledoux *et al.*, 2009). Moreover, absorption-line variability is witness to the influence of the afterglow radiation on the surrounding medium - bearing information on the distance between the absorbing gas and the GRB - (e.g., Vreeswijk *et al.*, 2007), but this typically requires a prompt response to be observed. RRM VLT/UVES observations of GRB afterglows therefore open unique insights on the properties of GRB-selected galaxies at $z = 2$–4, but the current sample is still limited to a handful of cases. Here we summarize remarkable findings from RRM VLT/UVES spectra of the GRB 080310 afterglow. A detailed study of the spectra is presented in De Cia *et al.* (2012), while photo-excitation/ionization modelling of absorption-line variability due to the GRB is performed in Vreeswijk *et al.*, in preparation.

2. Overview of the system

The consecutive four early UVES spectra of GRB 080310 show numerous absorption lines at the GRB host galaxy redshift ($z = 2.427$), including Lyα, typical resonance metal lines, Fe II fine-structure lines and a Fe III triplet of lines(UV34) arising from a highly-excited level, which has never been observed before in a GRB afterglow. Remarkably, Fe II and Fe III transitions from all (ground-state and excited) levels and the Cr II ground state show time variability. This is a clear indication of ongoing photo-ionization induced by the GRB radiation.

We estimated the ionic column densities with a four-component Voigt-profile fit of the absorption lines and measured a low $\log N(\text{H I}) = 18.7$ and a marked overabundance of Fe and Cr ([Si II/Fe II] $= -1.47 \pm 0.14$, [C II/Fe II] $= -1.74 \pm 0.17$ and [Si II/Cr II] $= -2.03 \pm 0.19$ in one of the absorption components labelled "b"). The pre-burst relative abundances, taking out the effect of ionization due to the GRB and the ambient radiation field, are even lower. We performed CLOUDY photo-ionization modelling of the expected pre-burst ionic column densities (derived in Vreeswijk *et al.*, in prep.) to estimate an intrinsic [Si/Fe] $= -1.4$ for the whole line profile, and even more extreme ratios for component "b" ([Si/Fe] $\leqslant$ [Si II/Fe II] and [C/Fe] $\leqslant$ [C II/Fe II]).

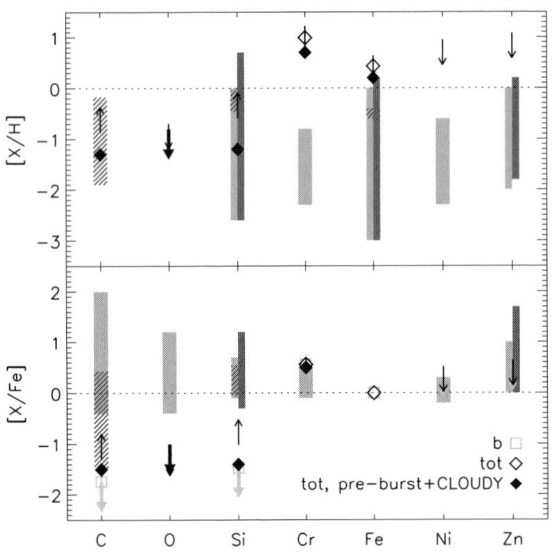

Figure 1 shows the abundances relative to H (top) and Fe (bottom) estimated for the GRB 080310 absorber (diamonds, squares and arrows). The thick arrows and filled symbols indicate ionization-corrected values, while thin arrows and open symbols refer to the observed ionic abundances. The typical abundances observed in QSO-damped Lyα absorbers (DLAs), GRB DLAs and Lyman limit systems are shown, as comparison, by the rectangular purple, orange and dashed-red regions, respectively. The GRB 080310 absorber clearly shows peculiar abundances produced by Fe and Cr overabundances, even more marked in component "b".

3. Implications

One possible explanation for the peculiar abundances in the GRB 080310 absorber is that specific Supernovae (SNe) have enriched the ISM with metals in the history of the GRB host galaxy. SN Ia yields can produce [Si/Fe] as low as -0.5 (e.g., Tanaka *et al.* 2011), while [Si/Fe] down to -2.5 are expected for PopIII star SNe, but with strong Cr depletion (Ohkubo *et al.* 2006). Thus, current models of SN yields cannot reproduce the observed abundances. An intriguing possibility is that destruction of drifted dust grains may have recycled a significant amount of refractory elements such as Fe and Cr into the ISM. This may be due to the GRB or to other processes such as SN shock waves in star forming regions. In any case, the extreme iron overabundance suggests i) a low amount of dust at the time of observing and ii) little recent star formation along the line-of-sight.

References

De Cia, A., Ledoux, C., Vreeswijk, P. M., Fox, A. J., Smette, A., *et al.* 2012, *A&A*, accepted
Ledoux, C., Vreeswijk, P. M., Smette, A., Fox, A. J., Petitjean, P., *et al.* 2009, *A&A*, 506, 661
Ohkubo, T., Umeda, H., Maeda, K., Nomoto, K., Suzuki, T., *et al.* 2006, *ApJ*, 645, 1352
Prochaska, J. X., Chen, H., Dessauges-Zavadsky, M., & Bloom, J. S. 2007, *ApJ*, 666, 267
Tanaka, M., Mazzali, P., Stanishev, V., Maurer, I., Kerzendorf, *et al.* 2011, *MNRAS*, 410, 1725
Vreeswijk, P. M., Ledoux, C., Smette, A., Ellison, S., Jaunsen, A., *et al.* 2007, *A&A*, 468, 83

Death of Massive Stars: Supernovae and Gamma-Ray Bursts
Proceedings IAU Symposium No. 279, 2012
P. Roming, N. Kawai & E. Pian, eds.

doi:10.1017/S1743921312013191

On the Origin of the 6.4 keV line from the GRXE

Romanus Eze[1,2], Kei Saitou[1,3] and Ken Ebisawa[1,3]

[1]Japan Aerospace Exploration Agency, Institute of Space and Astronautical Science, 3-1-1 Yoshinodai, Chuo-ku, Sagamihara, Kanagawa 252-5210
email: eze@ac.jaxa.jp

[2]Dept. of Astronomy & Astronomy, University of Nigeria,
email: romanus.eze@unn.edu.ng

[3]Dept. of Astronomy, Graduate School of Science, The University of Tokyo, 7-3-1 Hongo, Bunkyo-ku, Tokyo 113-0033

Abstract. The Galactic Ridge X-ray Emission (GRXE) spectrum has strong iron emission lines at 6.4, 6.7, and 7.0 keV, each corresponding to the neutral (or low-ionized), He-like, and H-like iron ions. The 6.4 keV fluorescence line is due to irradiation of neutral (or low ionized) material (iron) by hard X-ray sources, indicating uniform presence of the cold matter in the Galactic plane. In order to resolve origin of the cold fluorescent matter, we examined the contribution of the 6.4 keV line emission from white dwarf surfaces in the hard X-ray emitting symbiotic stars (hSSs) and magnetic cataclysmic variables (mCVs) to the GRXE. In our spectral analysis of 4 hSSs and 19 mCVs observed with Suzaku, we were able to resolve the three iron emission lines. We found that the equivalent-widths (EWs) of the 6.4 keV lines of hSSs are systematically higher than those of mCVs, such that the average EWs of hSSs and mCVs are 180^{+50}_{-10} eV and 93^{+20}_{-3} eV, respectively. The EW of hSSs compares favorably with the typical EWs of the 6.4 keV line in the GRXE of 90–300 eV depending on Galactic positions. Average 6.4 keV line luminosities of the hSSs and mCVs are 9.2×10^{39} and 1.6×10^{39} photons s^{-1}, respectively, indicating that hSSs are intrinsically more efficient 6.4 keV line emitters than mCVs. We estimated required space densities of hSSs and mCVs to account for all the GRXE 6.4 keV line emission flux to be 2×10^{-7} pc^{-3} and 1×10^{-6} pc^{-3}, respectively. We also estimated the actual 6.4 keV line contribution from the hSSs, which is as much as 30% of the observed GRXE flux, and that from the mCV is about 50%. We therefore conclude that the GRXE 6.4 keV line flux is primarily explained by hSSs and mCVs.

Keywords. Galaxy: disk — stars: binaries: symbiotic — stars: novae, cataclysmic variables — X-rays: stars

---

1. Introduction

Presence of the seemingly extended hard X-ray emission from the Galactic Ridge has been known since early 1980's (Galactic Ridge X-ray Emission; GRXE: Worrall *et al.* (1982), Warwick *et al.* (1985), Koyama *et al.* (1986). The nature and origin of this emission is yet to be settled. However, there is a strong argument in favor of collection of faint point sources as opposed to the diffused emission (e.g., Revnivtsev *et al.* (2006), Krivonos *et al.* (2007), Revnivtsev *et al.* (2009), Revnivtsev *et al.* (2010 and references therein), although the question remains "what are these Galactic point sources?"

We study origin of the 6.4 keV emission line in the GRXE using 4 hSSs and 19 mCVs observed with Suzaku, examining if this emission could be fully resolved by collection of point sources.

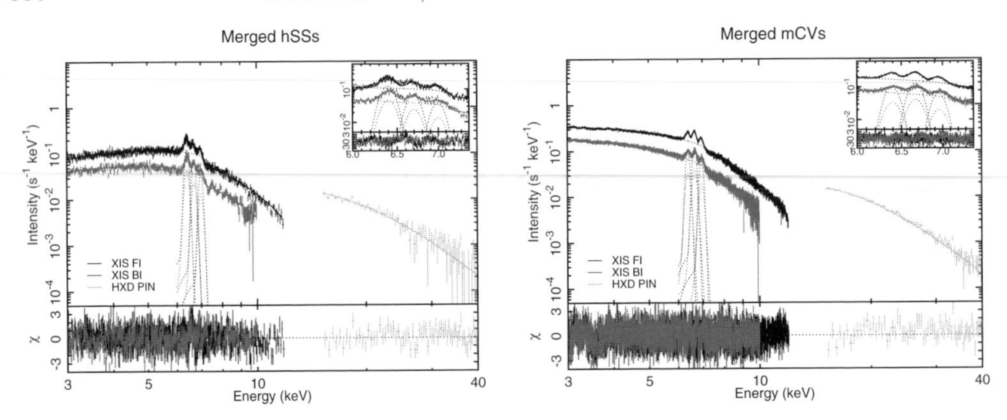

Figure 1. Spectra of the average symbiotic stars(left) and the average magnetic cataclysmic variables(right). In the upper panel, the data and the best-fit model are shown by crosses and solid lines, respectively. In the lower panel, the ratio of the data to the best-fit model is shown by crosses. The inset in the upper panel is an enlarged view for the Fe Kα complex lines.

2. Spectral Analysis and Results

Spectral analysis of all observations were performed using XSPEC version 12.7. We modeled the spectrum using absorbed bremsstralung model with three Gaussian lines for the three Fe Kα emission lines to measure the iron line fluxes. We assumed two types of absorption by full-covering and partial covering matter. The three Fe lines, neutral or low ionized (6.4 keV), He-like (6.7 keV), and H-like (7.0 keV) ions, were clearly resolved in all the sources except in one mCVs, where we were unable to detected the H-like (7.0 keV) significantly but the other two lines were detected.

We used `addascaspec` to average the spectra of hSSs and mCVs (as well as responses). The spectra for the average hSSs and average mCVs were presented in figure 1. We detected strong 6.4 keV iron line emission in the average hSSs spectrum with an equivalent width (EW) of 180^{+50}_{-10} eV and in the average mCVs spectrum with 93^{+20}_{-3} eV. We have found that the 6.4 keV line EW is much stronger in hSSs than in mCVs, which suggests that hSSs can be strong candidates of the GRXE 6.4 keV line emission. For comparison, 6.4 keV iron line EWs of the GRXE are of 90–390 eV, depending on the Galactic locations (Yamauchi *et al.* 2009).

In our rough estimate, as much as ~30% of the GRXE 6.4 keV line flux may be from hSSs, and ~50% from mCVs. We therefore conclude that the GRXE 6.4 keV line flux is primarily explained by hSSs and mCVs.

References

Koyama, K., Makishima, K., Tanaka, Y., & Tsunemi, H. 1986, *PASJ*, 38, 121

Krivonos, R., Revnivtsev, M., Churazov, E., Sazonov, S., Grebenev, S., & Sunyaev, R. 2007, *A&A*, 463, 957

Revnivtsev, M., Sazonov, S., Gilfanov, M., Churazov, E., & Sunyaev, R. 2006, *A&A*, 452, 169

Revnivtsev, M., Sazonov, S., Churazov, E., Froman, W., Vikhlinin, A., & Sunyaev, R. 2009, *Nature*, 458, 1142

Revnivtsev, M., van den Berg, M., Burenin, R., Grindlay, J. E., Karasev, D., & Forman, W. 2010, *A&A*, 515, A49

Warwick, R. S., Turner, M. J. L., Watson, M. G., & Willingale, R. 1985, *Nature*, 317, 218

Worrall, D. M., Marshall, F. E., Boldt, E. A., & Swank, J. H. 1982, *ApJ*, 255, 111

Yamauchi, S., *et al.* 2009, *PASJ*, 61, S225

Death of Massive Stars: Supernovae and Gamma-Ray Bursts
Proceedings IAU Symposium No. 279, 2012
P. Roming, N. Kawai & E. Pian, eds.

doi:10.1017/S1743921312013208

The Luminous Type Ibc Supernova 2010as

Gastón Folatelli

Kavli Institute for the Physics and Mathematics of the Universe,
Todai Institutes for Advanced Study, the University of Tokyo,
Kashiwa, Japan 277-8583 email: gaston.folatelli@ipmu.jp

On behalf of the MCSS, the CSP, and collaborators

Abstract. We present photometric and spectroscopic follow-up observations of SN 2010as carried out by the MCSS and CSP. The SN appears to be of the transitional type Ibc (SN Ibc) and is spectroscopically similar to the peculiar SN 2005bf. Based on distance and extinction estimates, a bolometric luminosity light curve is constructed showing that this was a relatively luminous SN Ibc. He I line expansion velocities are remarkably low and remain nearly constant with time, similarly to SN 2005bf. A preliminary model is presented with a progenitor ZAMS mass of 15 $M_{\odot}$ and a large yield of 0.35 $M_{\odot}$ of ^{56}Ni.

Keywords. supernovae: general, supernovae: individual (SN 2010as)

---

1. Photometry

SN 2010as was discovered in NGC 6000 on March 19.2 UT by the CHilean Automatic Supernova sEarch (CHASE; Maza *et al.* 2010). Follow-up in $BVRIg'r'i'z'$ bands was obtained with the 40-cm PROMPT 1, 3 and 4 telescopes at CTIO (Figure 1, left panel). The SN was caught 10 days before B-band maximum light and followed until ≈100 days after.

Observed colors corrected for Galactic reddening of $E(B-V)_{\rm MW} = 0.17$ mag (from NED†) were compared with those of a sample of SNe Ib and Ic from the CSP (Stritzinger *et al.* , in prep.) to derive a host-galaxy reddening of $E(B-V)_{\rm host} = 0.35$ mag‡. Adopting a distance of 32.7 ± 3.6 Mpc to the host (NED), reddening-free absolute peak magnitudes of $M_B = -18.1 \pm 0.5$ mag and $M_V = -18.5 \pm 0.4$ mag are obtained, indicating that SN 2010as was a relatively luminous SN Ibc.

2. Spectroscopy

A spectral time series between −10 and +309 days relative to B-band maximum was obtained by the MCSS and CSP with instruments from Las Campanas, ESO Paranal, SOAR and Gemini Observatories (see left panel of Figure 2). The pre-maximum spectra are of type Ic, dominated by Ca II and Fe II lines, with possible Hα or Si II at ≈6200 Å. Then He I lines develop marking the transition to a SN Ib. The right panel of Figure 2 compares the spectra with those of SNe 2005bf (Folatelli *et al.* 2006) and 2007Y (Stritzinger *et al.* 2009) at about one week before maximum light, around the time of maximum, and three weeks after. The three SNe are very similar before maximum, but then SN 2007Y develops strong He I lines. The other two SNe remain similar, with SN 2010as showing stronger Ca II features. An additional similarity with SN 2005bf is the peculiarly low He I line velocities observed at all times.

† NASA/IPAC Extragalactic Database
‡ This agrees with the value derived from a Na I D equivalent width of 2.1 Å measured on X-Shooter spectra.

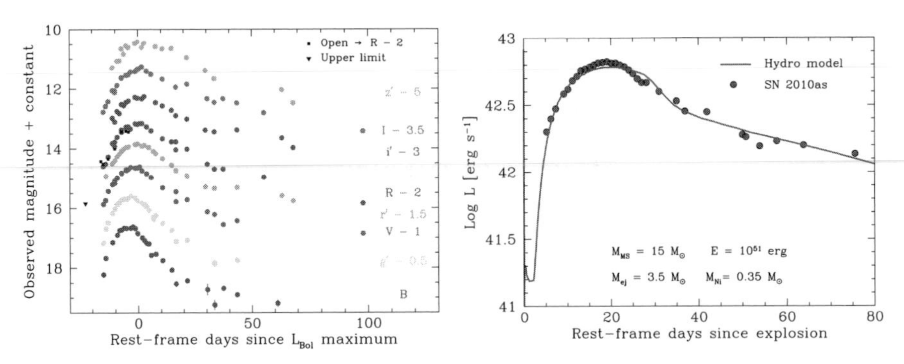

Figure 1. (Left) $BVRIg'r'i'z'$ light curves. (Right) Bolometric luminosity for a model (solid line) with the indicated physical parameters, compared with the data (dots).

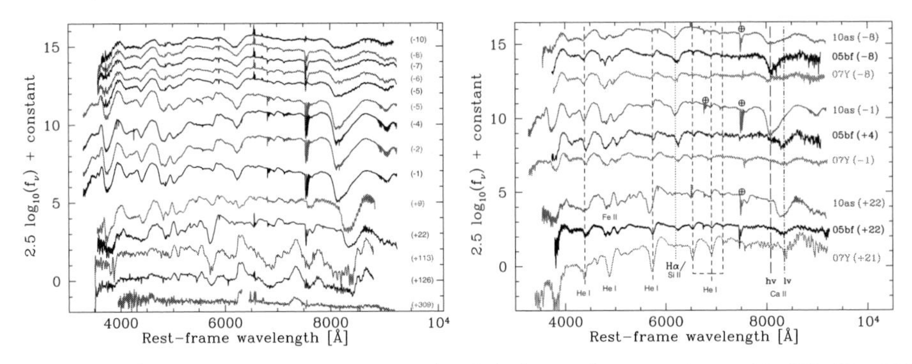

Figure 2. (Left) Spectroscopic time series. (Right) Spectral comparison with SNe Ibc 2005bf and 2007Y at three different epochs.

3. Physical parameters

A bolometric light curve was obtained by integrating the extinction-corrected optical flux and adding the extrapolation of a black-body fit toward the IR, and a straight line in the UV from the B-band point to zero flux at 2000 Å. The extrapolated UV+IR flux accounts for up to 50% of the total estimated flux. Assuming the extinction derived in Section 1, the resulting peak luminosity is $L_{\rm Bol} = 6.3 \times 10^{42}$ erg s^{-1}. This is comparable with the cases of the peculiar SNe 2005bf and 2009bb, which makes SN 2010as a relatively luminous Ibc event.

A preliminary hydrodynamical model (Bersten *et al.* 2011) was computed to reproduce the bolometric luminosity (see assumed physical parameters in the right panel of Figure 1). A relatively large amount of ^{56}Ni, $M_{\rm Ni} = 0.35\ M_{\odot}$, is required to explain the luminous peak. The explosion energy of $E = 10^{51}$ erg, however, overestimates the measured expansion velocities. This may be solved by assuming larger progenitor and ejected masses.

References

Bersten, M. C., *et al.* 2011, *ApJ*, 729, 61
Folatelli, G., *et al.* 2006, *ApJ*, 641, 1039
Maza, J., *et al.* 2010, *CBET*, 2215, 1
Stritzinger, M., *et al.* 2009, *ApJ*, 696, 713
Stritzinger, M., *et al.* 2012 in prep.

Death of Massive Stars: Supernovae and Gamma-Ray Bursts
Proceedings IAU Symposium No. 279, 2012
P. Roming, N. Kawai & E. Pian, eds.

doi:10.1017/S174392131201321X

A new equation of state Based on Nuclear Statistical Equilibrium for Core-Collapse Simulations

Shun Furusawa[1], Shoichi Yamada[1,2], Kohsuke Sumiyoshi[3] and Hideyuki Suzuki[4]

[1]Department of Science and Engineering, Waseda University, 3-4-1 Okubo, Shinjuku, Tokyo 169-8555, Japan
email: `furusawa@heap.phys.waseda.ac.jp`

[2]Advanced Research Institute for Science and Engineering, Waseda University, 3-4-1 Okubo, Shinjuku, Tokyo 169-8555, Japan

[3]Numazu College of Technology, Ooka 3600, Numazu, Shizuoka 410-8501, Japan

[4]Faculty of Science and Technology, Tokyo University of Science, Yamazaki 2641, Noda, Chiba 278-8510, Japan

Abstract. We calculate a new equation of state for baryons at sub-nuclear densities for the use in core-collapse simulations of massive stars. The formulation is the nuclear statistical equilibrium description and the liquid drop approximation of nuclei. The model free energy to minimize is calculated by relativistic mean field theory for nucleons and the mass formula for nuclei with atomic number up to ~ 1000. We have also taken into account the pasta phase. We find that the free energy and other thermodynamical quantities are not very different from those given in the standard EOSs that adopt the single nucleus approximation. On the other hand, the average mass is systematically different, which may have an important effect on the rates of electron captures and coherent neutrino scatterings on nuclei in supernova cores.

Keywords. equation of state, dense matter, nuclear abundances & supernovae

---

Equations of state (EOSs) of hot and dense matter play an important role in the dynamics of core-collapse supernovae both at sub- and supra-nuclear densities. In addition to thermodynamical quantities such as pressure, information on matter composition is provided by EOS. At sub-nuclear densities, the latter affects the rates of electron captures and coherent neutrino scatterings on nuclei, both of which in turn determine the electron fraction of collapsing cores, one of the most critical ingredients for the core dynamics.

At present, there are only two EOSs in wide use for the core-collapse simulation. Lattimer *et al.* (1991)'s EOS is based on Skyrme-type nuclear interactions and the so called compressible liquid drop model for nuclei surrounded by dripped nucleons. The EOS by H. Shen *et al.* (1998) *et al.* employs a relativistic mean field theory (RMF) to describe nuclear matter and the Thomas Fermi approximation for finite nuclei with dripped nucleons. It should be emphasized here that both EOSs adopt the so-called single nucleus approximation (SNA), in which only a single representative nucleus is included. In other words, the distribution of nuclei is ignored. There are a few EOSs assuming multi-nuclei (Botvina *et al.* (2010), Hempel *et al.* (2010) & G. Shen *et al.* (2011)). However, there is no model that takes into account both shell effects of nuclei at low densities and nuclear pasta phase transition to uniform matter with multi-nuclei at high densities.

The aim of our project is in a sense to merge all the previous EOSs to provide all the information needed for core-collapse simulations: thermodynamical quantities such as pressure as well as the information on the abundance of nuclei required for the calculation

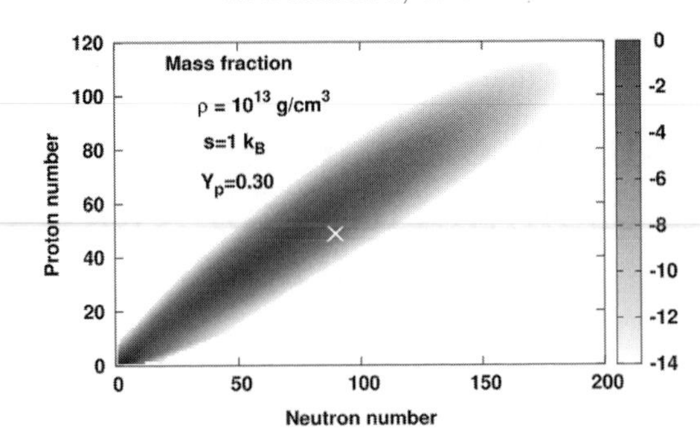

Figure 1. The mass fractions of nuclei in the (N, Z) plane for $\rho_B = 10^{11}\text{g/cm}^3$, $T = 1\text{MeV}$ and $Y_p = 0.3$. The cross indicates the representative nucleus for the H. Shen's EOS under the same condition.

of the weak interaction rates. The description of uniform nuclear matter is based on RMF in H. Shen's EOS and the approximate treatment of the pasta phase is similar to the one employed in Lattimer-Swesty's EOS. We solve a Saha-like equation to obtain the abundance of nuclei.

The formulation of this model is based on the NSE description using the mass descriptions for nuclei up to the atomic number of 1000 under the influence of surrounding nucleons and electrons. The mass data are taken from the experimental nuclear mass that allow us to take into account the nuclear shell effects. An extended liquid drop model is used to describe the nuclear mass whenever the experimental data are not available together with the medium effects, and in particular, the formulation of the pasta phases. Because of this combination of the mass data and the theoretical model, the free energy of multi-component system can reproduce the ordinary NSE results at low densities and make a continuous transition to the EOS for supra-nuclear densities. The details are given in Furusawa *et al.* (2011).

We find that the thermodynamical quantities such as free energy and pressure per baryon agree well among other EOSs. The matter compositions show noticeable differences, however. The average mass numbers, for example, are systematically smaller in our EOS than the mass numbers of the representative nuclei in the H.Shen EOS. We can see that the representative nucleus of H.Shen EOS is different from the abundance peak of our model in Fig.1. In other words, SNA can not reproduce the ensemble of nuclei.

We are currently constructing an EOS table, which will be available for supernova simulations and made available in the public domain. In so doing, electron capture rates, which are consistent with the abundance given by the EOS, should also be included.

References

Furusawa, S., Yamada, S., Sumiyoshi, K., & Suzuki, H. 2011, *ApJ*, 738, 178
Shen, G., Horowitz, C. J., & Teige, S. 2011, *Phys. Rev. C*, 83, 035802
Shen, H., Toki, H., Oyamatsu, K., & Sumiyoshi, K. 1998, *Nucl. Phys.*, A637, 435
Lattimer, J. M. & Swesty, F. D. 1991, *Nucl. Phys.*, A535, 331
Botvina, A. S. & Mishustin, I. N. 2010, *Nucl. Phys. A* 843, 98
Hempel, M. & Schaffner-Bielich, J. 2010, *Nucl. Phys. A* 837, 210

Death of Massive Stars: Supernovae and Gamma-Ray Bursts
Proceedings IAU Symposium No. 279, 2012
P. Roming, N. Kawai & E. Pian, eds.

doi:10.1017/S1743921312013221

Turbulent Magnetic Field Amplification behind Strong Shock Waves in GRB and SNR

Tsuyoshi Inoue

Department of Physics and Mathmatics, Aoyama Gakuin University,
Fuchinobe, Chuou-ku, Sagamihara 252-5258, Japan
email: inouety@phys.aoyama.ac.jp

Abstract. Using three-dimensional (special relativistic) magnetohydrodynamics simulations, the amplification of magnetic field behind strong shock wave is studied. In supernova remnants and gamma-ray bursts, strong shock waves propagate through an inhomogeneous density field. When the shock wave hit a density bump or density dent, the Richtmyer-Meshkov instability is induced that cause a deformation of the shock front. The deformed shock leaves vorticity behind the shock wave that amplifies the magnetic field due to the stretching of field lines.

Keywords. magnetic fields, turbulence, gamma rays: bursts, supernova remnants

1. Motivation and Results of Simulations

It is well known that magnetic fields play a very important role in high-energy astrophysical phenomena through particle acceleration and synchrotron emission. Strong magnetic fields around shock waves are often required to explain the observed emission. For example, at the shocks in supernova remnants (SNRs) and gamma-ray bursts (GRBs), magnetic field strengths orders of magnitude larger than their ambient fields are needed.

In this paper, we show that the magnetic field can be amplified far beyond the shock compression value behind the shock wave, if the shock wave propagates through an inhomogeneous density field (Giacalone & Jokipii 2007, Inoue *et al.* 2009, 2010, 2011, 2012). In the case of the SNR, observations have shown that the young SNR RX J1713.7-3946 is interacting with molecular clouds (e.g., Fukui *et al.* 2012) where there is strong evidence of magnetic field amplification up to $B \sim 1$ mG (Uchiyama *et al.* 2007).

The left panel of Fig. 1 shows the result of the three-dimensional (3D) magnetohydrodynamics (MHD) simulation of the shock propagation in a realistic cloudy ISM with $B_{\rm ini} = 5.0\ \mu$G (Inoue *et al.* 2012, see also Inoue & Inutsuka 2008, 2009 for the cloud formation). Because the interaction between the shock wave and cloud generates vorticity behind the shock through the induction of the Richtmyer-Meshkov instability, the shocked ISM that corresponds to the SNR becomes highly turbulent. The turbulence behind the shock wave amplifies magnetic field by stretching field lines. In the present case, the maximum field strength reaches $B_{\rm max} \sim 1$ mG (corresponding plasma $\beta \sim 1$) and the average field strength becomes $\langle |B| \rangle \sim 100\ \mu$G. The strongly magnetized regions with $B \sim 1$ mG are located at the interface regions of the clouds where strong shear flows are developed by the shock-cloud interaction. Their spatial scale, ~ 0.05 pc, which is essentially determined by the thickness of the interface due to the thermal conduction (Inoue *et al.* 2006, 2007), agrees pretty well with the scale of the observed amplified regions of $B \sim 1$ mG (Uchiyama *et al.* 2007).

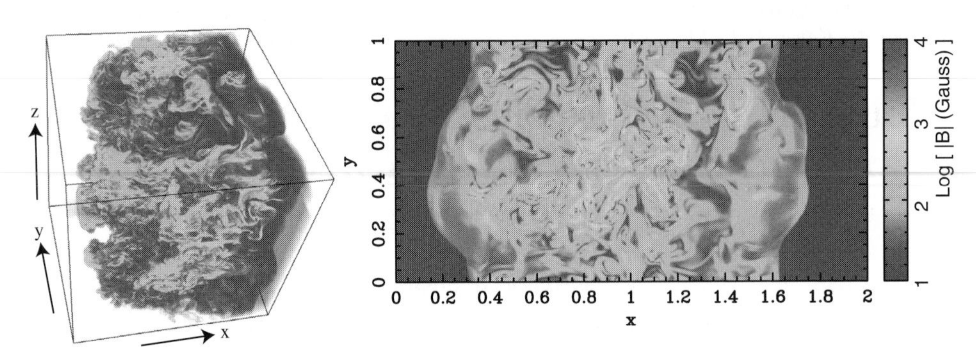

Figure 1. Left: map of magnetic field strength as a result of the MHD simulation of the shock propagation in inhomogeneous cloudy ISM. Regions in blue, green and red represent the regions with $B \lesssim 100\ \mu$G, $B \gtrsim 100\ \mu$G, and $B \gtrsim 500\ \mu$G, respectively. Right: 2D slice of magnetic field strength as a result of the 3D relativistic MHD simulation of inhomogeneous flow collision.

The GRBs are believed to be caused by the collisions of relativistic jets that are released when a black hole is formed. Because the jets are released very intermittently with time, it is reasonable to expect the inhomogeneous structure of the jets. Right panel of Fig. 1 shows the result of the 3D special relativistic MHD simulation of the relativistic jet-jet collision (Inoue *et al.* 2011). Owing to the density fluctuations imprinted in the colliding flows, the postshock layer becomes highly turbulent and the turbulence amplifies the magnetic field. We find that the velocity dispersion of turbulence induced behind the shock is proportional to the upstream density dispersion ($\Delta v \sim \Delta\rho$) and the velocity dispersion saturates when it reaches the postshock sound speed ($\sim c/\sqrt{3}$) at $\Delta\rho \sim \rho_0$. The magnetic field behind the shock is amplified until the magnetic energy becomes comparable to the kinetic energy of the decaying turbulence that is powered only at the shock front.

Acknowledgements

I would like to thank A. MacFadyen and J. Mao for fruitful discussions. Numerical computations were carried out on XT4 at the Center for Computational Astrophysics (CfCA) of National Astronomical Observatory of Japan. This work is supported by Grant-in-aids from the Ministry of Education, Culture, Sports, Science, and Technology (MEXT) of Japan, No.22·3369 and 23740154.

References

Fukui, Y. *et al.* 2012, *ApJ*, 746, 82
Giacalone, J. & Jokipii, J. R. 2007, *ApJ*, 663, L41
Inoue, T., Inutsuka, S., & Koyama, H. 2006, *ApJ*, 652, 1331
Inoue, T., Inutsuka, S., & Koyama, H. 2007, *ApJ*, 658, L99
Inoue, T. & Inutsuka, S. 2008, *ApJ*, 687, 303
Inoue, T. & Inutsuka, S. 2009, *ApJ*, 704, 161
Inoue, T., Yamazaki, R., & Inutsuka, S. 2009, *ApJ*, 695, 825
Inoue, T., Yamazaki, R., & Inutsuka, S. 2010, *ApJ*, 723, L108
Inoue, T., Asano, K., & Ioka, K. 2011, *ApJ*, 734, 77
Inoue, T., Yamazaki, R., Inutsuka, S., & Fukui, Y. 2012, *ApJ*, 744, 71
Uchiyama, Y. *et al.* 2007, *Nature*, 449, 576

Death of Massive Stars: Supernovae and Gamma-Ray Bursts
Proceedings IAU Symposium No. 279, 2012
P. Roming, N. Kawai & E. Pian, eds.

doi:10.1017/S1743921312013233

Neutron Star Kicks Affected by Standing Accretion Shock Instability for Core-Collapse Supernovae

Wakana Iwakami Nakano[1], Kei Kotake[2], Naofumi Ohnishi[3], Shoichi Yamada[4], and Keisuke Sawada[3]

[1]Institute of Fluid Science, Tohoku University, 2-1-1 Katahira, Aoba-ku, Sendai, Miyagi, 980-8577, Japan
email: wakana@dragon.ifs.tohoku.ac.jp

[2]Division of Theoretical Astronomy/Center for Computational Astrophysics, National Astronomical Observatory of Japan, 2-21-1, Osawa, Mitaka, Tokyo, 181-8588, Japan

[3]Department of Aerospace Engineering, Tohoku University, 6-6-01 Aramaki-Aza-Aoba, Aoba-ku, Sendai, 980-8579, Japan

[4]Advanced Research Institute for Science and Engineering, Waseda University, 3-4-1 Okubo, Shinjuku, Tokyo, 169-8555, Japan

Abstract. We investigate a proto-neutron star kick velocity estimated from kinetic momentum of a flow around the proto-neutron star after the standing accretion shock instability grows. In this study, ten different types of random perturbations are imposed on the initial flow for each neutrino luminosity. We found that the kick velocities of proto-neutron star are widely distributed from 40 km s^{-1} to 180 km s^{-1} when the shock wave reaches 2000 km away from the center of the star. The average value of kick velocity is 115 km s^{-1}, whose value is smaller than the observational ones. The kick velocities do not depend on the neutrino luminosity.

Keywords. supernovae, shock waves, hydrodynamics, instabilities.

---

1. Introduction

Pulsars are highly rotating and magnetized compact objects with periodic electromagnetic radiations. The velocities of their proper motions tend to be larger than those of other type of astronomical objects. The mean speeds of them are about 400 km s^{-1}, and the maximum velocity is around 1600 km s^{-1} (Hobbs (2005)). The mechanism of pulsar kicks has not been understood clearly yet. Pulsars are considered as neutron stars. Neutron stars are produced by core-collapse supernova explosions of massive stars. Some core-collapse supernova remnants have shown that non-spherically symmetric explosions occur (Winkler & Petre (2007)). Surrounding matters having nonuniformly distributed kinetic momenta might give a large kick velocity for a neutron star. Standing accretion shock instability (SASI) is regarded as playing a key role to the asymmetric explosions of core-collapse supernovae. The SASI makes the spherical shock wave deformed, and high entropy bubbles are generated behind the shock wave. The behavior of fluids inside and outside the bubbles produces the non-spherically symmetric distribution of kinetic momenta around the proto-neutron star (PNS). The kick velocity of PNS have been investigated with axisymmetric two-dimensional (2D) (Sheck (2006), Nordhaus (2010)) and non-axisymmetric three-dimensional (3D) (Wongwathanarat (2010)) simulations for core-collapse supernovae in recent years. These studies have obtained hundreds km s^{-1} of kick velocities.

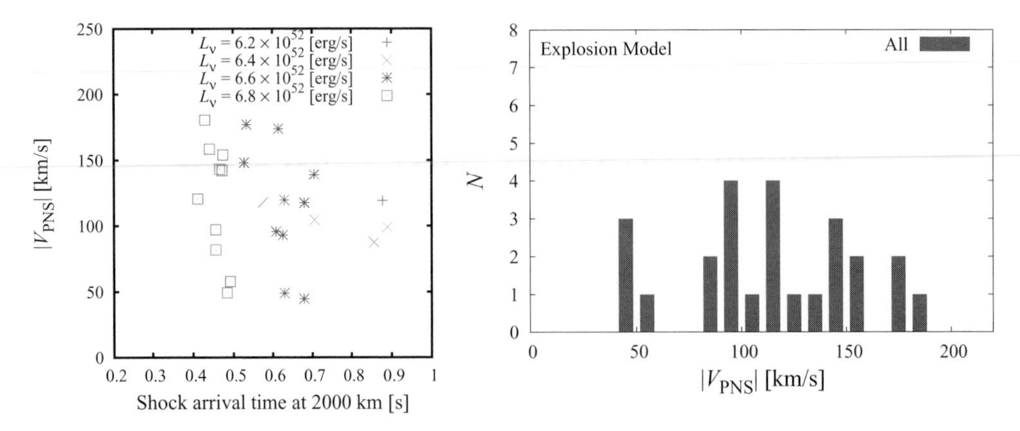

Figure 1. The kick velocities classified into neutrino luminosities as a function of shock arrival time at 2000 km (left panel) and the kick velocity distribution (right panel).

Our objective is to calculate the kick velocities from the results of three-dimensional simulations for core-collapse supernovae. We impose ten different types of random perturbations for each neutrino luminosity in the range 5.8 - 6.8$\times 10^{52}$ erg s^{-1}. The computational procedure is the same as explained in our previous paper (Iwakami (2008)).

2. Results and Discussions

We estimate the kick velocities from the equations described as

$$v_{\rm PNS}(t) = -\mathbf{P}_{\rm gas}(t)/M_{\rm PNS}(t) \qquad \mathbf{P}_{\rm gas}(t) = \int_{R_{\rm PNS}<r<\infty} \rho \mathbf{v}\, dV. \tag{2.1}$$

Fig. 1 shows in the left panel the shock arrival time dependence of the kick velocities. The shock arrival time is defined as the time when the shock wave reaches 2000 km. The kick velocity does not depend on the neutrino luminosity. The higher neutrino luminosity, the shorter the arrival time tends to be. So the kick velocity does not correlate with the speed of shock propagation.

Fig. 1 displays in the right panel the kick velocity distribution for 25 explosion models. The kick velocity is distributed from 40 km s^{-1} to 180 km s^{-1} Average value is 115 km s^{-1} which is smaller than the value of 400 km s^{-1} obtained by observation. The velocities are measured when the shock wave reaches 2000 km. If we use a larger computational region than 2000 km and calculate for longer time, the kick velocities would be larger.

3. Conclusions

The resulting kick velocities, distributed from 90 km s^{-1} to 180 km s^{-1} do not depend on the arrival time of the shock wave at 2000 km from the center of the star.

References

Hobbs, G., Lorimer, D. R., Lyne, A. G., & Kramer M. 2005, *MNRAS*, 360, 974

Winkler, P. F. & Petre, R. 2007, *ApJ*, 670, 635

Scheck, L., Kifonidis, K., Janka, H.-Th., & M'uller, E. 2006, *A&A*, 457, 963

Nordhaus, J., Brandt, T. D., Burrows, A., Livne, E., & Ott, C. D. 2010, *Phys. Rev. D*, 82, 103016

Wongwathanarat, A., Janka, H.-Th., & M'uller, E. 2010, *ApJ*, 725, L106

Iwakami, W., Kotake, K., Ohnishi, N., Yamada, S., & Sawada, K. 2008, *ApJ*, 678, 1207

Death of Massive Stars: Supernovae and Gamma-Ray Bursts
Proceedings IAU Symposium No. 279, 2012
P. Roming, N. Kawai & E. Pian, eds.

doi:10.1017/S1743921312013245

Supernova Nucleosynthesis with Neutrino Processes: Dependence of Fluorine abundance on Stellar Mass, Explosion Energy and Metallicity

Natsuko Izutani, Hideyuki Umeda, and Takashi Yoshida

Department of Astronomy, School of Science, University of Tokyo,
Bunkyo-ku, Tokyo, 113-0033
email: `izutani@astron.s.u-tokyo.ac.jp`

Abstract. We investigate the effects of neutrino-nucleus interactions on the production of Fluorine during normal supernovae and hypernovae, and discuss stellar mass, metallicity and explosion energy dependence of [F/Fe,Ne,O]. We find the clear trend of [F/Fe,O,Ne] with stellar mass and explosion energy, while no clear trend with metallicity. This trend of [F/O] can be used to constrain the contributed stellar mass by comparing with the observational abundance.

Keywords. neutrinos, nuclear reactions, nucleosynthesis, abundances, supernova: general

1. Introduction

The interaction of the neutrinos with matter and the effects on the nucleosynthesis have only been discussed for a few models (e.g., Woosley *et al.* 1990; Woosley & Weaver 1995; Yoshida *et al.* 2004; Heger *et al.* 2005; Yoshida *et al.* 2008; Nakamura *et al.* 2010). The ν-process does not affect the yields of major elements such as Fe and α elements, but it will increase those of some elements such as B, F, K, Sc, V, and Mn. In this paper, we focus on the effect of the ν-process on F during normal supernova (SN) and hypernova (HN) explosions, and discuss stellar mass, metallicity, and explosion energy dependence of [F/Fe,O,Ne].

2. Model & Method

We calculate the nucleosynthesis for core-collapse SNe with progenitor masses of $M =$ 15, 25, and 50 $M_\odot$ and initial metallicities of $Z = 0$, 0.004, and 0.02 for normal SNe and HNe. The explosion energy is set to be 1×10^{51} ergs for normal SNe, 10×10^{51} and 40×10^{51} ergs for HNe with $M_{\rm MS} = 25$ and 50 $M_\odot$, respectively. For normal SNe, the mass cut is set to meet the observed iron mass of 0.07 $M_\odot$. For HNe, the parameters of mixing fallback models are determined to get [O/Fe] = 0.5. The nuclear network includes 809 species up to ^{121}Pd (Izutani *et al.* 2009, Izutani & Umeda 2010). We adopt the ν-process up to ^{80}Kr as in Yoshida *et al.* (2008). The neutrino luminosity is assumed to be uniformly partitioned among the neutrino flavors, and decrease exponentially in time with a timescale of 3 s. The total neutrino energy is set to be $E_\nu = 3$ and 9×10^{53} ergs. The neutrino energy spectra are assumed to be Fermi-Dirac distributions with zero chemical potentials. The temperatures of $\nu_{\mu,\tau}$, $\bar{\nu}_{\mu,\tau}$ and $\nu_{\rm e}$, $\bar{\nu}_{\rm e}$ are set to be $T_\nu = 6$ and 4 MeV/k, respectively.

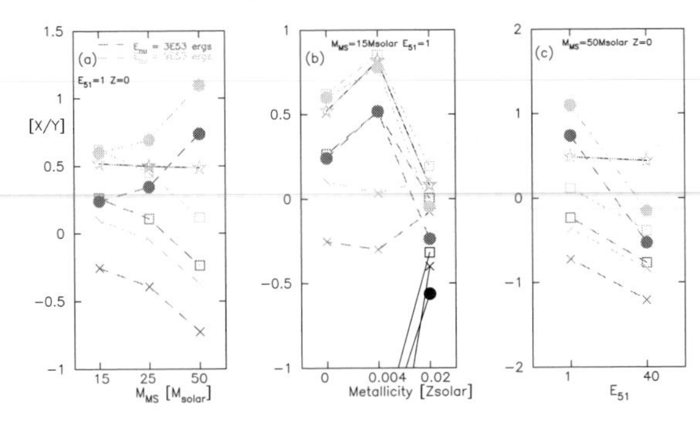

Figure 1. [F/Fe,O,Ne] (filled circles, open squares, and crosses) and [Ne/O] (stars) in the models without ν-processes (black solid lines), with ν-processes of $E_\nu = 3$ and 9×10^{53} ergs (magenta dashed lines and cyan dot-dashed lines). (a) Stellar mass dependence ([X/Y] in the models with $M_{\rm MS} = 15, 25, 50\ M_\odot$, $Z = 0$ and $E_{51} = 1$) (b) Metallicity dependence ([X/Y] in the models with $Z = 0, 0.004, 0.02$, $M_{\rm MS} = 15\ M_\odot$ and $E_{51} = 1$) (c) E_{51} dependence ([X/Y] in the models with $M_{\rm MS} = 50\ M_\odot$, $Z = 0$, $E_{51} = 1$ and 40.

3. Results and Discussion

With the ν-process, ^{19}F is produced in the O/Ne-enriched region through ^{20}Ne(ν, ν'p)^{19}F. Figure 1 (a) shows mass dependence of [F/Fe,O,Ne] and [Ne/O] in Z=0 star SNe. [Ne/O] is about 0.5 in these models. Without the ν-processes, [F/Fe,O,Ne] are ~ -5. With the ν-processes, [F/Fe,O,Ne] range from -1 to 1. [F/Fe] is higher for more massive stars because of the larger O/Ne-enriched region. By contrast, [F/O,Ne] are lower for more massive stars because the radius of the O/Ne-enriched region is larger, and the neutrino flux becomes smaller. Figure 1(b) shows metallicity dependence of [F/Fe,O,Ne] and [Ne/O] in 15 $M_\odot$ SNe. [Ne/O] is different between these models, though it is not clear whether this trend of [Ne/O] is due to metallicity or not. With the ν-processes, the F yield is increased by a factor of ~ 10 and 1000 for $Z = 0.02$ and 0, respectively. There is no clear trend of [F/Fe,O,Ne] with metallicity. Figure 1 (c) shows $E_{\rm exp}$ dependence of [F/Fe,O,Ne] and [Ne/O] in Z=0 50$M_\odot$ explosions. [Ne/O] is about 0.5 in these models. With the ν-processes, [F/Fe,O,Ne] range from -1 to 1. [F/O,Ne] are lower in the HN model. In the HN model, the shock wave reaches the O/Ne-enriched region earlier, and the region expands earlier, which causes smaller neutrino flux. [F/Fe] is also lower in the HN model, which is caused by both the larger mass of Fe and the smaller mass of F in the HN model.

It is true that ν-cross-sections contain some uncertainties. Nevertheless, the trend of [F/Fe,O,Ne] discussed above is robust for these uncertainties. For the galactic chemical evolution calculation using these yields, see Kobayashi *et al.* (2011).

References

Heger, A., Kolbe, E., Haxton, W. C., *et al.* 2005, *Phys. Lett. B.* 606, 258
Izutani, N., Umeda, H., & Tominaga, N. 2009, *ApJ*, 692, 1517
Izutani, N. & Umeda, H. 2010, *ApJ*, 720, L1
Kobayashi, C., Izutani, N., Karakas, A. I., Yoshida, T., Yong, D., & Umeda, H. 2011, *ApJ*, 739, L57
Nakamura, K., Yoshida, T., Shigeyama, T., & Kajino, T. 2010, *ApJ*, 718, L137
Umeda, H. & Nomoto, K., 2002, *ApJ*, 565, 385
Woosley, S. E., Hartmann, D. H., Hoffman, R. D., & Haxton, W. C. 1990, *ApJ*, 356, 272
Woosley, S. E. & Weaver, T. A. 1995, *ApJS*, 101, 181
Yoshida, T., Terasawa, M., Kajino, T., & Sumiyoshi, K. 2004 *ApJ*, 600, 204
Yoshida, T., Umeda, H., & Nomoto, K. 2008, *ApJ*, 672, 1043

Death of Massive Stars: Supernovae and Gamma-Ray Bursts
Proceedings IAU Symposium No. 279, 2012
P. Roming, N. Kawai & E. Pian, eds.

doi:10.1017/S1743921312013257

Progenitors of electron-capture supernovae

Samuel Jones[1,*], Raphael Hirschi[1,2], Falk Herwig[3], Bill Paxton[4], Francis X. Timmes[5,6] and Ken'ichi Nomoto[2]

[1]Astrophysics Group, Lennard Jones Building, Keele University ST5 5BG, UK
[2]Kavli IPMU, University of Tokyo, Kashiwa, Chiba 277-8583, Japan
[3]Department of Physics and Astronomy, Victoria, BC V8W 3P6, Canada
[4]KITP and Dept. of Physics, University of California, Santa Barbara, CA 93106 USA
[5]Joint Institute for Nuclear Astrophysics, University of Notre Dame, IN 46556, USA
[6]School of Earth and Space Exploration, University of Arizona, Tempe, AZ 85287, USA
*email: s.w.jones@epsam.keele.ac.uk

Abstract. We investigate the lowest mass stars that produce Type-II supernovae, motivated by recent results showing that a large fraction of type-II supernova progenitors for which there are direct detections display unexpectedly low luminosity (for a review see e.g. Smartt 2009). There are three potential evolutionary channels leading to this fate. Alongside the standard 'massive star' Fe-core collapse scenario we investigate the likelihood of electron capture supernovae (EC-SNe) from super-AGB (S-AGB) stars in their thermal pulse phase, from failed massive stars for which neon burning and other advanced burning stages fail to prevent the star from contracting to the critical densities required to initiate rapid electron-capture reactions and thus the star's collapse. We find it indeed possible that both of these relatively exotic evolutionary channels may be realised but it is currently unclear for what proportion of stars. Ultimately, the supernova light curves, explosion energies, remnant properties (see e.g. Knigge *et al.* 2011) and ejecta composition are the quantities desired to establish the role that these stars at the lower edge of the massive star mass range play.

Keywords. stars: evolution, supernovae: general

---

1. Preliminary models

For stars that develop degenerate ONe cores with $M_{\rm ONe} \gtrsim 1.37 M_\odot$, neon is ignited off-centre. Such an ignition was also found in the models of Nomoto (1984), but the subsequent evolution was not followed to a conclusion. As we discuss below, the nucleosynthesis in these neon-burning shells becomes rather complex and the speed and nature of their inward propagation determine the evolutionary outcome.

We computed a $9 M_\odot$ model from pre-main sequence using the MESA code (Paxton *et al.* 2011). Following the main sequence and core He-burning, our model ignites carbon non-degenerately at its centre where the stellar material becomes convectively unstable, which is typical of a massive star with $M_{\rm ini}/M_\odot \lesssim 20$. Following the ignition of convective secondary carbon burning shells, neon is ignited at the co-ordinate $\sim 0.75 M_\odot$ away from the centre. Ignition takes place where the maximum temperature now resides due to a temperature inversion in the core caused by the onset of degeneracy following the central extinction of carbon.

The centre contracts since there is no constant source of energy production there, only neutrino losses and URCA cooling. Meanwhile, energy production in the shell pushes the temperature there high enough to ignite oxygen, leaving behind mostly isotopes of Si and S. We find that the shell propagates by means of compressional heating, when gravitational energy released in contraction raises the temperature of the matter to the point of ignition. For that reason, in 1-dimension the propagation is seen as a series of flashes then contraction following extinction of the previous flash event.

There are two important timescales involved after neon is ignited: $\tau_{\rm flame}$, the timescale in which the propagating neon-oxygen burning shell will reach the centre and τ_ρ, the timescale in which the centre will contract to the threshold density critical for electrons to begin to capture on ^{20}Ne nuclei.

In our $9M_\odot$ model, $\tau_\rho < \tau_{\rm flame}$ and the central density surpasses the threshold for electron captures on ^{24}Mg to become energetic when the base of the flame is still $\sim 0.7M_\odot$ out from the centre. Assuming the Ledoux stability criterion for convection, the temperature peaks in the $\rho_{\rm c} - T_{\rm c}$ plane at locations corresponding to the threshold densities for electron captures on ^{24}Mg and ^{20}Ne as seen in Miyaji & Nomoto (1987) for example as the core contracts to higher densities. The central temperature begins to increase due to $^{20}{\rm Ne} + e^-$ at $\log_{10}(\rho_{\rm c}({\rm g\ cm^{-3}})) = 9.92$, eventually rising sufficiently to ignite oxygen.

The density at which oxygen is ignited (and thus, the critical density for electron captures on ^{20}Ne) determines whether the star will collapse or explode. For this reason it is crucial to include Coulomb corrections to the electron chemical potential, $\mu_{\rm e}$, and to the rate, both of which raise the effective threshold density (Gutierrez *et al.* 1996).

Lastly, while a S-AGB progenitor will have undergone a period of extended mass loss on the TP-SAGB, these low-mass massive stars for which $\tau_\rho < \tau_{\rm flame}$ will have the majority of their hydrogen-rich envelope intact at the point of collapse, perhaps with interesting consequences for the SN light curve.

2. Implications

Our models imply that there may be an additional evolutionary channel producing electron-capture supernovae. In order to have an idea of the statistical contribution of these stars to observed supernovae, numerical simulations with the most up-to-date microphysics must be computed. Since the neon-oxygen burning shell propagates by compressional heating in a degenerate regime, a large network involving many weak processes in addition to key URCA process reactions must be included in order to follow properly the $Y_{\rm e}$ profile.

We expect that the transition from S-AGB to massive star is not a finite jump, but instead is rather continuous, having some interesting nucleosynthetic and evolutionary consequences. What we would like to know is for what proportion of stars is neon ignited off-centre, for what proportion of those stars is the condition for EC-SN, $\tau_\rho < \tau_{\rm flame}$, satisfied; and if there are identifiable observational signatures that would enable us to differentiate between the different progenitor scenarios.

References

Gutierrez, J., Garcia-Berro, E., & Iben, Jr. *et al.* 1996, *ApJ*, 459, 701
Knigge, C., Coe, M. J., & Podsiadlowski, P. 2011, *Nature*, 479, 372
Miyaji, S. & Nomoto, K. 1987, *ApJ*, 318, 307
Nomoto, K. 1984, *ApJ*, 277, 791
Paxton, B., Bildsten, L., Dotter, A., Herwig, F., Lesaffre, P., & Timmes, F. 2011, *ApJS*, 192, 3
Smartt, S. J. 2009, *ARAA*, 47, 63

Death of Massive Stars: Supernovae and Gamma-Ray Bursts
Proceedings IAU Symposium No. 279, 2012
P. Roming, N. Kawai & E. Pian, eds.

doi:10.1017/S1743921312013269

Mass and metallicity constraints on supernova progenitors derived from integral field spectroscopy of the environment

Hanindyo Kuncarayakti[1], Mamoru Doi[1], Greg Aldering[2], Nobuo Arimoto[3], Keiichi Maeda[4], Tomoki Morokuma[1], Rui Pereira[5], Tomonori Usuda[6] and Yasuhito Hashiba[1]

[1]Institute of Astronomy, Graduate School of Science, the University of Tokyo
2-21-1 Osawa, Mitaka, Tokyo 181-0015, Japan
email: hanin@ioa.s.u-tokyo.ac.jp

[2]Physics Division, Lawrence Berkeley National Laboratory
1 Cyclotron Road, Berkeley, CA 94720, USA

[3]National Astronomical Observatory of Japan
2-21-1 Osawa, Mitaka, Tokyo 181-0015, Japan

[4]Kavli Institute for the Physics and Mathematics of the Universe, the University of Tokyo
5-1-5 Kashiwanoha, Kashiwa, Chiba 277-8583, Japan

[5]CNRS/IN2P3, Institut de Physique Nucléaire de Lyon
4 Rue Enrico Fermi, 69622 Villeurbanne Cedex, France

[6]Subaru Telescope, National Astronomical Observatory of Japan
650 North A'ohoku Place, Hilo, HI 96720, USA

Abstract. We have obtained optical integral field spectroscopy of the explosion sites of more than 25 nearby type-IIP/IIL/Ib/Ic supernovae using UH88/SNIFS, and additionally Gemini/GMOS IFU. This technique enables us to obtain both spatial and spectral information of the immediate environment of the supernovae. Using strong line method we measured the metallicity of the star cluster present at the explosion site, presumably the coeval parent stellar population of the supernova progenitor, and comparison with simple stellar population models gives age estimate of the cluster. With this method we were able to put constraints on the metallicity and age of the progenitor star. The age, i.e. lifetime, of the progenitor corresponds to the initial mass of the star. By far this is the most direct measurement of supernova progenitor metallicity and, if the cluster-progenitor association is confirmed, provides reliable determination of the initial mass of supernova progenitor stars.

Keywords. stars: supernovae

---

1. Introduction

Despite the large number of observed events, the current knowledge of supernova progenitors and of the final stages of massive stars still needs refinement. Theoretical predictions have suggested numerous possibilities on how a certain type of star produces which type of supernova. This poses the challenge of answering the question of which kind of star produces which supernova with observational evidences. It has been suggested that progenitor mass and metallicity are the most important parameters which drive the evolution of a massive star into supernova (e.g. Georgy *et al.* 2009). With integral field spectroscopy we were able to determine directly the metallicity of the explosion site, and derive the initial mass of the progenitor star via host cluster study. These constraints of mass and metallicity will give better understanding of the progenitor stars of supernovae.

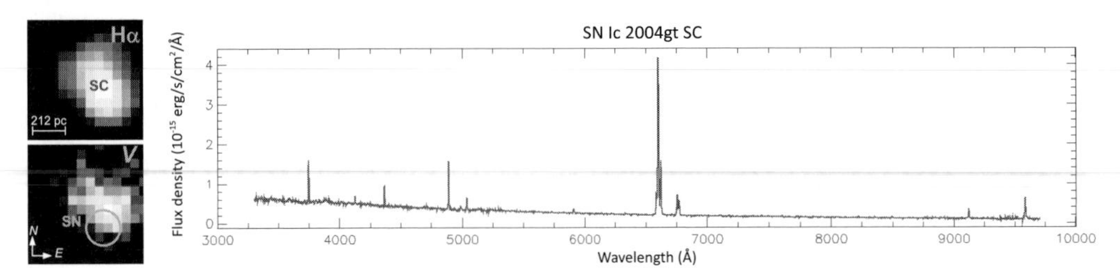

Figure 1. Reconstructed SNIFS IFU focal plane images of SN 2004gt explosion site in Hα and V-band (left panels), and the extracted spectrum of the host cluster (right).

2. Observation and data analysis

Using SNIFS integral field spectrograph (Aldering *et al.* 2002) attached to University of Hawaii 2.2 m telescope (UH88) at Mauna Kea and Gemini-N/GMOS-IFU (Hook *et al.* 2004; Allington-Smith *et al.* 2002), we observed 27 nearby type-IIP/IIL/Ib/Ic supernova sites in 2010–2011. Both SNIFS and GMOS datasets were analysed using IRAF. We performed aperture extraction to the flux-calibrated (x, y, λ) datacubes in the wavelength direction to obtain the spectra of the clusters in the field of view. We then measured the emission lines to determine metallicity using strong line method (Pettini & Pagel 2004). The age of the clusters was determined by comparison of Hα or near-infrared Ca II triplet equivalent widths with simple stellar population (SSP) models from Starburst99 (Leitherer *et al.* 1999) for the respective metallicities. Adopting the age of the parent stellar population as the supernova progenitor lifetime, we derived initial mass of the progenitor star using Padova stellar evolution models (Bressan *et al.* 1993).

3. Results

Figure 1 shows SN 2004gt site, one example of the results. SN 2004gt exploded in the outskirts of a cluster identified as the 5.8 Myr-old knot S in Whitmore *et al.* (2010). We derived the age of this cluster as 5.8 Myr. Metallicity was derived as 12 + log(O/H) = 8.72, consistent with Modjaz *et al.* (2011)'s determination of 12 + log(O/H) = 8.70. The derived age corresponds to the lifetime of a progenitor star with initial mass of 33.5 $M_\odot$.

We carried out our analysis on all of our samples and found that the result is mostly consistent with the currently accepted SN Ic > Ib > II progenitor mass sequence (Anderson & James 2008). Interestingly, we also found that the progenitors of all those types of core-collapse supernovae can appear on the same region in the mass-metallicity space. The full result of this work will be presented in Kuncarayakti *et al.* (2012, in preparation).

References

Aldering, G., *et al.* 2002, *Proc. SPIE*, 4836, 61
Allington-Smith, J., *et al.* 2002, *PASP*, 114, 892
Anderson, J. P. & James, P. A. 2008, *MNRAS*, 390, 1527
Bressan, A., *et al.* 1993, *A&AS*, 100, 647
Georgy, C., *et al.* 2009, *A&A*, 502, 611
Hook, I., *et al.* 2004, *PASP*, 116, 425
Leitherer, C., *et al.* 1999, *ApJS*, 123, 3
Modjaz, M., *et al.* 2011, *ApJ*, 731, 4
Pettini, M. & Pagel, B. E. J. 2004, *MNRAS*, 348, L59
Whitmore, B. C., *et al.* 2010, *AJ*, 140, 75

Death of Massive Stars: Supernovae and Gamma-Ray Bursts
Proceedings IAU Symposium No. 279, 2012
P. Roming, N. Kawai & E. Pian, eds.

doi:10.1017/S1743921312013270

Observing GRBs and Supernovae at Gemini Observatory as Target of Opportunity (ToO)

M. Lemoine-Busserolle[1], K. C. Roth[1], E. R. Carrasco[2], B. W. Miller[2], A. W. Stephens[1], I. Jorgensen[1] and B. Rodgers[2]

[1]Gemini North Observatory, 670 N. A'Ohoku St, Hilo, HI, USA 96720

[2]Gemini South Observatory, c/o AURA, Casilla 603, La Serena, CHILE

Abstract. The Gemini Observatories primarily operate a multi-instrument queue, with observers selecting observations that are best suited to weather and seeing conditions. The Target of Opportunity (ToO) observing mode is intended to allow observation of targets that cannot be specified in advance but which have a well defined external trigger such as distant supernovae or Gamma Ray bursts. In addition, the instrument and configuration best suited to observe the ToO may depend on properties of the event, such as brightness and redshift which again are impossible to know in advance. Queue observing naturally lends itself to Target of Opportunity (ToO) support since the time required to switch between programs and instruments is very short, and the staff observer is trained to operate all the available instruments and modes. Gemini Observatory has supported pre-approved ToO programs since beginning queue operations, and has implemented a rapid (less than 15 minutes response time) ToO mode since 2005. ToOs comprise a significant fraction of the queue (20–25% of the highest ranking band) nowadays. We discuss the ToO procedures, the statistics of rapid ToOs observing at Gemini North Observatory, the science related to GRBs and supernovae that this important mode has enabled.

Keywords. telescopes, supernovae, observations, gamma rays: bursts.

---

1. ToOs Trigger Types and Procedures

ToOs observations are a good match to queue observations. Because the science teams and template observations are approved in advance, the triggering of ToO observations involves no additional approvals, maximizing the efficiency and minimizing the time between detection of ToO targets and the collection of science data. This is particularly important for GRBs which often fade by 3-4 magnitudes in the first hour after their detection.In case of RAPID response trigger, the observation will be done within the next 24 hours. The minimum response time is about 20 minutes and these triggers can interrupt ongoing observations, both classical and queue. A STANDARD trigger will in general be executed more than 24 hours in the future. This is essentially identical to the normal queue mode except that the targets are not known in advance. Thus, observations will be placed in the queue based on science rank and observing conditions constraints (for more details see http://www.gemini.edu/?q=node/11014). All types of ToO observations can be prepared and submitted using the standard fetch/store operations of the Observing Tool (OT). The basics of ToO Phase II preparation are the same as for regular queue programs. However, in this case the PI needs to define template observations that will be used once the targets are known. Programs that observe more transient objects (e.g. SNe, GRBs) often use many instrument modes so they have to have more triggering options. All facility instruments that are offered for queue observing are available for ToO triggers, although GMOS MOS ToO programs are not accepted at this time. The instruments currently available on a given night, and details of the

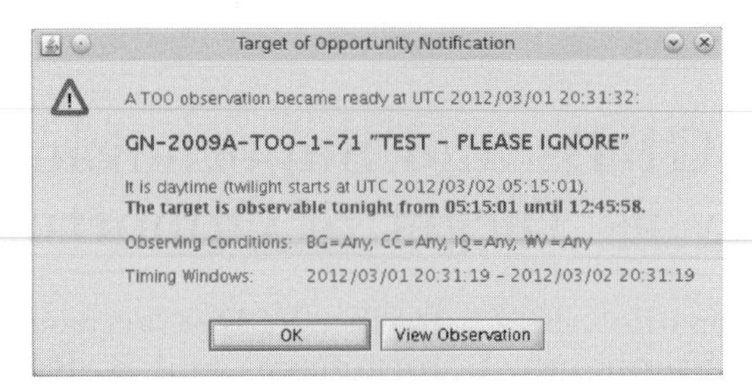

Figure 1. ToO notification in the Observing Tool.

GMOS North and South configurations, are kept up to date and are available for both the North and South telescopes. Standard ToO Gemini North laser guide star (LGS) observations are allowed; however, targets must be defined at least 8 days in advance of the start of the LGS observing block in which the observations are desired. Template observations should be made by the investigator for each instrument configuration that will be needed and should be stored in a folder called Templates. Once the investigator or the PI software triggered the observation a new active observation is generated in the observing database. In case of Standard ToO, an e-mail is automatically sent to the queue coordinators and the observation is given high priority in the queue generation for the next available night. For Rapid ToO, a windows pops up on the screen of the observer and any actives observing tools. These popup windows contain useful information on the target visibility and observing constraints. An audible alert in the control room occurs at the same time as the popup window and it only stops when the window is closed (Fig. 1). Once the notification is acknowledged by the observer the objective for imaging observations is an acquisition time of less than 6 minutes (including slew). For normal long-slit spectroscopy the objective is 15 minutes (currently the acquisition time is approximately 16 minutes and the slew time is 2 minutes.)

2. Science Highlight

Gemini Observatory rapid ToO follow-up observations of GRB optical have led to several breakthroughs in the understanding of these phenomena. In addition, standard ToO observations facilitate follow-up observations of supernovae as well as systematic monitoring of time variable targets. Some of science highlights that have resulted from GRBs and SNe observations can be found on the Gemini web site: The 2011 Nobel prize for physics citation recognizes two teams for the discovery of the accelerating expansion of the universe based on observations of SNe (http://www.gemini.edu/node/11688); the most distant GRB ever seen (http://www.gemini.edu/node/11634); and the first supernova to be discovered using laser guide star adaptive optics (http://www.gemini.edu/node/11226).

Acknowledgement

The Gemini Observatory is operated by the Association of Universities for Research in Astronomy, Inc., under a cooperative agreement with the NSF on behalf of the Gemini partnership: the National Science Foundation (United States), the Science and Technology Facilities Council (United Kingdom), the National Research Council (Canada), CONICYT (Chile), the Australian Research Council (Australia), Ministerio da Ciencia e Tecnologia (Brazil), and SECYT (Argentina).

Reference

Roth *et al.* 2010, *Proc. SPIE 7737*, 77370N (2010)

Death of Massive Stars: Supernovae and Gamma-Ray Bursts
Proceedings IAU Symposium No. 279, 2012
P. Roming, N. Kawai & E. Pian, eds.

doi:10.1017/S1743921312013282

Trigger Simulations for GRB Detection with the Swift Burst Alert Telescope

Amy Lien[1,4], Takanori Sakamoto[1,5], Neil Gehrels[1], David Palmer[2], and Carlo Graziani[3]

[1]NASA Goddard Space Flight Center,
Greenbelt, MD 20771, USA
email: amy.y.lien@nasa.gov; takanori@milkyway.gsfc.nasa.gov; neil.gehrels@nasa.gov

[2]Los Alamos National Lab,
B244, Los Alamos, NM 87545, USA
email: palmer@lanl.gov

[3]Department of Astronomy and Astrophysics, University of Chicago
5640 South Ellis Avenue, Chicago, IL 60637, USA
email: carlo@oddjob.uchicago.edu

[4]ORAU [5]CRESST/UMBC

Abstract. Understanding the intrinsic cosmic long gamma-ray burst (GRB) rate is essential in many aspects of astrophysics and cosmology, such as revealing the connection between GRBs, supernovae (SNe), and stellar evolution. *Swift*, a multi-wavelength space telescope, is quickly expanding the GRB category by observing hundreds of GRBs and their redshifts. However, it remains difficult to determine the intrinsic GRB rate due to the complex trigger algorithm adopted by *Swift*. Current studies of the GRB rate usually approximate the *Swift* trigger algorithm by a single detection threshold. Nevertheless, unlike the previously flown GRB instruments, *Swift* has over 500 trigger criteria based on count rates and additional thresholds for localization. To investigate possible systematic biases and further explore the intrinsic GRB rate as a function of redshift and the GRB luminosity function, we adopt a Monte Carlo approach by simulating all trigger criteria used by *Swift*. A precise estimation of the intrinsic GRB rate is important to reveal the GRB origins and their relation to the black-hole forming SNe. Additionally, the GRB rate at high redshifts provides a strong probe of the star formation history in the early universe, which is hard to measure directly through other methods.

Keywords. Gamma-ray Bursts

---

1. Introduction and Method

We adopt a Monte Carlo approach to explore the intrinsic cosmic long GRB rate, luminosity function, and other GRB characteristics. *Swift* triggers GRB detections mainly based on the photon count rates. In order to maximize the GRB observations, *Swift* has hundreds of trigger criteria to cover a wide range of possible burst durations and pulse shapes. Each criterion uses different time ranges for the background and foreground periods to calculate the corresponding signal-to-noise ratio (Fenimore *et al.* 2003; Graziani *et al.* 2003). If the signal-to-noise ratio passes the threshold adopted by the criterion, the event is "triggered" and will go through further procedures to confirm the nature of the event. In our program, we simulate all these criteria with different background/foreground time ranges, signal-to-noise thresholds, viewing angles related to the detector, and in different energy bands. Accurately mimicking the complex trigger algorithm adopted by *Swift* allows us to investigate possible systematic biases of *Swift*'s detections, and thus explore intrinsic GRB characteristics, such as their rate and luminosity function.

In order to preform a Monte Carlo simulation, we need to create a mock GRB sample based on some assumed GRB characteristics. We will then run the mock sample through the trigger code that simulates the *Swift* trigger algorithm. If the original assumptions of the GRB characteristics are accurate, the properties of the triggered events in the mock sample should match those of the actual GRB sample detected by *Swift*.

We generated the first mock GRB sample using the GRB rate and luminosity function in Wanderman & Piran (2010). These authors obtained the GRB rate and luminosity function by directly inverting the observed rate from *Swift*, with some empirical weighting factors to describe the probability of a burst being detected by *Swift* and acquiring redshift measurements. We also adopt the distribution of the GRB spectral index found in Sakamoto *et al.* (2009) to assign spectra to GRBs in our mock sample. Additionally, the GRB light curves (pulse shapes) in our mock sample are created based on Norris *et al.* (2005), which provides fitting functions for light curves of 23 GRBs detected by *Swift*.

2. Preliminary Results

Figure 1 shows our preliminary results based on the GRB characteristics discussed above. The red histogram plots the redshift distribution of the (normalized) number of the GRBs in the mock sample triggered by our code that simulates the *Swift* trigger algorithm. The blue histogram shows the redshift distribution of the actual *Swift*-detected GRBs for comparison (numbers are also normalized). Note that this *Swift*-detected GRB sample is a sub-sample of all GRBs observed by *Swift*. This sub-sample is selected by Fynbo *et al.* (2009), using criteria for choosing GRBs that produce an event sample that is relatively unaffected by selection biases in redshift measurements.

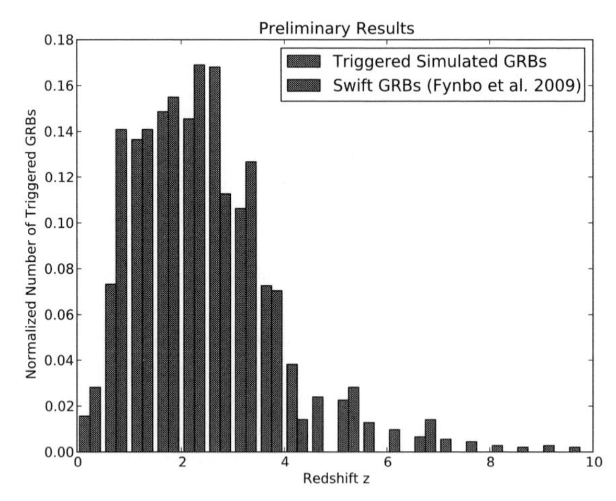

Figure 1. The normalized histogram of the triggered GRBs as a function of redshift. All numbers are normalized to the total number of triggered events. The triggered events from our simulation are plotted in red; the selected *Swift* GRBs sample (Fynbo *et al.* 2009) is plotted in blue for comparison.

References

Fynbo, J. P. U., Jakobsson, P., Prochaska, J. X., *et al.* 2009, *ApJS*, 185, 526

Graziani, C. 2003, *Gamma-Ray Burst and Afterglow Astronomy* 2001: 662, 79

Norris, J. P., Bonnell, J. T., Kazanas, D., *et al.* 2005, *ApJ*, 627, 324

Sakamoto, T., Sato, G., Barbier, L., *et al.* 2009, *ApJ*, 693, 922

Wanderman, D. & Piran, T. 2010, *mnras*, 406, 1944

Fenimore, E. E., Palmer, D., Galassi, M., *et al.* 2003, *Gamma-Ray Burst and Afterglow Astronomy* 2001: 662, 491

Death of Massive Stars: Supernovae and Gamma-Ray Bursts
Proceedings IAU Symposium No. 279, 2012
P. Roming, N. Kawai & E. Pian, eds.

doi:10.1017/S1743921312013294

The Ultra-Fast Flash Observatory's space GRB mission and science

H. Lim[1], S. Ahmad[2], P. Barrillon[2], S. Blin-Bondil[2], S. Brandt[3], C. Budtz-Jørgensen[3] A. J. Castro-Tirado[4], P. Chen[5], H. S. Choi[6], Y. J. Choi[7], P. Connell[8], S. Dagoret-Campagne[2], C. De La Taille[2], C. Eyles[8], B. Grossan[9], I. Hermann[7], M. -H. A. Huang[10], S. Jeong[1], A. Jung[1], J. E. Kim[1], S. -W. Kim[3], Y. W. Kim[1], J. Lee[1], E. V. Linder[1,9], T. -C. Liu[5], N. Lund[3], K. W. Min[7], G. W. Na[1], J. W. Nam[1], K. H. Nam[1], M. I. Panasyuk[12], I. H. Park[1], V. Reglero[8], J. Řípa[1], J. M. Rodrigo[8], G. F. Smoot[1,9], S. Svetilov[12], N. Vedenkin[12], and I. Yashin[12]

[1]Ewha Womans University, Seoul, Korea
email: heuijin.lim@gmail.com

[2]University of Paris-Sud 11, France

[3]National Space Institute, Denmark

[4]Instituto de Astrofisica de Andalucia, CSIC, Spain

[5]National Taiwan University, Taipei, Taiwan

[6]Korea Institute of Industrial Technology, Ansan, Korea

[7]Korea Advanced Institute of Science and Technology, Daejeon, Korea

[8]University of Valencia, Spain

[9]University of California, Berkeley, USA

[10]National United University, Miao-Li, Taiwan

[11]Yonsei University, Seoul, Korea

[12]Moscow State Univ., Moscow, Russia

Abstract. The Ultra-Fast Flash Observatory (UFFO) is a space mission to detect the early moments of an explosion from Gamma-ray bursts (GRBs), thus enhancing our understanding of the GRB mechanism. It consists of the UFFO Burst & Trigger telescope (UBAT) for the recognition of GRB positions using hard X-ray from GRBs. It also contains the Slewing Mirror Telescope (SMT) for the fast detection of UV-optical photons from GRBs. It is designed to begin the UV-optical observations in less than a few seconds after the trigger. The UBAT is based on a coded-mask X-ray camera with a wide field of view (FOV) and is composed of the coded mask, a hopper and a detector module. The SMT has a fast rotatable mirror which allows a fast UV-optical detection after the trigger. The telescope is a modified Ritchey-Chrétien telescope with the aperture size of 10 cm diameter, and an image intensifier readout by CCD. The UFFO pathfinder is scheduled to launch into orbit on 2012 June by the Lomonosov spacecraft. It is a scaled-down version of UFFO in order to make the first systematic study of early UV/optical light curves, including the rise phase of GRBs. We expect UBAT to trigger ~44 GRBs/yr and expect SMT to detect ~10 GRBs/yr.

Keywords. gamma rays: bursts, gamma rays: observations, instrumentation: detectors

---

1. Overview

The Ultra-Fast Flash Observatory (UFFO) aims to measure the early UV/optical observations of GRBs using the new approach of a beam steerer which can be implemented by a rotatable mirror in the UV/optical telescope. It allows UV/optical observations to

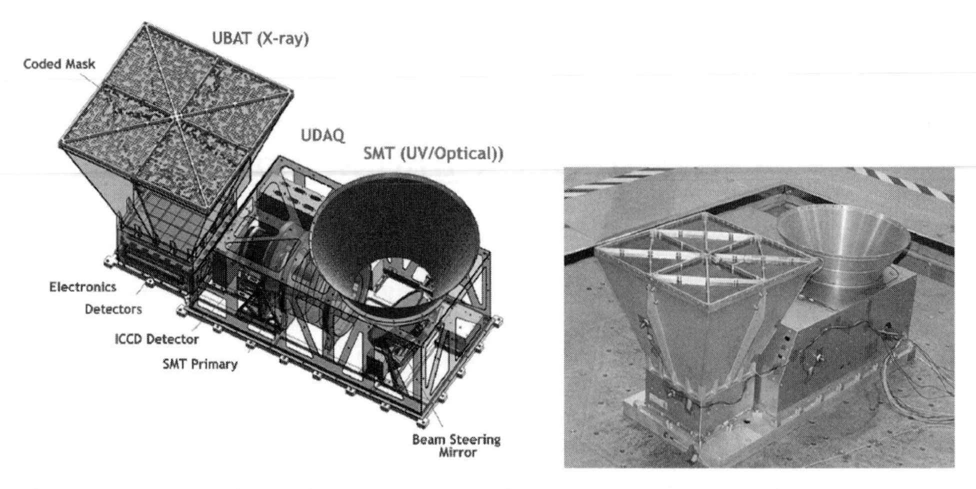

Figure 1. *Left*: UFFO pathfinder payload. *Right*: UFFO pathfinder pre-flight model which passed its space environment tests on 2011-July.

begin in less than a few seconds after the X-ray trigger Park *et al.* 2009). The UFFO will exploit this largely unexplored region of parameter space by providing a statistically significant sample of early UV/optical observations of GRBs.

The UFFO consists of a couple of wide Field-of-View (FOV) trigger telescopes, a narrow-FOV Slewing Mirror Telescope (SMT) for the fast measurement of the UV/optical photons from GRBs, and a gamma-ray monitor for energy measurement. The UFFO Burst Alert & Trigger Telescope (UBAT) will provide the primary trigger using X-rays from GRBs. It monitors the sky for GRBs and determines their position with sufficient accuracy (10 arcmin at 7.0σ) for follow-up in the UV/optical with the SMT. Whereas the fastest previous experiment, the *Swift* observatory, rarely observed GRBs in less than 60 seconds after the trigger, the UFFO is designed to begin UV/optical observations in less than a few seconds after the trigger. The SMT uses the novel approach of steering our telescope beam using a rotatable mirror, instead of re-orienting the instrument platform like *Swift* and other previous instruments. The UFFO pathfinder is the scaled-down version of UFFO with a physical size of 958.5(L)$\times$400(W)$\times$382.5(H) mm^3 (See Fig. 1(left)) and is scheduled to launch into orbit on 2012-June by the Lomonosov spacecraft. It will be on a sun-synchronous orbit at an altitude of $\sim$550 km. It successfully passed the thermal-vacuum test and the vibration-shock test on 2011-July at Taiwan NSPO (See Fig. 1(right)) and it is in the final stage for launch preparations.

Acknowledgements

This research was supported by the Basic Science Research Program through the National Research Foundation of Korea (NRF) that is funded by the Ministry of Education, Science and Technology (2010-0025056). This research is also supported by the World Class University (WCU) program through the NRF that is funded by the Ministry of Education, Science and Technology (R32-2008-000-10130-0) in Korea.

Reference

Park, I. H., *et al.* 2009, arXiv:0912.0773

Death of Massive Stars: Supernovae and Gamma-Ray Bursts
Proceedings IAU Symposium No. 279, 2012
P. Roming, N. Kawai & E. Pian, eds.

doi:10.1017/S1743921312013300

An X-ray study of mass-loss rate and wind acceleration of massive stars

Yoshitomo Maeda[1], Yasuharu Sugawara[2] and the WR140 collaborations

[1]Institute of Space and Astronautical Science, Japan Aerospace Exploration Agency 3-1-1 Yoshinodai, Sagamihara, Kanagawa 229-8510
email: ymaeda@astro.isas.jaxa.jp

[2]Department of Physics, Faculty of Science & Engineering, Chuo University, 1-13-27 Kasuga, Bunkyo, Tokyo 112-8551
email: sugawara@phys.chuo-u.ac.jp

Abstract. By monitoring WC7 and the O5.5 binary WR 140 with the *Suzaku* telescope, we demonstrate a new method to measure the mass loss rates of both stars. By using the absorption column density, we found a mass-loss rate for the WC7 component : $\dot{M}_{\rm WC7} \approx 1.2\times10^{-5}{\rm M}_{\odot}{\rm yr}^{-1}$. We also measured the mass-loss rate of the companion O component using a luminosity variation in phases: $\dot{M}_{\rm O5.5} \approx 5 \times 10^{-7}{\rm M}_{\odot}{\rm yr}^{-1}$.

Keywords. Massive star, GRB progenitor, mass-loss rate

1. Introduction

Evolution of massive stars is critical to understand the explosion of the gamma-ray burst (GRB). The stellar wind is also helpful to understand the afterglow of the GRB. However, fundamental parameters of massive stars such as stellar mass and mass-loss rate are not easy to be derived. Therefore, understanding of these parameters still remains an unresolved issue in astronomy.

WR 140 is a wide colliding wind binary system of WC7+O5.5. Its orbit has been well determined with $P_{\rm orb}$ = 2896.35 days, i = 119.6°, and e = 0.8964 by detailed optical monitoring (Monnier *et al.* 2011). We pursued this via *Suzaku* monitoring during the periastron passage in 2009, which includes broad band X-ray spectra. A full paper of the results will be published as Sugawara *et al.* (2012). In this proceeding, we introduce measurements of mass loss rates of both stars from Sugawara *et al.* (2012).

2. Wind acceleration and mass-loss rate

We estimate the mass-loss rate of the WC7 star using the observed column density $N_{\rm He}$. We found the column density obtained at around the periastron passage can be explained by $\dot{M}_{\rm WC7} \sim 1.2 \times 10^{-5}$ ${\rm M}_{\odot}{\rm yr}^{-1}$. Fahed *et al.* (2011) also estimated $\dot{M}_{\rm WC7}$ $\sim 3 \times 10^{-5}{\rm M}_{\odot}{\rm yr}^{-1}$ that is larger by a factor of three than our value.

According to Stevens *et al.* (1992), the X-ray luminosity of the colliding wind zone can be written as

$$L_{\rm X} \propto D^{-1}(1+A)/A^4 \qquad (2.1)$$

where the wind momentum flux ratio, $A = ((\dot{M}_{\rm WC7}\ v_{\rm WC7}(r))/(\dot{M}_{\rm O5.5}\ v_{\rm O5.5}(r)))^{1/2}$. We also adapted a simple beta law for the wind acceleration as

$$v(r) = v_{\infty}(1 - R/r)^{\beta}. \qquad (2.2)$$

Table 1. Summary of wind-acceleration parameter β and mass-loss rate $\dot{M}$.

$\dot{M}_{O5.5}$ / $\dot{M}_{WC7}$	$\beta_{O5.5}$	$\dot{M}_{O5.5}$ [$M_{\odot}yr^{-1}$]	β_{WC7}	$\dot{M}_{WC7}$ [$M_{\odot}yr^{-1}$]
0.2	3	2×10^{-6}	any	1.2×10^{-5}
0.1	2	1×10^{-6}	any	1.2×10^{-5}
0.04	1	5×10^{-7}	any	1.2×10^{-5}

Here, v_{∞} and R are the terminal wind-velocity and stellar radius, respectively. We used the value of $v_{\infty,WC7}$=2860 km s^{-1} (Williams *et al.* 1990), $v_{\infty,O5.5}$=3100 km s^{-1} (Setia Gunawan *et al.* 2001), $R_{WC7} = 2$ R$_{\odot}$ and $R_{O5.5} = 26$ R$_{\odot}$ (cf. Williams *et al.* (2009)). The mass-loss rate from the O5.5 star can be solved as a function of mass-loss rate ratio $\dot{M}_{O5.5}$ / $\dot{M}_{WC7}$ as summarized in Table 1. Since it is widely believed that β for O stars is about unity, we conclude $\dot{M}_{O5.5}$ to be $\sim 5 \times 10^{-7}$ M$_{\odot}$yr^{-1}.

3. Discussion

Using the direct imaging technique of the stars and the shocked cone combined with the optical spectroscopy, the mass of the O5.5 star was measured as 41 ± 6 M$_{\odot}$ (Fahed *et al.* 2011 and reference therein). Vink *et al.* (2001) theoretically gave a recipe (eq. 24) to calculate the mass-loss rate and predict for the O5.5 star ($\log L_{O5.5} = 6.18$, $T_{\rm eff,O5.5}$ =44,000 K) to be $\sim 6 \times 10^{-5}$ M$_{\odot}$yr^{-1} that is two orders of magnitude larger than our mass-loss number of $\dot{M}_{O5.5} \sim 5 \times 10^{-7}$ M$_{\odot}$yr^{-1}.

If we assume that the WC7 star had the same mass loss rate for $\sim$ 2 M yr, the initial mass of the WC7 star can be calculated as $\sim$40 M$_{\odot}$. The initial mass of the WC7 star should be much larger than the present mass of the companion O5.5 star if we assume that both stars were born at the same age. The WC7 star may then prefer a longer lifetime in the WR phase. A longer solution for the WR life in the solar metallicity is predicted for the case of fast-rotation (Meynet & Maeder 2005). If it is true, the WC7 star could be a progenitor of a future GRB event.

References

Fahed, R., *et al.* 2011, *MNRAS*, 418, 2
Meynet, G. & Maeder, A. 2005, *A&A*, 429, 581
Monnier, J. D., Zhao, M., Pedretti, E., *et al.* 2011, *ApJL*, 742, L1
Setia Gunawan, D. Y. A., van der Hucht, K. A., Williams, P. M., *et al.* 2001, *A&A*, 376, 460
Stevens, I. R., Blondin, J. M., & Pollock, A. M. T. 1992, *ApJL*, 386, 265
Sugawara, Y. *et al.* 2012, *PASJ*, submitted
Vink, J. S., de Koter, A., & Lamers, H. J. G. L. M. 2001, *A&A*, 369, 574
Williams, P. M., van der Hucht, K. A., Pollock, A. M. T., Florkowski, D. R., van der Woerd, H., & Wamsteker, W. M. 1990, *MNRAS*, 243, 662
Williams, P. M., *et al.* 2009, *MNRAS*, 395, 1749

Death of Massive Stars: Supernovae and Gamma-Ray Bursts
Proceedings IAU Symposium No. 279, 2012
P. Roming, N. Kawai & E. Pian, eds.

doi:10.1017/S1743921312013312

GRB host galaxies: theoretical investigation

Jirong Mao

Korea Astronomy and Space Science Institute 776, Daedeokdae-ro, Yuseong-gu, Daejeon, Republic of Korea 305-348
email: jirongmao@kasi.re.kr

Abstract. Long gamma-ray bursts (GRBs) can be linked to the massive stars and their host galaxies are assumed to be the star-forming galaxies within small dark matter halos. We apply a galaxy evolution model, in which the star formation process inside the virialized dark matter halo at a given redshift is achieved. The star formation rates (SFRs) in the GRB host galaxies at different redshifts can be derived from our model. The related stellar masses, luminosities, and metalicities of these GRB host galaxies are estimated. We also calculate the X-ray and optical absorption of GRB afterglow emission. At higher redshift, the SFR of host galaxy is stronger, and the absorption in the X-ray and optical bands of GRB afterglow is stronger, when the dust and metal components are locally released, surrounding the GRB environment. These model predictions are compared with some observational data as well.

Keywords. gamma rays: bursts, galaxies: high-redshift, dust, extinction.

1. Introduction

Long-duration gamma-ray burst (GRB) progenitors have been proposed as massive collapsing stars. Some oflong GRBs can be found in star formation galaxies which are dominated by young stellar populations. GRBs favor a metal-poor environment. Moreover, those high-redshift ($z > 2$) GRB host galaxies may trace the star formation of the universe. The high global star formation rate (SFR) history at redshifts larger than 6 indicate the possibility of high-redshift GRB production and the detection of host galaxies. On the other hand, the heavy attenuation shown in X-ray afterglow has been given, indicating a dense surrounding environment for those GRBs. It is also interesting to understand whether this kind of strong attenuation intrinsically evolves with redshift. The characteristics of the corresponding absorption shown in optical afterglow are still under debate. In order to have an explanation of GRB afterglow obscuration, the physical origin associated with star formation and galactic evolution should be studied in an unified scenario.

2. Results

We specify one physical model of star-forming and metal-poor galaxies as the hosts of long GRBs, exploiting the physical recipes from Granato *et al.* (2004). In the general scenario at each redshift bin, the SFR and galaxy mass in the given dark halo potential well have been calculated. In particular, Mao *et al.* (2007) have calculated the UV luminosities and the relative dust attenuation in the star-forming and metal-poor galaxies. The results are given in Fig. 1-Fig. 6. We invite readers to see the paper of mao (2010) for the physical processes and the denotations shown in the plots in detail. We note that our theoretical results from the model with dark matter halo mass of $5 \times 10^{11} M_{\odot}$ and the corresponding galactic evolutionary timescale of 5.0×10^{7} yr are consistent with those observational data well.

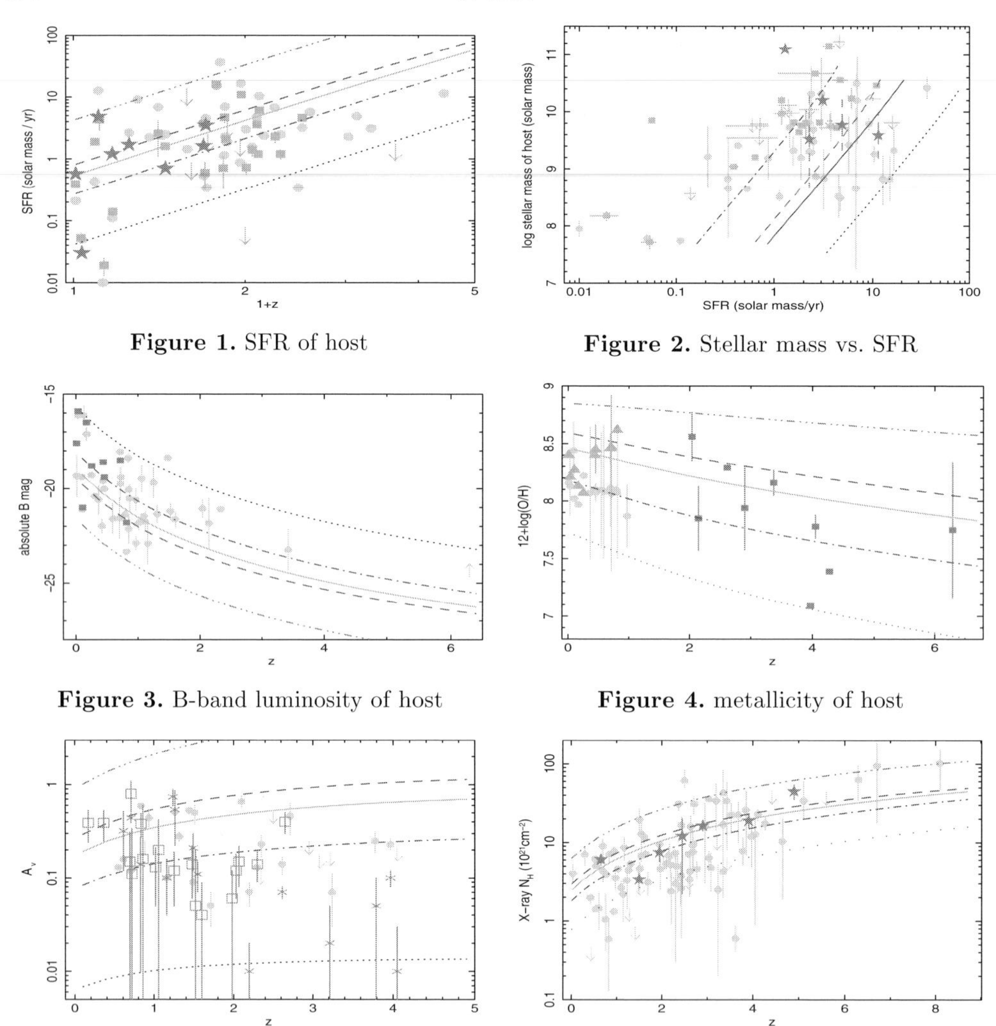

Figure 1. SFR of host

Figure 2. Stellar mass vs. SFR

Figure 3. B-band luminosity of host

Figure 4. metallicity of host

Figure 5. dust absorption of optical afterglow

Figure 6. attenuation of X-ray afterglow

3. Conclusion

We reveal that some properties from afterglow emissions and GRB hosts have shown a possible intrinsic cosmological evolution even some selection effects are considered. We speculate that improving the sensitivity of detectors on high-energy telescopes is not highly useful for catching more high-redshift but faint GRBs, since low-energy-released GRBs are almost absent in the high-redshift universe. We also caution that those high-redshift ($z > 5$) GRB host galaxies have not been significantly detected by space and ground-based telescopes (see Basa *et al.* 2012 and Tanvir *et al.* 2012 in detail).

References

Basa, S., *et al.* 2012, *arXiv*: 1201.6383
Granato, G. L., De Zotti, G., Silva, L., Bressan, A., & Danese, L. 2004, *ApJ*, 600, 580
Mao, J., Lapi, A., Granato, G. L., de Zotti, G., & Danese, L. 2007, *ApJ*, 667, 655
Mao, J. 2010, *ApJ*, 717, 140
Tanvir, N. R., *et al.* 2012, *arXiv*: 1201.6074

Death of Massive Stars: Supernovae and Gamma-Ray Bursts
Proceedings IAU Symposium No. 279, 2012
P. Roming, N. Kawai & E. Pian, eds.

doi:10.1017/S1743921312013324

Photospheric thermal radiation from GRB collapsar jets

Akira Mizuta[1] and Shigehiro Nagataki[2]

[1]Theory Center, Institute of Particle and Nuclear Studies, KEK (High Energy Accelerator Research Organization), 1-1 Oho, Tsukuba 305-0801, Japan
email: mizuta@post.kek.jp

[2]Yukawa Institute for Theoretical Physics, Kyoto University, Kitashirakawa Oiwake-cho, Sakyo-ku, Kyoto, 606-8502
email: nagataki@yukawa.kyoto-u.ac.jp

Abstract. Photospheric thermal radiation components from gamma-ray burst (GRB) jets are estimated based on relativistic hydrodynamic simulations of jet propagation. The light curves and spectra are derived, considering viewing angle effects. The light curves exhibit several seconds time variability and the luminosity is as large as that of GRB prompt emission. For observers at a viewing angle of several degrees the spectra below the peak energy are much softer than that of Planck distribution and close to typical GRB spectrum. Whereas the spectra for observers at small viewing angle are hard and close to Planck distribution. Numerical Amati and Yonetoku relations are reproduced.

Keywords. gamma-ray burst, hydrodynamics, numerical, radiation mechanisms: thermal

---

1. Introduction

The radiative mechanism of the GRB prompt emission is not fully understood. If the central engine produces a fireball, strong photospheric thermal radiation is expected during jet propagation into circum-stellar matter. The spectrum of the GRB prompt emission is a broken-power law. The νF_ν spectrum has a peak energy (E_p). The spectral index below the peak energy is typically unity which is much softer than that of single temperature Planck distribution. Recently thermal radiation components have been found in the spectrum of the GRB prompt emission, i.e., GRB090902B (Ryde *et al.* (2010), Zhang *et al.* (2011), and Ryde *et al.* (2011)), and GRB0909026B (Serino *et al.* (2011)).

The luminosity and spectrum of the photospheric thermal radiation from collapsar jets are derived theoretically by Pe'er & Ryde (2011), assuming steady and spherical outflow profile. Lazzati *et al.* (2009), Mizuta *et al.* (2011), Lazzati *et al.* (2011) and Nagakura *et al.* (2011) derived light curves of photospheric thermal radiation components from numerical relativistic hydrodynamic simulations. The luminosity is high enough to explain the prompt emission of typical GRBs. In this report, we show the results of photospheric thermal radiation, considering viewing angle effects.

2. Model, Methods, and Results

2D relativistic hydrodynamic simulations on jet propagation from collapsars have been performed, assuming axisymmetry and equatorial plane symmetry. Numerical methods and jet model are the same as those used in Mizuta *et al.* (2011), except the radial computational domain is extended to $r = 3\times10^{13}$ cm. The model 16TI which is developed by Woosley & Heger (2006) is employed as a progenitor star. Duration of the injection is 30 or 100 s.

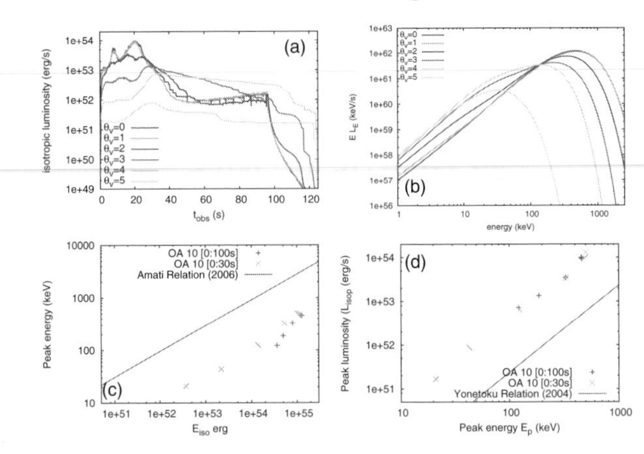

Figure 1. Light curves (a) and spectra (b) for different viewing angle (θ_v) observers (100 s injection model). $E_p - E_{\rm iso}$ (c) and $E_p - L_{{\rm iso}p}$ (d) plots with Amati and Yonetoku relations.

Figure 1(a) shows light curves of the photospheric thermal radiation for different viewing angles (θ_v) (100 s injection model). The luminosity quickly increases for small viewing angle and the light curves exhibit a few second time variability. On the other hand, the luminosity is not so high at viewing angles of several degrees. Fig. 1(b) shows spectra of photospheric thermal radiation for different viewing angles (100 s injection model). The spectrum below the peak energy is a power law and the indices for small viewing angles are $1 \sim 2.6$, while the indices for several degrees viewing angles are $1 \sim 1.5$, which is softer than that of single temperature Planck distribution. The isotropic radiation energy and peak energy are shown in Fig. 1(c). The isotopic peak luminosity and peak energy are shown in Fig. 1(d). Both plots show good correlations and reproduce the trend of Amati and Yonetoku relations, although absolute values do not match (numerical Amati and Yonetoku relations).

We would like to thank A. Heger for his kindness to allow us to use his progenitor model for this study. This work was carried out on SR16000 at YITP, Kyoto University, on the Space Science Simulator (NEC SX9) at JAXA, and on XT4 at CFCA at NAOJ. This work is partly supported by Grants-in-Aid from the MEXT Japan (20105005 (A.M.), 23105709 (S.N.)), Japan Society for the Promotion of Science (19104006 and 23340069 (S.N.)), and the Global COE Program 'The Next Generation of Physics, Spun from University and Emergence from MEXT of Japan' (S.N.).

References

Amati, L., *et al.* 2002, *A&A*, 390, 81
Lazzati, D., Morsony, B. J., & Begelman, M. C. 2009, *ApJ*, 700, L47
Lazzati, D., Morsony, B. J., & Begelman, M. C. 2011, *ApJ*, 732, 34
Mizuta, A., Nagataki, S., & Aoi, J. 2011, *ApJ*, 732, 26
Nagakura, H., Ito, H., Kiuchi, K., & Yamada, S. 2011, *ApJ*, 731, 80
Pe'er, A. & Ryde, F. 2011, *ApJ*, 732, 49
Ryde, F., *et al.* 2010, *ApJ*, 709, L172
Ryde, F., Pe'Er, A., Nymark, T., *et al.* 2011, *MNRAS*, 415, 3693
Serino, M., Yoshida, A., Kawai, N., *et al.* 2011, *PASJ*, 63, 1035
Woosley, S. E. & Heger, A. 2006, *ApJ*, 637, 914
Zhang, B.-B., Zhang, B., Liang, E.-W., *et al.* 2011, *ApJ*, 730, 141
Yonetoku, D., *et al.* 2004, *ApJ*, 609, 935

Death of Massive Stars: Supernovae and Gamma-Ray Bursts
Proceedings IAU Symposium No. 279, 2012
P. Roming, N. Kawai & E. Pian, eds.

doi:10.1017/S1743921312013336

Magnetorotational supernovae with different equations of state

Sergey G. Moiseenko and Gennady S. Bisnovatyi-Kogan

Space Research Institute,
Profsouznaya str. 84/32 117997 Moscow, Russia
email: moiseenko@iki.rssi.ru

Abstract. We present results of the simulation of a magneto-rotational supernova explosion. We show that, due to the differential rotation of the collapsing iron core, the magnetic field increases with time. The magnetic field transfers angular momentum and a MHD shock wave forms. This shock wave produces the supernova explosion. The explosion energy computed in our simulations is $0.5 - 2.5 \cdot 10^{51}$ erg. We used two different equations of state for the simulations. The results are rather similar.

Keywords. supernovae: general, methods: numerical

The important role of the rotation and the magnetic fields for core collapse supernova was suggested in the paper by Bisnovatyi-Kogan (1970). The idea of the magneto-rotational (MR) mechanism consists of angular momentum transfer outwards using the magnetic field, which is twisting due to the differential rotation. The toroidal component of the magnetic field is amplifying with time, which leads to increasing of the magnetic pressure and generating the supernova shock.

According to the results of two-dimensional modeling for the MR supernova mechanism, the initial stage of the linear growth of the toroidal magnetic field is followed by an exponential increase in both the toroidal and poloidal field components, accompanied by the development of MR instability.

For the simulations we used a specially developed implicit Lagrangian numerical scheme on a triangular grid of variable structure Ardeljan *et al.* (1996).

The results of our simulations show that the explosion energy depends weakly on the strength of the initial magnetic field. The 'shape' of the explosion depends qualitatively on the configuration of the magnetic field. For the quadrupole-like field the explosion develops preferably near the equatorial plane Ardeljan *et al.* (2005). For the dipole-like initial magnetic field the MR supernova explosion develops a mildly collimated jet Moiseenko *et al.* (2006).

In the paper by Bisnovatyi-Kogan *et al.* (2008) we simulated a MR supernova explosion for various initial core masses and rotational energies. The initial core mass was varied from 1.2 $M_\odot$ to 1.7 $M_\odot$. The specific rotational energy at the time when the magnetic field was switched on, E_{rot}/M_{core}, was varied from 0.19×10^{19} to 0.4×10^{19} erg/g. Fig. 1 presents the results of our calculations. It shows that the explosive energy of a MR supernova increases substantially with the mass of the iron core and the initial rotational energy (angular velocity; see Table). It is obvious that the energy released in a MR supernova is sufficient to explain supernovae with collapsing cores: $(0.5 - 2.6)10^{51}$ erg (Type II and Ib supernovae). The energies of Type Ic supernovae can be higher, probably due to the collapse of more massive cores, of the order of several tens of $M_\odot$.

For the simulations we used two different equations of state. The first one (Ardelyan *et al.* (1987)) is the approximation of tables from Ivanova *et al.* (1969). The second one

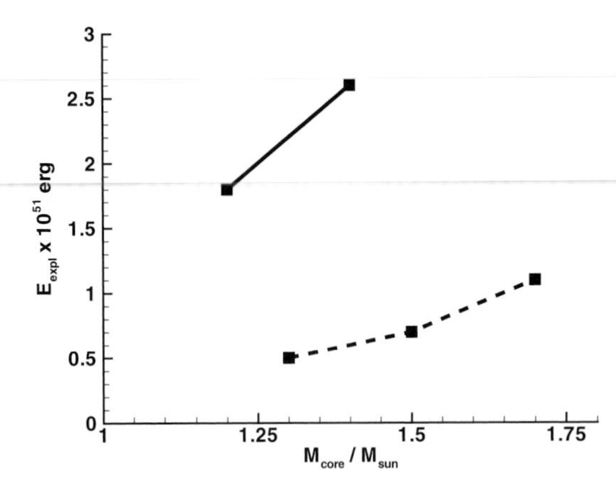

Figure 1. Energy of the explosion of a MR supernova as a function of the initial mass of the core for various specific rotational energies before the start of the evolution of the magnetic field (before beginning of the magnetic field evolution), $E_{rot}/M_{core} \approx 0.39 - 0.40 \cdot 10^{19}$ erg/g (solid line) and $E_{rot}/M_{core} \approx 0.19 - 0.23 \cdot 10^{19}$ erg/g (dashed line) (before collapse).

is from the paper by Shen *et al.* (1998). For the application of the second EOS we follow the paper by Kotake *et al.* (2003).

The results of simulations of the MR supernova explosion for both equations of state are in good agreement. The details of the simulations of the MR supernova with the EOS by Shen *et al.* (1998) will be published elsewhere.

Acknowledgements

This work was partially supported by RFBR grant 11-02-00602, grant NsH-5440.2012.2, program of the RAS 'Origin, structure and evolution of objects in the Universe'.

References

Ardeljan, N. V., Bisnovatyi-Kogan, G. S., Kosmachevskii, K. V., & Moiseenko, S. G., 1996, *A&AS*, 115, 573

Ardelyan, N. V., Bisnovatyi-Kogan, G. S., Popov, Yu. P., & Chernigovsky, S. V., 1987, *Astron. Zh.* 64, 761 (Soviet Astronomy, 1987, 31, 398)

Ardeljan, N. V., Bisnovatyi-Kogan, G. S., & Moiseenko, S. G., *Monthly Not. Roy. Astron. Soc.*, 359, 333 (2005).

Bisnovatyi-Kogan, G. S., 1970, *Astron. Zh.* 47, 813 (*Soviet Astronomy*, 1971, 14, 652

Bisnovatyi-Kogan G. S., Moiseenko S. G., & Ardelyan N. V., 2008, *Astronomy Reports*, 52,997

Ivanova, L. N., Imschennik, V. S., & Nadyozhin D. K., 1969 *Nauch. inf. Astron. Soveta*, 13,3

Kotake, K., Yamada, S., & Sato, K. 2003, *ApJ*, 595, 304

Moiseenko S. G., Bisnovatyi-Kogan G. S., & Ardeljan N. V., 2006, *Monthly Not. Roy. Astron. Soc.*, 370, 501

Shen, H., Toki, H., Oyamatsu, K., & Sumiyoshi, K. 1998, *Nucl. Phys. A*, 637, 435

Death of Massive Stars: Supernovae and Gamma-Ray Bursts
Proceedings IAU Symposium No. 279, 2012
P. Roming, N. Kawai & E. Pian, eds.

doi:10.1017/S1743921312013348

Influence of stellar oscillations on pulsar and magnetar magnetospheres

Viktoriya Morozova[1*], Bobomurat Ahmedov[2] and Olindo Zanotti[3]

[1]Institute of Nuclear Physics, Ulughbek, Tashkent 100214, Uzbekistan
[2]MPI for Gravitational Physics, D-14476 Potsdam, Germany
[3]Laboratory of Applied Mathematics, University of Trento, I-38100 Trento, Italy
[*]email: moroz_vs@yahoo.com

Abstract. We investigated influence of stellar oscillations on the electrodynamics of pulsars as well as magnetars magnetosphere. Besides finding noticeable modification of electromagnetic field and charge density in the polar cap vicinity of oscillating neutron stars we proposed qualitative hypotheses explaining phenomena of part time pulsars as well as sporadic radio emission from generally radio-quiet magnetars with the help of stellar oscillations.

Keywords. Stars: neutron, stars: oscillations, (stars:) pulsars: general.

Investigations of oscillating neutron stars are motivated by the detection of quasi-periodic oscillations (QPOs) in the spectra of soft gamma-ray repeaters (SGRs), which are thought to be neutron stars with very strong magnetic fields (Duncan & Thompson (1992)). The idea that stellar oscillations may induce high energy emission in neutron star magnetospheres was developed in a series of papers by Timokhin and collaborators (see Timokhin *et al.* (2000), Timokhin *et al.* (2008)), after the first pioneering investigations in McDermott *et al.* (1984), Muslimov & Tsygan (1986) and Rezzolla & Ahmedov (2004), where the case of an oscillating neutron star in a vacuum was considered. Starting from Abdikamalov *et al.* (2009), where the theoretical basis of our approach was developed, in a series of papers (Morozova *et al.* (2010), Morozova *et al.* (2012), Zanotti *et al.* (2012)) we have explored the influence of neutron star oscillations on such characteristics of pulsar and magnetar magnetosphere as charge density and electromagnetic field in the polar cap region of the magnetosphere, electromagnetic energy losses as well as conditions for the charged particles acceleration in the magnetosphere. In our research we used toroidal model of stellar oscillations described in Unno *et al.* (1989).

In Morozova *et al.* (2010) we explored the magnetosphere of a slowly rotating magnetized neutron star subject to toroidal oscillations in the relativistic regime. Under the assumption of a zero inclination angle between the magnetic moment and the angular momentum of the star, we analysed the Goldreich-Julian charge density and derived a second-order differential equation for the electrostatic potential. The analytical solution of this equation in the polar cap region of the magnetosphere revealed noticeable modification induced by oscillations on the accelerating electric field and on the charge density. We found that, after decomposing the oscillation velocity in terms of spherical harmonics, the first few modes with $m = 0, 1$ are responsible for energy losses that are almost linearly dependent on the amplitude of the oscillations and that, for the mode $(l, m) = (2, 1)$, can be a factor ~ 8 larger than the rotational energy losses, even for a velocity oscillation amplitude small in comparison with the linear velocity of stellar rotation. Based on these results we proposed a qualitative model for the explanation of the phenomenology of intermittent pulsars (Lyne (2009)). The idea is that stellar oscillations, periodically excited by star glitches, can create relativistic winds of charged particles because of the additional electric field. When the oscillations damp, the pulsar shifts below the death line in the $P - B$ diagram, thus entering the OFF invisible state of intermittent pulsars.

In Morozova *et al.* (2012) we investigated the conditions for radio emission in magnetospheres of rotating and oscillating magnetars. The activity of magnetars is observed in the form of bursts in X-ray and γ-ray bands, while there is no periodic radio emission from the majority of magnetars. We found that as soon as magnetar oscillations are taken into account, their death lines in the $P - B$ diagram shift downward and the conditions necessary for the generation of radio emission in the magnetosphere are met. Present observations (Malofeev *et al.* (2007), Malofeev *et al.* (2010)) showing a close connection between the burst activity of magnetars and sporadic detection of the radio emission from some magnetars are naturally accounted for within our interpretation.

In Zanotti *et al.* (2012) we explored he conditions for charged particle acceleration in the vicinity of the polar cap of pulsar magnetosphere in presence of stellar oscillations. We solved numerically the relativistic electrodynamics equations in the stationary regime, focusing on the computation of the Lorentz factor of a space-charge-limited electron flow accelerated in the polar cap region of a rotating and oscillating pulsar. We found that star oscillations may be responsible for a significant asymmetry in the pulse profile that depends on the orientation of the oscillations with respect to the pulsar magnetic field. In particular, significant enhancements of the Lorentz factor are produced by stellar oscillations in the super-GJ current density regime.

Some recent investigations have tried to connect the models of stellar oscillations with the observational data available for pulsars (see Rosen & Demorest (2011), Rosen *et al.* (2011)). The scenario that is emerging from these studies is that the presence of stellar oscillations creates different kinds of "noise" in the clock-like picture of pulses, i.e. changes in the pulse shape, changes in the spin-down rate, and the switching between different regimes of pulsar emission. As the next step in our future research we plan to thoroughly check the analytical results described above and the proposed hypotheses with the help of observational data on pulsar and magnetar radio emission.

References

Abdikamalov, E. B., Ahmedov, B. J., & Miller, J. C. 2009, *Mon. Not. R. Astron. Soc.*, 395, 443

Duncan, R. C. & Thompson, C. 1992, *ApJ*, 392, L9

Lyne A. G. 2009, in: W. Becker (ed.), Astrophys. Space Sci. Libr. Vol. 357, *Neutron Stars and Pulsars* (Berlin: Springer), p. 67

Malofeev, V. M., Malov, O. I., & Teplykh, D. A. 2007, *Astrophys. Space Sci*, 308, 211

Malofeev, V. M., Teplykh, D. A., & Malov, O. I. 2010, *Astron. Rep.*, 54, 995

McDermott, P. N., Savedoff, M. P., van Horn, H. M., Zweibel, E. G., & Hansen, C. J. 1984, *ApJ*, 281, 746

Morozova, V. S., Ahmedov, B. J., & Zanotti, O. 2010, *Mon. Not. R. Astron. Soc.*, 408, 490

Morozova, V. S., Ahmedov, B. J., & Zanotti, O. 2012, *Mon. Not. R. Astron. Soc.*, 419, 2147

Muslimov, A. G. & Tsygan, A. I. 1986, *Astrophys. Space Sci.*, 120, 27

Rezzolla, L. & Ahmedov, B. 2004, *Mon. Not. R. Astron. Soc.*, 352, 1161

Rosen, R. & Demorest, P. 2011, *ApJ*, 728, 156

Rosen, R., McLaughlin, M. A., & Thompson, S. E. 2011, *ApJ*, 728, L19

Timokhin, A. N., Bisnovatyi-Kogan, G. S., & Spruit, H. C. 2000, *Mon. Not. R. Astron. Soc.*, 316, 734

Timokhin, A. N., Eichler, D., & Lyubarsky, Y. 2008, *ApJ*, 680, 1398

Unno, W., Osaki, Y., Ando, H., Saio, H., & Shibahashi, H. 1989, *Nonradial Oscillations of Stars* (Tokyo: Univ. Tokyo Press)

Zanotti, O., Morozova, V. S., & Ahmedov, B. J. 2012, submitted to *Astron. Astrophys.*, DOI: 10.1051/0004-6361/201118380

Death of Massive Stars: Supernovae and Gamma-Ray Bursts
Proceedings IAU Symposium No. 279, 2012
P. Roming, N. Kawai & E. Pian, eds.

doi:10.1017/S174392131201335X

Carnegie Supernova Project: Spectroscopic Observations of Core Collapse Supernovae

Nidia I. Morrell†

Las Campanas Observatory, Carnegie Observatories, Casilla 601, La Serena, Chile
email: nmorrell@lco.cl

Abstract. The Carnegie Supernova Project (CSP) has performed, during the period 2004-2009, the optical and NIR follow up of 253 supernovae (SNe) of all types. Among those, 124 were core collapse events, comprising 93 SNe of type II and 31 of types Ib/Ic/IIb. Our follow up consisted of photometric observations suitable to build detailed light curves and a considerable amount of optical spectroscopy.
The bulk of our observations is carried out at Las Campanas Observatory, while access to other facilities is also provided thanks to our strong collaboration with the Millennium Center for Supernova Studies (MCSS).
Our spectroscopic observations were primarily aimed at typing possible new SNe, and follow-up the evolution of CSP targets. One of the goals of the follow-up of type II SNe is the application of independent distance indicators such as the Standard Candle (SCM) and the Expanding Photosphere (EPM) methods. Moreover, through the study of the spectroscopic evolution of these objects, from as early as possible after explosion to the nebular phases, we hope to contribute to their further understanding. Specific analysis of particular objects is underway by members of the CSP and an extended collaboration.

Keywords. (stars:) supernovae: general, techniques: spectroscopic

---

1. Introduction

The CSP (Hamuy *et al.* (2006)) conducted, between 2004 and 2009, 5 campaigns devoted to build detailed optical and NIR light curves of nearby ($z < 0.07$) SNe of all types.

Most of CSP targets came from the SN searches carried out at Lick (LOSS), Chile (CHASE) and those conducted by amateur astronomers such as Tim Puckett, Berto Monard, Tom Boles, Koichi Itagaki and more recently, the BOSS collaboration.

Optical spectroscopy was a key complement of the project. As soon as possible after a new SN was announced, a spectrum was obtained in order to determine its type, phase and redshift. This information was used to decide about which targets would be subject of dedicated follow-up, preference being given to young events, such as type Ia SNe before maximum brightness, and core collapse SNe soon after explosion.

Further spectral observations, in coincidence with the photometric follow up, were obtained as often as possible (always prioritizing the typing of new candidates) in order to study the spectroscopic evolution of CSP targets.

Spectral sequences of Type II SNe allow independent distance determinations via the Expansion Photosphere Method (Kirshner & Kwan (1974), Jones *et al.* (2009)) and the Standard Candle Method (Hamuy & Pinto (2002), Olivares (2008)).

† On behalf of the Carnegie Supernova Project

Multiple epoch spectroscopic observations of stripped core-collapse (Ib/Ic/IIb) SNe, combined with our complete sets of light curves, can be used to get further insight into the physical properties of these rare objects.

2. Observations

Most of the CSP spectroscopic observations have been carried out using the facilities available at LCO, i.e. the du Pont telescope (+WFCCD and Boller and Chivens spectrographs), as well as both Magellan telescopes (+LDSS3, IMACS, and MagE spectrographs).

Additional observations were obtained at CTIO, ESO and Gemini facilities.

Reductions have been carried out through IRAF † routines, usually via a set of specially designed scripts, including wavelength and flux callibration as well as telluric line removal.

Along the 5 CSP campaigns we obtained 770 spectra of 137 core collapse SNe (although not all of them were selected for CSP follow-up).

3. Results, work in prgress and CSP II

Some of the most conspicuous CSP targets have already been analysed in several papers (SN 2005bf: Folatelli *et al.* 2006; SN 2007Y: Stritzinger *et al.* 2009; SN 2009bb: Pignata *et al.* 2011). One of the best observed type IIP SNe ever, SN 2008bk, was the subject of extensive photometric and spectroscopic follow-up by CSP and other groups, from soon after discovery (Morrell & Stritzinger 2008) until well into the nebular phase. Its progenitor has been identified as a red supergiant (Van Dyk *et al.* 2012). Results from more than 3 years of observations by CSP and MCSS will be presented in a forthcoming paper (Pignata *et al.*, in preparation).

Detailed analysis of other CSP targets is underway by members of the CSP and their collaborators. After release, the CSP spectroscopic database will be publicly available at The Online Supernova Spectrum Archive (SUSPECT). Sample light curves can be found at the CSP website: http://csp1.lco.cl/~cspuser1/PUB/CSP.html

After 2009 we continued to obtain spectroscopic observations aimed at typing SNe candidates and follow-up selected targets in collaboration with MCSS.

In October 2011 we started the first CSP II (Carnegie Supernova Project II) campaign comprising optical and NIR imaging of our targets, as well as optical and NIR spectroscopy. While mostly focused on type Ia SNe, we still need to type new candidates, and we continue to do follow-up of selected SNe of different types.

References

Folatelli, G. *et al.* 2006, *ApJ*, 641, 1039
Hamuy, M. & Pinto, P. A. 2002, *ApJ*, 566, L63
Hamuy, M. *et al.* 2006, *PASP*, 118, 2
Jones, M. I. *et al.* 2009, *ApJ*, 696, 1176
Kirshner, R. P. & Kwan, J. 1974, *ApJ*, 193, 27
Morrell, N. & Stritzinger, M. 2008, *CBET*, 1335
Olivares, F. 2008, MSc Thesis, Universidad de Chile, 2008arXiv0810.55180
Pignata, G. *et al.* 2011, *ApJ*, 728, 14
Stritzinger, M. *et al.* 2009, *ApJ*, 696, 713
Van Dyk, S., *et al.* 2012, *AJ*, 143, 19

† IRAF is distributed by NOAO, operated by AURA Inc., under contract with NSF.

Death of Massive Stars: Supernovae and Gamma-Ray Bursts
Proceedings IAU Symposium No. 279, 2012
A.C. Editor, B.D. Editor & C.E. Editor, eds.

doi:10.1017/S1743921312013361

The Accretion-Powered Jet Propagations and Breakout Criteria for GRB Progenitors

Hiroki Nagakura[1,2], Yudai Suwa[1] and Kunihito Ioka[3]

[1]Yukawa Institute for Theoretical Physics, Kyoto University, Oiwake-cho, Kitashirakawa, Sakyo-ku, Kyoto, 606-8502, Japan
email: hirokin@yukawa.kyoto-u.ac.jp

[2]Department of Science and Engineering, Waseda University, 3-4-1 Okubo, Shinjuku, Tokyo 169-8555, Japan
email: hiroki@heap.phys.waseda.ac.jp

[3]KEK Theory Center and the Graduate University for Advanced Studies (Sokendai), 1-1 Oho, Tsukuba 305-0801, Japan

Abstract. We investigate the propagation of accretion-powered jets in various types of progenitor candidates of GRBs. We perform two dimensional axisymmetric simulations of relativistic hydrodynamics taking into account both the envelope collapse and the jet propagation. In our simulations, the accretion rate is estimated by the mass flux going through the inner boundary, and the jet is injected with a constant accretion-to-jet conversion efficiency η. By varying the efficiency η and opening angle θ_{op} for more than 30 models, we find that the jet can make a relativistic breakout from all types of progenitors for GRBs if a simple condition $\eta \gtrsim 10^{-3}(\theta_{op}/20^{\circ})^2$ is satisfied, that is consistent with analytical estimates, otherwise no explosion or some failed spherical explosions occur.

Keywords. gamma rays: burst, supernovae: general

---

1. Overview and Methods

Some populations of Gamma-Ray Bursts (GRBs) are thought to originate from the death of massive stars. It is widely believed that the jet is supposed to be launched due to interaction between black hole and accretion system. After the jet is launched in the vicinity of black hole, the jet head should propagate outward and break out from the stellar surface, otherwise the jet becomes non-relativistic ejecta and fails to create a GRB. Therefore, a relativistic jet breakout is a minimum requirement to produce GRBs. In this study, we investigate the jet break out criteria by simulating the jet propagation (see for more details in Nagakura *et al.* 2011b.)

We perform relativistic two-dimensional hydrodynamics simulations for envelope collapse and jet propagation. The numerical codes employed in this paper are essentially the same as those used in Nagakura *et al.* (2011a). In this study, the jet luminosity is determined by the mass accretion rate, which is a major difference from previous works (Almost every past study assumes a constant jet luminosity). We survey the parameter space of accretion-to-jet conversion efficiency (η) and opening angle of jet (θ_{op}), and discuss how these key quantities affect the jet dynamics. The numerical results are compared with the previous analytical work by Suwa & Ioka (2011).

2. Results

Fig. 1 shows the results of our study and comparison with analytical criteria. The upper left panel shows the Wolf-Rayet progenitor case (16TI model in Woosley & Heger

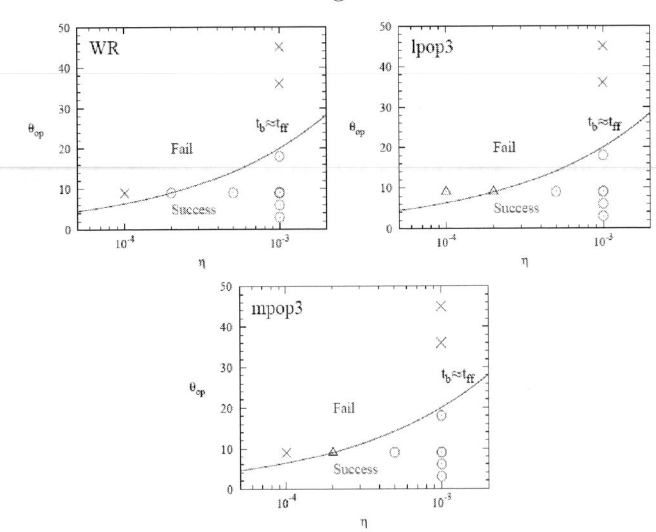

Figure 1. The score sheet of the shock break out. The x-axis denotes the conversion efficiency while the y-axis shows the opening angle.

(2006), hereafter SR model), while the upper right and lower panel show $40M_\odot$, which is the metal free pre-supernova model calculated by Woosley *et al.* (2002) (hereafter lpop3 model) and $1000M_\odot$, which is also the metal free presupernova model calculated by Ohkubo *et al.* (2009) (hereafter mpop3), respectively. The stellar radius for each model is 4×10^{10} cm, 1.5×10^{12} cm and 9×10^{12} cm, respectively. Last two models are investigated with Population III GRBs in mind. The circles correspond to the models which are conducive to GRB formation in our numerical simulations, while the crosses show the failed cases. The triangles correspond to mildly relativistic explosion models. As shown in this figure, if the accretion-to-jet conversion efficiency is high enough and opening angle of jet is small enough, the jet can propagate and break out relativistically even in massive progenitors such as lpop3 and mpop3 models. By comparing with analytical works (solid line in Fig 1), we obtain the simple break out criterion as

$$\eta \gtrsim 10^{-3} \left(\frac{\theta_{op}}{20^\circ}\right)^2 . \tag{2.1}$$

In summary, the central engine is not sufficient to produce GRBs by itself, and it should satisfy the condition of Eq. (2.1). It is interesting to note that the jet, whose conversion efficiency satisfies the above condition, succeeds to break out relativistically even if the progenitor has larger radius. It should be noted, however, that the inner part of the progenitor should also be compact in order to keep high mass accretion rate after the formation of the black hole, otherwise the jet luminosity becomes too weak to produce GRBs. Therefore Red Supergiant is not suitable for the production of GRBs.

References

Nagakura, H., Ito, H., Kiuchi, K., & Yamada, S. 2011a, *ApJ*, 731, 80
Nagakura, H., Suwa, Y., & Ioka, K. 2011b, arXiv:1104.5691
Ohkubo, T., Nomoto, K., Umeda, H., Yoshida, N., & Tsuruta, S. 2009, *ApJ*, 706, 1184
Suwa, Y. & Ioka, K. 2011, *ApJ*, 726, 107
Woosley, S. E., Heger, A., & Weaver, T. A. 2002, *Reviews of Modern Physics*, 74, 1015
Woosley, S. E. & Heger, A. 2006, *ApJ*, 637, 914

Death of Massive Stars: Supernovae and Gamma-Ray Bursts
Proceedings IAU Symposium No. 279, 2012
P. Roming, N. Kawai & E. Pian, eds.

doi:10.1017/S1743921312013373

Neutrino-driven supernova explosions powered by nuclear reactions

K. Nakamura[1], T. Takiwaki[1], K. Kotake[1,2], and N. Nishimura[3,4]

[1]Division of Theoretical Astronomy, National Astronomical Observatory of Japan,
Osawa 2-21-1, Mitaka, Tokyo, Japan 181-8588
email: nakamura.ko@nao.ac.jp

[2]Center for Computational Astrophysics, National Astronomical Observatory of Japan,
Osawa 2-21-1, Mitaka, Tokyo, Japan 181-8588

[3]Department Physik, Universität Basel,
Klingelbergstrasse 82, 4056 Basel, Switzerland

[4]GSI, Helmholtzzentrum für Schwerioneneforschung GmbH,
Planckstraße 1, 64291 Darmstadt, Germany

Abstract. We have investigated the revival of a shock wave by nuclear burning reactions at the central region of core-collapse supernovae. For this purpose, we performed hydrodynamic simulations of core collapse and bounce for 15 $M_{\odot}$ progenitor model, using ZEUS-MP code in axi-symmetric coordinates. Our numerical code is equipped with a simple nuclear reaction network including 13 α nuclei form ^{4}He to ^{56}Ni, and accounting for energy feedback from nuclear reactions as well as neutrino heating and cooling. We found that the energy released by nuclear reactions is significantly helpful in accelerating shock waves and is able to produce energetic explosion even if the input neutrino luminosity is low.

Keywords. hydrodynamics, neutrinos, nuclear reactions, nucleosynthesis, abundances, shock waves, supernovae: general

---

1. Introduction

In this work we argue that the energy released by nuclear reactions may be able to realize the revival of a shock wave which stalls inside the iron core of a supernova progenitor because of endothermic photodisintegration reaction. The exact mechanism of the explosion and the crucial ingredients are still controversial, and neutrino-driven explosion model is one of the mechanisms suggested for core-collapse supernova explosion (Scheck *et al.* 2008; Marek & Janka 2009). Moreover, nuclear energy is a possible propellant of shock into energetic supernova explosions. We present the numerical simulations of core-collapse supernovae by means of multi-dimensional hydrodynamic code including a simple nuclear reaction network including 13 alpha nuclei.

2. Numerical Scheme and Results

A 15 $M_{\odot}$ star (Limongi & Chieffi 2006) is employed for the progenitor star. Details regarding numerical codes and results for the other models are summarized in our forthcoming paper (Nakamura *et al.* in preparation).

In Figure 1 we present a snapshot of the distributions of entropy and some representative elements for neutrino luminosity $L_{\nu,0} = 3.0\times10^{52}$ erg s^{-1} which decays exponentially in a time scale $t_d = 1.1$ s. We can see that oxygen is completely burned out and heavier nuclei like silicon are produced at a weak shock in accreting gas, followed by a strong shock wave where a fraction of silicon is converted into nickel. Soon the strong shock

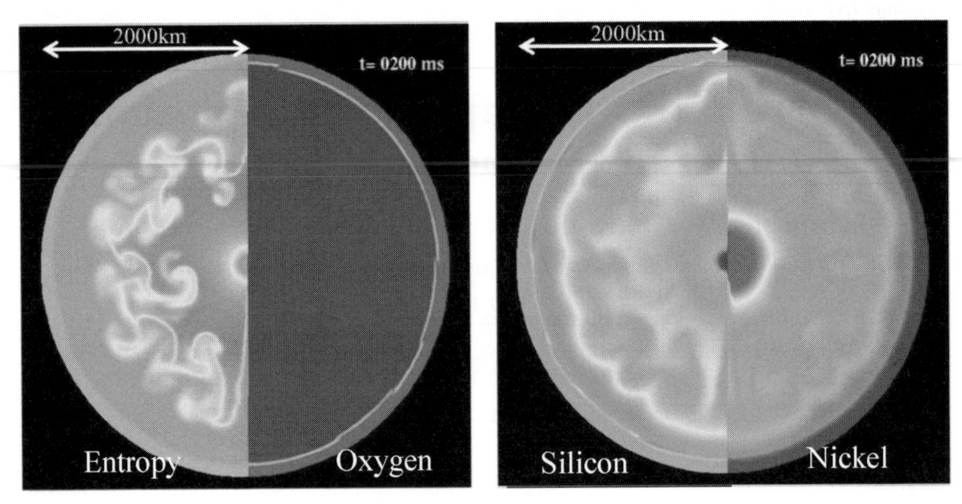

Figure 1. A snapshot of the distributions of entropy and some representative elements at $t_{pb} = 200$ ms within the radius $r = 2000$ km. Shown is the result from 2-dimensional simulation for LC15 model with $L_{\nu,0} = 3.0 \times 10^{52}$ erg s^{-1} and $t_d = 1.1$ s.

wave catches up with the weak shock front and at this phase the energy deposition rate through nuclear reactions shows a rapid rise and heats up the region behind the shock front, resulting in the shock acceleration and increasing explosion energy. The entropy profile behind the strong shock shows hydrodynamic instability which well mixes materials behind the shock front. In general such kinds of hydrodynamic instabilities make accreting materials remain at so-called "gain region" and help neutrino heating become effective. Compared to spherical explosions, we confirmed that the effects of spatial dimensions enhance the efficiency of neutrino heating and 2-dimensional models explode more easily in terms of shock velocity and also the minimum neutrino luminosity for explosion. For example, with a fixed t_d of 1.1 s and without nuclear energy, spherical explosions need initial neutrino luminosity more than 2.5×10^{52} erg s^{-1}. On the other hand, for the 2-dimensional explosion case, the minimum neutrino luminosity is reduced to 2.2×10^{52} erg s^{-1}. The effect of nuclear reactions becomes outstanding when the input neutrino luminosity is low. Taking the case of $L_{\nu,0} = 2.4 \times 10^{52}$ erg s^{-1} and $t_d = 3.0$ s as an example, we obtained final explosion energy of 8.1×10^{50} erg with the aid of nuclear reactions. This parameter set holds itself explosive even if a hydrodynamic energy equation does not include the term of the energy released via nuclear reactions. In the case without nuclear reactions, however, the final explosion energy is 5.9×10^{50} erg, more than 25 % less than the case with nuclear reactions. This shortage corresponds to the net energy released via nuclear reactions.

We conclude that nuclear reactions obviously contribute to shock acceleration and bulking up of explosion energy of core-collapse supernovae. In order to make this process effective a shock front needs to break out of the iron core and reach the oxygen-rich layer, which should be driven by neutrino heating or other energy sources.

References

Limongi, M. & Chieffi, A. 2006, *ApJ*, 647, 483
Marek, A. & Janka, H.-T. 2009, *ApJ*, 694, 664
Nakamura, K., Takiwaki, T., Kotake, K., & Nishimura, N. in preparation
Scheck, L., Janka, H.-T., Foglizzo, T., & Kifonidis, K. 2008, *A&A*, 477, 931

Death of Massive Stars: Supernovae and Gamma-Ray Bursts
Proceedings IAU Symposium No. 279, 2012
P. Roming, N. Kawai & E. Pian, eds.

doi:10.1017/S1743921312013385

Stellar Core Collapse and Exotic Matter

Ken'ichiro Nakazato[1] and Kohsuke Sumiyoshi[2]

[1]Department of Physics, Faculty of Science & Technology, Tokyo University of Science, Yamazaki 2641, Noda, Chiba 278-8510, Japan
email: nakazato@rs.tus.ac.jp

[2]Numazu College of Technology, Ooka 3600, Numazu, Shizuoka 410-8501, Japan

Abstract. Some supernovae and gamma-ray bursts are thought to accompany a black hole formation. In the process of a black hole formation, a central core becomes hot and dense enough for hyperons and quarks to appear. In this study, we perform neutrino-radiation hydrodynamical simulations of a stellar core collapse and black hole formation taking into account such exotic components. In our computation, general relativity is fully considered under spherical symmetry. As a result, we find that the additional degrees of freedom soften the equation of state of matter and promote the black hole formation. Furthermore, their effects are detectable as a neutrino signal. We believe that the properties of hot and dense matter at extreme conditions are essential for the studies on the astrophysical black hole formation. This study will be hopefully a first step toward a physics of the central engine of gamma-ray bursts.

Keywords. black hole physics, dense matter, equation of state, hydrodynamics.

---

1. Introduction

The fate of stars with $M \gtrsim 25M_\odot$ can be observationally split into two branches, namely, a hypernova branch and a faint-supernova branch, and they are both thought to form black holes eventually (Nomoto *et al.* 2006). The progenitors of the hypernova branch are thought to rotate strongly and associate with the gamma-ray bursts. On the other hand, nonrotating and weakly rotating massive stars are constituents of the faint-supernova branch and more massive stars are thought to result in so-called failed supernovae. In this study, we perform neutrino-radiation hydrodynamical simulations of a stellar core collapse and black hole formation taking into account hyperons and quarks. As a result, we find that the additional degrees of freedom soften the equation of state (EOS) of matter and promote the black hole formation. Further details of this study can also be found in our recent paper (Nakazato *et al.* 2012).

2. Setups

In the process of a black hole formation, a central core becomes hot and dense enough for hyperons and quarks to appear. In order to compute the stellar core collapse and black hole formation, we use the numerical code of general relativistic ν-radiation hydrodynamics which solves the Boltzmann equation for neutrinos together with Lagrangian hydrodynamics under spherical symmetry. As an initial condition, the stellar model with $40M_\odot$ and solar metallicity from the evolutionary calculation Woosley & Weaver (1995) is adopted. To examine the effects of hyperons, we utilize the tables of EOS by Ishizuka *et al.* (2008), which are based on an $SU_f(3)$ extended relativistic mean field model and constructed as an extension of the EOS by Shen *et al.* (1998). Since it is undetermined whether the Σ-N interaction is attractive or repulsive, we adopt both EOS sets with potential depths $(U_\Lambda, U_\Sigma, U_\Xi) = (-30$ MeV, $+30$ MeV, -15 MeV) for the repulsive case

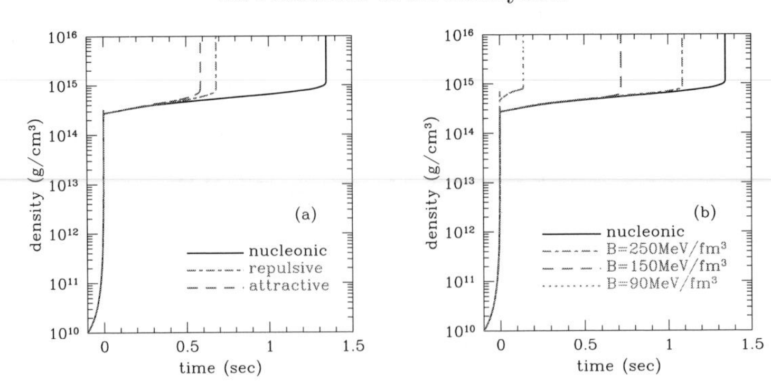

Figure 1. Time evolutions of the central baryon mass density for the collapse of models with *(a)* hyperons and *(b)* quarks. In both panels, the solid lines show the result for the model without hyperons and quarks.

and (−30 MeV, −30 MeV, −15 MeV) for the attractive case. As for the quark effects, we investigate the MIT bag model with the EOS by Shen *et al.* (1998) as in Nakazato *et al.* (2008). In this model, the ambiguities of the interaction are encapsulated in one parameter called bag constant, B, and we examine $B = 90$, 150 and 250 MeV/fm^3. Note that, for the EOS's with hyperons or quarks, the maximum masses of neutron stars are smaller than $1.97 \pm 0.04 M_\odot$, the mass of the binary millisecond pulsar J1614-2230 Demorest *et al.* (2010). While our focus is on investigating systematic differences due to the interactions of hyperons and quarks, further studies on the EOS are also important.

3. Results

In Figure 1, we show the time profiles of the central baryon mass density. The core is bounced once and then recollapses to a black hole. We can recognize that the time interval between the bounce and black hole formation gets shorter as we put additional degrees of freedom, hyperons or quarks. As for the hyperonic models, the black hole is formed earlier for the attractive case because Σ hyperons appear more easily and soften the EOS. On the other hand, for the quark models, the time interval of the lower bag constant case is shorter because the phase transition occurs at lower density and triggers the black hole formation. Note that the density at the bounce of the model with $B = 90$ MeV/fm^3 differs from those of other models because the transition occurs already at the bounce. Since the duration of neutrino emission corresponds to the time interval between the bounce and black hole formation, we may be able to probe observationally the effects of hyperons and quarks using the difference of the time interval in future.

References

Demorest, P. B., Pennucci, T., Ransom, S. M., Roberts, M. S. E., & Hesseles, J. W. T. 2010, *Nature*, 467, 1081

Ishizuka, C., Ohnishi, A., Tsubakihara, K., Sumiyoshi, K., & Yamada, S. 2008, *J. of Phys. G*, 35, 085201

Nakazato, K., Furusawa, S., Sumiyoshi, K., Ohnishi, A., Yamada, S., & Suzuki, H. 2012, *ApJ*, 745, 197

Nakazato, K., Sumiyoshi, K., & Yamada, S. 2008, *Phys. Rev. D*, 77, 103006

Nomoto, K., Tominaga, N., Umeda, H., Kobayashi, C., & Maeda, K. 2006, *Nucl. Phys., A*, 777, 424

Shen, H., Toki, H., Oyamatsu, K., & Sumiyoshi, K. 1998, *Prog. Theor. Phys.*, 100, 1013

Woosley, S. E. & Weaver, T. A. 1995, *ApJS*, 101, 181

Death of Massive Stars: Supernovae and Gamma-Ray Bursts
Proceedings IAU Symposium No. 279, 2012
P. Roming, N. Kawai & E. Pian, eds.

doi:10.1017/S1743921312013397

Revisiting Metallicity of Long Gamma-Ray Burst Host Galaxies: The Role of Chemical Inhomogeneities in Galaxies

Yuu Niino

Division of Optical and NIR Astronomy, National Astronomical Observatory of Japan
2-21-1 Osawa, Mitaka, Tokyo, Japan email: yuu.niino@nao.ac.jp

Abstract. Some theoretical studies on the origin of long gamma-ray bursts (GRBs) using stellar evolution models suggest that a low metallicity environment may be a necessary condition for a GRB to occur. However, recent discoveries of high-metallicity host galaxies of some GRBs cast doubt on the requirement of low-metallicity in GRB occurrence. In this study, we predict the metallicity distribution of GRB host galaxies, assuming empirical formulations of galaxy properties. We take internal dispersion of metallicity within each galaxy into account. Assuming GRBs trace low-metallicity star formation 12+log(O/H) < 8.2, we find that $\gtrsim 10\%$ of GRB host galaxies may have $Z > Z_{\odot}$, depending on the internal dispersion of metallicity within galaxies.

Keywords. gamma rays: bursts, galaxies: abundances

1. Models of Galaxies

We assume following empirical formulations of galaxy properties at $z \sim 0.1$, to compute expected $M_{\star}$ and metallicity distribution of GRB host galaxies.

• Stellar mass function (Bell *et al.* 2003; Drory & Alverez 2008)

• Star formation rate (SFR) as a function of $M_{\star}$ (Brinchmann *et al.* 2004; Stanek *et al.* 2006; Drory & Alverez 2008)

• $M_{\star}$–Z relation (Savaglio *et al.* 2005, recalibration of SDSS sample with R23 method)

Dispersion of SFR and $M_{\star} - Z$ relation is $\sigma = 0.3$ & 0.1 dex, respectively.

Metallicity of GRB host galaxies is an important clue to study the metallicity of GRB progenitors. However metallicity of a galaxy is not necessarily equal to metallicity of young stars formed in the galaxy. The Milky Way (MW) has mean metallicity 12+log(O/H) ~ 8.9, but some young B-type stars and HII regions in the Milky Way have 12+log(O/H) ~ 8.2 (e.g. Afflerbach *et al.* 1997; Rolleston *et al.* 2000).

Motivated by the observations of galactic B-type main sequence stars and HII regions, we consider internal metallicity distribution of young stars within each galaxy with log-normal tail ($\sigma_{Z,\rm int} = 0.1$, 0.3 & 0.5). In each galaxy, we assume $R_{\rm GRB} \propto \mathrm{SFR}_{12+\log(\mathrm{O/H})<8.2}$.

2. Results

The computed metallicity distribution is shown in figure 1. No metallicity preference model contradicts observations as suggested by Stanek *et al.* (2006). In $\sigma_{Z,\rm int} = 0.3$, 0.5 models, high-metallicity galaxies with 12+log(O/H) > 8.8 contribute 5–25% of cosmic low-metal star formation [12+log(O/H) < 8.2], or GRB rate. Double peak distribution of host metallicity appears when $\sigma_{Z,\rm int}$ positively correlates with $M_{\star}$.

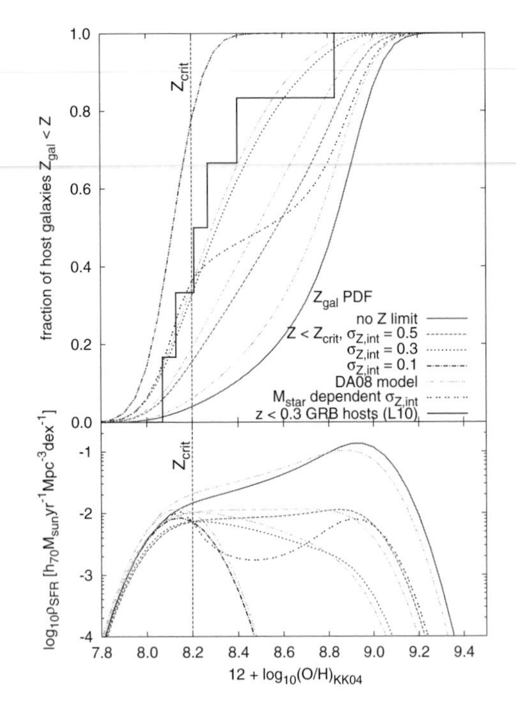

Figure 1. Metallicity distribution of GRB host galaxies predicted in our model compared to current sample collected by Levesque *et al.* (2010).

3. Discussion

We have computed the expected metallicity distribution of GRB host galaxies assuming empirical formulations of galaxy properties at $z \sim 0.1$ and the model of GRB rate in which GRBs occur only from low-metallicity stars. Our results show that high-metallicity galaxies [12+log(O/H) $>$ 8.8] may have significant contribution to cosmic GRB rate. This means that metallicities of GRB host galaxies may be systematically different from those of GRB progenitors, and the low-metallicity scenario can be reconciled with the observations of high-metallicity host galaxies of GRBs. More detailed discussion is in Niino (2011).

References

Afflerbach, A., Churchwell, E., & Werner, M. W. 1997, *ApJ*, 478, 190

Bell, E. F., McIntosh, D. H., Katz, N., & Weinberg, M. D. 2003, *ApJS*, 149, 289

Brinchmann, J., Charlot, S., White, S. D. M., Tremonti, C., Kauffmann, G., Heckman, T., & Brinkmann, J. 2004, *MNRAS*, 351, 1151

Drory, N. & Alvarez, M. 2008, *ApJ*, 680, 41

Levesque, E. M., Kewley, L. J., Berger, E., & Jabran Zahid, H. 2010, *AJ*, 140, 1557

Niino, Y. 2011, *MNRAS*, 417, 567

Rolleston, W. R. J., Smartt, S. J., Dufton, P. L., & Ryans, R. S. I. 2000, *A&A*, 363, 537

Savaglio, S. *et al.* 2005, *ApJ*, 635, 260

Stanek, K. Z. *et al.* 2006, *AcA.*, 56, 333

Death of Massive Stars: Supernovae and Gamma-Ray Bursts
Proceedings IAU Symposium No. 279, 2012
P. Roming, N. Kawai & E. Pian, eds.

doi:10.1017/S1743921312013403

Radiation from accelerated particles in shocks

K.-I. Nishikawa[1], B. Zhang[2], E. J. Choi[3], K. W. Min[3], J. Niemiec[4], M. Medvedev[5], P. Hardee[6], Y. Mizuno[7], A. Nordlund[8], J. Frederiksen[8], H. Sol[9], M. Pohl[10], D. H. Hartmann[11], and G.J. Fishman[12]

[1]Center for Space Plasma and Aeronomic Research, University of Alabama in Huntsville, 320 Sparkman Drive, Huntsville, AL 35805, USA
email: ken-ichi.nishikawa-1@nasa.gov

[2]Department of Physics and Astronomy, University of Nevada, Las Vegas, NV 89154, USA
[3]Korea Advanced Institute of Science and Technology, Daejeon 305-701, South Korea
[4]Institute of Nuclear Physics PAN, ul. Radzikowskiego 152, 31-342 Krakow, Poland
[5]Department of Physics and Astronomy, University of Kansas, KS 66045, USA
[6]Department of Physics and Astronomy, The University of Alabama, Tuscaloosa, AL 35487, USA
[7]Institute of Astronomy, National Tsing-Hua University , No. 101, Sec. 2, Kuang-Fu Road., Hsinchu, Taiwan 30013, R.O.C.
[8]Niels Bohr Institute, Juliane Maries Vej 30, 2100 Kbenhavn, Denmark
[9]LUTH, Observatore de Paris-Meudon, 5 place Jules Jansen, 92195 Meudon Cedex, France
[10]Institut fuer Physik und Astronomie, Universitaet Potsdam, 14476 Potsdam-Golm, Germany
[11]Department of Physics and Astronomy, Clemson University, Clemson, SC 29634, USA
[12]NASA/MSFC, 320 Sparkman Drive, Huntsville, AL 35805, USA

Abstract. Recent PIC simulations of relativistic electron-positron (electron-ion) jets injected into a stationary medium show that particle acceleration occurs in the shocked regions. Simulations show that the Weibel instability is responsible for generating and amplifying highly nonuniform, small-scale magnetic fields and for particle acceleration. These magnetic fields contribute to the electron's transverse deflection behind the shock. The "jitter" radiation from deflected electrons in turbulent magnetic fields has properties different from synchrotron radiation calculated in a uniform magnetic field. This jitter radiation may be important for understanding the complex time evolution and/or spectral structure of gamma-ray bursts, relativistic jets in general, and supernova remnants. In order to calculate radiation from first principles and go beyond the standard synchrotron model, we have used PIC simulations. We present synthetic spectra to compare with the spectra obtained from Fermi observations.

Keywords. Relativistic jets, Weibel instability, magnetic field generation, particle acceleration, radiation.

We have calculated the radiation spectra directly from our simulations by integrating the expression for the retarded power, derived from the Liénard-Wiechert potentials for a large number of representative particles in the PIC representation of the plasma (Nishikawa *et al.* 2009, 2011). In order to obtain the spectrum of the synchrotron/jitter emission, we consider an ensemble of electrons selected in the region where the Weibel instability has fully grown and where the electrons are accelerated in the self-consistently generated magnetic fields.

Figure 1 shows how our synthetic spectrum matches with spectra obtained from Fermi observations. Figure 1a shows the observed spectra in νF_ν as modeled by Abdo *et al.* (2009) at five different time intervals.

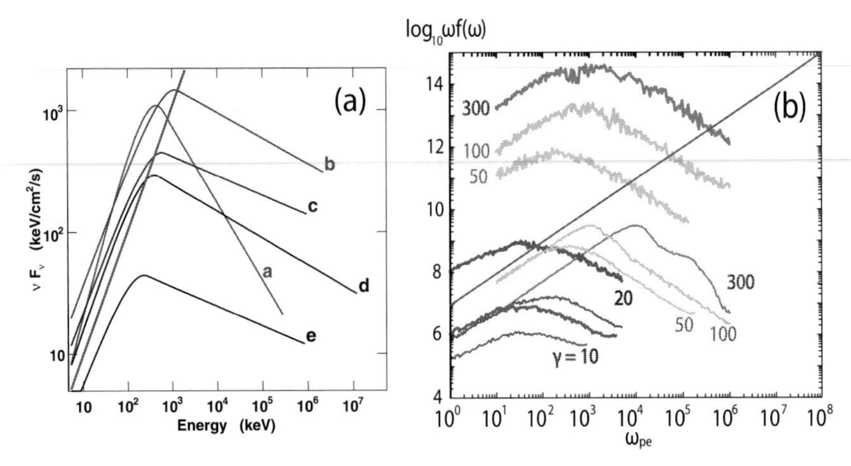

Figure 1. Comparison of a synthetic spectrum with spectra obtained from Fermi observations. Figure 1a shows the modeled Fermi spectra in νF_ν units for five time intervals. A flat spectrum would indicate equal energy per decade in photon energy. The changing shapes show the evolution of the spectrum over time. Figure 1b shows the spectra for the cases of γ = 10, 20, 50,100, and 300 with cold (thin lines) and warm (thick lines) electron jets. The low frequency slope is approximately 1.

The red line in Fig. 1a indicates a slope of one, and except for the spectrum at time "a" the low frequency slopes are all approximately one. This is similar to a Bremsstrahlung-like spectrum at least for the low frequency side. As shown in Fig. 1b the slope at low frequency is very similar to the observed spectra. The peaks and slopes at high frequencies change over time.

Emission computed using the method described above is obtained self-consistently, and automatically accounts for magnetic field structures on the small scales responsible for jitter emission. By performing such calculations for simulations using different parameters, we can investigate and compare the different regimes of jitter- and synchrotron-type emission (Medvedev 2006). Thus, we should be able to address the low frequency GRB spectral index violation of the synchrotron spectrum line of death (Medvedev 2006).

Acknowledgements

This work is supported by NSF-AST-0506719, AST-0506666, AST-0908040, AST-0908010, NASA-NNG05GK73G, NNX07AJ88G, NNX08AG83G, NNX08AL39G, and NNX09AD 16G. JN was supported by MNiSW research project N N203 393034, and The Foundation for Polish Science through the HOMING program, which is supported through the EEA Financial Mechanism. Simulations were performed at the Columbia facility at the NASA Advanced Supercomputing (NAS), and on the IBM p690 (Copper) at the National Center for Supercomputing Applications (NCSA) which is supported by the NSF. Part of this work was done while K.-I. N. was visiting the Niels Bohr Institute. Support from the Danish Natural Science Research Council is gratefully acknowledged. This report was finalized during the program "Particle Acceleration in Astrophysical Plasmas" at the Kavli Institute for Theoretical Physics which is supported by the National Science Foundation under Grant No. PHY05-51164.

References

Abdo, A. A., *et al.* 2009, *Science*, 323, 1688
Medvedev, M. V. 2006, *ApJ*, 637, 869
Nishikawa, K.-I., *et al.* 2009b, *ApJ*, 689, L10
Nishikawa, K.-I., *et al.* 2011, *AdvSR*, 47, 1434

Death of Massive Stars: Supernovae and Gamma-Ray Bursts
Proceedings IAU Symposium No. 279, 2012
P. Roming, N. Kawai & E. Pian, eds.

doi:10.1017/S1743921312013415

Black-Hole Formation in Potential γ-Ray Burst Progenitors

Evan O'Connor[1], Luc Dessart[1,2], and Christian D. Ott[1]

[1]TAPIR, California Institute of Technology
email: evanoc@tapir.caltech.edu cott@tapir.caltech.edu

[2]Laboratorie d'Astrophysique de Marseille, Universitè Aix-Marseille & CNRS
email: luc.dessart@oamp.fr

Abstract. We present the results of a study by Dessart *et al.* (2012), where we performed stellar collapse simulations of proposed long-duration γ-ray burst (LGRB) progenitor models and assessed the prospects for black hole formation. We find that many of the proposed LGRB candidates in Woosley & Heger (2006) have core structures similar to garden-variety core-collapse supernova progenitors and thus are not expected to form black holes, which is a key ingredient of the collapsar model of LGRBs. The small fraction of proposed progenitors that are compact enough to form black holes have fast rotating iron cores, making them prone to a magneto-rotational explosion and the formation of a proto-magnetar rather than a black hole. This leads us to our take-home message, that one must consider the iron-core structure (eg. $\rho(r), \Omega(r)$) of evolved massive stars before making assumptions on the central engine of LGRBs.

Keywords. (stars:) supernovae: general, gamma rays: bursts

---

1. Introduction

A great puzzle in massive star evolution is to understand the necessary departures from the general core collapse scenario to produce a LGRB in addition to a SN explosion, as spectroscopically confirmed in, to date, at least six LGRB/SNe pairs. The very low occurrence rate of LGRB/SN per CCSN calls for progenitor properties that are rarely encountered in star formation/evolution.

Two LGRB central-engine models are currently favored. They suggest the key components for a successful LGRB/SN are a compact progenitor with a short light-crossing time of $\sim$1 s and fast rotation at the time of collapse. One is the collapsar model (Woosley 1993), in which a fast-rotating progenitor fails to explode in its early post-bounce phase and instead forms a black hole, while the in-falling envelope eventually forms a Keplerian disk feeding the hole on an accretion timescale comparable to that of the LGRB. The other model involves a proto-magnetar (Wheeler *et al.* 2000), in which the LGRB is born after a successful SN explosion, either by the neutrino or the magneto-rotational mechanism, although the latter seems more likely given the rapid rotation required for the magnetar (Dessart *et al.* 2008).

The only stellar-evolutionary models for LGRB progenitors that are evolved until the onset of core collapse are those of Woosley & Heger (2006). We thus focus on their model set for our investigation on the dynamics of the CCSN engine and the potential formation of a black hole in the collapsar context. In this work, we use the open-source, spherically symmetric, general relativistic, Eulerian hydrodynamics code `GR1D` (O'Connor & Ott 2010). Rotation is included through a centrifugal-acceleration term in the momentum equation.

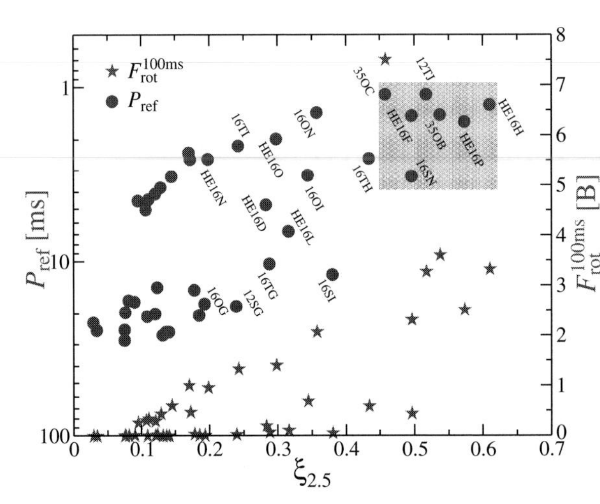

Figure 1. Reference proto-neutron star spin period, $P_{\rm ref}$, taken at the onset of explosion (left axis, blue dots) and the free energy stored in differential rotation 100 ms after bounce $F^{\rm rot}_{100\,\rm ms}$ (right axis, red stars) vs. bounce compactness $\xi_{2.5}$. Most models have a low bounce compactness and are unlikely to form a black hole. Models with a high bounce compactness (dark shaded box) systematically have short spin periods and a large budget of free energy stored in the differential rotation. Comparing the reference spin periods of this figure to Metzger *et al.* (2011), one would predict that none of these models form black holes.

2. Results

Our simulations first demonstrate that most of the Woosley & Heger (2006) models have a small bounce compactness $\xi_{2.5}$, where $\xi_M = (M/2.5\,M_\odot)/(R_M/10^8\,\rm cm)$ (Fig. 1). O'Connor & Ott (2011) argue that a bounce compactness of $\sim$0.45 represents a threshold value for the neutrino mechanism since above it an unrealistic neutrino-heating efficiency is required to prevent black hole formation. The progenitors in Fig. 1 that are below this threshold value are, in terms of compactness, similar to garden variety, low-mass, non-rotating, progenitors and do not seem to have any more reason to form a black hole than, e.g., the RSG progenitors expected to produce SNe II-Plateau. However, if we assume that the magneto-rotational instability is able to generate large scale magnetic fields on short timescales, the fast rotating progenitors in Fig. 1 ($P_{\rm ref} \lesssim 10\,\rm ms$) may be ideal progenitors of a proto-magnetar-powered LGRB.

There are seven models in Woosley & Heger (2006) that have bounce compactnesses larger than the threshold value of O'Connor & Ott (2011); they are highlighted with a shaded box in Fig. 1. If these progenitors were non-rotating one might expect a failed supernova and black hole formation, however each of the progenitors has significant rotation ($P_{\rm ref} \lesssim 3\,\rm ms$) and most have a significant amount of free energy in differential rotation, which can be tapped by the magneto-rotational instability. Dessart et *al.* (2008) evolved the 35OC model with magnetic fields in 2D and found a strong magneto-rotational explosion. This leads us to predict that the other models in this category will have a similar fate. If these models explode, they may also make excellent candidates for progenitors of the proto-magnetar-powered LGRB.

References

Dessart, L., Burrows, A., Livne, E., & Ott, C. D. 2008, *Astrophys. J. Lett.*, **673** L43

Dessart, L., O'Connor, E., & Ott, C. D. 2012, *submitted to Astrophys. J.*, arXiv:1203:1926

Metzger, B. D., Giannios, D., Thompson, T. A., Bucciantini, N., & Quataert, E. 2011, *Mon. Not. Roy. Astron. Soc.*, **413** 2031

O'Connor, E. & Ott, C. D. 2010, *Class. Quantum Grav.*, **27** 114103

O'Connor, E. & Ott, C. D. 2011, *Astrophys. J.*, **730** 70

Wheeler, J. C., Yi, L., Höflich, P., & Wang, L. 2000, *Astrophys. J.*, **537** 810

Woosley, S. 1993, *Astrophys. J.*, **405** 273

Woosley, S. & Heger, A. 2006, *Astrophys. J.*, **637** 914

Woosley, S. & Heger, A. 2007, *Phys. Rep.*, **442** 269

Death of Massive Stars: Supernovae and Gamma-Ray Bursts
Proceedings IAU Symposium No. 279, 2012
P. Roming, N. Kawai & E. Pian, eds.

doi:10.1017/S1743921312013427

The Fast Evolution of SN 2010bh associated with GRB 100316D

Felipe Olivares E.[1], Jochen Greiner[1], Patricia Schady[1], Arne Rau[1], Sylvio Klose[2], and Thomas Krühler[3] for the GROND team

[1]Max-Planck-Institute für extraterrestrische Physik,
Giessenbachstraße 1, 85748, Garching, Germany, email: foe@mpe.mpg.de

[2]Thüringer Landessternwarte Tautenburg, Sternwarte 5, 07778, Tautenburg, Germany

[3]Dark Cosmology Centre, Niels Bohr Institute, University of Copenhagen,
Juliane Maries Vej 30, 2100, Copenhagen, Denmark

Abstract. We report on the type-Ic SN 2010bh associated with XRF 100316D at $z = 0.059$, which is among the latest spectroscopically confirmed GRB-SNe (Bufano *et al.* 2012). This supernova proves to be the most rapidly evolving GRB-SN to date.

Keywords. supernovae: individual (SN 2010bh), gamma-rays: burts

1. Introduction

So far only a handful of supernovae (SNe) associated with gamma-ray bursts (GRBs) have been spectroscopically confirmed; see Woosley & Bloom (2006) and Hjorth & Bloom (2011) for reviews. Out of those only SN 2006aj associated with XRF 060218 has shown signatures of the cooling of the shock breakout (Campana *et al.* 2006).

For the case of the association SN 2010bh/XRF 100316D, GROND provided data from 0.5 to 80 days after the burst covering a wavelength range from 350 to 1800 nm, significantly expanding the pre-existing data set for this event (e.g., Cano *et al.* 2011).

2. Results

Broad Band Spectral Energy Distribution: Detections at 50 ks in $g'r'i'z'J$ are combined with a UVOT $uvw1$-band detection at 33 ks and an interpolated XRT spectrum. We model the SED (left panel of Fig. 1) with a blackbody associated with the cooling of the shock breakout and a power-law representing the afterglow. We obtain a host-galaxy extinction of $A_{V,\mathrm{host}} = 1.2 \pm 0.1$ mag and a metal absorption equivalent to a hydrogen column density of $N_{H,\mathrm{host}} = (4.4 \pm 0.4) \times 10^{22}$ cm^{-2}.

Evolution of the Thermal Component: Early X-ray measurements from Starling *et al.* (2011) complement our results for the temperature and radius of the blackbody component (Fig. 1, right panel). Assuming a linear growth, we obtain $v = 8000$ km/s. The best model is a power law, which yields an initial apparent radius of $R_0 = (7.0 \pm 0.9) \times 10^{11}$ cm. As seen in the upper right panel of Fig. 1, adiabatic expansion fails to reproduce $T_{BB}(t)$.

Pseudo Bolometric Light Curve: The flux integrated from the g' to the H bands was fitted using the two component model from Maeda *et al.* (2003), which consists of a dense inner core and an outer core with lower opacity (e.g., Valenti *et al.* 2008). The transition from optically thick to thin occurs at ≈ 33 days after the burst for SN 2010bh. The physical parameters of the explosion are $M_{\mathrm{Ni}} = (0.21 \pm 0.03) M_\odot$, $M_{\mathrm{ej}} = (2.60 \pm 0.23) M_\odot$, and $E_{\mathrm{k}} = (2.4 \pm 0.7) \times 10^{52}$ erg.

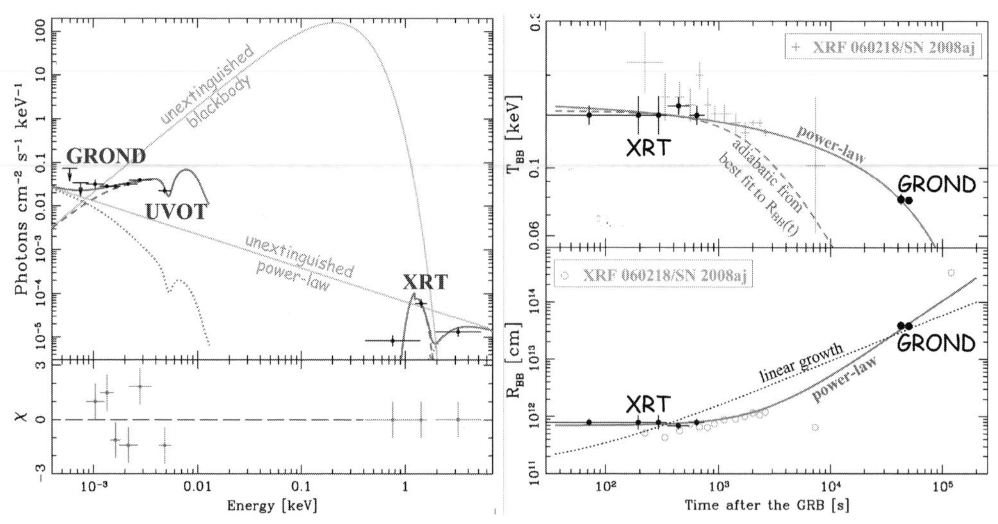

Figure 1. ***Left:*** Broad-band spectral energy distribution. The afterglow spectral slope is $\Gamma = 1.8 \pm 0.1$. ***Right:*** Temporal evolution of the blackbody component. Temperature is shown in the upper panel and blackbody radius in the lower panel.

3. Conclusions

Combining GROND and *Swift* data, the early broad-band SED is modeled with a blackbody and power-law component attenuated by ISM absorption in the host galaxy. The evolution of the thermal component reveals a cooling envelope at an apparent initial radius, which is compatible with a dense wind surrounding a WR star. Moreover, the early-time expansion velocities are compatible with the SN nature. Multicolor templates of SN 1998bw show that SN 2010bh is on average 70% as bright as SN 1998bw (see Olivares E. *et al.* 2012). Reaching maximum brightness at 8 − 9 days after the burst in the blue bands, SN 2010bh is the most rapidly evolving GRB-SNe to date. A two-component parametrized model fitted to the pseudo bolometric light curve delivered physical parameters of the explosion. The kinetic energy makes this SN the second most energetic GRB-SN after SN 1998bw. SN 2010bh also shows one of the earliest peaks ever recorded and it fades more rapidly than any other GRB-SN or type-Ic SN. Further analysis can be found in Olivares E. *et al.* (2012).

Acknowledgements

Part of the funding for GROND (both hardware as well as personnel) was generously granted from the Leibniz-Prize to Prof. G. Hasinger (DFG grant HA 1850/28-1). This work made use of data supplied by the UK *Swift* Science Data Centre at the University of Leicester.

References

Bufano, F., Pian, E., Sollerman, J., *et al.* 2012, *ApJ* submitted, arXiv:1111.4527
Campana, S., Mangano, V., Blustin, A. J., *et al.* 2006, *Nature*, 442, 1008
Cano, Z., Bersier, D., Guidorzi, C., *et al.* 2011, *ApJ*, 740, 41
Hjorth, J. & Bloom, J. S. 2011, in *Gamma-Ray Bursts*, ed. C. Kouveliotou, R. A. M. J. Wijers, & S. E. Woosley (Cambridge: Cambridge Univ. Press), arXiv:1104.2274
Maeda, K., Mazzali, P. A., Deng, J., *et al.* 2003, *ApJ*, 593, 931
Olivares E., F., Greiner, J., Schady, P., *et al.* 2012, *A&A*, 539, A76
Starling, R. L. C., Wiersema, K., Levan, A. J., *et al.* 2011, *MNRAS*, 411, 2792
Valenti, S., Benetti, S., Cappellaro, E., *et al.* 2008, *MNRAS*, 383, 1485
Woosley, S. E. & Bloom, J. S. 2006, *ARAA*, 44, 507

Death of Massive Stars: Supernovae and Gamma-Ray Bursts
Proceedings IAU Symposium No. 279, 2012
P. Roming, N. Kawai & E. Pian, eds.

doi:10.1017/S1743921312013439

Exploding SNe with jets: time-scales

Oded Papish and Noam Soker

Dept. of Physics, Technion, Haifa 32000, Israel; papish@physics.technion.ac.il;
soker@physics.technion.ac.il.

Abstract. We perform hydrodynamical simulations of core collapse supernovae (CCSNe) with a cylindrically-symmetrical numerical code (FLASH) to study the inflation of bubbles and the initiation of the explosion within the frame of the jittering-jets model. We study the typical time-scale of the model and compare it to the typical time-scale of the delayed neutrino mechanism. Our analysis shows that the explosion energy of the delayed neutrino mechanism is an order of magnitude less than the required 10^{51} erg.

Keywords. supernovae: general

1. The Jittering-Jets Model

Accretion-outflow systems are commonly observed in astrophysics when compact objects accrete mass via an accretion disk that launches jets. This is the basic engine assumed in jet-based CCSN models (Papish & Soker 2011, Lazzati *et al.* 2011). Our *Jittering-Jet Model* for CCSN explosions is based on the following ingredients. (1) We do not try to revive the stalled shock. To the contrary, our model requires the material near the stalled-shock to fall inward and form an accretion disk around the newly born neutron star (NS) or black hole (BH). (2) We conjecture that due to stochastic processes and the stationary accretion shock instability (SASI; e.g., Blondin & Shaw 2007, Fernández 2010) segments of the post-shock accreted gas possess local angular momentum (see also Foglizzo *et al.* (2012) for an experimental demonstration). (3) We assume that the accretion disk launches two opposite jets. Due to the rapid change in the disk's axis, the jets can be intermittent and their direction rapidly varying. These are termed jittering jets. (4) The jets penetrate the infalling gas up to a distance of few$\times$1000 km, i.e., beyond the stalled-shock. The jets deposit their energy inside the star via shock waves, and form hot bubbles. (5) The jets are launched only in the last phase of accretion onto the NS.

We perform 2.5D numerical simulations using the FLASH code (Fryxell *et al.* 2000). We use 2D cylindrical coordinates on a grid of size 1.5×10^9 cm in each direction. We use 10 mesh grid refinements which gives us a resolution of 3 km at the inner boundary of the grid of $r = 70$ km. For the initial conditions of the ambient gas in the core we used the $15 M_\odot$ model of Liebendörfer *et al.* (2005) who made a 1D simulations of the core bounce. We start the simulation by launching jets at 0.25 s after bounce. We inject a jet (only one jet as only one half of space is simulated) at 75 km from the center with a full opening angle of 10 degrees. To simulate the jittering effect we inject the jet for a time interval of $\Delta t_j = 0.05$ s with a constant angle θ_n relative to the z axis. After this time interval, we stop the jets for a period of $\Delta t_p = 0.05$ s. We then continue with a jet at a different angle of θ_{n+1}, for the same $\Delta t_j = 0.05$ s. We repeat this process for several times. The velocity of the jet is taken to be 10^{10} cm s^{-1} with a mass outflow rate of $\dot{M}_{2j} = 4 \times 10^{31}$ g s^{-1}. The total energy carried by the two opposite jets combined over all episodes is $E_{2j} = 2 \times 10^{51}$ erg. Our preliminary results are presented in Fig. 1.

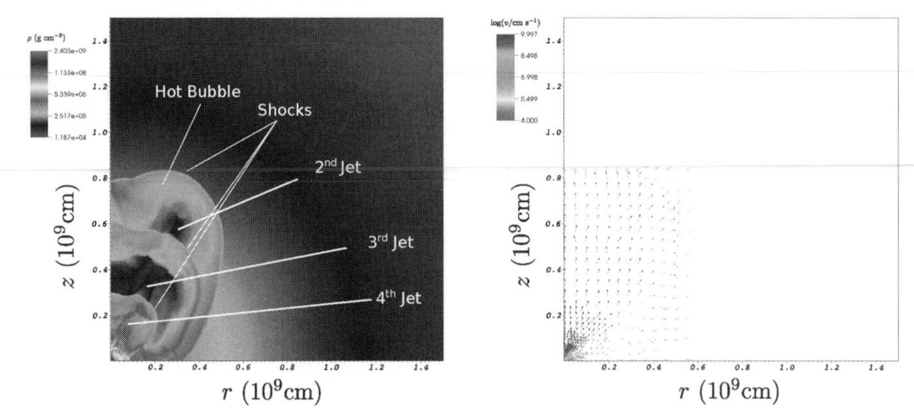

Figure 1. Time $t = 0.61$ s after bounce with the first jets episode starting at $t = 0.25$ s after bounce. 4 episodes of jets were launched up to this time, in directions $\theta = 10°, 30°, 50°$, and $35°$. The first two episodes are already mixed together; the 4th jet's episode was turned off 0.01 seconds ago. Note that accretion from the equatorial plane (the horizontal plane in the figure) continues alongside with the outflow induced by the jets. The left plot shows the density, the right plot shows the log of the velocity. All values are in cgs units. The size of the grid is 1.5×10^9 cm in each direction.

2. Time-scale Considerations

The typical parameters of the model are the following. The active phase of the jets lasts for a typical time equal to few times the free fall time from the region where the baryonic mass is ~1.5-1.6 $M_\odot$. This radius is ~ 3000 km and the free fall time is 0.4 s. The jets deposit their energy in a typical radius of $r_d \simeq 10^4$ km. The average power of the jets is taken to be $P_{2j} \simeq 10^{51}$ erg s^{-1}. The jets are active for $t_j \simeq 1-2$ s. This is also the dynamical time at r_d

$$t_d \equiv \frac{r_d}{v_{\rm esc}(r_d)} = 1.6 \left(\frac{r_d}{10^4 \text{ km}}\right)^{3/2} \text{ s}, \tag{2.1}$$

where $v_{\rm esc}$ is the escape velocity at r_d, and where the central mass is taken to be $1.5 M_\odot$. The jets' power and the interaction time at $r_d \sim 10^4$ km are consistent with the energy required to explode the star. ¿From this time on, the hot bubbles (which might merge to one bubble) explode the star.

We find that the same consideration of interaction time and explosion power is problematic for neutrino driven CCSN explosion models. In recent numerical results (e.g. Brandt *et al.* 2011, Hanke *et al.* 2011, Kuroda *et al.* 2012) the mechanical energy achieved by neutrino driven models is still significantly short of the desired ~10^{51} erg required to explode the star. In the recent simulations of Mueller *et al.* (2012) the sum of the kinetic, thermal, and nuclear energy of the expanding gas in the core is a factor of ~ 4 smaller than the observed energy of CCSNe. We note that in Nordhaus *et al.* (2011) and Scheck *et al.* (2006) the explosion is achieved mainly by a continuous wind. Here we refer to the delayed neutrino mechanism models where the energy of the wind is negligible.

Let us apply the interaction time considerations to the delayed neutrino mechanism. The gain region, where neutrino heating is efficient, occurs in the region $r \simeq 200-700$ km (Janka 2001). If energy becomes significant the gas will be accelerated and escape within a time of $\sim r/[0.5 v_{\rm esc}(r)]$. From Janka (2001) we find the neutrino "optical depth" from r to infinity to be $\tau \sim (r/100 \text{ km})^{-3}$. The typical electron (and positron) neutrino luminosity is $L_\nu \simeq 5 \times 10^{52}$ erg s^{-1} (Mueller *et al.* 2012). Over all, if the interaction occurs near a radius r in the gain region, the energy that can be acquired by the expanding gas is

$$E_{\rm shell} \simeq \frac{2r}{v_{\rm esc}(r)} \tau L_\nu \simeq 10^{50} \left(\frac{r}{200 \text{ km}}\right)^{-1.5} \left(\frac{L_\nu}{5 \times 10^{52} \text{ erg s}^{-1}}\right) \text{erg}. \tag{2.2}$$

We claim, therefore, that the total energy that can be used to revive the shock is limited to a typical value of $\sim 10^{50}$ erg. This is along the recent results of numerical simulations of the delayed neutrino mechanism cited above.

We conclude that the delayed neutrino explosion mechanism, where the explosion is due to neutrino heating in the gain region, as proposed by Wilson (Bethe & Wilson 1985), cannot work. It might lead to the reviving of the stalled shock under some circumstances, but it cannot lead to an explosion with an energy of 10^{51} erg.

Our basic conclusion is that if no ingredient is added to the neutrino delayed mechanism, it falls short of the required energy to explode the star by an order of magnitude. Such an ingredient can be a strong wind, such as was applied by artificial energy deposition (Nordhaus *et al.* 2011). In their 2.5D simulations Scheck *et al.* (2006) achieved explosion that was mainly driven by a continuous wind. The problem we see with winds is that they are less efficient than jets. Indeed, in order to obtain an explosion the winds in the simulations of Scheck *et al.* (2006) had to be massive. For that, in cases where they obtained energetic enough explosions the final mass of the NS was low ($M_{\rm NS} < 1.3 M_\odot$). The problem we find is that such a wind must be active while accretion takes place; the accretion is required for the energy source.

An inflow-outflow situation naturally occurs with jets launched by accretion disks. For that, we propose the jittering-jet mechanism. Namely, we require the accretion process to continue for ~ 1 s. In our model there is no need to revive the accretion shock.

Acknowledgements

We thank Thomas Janka, Kei Kotake, and Jason Nordhaus for useful comments. This research was supported by the Asher Fund for Space Research at the Technion, and by the Israel Science Foundation.

References

Bethe, H. A. & Wilson, J. R. 1985, *ApJ*, 295, 14
Blondin, J. M. & Shaw, S. 2007, *ApJ*, 656, 366
Brandt, T. D., Burrows, A., Ott, C. D., & Livne, E. 2011, *ApJ*, 728, 8
Fernández, R. 2010, *ApJ*, 725, 1563
Foglizzo, T., Masset, F., Guilet, J., & Durand, G. 2012, *Physical Review Letters*, 108, 051103
Fryxell, B., Olson, K., Ricker, P., *et al.* 2000, *ApJS*, 131, 273
Hanke, F., Marek, A., Mueller, B., & Janka, H.-T. 2011, arXiv:1108.4355
Janka, H.-T. 2001, *A&A*, 368, 527
Lazzati, D., Morsony, B. J., Blackwell, C. H., & Begelman, M. C. 2011, arXiv:1111.0970
Liebendörfer, M., Rampp, M., Janka, H.-T., & Mezzacappa, A. 2005, *ApJ*, 620, 840
Kuroda, T., Kotake, K., & Takiwaki, T. 2012, arXiv:1202.2487
Mueller, B., Janka, H.-T., & Marek, A. 2012, arXiv:1202.0815
Nordhaus, J., Brandt, T., Burrows, A., & Almgren, A. 2011, arXiv:1112.3342
Papish, O. & Soker, N. 2011, *MNRAS*, 416, 1697
Scheck, L., Kifonidis, K., Janka, H.-T., & Mueller, E. 2006, *A&A*, 457, 963

Death of Massive Stars: Supernovae and Gamma-Ray Bursts
Proceedings IAU Symposium No. 279, 2012
P. Roming, N. Kawai & E. Pian, eds.

doi:10.1017/S1743921312013440

Observations of GRBs in the mm/submm range at the dawn of the ALMA era

A. de Ugarte Postigo[1,2], A. Lundgren[3,4], S. Martín[3], D. García-Appadoo[3,4], I. de Gregorio Monsalvo[3,4], C.C. Thöne[1], J. Gorosabel[1], A. J. Castro-Tirado[1], R. Sánchez-Ramírez[1], and J. C. Tello[1] on behalf of a larger collaboration

[1]Instituto de Astrofísica de Andalucía (IAA-CSIC), Spain;
email: deugarte@iaa.es

[2]Dark Cosmology Centre, Niels Bohr Institute, Univ. of Copenhagen, Denmark [3]European Southern Observatory, Chile [4]Joint ALMA Observatory, Chile

Abstract. Gamma-ray bursts (GRBs) generate an afterglow with an emission peaking in the millimetre and submillimeter (mm/submm) range during the first hours to days, making the study in these wavelengths of great importance. Here we give an overview of the data that has been collected for GRB observations in this wavelengths until September 2011. The total sample includes 102 GRBs, of which 88 have afterglow observations, and the rest are only host galaxy searches. The 22 detections cover the redshift range between 0.168 and 8.2 and have peak luminosities that span 2.5 orders of magnitude. With the start of the operations at ALMA, the sensitivity with respect to previous facilities has already improved by over an order of magnitude. We estimate that, once completed, ALMA will be able to detect ~98 % of the afterglows.

This proceeding is based on the work published by de Ugarte Postigo *et al.* (2012).

Keywords. gamma rays: bursts, submillimeter

1. Observations and sample

During the last 3 years we have followed-up 11 GRBs (plus an additional Galactic X-ray binary, initially identified as a GRB) and discovered 2 counterparts in submm with observing programmes at APEX and SMA. This is put into context with the most complete sample of continuum observations that have been published to date of GRB afterglows and their host galaxies in the mm/submm wavelength range, covering from early 1997 until the 30th of September 2011.

The complete sample includes observations of 102 bursts, of which 88 are searches for GRB afterglows, with 22 detections. There have been specific host galaxy searches for 36 cases, although limits can be provided for the 102 bursts that have been followed. Host galaxy detections have only been achieved in four cases: GRB 000210, GRB 000418, GRB 010222 and XT 080109.

2. Afterglow models

GRB afterglows can be described, in the simplest case, using the fireball model (Sari *et al.* 1998). According to it, material is ejected at ultrarelativistic velocities through collimated jets. When this material interacts with the medium surrounding the progenitor, the accelerated particles emit a synchrotron spectrum that is characterised by three break frequencies: ν_m is the characteristic synchrotron frequency and is the maximum of the emission, ν_c is the cooling frequency, above which radiative cooling is significant, and ν_a is the synchrotron self-absorption frequency.

A reverse shock, produced inside the ejecta, can generate an additional early emission (Piran *et al.* 1999). This has been rarely observed in the optical wavelengths but is expected to have a significant contribution in the early mm/submm emission. For example, the mm detection of GRB 090423, at a redshift of 8.2, seems to show excess emission possibly due to a reverse shock.

3. Observations of individual GRBs

Multiwavelength observations of GRB afterglows are the only way to determine with precision the physics involved in the GRB and learn about the environment that surrounds it. Studies in mm/submm are especially interesting in the case of optically-dark bursts, as they allow us to access what optical observations cannot.

Optical samples of GRBs are limited by the extinction in the host galaxy that, if large, make the optical emission undetectable. The negligible effect of dust extinction in the mm/submm bands allows us to study a more complete sample. As an example we can look at GRB 051022, one of the darkest bursts detected to date, for which an optical counterpart was not found. Observations in mm wavelengths allowed us to localise and study the afterglow and host galaxy (Castro-Tirado *et al.* 2007). The other main cause for optically dark GRBs is their high redshift. In these cases the absorption produced at frequencies higher than the Lyman limit does not allow us to obtain optical detections of GRB afterglows beyond redshifts of 6. These events are important to understand the formation of the first stars in the Universe. Proof that they can be detected in the mm/submm range is the fact that, out of the three GRB afterglows observed at $z > 6$, two have been detected (GRB 050904 at $z = 6.3$, Tagliaferri *et al.* 2005, Haislip *et al.* 2006, and GRB 090423 at $z = 8.2$, Tanvir *et al.* 2009, Salvaterra *et al.* 2009).

4. GRBs in the ALMA era

Using data from the sample and assumptions based on samples at other wavelengths, we estimate the real peak flux density distribution of GRBs, from which an average peak flux density value of 0.33 mJy can be expected. Using the detection limits calculated for ALMA, we can expect that the completed observatory should be able to detect 98%. In the case of bright GRB afterglows, ALMA will be able to study spectral features and perform polarimetric studies of the afterglow, which have been out of reach until now. With ALMA we will be, for the first time, in position to undertake studies of samples of GRB host galaxies. We will be able to perform studies of the continuum emission to characterise the dust content and determine the unextinguished star formation rate of the hosts. Through the study of emission features from the host, we will be able to understand the molecular content and the chemical enrichment of the strong star-forming regions in which GRBs are found, at redshifts that go back to the epoch in which the first stars were formed.

References

Castro-Tirado, A. J., Bremer, M., McBreen, S., *et al.* 2007, *A&A*, 475, 101
Haislip, J. B., Nysewander, M. C., Reichart, D. E., *et al.* 2006, *Nature*, 440, 181
Piran, T. 1999, *Physics Reports*, 314, 575

Salvaterra, R., Della Valle, M., Campana, S., *et al.* 2009, *Nature*, 461, 1258
Sari, R., Piran, T., & Narayan, R. 1998, *ApJL*, 497, L17
Tagliaferri, G., Antonelli, L. A., Chincarini, G., *et al.* 2005, *A&A*, 443, L1
Tanvir, N. R., Fox, D. B., Levan, A. J., *et al.* 2009, *Nature*, 461, 1254
de Ugarte Postigo, A., Lundgren, A., Martín, S., *et al.* 2012, *A&A*, 538, A44

Death of Massive Stars: Supernovae and Gamma-Ray Bursts
Proceedings IAU Symposium No. 279, 2012
P. Roming, N. Kawai & E. Pian, eds.

doi:10.1017/S1743921312013452

Early Time Bolometric Light Curves of Type–II Supernovae Observed by *Swift*

T. A. Pritchard[1] **and P. W. A. Roming**[2,1]

[1]Dept. of Astronomy & Astrophysics, The Pennsylvania State University,
525 Davey Lab, University Park, PA 16802, USA
email: tapritchard@astro.psu.edu

[2]Southwest Research Institute, Department of Space Science
6220 Culebra Rd, San Antonio, Texas, 78238 USA
email: proming@swri.edu

Abstract. We present early time (~0-50 days) bolometric light curves of UV-bright Core Collapse Supernovae observed with the *Swift* UV/Optical Telescope. We also generate pseudo-bolometric light curves from Swift UV and optical data and examine these by subtype as well as the observed and interpolated UV and IR flux contributions by epoch and bolometric corrections at early times from UV data.

Keywords. supernovae:general, ultraviolet: general, catalogs

1. Introduction

We present early time UV and optical observations of Core-Collapse SNe (CCSNe) obtained by *Swift* since its launch in 2005. UV observations of these objects at early times are rare, and for UV-bright objects (II, IIP, IIn, IIL SNe) the UV may account for upwards of 80% of the total luminosity during this early portion. Due to the paucity of UV instruments available this large fraction of a SN's light often goes unaccounted for in the first several weeks of a SN. To help rectify this situation we present a collection of observations by the *Swift* satellite in the UV and optical bands. We generate pseudo-bolometric light curves from this data, and using blackbody fits to approximate the UV and IR tails outside of Swift coverage compute a bolometric correction to arrive at a bolometric light curve. We are working to incorporate observations from other telescopes currently so that we may increase the observed coverage of these SNe and increase the quality of our bolometric light curves.

2. Overview

We use synthetic photometry from reddened model blackbodies fit to the *Swift* light curve data, and UVOT count rate to flux conversions interpolated from data in Appendix A of Brown *et al.* 2010 using our best fit blackbody temperature, to generate a pseudo-bolometric light curve for each CCSN. At each epoch we then use the integrated flux of the best-fit reddened model blackbody at our unobserved wavelengths to calculate UV (<1600 Angstrom) and IR (>6000 Angstrom) corrections. These correction factors are then added to the pseudo-bolometric light curve to arrive at a bolometric light curve. Due to the nature of the CCSNe spectra, and the *Swift* filter selection we find that this works best for SNe with a thick hydrogen shell as these SNe are both typically UV-bright at early times as well as the SNe spectra more closely resembling our blackbody model.

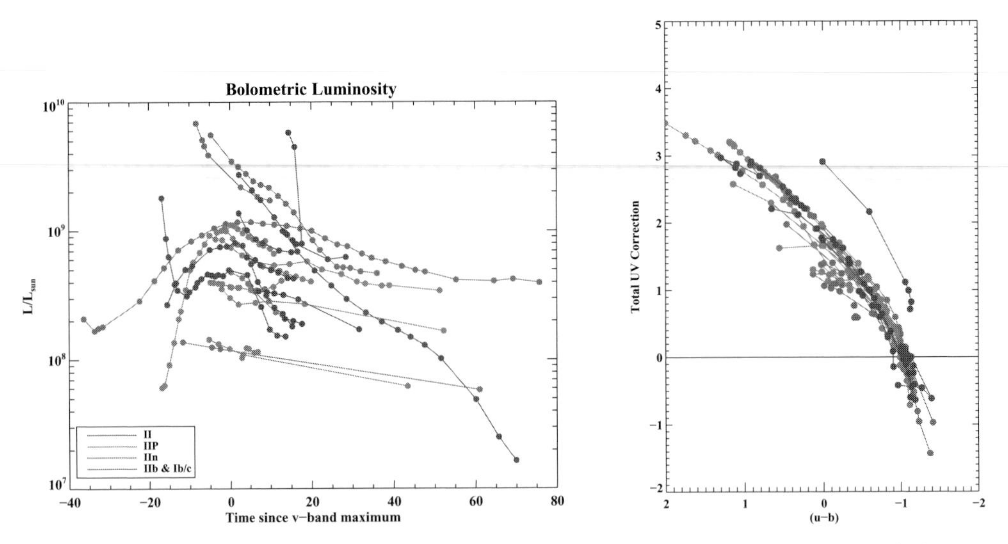

Figure 1. *Left*: UVOT Bolometric lightcurves from the *Swift* sample centered around the time of maximum light in the visual band. *Right*: Correction factor in magnitudes as a function of CCSNe color

3. Implications

Using the *Swift* only data we find a maximum unobserved UV correction of only $\sim 25\%$ of the total flux. This is a factor of > 3 better than can be done with ground based optical and IR data only at similar epochs, as seen in Bersten & Hamuy 2009. This allows us to calculate an accurate distance-independent UV correction based on optical colors as visible in Figure 1) (*Right*). We perform this calculation in a similar manner as normally done for a bolometric correction; but we only include *Swifts* observed UV flux (filters u, uvw1, uvm2, and uvw2 filters) plus the calculated UV correction factor. We find that all CCSNe subtypes behave similarly, however $u - b$ color appears to be a much stronger indicator than $b - v$. This is expected since u-band traces the UV-flux much more closely than b.

In Figure 1 (*Left*) we plot our sample's bolometric light curves. These UV-bright CC-SNe reach maximum light very rapidly, which is seen in our sample in both the lack of a clear UV-rise in most SNe UV light curves as well as the few clear peaks we see in the bolometric light curves. While our IR-correction at early times is quite small, as low as 5% for some CCSNe, after $\sim$ 20 days post v-band maximum this has grown to be significant. At the end of *Swift* observations, due to the SNe spectrum cooling, we have found this correction to be as high as 80% of the total SNe flux. Integration of optical and IR ground based observations red-ward of the UVOT bandpass are underway, and are necessary to enhance the quality of the bolometric light curves at these times. Further observations of very early supernovae to observe the UV rise will continue to improve these early time bolometric light curves as well.

References

Bersten, Melina C. & Hamuy, Mario 2009, *ApJ*, 701, 200
Brown, Peter J., *et al.* 2010, *ApJ* 721, 1608

Death of Massive Stars: Supernovae and Gamma-Ray Bursts
Proceedings IAU Symposium No. 279, 2012
P. Roming, N. Kawai & E. Pian, eds.

doi:10.1017/S1743921312013464

Cosmological effects on the observed flux and fluence distributions of gamma-ray bursts

Jakub Řípa[1], Attila Mészáros[2] and Felix Ryde[3]

[1]Institute for the Early Universe, Ewha Womans University, 120-750 Seoul, Korea
email: ripa@ewha.ac.kr; sirrah.troja.mff.cuni.cz

[2]Astronomical Institute, Charles University, 180 00 Prague, Czech Republic
email: meszaros@cesnet.cz

[3]Department of Physics, Royal Institute of Technology, AlbaNova University Center, SE-106 91 Stockholm, Sweden
email: felix@particle.kth.se

Abstract. Several claims have been put forward that an essential fraction of long-duration BATSE gamma-ray bursts should lie at redshifts larger than 5. This point-of-view follows from the natural assumption that fainter objects should, on average, lie at larger redshifts. However, redshifts larger than 5 are rare for bursts observed by Swift. The purpose of this article is to show that the most distant bursts in general need not be the faintest ones. We derive the cosmological relationships between the observed and emitted quantities, and arrive at a prediction that is tested on the ensembles of BATSE, Swift and Fermi bursts. This analysis is independent on the assumed cosmology, on the observational biases, as well as on any gamma-ray burst model. We arrive to the conclusion that apparently fainter bursts need not, in general, lie at large redshifts. Such a behaviour is possible, when the luminosities (or emitted energies) in a sample of bursts increase more than the dimming of the observed values with redshift. In such a case dP(z)/dz > 0 can hold, where P(z) is either the peak-flux or the fluence. This also means that the hundreds of faint, long-duration BATSE bursts need not lie at high redshifts, and that the observed redshift distribution of long Swift bursts might actually represent the actual distribution.

Keywords. gamma rays: bursts, cosmology: observations

---

1. Introduction

This article briefly summarizes works published by Mészáros *et al.* (2011a) and Mészáros *et al.* (2011b). It can be shown that for the observed peak-flux or fluence $P(z)$ of gamma-ray bursts (GRBs) it holds: $P(z) = \frac{(1+z)^N \tilde{L}(z)}{4\pi d_l^2(z)}$, z is the redshift; $d_l(z)$ is the luminosity distance; $\tilde{L}(z)$ is isotropic peak-luminosity or emitted energy. $N = 0$; 1; 2 depending on the units of the observables: $N = 0$ if $P(z)$ is peak-flux in units erg/(cm^{2}s) and then $\tilde{L}(z)$ in units erg/s; $N = 1$ if $P(z)$ is fluence in units erg/cm^2 and then $\tilde{L}(z)$ in ergs or $N = 1$ if $P(z)$ is peak-flux in units ph/(cm^{2}s) and then $\tilde{L}(z)$ in units ph/s; $N = 2$ if $P(z)$ is fluence in units ph/cm^2 and then $\tilde{L}(z)$ is in photons. $P(z)$ is given by photons with energies $E_1 \leqslant E \leqslant E_2$, where $E_{1,2}$ is the detector range. $\tilde{L}_{\rm ph}(z)$ is from energy range $E_1(1+z) \leqslant E \leqslant E_2(1+z)$.

It is a standard cosmology that for small redshifts ($z \ll 0.1$) $d_l(z) \propto z$, and for high redshifts $\lim_{z\to\infty} \frac{d_l(z)}{1+z}$ = finite and positive number for any H_o, Ω_M, Ω_Λ. If we assume that $\tilde{L}(z) \propto (1+z)^q; N + q > 2$ for $z \to \infty$ then $\frac{dP(z)}{dz} > 0$, for $z \to \infty$. If $\tilde{L}(z)$ increases with z faster than $\propto (1+z)^{2-N}$ for high redshifts, then inverse behaviour can happen, and the apparently brighter GRBs can be at higher redshifts; $dP(z)/dz > 0$ can occur.

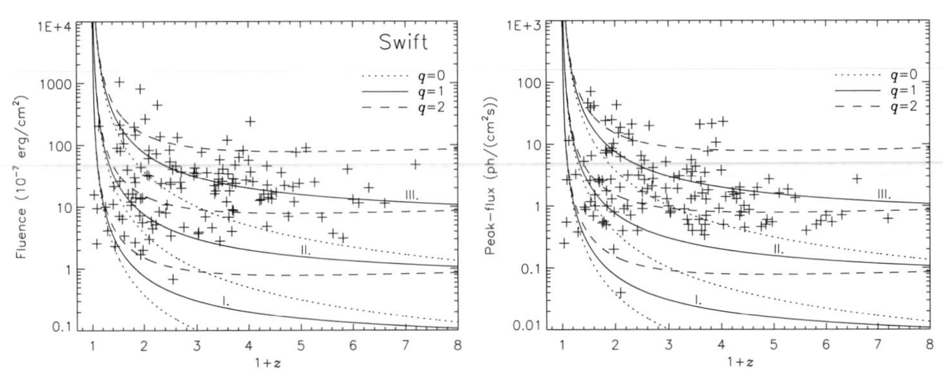

Figure 1. Distributions of fluences (left panel) and peak-fluxes (right panel) Swift long GRBs. On the left panel the curves denote the values of fluences for $\tilde{E}_{\rm iso} = \tilde{E}_o(1+z)^q$ (constants $\tilde{E}_o$ are in 10^{51} erg: I. 0.1; II. 1.0; III. 10.0). On the right panel the curves denote the values of peak-fluxes for $\tilde{L}_{\rm iso} = \tilde{L}_o(1+z)^q$ (constants $\tilde{L}_o$ are in 10^{58} ph/s: I. 0.01; II. 0.1; III. 1.0). Here $N = 1$ thus value $q = 1$ is the limiting case.

For small redshifts z, $P(z)$ decreases as z^{-2}, but if $\tilde{L}(z) \propto (1+z)^q$ and $N+q > 2$ then $P(z)$ starts to increase as z^{N+q-2} for high z. To answer the question where the $z_{\rm turn}$ is, i.e. where $dP(z)/dz = 0$, one can simply search for the minimum of $Q(z) = (1+z)^{N+q}/d_l(z)^2$, i.e., when $dQ(z)/dz = 0$.

2. Samples of long GRBs

We studied 134 long Swift GRBs with know redshifts. The distribution of observed flunces/peak fluxes together with the curves of constant $\tilde{E}_{\rm iso}$ and $\tilde{L}_{\rm iso}$ are shown in Fig. 1. We also studied 6 Fermi GRBs, 9 BATSE GRBs with measured redshifts and 13 BATSE GRBs with calculated pseudo-redshifts. These samples also demonstrate that fainter bursts can well have smaller redshifts (for details see Mészáros *et al.* (2011a)).

3. Conclusion

The theoretical study of the z-dependence of the observed fluences and peak-fluxes of GRBs have shown that fainter bursts could well have smaller redshifts. This is really fulfilled for the four different samples of long GRBs. These results do not depend on the cosmological parameters and on GRB models.

Acknowledgements

This study was supported by the OTKA grant K77795, by the Grant Agency of the Czech Republic grants No. P209/10/0734, by the Research Program MSM0021620860 of the Ministry of Education of the Czech Republic, and by Creative Research Initiatives (RCMST) of MEST/NRF and the World Class University grant no R32-2008-000-101300.

References

Mészáros, A., Řípa, J., & Ryde, F. 2011a, *Astronomy & Astrophysics*, 529, A55
Mészáros, A., Řípa, J., & Ryde, F. 2011b, *Acta Polytechnica*, 51, 45

Death of Massive Stars: Supernovae and Gamma-Ray Bursts
Proceedings IAU Symposium No. 279, 2012
P. Roming, N. Kawai & E. Pian, eds.

doi:10.1017/S1743921312013476

Study of very early phase GRB afterglows with MITSuME

Yoshihiko Saito[1], Yoichi Yatsu[1], Hideya Nakajima[1], Nobuyuki Kawai[1], Katsuaki Asano[1], Yu Aoki[1], Mayumi Hayashi[1], Seongdeng Song[1], Kosuke Kawakami[1], Kazuki Tokoyoda[1], Takahiro Enomoto[1], Ryuichi Usui[1], Daisuke Kuroda[2], Kenshi Yanagisawa[2], Hiroyasu Shimizu[2], Hiroyuki Toda[2], Shogo Nagayama[3], Hidekazu Hanayama[4], Michitoshi Yoshida[5] and Koji Ohta[6]

[1]Dept. of Physics, Tokyo Institute of Technology,
2-12-1, Oookayama, Meguro, Tokyo, 152-8551, Japan
email: `saitoys@hp.phys.titech.ac.jp`

[2]Okayama Astrophysical Observatory, National Astronomical Observatory of Japan,
Kamogata, Okayama 719-0232, Japan

[3]National Astronomical Observatory of Japan,
2-21-1, Ohsawa, Mitaka, Tokyo, 181-8588, Japan

[4]Ishigaki Astronomical Observatory, National Astronomical Observatory of Japan,
1024-1, Arakawa, Ishigaki, Okinawa, 907-0024, Japan

[5]Hiroshima Astrophysical Science Center, Hiroshima University
1-3-1, Kagamiyama, Higashi-Hiroshima, Hiroshima, 739-8526, Japan

[6]Dept. of Astronomy, Kyoto University,
Kitashirakawa-Oiwake, Sakyo, Kyoto, 606-8502, Japan

Abstract. We review the results of very early phase optical follow-up observations of recent gamma-ray bursts (GRBs) with the multi-color optical telescopes "MITSuME". The MITSuME telescopes were designed to perform "real time" and "automatic" follow-up observations prompted by the GCN alerts via the Internet. The rapidly slewing equatorial mounts allow MITSuME to start photometric observations within 100 seconds after the trigger for several GRBs. In particular, we detected a brightening just after the trigger for two GRBs. These phenomena could be interpreted as the "on-set" of afterglow. In this paper we summarize these optical observations with a brief interpretation.

Keywords. GRBs, Afterglows, Optical

---

1. Introduction

Early phase variability in the optical afterglow of GRBs should provide some important information about the circumburst medium and, therefore, on the origin of GRBs.

The robotic telescope system MITSuME (Multicolor Imaging Telescope for Survey and Monstrous Explosions), prompted by the GCN alerts, can start observations within about a minute after the triggers. We have three optical telescopes in different regions in Japan (Akeno 0.5m, Okayama 0.5m, and Ishigakijima 1.05m). Each telescope has a tricolor camera that can take images in three different color ($g\prime$(SDSS), R_C, I_C (Johnson-Cousins)) at the same time.

In the next section we report the results of the observation in which we could successfully monitor the light curve of the GRB afterglow.

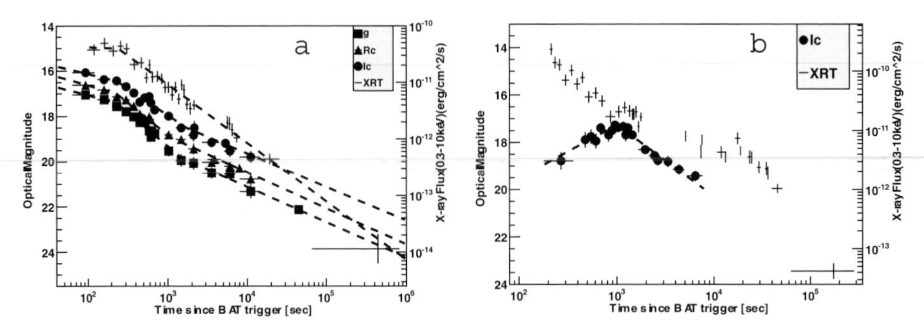

Figure 1. (a) The light curve of GRB090426. (b) The light curve of GRB100219A.

2. Results of Observation

In the last three years, we have successfully observed six GRBs. Brief reports for four of those GRBs are described as follows.

GRB090426. This GRB should be the most distant (z=2.61) short burst (Antonelli *et al.* 2009). We can find two breaks in the optical light curve as shown in Fig. 1a. This result implies two possible scenarios that are "Two component jet scenario" (Racusin *et al.* 2008) and "Multiple energy injection scenario" (Xin *et al.* 2011).

GRB091208B. This is a long GRB at z=1.06. We find that the optical light curve has a simple power law decay until $T_0 \sim 2000$ sec. The energy spectral index of electrons could be estimated from the index of this power law. This result is consistent with the one estimated from the decay index of X-ray.

GRB100219A. Because this GRB has a large redshift (z=4.67), we could identify the optical counterpart only in I_c band. We can follow the brightening phase in the light curve until $T_0 \sim 1000$ sec, while X-ray light curve appears to have a small bump around this time (Fig. 1b). The simultaneous optical and X-ray rebrightening could be explained as the onset of the afterglow. From the peak of onset, we derived that the initial Lorentz factor should be $\Gamma_0 \sim 84$. Based on the optical brightening index, we also find that the ISM around the GRB should have an intermediate density profile between the ISM model and the wind model.

GRB100906A. Because the Akeno telescope of MITSuME could start the observation of this GRB 26 seconds after the trigger, we could obtain a very early phase light curve. This light curve shows a brightening phase until $T_0 \sim 150$ sec and a power law decay thereafter. This optical onset suggests that the ISM around the GRB should be explained by a wind model and the initial Lorentz factor should be $\Gamma_0 \sim 160$.

References

Antonelli, L. A., D'Avanzo, P., Perna, R., *et al.* 2009, *A&A*, 507L, 45
Racusin, J. L., Karpov, S. V., Sokolowski, M., *et al.* 2008, *Nature*, 455, 183
Xin, L., Liang, E., Wei, J., *et al.* 2011, *MNRAS*, 410, 27

Death of Massive Stars: Supernovae and Gamma-Ray Bursts
Proceedings IAU Symposium No. 279, 2012
P. Roming, N. Kawai & E. Pian, eds.

doi:10.1017/S1743921312013488

Origin of Ultra-High Energy Cosmic Rays: Nuclear Composition of Gamma-Ray Burst Jets

Sanshiro Shibata[1] and Nozomu Tominaga[1,2]

[1]Department of physics, Konan University, 8-9-1 Okamoto, Kobe, Hyogo 658-8501, Japan
Email: d1221001@center.konan-u.ac.jp

[2]Kavli Institute for the Physics and Mathematics of the Universe, University of Tokyo, 5-1-5 Kashiwanoha, Kashiwa, Chiba 277-8583, Japan

Abstract. Ultra-high energy cosmic rays (UHECRs) are the most energetic particles flying from space and their source is not clarified yet. Recently, the Pierre Auger Observatory (PAO) suggests that UHECRs involve heavy nuclei. The PAO results require that a considerable fraction of metal nuclei must exist in the accelerating site, which can be realized only in the stellar interior. This puts strong constraints on the origin of UHECRs. In order to definitize the constraints from PAO results, we investigate the fraction of metal nuclei in a relativistic jet in gamma-ray burst associated with core-collapse supernova. If the jet is initially dominated by radiation field, quasi-statistical equilibrium (QSE) is established and heavy nuclei are dissociated to light particles such as ^{4}He during the acceleration and expansion. On the other hand, if the jet is mainly accelerated by magnetic field heavy or intermediate mass nuclei can survive. The criterion to contain the metal nuclei is that the temperature at the launch site is below 4.5×10^9K. Therefore, if the composition of UHECRs is dominated by metal nuclei, a GRB with the magnetized jet is the most plausible candidate of the accelerating site.

Keywords. gamma rays: bursts, nuclear reactions, nucleosynthesis, abundances, MHD,

1. Introduction

The most energetic cosmic rays with $E \geqslant 10^{18}$ eV are called ultra-high energy cosmic rays (UHECRs). Their accelerating site is not clarified yet, although a few astronomical objects have been proposed as the candidate of the accelerating source.

Recent experiments clarify nuclear composition of UHECRs by measuring depth of air shower maximum. In particular, results of the Pierre Auger Observatory indicate that UHECRs are dominated by heavy nuclei at high energy (Abraham *et al.* 2010). This indicate the metallicity at the accelerating site is extremely high.

Therefore, we focus on gamma-ray bursts (GRB) as the candidate of the accelerating site, since GRB is associated with a death of massive star and there are alot of metals in the stellar interior. We investigate whether the jets can possess metal nuclei with high metallicities.

2. Method & Model

We treat a GRB jet as a steady, spherically symmetric, magnetized outflow with efficient magnetic dissipation. We assume that magnetic fields in the outflow are dominated by a toroidal component and that the field efficiently dissipates via magnetic reconnection as the outflow expands. Such efficient dissipation of magnetic fields creates strong magnetic pressure gradient that makes possible a direct conversion of magnetic field energy

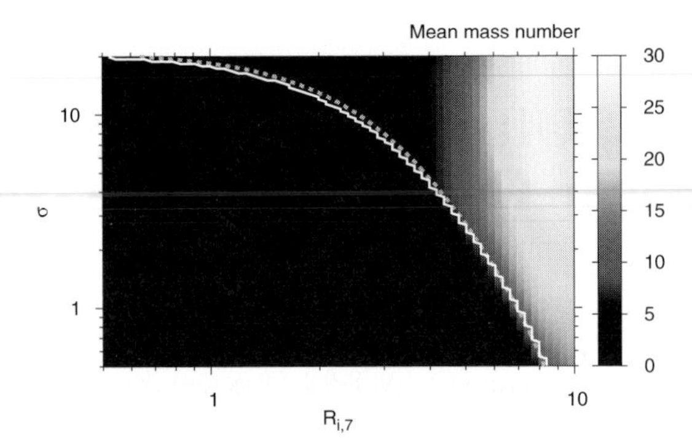

Figure 1. Mean mass number of final composition of the jet.

into kinetic energy (Drenkhahn 2002). Solving the equations for such outflow numerically, time evolution of density and temperature are obtained and then nuclear reactions in the outflow are calculated as a postprocessing. The nuclear reaction network includes 281 isotopes up to ^{79}Br.

We parameterize the GRB outflow with six parameters: isotropic energy deposition rate L_j^{iso}, initial radius of the outflow R_i, initial Lorentz factor Γ_i, maximum Lorentz factor Γ_{max}, angular frequency of central object (including dimensionless factor) $\epsilon\Omega$, and initial magnetization parameter σ_i defined by $\sigma_i \equiv b_i^2/w_i$, where b_i and w_i is the initial magnetic four-vector and the initial enthalpy, respectively.

The initial composion of the outflow is set to be an integration of accreted matter which is obtained by a calculation of relativistic jet-induced explosions of C+O stars with the use of a two-dimensional relativistic Eulerian hydrodynamic code with the Newtonian self-gravity (Tominaga 2009).

3. Results & Conclusion

Figure 1 shows mean mass number of final nuclear composition as functions of R_i and σ_i for the models with $L_j^{iso} = 4 \times 10^{52}$ ergs^{-1}, $\Gamma_i = 1.22$, $\Gamma_{max} = 100$, and $\epsilon\Omega = 10^3$ s^{-1}. The white solid line represents the criterion for establishment of quasi statistical equilibrium (QSE; e.g., Woosley *et al.* 1973). This criterion is well fitted by a contour of initial temperature $T_{i,9} = T_i/10^9\mathrm{K} = 4.5$ (a dotted line in Figure 1). In the models with lower σ_i and smaller R_i than the white solid line, the composition is described by QSE and the metal nuclei are almost destroyed due to the high entropy. On the other hand, the condition for QSE is avoidable in the models with appropriate R_i and σ_i which realize $T_{i,9} < 4.5$.

Therefore, we conclude that the metal nuclei-dominated UHECRs can be originated by GRBs if the relativistic GRB jets are accelerated by the magnetic field and initial temperatures of the jets are smaller than 4.5×10^9 K.

References

Abraham, J., *et al.* 2010, *Phys. Rev. Lett.*, 104, 091101

Drenkhahn, G. 2002, *A&A*, 387, 714

Tominaga, N. 2009, *ApJ*, 690, 526

Woosley, S. E., Arnett, W. D., & Clayton D. D., 1973, *ApJS*, 26, 231

Death of Massive Stars: Supernovae and Gamma-Ray Bursts
Proceedings IAU Symposium No. 279, 2012
P. Roming, N. Kawai & E. Pian, eds.

doi:10.1017/S174392131201349X

Wave-Driven Mass Loss: A mechanism for late-stage stellar eruptions

Josh Shiode and Eliot Quataert

UC Berkeley, Department of Astronomy,
601 Campbell Hall #3411, UC Berkeley,
Berkeley, CA, USA 94720-3411

Abstract. During the late stages of stellar evolution in massive stars (carbon fusion and later), the fusion and neutrino luminosities in the core of the star exceed the Eddington luminosity. This can drive vigorous convective motions which in turn excite a super-Eddington flux in internal gravity waves. We show that an interesting fraction of the energy in excited gravity waves can, in some cases, convert into sound waves as the gravity waves propagate (tunnel) towards the stellar surface. The subsequent dissipation of the sound waves can unbind up to several $M_\odot$ of the stellar envelope. This wave-driven mass loss can explain the existence of extremely large stellar mass loss rates just prior to core-collapse, which are inferred via circumstellar interaction in some core-collapse supernovae (e.g., SNe 2006gy and PTF 09uj).

Keywords. stars: mass loss, winds, outflows; supernovae: general

---

1. Introduction

The interaction between an outgoing supernova shock and $\sim$1 – 10 $M_\odot$ of material ejected in the $\sim$ year prior to core-collapse, and thus ending up $\sim$100 AU from the progenitor, can explain the most optically luminous supernovae (SNe) yet detected (e.g., Smith & McCray 2007), including, e.g., SN 2006gy (Smith *et al.* 2007) and perhaps the emerging class of hydrogen-poor superluminous SNe (SLSNe) discussed by both Quimby and Chornock in these proceedings. However, the giant eruptive mass-loss events preceding these SNe lack a plausible mechanism.

During this last $\sim$ year of evolution, the fusion and neutrino luminosities in the core of a massive star significantly exceed the Eddington luminosity (see, e.g. Woosley *et al.* 2002, for a review). We propose that in some SLSNe progenitors, gravity waves, excited by strong convection in the burning core, can tunnel out to the stellar envelope, and deposit enough energy to drive the inferred prodigious pre-supernova mass-loss (Quataert & Shiode 2012).

2. Convection and Wave Excitation During Late Stages of Massive Star Evolution

Convection excites a gravity-wave (g-mode) luminosity given by $L_{wave} \sim \mathcal{M}_{conv}\, L_{conv}$, where $\mathcal{M}_{conv}$ is the convective Mach number (v_{conv}/c_{sound}) and L_{conv} is the convective luminosity (Goldreich & Kumar 1990). This energy is predominantly in g-modes with spherical harmonic degree, ℓ, and frequency, ω, determined by the convective mixing length and timescales, so that the spherical degree $\ell \sim r/H$ and frequency $\omega \sim v_{conv}/H$, where r is the local radius and H the local pressure scale height (Kumar et al. 1999). During core O/Ne burning, $\sim$ 1 yr before core-collapse, for a typical massive star, waves will be excited in a buoyantly-stable layer overlying the convective core, with $L_{wave} \sim 0.01\, L_{conv} \sim 10^{-3} L_{nuc} \gg L_{Eddington}$.

3. Tunneling vs. Local Damping

Convectively excited gravity-waves can propagate as acoustic waves (p-modes) in the stellar envelope if they can tunnel through the evanescent region that separates the g- and p-mode propagation cavities. The width of this region and effective strength of the potential barrier are defined by the ℓ and ω of the mode. Waves can tunnel if they do not reach amplitudes comparable to their wavelength in the cavity where they are excited (i.e., break) and their time to damp by neutrino radiation is longer than the time to tunnel. These conditions are given by (respectively):

$$k_r \xi_r \sim \Lambda^{3/2} \left(\frac{N}{\omega}\right)^{3/2} \left(\frac{\mathrm{F}_{\mathrm{wave}}}{\rho r^3 \omega^3}\right)^{1/2} \lesssim 1, \tag{3.1}$$

$$t_\nu \sim \frac{t_{\mathrm{nuc}}}{100} > t_{\mathrm{tunnel}} \sim \frac{r_{in}}{v_g} \left(\frac{r_{out}}{r_{in}}\right)^{2\Lambda}, \tag{3.2}$$

where $\mathrm{F}_{\mathrm{wave}}$ is the wave energy flux, $\Lambda^2 = \ell(\ell+1)$, t_{nuc} the nuclear burning timescale ($\sim$ year), ρ the density, r the radius, k_r the radial wavenumber, ξ_r the wave's radial amplitude, N the Brunt-Väisälä frequency, v_g the g-mode group velocity, and r_{in} and r_{out} the inner and outer radii of the tunneling region. Waves that tunnel will deposit their energy in the envelope when the radiative diffusion time across a wavelength is comparable to the group travel time over a scale height, which can be written as

$$L_{\mathrm{rad}} \gtrsim L_{\mathrm{damp}} \equiv \frac{4\pi r^2 \rho c_{\mathrm{sound}}^3}{(k_r H)^2}, \tag{3.3}$$

where the symbols are as given above.

4. Conclusions

We find that high frequency modes, $\omega \gtrsim 3\, v_{\mathrm{conv}}/H$, of low spherical degree, $\ell \lesssim$ few, are most likely to tunnel out to the stellar envelope. Modes of lower frequency likely do not satisfy condition 3.1, and are thus more likely to induce mixing locally above the convective core. If most of the super-Eddington flux in gravity waves meets both criteria 3.1 and 3.2, they will tunnel out to the stellar envelope and deposit their energy at a radius given by 3.3, where it is capable of unbinding up to several solar masses of material during the year prior to core-collapse. If instead, the excitation is preferentially towards modes of lower frequency, these waves will likely break locally and induce mixing above the convective core (see simulations by Meakin & Arnett 2006, 2011 for possible examples of this). An outstanding question is understanding what stellar parameters (mass, rotation, metallicity, age) are the most susceptible to wave-driven mass loss or mixing. This depends on both the precise internal structure of massive stars and the power-spectrum of internal gravity waves excited by stellar convection.

References

Arnett, W. D. & Meakin, C. 2011, *ApJ*, 733, 78
Goldreich, P. & Kumar, P. 1990, *ApJ*, 363, 694
Kumar, P., Talon, S., & Zahn, J.-P. 1999, *ApJ*, 520, 859
Meakin, C. A. & Arnett, D. 2006, *ApJL*, 637, L53
Quataert, E. & Shiode, J. 2012, *MNRAS*, in press
Smith, N. & McCray, R. 2007, *ApJL*, 671, L17
Smith, N., *et al.* 2007, *ApJ*, 666, 1116
Woosley, S. E., Heger, A., & Weaver, T. A. 2002, *Reviews of Modern Physics*, 74, 1015

Death of Massive Stars: Supernovae and Gamma-Ray Bursts
Proceedings IAU Symposium No. 279, 2012
P. Roming, N. Kawai & E. Pian, eds.

doi:10.1017/S1743921312013506

Radio Insight into the Nature of Type IIb Progenitors

Christopher J. Stockdale[1], S. D. Ryder[2], A. Horesh[3], K. W. Weiler[4], N. Panagia[5], S. D. Van Dyk[6], F. E. Bauer[7], S. Immler[8], R. A. Sramek[9], D. Pooley[10], J. M. Marcaide[11], and N. Kassim[12]

[1]Marquette University
Dept. of Physics, P.O. Box 1811, Milwaukee, WI 53201, USA
email: chris.stockdale@mu.edu

[2]Australian Astronomical Observatory
P.O. Box 296, Epping, NSW 1710, Australia

[3]California Institute of Technology
Cahill Center for Astrophysics, Pasadena, CA, 91125, USA

[4]Computational Physics, Inc.
8001 Braddock Rd., Suite 210, Springfield, VA 22151, USA

[5]Space Telescope Science Institute
3700 San Martin Drive, Baltimore, MD 21218, USA

[6]Spitzer Science Center/Caltech
Mailcode 220-6, Pasadena, CA 91125, USA

[7]Pontificia Universidad Católica de Chile
Departamento de Astronomía y Astrofísica, Casilla 306, Santiago 22, Chile

[8]Center for Research and Exploration in Space Science and Technology
NASA Goddard Space Flight Center, Greenbelt, MD 20771, USA

[9]National Radio Astronomy Observatory
P.O. Box O, Socorro, NM 87801 USA

[10]Sam Houston State University
Department of Physics, Farrington Building, Suite 204, Box 2267, Huntsville, TX 77341 USA

[11]Universidad de Valencia
Departamento de Astronomía, 46100 Burjassot, Spain

[12]Naval Research Laboratory
Code 7210, Washington, D.C. 20375, USA

Abstract. We present the results of over two decades of radio observations of type IIb Supernovae with the Very Large Array and the Australia Telescope Compact Array. These radio studies illustrate the need for multi-wavelength follow-up to determine the progenitor scenario for type IIb events.

---

1. Overview

Type IIb Supernovae often have very similar optical spectra initially and frequently evolve into type Ib Supernovae. The optical classification IIb was invented to describe the spectral characteristics of type II SN 1987K, distinguished by its weak hydrogen absorption lines and a weak HeI emission line (Filippenko 1988). The recent SNe 2011dh (Horesh *et al.* 2011a; Horesh *et al.* 2011b; Soderberg *et al.* 2011; Krauss *et al.* 2012) and 2011hs (Ryder *et al.* 2011) bring the total type IIb SNe detected in the radio to eight, including SNe 1993J (Weiler *et al.* 2007), 2001gd (Stockdale *et al.* 2007), 2001ig (Ryder *et al.* 2004), 2003bg (Soderberg *et al.* 2006), 2008ax (Stockdale *et al.* 2008a; Roming *et al.* 2009), and 2008bo (Stockdale *et al.* 2008b).

SN 1993J is the best studied case of a radio-loud supernova, (Weiler *et al.* 2007; Marti-Vidal *et al.* 2011; and references therein) reaching maximum radio light at 6 cm near 150 days after explosion and declining with a power-law time decay index of -0.73 and a spectral index of -0.81. SN 2001gd had a similar radio evolution to SN 1993J with both SNe showing no clear signs of short-term variability (Stockdale *et al.* 2007). However both show unexpected declines in their radio light curves that are interpreted as abrupt changes in the mass-loss rates of their progenitors. For SN 1993J, this change implies that its progenitor star underwent an abrupt increase in its mass-loss rate from 3×10^{-6} up to 7×10^{-6} solar masses yr^{-1} between 10,000 and 7,000 yrs prior to explosion, with a nearly linear decline in mass-loss rate to $\sim 10^{-6}$ solar masses yr^{-1} just prior to explosion.

Radio emission from another excellent example of this type IIb class, SN 2001ig, shows a periodic oscillation in the radio light curve of 150 days, while SN 2008ax exhibits an approximate periodicity of 80 days in its radio emission. The periodicity observed in SNe 2001ig (Ryder *et al.* 2004), 2003bg (Soderberg *et al.* 2006) and 2008ax (Roming *et al.* 2009) can be explained by a binary companion that influenced the wind-established CSM of the SN progenitor star to create spiral density fluctuations that the SN blastwave later shocks (e.g. SN 1979C Weiler *et al.* 1992; Schwarz & Pringle 1996).

2. Implications

These studies demonstrate that the radio properties can effectively allow us to distinguish between two possible progenitor scenarios which have nearly indistinguishable optical spectra: 1) a single massive star which has shed most of its hydrogen as a function of normal evolution and 2) a less massive star whose hydrogen envelope is being influenced and depleted by a binary companion. The prevailing evidence appears to lean toward most type IIb SNe being binary systems with varying degrees of interaction (Van Dyk *et al.* 2011; Chevalier & Soderberg 2010).

References

Chevalier & Soderberg 2010, *ApJ*, 711, L40
Horesh, A. *et al.* 2011, *ATEL*, 3405
Horesh, A. *et al.* 2011, *ATEL*, 3411
Filippenko, A. 1988, *AJ* 96, 1941
Krauss, M., *et al.* 2012, *arXiv*, 1201.0770
Marti-Vidal *et al.* 2011, *A&A*, 526A, 143
Roming, P., *et al.* 2009, *ApJ*, 704, L118
Ryder, S. D., *et al.* 2004, *MNRAS*, 349, 1093
Ryder, S. D. *et al.* 2011, *ATEL*, 3789
Schwarz & Pringle 1996, *MNRAS*, 282, 1018
Soderberg, A. M., *et al.* 2006, *ApJ*, 651, 1005
Soderberg, A. M., *et al.* 2011, *arXiv*, 1107.1876
Stockdale, C. J., *et al.* 2007, *ApJ*, 671, 689
Stockdale, C. J., *et al.* 2008, *CBET*, 1299
Stockdale, C. J., *et al.* 2008, *IAUC*, 8939
Van Dyk, S. D., *et al.* 2011, *ApJ*, 741, L28
Weiler, K. W., *et al.* 2007, *ApJ*, 671, 1959
Weiler *et al.* 1992, *ApJ*, 399, 672

Death of Massive Stars: Supernovae and Gamma-Ray Bursts
Proceedings IAU Symposium No. 279, 2012
P. Roming, N. Kawai & E. Pian, eds.

doi:10.1017/S1743921312013518

Numerical code of the neutrino-transfer in three dimensions for core-collapse supernovae

Kohsuke Sumiyoshi[1] **and Shoichi Yamada**[2]

[1]Numazu College of Technology, Ooka 3600, Numazu, Shizuoka 410-8501, Japan
email: sumi@numazu-ct.ac.jp

[2]Science and Engineering & Advanced Research Institute for Science and Engineering, Waseda University, Okubo, 3-4-1, Shinjuku, Tokyo 169-8555, Japan
email: shoichi@heap.phys.waseda.ac.jp

Abstract. We develop a new numerical code of the *multi-energy and multi-angle* neutrino-radiation transfer in three dimensions (3D) for core-collapse supernovae. Our 3D code to solve the Boltzmann equations is based on the discretized-ordinate (S_N) method with a fully implicit differencing for time advance. A basic set of neutrino reactions is implemented in the collision terms together with a realistic equation of state. By following the time evolution of neutrino distributions in six dimensions (3 spatial and 3 momentum-space) by the 3D Boltzmann solver, we study the 3D feature of neutrino transfer for given background models of supernova cores in order to understand the explosion mechanism through neutrino heating in multi dimensions.

Keywords. methods: numerical, neutrinos, radiative transfer, stars: massive, stars: neutron, supernovae: general

1. Challenges to compute the neutrino transfer in supernovae

Computation of the neutrino-radiation transfer together with hydrodynamics is mandatory for the study of gravitational collapse of massive stars. The neutrino heating mechanism in non-spherical configurations induced by hydrodynamical instabilities is the essential key to clarify the elusive mechanism of supernova explosion. Simplifications of the neutrino transfer by dropping energy or angle dependence are not reliable in principle since neutrino energy- and angle-dependent interactions determine the transfer of energy in the supernova dynamics.

To pin down the crucial part of the explosion mechanism, full calculations of the neutrino-radiation hydrodynamics in 3D without invoking approximations of neutrino transfer are required. In two dimensions, the approximate treatments of neutrino transfer have been adopted in most of the state-of-the-art calculations (Kotake (2012)). Only recently the Boltzmann equations in axial symmetry were directly solved by Ott *et al.* (2008). The 3D numerical simulations with an approximate neutrino transfer have been reported more recently (Takiwaki, Kotake & Suwa (2012)) in the ray-by-ray approach.

2. Applications of the new 3D Boltzmann solver

Rapid growth of supercomputing power enables us to compute the neutrino transfer in 3D (Sumiyoshi & Yamada (2012)). It is a challenging task to describe the time evolution of neutrino distributions in six dimensions (three spatial coordinates with one energy and two angles). We have developed a numerical code to solve the Boltzmann equations for multi-energy and multi-angle group in 3D spatial coordinates. We solve the time

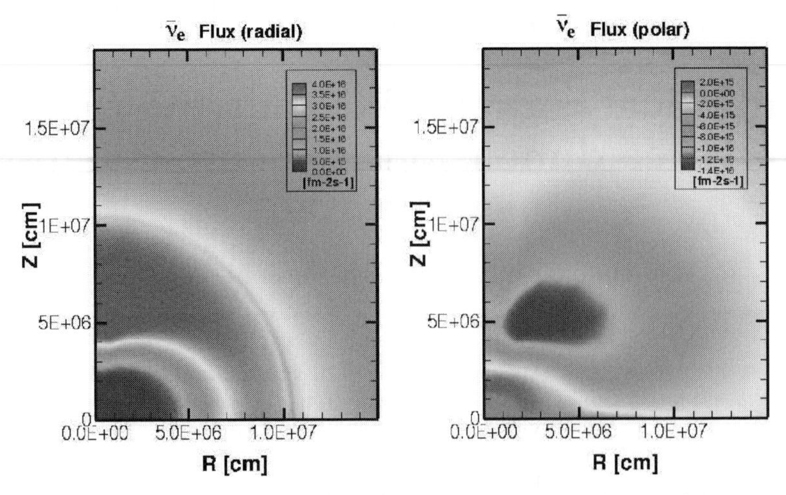

Figure 1. Radial (left) and polar (right) components of the fluxes of electron-type anti-neutrinos in the 2D deformed supernova core on a meridian slice.

evolution of neutrino distributions by the discrete-ordinate (S_n) method with a fully implicit differencing for time advance. The basic set of neutrino reactions (including pair processes) for the neutrinos of three species (ν_e, $\bar{\nu}_e$, $\nu_{\mu/\tau}$) is implemented with a set of EOS table (Sumiyoshi *et al.* (2005), Shen *et al.* (2011)).

The 3D Boltzmann solver is applied to examine the neutrino transfer for given background profiles in 2D/3D from Takiwaki, Kotake & Suwa (2012) and Sekiguchi & Shibata (2011). In Figure 1, we show an exemplar model of axially deformed supernova cores after the bounce. Due to the oblate shape of the proto-neutron star at center, the radial flux is enhanced near the polar direction. The contribution of the polar flux (θ-direction) is substantial between the pole and the equatorial plane. We note that this feature can be properly described by the 3D Boltzmann solver, being different from the ray-by-ray methods, in which only the radial flux can be described. Developments of the numerical code for the 3D neutrino-radiation hydrodynamics are in progress in order to explore the 3D supernova dynamics with the exact solution of neutrino transfer.

This work is based on the collaboration with H. Nagakura, S. Furusawa, H. Matsufuru, A. Imakura, T. Sakurai, Y. Sekiguchi, T. Takiwaki and K. Kotake. The numerical computations were performed on the supercomputers at YITP, KEK, RCNP, UT, Japan. This work is supported by the Grant-in-Aid for Scientific Research (Nos. 19104006, 20105004, 20105005, 21540281, 22540296) and the HPCI Strategic Program by MEXT, Japan.

References

Kotake, K. 2012, *Comptes Rendus Physique* in press; arXiv:1110.5107
Ott, C. D., Burrows, A., Dessart, L., & Livne, E. 2008, *ApJ*, 685, 1069
Sekiguchi Y. & Shibata M. 2011, *ApJ*, 737, 6
Shen, H., Toki, H., Oyamatsu, K., & Sumiyoshi, K. 2011, *ApJS*, 197, 20
Sumiyoshi, K., Yamada, S., Suzuki, H., Shen, H., Chiba, S., & Toki, H. 2005, *ApJ*, 629, 922
Sumiyoshi, K. & Yamada, S. 2012, *ApJS*, 199, 17
Takiwaki, T., Kotake, K., & Suwa, Y. 2012, *ApJ*, 749, 98

Death of Massive Stars: Supernovae and Gamma-Ray Bursts
Proceedings IAU Symposium No. 279, 2012
P. Roming, N. Kawai & E. Pian, eds.

doi:10.1017/S174392131201352X

On the importance of the equation of state for the neutrino-driven supernova explosion mechanism

Yudai Suwa

Yukawa Institute for Theoretical Physics, Kyoto University, Oiwake-cho, Kitashirakawa, Sakyo-ku, Kyoto, 606-8502, Japan; email: suwa@yukawa.kyoto-u.ac.jp

Abstract. We present two-dimensional numerical simulations of core-collapse supernova including multi-energy neutrino radiative transfer. We aim to examine the influence of the equation of state (EOS) for the dense nuclear matter. We employ four sets of EOSs, namely, those by Lattimer and Swesty (LS) and Shen *et al.*, which became standard EOSs in the core-collapse supernova community. We reconfirm that not every EOS produces an explosion in spherical symmetry, which is consistent with previous works. In two-dimensional simulations, we find that the structure of the accretion flow is significantly different between LS EOS and Shen EOS, inducing an even qualitatively different evolution of the shock wave, namely, the LS EOS leads to shock propagation beyond 2000 km from the center, while the Shen EOS shows only oscillations within 500 km. The possible origins of the difference are discussed.

Keywords. equation of state — hydrodynamics — neutrinos — stars: neutron — supernovae: general

---

1. Introduction

Core-collapse supernova explosions are triggered by the gravitational energy released during the transition from a stellar core into a protoneutron star (PNS), which has a temperature on the order of tens of MeV and densities on the order of normal nuclear matter density (3×10^{14} g cm^{-3}). There are only few equations of state (EOS) valid for these conditions. The most commonly used nuclear EOS in supernova simulations are the EOS from Lattimer & Swesty (1991) (hereafter LS), based on the incompressible liquid-drop model including surface effects, and from Shen *et al.* (1998) (SHEN). The latter is based on relativistic mean field (RMF) theory and Thomas-Fermi approximation.

There are several studies, which investigated EOS dependences on the hydrodynamical features in spherical symmetry, while a similar study in multi-dimensional simulation is not performed so far. Recently, several simulations that successfully produced the explosion are reported (Marek & Janka (2009), Suwa *et al.* (2010)). Therefore, we can now investigate how the EOS could affect the supernova dynamics.

In this paper we present results of numerical simulation of core-collapse supernovae of massive iron-core progenitors. Our model is based on neutrino radiation hydrodynamics and includes multi-energy radiative transfer. We employ four EOS, LS (with three values of incompressibility) and SHEN. We focus on the shock formation and evolution on a long timescale (more than 500 ms after core bounce) and investigate whether the shock finally obtains enough energy via neutrino heating to expand outward.

2. Numerical Method and Results

Methods. Our 2D simulations are performed using a code which is based on the spectral neutrino transport scheme IDSA, developed by Liebendörfer *et al.* (2009), and the

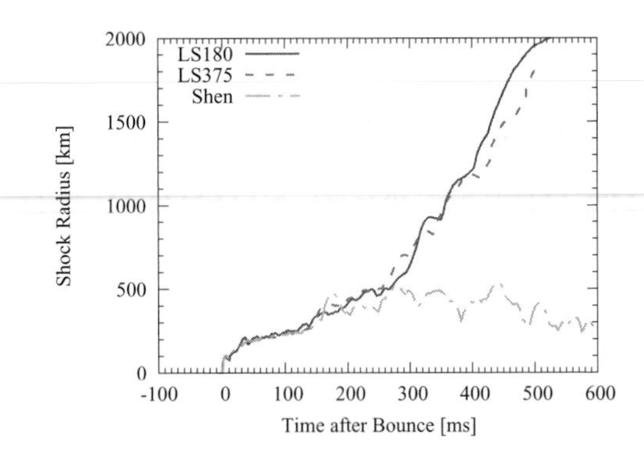

Figure 1. The time evolution of the maximum radius of shock wave.

ZEUS-2D code (Stone & Norman (1992), Suwa *et al.* (2010), Suwa *et al.* (2011)). In this study, we employ the two standard EOS for matter in NSE, LS and SHEN, in supernova simulations. As for the LS, there are three different versions available, for three different values of the incompressibility, $K = 180$ MeV (LS180), 220 MeV (LS200), and 375 MeV (LS375). Here we present results of LS180 and LS375 because LS220 is similar to LS180 in our simulation. We use the 15 $M_\odot$ progenitor from Woosley & Weaver (1995).

Results. We compare the results obtained in core-collapse supernova simulations of the 15 $M_\odot$ progenitor using the different EOS, i.e., LS180, LS375, and SHEN. Fig. 1 shows the shock-radius evolution for each model. Each of them has three lines, corresponding to maximum, angular average and minimum shock radii (from top to bottom). From Fig. 1, we can see that LS180 and LS375 (red solid lines and green dotted lines) show similar trajectories. There is a gradual expanding phase for the postbounce time $t_{\rm pb} \lesssim 200$–300 ms where the maximum shock radii reach about 500 km, after which the shock radii start to grow more rapidly for both models. Later, no further contraction is observed. The maximum shock radii reach about 2000 km for LS180 and 1800 km for LS375 at 500 ms post bounce. On the other hand, for the 2D simulation using SHEN the shock wave does not continue to grow to increasingly larger radii, even at late times. Although conclusions about possible explosions using LS180 and LS375 are still weak at the post bounce times when the 2D simulations were stopped (the explosion energy remains $\sim 10^{50}$ erg in both cases), explosions for the 2D simulations using SHEN are highly unlikely to occur.

Numerical computations were in part carried on Cray XT4 and medium-scale clusters at CfCA of the National Astronomical Observatory of Japan, and on SR16000 at YITP in Kyoto University. This study was supported in part by the Grants-in-Aid for the Scientific Research from the Ministry of Education, Science and Culture of Japan (Nos. 19540309, 20740150 and 23840023) and MEXT HPCI STRATEGIC PROGRAM.

References

Lattimer, J. M. & Swesty, F. D. 1991, *Nuclear Physics A*, 535, 331
Liebendörfer, M., Whitehouse, S. C., & Fischer, T. 2009, *Astrophys. J.*, 698, 1174
Marek, A. & Janka, H. 2009, *Astrophys. J.*, 694, 664
Shen, H., Toki, H., Oyamatsu, K., & Sumiyoshi, K. 1998, *Nucl. Phys.*, A637, 435
Stone, J. M. & Norman, M. L. 1992, *Astrophys. J.*, 80, 753
Suwa, Y. *et al.* 2010, *PASJ*, 62, L49
Suwa, Y., Kotake, K., Takiwaki, T., Liebendörfer, M., & Sato, K. 2011, *Astrophys. J.*, 738, 165
Woosley, S. E. & Weaver, T. A. 1995, *Astrophys. J.*, 101, 181

Death of Massive Stars: Supernovae and Gamma-Ray Bursts
Proceedings IAU Symposium No. 279, 2012
P. Roming, N. Kawai & E. Pian, eds.

doi:10.1017/S1743921312013531

Optical to X-rays SNe light curves following shock breakout through a thick wind

Gilad Svirski[1], Ehud Nakar[1] and Re'em Sari[2]

[1]Raymond and Beverly Sackler School of Physics & Astronomy, Tel Aviv University, Tel Aviv 69978, Israel

[2]Racah Institute for Physics, The Hebrew University, Jerusalem 91904, Israel

Abstract. We present luminosity and temperature light curves following supernova shock breakouts through a thick wind. These events are very luminous and their spectrum may contain both an X-ray and a UV/optical component. For breakout pulse durations between a week and a month, the X-ray component luminosity peaks $100 - 500$ days after the explosion, respectively.

---

1. A shock breakout through a thick wind

Supernovae (SNe) shock breakouts through a thick wind last much longer than breakouts from a stellar surface, and are thus easier to detect. If a star is surrounded by a wind with an optical depth $\tau_w > c/v$, where v is the shock speed, then the shock continues into the wind without releasing photons to the observer. Once the wind optical depth drops to c/v the shock is breaking out of the wind, releasing all its energy as an intense pulse. Following breakout, the radiation mediated shock is replaced by a collisionless shock. We examine the light curve and the observed spectrum at and following the breakout (see also Chevalier & Irwin 2012).

2. Main conclusions

Wind breakouts are very luminous, 10^{43-44} erg/s for breakout times of days or longer. Two spectral components are observed: a UV soft component, dominating the luminosity at breakout with temperatures of $10^4 - 10^6$ K, depending on the shock breakout velocity. A hard component (X-rays/γ-rays), contributing a fraction of 10^{-4} of the soft component luminosity at breakout, rising quickly to domination at $\sim 10 - 50$ breakout times, with a breakout temperature of ~ 0.1 keV, rising to $1 - 100$ keV at peak. The light curve and spectrum evolution details depend mostly on the breakout time. In terms of prospects for X-rays and soft γ-rays detection, it is best to observe $100 - 500$ days after explosions with breakout timescales between a week and a month.

3. Processes, timescales and light curves

All the shock internal energy is radiated away (i.e, fast cooling). The bolometric luminosity depends only on the shock velocity and external density, independent of the spectrum. Electrons are heated behind the collisionless shock to 60 keV (Katz *et al.* 2011). These electrons are the energy source of the emission. The soft component is free-free emission of the unshocked wind, heated by the radiation of the post-shock hot electrons. The hard component photons are free-free emission of the 60 keV post-shock electrons, softened during diffusion through the unshocked wind. The hard component luminosity is suppressed by inverse Compton (IC) cooling of the hot electrons over the

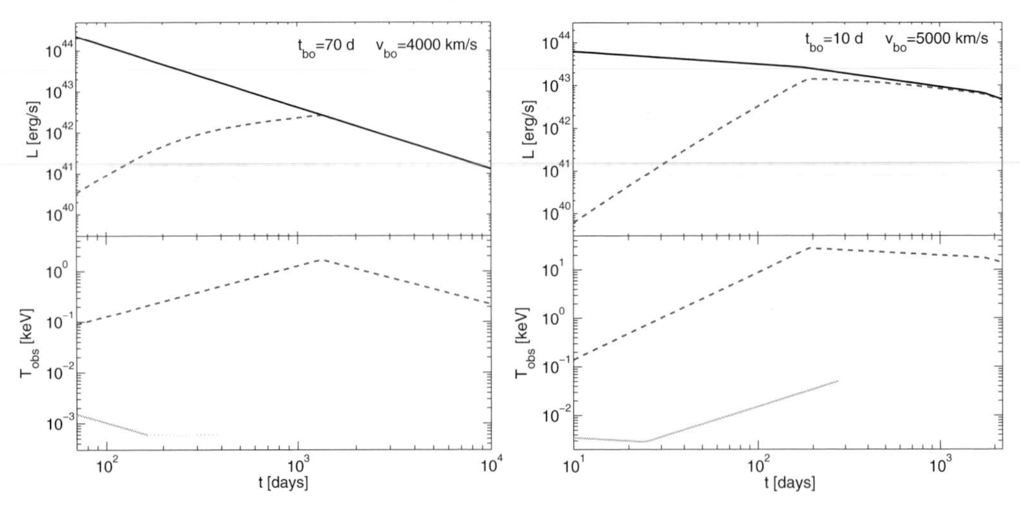

Figure 1. *Left*: Late breakout ($t_{SP} < t_{bo}$). Breakout after the entire ejecta is shocked. Brightest emission, released as optical-UV flash. X-rays observed only long after breakout, if the wind is not terminated by then. Figure parameters chosen to fit SN 2006gy. Solid lines are bolometric luminosity and temperature of the soft component. Dashed lines are the hard component luminosity and temperature. *Right*: Early breakout ($t_{bo} < t_{\rm hard} < t_{SP}$). Soft component temperature roughly constant while in equilibrium, climbs quickly afterwards. X-rays/γ-rays peak luminosity, at 100-300 d after the SN explosion, is similar to breakout bolometric luminosity. A larger t_{bo} yields a brighter X-ray event. May explain SNe like PTF 09uj (Ofek *et al.* 2010).

soft component and by Compton losses during diffusion. Following breakout, as the wind opacity decreases, both the IC cooling and the Compton loses decrease. As a result the hard component gains dominance over the soft component with time.

Depending on the shock velocity, the breakout radiation may deviate from thermal equilibrium. In such case the observed breakout temperature is high and it depends strongly on the shock velocity. For further discussion see Weaver (1976), Katz *et al.* (2010) and Nakar & Sari (2010).

Three timescales are involved: (1) the breakout time, (2) the time when the hard component luminosity becomes dominant, and (3) the transition of the forward shock to a snowplow expansion (when mass of the swept wind becomes comparable to the ejecta mass). The temporal order of these timescales is determined by the breakout time and dictates three evolution scenarios which we denote as early, intermediate and late breakouts. The early and late breakout scenarios are depicted and described in Figure 1. This poster is a summary of Svirski *et al.* (2012)

References

Chevalier, R. A. & Irwin, C. M. 2012, *ApJL*, 747, L17
Katz, B., Budnik, R., & Waxman, E. 2010, *ApJ*, 716, 781
Katz, B., Sapir, N., & Waxman, E. 2011, ArXiv e-prints
Nakar, E. & Sari, R. 2010, *ApJ*, 725, 904
Ofek, E. O., *et al.* 2010, *ApJ*, 724, 1396
Svirski, G., Nakar, E., & Sari, R. 2012, ArXiv e-prints
Weaver, T. A. 1976, *ApJS*, 32, 233

Death of Massive Stars: Supernovae and Gamma-Ray Bursts
Proceedings IAU Symposium No. 279, 2012
P. Roming, N. Kawai & E. Pian, eds.

doi:10.1017/S1743921312013543

Type II-P supernovae in the mid-infrared

Tamás Szalai and József Vinkó

Department of Optics and Quantum Electronics, University of Szeged,
Dóm tér 9., Szeged H-6720, Hungary; email: szaszi@titan.physx.u-szeged.hu

Abstract. We present detailed analysis of mid-infrared (MIR) data for 9 type II-P supernovae from the public *Spitzer* database. Spectral energy distributions (SEDs) from observed fluxes are fitted with simple models to get basic information about the dust as the presumed source of MIR radiation. We found two SNe, 2005ad and 2005af, which likely have newly-formed dust in their environment, while in the other seven cases the observed MIR flux may originate from pre-existing circumstellar or interstellar dust.

Keywords. supernova, circumstellar dust, infrared excess

1. Unpublished supernova data in the Spitzer database

Core-collapse supernovae (CC SNe), especially those of type II-plateau (II-P), are thought to be important contributors to cosmic dust production. The most obvious indicator of the presence of newly-formed and/or pre-existing dust is the time-dependent MIR excess coming from the environment of SNe. While in the past years several CC SNe were monitored by the *Spitzer Space Telescope*, there have been only a few of these objects analyzed and published up to now.

We studied the public *Spitzer* data of twelve type II-P (or peculiar II-P) SNe. In nine cases (SNe 2003J, 2003ie, 2004A, 2005ad, 2005af, 2006bp, 2006my, 2006ov, 2007oc), we could identify a mid-IR point source at the SN position, and carried out a complete analysis.

2. Analysis of mid-IR data, model fitting of SEDs

We computed simple aperture photometry on the post-BCD frames taken with Infrared Array Camera (IRAC), MIPS (Multiband Imaging Spectrometer) 24.0 μm channel and IRS PUI (Infrared Spectrograph, peak-up imaging mode) with IRAF and MOPEX, taking into account all device-specific corrections. The available IRS spectra were processed using SPICE. MIR SEDs were calculated from observed fluxes. The continuum fluxes of reduced IRS spectra match well with the SED flux levels. Early-time MIR fluxes of SN 2005af, which are the only data already published elsewhere (Kotak *et al.* 2006), are also in good agreement with our values.

Assuming that the radiation is purely thermal, the main source of MIR flux is most likely warm dust. We fitted blackbodies (BBs) and analytic dust models to the observed, dereddened SED points to get information about physical properties and total amount of dust. Analytic models are based on Eq.1 in Meikle *et al.* (2007) assuming a homogeneous (constant-density) and uniform dust distribution within a sphere, using a power-law grain-size distribution. For most cases we used amorphous carbon (AC) grains (Colangeli *et al.* 1995), except for the cases of SNe 2005af and 2006my, where adequate solutions were possible only by applying a C-Si-PAH mixture (Weingartner & Draine 2001).

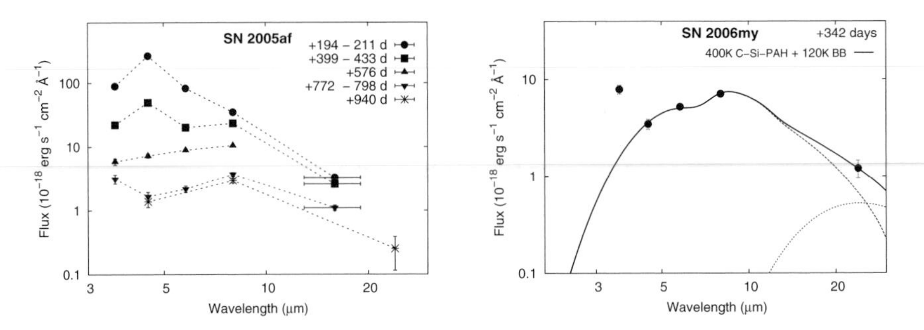

Figure 1. *Left*: Mid-IR SEDs of SN 2005af at different epochs. *Right*: An example of SED models: C-Si-PAH warm + cold BB components, SN 2006my (+342 days).

3. Conclusions

With the nine studied objects, we almost doubled the number of type II-P SNe having detailed MIR data based on *Spitzer* measurements. In two cases, SNe 2005ad and 2005af, we found cooling temperatures and decreasing luminosities of the warm component which are similar to the values found in other SNe that are thought to have newly-formed dust in their environment (e.g. SN 2004et – Kotak *et al.* 2009; SN 2004dj – Szalai *et al.* 2011, Meikle *et al.* 2011).

The calculated temperatures for the other SNe do not show strong temporal variation, while the derived luminosities as well as radii, are too high to be compatible with local dust. Also, the calculated dust masses in these cases are orders of magnitudes higher than the observed amount of dust around well-studied SNe listed above. The large radius of the warm component may suggest pre-existing dust in these cases, thereby making it unclear if there was new dust formed around these SNe.

Nevertheless, theoretical models predict orders of magnitude more newly-formed dust in CC SNe. Our conclusions support the previous observational results that warm new dust in the environment of SNe contributes only slightly to cosmic dust content. A more important contributor, as latest results suggest, may be the colder (<50 K) dust which could be found in older SN remnants (see e.g. Matsuura *et al.* 2011).

Acknowledgments

This work is supported by the Hungarian OTKA Grant K76816, by the European Union and co-funded by the European Social Fund through the TÁMOP 4.2.2/B-10/1-2010-0012 grant.

References

Colangeli, L., Mennella, V., Palumbo, P., *et al.* 1995, *A&AS*, 113, 561
Kotak, R., Meikle, W. P. S., Pozzo, M., *et al.* 2006, *ApJ* (Letters), 651, L117
Kotak, R., Meikle, W. P. S., Farrah, D., *et al.* 2009, *ApJ*, 704, 306
Matsuura, M., Dwek, E., Meixner, M., *et al.* 2011, *Science*, 333, 1258
Meikle, W. P. S., Mattila, S., Pastorello, A., *et al.* 2007, *ApJ*, 665, 608
Meikle, W. P. S., Kotak, R., Farrah, D., *et al.* 2011, *ApJ*, 732, 109
Szalai, T., Vinkó, J., Balog, Z., *et al.* 2011, *A&A*, 527, A61
Weingartner, J. C. & Draine, B. T. 2001, *ApJ*, 548, 296

Death of Massive Stars: Supernovae and Gamma-Ray Bursts
Proceedings IAU Symposium No. 279, 2012
P. Roming, N. Kawai & E. Pian, eds.

doi:10.1017/S1743921312013555

The Type II supernovae 2006V and 2006au: two SN 1987A-like events

Francesco Taddia

Department of Astronomy, The Oskar Klein Center, Stockholm University, AlbaNova, 10691 Stockholm, Sweden
email: ftadd@astro.su.se

Abstract. We studied optical and near-infrared (NIR) light curves, and optical spectra of Supernovae (SNe) 2006V and 2006au, two objects monitored by the Carnegie Supernova Project (CSP) and displaying remarkable similarity to SN 1987A, although they were brighter, bluer and with higher expansion velocities. SN 2006au also shows an initial dip in the light curve, which we have interpreted as the cooling tail of the shock break-out. By fitting semi-analytic models to the UVOIR light curve of each object, we derive the physical properties of the progenitors and we conclude that SNe 2006V and 2006au were most likely Blue Supergiant (BSG) stars that exploded with larger energies as compared to that of SN 1987A. We are currently investigating the host galaxies of a few BSG SNe, in order to understand the role played by the metallicity in the production of these rare exploding BSG stars.

Keywords. supernovae: general, supernovae: individual (SN 2006V, SN 2006au, SN 1987A)

Supernova (SN) 1987A, a milestone in the history of supernova observations, was the explosion of a compact blue supergiant (BSG) star. Due to the small progenitor radius, the explosion energy was mainly spent to adiabatically expand the SN and therefore its light curve exhibited a long rise to maximum and a low luminosity at early times. Objects similar to SN 1987A (87A-like or BSG SNe) are particularly rare. Pastorello *et al.* (2005) presented SN 1998A, the first BSG SN after SN 1987A. SN 2000cb was recently presented by Kleiser *et al.* (2011) and modelled by Utrobin & Chugai (2011). In Pastorello *et al.* (2012) we have recently drawn up a list of 12 87A-like events. Among them, only 6 have an exhaustive dataset, including the aforementioned SNe and our objects from Taddia *et al.* (2012) that we present here. SN 2006V and SN 2006au were respectively located in the spiral galaxies UGC 6510 and UGC 11057, at 72.7±5.0 and 46.2±3.2 Mpc. They were observed by the CSP, which obtained optical+NIR photometry and optical spectroscopy, thanks to the Swope and the Du Pont telescopes at Las Campanas Observatory (Chile). The objects were observed for ~130 and ~90 days, so that the light curve tail powered by the ^{56}Co decay was detected for SN 2006V but not for SN 2006au. As we can see in Fig. 1, both SNe follow the light-curve shape of SN 1987A. Additionally, SN 2006au also presents an initial slowly declining phase, which we have interpreted as the cooling tail after the shock break-out. This feature has been observed for a small sample of SNe, including SNe 1987A, 1993J, 1999ex, and 2008D. From the comparison of their absolute magnitudes, both SNe 2006V and 2006au turn out to be brighter than SN 1987A. We notice that for SN 2006V we assumed negligible host galaxy extinction whereas for SN 2006au we assumed $E(B-V)_{host} = 0.312$ mag (from the equivalent width of the Na I D absorption line). The colors computed from the photometry show that both SNe are bluer than SN 1987A, suggesting higher photospheric temperature. This is obviously confirmed by the spectral comparison, since our objects suffer low suppression in the blue part of the spectrum if compared to SN 1987A. The P-Cygni profiles which characterize

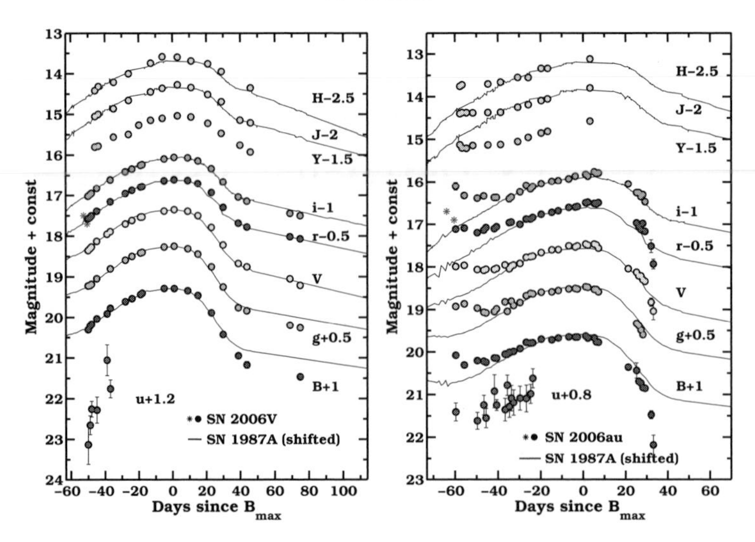

Figure 1. Light curves of SNe 2006V and 2006au compared to those of SN 1987A.

the spectral lines of our SNe reveal high expansion velocity, in particular for SN 2006au. Overall the spectra exhibit typical Type II features. The complete photometric datasets enabled us to build bolometric light curves (BLCs) for both SNe, by converting the magnitudes into fluxes at the proper effective wavelength and then integrating the resulting spectral energy distributions. We also estimated the explosion day of our objects by applying the expanding photospheric method, e.g. Dessart & Hillier (2005). This allowed us to properly model the BLCs, using the semi-analytic model of Imshennik & Popov (1992). We also modelled the early BLC of SN 2006au with the analytic model of Chevalier (1992). The best model fit gave us estimates for the ejecta mass (M_{ej}), the explosion energy (E), the progenitor radius (R) and the ^{56}Ni mass of both events. We obtained $M_{ej} \approx 20\ M_{\odot}$, $E \approx 2-3 \times 10^{51}$ erg and $R \approx 75-100\ R_{\odot}$. The ^{56}Ni mass was estimated to be 0.127 $M_{\odot}$ for SN 2006V and $\leqslant$0.073 $M_{\odot}$ for SN 2006au. The physical parameters we derived for the progenitors, in particular the small radii, are consistent with a scenario where both SNe 2006V and 2006au were BSGs which exploded with larger energies as compared to that of SN 1987A. In order to understand why these rare 87A-like SNe have BSG progenitors, we are measuring the metallicity at the explosion site of each event in the sample by Pastorello *et al.* (2012). We will be able to test if the sub-solar metallicity is a necessary ingredient to produce exploding BSG stars, as it has been thought for SN 1987A (see Podsiadlowski 1992). Preliminary results suggest some of these SNe were produced at solar metallicity, opening interesting questions on their origin.

References

Chevalier, R. A. 1992, *ApJ*, 394, 599
Dessart, L. & Hillier, D. J. 2005, *A&A*, 439, 671
Imshennik, V. S. & Popov, D. V. 1992, *Astron. Zh.*, 69, 497
Kleiser, I. K. W., Poznanski, D., Kasen, D., *et al.* 2011, *MNRAS*, 415, 372
Pastorello, A., Baron, E., Branch, D., *et al.* 2005, *MNRAS*, 360, 950
Pastorello, A., Pumo , M. L., Havasardyan, H., *et al.* 2012, *A&A*, 537, A141
Podsiadlowski, P. 1992, *PASP*, 104, 717
Taddia, F., Stritzinger, M. D., Sollerman, J., *et al.* 2012, *A&A*, 537, A140
Utrobin, V. P. & Chugai, N. N. 2011, *A&A*, 532, A100

Death of Massive Stars: Supernovae and Gamma-Ray Bursts
Proceedings IAU Symposium No. 279, 2012
P. Roming, N. Kawai & E. Pian, eds.

doi:10.1017/S1743921312013567

Magnetic Energy Release in Relativistic Plasma

Hiroyuki R. Takahashi[1] **and Ken Ohsuga**[2]

[1]Center for Computational Astrophysics, National Astronomical Observatory of Japan,
Osawa, Mitaka, Tokyo 181-8588, Japan
email: takahshi@cfca.jp

[2]National Astronomical Observatory of Japan,
Osawa, Mitaka, Tokyo 181-8588, Japan

Abstract. The efficiency of the energy conversion rate in the relativistic magnetic reconnection is investigated by means of Relativistic Resistive Magnetohydrodynamic (R2MHD) simulations. We confirmed that the simple Sweet-Parker type magnetic reconnection is a slow process for the energy conversion as theoretically predicted by Lyubarsky (2005). After the Sweet-Parker regime, we found a growth of the secondary tearing instability in the elongated current sheet. Then the energy conversion rate and the outflow velocity of reconnection jet increase rapidly. Such a rapid energy conversion would explain the time variations observed in many astrophysical flaring events.

To construct a more realistic model of relativistic reconnection, we extend our R2MHD code to R3MHD code by including the radiation effects (Relativistic Resistive Radiation Magnetohydrodynamics R3MHD). The radiation field is described by the 0th and 1st moments of the radiation intensity (Farris *et al.* 2008, Shibata *et al.* 2011). The code has already passed some one-dimensional and multi-dimensional numerical problems. We demonstrate the first results of magnetic reconnection in the radiation dominated current sheet.

Keywords. magnetohydrodynamics, radiation, relativity

1. Introduction

Magnetic reconnection is one of the most important subjects in the studies of space, laboratory, and astrophysical plasmas. Recently, it has been recognized that the magnetic reconnection plays an essential role for energy conversion in astronomical compact objects, such as neutron stars (Kennel & Coroniti 1984), soft gamma-ray repeaters (Masada *et al.* 2010), and gamma-ray bursts (Drenkhahn 2002, McKinney & Uzdensky 2010, Zhang & Yan 2011). However, there are a few theoretical and numerical studies on the relativistic magnetic reconnection (Lyutikov & Uzdensky 2003, Watanabe & Yokoyama 2006, Zenitani *et al.* 2010, Zanotti & Dumbser 2011).

We performed 2-dimensional Relativistic Resistive Magnetohydrodynamic (R2MHD) simulations to study the energy conversion in the magnetic reconnection. Then, we developed the Relativistic Resistive Radiation Magnetohydrodynamic (R3MHD) code to study the radiation effects. In this paper, we show the first results of the R3MHD simulation.

2. Results

We first perform R2MHD simulations of the magnetic reconnection with a uniform resistivity model. Then, Sweet-Parker type magnetic reconnection is realized in the relativistic plasma. In this case, most of the magnetic energy is converted into the thermal

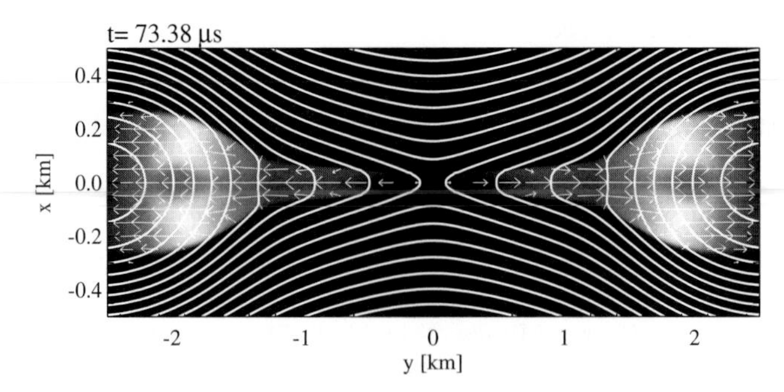

Figure 1. First numerical results of R3MHD simulation on the Petschek type magnetic reconnection. Color and arrows show the radiation energy density and flux in the observer frame, while the curves show the magnetic field lines.

energy by the Ohmic dissipation, resulting in the formation of a hot ($k_B T > mc^2$) outflow. Since the excess of the thermal energy leads to increasing the plasma inertia, the outflow speed is only mildly relativistic (Lorentz factor ~ 1). The reconnection rate strongly depends on the magnetic Reynolds number R_M as $\simeq R_M^{-0.5}$, as well as that in the non-relativistic magnetic reconnection (Komissarov 2005, Takahashi *et al.* 2011).

When the localized resistivity model is adopted, the Petschek type magnetic reconnection is realized in the relativistic plasma (Watanabe & Yokoyama 2006, Zenitani *et al.* 2010, Zanotti & Dumbser 2011). Figure 1 shows the first results of R3MHD simulation on the relativistic magnetic reconnection. We found that the radiation energy density is enhanced inside the reconnection outflow in our model.

Acknowledgement

Numerical computations were carried out on Cray XT4 at the Center for Computational Astrophysics, CfCA, at the National Astronomical Observatory of Japan, on Fujitsu FX-1 at the JAXA Supercomputer System (JSS) at the Japan Aerospace Exploration Agency (JAXA), and on T2K at the University of Tokyo. A part of this research has been funded by MEXT HPCI STRATEGIC PROGRAM and supported by Ministry of Education, Culture, Sports, Science, and Technology (MEXT) for Research Activity Start-up (HRT) 23840045, and for Young Scientist (KO) 20740115, 23340040.

References

Drenkhahn, G. 2002, *A&A*, 387, 714
Farris, B. D., Li, T. K., Liu, Y. T., & Shapiro, S. L. 2008, *Phys. Rev.* D, 78, 024023
Kennel, C. F. & Coroniti, F. V. 1984, *ApJ*, 283, 710
Lyubarsky, Y. E. 2005, *MNRAS*, 358, 113
Lyutikov, M. & Uzdensky, D. 2003, *ApJ*, 589, 893
Masada, Y., Nagataki, S., Shibata, K., & Terasawa, T. 2010, *PAJS*, 62, 1093
McKinney, J. C. & Uzdensky, D. A. 2010, ArXiv e-prints
Shibata, M., Kiuchi, K., Sekiguchi, Y., & Suwa, Y. 2011, *Progress of Theoretical Physics*, 125, 1255
Takahashi, H. R., Kudoh, T., Masada, Y., & Matsumoto, J. 2011, *ApJ*, 739, L53
Watanabe, N. & Yokoyama, T. 2006, *ApJ*, 647, L123
Zanotti, O. & Dumbser, M. 2011, *MNRAS*, 418, 1004
Zenitani, S., Hesse, M., & Klimas, A. 2010, *ApJ*, 716, L214
Zhang, B. & Yan, H. 2011, *ApJ* 726, 90

Death of Massive Stars: Supernovae and Gamma-Ray Bursts
Proceedings IAU Symposium No. 279, 2012
P. Roming, N. Kawai & E. Pian, eds.

doi:10.1017/S1743921312013579

Evolution of stars just below the critical mass for iron core formation

Koh Takahashi, Hideyuki Umeda and Takashi Yoshida

Department of Astronomy, Graduate school of Science, University of Tokyo,
7-3-1 Hongo, Bunkyo-ku, Tokyo 113-0033, Japan
email: ktakahashi@astron.s.u-tokyo.ac.jp

Abstract. We calculate the evolution of stars with their initial mass of 9-11$M_\odot$ under very fine initial mass grid of 0.01$M_\odot$. We determine the lower critical mass for Ne ignition in an ONe core that has not undergone the thermal pulse episode. The values are 9.83$M_\odot$ for the initial mass and 1.365$M_\odot$ for the CO core mass. A star with an initial mass slightly larger than the critical, undergoes an off-center Ne+O ignition. Since the energy production rate of Ne+O burning and lasting electron capture reactions is sufficiently large to ignite Si, an Fe core forms as a result of shell Si burning. For a star just below the critical mass, an ONe core continues to contract. In such a high density core, electron capture by nuclei produced through C burning affects the core evolution. After ^{20}Ne starts to capture electrons, the core may ignite O and undergo O detonation. The fate may be an Electron Capture Supernova.

Keywords. stars: evolution, supernovae: general, nucleosynthesis

---

1. Introduction

A star with initial mass of 9-11 $M_\odot$ forms a CO core around the Chandrasekhar limit $M_{Ch} \simeq 1.38M_\odot$. While it is well known that a star with a CO core sufficiently larger than M_{Ch} forms an Fe core after non-degenerate Si ignition and collapses (Woosley *et al.* 1986), in the case for a star with a CO core slightly less than the limit, the evolution after an ONe core formation is still poorly known. This is due to complicated aspects of their evolution such as thermal pulse (Poelarends *et al.* 2008) or convective URCA process (Ritossa *et al.* 2008).

However, the importance for their understanding is apparent. Until now, the only model for a supernova explosion that can be achieved in first-principle calculation is Electron Capture Supernova (ECSN) which have a progenitor of the Chandrasekhar limit mass ONe core (Nomoto 1987, Kitaura *et al.* 2006). We investigate detailed evolution of 9-11$M_\odot$ stars which form degenerate ONe cores.

2. The critical mass for Ne ignition

Figure 1 shows evolution tracks in terms of the central temperature and density after C burning. We see clear differences between less massive 9.82, 9.83$M_\odot$ stars and more massive 9.9, 10$M_\odot$ stars in the tracks. For less massive stars of 9.82, 9.83$M_\odot$, as the density at the center reaches the value of log $\rho \sim 9$, the temperature starts to decrease. This is owing to URCA cooling by ^{25}Mg and ^{23}Na. On the other hand, in the tracks for more massive stars, spikes appear at log $\rho \sim 8$ in which the core expands and contracts adiabatically due to Ne+O or Ne+O+Si shell burning. We determine 9.83$M_\odot$ as the lower critical initial mass for Ne ignition. The corresponding CO core mass is 1.365$M_\odot$. Hereafter we describe some details of the evolution.

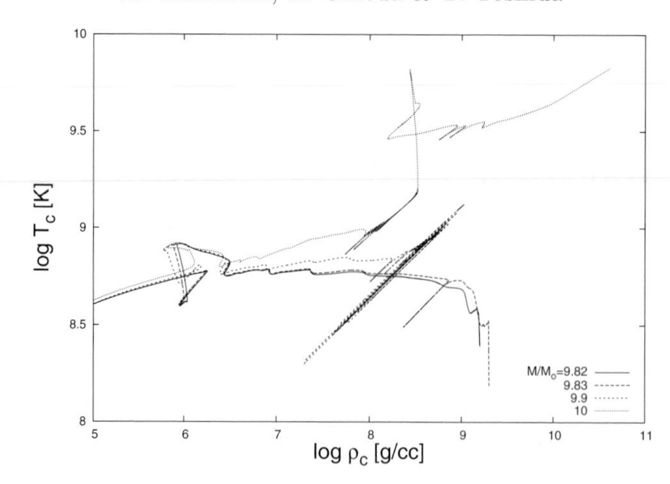

Figure 1. Evolution tracks in terms of central temperature and density.

9.82M$_\odot$ model. The star with the initial mass of 9.82M$_\odot$ forms a CO core of 1.360M$_\odot$ and later an ONe core forms after semi-degenerate shell C burning. As the core contracts, the inverse temperature distribution comes out and the highest temperature region eventually moves outward to the edge of the core. This is the general feature which appears in a high density core supported by degenerate electrons. Finally, the core keeps contracting without Ne ignition. In the contracting ONe core, reactions of electron capture by C burning products occur in succession. After ^{20}Ne start to capture electrons, temperature at the center will increase rapidly due to a decrease in the number of electrons which support the core. As the temperature reaches the critical value to ignite Ne and O, O detonation may take place at the center and the fate may be an ECSN.

10M$_\odot$ model. A star with a slightly larger core undergoes shell Ne ignition and the fate will be CCSN. In the less-degenerate CO core, C ignites at the center of the 10M$_\odot$ star, while for the 9.9M$_\odot$ star, the formation of the ONe core occurs in the same way as the case of 9.82M$_\odot$ star. Since it has a larger CO core, 1.49M$_\odot$, and electrons at the center are less degenerate, off-center Ne ignition takes place. The Ne burning becomes a flash due to the high degeneracy of the core and is followed by shell Si burning, as well as by shell O burning. A strong shell burning occurs and Ne, O and Si simultaneously burn in the shell. The Fe core forms after the burning front reaches the center, and the core contracts and collapses.

Shell Ne+O+Si burning will be the general aspects for stars which have slightly larger core mass than the critical of 1.365M$_\odot$ for the CO core. The star of 9.9M$_\odot$ for its initial and 1.404M$_\odot$ for the CO core also shows Ne+O+Si burning. In the case of the star of 9.83M$_\odot$, a contracting phase exists between shell Ne+O burning and shell Si burning. So far, the critical mass for Ne ignition is 9.83M$_\odot$ for the initial mass, 1.365M$_\odot$ for the CO core.

References

Kitaura, F. S., Janka, H.-Th., & Hillebrandt, W. 2006, *A&A*, 450, 345
Nomoto, K. 1987, *ApJ*, 322, 206
Poelarends, A. J., Herwig, F., Langer, N., & Heger, A. 2008, *ApJ*, 675, 614
Rittosa, C., Garcia-Berro, E., & Iben, I. J. 1999, *ApJ*, 525, 381
Woosley, S. E. & Weaver, T. A. 1986, *ARAA*, 24, 205

Death of Massive Stars: Supernovae and Gamma-Ray Bursts
Proceedings IAU Symposium No. 279, 2012
P. Roming, N. Kawai & E. Pian, eds.

doi:10.1017/S1743921312013580

3D hydrodynamic core-collapse SN simulations for an 11.2 $M_{\odot}$ star with spectral neutrino transport

Tomoya Takiwaki[1], **Kei Kotake**[2] **and Yudai Suwa**[3]

[1]Center for Computational Astrophysics, National Astronomical Observatory of Japan, 2-21-1, Osawa, Mitaka, Tokyo, 181-8588, Japan
email: takiwaki.tomoya@nao.ac.jp

[2]Division of Theoretical Astronomy, National Astronomical Observatory of Japan, 2-21-1,Osawa, Mitaka, Tokyo, 181-8588, Japan

[3]Yukawa Institute for Theoretical Physics, Kyoto University, Oiwake-cho, Kitashirakawa, Skyo-ku, Kyoto, 606-8502, Japan

Abstract. We have performed three-dimensional (3D) hydrodynamical simulations of core-collapse supernovae (SNe) with multigroup neutrino transport to study non-axisymmetric effects in the context of neutrino heating explosion mechanism. By comparing one- (1D) and two dimensional (2D) results with those of 3D, we study how the increasing spatial multi-dimensionality affects the postbounce SN dynamics. The calculations were performed with an energy-dependent treatment of the neutrino transport that is solved by the isotropic diffusion source approximation scheme. In agreement with previous studies, our 1D model does not produce explosions for the $11.2M_{\odot}$ star, while the neutrino-driven revival of the stalled bounce shock is obtained both in the 2D and 3D models. Our results show that convective matter motions below the gain radius become much more violent in 3D than 2D, making the neutrino luminosity larger for 3D. Enhanced by the large neutrino luminosity, the shock of the 3D model expands faster than that of the 2D. Our results show that the evolution of the shock is sensitive to the employed numerical resolutions. To draw a robust conclusion, 3D simulations with much higher numerical resolution and more advanced treatment of neutrino transport and gravity is needed.

Keywords. Supernovae:general, neutrinos, hydrodynamics

1. Introduction

Pushed by mounting observations of the SN blast morphology (e.g., Wang *et al.* 2001, Maeda *et al.* 2008), it is now almost certain that the breaking of the spherical symmetry is the key to solve the SN problem. So far, a number of 2D (axi-symmetric) hydrodynamic simulations have shown that hydrodynamic motions associated with convective overturn (e.g., Burrows *et al.* 1995) and the Standing-Accretion-Shock-Instability (SASI, e.g., Blondin *et al.* 2003) can help the onset of the neutrino-driven explosion.

It is natural to wonder how 3D convection affects the explosion. Nordhaus *et al.* (2010) argued that the critical neutrino luminosity for producing neutrino-driven explosions becomes smaller in 3D than 2D (an objection was proposed by Hanke *et al.* 2011). Encouraged by that work, we explore in this study possible 3D effects in the SN mechanism by performing 3D, multigroup, radiation hydrodynamic core collapse simulations.

2. Result

We focus here on the evolution of an $11.2M_{\odot}$ star of Woosley *et al.* (2002). Figure 1 shows the blast morphology for our 1D (left), 2D (middle), and 3D (right) model,

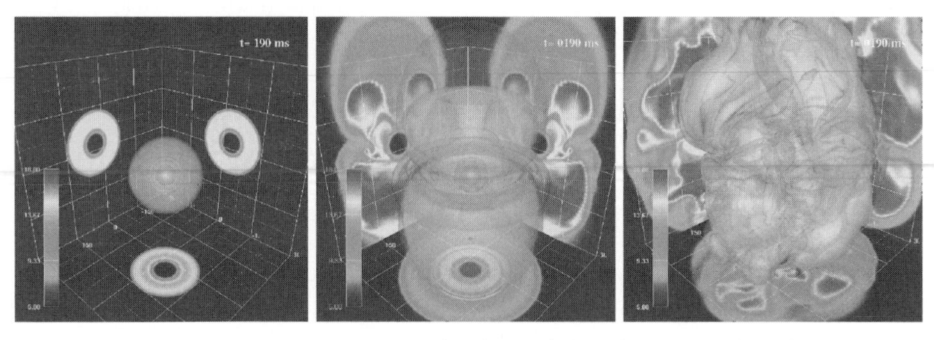

Figure 1. Blast morphology for 1D (left), 2D (middle), and 3D (right) model.

respectively. The shock of the 1D has already stalled but that of the 2D and 3D has not. The matter in the 2D and 3D receives neutrino heating promoted by the convective overturn and SASI.

We compare the evolution of the shocks in the left panel of Figure 2. The 1D and the other models are completely different. The convection significantly affects the shock. The shock of the 3D model goes faster than that of the 2D since the neutrino luminosity of the 3D is larger than the 2D as shown in the right panel of Figure 2.

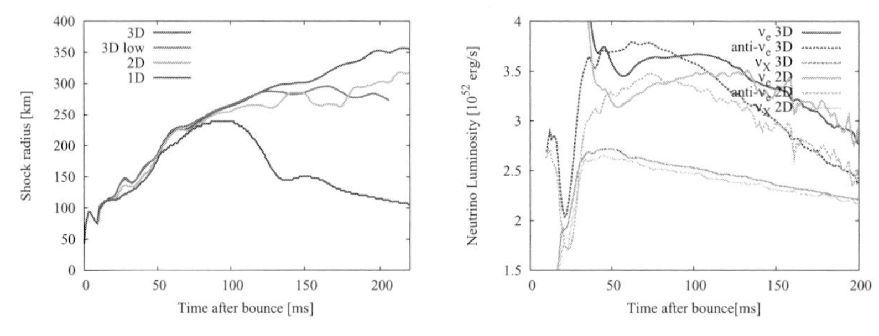

Figure 2. *Left*: Evolution of the average shock radius. *Right*: Evolution of the neutrino luminosity. While the red line is the 3D model with the zone of $320(r)$x$64(\theta)$x$128(\phi)$, the green line is the 2D axi-symmetric model with the zone of $320(r)$x$64(\theta)$ and the blue line is the 1D spherical model with a radial zone of $320(r)$. The purple line corresponds to the 3D low-resolution model of $200(r)$x$32(\theta)$x$64(\phi)$

Our results show that the evolution of the shock is sensitive to the employed numerical resolution. To draw a robust conclusion, 3D simulations with much higher numerical resolution are necessary.

References

Blondin, J. M., Mezzacappa, A., & DeMarino, C. 2003, *ApJ*, 584, 971
Burrows, A., Hayes, J., & Fryxell, B. A. 1995, *ApJ*, 450, 830
Hanke, F., Marek, A., Mueller, B., & Janka, H.-T., arXiv:1108.4355
Maeda, K., Kawabata, K., Mazzali, P. A., *et al.* 2008, *Science*, 319, 1220
Nordhaus, J., Burrows, A., Almgren, A., & Bell, J. 2010, *ApJ*, 720, 694
Wang, L., Howell, D. A., Höflich, P., & Wheeler, J. C. 2001, *ApJ*, 550, 1030
Woosley, S. E., Heger, A., & Weaver, T. A. 2002, *Reviews of Modern Physics*, 74, 1015

Death of Massive Stars: Supernovae and Gamma-Ray Bursts
Proceedings IAU Symposium No. 279, 2012
P. Roming, N. Kawai & E. Pian, eds.

doi:10.1017/S1743921312013592

Spectral Evolutions Study of Gamma-Ray Burst Exponential Decays with Suzaku-WAM

Makoto S. Tashiro[1], Kaori Onda[1], Kazutaka Yamaoka[2], Masahiro Ohno[3], Satoshi Sugita[4], Takeshi Uehara[5], and Hiromi Seta[6]

[1]Department of Physics, Saitama University, Sakura, Saitama, Japan,
email: tashiro@phy.saitama-u.ac.jp

[2]Research Institute for Science and Engineering, Waseda University, Shinjuku, Tokyo, Japan

[3]JAXA/Institute of Space and Astronautical Science, Sagamihara, Japan

[4]EcoTopia Science Institute, Nagoya University, Chikusa-ku, Nagoya, Japan

[5]Department of Physics, Hiroshima University, Kagami-Yama, Higashi-Hiroshima, Japan

[6]Research Center for Measurement in Advanced Science, Rikkyo University, Tokyo, Japan

Abstract. An observational study is presented of the spectral evolution of gamma-ray burst (GRB) prompt emissions with the Suzaku Wide-band All-sky Monitor (WAM). We selected 6 bright GRBs exhibiting 7 well-separated fast-rise-exponential-decay (FRED) shaped light curves to investigate spectral changes by evaluating exponential decay time constants of the energy-resolved light curves. In addition, we carried out time-resolved spectroscopy of two of them which were located with accuracy sufficient to evaluate the time-resolved spectra with precise energy response matrices. The two imply different emission mechanisms; the one is well reproduced with a cooling blackbody radiation model with a power-law component, while the other prefers non-thermal emission model with a decaying turn over energy.

Keywords. gamma rays: bursts, radiation mechanisms: nonthermal, thermal

---

1. Introduction

Gamma-ray burst (GRB) prompt emission spectra are often described with a power law, an exponentially cut-off power law, or a smoothly connected broken power law (GRBM; Band *et al.* (1993)). They are consistent with the energy spectral distributions from optically thin synchrotron or synchrotron self-Compton radiation produced by relativistic electrons accelerated through shock fronts in the outflows . Alternatively, an optically thick thermal (blackbody) component has been proposed in addition to the pure non-thermal (power-law) model (e.g. Meszaros & Rees (2000)).

In this paper, we study luminosity-spectrum evolution to investigate the emission mechanism by using data collected with the Suzaku Wide-band All-sky Monitor (WAM; Yamaoka *et al.* (2009)). The WAM has a good advantage in the effective area at the energy band in which we often observe turn-over frequency of the GRB prompt emission. To reduce the number of parameters affecting the time evolution, here we focus on fast-rise-and-exponential-decay (FRED) light curves, which are the most promising for investigating the radiation process separately from the geometrical effects of the emission regions. The relatively fast rise implies that the time scale of geometrical variation is sufficiently short not to dominate the longer decaying process. The exponential decay time scale is thus expected to reflect the state evolution of the emission region. Therefore, we selected from WAM archive those event which have: bright (>1000 c s^{-1} at peak, asymmetric (fast rise-slow decay), and no overlapped peak in decay phase. Consequently we study 7 FRED peaks from 6 GRBs in this paper. Detailed description of this study is seen in Tashiro *et al.* (2012).

2. Results

2.1. *Energy resolved lightcurve*

We evaluated each decaying portion in three bands of WAM lightcurve data (nominally 50-110keV, 110-240 keV, and 240-520 keV). We confirmed that every decay is exponential and no power-law decay model was accepted. In general, the shorter time constant is observed in the higher energy band light curve. In order to evaluate this trend quantitatively, we fitted the derived time constants of decays with a power-law function, $\tau(E) \propto E^{-\gamma}$. All of the peaks shows clear spectral variation ($\gamma < 0$) in the accuracy of 90 % confidence level, an the derived average $\tau_{\rm ave} = -0.34 \pm 0.12$. Five of 7 peaks accept simple synchrotron/IC cooling interpretation ($\gamma = -1/2$), though we cannot reject thermal interpretation only with this energy resolved lightcurve study.

2.2. *Time resolved spectrum*

The incident angle of gamma-rays only from GRB 081224 and GRB 100707A were determined with an accuracy sufficient to generate reliable energy response matrices.

We examined four overlaid 1-s time-resolved spectra from GRB 081224 in the decay phase. A blackbody radiation with a power-law model (BBPL) succeeded to describe the time resolved spectra with naturally lowering kT and decreasing normalization, which implies a relatively constant power-law component with a gradually cooling blackbody component. The derived behavior of kT is well reproduced by a power-law function of time with an index of $-(0.43^{+0.27}_{-0.28})$.

On the contrary, the four 1-s time-resolved spectra from the decay phase in GRB 100707A exhibit a statistically significant preference for GRBM (χ^2/d.o.f.= 64.7/57) over BBPL (χ^2/d.o.f. = 131.4/39). In order to evaluate the spectral softening in the best-fit value of the turn-over energy E_0, we tied the first and second spectral indices. The parameter of E_0 exhibits a power-law-type decrease in time with an index of $-(1.1^{+0.6}_{-0.8})$, which is consistent with the expected non-thermal (synchrotron/IC) cooling model.

3. Conclusion

We studied 7 well-separated bright FRED peaks with exponential decay but power-law decays are not accepted. They clearly show spectral evolution and the energy indices of the time constants are concentrated around -0.3. We also showed time resolved spectroscopy for GRB 081224 and GRB 100707A. These behaviors observed in the light curves and the spectral evolution analysis suggest that the emission mechanism in GRBs consists of at least two different components, such as the thermal and non-thermal processes.

References

Band, D. *et al.* 1993, *ApJ* 413, 281
Meszaros, P. & Rees, M. J. 2000, *ApJ*, 530, 292
Tashiro, M. *et al.* 2012, *PASJ*, 64, 26
Yamaoka, K. *et al.* 2009, *PASJ*, 61, S35

Death of Massive Stars: Supernovae and Gamma-Ray Bursts
Proceedings IAU Symposium No. 279, 2012
P. Roming, N. Kawai & E. Pian, eds.

doi:10.1017/S1743921312013609

Shock Breakout of Type II Plateau Supernova

Nozomu Tominaga[1,2], **Tomoki Morokuma**[3] **and Sergei I. Blinnikov**[4,2]

[1]Department of Physics, Faculty of Science and Engineering, Konan University,
8-9-1 Okamoto, Kobe, Hyogo 658-8501, Japan
email: `tominaga@konan-u.ac.jp`

[2]Kavli Institute for the Physics and Mathematics of the Universe, University of Tokyo,
5-1-5 Kashiwanoha, Kashiwa, Chiba 277-8583, Japan

[3]Institute of Astronomy, University of Tokyo, Mitaka, Tokyo 181-0015, Japan

[4]Institute for Theoretical and Experimental Physics (ITEP), Moscow 117218, Russia

Abstract. Type II-plateau supernovae (SNe II-P) are fainter than Type Ia SNe and thus have so far been observed only at $z < 1$. We introduce shock breakout and propose a distant SN II-P survey at $z > 1$ with shock breakout. The first observation of shock breakout from the rising phase is reported in 2008. We first construct a theoretical model reproducing the UV-optical light curves (LCs) of the first example and demonstrate that the peak apparent g-band magnitude of the shock breakout would be $m_{\rm g} \sim 26.4$ mag if an identical SN occurs at a redshift $z = 1$, which can be reached by 8m-class telescopes. Furthermore, we present LCs of shock breakout of SN explosions with various main-sequence masses, metallicities, and explosion energies and derive the observable SN rate and reachable redshift as functions of filter and limiting magnitude by taking into account an initial mass function, cosmic star formation history, intergalactic absorption, and host galaxy extinction. The g-band observable SN rate with limiting magnitude 27.5 mag is 3.3 SNe deg^{-2} day^{-1} and half of them are located at $z > 1.2$.

Keywords. shock waves, radiative transfer, supernovae: general, supernovae: individual (SNLS-04D2dc), stars: evolution, surveys

---

1. Introduction

Shock breakout is the bolometrically brightest phenomenon in supernovae (SNe) and thus it is suggested to be observable even if it takes place at $z \gtrsim 1$ (e.g. Chugai *et al.* 2000). Recently, shock breakout of Type II plateau SNe (SNe II-P) was detected for SNLS-04D2dc (Schawinski *et al.* 2008; Gezari *et al.* 2008) and SNLS-06D1jd (Gezari *et al.* 2008) by the *GALEX* satellite and thus it is coming under the spotlight to probe high-z CCSNe by the shock breakout of SNe II-P. However, the observable properties are poorly understood. Therefore, in this contribution, we present theoretical predictions of observables of shock breakout of SNe II-P calculated by a multigroup radiation hydrodynamics code STELLA (Blinnikov *et al.* 1998; 2000; 2006).

2. Results

First, we focus on SNLS-04D2dc because SNLS-06D1jd has sparse UV observations with relatively low signal-to-noise ratio and compare the synthetic LCs with the multicolor observations of SN IIP SNLS-04D2dc. We successfully construct a SN II-P model reproducing well the shock breakout and plateau consistently and constrain SN and progenitor properties (left panel of Fig. 1, Tominaga *et al.* 2009). Furthermore, based on the multicolor LC model, we demonstrate that the peak apparent g-band magnitude of the

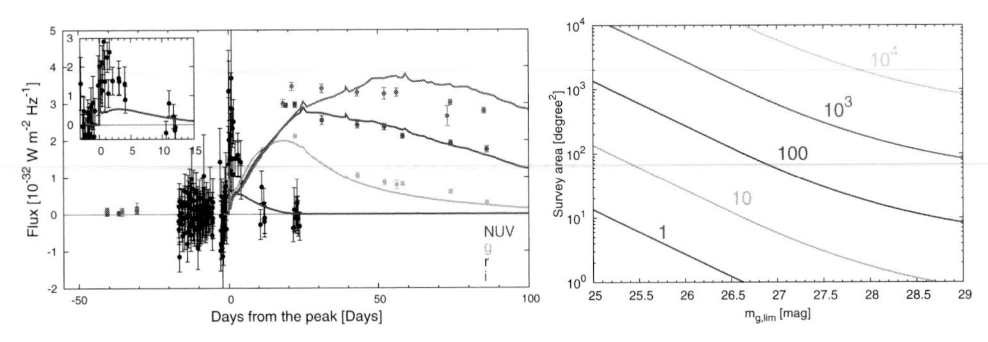

Figure 1. (Left)Comparison between the SNLS-04D2dc observations (*points*) and a SN IIP model reddened for the host galaxy extinction (*lines*) (*black* and *red*: near UV, *green*: g band, *blue*: r band, *magenta*: i band). The inset enlarges the phase when the SN emitted UV light. (Right) Observable SN rate per day in g' band. The lines show different observable SN rate: 1 (red), 10 (green), 100 (blue), 10^3 (magenta), and 10^4 (cyan).

shock breakout would be $m_{\rm g} \sim 26.4$ mag if a SN being identical to SNLS-04D2dc occurs at a redshift $z = 1$, which can be reached by 8m-class telescopes. The result evidences that the shock breakout has a great potential to detect SNe IIP at $z \gtrsim 1$.

Next, we adopt progenitor models with various main-sequence masses which are taken from Umeda & Nomoto (2005) and present multicolor light curves of shock breakout of SNe II-P (Tominaga *et al.* 2011). As a result, we predict apparent multicolor light curves of shock breakout at various redshifts z and derive the observable SN rate and reachable redshift as functions of filter x and limiting magnitude $m_{x,\rm lim}$ by taking into account an initial mass function, cosmic star formation history, intergalactic absorption, and host galaxy extinction. We propose a realistic survey strategy optimized for shock breakout. For example, the g'-band observable SN rate for $m_{g',\rm lim} = 27.5$ mag is 3.3 SNe degree^{-2} day^{-1} and half of them are located at $z \geqslant 1.2$ (right panel of Fig. 1).

Although the survey parameters should be customized to observation purposes and telescope/instrument, future/ongoing wide and/or deep surveys, e.g. Palomar Transient Factory (PTF), Lick Observatory Supernova Search (LOSS), Catalina Real-Time Transient Survey (CRTS), Kiso/Kiso Wide Field Camera (KWFC), Skymapper, Dark Energy Survey (DES), Panoramic Survey Telescope and Rapid Response System (Pan-STARRS), Subaru/Hyper Suprime-Cam (HSC), and Large Synoptic Survey Telescope (LSST), will find a large number of shock breakout events. We emphasize that the multicolor observations in blue optical bands with $\sim$ hour intervals, preferably over $\geqslant 2$ continuous nights, are essential to efficiently detect, identify, and interpret shock breakout and conclude that the most essential observation is the multicolor photometry with short intervals less than 1 day and that the observation over $\geqslant 2$ continuous nights is favorable.

References

Blinnikov, S., Lundqvist, P., Bartunov, O., Nomoto, K., & Iwamoto, K. 2000, *ApJ*, 532, 1132
Blinnikov, S. I., Eastman, R., Bartunov, O. S., *et al.*, 1998, *ApJ*, 496, 454
Blinnikov, S. I., Röpke, F. K., Sorokina, E. I., *et al.*, 2006, *A&A*, 453, 229
Chugai, N. N., Blinnikov, S. I., & Lundqvist, P. 2000, *Mem. della Soc. Astro. Ita.*, 71, 383
Gezari, S., Dessart, L., Basa, S., *et al.* 2008, *ApJL*, 683, L131
Schawinski, K., Justham, S., Wolf, C., *et al.* 2008, *Science*, 321, 223
Tominaga, N., Blinnikov, S., Baklanov, P., *et al.*, 2009, *ApJL*, 705, L10
Tominaga, N., Morokuma, T., Blinnikov, S., *et al.*, 2011, *ApJS*, 193, 20
Umeda, H. & Nomoto, K. 2005, *ApJ*, 619, 427

Death of Massive Stars: Supernovae and Gamma-Ray Bursts
Proceedings IAU Symposium No. 279, 2012
P. Roming, N. Kawai & E. Pian, eds.

doi:10.1017/S1743921312013610

GRB 100816A and the nature of intermediate duration gamma-ray bursts

Rachel L. Tunnicliffe and Andrew Levan

Department of Physics, University of Warwick, Coventry, CV4 7AL, UK
email: r.l.tunnicliffe@warwick.ac.uk

Abstract. Gamma-ray bursts are normally split into two classes, primarily determined by their observed duration, so called long (> 2 s) and short (< 2 s) GRBs. There have been many claims of a third duration class, with emission lasting for intermediate periods between $2-5$ s, although the reality of this class remains controversial. Here, we investigate this further utilising the 2.9 s duration, spectrally hard GRB 100816A. This burst lies well offset from its host galaxy, has no evidence for an associated supernova (albeit to only moderately constraining limits), and has properties which appear to be genuinely intermediate between long- and short-population bursts. We extend this analysis by comparing the physical locations of a population of intermediate duration GRBs with those of short-GRBs and long-GRBs, concluding that the intermediate sample is indistinguishable from the long-GRB population, whose locations are very different from other transients.

Keywords.

1. Introduction

Studies of BATSE and Swift GRBs have shown that their duration distribution can be best fit by the addition of a third class of burst, of intermediate duration, but typically softer spectra. In principle these GRBs could arise from progenitors distinct from those thought to form LGRBs and SGRBs, but they have proved extremely difficult to pinpoint (see e.g. de Ugarte Postigo *et al.* (2010)).

Here, we consider GRB 100816A, whose properties appear genuinely intermediate between LGRBs and SGRBs. The burst is formally long (2.9 s), but spectrally hard, and lies well offset from its host galaxy. No clear SN is detected to limits a factor of $\sim$10 fainter than SN 1998bw.

2. Properties of intermediate GRBs

2.1. *Host galaxy properties*

The locations of transient objects are highly diagnostic of their progenitors. Fruchter *et al.* (2006) have shown that LGRBs are highly concentrated on their host light, much more so than core collapse supernovae generally. Further, Fong *et al.* (2009) have similarly shown that SGRBs are highly scattered on the light of their host galaxies. Using a sample of intermediate bursts figure 1(a) shows that the intermediate GRBs seem to trace the same locations as LGRBs, ruling out a location similar to SGRBs, or a linear proportionality to the light at the 98% and 90% confidence levels. This suggests that most intermediate GRBs are in fact arising from collapsars similar to LGRB progenitors.

We note that GRB 100816A itself does reside within a fainter region of its host but this is not unprecedented in the long GRB population.

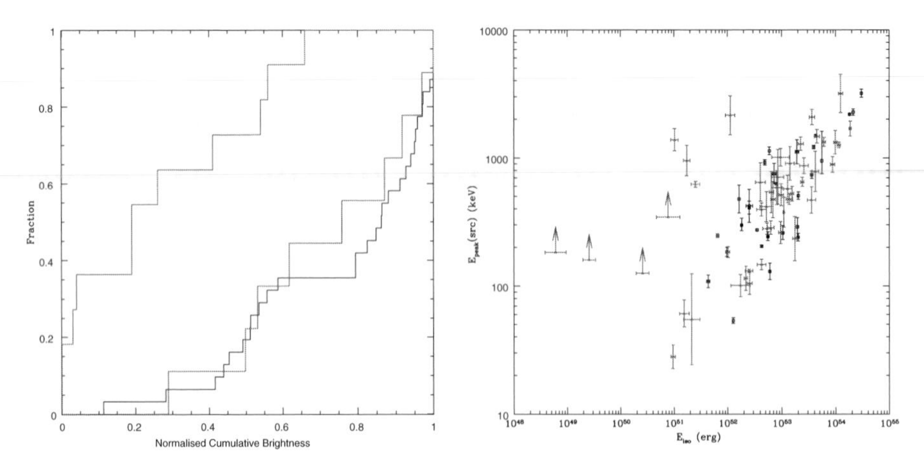

Figure 1. The left hand panel shows the distribution of short (blue), intermediate (red) and long (black) GRBs on the light of their host galaxies. Intermediate and long GRBs are highly concentrated on their host light and are clearly distinct from short GRBs. The right hand panel shows the energy characteristics of short (blue), intermediate (red) and long (black) GRBs in the rest frame of the burst. Long and intermediate GRBs follow the Amati relation whereas short GRBs clearly lie offset from this distribution.

2.2. *Prompt and afterglow characteristics*

Compared to other intermediate and long GRBs, GRB 100816A is spectrally hard. However, figure 1(b) shows that when looking at their intrinsic properties all intermediate GRBs, including GRB 100816A, follow the relation between the isotropic equivalent energy, E_{iso}, and E_{peak} in the rest frame of the GRB, first identfied by Amati *et al.* (2002). All short GRBs are well offset from this relation.

Comparing the optical afterglow magnitudes to values from Kann *et al.* (2010) and Kann *et al.* (2011) we find GRB 100816A to be intermediate between the long and short populations. For optical and X-ray afterglows de Ugarte Postigo *et al.* (2010) found this to be general property of intermediate GRBs.

3. Conclusions

We have shown that the majority of intermediate duration GRBs are distributed in host galaxies identically to the LGRBs, suggesting a very similar progenitor mechanism. However, some events such as GRB 100816A appear to lie in regions intermediate between LGRBs and SGRBs in many observational parameters. This suggests that, even with excellent observational data the task of distinguishing the origin of bursts around the 2 s divide is extremely challenging.

References

de Ugarte Postigo, A. *et al.* 2010, *A&A*, 525, 109
Amati, L. *et al.* 2002, *A&A*, 390, 81
Kann, D. A. *et al.* 2010, *ApJ*, 720, 1513
Kann, D. A. *et al.* 2011, *ApJ*, 734, 96
Fruchter, A. S. *et al.* 2006, *Nature*, 441, 463
Fong, W., Berger, E., & Fox, D. B. 2009, *ApJ*, 708, 9

Death of Massive Stars: Supernovae and Gamma-Ray Bursts
Proceedings IAU Symposium No. 279, 2012
P. Roming, N. Kawai & E. Pian, eds.

doi:10.1017/S1743921312013622

Classification of long Gamma Ray Bursts using cosmologically corrected temporal estimators

Nicolas A. Vasquez[1] **and Christian Vasconez**[2]

[1]Observatorio Astronomico de Quito
Av Gran Colombia S/N, Parque de la Alameda, Quito, Ecuador
email: nvasquez.observatorio@epn.edu.ec

[2]Observatorio Astronomico de Quito
Av Gran Colombia S/N, Parque de la Alameda, Quito
email: c.vasconez@epn.edu.ec

Abstract. The canonical classification of GRBs establishes two types of bursts, long and short. Although an intermediate class of GRBs was suggested, its existence is not yet conclusive. In the present work, we explore the temporal classification of GRBs in the burst frame, because in recent years the statistics of bursts with known redshifts has increased. We studied a sample of *Swift* GRBs with known redshifts to determine three different time estimators: autocorrelation functions, emission times and duration times. In order to look for a subclass in long GRBs , we studied the distribution of the cosmologically corrected time estimators. The distribution of time estimators of the sample suggests an internal division of long GRBs. The proposed bimodality is also supported in the isotropic luminosity - time estimator planes and we discuss some possible implications of the classification of GRBs in the burst frame.

Keywords. gamma rays: bursts, gamma rays: observations

1. Introduction

The introduction of a third class of bursts (Horvath *et al.* (2010)) within the canonical classification of GRBs (Kouveliotou *et al.* (1993)) has improved our understanding of the progenitors and radiation mechanisms of these explosions. Nevertheless, classifications of bursts has been done in the observer frame which includes some of the underlying effects of the cosmological expansion of the Universe. In the present study we performed a temporal classification in the burst frame as a extension of a previous work (Vasquez & Kawai (2010)).

2. Data analysis

Our sample is composed of 15 GRBs detected simultaneously by the *Swift* and *Suzaku* missions with measured redshifts. We compute the duration time of the 50 percent of the total fluence (T_{50}), the emission time of the 50 percent of the total fluence (t_{50}; Mitrofanov *et al.* (1998)), and the autocorrelation function (ACF) of the light curve evaluated at FWHM (Link *et al.* (1995)) based on Monte Carlo simulations of the light curves. We found that the emission time is a better temporal estimator as is shown in Figure 1. The correlation coefficient between the isotropic luminosity improves from -0.28 to 0.78 when the four long-dim bursts are taken as different type of GRBs. This behavior is recurrent with the other temporal estimators.

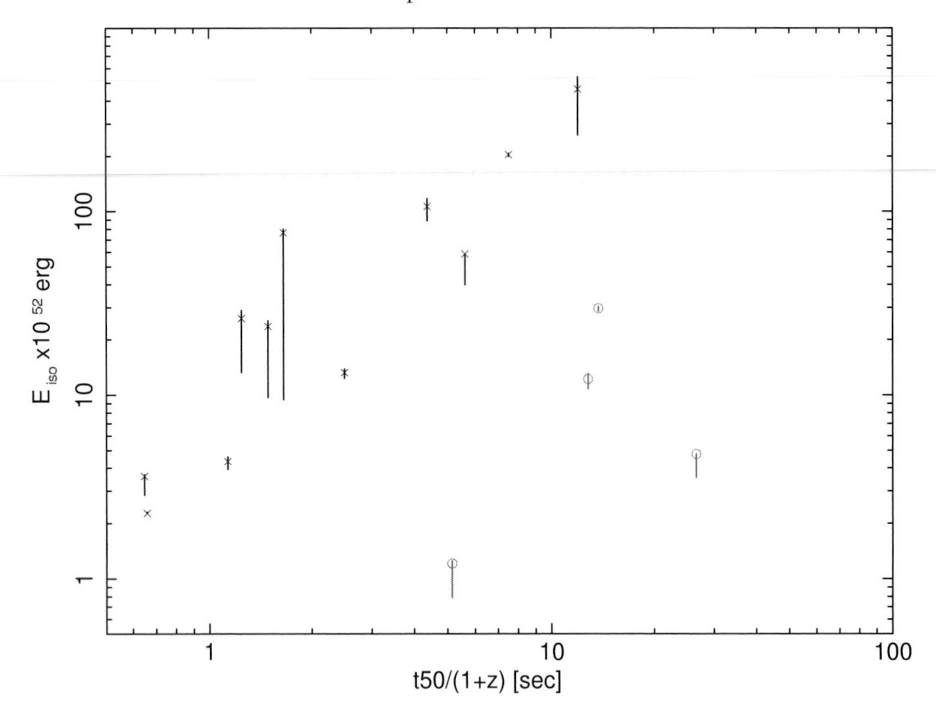

Figure 1. Isotropic luminosity as a function of the emission time (t_{50}) in the 50–150 keV band. In the right bottom corner, the four long-dim bursts are localized.

3. Implications

One of the most intriguing questions is to know if the classifications of bursts in the observer frame are hiding some intrinsic properties due to cosmological effects. Borgonovo *et al.* (2007) show that long GRBs have a bimodal distribution when the ACF is cosmologically corrected (Borgonovo *et al.* (2007)). Following this idea, we also found a similar bimodal distribution of long GRBs when the ACF is computed in the rest frame. Since we expect that cosmological dilation effects are the same for all timescales, we also extend the bimodality for other time estimators: duration times of outburst activity (T_{50}) and the time intervals where the engine is in the most active phase (t_{50}). There are four GRBs, that in the time-distance plane, form a subgroup. Long-near bursts and long-distant bursts, seem to have different origin and behavior. In the light curves, we also observed that there are some common trends: the time history is composed of broad pulses that underlie a longer active phase. Evidence for a subclass of GRBs is also present in the time-energy planes, which reinforces the possibility of different engines or progenitor for GRBs that have longer active phases, are dimmer and have smaller redshifts.

References

Borgonovo L., *et al.* 2007, *A&A*, 465, 765
Horvath Z., *et al.* 2010, *ApJ*, 713, 552
Kouveliotou C., *et al.* 1993, *ApJL*, 413, 101
Link, B., Epstein, R. I., & Priedhorsky, W. C. 1995, *ApJ*, 631, 976
Mitrofanov I., *et al.* 1998, *ApJ*, 522, 1069
Vasquez, N. & Kawai, N. 2010, *Inter. J. of Modern Physics D*, 19, 997

Death of Massive Stars: Supernovae and Gamma-Ray Bursts
Proceedings IAU Symposium No. 279, 2012
P. Roming, N. Kawai & E. Pian, eds.

doi:10.1017/S1743921312013634

The Red Supergiant Problem: Circumstellar dust as a solution

Joe Walmswell[1] and John Eldridge[2]

[1]Institute of Astronomy, The Observatories, University of Cambridge
Madingley Road, Cambridge CB3 0HA, United Kingdom
email: jjw49@ast.cam.ac.uk

[2]The Department of Physics, The University of Auckland
Private Bag 92019, Auckland, New Zealand
email: j.eldridge@auckland.ac.nz

Abstract. We investigate the red supergiant problem: the apparent dearth of Type IIP supernova progenitors with masses between 16 and 30 $M_{\odot}$. Although red supergiants with masses in this range have been observed, none have been identified as progenitors in pre–explosion images. We show that, by failing to take into account the additional extinction resulting from the dust produced in the red supergiant winds, the luminosity of the most massive red supergiants at the end of their lives is underestimated. We re–estimate the initial masses of all Type IIP progenitors for which observations exist and analyse the resulting population. We find that the most likely maximum mass for a Type IIP progenitor is 21^{+2}_{-1} $M_{\odot}$. This is in closer agreement with the limit predicted from single star evolution models.

Keywords. .stars: evolution – supernovae: general – stars: supergiants

---

1. The problem

It is generally acknowledged that all Type IIP supernovae are the result of core–collapse in either asymptotic giant branch (AGB) stars or red supergiants (RSGs). The strong hydrogen lines and the plateau in the light curve imply a considerable mass of hydrogen in these progenitors. We know from observations that stars with masses between about 12 and 25–30 $M_{\odot}$ will evolve into RSGs. However surveys of Type IIP progenitors have failed to identify high–mass RSGs among them. This is the red supergiant problem: how do massive RSGs end their lives?

2. Our solution

Most of the proposed solutions involve alternative fates for the massive RSGs, such as sudden mass-loss episodes or direct collapse to a black hole. We suggest that because mass–loss increases with stellar mass these stars explode whilst surrounded by large amounts of wind material of their own creation. This causes a considerable amount of additional extinction in observations of the progenitors, resulting in underestimates for the luminosities and hence the deduced masses. Evidence for this includes the abnormally high observed extinction values for certain massive RSGs, reaching several A_V in some cases (Massey *et al.* 2005).

We do not try to model the dusty wind behavior. It is a complex and variable 3D process and we lack the information to calibrate the inputs. Instead we use the work of Massey *et al.* (2005), who measured the dust production rates for a population of RSGs and plotted them against their bolometric luminosity. This gave a least-squares fit, allowing us to calculate an empirical average dust mass rate for a star.

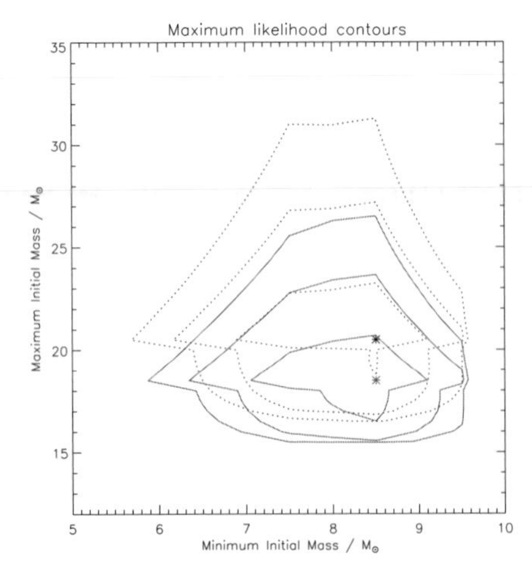

Figure 1. The optimal solutions and the maximum-likelihood contours for the 68, 90 and 95 per cent confidence regions. The solid lines are for the original dust–free models and the dotted lines are for the models with dust included.

We used this relation to modify a grid of stellar models produced with the Cambridge STARS code (Eldridge & Tout 2004). This gave the magnitudes in various pass-bands, both with and without the effect of extra dust. We deduced the dust mass rate at each time step and interpolated to get it as a function of time. We then integrated this to get the dust mass in shells around the star and then found the total extinction due to these shells.

3. Our results

We analysed observations of six stars known to be progenitors of Type IIP SNe. We also considered twelve instances where the progenitor was not identified but where the sensitivity of the instrument gave an upper limit to its luminosity. We use the compilation of SN detections and non-detections of Smartt *et al.* (2009). All progenitor information can be found in that paper and the references therein. We supplement this with SN 2009md (Fraser *et al.* 2011).

We deduced mass ranges for these 18 SNe and performed maximum-likelihood calculations to find the most probable upper and lower progenitor mass limits (Figure 1). We find that when the effect of circumstellar dust is ignored, the most probable upper limit is 18 $M_\odot$ and the 90 per cent confidence limit is 23 $M_\odot$. However, the dust models give 21 $M_\odot$ for the upper limit and, more significantly, a 90 per cent limit of 27 $M_\odot$. This means that, with the current data, it is difficult to argue that the red supergiant problem really exists. Without accurate measures of the stellar, as opposed to the local, extinction, the increase of dust production with mass leads to aliasing at higher masses.

References

Eldridge, J. J. & Tout, C. A., 2004, *MNRAS*, 353, 87

Fraser, M., Ergon, M., Eldridge, J., Valenti, S., Pastorello, A., Sollerman, J., Smartt, S. J., Agnoletto, I., & Arcavi, I., etc., 2011, *MNRAS*, 417, 1417

Massey, P., Plez, B., Levesque, E. M., Olsen, K. A. G., Clayton, G. C., & Josselin, E., 2005, *ApJ*, 634, 1286

Smartt, S. J., Eldridge, J. J., Crockett, R. M., & Maund, J. R., 2009, *MNRAS*, 395, 1409

Death of Massive Stars: Supernovae and Gamma-Ray Bursts
Proceedings IAU Symposium No. 279, 2012
P. Roming, N. Kawai & E. Pian, eds.

doi:10.1017/S1743921312013646

SVOM Visible Telescope: Performance and Data Process Scheme

C. Wu, Y. L. Qiu and H. B. Cai
on behalf of SVOM team

National Astronomical Observatories, Chinese Academy of Sciences,
Postbus 100012, Beijing, China
email: cwu@bao.ac.cn

Abstract. VT (Visible Telescope) is an instrument onboard SVOM (Space-based multi-band astronomical Variable Objects Monitor) satellite working in the visible band, which will play an important role in follow-up of two categories of GRBs: very distant events at higher redshift and faint/soft nearby events in SVOM mission. To fulfill these primary science requirements, decent sensitivity and wavelength coverage are fundamental for VT design. VT performance and data process strategy were successfully studied on its feasibility in Phase A, which is presented in this poster. Additionally, preliminary VT image simulator is also introduced here.

Keywords. space vehicles: instruments, telescope, gamma rays: bursts.

1. Introduction

The SVOM mission is dedicated to observe gamma-ray bursts (GRB) in multi-bands from visible to MeV. The multi-bands observation is provided by four instruments on SVOM satellite, i.e. VT, MXT, GRM, and ECLAIR. VT is the only one working in the visible band, which serves the purpose of improving the GRB localizations obtained by the higher energy detectors ECLAIR and MXT to sub-arcsec precision through the observation of the optical afterglow. It will thus ensure a deep and uniform sample of optical-afterglow light-curves, and make a preliminary selection of optically dark GRBs and high-redshift GRB candidates (for detail, see Paul *et. al.* 2011).

2. VT performance

The general VT performance is presented in Table 1. The primary points in VT design are: (i) decent sensitivity and wavelength coverage: It is required by the scientific goal to detect faint and high-z GRBs. Therefore, VT will have two simultaneous channels with high EQ CCD and wavelength coverage from 400 nm to 950 nm. Detector parameters, such as read noise, working temperature, pixel size are also taken into account. (ii) The Field Of View (FOV) of VT should cover the entire ECLAIR localization error box of the GRB. Currently, the FOV of VT, modified to $26' \times 26'$, is just based on ECLAIR localization strategy update. (iii) Effectiveness and reliability: VT shall have the following operation modes: OFF, OBSERVATION MODE, STAND-BY, CALIBRATION, CCD CLEANING, FOCUS ADJUSTING, RESET, AND CCD BAKE-OUT, to make it working effectively and reliably. Furthermore, CCD shielding and internal shutter mechanism are also considered in VT design.

Table 1. Overview of VT performance.

Parameter	Value	Parameter	Value
Aperture	450 mm (diameter)	Focal Length	3600 mm
FOV	$26' \times 26'$ (new modification)	Trans. efficiency	$\geqq 0.7$ (optical path)
Channels	400–650 nm, 650–950 nm	Read noise	$\leqq 6e$ (100 s/300 s)
m_{Limit}	V = 23 (300 s exp; SNR$\geqslant$5)	T_{CCD}	$\leqq -65°C$

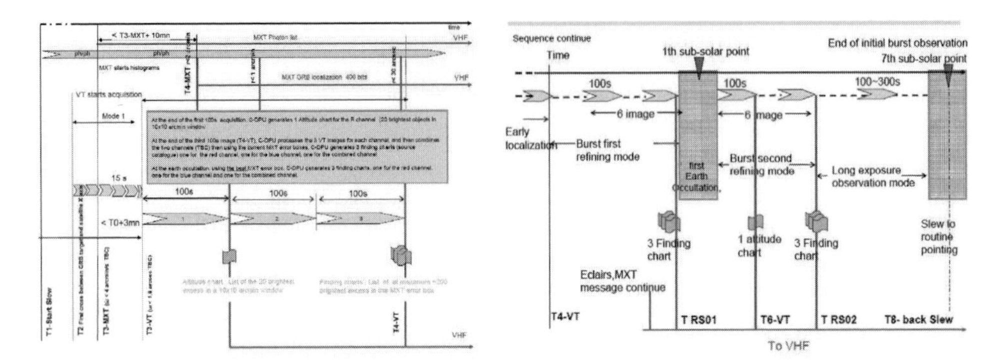

Figure 1. VT observation strategy.

3. Observation & data process strategy

Observation strategy. The on-board localization sequence is performed during the first and the second orbit of the initial burst observation, which is shown in Figure 1.

Data process strategy. On-board data process includes attitude chart and finding chart process for each 300s (see Figure 1). Ground data process comprises three pipelines: (i) Pipeline I: Astrometry calibration for GRB position of MXT with VT Attitude. (ii) Pipeline II: To Search GRB candidate in finding chart and do astrometry calibration. The output is GRB position in VT field. (ii) Pipeline III: Search of GRB fainter candidate in subimage and also do astrometry calibration. The output is GRB position in VT field.

4. VT image simulator

Purpose. (i) Pipeline: To provide input for pipeline development. (ii)Design: To validate the performance of VT design. (iii) Calibration: To understand SVOM photometry system. (iv) Observation strategy: To estimate GRB detection efficiency.

Methods & tools. Input of the simulator is the telescope optical system efficiency, CCDs performance parameters: gain, read noise, response curve, jitter stability, sky background, Space radiation environments, GRB knowledge (number, types, spatial distribution etc.), survey catalogues, SVOM observation strategy. Its output is VT image with GRB. Used tools are: skymaker,Pysyphot,IRAF artdata package, and HST handbook.

C.W. acknowledges support from the National Natural Science Foundation of China grant 10903010.

References

Paul, J., Wei, J. Y., Basa, S., Zhang, S. N., & Anders, E. 2011, *CRPhy*, 12, 298

Death of Massive Stars: Supernovae and Gamma-Ray Bursts
Proceedings IAU Symposium No. 279, 2012
P. Roming, N. Kawai & E. Pian, eds.

doi:10.1017/S1743921312013658

Development of a micro-satellite TSUBAME for X-ray polarimetry of GRBs

Yoichi Yatsu[1], Mayumi Hayashi[1], Kousuke Kawakami[1], Kazuki Tokoyoda[1], Takahiro Enomoto[1], Takahiro Toizumi[1], Nobuyuki Kawai[1], Kazuya Ishizaka[2], Azusa Muta[2], Hiroyuki Morishita[2], Saburo Matsunaga[2,3], Takeshi Nakamori[4], Jun Kataoka[4], and Shin Kubo[5]

[1]Dept. of Physics, Tokyo Institute of Technology,
2-12-1 Oookayama Meguro Tokyo, 152-8551, Japan
email: yatsu@hp.phys.titech.ac.jp

[2]Dept. of Mechanical and Aerospace Engineering, Tokyo Institute of Technology, Japan

[3]Dept. of Space Structure and Materials, Institute of Space and Astronautical Science, Japan

[4]Research Institute for Science and Engineering, Waseda University, Japan

[5]CLEAR PULSE CO., LTD., Japan

Abstract. Hard X-ray polarization is believed to be one of the most promising methods to investigate the physical processes just around the central engines by constraining the magnetic environment. For this purpose we are now developing a compact and highly sensitive hard X-ray polarimeter aboard a university class micro-satellite "TSUBAME". We are now developing the flight model of the satellite aiming for the launch in late 2012 from Russia.

Keywords. GRBs, Polarimetry, micro-satellite.

---

1. Introduction

The "fireball model" in which gamma-rays originate in a relativistic outflow driven by some energy source is commonly accepted for explaining gamma-ray bursts (GRBs). However the physical connection between the "fireballs" and the central engines is still unclear. It is often claimed that a magnetic field plays an important role in generating the relativistic outflow from the gravitational energy and, consequently, the gamma-ray emission. For this issue, X-ray/gamma-ray polarimetry is believed to be one of the most promising methods for providing information on the magnetic environment around the emitting region.

GRBs become much brighter than the Sun in the gamma-ray sky and they occur at a rate of about one per day in the universe. Therefore, if we can start observations soon after the detection, even a small detector mounted on a micro-satellite can provide enough information to constrain the physical mechanism in GRBs, and this fact was already clarified by the GAP aboard IKAROS (Yonetoku *et al.* 2011).

2. Satellite and Detector System

For this science mission we chose a 50 kg micro-satellite bus as a platform (Fig. 1). Although the mission payload is strictly limited in size, mass, and power supply the micro-satellite has a quite attractive advantage: launch opportunity. In order to perform "on-axis" pointing observations the satellite bus possesses a high speed attitude control system using compact and high-torque actuators, so-called "control moment gyroscopes".

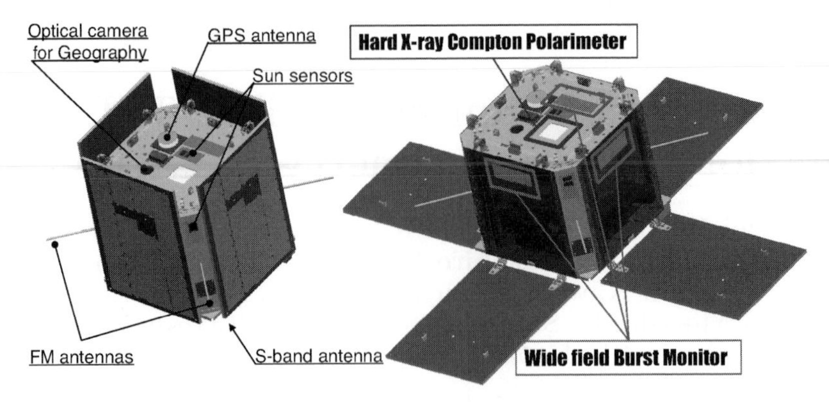

Figure 1. Overview of the micro-satellite TSUBAME (3D-CAD model).

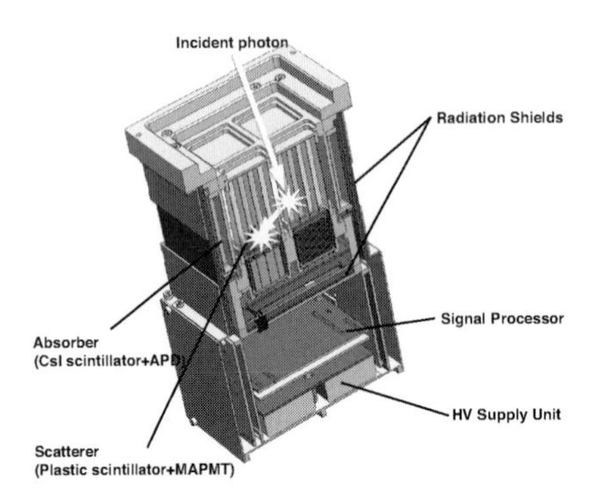

Figure 2. Schematic view of the hard X-ray Compton polarimeter.

Table 1. Expected performance of HXCP.

Energy band	30–200 keV
Field of view	15° × 15°
Effective area	7.1 cm^2 at 100 keV
Modulation factor	54 % at 100 keV
MDP[a]	6.2 % for GRB021206

[a] MDP is the minimum detectable polarization with 3σ confidence level. Detailed information can be found in Yatsu *et al.* (2011).

These that enable high speed attitude control faster than 6° s^{-1}. Thanks to a wide field burst monitor for real time position determination of GRBs, TSUBAME can start a pointing observation within ~15 s after the detection for any GRB in the half-sky field of view of the burst monitor.

The main detector, the Hard X-ray Polarimeter (Fig. 2) measures X-ray polarization utilizing anisotropy of the Compton scattering. The expected performance of the HXCP is summarized in Table. 1. While the WBM always monitors half the sky and if a GRB detected WBM calculate the coordinate with on-board MPU.

References

Yatsu, Y., Enomoto, T., Kawakami, K., Tokoyoda, K., Toizumi, T., Kawaia, N., Ishizaka, K., Matsunaga,S., Nakamori, T., Kataoka, J., & Kubo, S. 2011, *SPIE*, Volume 8145, pp. 814508-814508-11

Yonetoku, D., Murakami, T., Gunji, S., Mihara, T., Toma, K., Sakashita, T., Morihara, Y., Takahashi, T., Toukairin, N., Fujimoto, H., Kodama, Y., Kubo, S. *et al.* 2011, *ApJ*, 743, Issue2, L30

Death of Massive Stars: Supernovae and Gamma-Ray Bursts
Proceedings IAU Symposium No. 279, 2012
P. Roming, N. Kawai & E. Pian, eds.

doi:10.1017/S174392131201366X

Study of emission mechanism of GRBs probed by the gamma-ray polarization with IKAROS-GAP

Daisuke Yonetoku[1], Toshio Murakami[1], Tomonori Sakashita[1], Yoshiyuki Morihara[1], Shuichi Gunji[2], Tatehiro Mihara[3], Kenji Toma[4], and GAP team

[1]College of Science and Engineering, School of Mathematics and Physics, Kanazawa University, Kakuma, Kanazawa, Ishikawa 920-1192, Japan,
email: (DY) yonetoku@astro.s.kanazawa-u.ac.jp

[2]Department of Physics, Faculty of Science, Yamagata University, 1-4-12, Koshirakawa, Yamagata, Yamagata 990-8560, Japan

[3]Cosmic Radiation Laboratory, RIKEN, 2-1, Hirosawa, Wako City, Saitama 351-0198, Japan

[4]Department of Earth and Space Science, Osaka University, Toyonaka 560-0043, Japan

Abstract. We report a polarization measurement in prompt γ-ray emission of GRB 100826A with the Gamma-Ray Burst Polarimeter (GAP) aboard the small solar power sail demonstrator IKAROS. We detected a firm change of polarization angle (PA) during the prompt emission with 99.9 % (3.5 σ) confidence level, and an average polarization degree (Π) of 27 ± 11 % with 99.4 % (2.9 σ) confidence level. Here the quoted errors are given at 1 σ confidence level for two parameters of interest. Non-axisymmetric (e.g., patchy) structures of the magnetic fields and/or brightness inside the relativistic jet are therefore required within the observable angular scale of $\sim \Gamma^{-1}$. Our observation strongly indicates that the polarization measurement is a powerful tool to constrain the GRB production mechanism, and more theoretical works are needed to discuss the data in more details.

Keywords. Gamma-Ray Bursts, polarization, emission mechanism

---

1. Introduction

Gamma-ray bursts (GRBs) are the most energetic explosions in the universe. Since the discovery of X-ray afterglow of GRBs by BeppoSAX Costa *et al.* (1997) and the identification of optical counterparts at cosmological distance, many observational facts and theoretical works have led us to understand the nature of GRBs. However, a crucial issue still remaining to answer is how to release such a huge energy as γ-ray photons. In spite of extensive discussions with the spectral and lightcurve information being collected, there are still several possible models of GRBs Mészáros (2006). The polarimetric observations provide us completely different information. A firm detection of linear polarization of γ-ray photons will further constrain the emission models Lazzati (2006), Toma *et al.* (2009).

2. Data Analyses

The GAP detected GRB 100826A on 26 August 2010 at 22:57:20.8 (UT). The lightcurve of the prompt emission is shown in Fig. 1 (left). First of all, we performed the polarization analysis for the entire dataset, but we set only an upper limit of $\Pi < 30$ % (2 σ confidence level). Therefore we divided the entire data into two time intervals, Interval-1 and -2, as labeled in the lightcurve in Figure 1 (right). (Although Interval-2 has several spikes, we

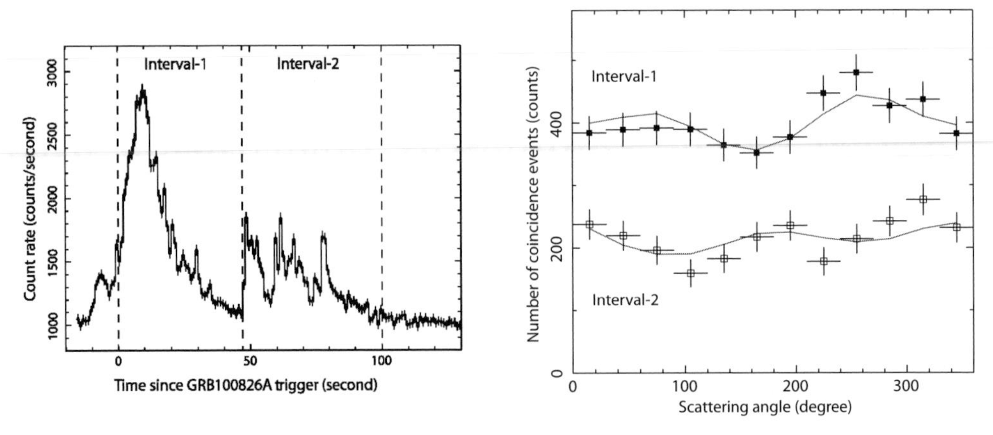

Figure 1. (Left) The lightcurve of GRB 100826A. (Right) The modulation curve of each interval shown in the right panel.

combined all of them to keep photon statistics.) We investigated the polarization degrees, Π, and the polarization angle (ϕ_p) separately for Interval-1 and -2.

The observed modulation curve is shown in Fig. 1 (right). The response of GAP for irradiation from 20.0 degree off-axis is modeled by the Monte-Carlo method with Geant 4 simulator. The gray solid lines in Fig. 1 (right) are the best-fit functions for Interval-1 and -2. The best values are $\Pi_1 = 25 \pm 15$ % with $\phi_{p1} = 159 \pm 18$ degrees for Interval-1 and $\Pi_2 = 31 \pm 21$ % with $\phi_{p2} = 75 \pm 20$ degrees for Interval-2, respectively. Hereafter the quoted errors are 1 σ confidence for two parameters of interest. The significance of polarization detection is rather low: 95.4 % and 89.0 % for Interval-1 and -2, while the difference of polarization angles is significant with 99.9 % (3.5 σ) level.

In the next step, we performed a combined fit for the two intervals, assuming that the polarization degree for Interval-2 is the same as that for Interval-1. This means that we treat Π as one free parameter to improve the statistics with the reduction of model parameters. Here the two polarization angles were still free parameters for both intervals, because a change of angle is apparent. The best-fit polarization degree is $\Pi = 27 \pm 11$ % with $\chi^2 = 21.8$ for 19 degrees of freedom. The significance of detection of polarization is 99.4 % (2.9 σ) confidence level.

We detected significant polarization from two more events, GRB 110301A and GRB 110721A. Detailed discussion can be found in Yonetoku *et al.* (2011b). We consider that the emission mechanism of GRBs may be synchrotron radiation in relatively coherent magnetic fields, and the existence of patchy emission regions can explain the observed rapid change of polarization angle.

References

Costa, E. *et al.*, 1997, *Nature*, 387, 783-785
Lazzati, D. *et al.*, 2006, *New Journal of Physics*, 8, 131
Mészáros, P. 2006, *Rep. Prog. Phys.*, 69, 2259
Toma, K., Sakamoto, T., Zhang, B., *et al.* 2009, *ApJ*, 698, 1042
Yonetoku, D., Murakami, T., Gunji, S., Mihara, T., *et al.*, 2011, *PASJ*, 63, 3
Yonetoku, D., Murakami, T., Gunji, S., Mihara, T., Toma, K., *et al.*, 2011, *ApJ*, 743, L30